CWW1314966

Cereal Processing Technologies

Cereals are the principal dietary components of the human diet and have been for several thousand years. Whole grain cereals are not only an excellent source of energy, but also enrich the diet. The processing of cereals prior to consumption is a necessary step in the production chain to make them palatable and enhance bio- and techno-functional performance.

Cereal Processing Technologies: Impact on Nutritional, Functional, and Biological Properties reviews cereal processing technologies and their impact on quality attributes of cereals, detailing the processing techniques of cereals with recent advancements followed by their impact on nutritional, functional, and biological potential. Each chapter covers three major components as a) technological details for the processing treatment, b) impact on nutritive, functional and biological properties, and c) characterization of processed products.

Key Features:

- Focuses on different cereals for nutritive and functional characteristics
- Explores mechanical, biological, thermal and non-thermal processing treatments of cereals
- Presents impact of different treatments on biological and techno-functional properties of cereals
- Discusses characteristics of the processed products

The contents of ***Cereal Processing Technologies*** are an asset for researchers, students, and professionals, and can be potentially used as reference and important resource for academia and future investigations. This book helps readers identify how different techniques for processing of cereal grains enhance the targeted nutritional and functional quality.

Cereal Processing Technologies

Impact on Nutritional, Functional, and Biological Properties

Edited by
Rajan Sharma, B.N. Dar, and Savita Sharma

CRC Press
Taylor & Francis Group
Boca Raton London New York

CRC Press is an imprint of the
Taylor & Francis Group, an **informa** business

First edition published 2024
by CRC Press
6000 Broken Sound Parkway NW, Suite 300, Boca Raton, FL 33487-2742

and by CRC Press
4 Park Square, Milton Park, Abingdon, Oxon, OX14 4RN

CRC Press is an imprint of Taylor & Francis Group, LLC

Library of Congress Cataloging-in-Publication Data
Names: Sharma, Rajan, editor. | Dar, Basharat Nabi, editor. | Sharma, Savita (Food technologist), editor.
Title: Cereal processing technologies : impact on nutritional, functional,
and biological properties / edited by Rajan Sharma, B. N. Dar, Savita Sharma.
Other titles: Impact on nutritional, functional, and biological properties
Description: First edition | Boca Raton, FL : Taylor and Francis, 2024 |
Includes bibliographical references and index.
Identifiers: LCCN 2023003280 (print) | LCCN 2023003281 (ebook) |
ISBN 9781032120805 (hardback) | ISBN 9781032150338 (paperback) |
ISBN 9781003242192 (ebook)
Subjects: LCSH: Cereal products–Processing. | Agricultural innovations.
Classification: LCC TP434 .C45 2024 (print) |
LCC TP434 (ebook) | DDC 664/.7–dc23/eng/20230712
LC record available at https://lccn.loc.gov/2023003280
LC ebook record available at https://lccn.loc.gov/2023003281

ISBN: 978-1-032-12080-5 (hbk)
ISBN: 978-1-032-15033-8 (pbk)
ISBN: 978-1-003-24219-2 (ebk)

DOI: 10.1201/9781003242192

Typeset in Times
by Newgen Publishing UK

Contents

Preface

Cereals have been the principal dietary components of the human diet for several thousand years. Whole grain cereals are not only an excellent source of energy but also enrich the diet with fibres, minerals, vitamins, and polyphenolic constituents comprising phenolic acids, flavonoids, phytosterols, and carotenoids. These bioactive components have a positive pharmacological impact on human health via a variety of mechanisms such as antioxidative, anti-inflammatory, antidepressant, and antiproliferative pathways. Processing of cereals prior to consumption is a necessary step in the production chain to make them palatable and to enhance bio- and techno-functional performance. Recent decades have witnessed significant variation in the conventional processing techniques of cereals and several novel technologies have been introduced to augment the functionality of cereals without compromise to the aesthetic appeal, nutritive excellence, and biological potential.

The present book aims to critically review the updated processing technologies and their impact on functionality of cereals followed by characterization of processed products. These techniques have been categorized into four major sections as mechanical, biological, thermal and non-thermal methods. Different principles of cereal processing operations result in diverse effects on their performance. Milling and related mechanical treatments cause significant reduction in biological potential of grains due to elimination of nutri-dense bran and aleurone layers but enhance the techno-functional characteristics. Similarly, biological treatments such as soaking and germination cause considerable reduction in anti-nutritional factors, which restrains the absorption and metabolism of minerals and proteins. On the same note, thermally processed cereals exhibit varied functional properties due to heat-induced denaturation of proteins and starch gelatinization depending upon the severity of the treatment. Lately, non-thermal treatments to cereals such as cold plasma, ozonation, and ultrasonic processing have been explored to modulate the biochemical composition and functional characteristics to valorize them for potential food industry needs.

The objective of the present book is to detail the processing techniques of cereals with recent advancements followed by their impact on nutritional, functional, and biological potential. This book will benefit students, academicians, and researchers and serve as an asset for future investigations in the domain of cereal science and technology.

About the Editors

Rajan Sharma

Rajan Sharma, PhD is Teaching Assistant at Punjab Agricultural University, Ludhiana, India. Dr Sharma holds specialization in the domain of cereal technology. His doctoral dissertation focused on characterization, processing, and value addition of nutri-cereals. He earned a B.Tech. (Hons) in Food Technology and M.E. in Food Technology in merit with University Gold Medals. He has authored more than 25 peer-reviewed articles, five book chapters, and two feature articles in reputed international and national journals. He is an active participant in international conferences, seminars, and workshops. To his credit, he has won two prestigious awards: IFI Best Feature Article Award 2016 by the Association of Food Scientists and Technologists (India) and Outstanding Paper 2019 in Emerald Literati Awards.

B.N. Dar

Basharat Nabi Dar, PhD, is Assistant Professor in the Department of Food Technology at Islamic University of Science & Technology, Awantipora, JK, India. He earned his PhD in Food Technology with a specialization in Cereal Technology from Punjab Agricultural University. He is a recipient of CV Raman Fellowship and UGC (University Grants Commission) Research Award (2014–2016) in the field of agricultural sciences. He has worked on multimillion-dollar research projects with the Department of Biotechnology, Government of India, and MOFPI, GOI. He is also associated with several other major research projects as Co-PI/member. Dr Dar is a member of several professional Food Science and Technology associations, including IFT, AFSTI, ASFFBC, WASET, ISEKI, and FSSAI. He has published more than 70 research and review papers and 10 book chapters, edited two books in his field in various prestigious national and international journals. Dr Dar is a visiting scientist at the Department of Food Science, CALS, Cornell University, USA.

Savita Sharma

Prof. Savita Sharma, PhD, is Principal Food Technologist (Dough Rheology), as well as Head of the Department of Food Science and Technology at Punjab Agricultural University, Ludhiana, India. She has contributed more than 32 years to research, teaching, and extension activities related to cereal science and technology. She has been working to develop novel technologies for cereal products such as functional foods and convenience products in addition to extensive characterization and utilization of bioprocessed grains including cereals and pulses. She has supervised more than 30 MSc and 13 PhD scholars. She has completed eight research projects funded by national agencies. Prof. Sharma has authored six books and more than 150 peer-reviewed articles in renowned journals. She has attended and chaired several technical sessions in international conferences, conventions, and seminars and is the recipient of several prestigious awards and distinctions including Fellow of National Academy of Dairy Science (India).

Contributors

Farhana Mehraj Allai
Islamic University of Science & Technology
Awantipora, Jammu and Kashmir, India

Anupama Anand
Dr YS Parmar University of Horticulture and Forestry
Nauni, Himachai Pradesh, India

Uday S. Annapure
Institute of Chemical Technology
Mumbai, Maharashtra, India
Institute of Chemical Technology, Marathwada Campus,
Jalna, Maharashtra, India

Z.R. Azaz Ahmad Azad
Aligarh Muslim University
Aligarh, Uttar Pradesh, India

Manisha Bhandari
Punjab Agricultural University
Ludhiana, Punjab, India

Farhan M. Bhat
Himachal Pradesh Agriculture University
Palampur, Himachal Pradesh, India

Fatma Boukid
ClonBio Group LTD
Dublin, Ireland

Zhenui Cao
COFCO Nutrition & Health Research Institute
Beijing, China

Purba Chakraborty
Dr. SSB UICET, Panjab University
Chandigarh, India

Narender Kumar Chandla
Guru Angad Dev Veterinary and Animal Sciences University (GADVASU)
Ludhiana, Punjab, India

Navnidhi Chhikara
Guru Jambheshwar University of Science & Technology
Hisar, Haryana, India

B.N. Dar
Islamic University of Science & Technology
Awantipora, Jammu and Kashmir, India

Arup Jyoti Das
Tezpur University
Assam, India

Gunjana Deka
Mizoram University
Aizawl, India

Sankar Chandra Deka
Tezpur University
Assam, India

Gunaseelan Eazhumalai
Institute of Chemical Technology
Mumbai, Maharashtra, India

P.S. Gaikwad
National Institute of Food Technology, Entrepreneurship and Management – Thanjavur
Thanjavur, Tamil Nadu, India

Gargi Ghoshal
Dr. SSB UICET, Panjab University
Chandigarh, India

Balmeet Singh Gill
Guru Nanak Dev University
Amritsar, Punjab, India

Khalid Gul
National Institute of Technology
Rourkela, India

Anil Gupta
Dr YS Parmar University of Horticulture and Forestry
Nauni, Himachai Pradesh, India

Mohamad Mazen Hamoud-Agha
Le moulin de Cadillac
Noyal-Muzillac, France

Sundaramoorthy Haripriya
Pondicherry University
Kalapet, India

Swati Joshi
Punjab Agricultural University
Ludhiana, Punjab, India

Ranjitha Gracy T. Kalaivendan
Institute of Chemical Technology
Mumbai, Maharashtra, India

Harpreet Kaur
Guru Nanak Dev University
Amritsar, Punjab, India

Maninder Kaur
Guru Nanak Dev University
Amritsar, Punjab, India

Manisha Kaushal
Dr YS Parmar University of Horticulture and Forestry
Nauni, Himachai Pradesh, India

Anju Boora Khatkar
Guru Angad Dev veterinary and Animal Sciences University (GADVASU)
Ludhiana, Punjab, India

Sunil Kumar Khatkar
Guru Angad Dev veterinary and Animal Sciences University (GADVASU)
Ludhiana, Punjab, India

Tapasya Kumari
Tezpur University
Assam, India

Nisar Ahmad Mir
Chandigarh University
Mohali, India

Anusha Mishra
Institute of Chemical Technology, Mumbai
Mumbai, Maharashtra, India

Vikas Nanda
Sant Longowal Institute of Engineering and Technology,
Longowal, Punjab, India

Pratik Nayi
National Pingtung University of Science and Technology
Neipu, Taiwan

Monica Ostwal
National Institute of Food Technology, Entrepreneurship and Management – Thanjavur
Thanjavur, Tamil Nadu, India

Jessica Pandohee
Curtin University
Bentley, Western Australia, Australia

Anil Panghal
Chaudhary Charan Singh Haryana Agricultural University
Hisar, Haryana, India

Pichan Prabhasankar
CSIR – Central Food Technological Research Institute,
Mysore, India

Emi Grace Mary Gowshika R.
Women's Christian College
Chennai, Tamil Nadu, India

Chandrakala Ravichandran
Rajalakshmi Engineering College
Chennai, Tamil Nadu, India
National Institute of Food Technology Entrepreneurship and Management
Kundli, Haryana, India

Nalla Bhanu Prakash Reddy
National Institute of Food Technology, Entrepreneurship and Management – Thanjavur
Thanjavur, Tamil Nadu, India

Reetu
Maharaja Ranjit Singh Punjab Technical University
Bathinda, Punjab, India

Arya S.S.
Institute of Chemical Technology, NM
Matunga, Mumbai, India

Kawaljit Singh Sandhu
Maharaja Ranjit Singh Punjab Technical University
Bathinda, Punjab, India

Nilakshi Chauhan
Himachal Pradesh Agriculture University
Palampur, Himachal Pradesh, India

Arashdeep Singh
Punjab Agricultural University
Ludhiana, Punjab, India

Dilpreet Singh
Guru Nanak Dev University
Amritsar, Punjab, India

Sukriti Singh
Uttaranchal University
Dehradun, Uttarakhand, India

Rajat Suhag
Free University of Bozen-Bolzano
Bolzano, Italy
National Institute of Food Technology, Entrepreneurship and Management
Kundli, Haryana, India

Mamta Thakur
ITM University
Gwalior, India

Lisa F.M. Lee Nen That
RMIT University
Bundoora, Victoria, Australia

Pinchu Elizabath Thomas
CSIR – Central Food Technological Research Institute,
Mysore, India

Amit Kumar Tiwari
Maharaja Ranjit Singh Punjab Technical University
Bathinda, Punjab, India

Ashutosh Upadhyay
National Institute of Food Technology Entrepreneurship and Management
Kundli, Haryana, India

Devina Vaidya
Dr YS Parmar University of Horticulture and Forestry
Nauni, Himachai Pradesh, India

Vandana Yalakki
Institute of Chemical Technology, NM Parikh Marg
Matunga, Mumbai, India

Ranjana Verma
Himachal Pradesh Agriculture University
Palampur, Himachal Pradesh, India

Bo Wang
Australian Catholic University
New South Wales, Australia

Xiaoming Wei
COFCO Nutrition & Health Research Institute
Beijing, China

B.K. Yadav
National Institute of Food Technology, Entrepreneurship and Management – Thanjavur
Thanjavur, Tamil Nadu, India

D.N. Yadav
Central Institute of Post-Harvest Engineering & Technology
Ludhiana, Punjab, India

Shulin Yang
COFCO Grains Holdings Limited
Beijing, China

Xin Ying
COFCO Nutrition & Health Research Institute
Beijing, China

PART I

Introduction

CHAPTER 1

Cereals as Value-Added Food Components

Fatma Boukid[1] and B.N. Dar[2]
[1]ClonBio Group LTD, Dublin, Ireland
[2]Department of Food Technology, Islamic University of Science & Technology, Awantipora, Jammu and Kashmir, India

CONTENTS

1.1 INTRODUCTION

The family of grains known as the Gramineae, which has nine species, is referred to as "cereals." These species include *Triticum* (Wheat), *Zea* (Corn), *sorghum* (Sorghum), *Pennisetum* (millet), *Secale* (rye), *Hordeum* (barley), *Avena* (oat), *Oryza* (rice), and *Triticosecale* (triticale) (Boukid, 2021). Cereals have a long history as staple foods for human and livestock nutrition. Prior to domestication and agriculture, the ancient world had wild species of wheat that were the ancestors of modern wheat (Boukid et al., 2018). As proof of the change from the hunter-gatherer living to the emergence of agricultural cultures, the very first fossil of domesticated grains was found at several archaeological sites between 8,000 to 10,000 years ago (Pankin & von Korff, 2017). Ancient agricultural cultures in the Fertile Crescent region cultivated the earliest cereal grains (Cooper, 2015). The first species to be domesticated from their wild relatives – *Triticum Monococcum, Triticum Turgidum subsp. Dicoccoides,* and *Hordeum vulgare L. subsp. spontaneum* (C. Koch) Thell – were the cereal crops einkorn (*Triticum Monococcum subsp. Monococcum*), emmer (*Triticum Turgidum subsp. Dicoccum*), and barley (*Hordeum Vulgare subsp. Spontaneum*) (Royo et al., 2017). The shift from wild to domesticated species, referred to as the "domestication syndrome," consisted of

DOI: 10.1201/9781003242192-2

morphological, physiological, and genetic changes (Hammer, 1984). The domestication syndrome included two main changes, the non-brittle rachis mutation resulting in non-shattering domesticated cereals and the non-hulled mutation resulting in free-threshing domesticated cereal. Domestication also included other relevant changes such as increasing seed size and number, increasing carbohydrates contents, and reducing protein and mineral contents (Fuller, 2007; Harlan & Zohary, 1966; Nevo et al., 1986).

The five principal cereals cultivated in the world are wheat, rice, maize, sorghum, and barley. Total production of cereal crops reached 2711 million tons being harvested globally in 2020, comprising 761 million tons of wheat, 503 million tons of rice, and 520 million tons of maize, whereas the production of coarse grains (barley, oats, rye, sorghum, and millets) reached 1447 million tons (FAO, 2021b). Major cereal crops, including wheat, maize, and rice, have a global cropping area of almost 700 million hectares (FAO, 2021c). These cereals provide essential nutrients and supply about 52.5% (around 38% in developed countries and 58% in developing countries) of total calories available for human consumption (FAO, 2021b). The main reasons behind the widespread popularity of these cereals are their adaptability to harsh environments, high productivity, and ease of culture, harvesting, transport, and storing, as well as versatility and nutritional value enabling their use as raw ingredients to make several products such as bread, baked goods, noodles, and pasta (DeFries et al., 2018; Lafiandra et al., 2014; Mefleh, 2021; Papageorgiou & Skendi, 2018). Coarse cereals, or called "minor cereals," are used primarily for animal feed or brewing, while in African countries they are used in a direct way for food production such as in porridge, bread, stew, burghul, and freekeh (Boukid, 2021a; Grote et al., 2021). Nutritionally, these crops provide macro- (e.g., carbohydrates, fiber, and proteins) and micro-nutrients (e.g., minerals and vitamins). Here, in the present chapter, we provide a comprehensive overview of the five principal cereals used globally with a focus on their characteristics, nutritional composition, and applications relying on a compilation of recent literature.

1.2 AN OVERVIEW OF PRINCIPAL CEREAL CROPS

1.2.1 Wheat

In southeast Turkey, 9000 years ago, hexaploid *Triticum Aestivum subsp. Spelt*, the ancestor of common wheat (*Triticum turgidum subsp. Aestivum*), was developed through natural hybridization between domesticated Emmer and diploid Aegilops (Mefleh, 2021; Shewry et al., 2013). Over the centuries, the cropping area and the production of einkorn, emmer, and spelt reduced due to several reasons mainly lower yields (Shewry & Hey, 2015). In recent years, these crops have been reintroduced to enlarge the market of cereal-based products. Currently, the major grown wheat species is hexaploid bread wheat, with over 25,000 varieties adapted to different environments (Shewry et al., 2013). Globally, durum wheat is the 10th most important cereal with an annual production of ≈40 million tons (16 million hectares of planting area) (Beres et al., 2020), which represents ≈5% of total wheat production (International Grains Council, 2020). *T. durum* is characterized by high adaptability to high-temperature conditions and dry environments. Durum wheat is a main ingredient in making pasta, couscous, and bulgur (Boukid, 2021a; Shewry & Hey, 2015). Wheat plays a crucial role in the global economy (Shewry & Hey, 2015). This also implies finding solutions to respond to the increasing global demand to achieve food security. Modern breeding has focused on producing varieties with high yields that gradually replaced old varieties having lower yield. Varieties resistant to biotic and abiotic stress are also one of the objectives of modern breeding. The use of fertilizers and large commercial farms also are boosting wheat production. Global wheat production is expected to reach 839 million tons by 2029 (FAO, 2020).

On the other hand, there are some limiting factors for wheat demand. Indeed, wheat has been increasingly reported to trigger immune response (Boukid, 2021b). Wheat proteins were reported to be triggers of celiac disease (CD) and wheat allergy (WA) (Brouns et al., 2019). Those with a genetic predisposition to CD are said to have a chronic small intestine immune-mediated enteropathy caused by dietary gluten intake (Barilli et al., 2018; Graziano et al., 2019). Globally, ≈25 to 40% of the population carry HLA-DQ2 and/or DQ8, but only around 1% are celiac patients. Proteins in wheat (gliadins and glutenins), rye (secalin), and barley (hordein) are the main triggers of CD (Boukid, Prandi, Sforza, et al., 2017b; Gaiani et al., 2020). Several attempts have been made to reduce wheat epitopes (Boukid, Mejri, et al., 2017). Regarding advanced breeding approaches, untargeted methods such as mutagenesis and transgenesis have shown limitations in silencing specific genes coding gluten epitopes. However, targeted methods have down-regulated the expression of gluten epitopes and provided transgenic wheat with low-gluten content (Barro et al., 2016). Even though there was an assumption that modern wheat breeding increased adverse reactions to wheat, scientific evidence has shown no relatedness (Boukid, Mejri, et al., 2017). Research experiments have been conducted and results varied depending on the severity of the treatment (Boukid et al., 2018; Boukid, Prandi, Buhler, et al., 2017; Francavilla et al., 2017). The disadvantages of such techniques include the possibility of having a significant negative influence on the performance of wheat when used as an ingredient in food making. However, developing wheat strains with a gluten concentration less than 20 ppm by breeding or processing could find a market as preventative and/or therapeutic gluten-free solutions. Indeed, beside celiac patients, there is a growing population (around 1% of the population) adopting a gluten-free diet as a healthier lifestyle (HE et al., 2015). Besides CD, wheat allergy is an allergic reaction following eating foods containing wheat (Cianferoni, 2016). A significant occupational obstructive airway disorder that affects 4–25% of bakery workers globally is baker's asthma caused by allergic reactions due to inhaling wheat flour (Boukid, Prandi, Sforza, et al., 2017a). Non-celiac gluten sensitivity (NCGS) is another wheat-related disorder, but it is complicated to distinguish between CD and NCGS due to similarities in symptoms (Leffler et al., 2013). Other wheat ingredients including gluten may cause NCGS (Barbaro et al., 2018). There is conflicting information on the cause of NCGS. Social media also plays a role in depicting wheat as the enemy, which is not scientifically sustained. This might have an impact on the wheat market especially in developed countries.

1.2.2 Rice

Oryza sativa L. in Asia and Oryza glaberrima Steud in Africa are the two species that are cultivated the most often. Wild *O. rufipogon* was used to domesticated *O. sativa* in China 8000 years ago, and is currently being cultivated worldwide, while *O. glaberrima* was domesticated from *O. barthii* in West Africa where it has been cultivated for 3500 years (Rosell & Marco, 2008; Win et al., 2017). Japonica (short grain) and Indica (long grain) are the two major domestic types of *O. sativa* (Mefleh, 2021). Owing to its high biodiversity, there are thousands of rice varieties grown worldwide that can be classified as short, medium, or long grain (Carcea, 2021). Rice genotypes have different colors including black, red, and purple bran layers (Oikawa et al., 2015). This pigmentation is controlled by Ra, Rc, and Rd genes (Ahmad et al., 2015). Modern rice breeding practices have contributed to the increase of disease resistance, adaptability, eating quality, and productivity (Guzman et al., 2017). Currently, rice is a staple food in the human diet for more than two thirds of the world's population. Global rice production is projected to reach 582 million tons in 2029, in which Asia is expected to have the largest share accounting for 61 million tons (FAO, 2020). In the last decade, rice has also played an important role as a main ingredient in gluten-free products (Morreale et al., 2019). The main cropping areas are located in Asia (90% of the total world production) (Bandumula, 2018).

1.2.3 Maize

Maize originated from a single domestication from teosinte (*Zea mays* ssp. *parviglumis*) in southern Mexico about 9000 years ago (Orozco-Ramírez et al., 2016; Yang et al., 2019). The domestication of maize maintained the wind-born pollen of teosinte, but changed other traits to make it more suitable for human consumption (Hake & Ross-Ibarra, 2015). Maize is a staple food in Latin America and sub-Saharan Africa owing to its long history of consumption and cultivation (Palacios-Rojas et al., 2020). Maize was introduced into China nearly 500 years ago (Shu et al., 2021). Maize is currently the second most important cereal crop after rice in Asia. Global maize production reached 1315 million tons in 2020 with China, the United States, Brazil, Argentina, and Ukraine being the largest producers (FAO, 2020). Recently, breeders developed a provitamin A biofortified maize (15 μg/g DW) to increase vitamin A intake in comparison to conventional maize (0.25 and 2.5 μg/g DW) (Awobusuyi et al., 2021).

1.2.4 Barley

Hordeum vulgare L. was domesticated from the wild, *Hordeum spontaneum* C. Koch. Barley has been a staple crop since the old world Neolithic agriculture (Wang et al., 2015). Domesticated barley spread globally from the Near Eastern region to be cultivated in several countries such as Morocco, Algeria, Crete, Ethiopia, and Tibet (Pankin & von Korff, 2017), which explains the difficulty in identifying the origin of barley (Bothmer et al., 2003; Molina-Cano et al., 1987; Von Bothmer & Komatsuda, 2010). Barley has a short growing season, and can grow under different climates, irrigated and dry land production areas. The global production of barley reached 159.74 million tons during the 2020–2021 crop season (Statista, 2021). China is the largest producer with 440.15 million hectoliters followed by the United States and Brazil.

1.2.5 Sorghum

Archeological evidence has identified the origin of sorghum in Sudan, Ethiopia, and West Africa, around 7500 BC (Winchell et al., 2018). The origin of sorghum is not completely clear, but it has been reported that the first domesticated sorghums were found in eastern Sudan around 3000 BC (Fuller & Stevens, 2018; Winchell et al., 2017). Like other cereals, wild sorghum was used to increase the biodiversity in modern selection programs (Cowan et al., 2020). *Sorghum bicolor* (L.) Moench was derived from the wild progenitor *S. bicolor* subsp. *verticilliflorum*, which was commonly distributed in Africa (Cowan et al., 2020) and then spread to India and China. Sorghum spread from China to the United States (Venkateswaran et al., 2014). It reached Australia in the 19th century and currently it is a major summer crop contributing 5% of the global export of sorghum (Ananda et al., 2020). Currently, sorghum is grown throughout the world, with the majority produced in Asia and Africa. This crop has high resistance to pests and diseases with the ability to grow in arid areas (Ficco et al., 2016; Palavecino et al., 2016).

1.3 CONTRIBUTION OF CEREALS TO DAILY INTAKE OF NUTRIENTS

Cereals are sources of energy, proteins, and fibers. Nevertheless, their chemical composition can drastically vary depending on several factors related to the cereal/variety and/or cultivation conditions. Table 1.1 summarizes the chemical composition of the principal cereals retrieved from the Food Composition Databases of the United States Department of Agriculture (USDA).

Table 1.1 Nutrient Composition of Cereals (USDA, 2021)

	Unit	Wheat[a]	Rice[b]	Maize[c]	Barley[d]	Sorghum[e]
Proximates:						
Water	G	9.44	11.6	10.4	9.44	10.3
Energy	kcal	370	369	365	354	359
Protein	G	15.1	6.94	9.42	12.5	8.43
Total lipid (fat)	G	2.73	1.3	4.74	2.3	3.34
Ash	G	1.56	0.35	1.2	2.29	1.32
Carbohydrates:						
Carbohydrate, by difference	G	71.2	79.8	74.3	73.5	76.6
Fiber, total dietary	G	10.6	0.5	7.3	17.3	6.6
Minerals:						
Calcium	mg	38	6	7	33	12
Iron	mg	3.86	0.22	2.71	3.6	3.14
Magnesium	mg	136	22.9	127	133	123
Phosphorus	mg	352	94	210	264	278
Potassium	mg	376	75	287	452	324
Sodium	mg	3	5	35	12	3
Zinc	mg	3.24	1.19	2.21	2.77	1.63
Copper	mg	0.452	0.209	0.314	0.498	0.253
Manganese	mg	3.56	0.892	0.485	1.94	1.26
Selenium	µg	23.6	5.3	15.5	37.7	12.2
Molybdenum	µg	58.5	45.5	nr	nr	nr
Vitamins and Other Components:						
Thiamin	mg	0.504	0.09	0.385	0.646	0.329
Riboflavin	mg	0.128	0	0.201	0.285	0.061
Niacin	mg	5.55	1.25	3.63	4.6	4.5
Vitamin B-6	mg	0.268	0.052	0.622	0.318	0.325
Folate, total	µg	39	16	nr	19	25

[a] whole wheat, unenriched; [b] Flour, rice, white, unenriched, Flour; [c] Corn grain, white; [d] Barley, hulled; [e] Sorghum flour, whole grain

The major components of wheat flour are protein (10%–12%) and starch (70–75%), followed by minerals (2–3%) and lipids (2%). Carbohydrates consist of sugars and dietary fibers. Starch is the predominant source of energy. Dietary fibers (insoluble and insoluble forms) have several health benefits such as improved digestive health and reduced risk of several chronic diseases (Mitchell & Shewry, 2015; Qiu et al., 2017). Wheat protein content is a key quality factor for all uses of wheat including bread baking to pasta, cakes, noodles and biscuits. Wheat protein is rich in proline and glutamine, but deficient in certain essential amino acids, namely lysine, threonine, tryptophan, histidine, and methionine. (Shewry, 2019). Wheat flour contains 1–2% total fats, in which 78.62% are unsaturated fatty acids consisting mainly of oleic (20.28%) and linoleic acids (57.67%) (Nikolić et al., 2008). Fats contribute to the quality of cereal products, for instance in bread fermentation process (Pareyt et al., 2011). Wheat provides significant amounts of vitamins such as B and E, and minerals such as iron (Papageorgiou & Skendi, 2018). Wheat is also a high source of betaine and choline, two nutrients that aid in the prevention of chronic illnesses and may improve general health (Filipčev et al., 2018). Wheat is also a substantial source of free, bound, or conjugated phenolic acids, including ferulic acid, p-coumaric acid, vanillic acid, caffeic acid, syringic acid, and p-hydroxybenzoic acid (Ramrez-Maganda et al., 2015; Boukid et al., 2019).

All types of rice have starch as their main source of carbohydrates (Zhu, 2015). Non-glutinous rice's starch, which makes up the majority of the grain, is made up of 10–30% amylose and 70–90% amylopectin. Due to its greater amounts of lysine and sulfur-containing amino acids than wheat (AAS of 43) and corn (AAS of 35), rice protein is about 7–10% and is distinguished by a balanced amino acid profile (Frank et al., 2012). Rice contains peptides with biological activities. Rice is also classified as a hypoallergenic cereal due to the absence of gluten. Rice is characterized by low fat content. Rice also contains antioxidant compounds such as phenolic acid flavonoids and anthocyanin depending on the variety. Phenolic compounds are mainly cinnamic, protocatechuic, gallic acid, ferulic, coumaric, and caffeic acids (Al-Taher et al., 2017; Alves et al., 2016). Flavonoids and anthocyanins are mostly found in black rice (Ito & Lacerda, 2019; Kushwaha, 2016).

The main chemical component in maize is starch followed by protein (Preciado-Ortíz et al., 2018). Depending on environment and genotype, protein content in maize ranges from 7–14%. Maize is considered a good oil source having low level of saturated fatty acids (Chávez-Santoscoy et al., 2016). Maize is rich in phenolic acids where hydroxycinnamic acid is the most abundant (Cai & Shi, 2010). The main components of barley are starch, protein, and fiber (Panizo-Casado et al., 2020; Zhu, 2017). Furthermore, barley is characterized by its relevant amounts of bioactive peptides, polysaccharides, and phenolic compounds (Sharma, 2010; Madhujith, 2006). The main phenolics in barley are ferulic acid, p-coumaric acid, caffeic acid, protocatechuic acid, flavan-3-ols, and flavonols (Quan et al., 2018). Sorghum is a grain rich in starch, protein, oil, fiber, and polyphenolic compounds (Palavecino et al., 2016; Sanjari et al., 2021). Sorghum polyphenols are indicated as health-beneficial crops playing a role in cancer prevention, glycemic response improvement, and inflammation inhibition (Ficco et al., 2016).

1.4 CONSUMPTION PATTERN

Cereal grains are an important source of energy (calories) especially in developing countries because of their high starch content and cheap cost. For more than 7.5 billion people, cereals and diets based on cereals constitute the primary source of carbohydrates, protein, vitamins, notably B-group vitamins, and minerals. The food use of cereals and cereal-based products is increasing but at a diminishing rate. According to the Food and Agriculture Organization, world cereal consumption in 2021–2022 is forecasted at 2811 million tons, 49 million tons (1.8% perent) higher than in 2020–2021. The average annual consumption in developing nations is currently 173 kg (Fig. 1.1), accounting for almost 56% of all calories (Bruinsma, 2017). Due to agro-ecological limitations, wheat has historically experienced the fastest rate of consumption growth among the major cereals and will likely continue to do so in the future, especially in nations with low or no production levels. Contrarily, it is anticipated that the per capita consumption of rice will maintain its recent trend toward stabilization with a slight decline, reflecting developments, primarily in the East Asia region. Cereals are a major dietary staple in the majority of nations. The three most commonly used grains, which account for 93% of all cereal calories, are wheat, rice, and maize. The primary food source for Europeans, Asians, and Americans, respectively, is one of these grains. In Africa and India, sorghum and millets are widely farmed and consumed. Around 50% of the dietary energy consumed worldwide comes from cereals, a proportion that has remained remarkably steady throughout time. However, modifications are being made. A more thorough examination of dietary energy consumption reveals a decline in emerging nations, where the proportion of energy coming from grains has dropped from 60% to 54% (Bruinsma, 2017). Wheat and rice are cited for the falling trend since

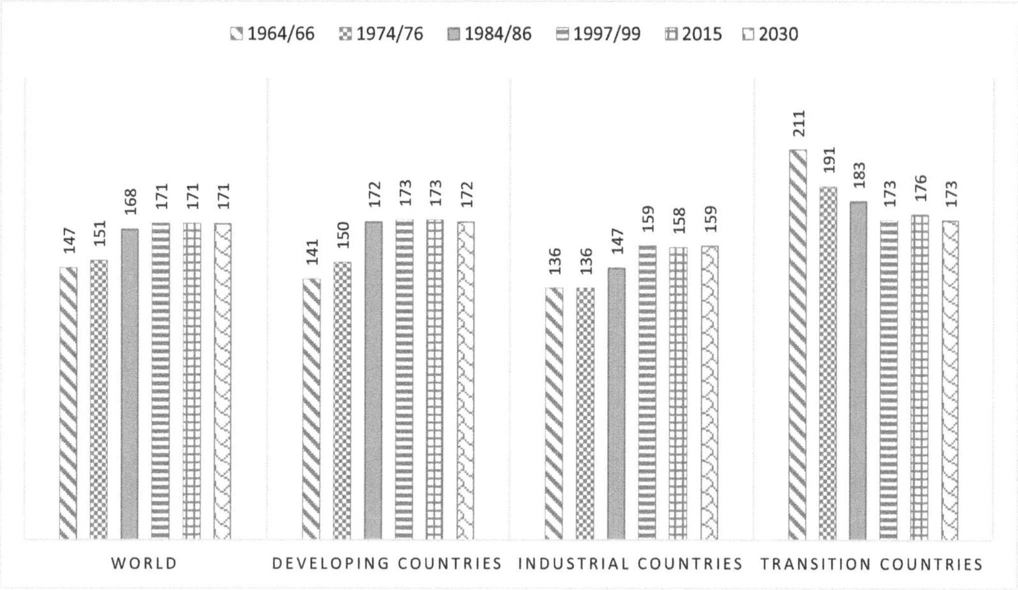

Figure 1.1 Consumption pattern of cereals (Kg/capita/year) (Data adopted from World agriculture: towards 2015/2030, AN FAO PERSPECTIVE Edited by Jelle Bruinsma).

they are now less popular meals in middle-income nations like Brazil and China. This trend is expected to continue for the next 30 years or more. The world's total cereal production has reached 2788 million tonnes in 2021–2022 and the world has also seen expected growth in their utilization. Wheat consumption is increasing globally along with population increase, resulting in a per capita level that is largely steady. The expected increase in maize use over the next year is 2.5%. The usage of sorghum is anticipated to rise in 2021–2022 due to rising food and feed demand. In 2021–2022, global rice consumption is still anticipated to increase by 1.6% year to a new high of 518.8 million tons. Although it is expected that feed and industrial applications of rice will continue to rise, food consumption is thought to be the key driver of annual worldwide usage expansion, growing at a slightly higher rate than population growth (FAO, 2021a).

Cereals and cereal-based products are staple foods in most human diets (McKevith,2004; Kushi, 1999; McIntoshin, 2001) of both developed and developing countries, contributing a major proportion of dietary energy and nutrients, contributing in global terms more than 50% of energy supply (WHO, 2003).

1.5 CONCLUSION

For both humans and animals, cereal and cereal products are important dietary sources. Cereal grains provide nutrition and energy in the form of carbohydrates, proteins, fats, fiber, minerals, and vitamins. Cereal grains are a significant source of energy in underdeveloped nations due to high starch content and cheap cost. The amount of cereal consumed in these regions is sufficient to add significant levels of protein to both children's and adults' diets. In order to fulfill consumer demands for high nutritional content and high production and to assure global food security, modern breeding efforts continue to concentrate on choosing new varieties.

REFERENCES

Ahmad, F., Hanafi, M. M., Hakim, M. A., Rafii, M. Y., Arolu, I. W., & Abdullah, S. N. A. (2015). Genetic Divergence and Heritability of 42 Coloured Upland Rice Genotypes (Oryzasativa) as Revealed by Microsatellites Marker and Agro-Morphological Traits. *PLOS ONE, 10*(9), e0138246. https://doi.org/10.1371/JOURNAL.PONE.0138246

Al-Taher, F., Cappozzo, J., Zweigenbaum, J., Lee, H. J., Jackson, L., & Ryu, D. (2017). Detection and quantitation of mycotoxins in infant cereals in the U.S. market by LC-MS/MS using a stable isotope dilution assay. *Food Control, 72*, 27–35. https://doi.org/10.1016/j.foodcont.2016.07.027

Alves, G. H., Ferreira, C. D., Vivian, P. G., Monks, J. L. F., Elias, M. C., Vanier, N. L., & De Oliveira, M. (2016). The revisited levels of free and bound phenolics in rice: Effects of the extraction procedure. *Food Chemistry, 208*, 116–123.

Ananda, G. K. S., Myrans, H., Norton, S. L., Gleadow, R., Furtado, A., & Henry, R. J. (2020). Wild Sorghum as a Promising Resource for Crop Improvement. *Frontiers in Plant Science, 11*. https://doi.org/10.3389/FPLS.2020.01108/ENDNOTE

Awobusuyi, T. D., Oyeyinka, S. A., Siwela, M., & Amonsou, E. O. (2021). Nutritional properties of provitamin A-biofortified maize amahewu prepared using different inocula. *Food Bioscience, 42*, 101217. https://doi.org/10.1016/J.FBIO.2021.101217

Bandumula, N. (2018). Rice Production in Asia: Key to Global Food Security. *Proceedings of the National Academy of Sciences India Section B – Biological Sciences, 88*(4), 1323–1328. https://doi.org/10.1007/S40011-017-0867-7/TABLES/3

Barbaro, M. R., Cremon, C., Stanghellini, V., & Barbara, G. (2018). Recent advances in understanding non-celiac gluten sensitivity. *F1000Research, 7*. https://doi.org/10.12688/F1000RESEARCH.15849.1

Barilli, A., Gaiani, F., Prandi, B., Cirlini, M., Ingoglia, F., Visigalli, R., Rotoli, B. M., de'Angelis, N., Sforza, S., de'Angelis, G. L., & Dall'Asta, V. (2018). Gluten peptides drive healthy and celiac monocytes toward an M2-like polarization. *Journal of Nutritional Biochemistry, 54*, 11–17. https://doi.org/10.1016/j.jnutbio.2017.10.017

Barro, F., Iehisa, J. C. M., Giménez, M. J., García-Molina, M. D., Ozuna, C. V., Comino, I., Sousa, C., & Gil-Humanes, J. (2016). Targeting of prolamins by RNAi in bread wheat: effectiveness of seven silencing-fragment combinations for obtaining lines devoid of coeliac disease epitopes from highly immunogenic gliadins. *Plant Biotechnology Journal, 14*(3), 986–996. https://doi.org/10.1111/pbi.12455

Beres, B. L., Rahmani, E., Clarke, J. M., Grassini, P., Pozniak, C. J., Geddes, C. M., Porker, K. D., May, W. E., & Ransom, J. K. (2020). A Systematic Review of Durum Wheat: Enhancing Production Systems by Exploring Genotype, Environment, and Management (G × E × M) Synergies. *Frontiers in Plant Science, 11*. https://doi.org/10.3389/FPLS.2020.568657/BIBTEX

Bothmer, R. von, Sato, K., Komatsuda, T., Yasuda, S., & Fischbeck, G. (2003). Chapter 2 The domestication of cultivated barley. *Developments in Plant Genetics and Breeding, 7*(C), 9–27. https://doi.org/10.1016/S0168-7972(03)80004-X

Boukid, F., Mejri, M., Pellegrini, N., Sforza, S., & Prandi, B. (2017). How Looking for Celiac-Safe Wheat Can Influence Its Technological Properties. *Comprehensive Reviews in Food Science and Food Safety, 16*(5). https://doi.org/10.1111/1541-4337.12288

Boukid, F., Prandi, B., Buhler, S., & Sforza, S. (2017). Effectiveness of Germination on Protein Hydrolysis as a Way to Reduce Adverse Reactions to Wheat. *Journal of Agricultural and Food Chemistry, 65*(45). https://doi.org/10.1021/acs.jafc.7b03175

Boukid, F., Prandi, B., Sforza, S., Sayar, R., Seo, Y. W., Mejri, M., & Yacoubi, I. (2017a). Understanding the Effects of Genotype, Growing Year, and Breeding on Tunisian Durum Wheat Allergenicity. 1. the Baker's Asthma Case. *Journal of Agricultural and Food Chemistry, 65*(28). https://doi.org/10.1021/acs.jafc.7b02040

Boukid, F., Prandi, B., Sforza, S., Sayar, R., Seo, Y. W., Mejri, M., & Yacoubi, I. (2017b). Understanding the Effects of Genotype, Growing Year, and Breeding on Tunisian Durum Wheat Allergenicity. 2. the Celiac Disease Case. *Journal of Agricultural and Food Chemistry, 65*(28). https://doi.org/10.1021/acs.jafc.7b02041

Boukid, F., Prandi, B., Vittadini, E., Francia, E., & Sforza, S. (2018). Tracking celiac disease-triggering peptides and whole wheat flour quality as function of germination kinetics. *Food Research International, 112*. https://doi.org/10.1016/j.foodres.2018.06.055

Boukid, F. (2021). *Cereal-Based Foodstuffs: The Backbone of Mediterranean Cuisine*. Springer International Publishing. https://doi.org/10.1007/978-3-030-69228-5

Boukid, Fatma. (2021a). Cereal-Derived Foodstuffs from North African-Mediterranean: From Tradition to Innovation. In Fatma Boukid (Ed.), *Cereal-Based Foodstuffs: The Backbone of Mediterranean Cuisine* (pp. 117–150). Springer International Publishing. https://doi.org/10.1007/978-3-030-69228-5_5

Boukid, Fatma. (2021b). The Bright and Dark Sides of Wheat. In *Cereal-Based Foodstuffs: The Backbone of Mediterranean Cuisine* (pp. 231–246). Springer International Publishing. https://doi.org/10.1007/978-3-030-69228-5_9

Boukid, Fatma, Dall'Asta, M., Bresciani, L., Mena, P., Del Rio, D., Calani, L., Sayar, R., Seo, Y. W., Yacoubi, I., & Mejri, M. (2019). Phenolic profile and antioxidant capacity of landraces, old and modern Tunisian durum wheat. *European Food Research and Technology*, *245*(1), 73–82. https://doi.org/10.1007/s00217-018-3141-1

Boukid, Fatma, Folloni, S., Sforza, S., Vittadini, E., & Prandi, B. (2018). Current Trends in Ancient Grains-Based Foodstuffs: Insights into Nutritional Aspects and Technological Applications. *Comprehensive Reviews in Food Science and Food Safety*, *17*(1), 123–136. https://doi.org/10.1111/1541-4337.12315

Brouns, F., Rooy, G. van, Shewry, P., Rustgi, S., & Jonkers, D. (2019). Adverse Reactions to Wheat or Wheat Components. *Comprehensive Reviews in Food Science and Food Safety*, *18*(5), 1437–1452. https://doi.org/10.1111/1541-4337.12475

Bruinsma, J. (2017). *World agriculture: towards 2015/2030: an FAO study*. Routledge.

Cai, L., & Shi, Y. C. (2010). Structure and digestibility of crystalline short-chain amylose from debranched waxy wheat, waxy maize, and waxy potato starches. *Carbohydrate Polymers*, *79*(4), 1117–1123. https://doi.org/10.1016/j.carbpol.2009.10.057

Carcea, M. (2021). Value of Wholegrain Rice in a Healthy Human Nutrition. *Agriculture 2021, Vol. 11, Page 720*, *11*(8), 720. https://doi.org/10.3390/AGRICULTURE11080720

Chávez-Santoscoy, R. A., Gutiérrez-Uribe, J. A., Serna-Saldivar, S. O., & Perez-Carrillo, E. (2016). Production of maize tortillas and cookies from nixtamalized flour enriched with anthocyanins, flavonoids and saponins extracted from black bean (Phaseolus vulgaris) seed coats. *Food Chemistry*, *192*, 90–97. https://doi.org/10.1016/j.foodchem.2015.06.113

Cianferoni, A. (2016). Wheat allergy: diagnosis and management. *Journal of Asthma and Allergy*, *9*, 13–25. https://doi.org/10.2147/JAA.S81550

Cooper, R. (2015). Re-discovering ancient wheat varieties as functional foods. *Journal of Traditional and Complementary Medicine*, *5*(3), 138–143. https://doi.org/10.1016/J.JTCME.2015.02.004

Cowan, M. F., Blomstedt, C. K., Norton, S. L., Henry, R. J., Møller, B. L., & Gleadow, R. (2020). Crop wild relatives as a genetic resource for generating low-cyanide, drought-tolerant Sorghum. *Environmental and Experimental Botany*, *169*. https://doi.org/10.1016/J.ENVEXPBOT.2019.103884

DeFries, R., Chhatre, A., Davis, K. F., Dutta, A., Fanzo, J., Ghosh-Jerath, S., Myers, S., Rao, N. D., & Smith, M. R. (2018). Impact of Historical Changes in Coarse Cereals Consumption in India on Micronutrient Intake and Anemia Prevalence. *Food and Nutrition Bulletin*, *39*(3), 377–392.

FAO. (2020). *OECD-FAO Agricultural Outlook 2020–2029*. https://doi.org/10.1787/1112C23B-EN

FAO. (2021a). World Food and Agriculture - Statistical Yearbook 2021. Rome.

FAO. (2021b). *FAO Cereal Supply and Demand Brief | World Food Situation | Food and Agriculture Organization of the United Nations*. www.fao.org/worldfoodsituation/csdb/en/

FAO. (2021c). *World cropped area, yield and production of Cereals*. www.fao.org/economic/the-statistics-division-ess/chartroom-and-factoids/chartroom/35-world-cropped-area-yield-and-production-of-cereals/en/

Ficco, D. B. M., De Simone, V., De Leonardis, A. M., Giovanniello, V., Del Nobile, M. A., Padalino, L., Lecce, L., Borrelli, G. M., & De Vita, P. (2016). Use of purple durum wheat to produce naturally functional fresh and dry pasta. *Food Chemistry*, *205*, 187–195. https://doi.org/10.1016/j.foodchem.2016.03.014

Filipčev, B., Kojić, J., Krulj, J., Bodroža-Solarov, M., & Ilić, N. (2018). Betaine in Cereal Grains and Grain-Based Products. *Foods*, *7*(4). https://doi.org/10.3390/FOODS7040049

Francavilla, R., De Angelis, M., Rizzello, C. G., Cavallo, N., Dal Bello, F., & Gobbetti, M. (2017). Selected Probiotic Lactobacilli Have the Capacity To Hydrolyze Gluten Peptides during Simulated Gastrointestinal Digestion. *Applied and Environmental Microbiology*, *83*(14), e00376-17. https://doi.org/10.1128/AEM.00376-17

Frank, T., Reichardt, B., Shu, Q., & Engel, K. H. (2012). Metabolite profiling of colored rice (Oryza sativa L.) grains. *Journal of Cereal Science*, *55*(2), 112–119.

Fuller, D. Q. (2007). Contrasting patterns in crop domestication and domestication rates: Recent archaeobotanical insights from the old world. *Annals of Botany*, *100*(5), 903–924. https://doi.org/10.1093/AOB/MCM048

Fuller, D. Q., & Stevens, C. J. (2018). Sorghum domestication and diversification: A current archaeobotanical perspective. *Plants and People in the African Past: Progress in African Archaeobotany*, 427–452. https://doi.org/10.1007/978-3-319-89839-1_19

Gaiani, F., Graziano, S., Boukid, F., Prandi, B., Bottarelli, L., Barilli, A., Dossena, A., Marmiroli, N., Gullì, M., de'Angelis, G.L., Sforza, S. (2020). The Diverse Potential of Gluten from Different Durum Wheat Varieties in Triggering Celiac Disease: A Multilevel In Vitro, Ex Vivo and In Vivo Approach. *Nutrients*, *12*(11), 1–16. https://doi.org/10.3390/NU12113566

Graziano, S., Marando, S., Prandi, B., Boukid, F., Marmiroli, N., Francia, E., Pecchioni, N., Sforza, S., Visioli, G., & Gullì, M. (2019). Technological Quality and Nutritional Value of Two Durum Wheat Varieties Depend on Both Genetic and Environmental Factors. *Journal of Agricultural and Food Chemistry*, *67*(8). https://doi.org/10.1021/acs.jafc.8b06621

Grote, U., Fasse, A., Nguyen, T. T., & Erenstein, O. (2021). Food Security and the Dynamics of Wheat and Maize Value Chains in Africa and Asia. *Frontiers in Sustainable Food Systems*, *4*. https://doi.org/10.3389/FSUFS.2020.617009/FULL

Guzman, M. K. de, Parween, S., Butardo, V. M., Alhambra, C. M., Anacleto, R., Seiler, C., Bird, A. R., Chow, C.-P., & Sreenivasulu, N. (2017). Investigating glycemic potential of rice by unraveling compositional variations in mature grain and starch mobilization patterns during seed germination. *Scientific Reports*, *7*(1), 5854. https://doi.org/10.1038/s41598-017-06026-0

Hake, S., & Ross-Ibarra, J. (2015). Genetic, evolutionary and plant breeding insights from the domestication of maize. *ELife*, *4*(0), e05861–e05861. https://doi.org/10.7554/elife.05861

Hammer, K. (1984). Das Domestikationssyndrom. *Die Kulturpflanze*, *32*(1), 11–34. https://doi.org/10.1007/BF02098682

Harlan, J. R., & Zohary, D. (1966). Distribution of wild wheats and barley. *Science*, *153*(3740), 1074–1080. https://doi.org/10.1126/SCIENCE.153.3740.1074

HE, M., P, W., & AY, G. (2015). Racial Differences in the Prevalence of Celiac Disease in the US Population: National Health and Nutrition Examination Survey (NHANES) 2009–2012. *Digestive Diseases and Sciences*, *60*(6), 1738–1742. https://doi.org/10.1007/S10620-014-3514-7

InternationalGrainsCouncil. (2020). *World Grain Statistics 2016.* available: www.igc.int/en/subscriptions/subscription.aspx (accessed Jully 21, 2021).

Ito, V. C., & Lacerda, L. G. (2019). Black rice (Oryza sativa L.): A review of its historical aspects, chemical composition, nutritional and functional properties, and applications and processing technologies. *Food Chemistry*, *301*, 125304. https://doi.org/10.1016/J.FOODCHEM.2019.125304

Kushi, L. H., Meyer, K. A., & Jacobs Jr, D. R. (1999). Cereals, legumes, and chronic disease risk reduction: evidence from epidemiologic studies. *The American journal of clinical nutrition*, *70*(3), 451s-458s.

Kushwaha, U. K. S. (2016). Black Rice. *Black Rice*, 21–47. https://doi.org/10.1007/978-3-319-30153-2_2

Lafiandra, D., Riccardi, G., & Shewry, P. R. (2014). Improving cereal grain carbohydrates for diet and health. In *Journal of Cereal Science* (Vol. 59, Issue 3, pp. 312–326). Academic Press. https://doi.org/10.1016/j.jcs.2014.01.001

Leffler, D., Schuppan, D., Pallav, K., Najarian, R., Goldsmith, J. D., Hansen, J., Kabbani, T., Dennis, M., & Kelly, C. P. (2013). Kinetics of the histological, serological and symptomatic responses to gluten challenge in adults with coeliac disease. *Gut*, *62*(7), 996–1004. https://doi.org/10.1136/gutjnl-2012-302196

McIntosh, G. H. (2001). Cereal foods, fibres and the prevention of cancers. *Australian journal of nutrition and dietetics*, *58*(4), S35–S35.

McKevith, B. (2004). Nutritional aspects of cereals. *Nutrition Bulletin*, *29*(2), 111–142.

Mefleh, M. (2021). Cereals of the Mediterranean Region: Their Origin, Breeding History and Grain Quality Traits. In *Cereal-Based Foodstuffs: The Backbone of Mediterranean Cuisine* (pp. 1–18). Springer International Publishing. https://doi.org/10.1007/978-3-030-69228-5_1

Mitchell, R. A. C., & Shewry, P. R. (2015). Dietary Fibre: Wheat Genes for Enhanced Human Health. *Advances in Wheat Genetics: From Genome to Field*, 411–419. https://doi.org/10.1007/978-4-431-55675-6_46

Molina-Cano, J. L., Fra-Mon, P., Salcedo, G., Aragoncillo, C., de Togores, F. R., & García-Olmedo, F. (1987). Morocco as a possible domestication center for barley: biochemical and agromorphological evidence. *Theoretical and Applied Genetics*, *73*(4), 531–536. https://doi.org/10.1007/BF00289190

Morreale, F., Boukid, F., Carini, E., Federici, E., Vittadini, E., & Pellegrini, N. (2019). An overview of the Italian market for 2015: cooking quality and nutritional value of gluten-free pasta. *International Journal of Food Science and Technology*, *54*(3). https://doi.org/10.1111/ijfs.13995

Nevo, E., Beiles, A., & Zohary, D. (1986). Genetic resources of wild barley in the Near East: structure, evolution and application in breeding. *Biological Journal of the Linnean Society*, *27*(4), 355–380. https://doi.org/10.1111/J.1095-8312.1986.TB01742.X

Nikolić, N., Radulović, N., Momcilović, B., Nikolić, G., Lazić, M., & Todorovic, Z. (2008). Fatty acids composition and rheology properties of wheat and wheat and white or brown rice flour mixture. *European Food Research and Technology 2008 227:5*, *227*(5), 1543–1548. https://doi.org/10.1007/S00217-008-0877-Z

Oikawa, T., Maeda, H., Oguchi, T., Yamaguchi, T., Tanabe, N., Ebana, K., Yano, M., Ebitani, T., & Izawa, T. (2015). The birth of a black rice gene and its local spread by introgression. *Plant Cell*, *27*(9), 2401–2414. https://doi.org/10.1105/TPC.15.00310

Orozco-Ramírez, Q., Ross-Ibarra, J., Santacruz-Varela, A., & Brush, S. (2016). Maize diversity associated with social origin and environmental variation in Southern Mexico. *Heredity 2016 116:5*, *116*(5), 477–484. https://doi.org/10.1038/hdy.2016.10

P Sharma, H. G. (2010). Antioxidant and polyphenol oxidase activity of germinated barley and its milling fractions. *Food Chem.*, *120*(3), 673–678. https://doi.org/10.1016/j.foodchem.2009.10.059

Palacios-Rojas, N., McCulley, L., Kaeppler, M., Titcomb, T. J., Gunaratna, N. S., Lopez-Ridaura, S., & Tanumihardjo, S. A. (2020). Mining maize diversity and improving its nutritional aspects within agro-food systems. *Comprehensive Reviews in Food Science and Food Safety*, *19*(4), 1809–1834. https://doi.org/10.1111/1541-4337.12552

Palavecino, P. M., Penci, M. C., Calderón-Domínguez, G., & Ribotta, P. D. (2016). Chemical composition and physical properties of sorghum flour prepared from different sorghum hybrids grown in Argentina. *Starch – Stärke*, *68*(11–12), 1055–1064. https://doi.org/10.1002/STAR.201600111

Panizo-Casado, M., Déniz-Expósito, P., Rodríguez-Gald0n, B., Afonso-Morales, D., Ríos-Mesa, D., Díaz-Romero, C., & Rodríguez-Rodríguez, E. M. (2020). The chemical composition of barley grain (Hordeum vulgare L.) landraces from the Canary Islands. *Journal of Food Science*, *85*(6), 1725–1734. https://doi.org/10.1111/1750-3841.15144

Pankin, A., & von Korff, M. (2017). Co-evolution of methods and thoughts in cereal domestication studies: a tale of barley (Hordeum vulgare). *Current Opinion in Plant Biology*, *36*, 15–21. https://doi.org/10.1016/J.PBI.2016.12.001

Papageorgiou, M., & Skendi, A. (2018). 1 – Introduction to cereal processing and by-products. In *Sustainable Recovery and Reutilization of Cereal Processing By-Products* (pp. 1–25). https://doi.org/10.1016/B978-0-08-102162-0.00001-0

Pareyt, B., Finnie, S. M., Putseys, J. A., & Delcour, J. A. (2011). Lipids in bread making: Sources, interactions, and impact on bread quality. *Journal of Cereal Science*, *54*(3), 266–279. https://doi.org/10.1016/J.JCS.2011.08.011

Preciado-Ortíz, R. E., Vázquez-Carrillo, M. G., Figueroa-Cárdenas, J. de D., Guzmán-Maldonado, S. H., Santiago-Ramos, D., & Topete-Betancourt, A. (2018). Fatty acids and starch properties of high-oil maize hybrids during nixtamalization and tortilla-making process. *Journal of Cereal Science*, *83*, 171–179. https://doi.org/10.1016/J.JCS.2018.08.015

Qiu, S., Yadav, M. P., & Yin, L. (2017). Characterization and functionalities study of hemicellulose and cellulose components isolated from sorghum bran, bagasse and biomass. *Food Chemistry*, *230*, 225–233. https://doi.org/10.1016/j.foodchem.2017.03.028

Quan, M., Li, Q., Zhao, P., & Tian, C. (2018). Chemical composition and hepatoprotective effect of free phenolic extract from barley during malting process. *Scientific Reports 2018 8:1*, *8*(1), 1–9. https://doi.org/10.1038/s41598-018-22808-6

Ramírez-Maganda, J., Blancas-Benítez, F. J., Zamora-Gasga, V. M., García-Magaña, M. de L., Bello-Pérez, L. A., Tovar, J., & Sáyago-Ayerdi, S. G. (2015). Nutritional properties and phenolic content of a bakery product substituted with a mango (Mangifera indica) 'Ataulfo' processing by-product. *Food Research International, 73*, 117–123. https://doi.org/10.1016/j.foodres.2015.03.004

Rosell, C. M., & Marco, C. (2008). Rice. *Gluten-Free Cereal Products and Beverages*, 81–III. https://doi.org/10.1016/B978-012373739-7.50006-X

Royo, C., Soriano, J. M., & Alvaro, F. (2017). Wheat: A Crop in the Bottom of the Mediterranean Diet Pyramid. *Mediterranean Identities – Environment, Society, Culture*. https://doi.org/10.5772/intechopen.69184

Sanjari, S., Shobbar, Z. S., Ghanati, F., Afshari-Behbahanizadeh, S., Farajpour, M., Jokar, M., Khazaei, A., & Shahbazi, M. (2021). Molecular, chemical, and physiological analyses of sorghum leaf wax under post-flowering drought stress. *Plant Physiology and Biochemistry, 159*, 383–391. https://doi.org/10.1016/J.PLAPHY.2021.01.001

Shewry, P. (2019). What is gluten—Why is it special? In *Frontiers in Nutrition* (Vol. 6). Frontiers Media S.A. https://doi.org/10.3389/fnut.2019.00101

Shewry, P. R., Hawkesford, M. J., Piironen, V., Lampi, A.-M., Gebruers, K., Boros, D., Andersson, A. A. M., Åman, P., Rakszegi, M., Bedo, Z., & Ward, J. L. (2013). Natural Variation in Grain Composition of Wheat and Related Cereals. *Journal of Agricultural and Food Chemistry, 61*(35), 8295–8303. https://doi.org/10.1021/JF3054092

Shewry, P. R., & Hey, S. J. (2015). The contribution of wheat to human diet and health. *Food and Energy Security, 4*(3), 178. https://doi.org/10.1002/FES3.64

Shu, G., Cao, G., Li, N., Wang, A., Wei, F., Li, T., Yi, L., Xu, Y., & Wang, Y. (2021). Genetic variation and population structure in China summer maize germplasm. *Scientific Reports 2021 11:1, 11*(1), 1–13. https://doi.org/10.1038/s41598-021-84732-6

Statista. (2021). *World barley production*. https://www.statista.com/statistics/271973/world-barley-production-since-2008/

T Madhujith, M. I. F. S. (2006). Antioxidant properties of pearled barley fractions. *J. Agric. Food Chem., 54*(9), 3283–3289. https://doi.org/10.1021/jf0527504

USDA. (2021). *Food Composition Databases Show Foods List*. https://ndb.nal.usda.gov/ndb/search/list

Venkateswaran, K., Muraya, M., Dwivedi, S. L., & Upadhyaya, H. D. (2014). *Wild Sorghums-Their Potential Use in Crop Improvement*.

Von Bothmer, R., & Komatsuda, T. (2010). Barley origin and related species. *Barley: Production, Improvement, and Uses*, 14–62. https://doi.org/10.1002/9780470958636.CH2

Wang, Y., Ren, X., Sun, D., & Sun, G. (2015). Origin of worldwide cultivated barley revealed by NAM-1 gene and grain protein content. *Frontiers in Plant Science, 6*(SEPTEMBER). https://doi.org/10.3389/FPLS.2015.00803/REFERENCE

Win, K. T., Yamagata, Y., Doi, K., Uyama, K., Nagai, Y., Toda, Y., Kani, T., Ashikari, M., Yasui, H., & Yoshimura, A. (2017). A single base change explains the independent origin of and selection for the nonshattering gene in African rice domestication. *New Phytologist, 213*(4), 1925–1935. https://doi.org/10.1111/NPH.14290

Winchell, F., Brass, M., Manzo, A., Beldados, A., Perna, V., Murphy, C., Stevens, C., & Fuller, D. Q. (2018). On the Origins and Dissemination of Domesticated Sorghum and Pearl Millet across Africa and into India: a View from the Butana Group of the Far Eastern Sahel. *African Archaeological Review, 35*(4), 483–505. https://doi.org/10.1007/S10437-018-9314-2/EMAIL/CORRESPONDENT/C1/NEW

Winchell, F., Stevens, C. J., Murphy, C., Champion, L., & Fuller, D. Q. (2017). Evidence for sorghum domestication in fourth millennium BC Eastern Sudan: Spikelet morphology from ceramic impressions of the Butan group. *Current Anthropology, 58*(5), 673–683. https://doi.org/10.1086/693898

World Health Organization. (2003). *Diet, nutrition, and the prevention of chronic diseases: report of a joint WHO/FAO expert consultation* (Vol. 916). World Health Organization.

Yang, C. J., Samayoa, L. F., Bradbury, P. J., Olukolu, B. A., Xue, W., York, A. M., Tuholski, M. R., Wang, W., Daskalska, L. L., Neumeyer, M. A., Sanchez-Gonzalez, J. de J., Romay, M. C., Glaubitz, J. C., Sun, Q.,

Buckler, E. S., Holland, J. B., & Doebley, J. F. (2019). The genetic architecture of teosinte catalyzed and constrained maize domestication. *Proceedings of the National Academy of Sciences*, *116*(12), 5643–5652. https://doi.org/10.1073/PNAS.1820997116

Zhu, F. (2015). Interactions between starch and phenolic compound. *Trends in Food Science and Technology*, *43*(2), 129–143.

Zhu, F. (2017). Barley Starch: Composition, Structure, Properties, and Modifications. *Comprehensive Reviews in Food Science and Food Safety*, *16*(4), 558–579. https://doi.org/10.1111/1541-4337.12265

CHAPTER 2

Nutritional Potential of Cereals

Pinchu Elizabeth Thomas and Pichan Prabhasankar
Flour Milling Baking and Confectionery Technology Department,
CSIR – Central Food Technological Research Institute, Mysore, Karnataka, India

CONTENTS

DOI: 10.1201/9781003242192-3

2.1 INTRODUCTION

Wheat (*Triticum*), rice (*Oryza*), corn (*Zea*), rye (*Secale*), barley (*Hordeum*), oat (*Avena*), millets, sorghum (*Sorghum*), and triticale (*Triticosecale*) are the principal cereals that belong to the grass family Gramineae. They are grown for their edible seeds or grains, which are highly nutritious. Rice and wheat are the most common cereals in human diets. Cereals are the cheapest source of energy, providing over half of the world's food calories. In addition, they contribute an excellent supply of B vitamins, fibre, traces of minerals and pigments, as well as bioactive compounds that are beneficial to humans and animals. These grains are eaten as either partially or fully processed. Moreover, cereals required optimum post-harvest conditions and processing techniques to produce desirable products with extended shelf life. In cereals, a moisture level of less than 14% is safe to avoid spoilage. They are non-perishable and easy to pack, store, and transport.

Bran, endosperm, and embryo or germ make up the grain. Cereals typically contain 70–80% carbohydrate, primarily starch, 6.5–14% protein, and 2–4% fat, except for oats (includes 7.5% fat). Moreover, they are high in vitamins, including the B complex vitamins and vitamin E, and minerals like potassium, calcium, magnesium, phosphorus, iron, and zinc (Manay & Shadaksharaswamy, 2001). Cereals and cereal products can be fortified and enriched with bioactive components to improve or restore the nutritional value. The by-products of cereals are also used to produce economically or industrially important enzymes. Recent cereal processing trends focus on developing value-added food products by combining traditional and innovative cereal processing methods. Therefore, this chapter provides detailed information about nutritional aspects of cereals and cereal processing techniques and their effect on nutritional, functional, and bioactive characteristics.

2.2 NUTRITIONAL FACTS OF WHOLE-GRAIN CEREALS

The whole-grain cereals possess a healthy nutritional profile because they contain a significant amount of vitamins, minerals, fibre, resistant starch, antioxidants, and phytochemicals rather than carbohydrate, protein, and fat. They are considered as the primary carbohydrate resources for humans. Among the cereals, wheat, rice, corn, and a fair quantity of sorghum and millets are considered as part of the staple diet.

2.2.1 Wheat

Wheat is one of the most widely eaten grain crops in the world. Common or bread wheat (*Triticum aestivum*) and durum wheat (*Triticum durum*) are the two kinds of wheat grown for human consumption. The nutritional quality of wheat is influenced by both internal and external variables, such as variety and chemical composition, as well as the growing environment and storage. Generally, wheat contains 13–17% bran, 80–85% endosperm, and 2–3% germ. Around 53% of the bran is contributed by fibre components that protect the germ and endosperm. The starchy endosperm contains 1.5% fat, 13% protein, 0.5% minerals, and 1.5% fibre. The germ is made of 25% protein, 8–13% fat, and 4–5% minerals. The germ also contains vitamin E and amino acids such as alanine, arginine, asparagine, glycine, lysine, and threonine (Sramkova et al., 2009). Wheat contains around 10–18% of protein, mainly albumin, globulin, gliadin, and glutenin fractions. Gluten is a specific wheat protein made up of gliadins and glutenins. Gluten accounts for 75% of wheat's total proteins and is responsible for dough's rheological characteristics. Wheat is divided into two types based on protein: hard wheat (11–13%) and soft wheat (8–10%).

Cereals store energy in the form of starch granules, which are made up of amylose (20–30%) and amylopectin (70–80%) (Konik–Rose et al., 2007). Cellulose and pentosans are the most important non-starch polysaccharides found in wheat. Non-starch polysaccharides are mostly concentrated in the wheat bran fraction, containing around 9% cellulose and 29% non-starch polysaccharides. Polysaccharides are found in more significant amounts in whole flour than in processed flour. Pentosans, also known as hemicelluloses, assist in reducing fat and cholesterol absorption. Pentosans improve the ability of flour to absorb water and the viscosity of dough. Cereals include a small amount of lipids, yet they have a great impact on the quality and texture of the food. Dietary fibre is another wheat constituent with numerous health benefits, including protection against heart disease and cancer, regular bowel movements, constipation prevention, normalization of blood lipids, glucose absorption, and insulin secretion regulation. Arabinoxylan is an insoluble fibre found in the cell wall of the endosperm. It is an ideal substrate for the fermentative generation of short-chain fatty acids in the colon, especially butyrate, which aids in bowel health and cancer prevention (Philippe et al., 2006). The aleurone and neighbouring sub-aleurone regions of the grain contain the soluble fibre, β-glucan. Wheat contains minerals. However, the mineral concentration varies depending on the fraction of the grain. Wheat grains have a high mineral content in the aleurone layer (particularly P, K, and Mg), but the endosperm layer has a low mineral content. The milling process has a significant impact on wheat flour's mineral content, so it is necessary to minimize the loss of the aleurone layer during milling (Sramkova et al., 2009).

2.2.2 Rice

Rice (*Oryza sativa* L.) is a staple food for more than half of the world's population, mainly in Asia. Generally, the grain contains the edible caryopsis enclosed in a protective outer covering known as a hull or husk. The general structure of rice includes husk, pericarp, seed coat, embryo, and endosperm. The composition of the rice varies with variety and climatic conditions. Rice contains 73.4–80.8% carbohydrate, mainly starch. In addition to starch, it contains free sugars and fibre. The nutritional value of rice germ, pericarp, and aleurone layer is higher than that of endosperm because most of the vitamins, minerals, and pigments are present in those fractions. While comparing with wheat, rice contains less protein. The major protein of rice is glutelin, known as oryzenin. Other minor proteins include albumin, globulin, and prolamins. The polishing of rice decreases its biological value and improves protein digestibility. Rice contains a significant amount of semi-essential amino acid, arginine. Like other cereal grains, most of the minerals (calcium, iron, and phosphorous) are present in the germ and pericarp of rice. The important rice enzymes are amylases, proteases, lipases oxidases, peroxidases, and phenolases. The activity of amylase, lipase, and peroxidase declines during storage. Anthocyanins and carotenoids are the pigments responsible for the characteristic colour of some varieties of rice (Manay & Shadaksharaswamy, 2001). Rice contributes high calories, which may lead to diabetes mellitus due to the high glycemic index. As a staple food, rice can be used for nutritional supplement programmes to prevent micronutrient deficiency.

2.2.3 Corn

Corn, often known as maize (*Zea mays* L.), is a widely used grain in the food processing industry. Dent maize (has depression or dent at the crown of the kernel), flint maize (hard, spherical kernels), sweet corn (dent-type maize high in carbohydrate used to make flour), and popcorn are the four types of corn (flint-type maize, which expands while heating). The kernel consists of endosperm (83%), germ (11%), pericarp (5%), and tip cap (1%) (Gwirtz & Garcia-casal, 2014). A protein matrix surrounds the starchy endosperm. The starch may be hard or vitreous and soft or opaque. The starch fraction has a major influence on its digestibility and degradation. The embryo is rich in fat (around 33.3%), B complex vitamins, and antioxidants (vitamin E). Corn oil is rich in polyunsaturated fatty acids (54.7%). The pericarp of corn has high fibre (8.8%) content, B vitamins, and minerals (Gwirtz & Garcia-casal, 2014). The predominant maize protein is the prolamine known as zein.

2.2.4 Barley

Barley (*Hordeum Vulgare*) is the world's fourth most important cereal after wheat, rice, and maize and one of the oldest cultivated cereals. It is used as a bread grain and for animal feed, malting, and brewing. Today, the use of barley as food is increased due to its nutritional potential. Barley contains the healthy fibre β-glucan that helps to reduce serum cholesterol. Like other cereals, the principal constituent of barley is a carbohydrate, mainly starch, followed by protein and fat. It contains around 10–20% dietary fibre, which helps to reduce cholesterol, glycemic index, and colon cancer. In addition to β-glucan, barley contains cellulose, fructans, arabinoxylans, galactomannans, and arabinogalactans. Barley contains tocopherols and tocotrienols, which are lipids with antioxidant properties. α-D-tocotrienol and α-linoleic acid found in barley help to inhibit cholesterol synthesis. Barley bioactives such as phytin, phenolic acids, vitamin E, proanthocyanidins, and catechins help prevent the formation of carcinogens. It is also a good source of B vitamins and minerals. The main part of barley consists of husk, pericarp, testa, aleurone layer, endosperm, and embryo. The husk and pericarp mainly consist of dietary fibre that protects the grain. Antioxidants such as proanthocyanidins and catechins are present in the testa. The embryo is rich in starch, protein, lipids, and minerals. The scutellum is made of hemicelluloses and contains ferulic acid. The aleurone layer contains arabinoxylans, β-glucans and phenolic acids. Barley contains around 75–80% endosperm. The endosperm is rich in starch and contains β-glucan, polysaccharides, and uronic acid (Izydorczyk et al., 2016).

2.2.5 Oats

Among the different species of *Avena sativa* L., white oats are the most important cultivated form. Oats are covered with numerous trichomes or hair-like protuberance. The kernel is enclosed by a hull, which is made of lemma and palea. The hull is loosely attached to the groat. The groat is enveloped by bran layers pericarp, seed coat, and aleurone cells. Oats have a sweetish flavour, which makes them a favourite among breakfast cereals. The unique property of oat is contributed by globulin protein, phenolic compound, dietary fibre, vitamins, and minerals. Oats contain a significant amount of soluble fibre, β-glucan having cholesterol-lowering properties. Avenanthramides, a unique antioxidant present in oats, helps protect blood vessels from the damaging effects of LDL (low-density lipoprotein) cholesterol. Oat protein's peculiar amino acid sequence makes it ideal for producing gluten-free food products for celiac patients (Biel et al., 2020).

2.2.6 Sorghum

Sorghum (*Sorghum bicolor* L. Moench) is a staple food of many parts of Asia, Africa, and the Middle East. The quality of sorghum mainly depends on the nutritional value, organoleptic properties, anti-nutritional factors, and processing characteristics. The protein quality of the cereal depends on the composition of essential amino acids. It contains a significant amount of glutamic acid, leucine, alanine, proline, and aspartic acid. Starch is the main carbohydrate of sorghum and accounts for around 75%. The amylose content of starch is an important factor that affects the physicochemical properties of starch. Sorghum contains around 2.1–7.6 % fat, 1.0–3.4% crude fibre, and 1.3–3.3% minerals. Sorghum is rich in minerals, mainly phosphorus, potassium, and magnesium. In addition, sorghum contains polyphenols, called tannins, which are responsible for grain pigmentation and interfere with the bioavailability of nutrients (Subramanian & Jambunathan, 1984).

2.2.7 Rye

Rye *(Secale cereale)* is one of the main crops in Eastern Europe. It is also known as the bread grain of poor people. The rye grain is protected by husk and contains pericarp, testa, embryo, and endosperm. Rye has soluble fibre arabinoxylan and β-glucan, which possess health benefits. The aleurone layer contains a significant amount of proteins, B vitamins, and minerals (manganese, iron, copper, zinc, selenium, magnesium, and fluoride). It also contains lignans that are converted to enterodiol and enterolactone by intestinal microflora and positively affect health. The polyphenols present in the rye, such as tannins, phytic acid, ferulic acid, etc., provide antioxidant and anticancerous properties.

2.2.8 Millets

Millets are one of the major crops of developing countries because they can grow in any adverse environment. Millets are known as coarse cereals, and include pearl millet (*Pennisetum glaucum*), finger millet (*Eleusine coracana*), proso millet (*Penicum miliaceum*), foxtail millet (*Setaria italic*), kodo millet (*Paspalum setaceum*), little millet (*Panicum sumatrense*), and barnyard millet (*Echinochloa utilis*). The four major millets include pearl millet, foxtail millet, proso millet, andfinger millet. Whereas barnyard millet, kodo millet, little millet, and guinea millet are considered as minor millets. Significant calcium, dietary fibre, polyphenols, and protein concentration makes the millets unique among the cereals. The demand for millets as one of the major ingredients in multigrain and gluten-free food formulations has increased tremendously due to their positive health effects. In addition, they are rich in essential amino acids, particularly sulphur containing amino acids. Millets also have a good amount of magnesium and phosphorus. Magnesium helps to reduce the effect of migraine and heart attacks, whereas phosphorus is the main part of adenosine triphosphate (ATP), a precursor of energy. They are rich in fibre and polyphenols. They act as an antioxidant and play a significant role in the defence mechanism of the body (Devi et al., 2014). The free radical scavenging activity of finger millet, kodo millet, foxtail millet, barnyard millet, and little millet were reported by Devi et al. and Mohamed et al., which confirms their antioxidant potential (Mohamed et al., 2012). The fatty acid profiling of millets shows that they are rich in linoleic, oleic, and palmitic acid (Ibrahima et al., 2004).

2.2.9 Triticale

Triticale is a hybrid combination of wheat and rye possessing the combined quality of both. The qualities include better yield, grain quality of wheat, and disease-resistant properties of rye. Triticale is also known to play a vital role in the development of novel healthy cereal products. Triticale has a chemical makeup that is nearly identical to wheat, however, it has more nutritious constituents when compared to wheat in terms of nutritional value. Triticale has more lysine, arginine, aspartic acid, and alanine than wheat and has a greater crude protein content. Because of its high lysine content, triticale is a better alternative to other cereals in terms of protein digestibility and mineral balance (Salmon, 1984). The endosperm contains non-starch polysaccharides such as pentosans and β-glucan, as in wheat and rye. Triticale flour may be used to produce bakery products, particularly bread, even though it has less gluten than wheat. Triticale has the same starch content as wheat and is greater than rye (McGoverin et al., 2011). The starch composition affects the digestibility and bread-making property of triticale. It is also rich in essential amino acids, making it superior in nutritional value to wheat (Biel et al., 2020).

2.3 PROCESSING OF CEREALS

To convert whole grain cereals into a product, they are exposed to various primary and secondary processing methods. The primary processing of cereal includes cleaning, grading, milling, and parboiling. At the same time, secondary processing includes fermentation, extrusion, puffing, flaking, shredding, baking, and roasting, which convert them to edible products. Milling is the most significant processing method resulting in flour production, a value-added product from grains (Kanojia et al., 2018). Dry milling and wet milling are the two forms of milling. Dry milling is an abrasive processing method that produces polished grain by progressively separating the seed coat, aleurone layer, subaleurone layer, and germ from the endosperm. The milling of wheat, rice, and oats are examples of dry milling. The by-products generated after dry milling are high in bioactive substances. Generally, wet milling is used to make starch, sugar, gluten, and other products. Corn starch production is an example of wet milling (Papageorgiou et al., 2020). Parboiling is a processing technique given to paddy before milling. Parboiling helps to improve the nutritional profile, storage life, and head rice yield during milling. Whereas malting is a controlled germination process in which carbohydrates and protein are degraded enzymatically (e.g., barley and sorghum). Malting also assists the brewing process by increasing the quantity of free simple sugars and amino acids in the grain (Taylor et al., 2005). Extrusion cooking is a high-temperature and high-pressure processing method that converts cereals into value-added food products, including breakfast cereals, snacks, ready-to-cook and ready-to-eat products. In addition to these processing techniques, fermentation, puffing, flaking, shredding, and roasting are employed to create a wide range of cereal products.

2.3.1 Milling

Milling is a preliminary processing treatment given to cereals to get the edible fraction, endosperm, by removing the husk or bran layers. The basic steps in the whole milling process include cleaning, tempering, conditioning, breaking and reduction, sifting, and purifying. Milling can be done either by traditional or modern milling methods. The classical or conventional milling methods are hand grinding, stone grinding, and mortar and pestle grinding. At the same time, modern mills use rollers to grind the cereals. Usually, after every stage of grinding, the material is sifted to separate uniformly sized particles. In the purifying process, the coarser fractions are separated and graded based on their size and quality.

2.3.1.1 Wheat Milling

Wheat is milled either by traditional or mechanized methods to separate the endosperm from the husk. Traditional wheat milling processes produce whole wheat flour or atta. The major disadvantage of this method is that indigestible components like cellulose and phytate are not eliminated during milling. But modern or mechanized flour mills use metal cylinders or rollers to mill cereals. Before milling, the grains are cleaned to remove physical impurities such as sticks, stones, sand, dust, foreign seeds, and metal bits. The cleaning house operation uses sieves, spiral seed separators, magnetic separators, electrostatic separators, aspirators (air currents), scorers (by friction), de-stoners, and water washes. Subsequently, the grains are subjected to a conditioning and tempering process, which regulates moisture levels and improves grain quality during milling. Tempering (adding water to the grain) and conditioning (heating) help fast diffusion of water into the bran and endosperm. It hardens (firm and rubbery) the bran and mellows the inner endosperm (Mckevith, 2004). After this treatment, the moisture content of hard wheat changes to 15 to 19% and soft wheat to 14.5 to 17%. Both conditioning and tempering are influenced by temperature, processing time, and the amount of water used. The optimum temperature is typically 115^{0}F because higher temperatures may alter

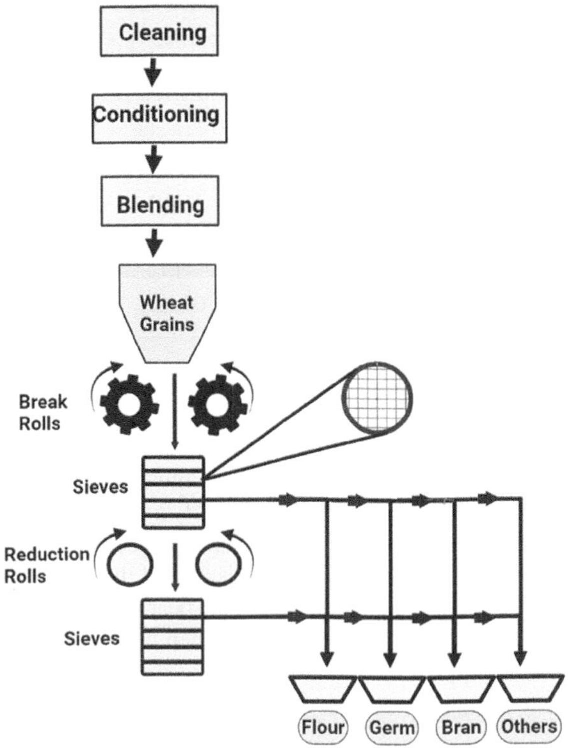

Figure 2.1 General milling process of wheat.

the starch and protein properties. Wheat grains are then milled by passing through roller mills, namely, break rolls and reduction rolls. The surface of reduction rolls is smooth, but the break rolls have a corrugated surface. The break rollers separate the endosperm and germ from the pericarp by breaking the kernel. The break fraction comprises bran, the coarse component of the endosperm (sizings), the smaller endosperm particles (middlings), and break flour fractions. These fractions are then passed through sifters and purifiers to eliminate the bran from the endosperm particles. The sizings and middlings are reduced to flour using the reduction rollers. Atta (whole wheat flour), suji or rava (semolina), and maida (endosperm flour) are the principal products obtained from wheat milling. The freshly milled flour is next bleached to remove the yellow color resulting from the pigment xanthophyll. Bleaching can be done by exposing the flour to air or oxidizing agents such as chlorine, chlorine dioxide, and benzoyl peroxide. The bread-making quality of freshly milled flour can be improved by storing the flour for 1–2 months. Enzymatic modifications known as maturation or aging occur during storage. During storage, the elasticity of gluten is maintained by a reduction in disulphide bonds and an increase in sulfhydryl bonds. Improvers such as potassium bromate, calcium phosphate, and ascorbic acid are employed to speed up the maturation process (Manay & Shadaksharaswamy, 2001). The general steps involved in wheat milling are given in Figure 2.1.

2.3.1.2 Rice Milling

Rice is commonly consumed as whole grain or processed into flour. Home pounding or automated rice mills are used to mill paddy. Generally, paddy is dried to a moisture content of 12–13% before processing. Paddy is cleaned and then passed through screens of varying perforations to eliminate

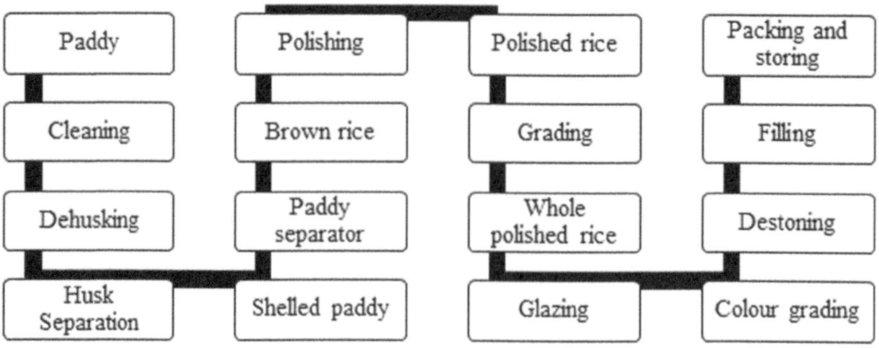

Figure 2.2 Rice milling process.

small, large, and heavy impurities. Lighter impurities are removed by air currents and sent through a magnetic separator to remove metal particles. Cleaned paddy is then put through a husking machine consisting of two stones or rubber discs revolving at various speeds. The husk is then removed by aspiration and brown rice is obtained. The hulled rice is passed through a series of cone-shaped or horizontal-shaped machines known as pearling cones. The pearling technique removes the bran and germ from brown rice. The unpolished rice obtained is then put through a brush machine, which polishes it by removing the aleurone layer. Finally, talc or sugar coating is applied to the polished milled rice to give it a glossy appearance. The different stages of the rice milling process are given in Figure 2.2.

2.3.1.3 Corn Milling

Dry or wet milling can be used to process corn. Wet milling produces starch, whereas dry milling produces grits. After cleaning, the moisture content of the grains is raised to 21% by conditioning and tempering. The grain is then passed through a huller and degermer to remove the hull and germ from the kernel. This can be done either in a de-germer and corn huller/a roller mill and separator/entoleters and gravity separators. The endosperm is then ground to grits by passing through rollers. The material is subsequently pushed through further rollers to get required particle size. Corn is wet milled to produce starch, germ, fibre, and protein. Wet milling starts with steeping, in which grain is soaked in water for 30–40 hours (Mckevith, 2004) to absorb moisture and swell. Steeping activates enzymes and aids in the breakdown of the structure. Additionally, it helps in increasing protein solubility and reducing starch–protein interactions. After steeping, corn is coarsely ground in disc mill to separate germ from the kernel. The pulverized slurry is then passed through hydroclones to remove lighter-weight germs. Further, the germs are dried and ground into meal and oil. The resulting fractions are screened, and larger particles are finely ground to separate the starch, protein, and fibre. Fibre is collected and washed. The remaining fractions are centrifuged with disc nozzles to separate the heavier starch from the gluten. After separation, both the starch and gluten are dewatered using additional centrifuges and vacuum filters.

2.3.1.4 Barley Milling

Barley is dehulled in a pearling machine or compressed air dehullers that use pressurized air to give a mechanical shock to the grain that removes the hull from the kernel. The aspiration process later separates the hull. The pearling process includes abrasive scouring that progressively eliminates hull, bran layers, and the embryo. The pearling system comprises a rotating carborundum or grinding stones, which rotate immediately within a perforated cylinder. The hull and adjacent portions are scraped off from the kernel by rubbing action. The degree of the pearling process depends on: (i) the rate of feeding, (ii) the distance between the abrasive stone and screen, (iii) the roughness of the stone, and (iv) the duration of processing. Based on the degree of pearling, the process produces dehulled or blocked barley, pot barley, and pearled barley. The blocking process eliminates the husk only, whereas pearling alters the composition of the grain by removing its outer fractions. Additionally, pearling improves the palatability and the overall acceptability of the grain. The dehulled barley is then passed through roller mills to fractionate. The important bioactive compounds like β-glucans, dietary fibre, vitamins, minerals, and phenolic compounds are mainly distributed in the coarsely milled fractions. The other methods of barley processing, such as cooking, hydrothermal processing, and extrusion, help to manufacture convenient foods with improved nutritional profile and shelf life (Izydorczyk et al., 2016).

2.3.1.5 Oats Milling

Oats are generally milled either by a traditional dry-shelling system or the modern green shelling system. The hull is removed from the oat, then steamed and roasted to get oat groat during processing. Generally, most of the oats are steamed and flattened to make old-fashioned or regular oats, quick oats, and instant oats. The old-fashioned oat is a whole grain cereal as it includes the bran, germ, and endosperm. Instant oats are prepared by cutting the whole groats into thirds, followed by steaming and rolling. The high roasting temperatures and thinness of the flake make them cook instantly while adding boiling water. Steel cut oats, also known as Irish oatmeal, are steamed and roasted to inactivate the enzymes. This helps to prevent rancidity and provides a toasted flavor to the final product. The whole groats are cut, but they are not rolled.

2.3.1.6 Rye Milling

Rye grain is mainly used to make flour, rye bread, alcoholic beverages such as rye beer, whisky, vodka, and animal feed. It contains a significant amount of non-cellulose polysaccharides, which have the high water-binding capacity and immediately give a satiety feeling. The processing of rye involves cleaning, tempering (14.5 to 15.5% moisture), break rolling, and sifting to get rye meal. When rye is milled, it yields rye flour with varying extraction rates. Rye flour is used to prepare soft bread and coarse flour for hard bread. Rye flakes are mainly used to make breakfast cereals. Most importantly, the fibre loss is lower during milling because the bran attaches tenaciously to the endosperm (Slavin et al., 2001).

2.3.1.7 Sorghum Milling

Sorghum is usually milled in stone mills to separates endosperm, germ, and bran. Little sorghum is milled in commercial mills. But very small sorghum uses traditional mortar and pestle methods for decortication and saddle stones for milling. Sorghum can also be ground by hammer milling or roller milling after the conditioning process (Abdelrahim & Mudawi, 2014). Dry milling of sorghum produces grits, whereas wet milling yields starch. The milling quality of sorghum depends on the processing method. The decortication process removes the bran from the grain and reduces the nutritional value of sorghum. Moreover, it enhances the protein quality and palatability of sorghum.

2.3.1.8 *Millet Milling*

The mechanical processing methods of millets include decortication, milling, and sieving. In addition, bioprocessing methods like germination, malting, and fermentation also improve the nutritional characteristics of millets. Traditional processing methods such as puffing or popping and advanced procedures like roller-drying and extrusion-cooking are used to make ready-to-eat millet products. The soaking and cooking process reduces the anti-nutritional factors present in them.

2.3.1.9 *Triticale Milling*

Triticale (*Triticosecale*) is a hybrid of wheat and rye. Triticale possesses the winter hardiness of rye and the baking properties of wheat. It is mainly used as a feed crop. The flour obtained after milling of triticale can be used to prepare bread.

2.3.2 Parboiling

Parboiling helps to improve the nutritional, organoleptic, textural, and cooking quality of rice. Parboiling includes soaking paddy in water for a definite period, followed by heating it once or twice in steam and drying before milling. Soaking and heating cause the starch present in the rice to gelatinize. Also, it aids the movement of water-soluble vitamins present in the bran to endosperm. As a result, it alters the physico-chemical and nutritional profile of rice. During the parboiling process, the paddy grains absorb water and cause the starch molecules to swell. Amylose fraction of the starch is an important factor that decides the gelatinization character, pasting properties, cooking quality, and glycemic index of rice. Single parboiled rice has been heated once, whereas double parboiled rice has been heated twice. Parboiling can be done in traditional, modern, or industrial ways, as shown in Figure 2.3.

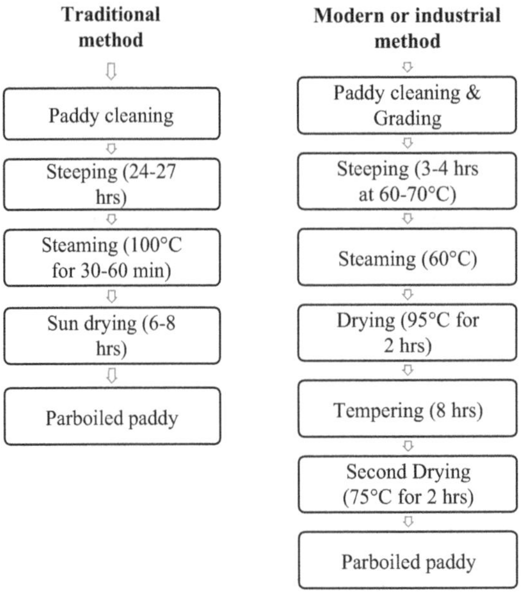

Figure 2.3 Parboiling methods of rice.

The advantages of parboiling are that dehusking becomes easier, grain gets tougher, and reduction in excessive breaking during milling occurs. Parboiling also results in higher head rice recovery, increased nutritional value, and improved resistance to insects and fungus infection. While cooking, parboiled rice does not become a sticky mass. The drawbacks of the parboiling process are the formation of an unpleasant odor and the occurrence of color change owing to fermentation.

2.3.3 Malting Process

Barley is the most popular grain used for malting because of its high starch-to-protein ratio and adherent husk. Other cereals, such as wheat, oats, rye, sorghum, millets, and triticale, are malted and used in brewing, distilling, and food manufacturing besides barley (MacLeod & Evans, 2016). The malting process involves steeping the barley in water for germination, followed by drying and curing in a kiln. The barley is steeped in water at 12°C for 36 hours with regular aeration to achieve a moisture level sufficient to activate hydrolytic enzymes. It is then allowed to germinate at 14°C for 4–6 days. The starch granules are degraded and exposed by the enzymes that move across the starchy endosperm during germination. The "green malt" obtained after germination is kilned at a temperature of not more than 85°C to reduce the moisture content (less than 5%). The kilning process also ceases the germination process and stabilizes the malt. Moreover, it develops color, aroma, and flavour in the malt. The changes that take place during malting are shown in Figure 2.4.

2.3.4 Extrusion

Extrusion is a high-temperature, a high-pressure cooking process that converts raw materials into various products. Extrusion technology is widely used in the manufacture of breakfast cereals, ready-to-eat snacks, confections, pasta, noodles, etc. The extruder consists of one (single-screw extruder) or two (twin-screw extruder) screws placed on a shaft that spins in a fixed barrel. The most important components of an extruder are the feeding system, preconditioner, barrel, and knife assembly. The feeding system is the basic part that uniformly delivers the ingredients into the next component, the preconditioner. The extruder barrel consists of screws, sleeves, barrel heads, and dies. The final component, knife assembly, functions by cutting the product into desired size and shape. Extruder performs unit operations such as conveying, compressing, mixing, kneading, forming, cooking, shearing, heating, puffing, cooling, and shaping the materials. The rheological properties of the dough during processing determine the quality of the fined product. The process is influenced by the nature of the raw material, type of extruder, and processing conditions. The extrusion processes may be cold, hot, steam-induced, and co-extrusion. Cold extrusion produces pasta, noodles, and other foods that combine and mold the ingredients without applying heat or cooking.

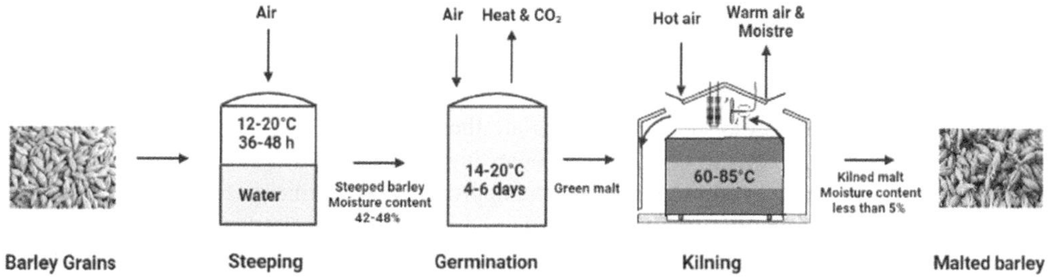

Figure 2.4 Changes take place during the malting process.

Hot extrusion produces textured food items by high-temperature heating of ingredients for a short period under pressure. Steam-induced expansion is the melt expansion resulting in highly expanded products due to water flashing off at the die exit. For example, expanded snacks and breakfast cereals are prepared by the steam-induced expansion method. Co-extrusion is a combined process of steam-induced expansion and filling injection, which is mainly used to produce expanded products with dual textures (products with crispy shells and soft filling). Wheat, rice, corn, sorghum, and oats are the most common cereal grains used to make extruded products. Whole grain flour, refined flour, or their fractions are used as raw material for extrusion. The chemical composition of the flour significantly influences the behaviour of the flour during extrusion processing. Refined flour is mainly used for extrusion because of their high starch, low protein and low fibre content. This gives better expansion and sensory attributes than whole grain flour. The texture is the critical quality parameter that must be considered in extrusion (Pichmony et al., 2020).

2.3.5 Fermentation

Fermentation is an age-old cereal processing method that improves the organoleptic properties, textural characteristics, and shelf life of cereal-based foods. It is defined as the acceptable biochemical modification of complex food constituents by microorganisms and their enzymes (Kohajdov & Karovi, 2007). Fermented cereals are used to make baked products like bread and alcoholic beverages. In bread making, carbon dioxide is produced by the action of yeast enzymes on starch fractions and result in increased loaf volume. Alcoholic fermentation of cereals produces ethyl alcohol and carbon dioxide. In addition to nutrients, cereals contain some anti-nutritional compounds. Mostly, these anti-nutritional compounds are bonded to nutrients and prevent them from being released into the food. The fermentation and germination processes break down these interactions, making nutrients available to digestive enzymes. Fermentation enhances the protein quality due to the synthesis of amino acids and improves protein digestibility.

2.3.6 Other Processing Methods

In addition to these methods, other cereal processing methods such as puffing, flaking, and shredding are also used to produce breakfast cereals and ready-to-eat snacks. During popping or puffing, the grains are subjected to high-temperature, short-time heating that causes the internal moisture to expand and pop out. The puffing process yields a light or crispy structure to food products. Puffing of cereals can be attained by using the oven, hot sand, hot oil frying, microwave heating, or a gun. Oven puffing works by applying a burst of heat at atmospheric pressure to the grain, which causes the water in the grain to vaporize and expand. Gun puffing transfers the superheated steam held by the cereal grain from high pressure to low pressure, thus enabling the water to vaporize and expand (Kent & Evers, 1994). Generally, wheat, rice, corn sorghum, and ragi are puffed. The popping quality of cereals depends on the variety, grain composition, physical properties, and the popping method. The amylose content of cereal affects the puffing of cereal especially, rice. Corn with a large kernel size results in less puffing due to soft endosperm.

Flaked cereals are made from whole cereal grains of wheat, rice, and maize. The grains are conditioned and tempered after cleaning to regulate the moisture content. After that, the material is passed through smooth rolls to break up the outer layer. The rollers are arranged in such a way that the gap between them is adjustable to get the appropriate flake thickness. The grains are then cooked under pressure and dried. Further, the material is flaked between heavy rollers and subsequently roasted. For making shredded products, whole grain wheat is generally used. The wheat is first cooked in water at atmospheric pressure to gelatinize the starch. It is then cooled to room temperature and allowed to rest for some time, known as conditioning. The conditioned grain is fed into shredders, consisting of two metal shredding rolls that rotate in opposite directions. The surface of one roll is smooth, but the other is grooved. During the process, the rolls are cooled with water to regulate the surface temperature. The material emerges from the roll as long parallel shreds, and that forms a thick mat. It is then dried or baked.

2.4 IMPACT OF PROCESSING TREATMENT ON NUTRITIONAL CHARACTERISTICS

Generally, whole grain cereal or whole grain flour possess a healthier profile than processed or refined cereal products. Cereals are subjected to various processing treatments to convert them into products. Carbohydrate constitutes a significant part of cereals. Processing treatments such as dehulling and milling increase the carbohydrate content of grains. The fibre content of cereals decreases during processing because it is difficult to grind and remove as coarse material during the screening process. Most of the anti-nutritional factors present in the cereal are reduced during fermentation or malting, which further improves the bioaccessibility of minerals. Dehulling is a mechanical process that removes the bran fractions from the grain. Dehulling has a major impact on the loss of bioactives from cereals because most of them are concentrated in the outer portion of the grain. Milling is another primary treatment method of cereals that rearranges minerals, enzymes, and anti-nutritional factors. Soaking helps to increase the nutritional value and reduce the cooking time of cereals. During soaking, cereals absorb water and activate the enzymes, which result in softening of the cell wall and leaching of water-soluble nutrients from the grains. The effect of soaking depends on the duration, temperature, and pH of water (Raes et al., 2014).

Among the various cereal processing techniques, milling is widely used to convert grains into different value-added products. It is a grinding process that converts the whole grain into flour. The milling process has a significant influence on various cereal constituents. The nutritional profile of the finished product depends on the extend to which the outer layers are removed by milling. Most of the significant food constituents such as vitamins, minerals, pigments, dietary fibre, antioxidants, and some enzyme inhibitors are present on the outer layers of the cereal grain. The milling produces refined products that may lead to decreases in nutrient content. It also reduces the anti-nutritional factors present in the cereals. However, reduction in phytic acid, tannin, and polyphenol improve the bioaccessibility of protein, carbohydrates, and minerals. Carbohydrates and protein are mainly found in the endosperm of the grain, and therefore, they are less impacted by the milling process. Pearling technique improves the overall acceptability of barley grains. The amount of protein, fat, tannin, and soluble fibre increases during the initial stages of the pearling process, but it declines after decortication. It also reduces the insoluble fibre content. The by-products of pearling are rich in bioactive compounds such as phytate, tocols, phenolic compounds, and insoluble dietary fibre and have antioxidant properties (Izydorczyk et al., 2016).

Parboiling is a unit operation given to paddy to improve its nutritional, organoleptic, and physico-chemical qualities. Parboiling causes the starch to gelatinize and thereby reduces the breakage of rice during milling. It also enhances the cooking quality of rice by reducing the stickiness of the cooked rice. Usually, parboiled rice takes a longer time to cook. The starch granules of parboiled rice hydrate slowly due to the strong cohesion between endosperm cells and hence longer cooking time. Cooked parboiled rice is much more intact and keeps their natural structure when compared to non-parboiled rice. The heat treatment during parboiling inactivates the enzyme lipase found in the bran layer of brown rice. This extends the storage life of parboiled brown rice by lowering the risk of oxidative rancidity. Moreover, the moisture content of rice experiences a drop after parboiling because of the drying process. The mineral concentration of rice is significantly affected by parboiling. It is increased gradually during the first stages of parboiling and later decreases as the temperature of parboiling increased. The crude fibre content remains constant and steady with an increase in temperature. When compared to non-parboiled rice, parboiled rice had a lower protein level. The protein loss is due to leaching during soaking and molecule rupturing during steaming. After parboiling, the protein sinks into the compact mass of gelatinized starch. This makes the protein less extractable resulting in a lower protein concentration. Because of the leaching and breaking of fat molecules during the steaming process, parboiled rice has a greater fat content than non-parboiled rice (Rhaghavendra Rao & Juliano, 1970). The gelatinization and partial dextrinization of starch enhance the carbohydrate content of parboiled rice. A change in vitamin content during steaming was also observed due to leaching and heating occurs at the time of parboiling (Chukwu & Oseh, 2009).

Traditional cooking methods use high temperatures for extended periods leading to the loss of heat-sensitive nutritional components. Extrusion is a high-temperature, short-time procedure that helps to retain most of the nutrients. While the digestibility of starch is affected by extrusion cooking, it still depends on the starch fragmentation, whether it is normal, waxy, or dextrinized (Robin et al., 2016). Extrusion causes amylose and amylopectin molecules in the starch to melt and gelatinize. This results in structural and functional changes of starch molecules (Guy, 2001). Retrogradation of starch and the formation of the starch–lipid complex during extrusion cooking result in resistant starch formation (Kim et al., 2006). Resistant starch plays a vital role in food processing industry as an anti-stailing agent and dough conditioner. It also aids in the slowing of glucose release, lowering the glycemic index and insulin responses. Additionally, extrusion influences the solubility and molecular weight of other non-starch polysaccharides (e.g., β-glucan) and affects starch digestibility. The metabolites produced by intestinal bacteria after fermenting non-digestible carbohydrates influences the regulation of hunger, insulin secretion, lipid metabolism, and inflammation. Extrusion cooking also improves insoluble fibre accessibility through gut microbiota fermentation (Kasubuchi et al., 2015). Moreover, extrusion enhances dietary fibre content by the production of resistant starch and the transfer of insoluble to soluble dietary fibre. Extrusion has minimal effect on protein quality, even though it alters essential amino acid bioavailability and digestibility. Denaturation of protein and inactivation of anti-nutritional substances are responsible for improved protein digestibility. Extrusion of wheat, rice, and maize flour result in a substantial reduction in lysine concentration (Pastor-Cavada et al., 2011). Thermal degradation of important amino acids such as threonine, valine, leucine, and isoleucine has also been observed during extrusion cooking (Arrage et al., 1992). The solubility of protein also decreases by extrusion because of aggregation. The modifications that occur during extrusion are dependent on the type of protein. According to Fapojuwo et al., extrusion cooking improved the protein digestibility of sorghum flour by disrupting disulfide bonds at high screw speeds (Fapojuwo et al., 1987). Extruded food contains very little lipids, but they interact with carbohydrates and protein during the extrusion process. The binding of fat with other constituents helps to preserve the nutritional stability of extruded foods. The stability of vitamins during extrusion is affected by processing variables such as time, temperature, screw speed, pressure, feed rate, vitamin source, and structure. In the case of vitamins, there is a destruction of vitamin C and B complex vitamins. High extrusion temperatures are also harmful to vitamin A and E (Gulati et al., 2017). Bioaccessibility of carotenoids, a precursor to vitamin A, also improves through extrusion. The loss of carotenoids during corn processing can be reduced by combining it with other carotenoid-rich cereals having antioxidant potential (Cueto et al., 2017). The presence of mineral binding substances such as phytic acid, phenolic compounds, dietary fibres, and proteins causes a change in mineral bioavailability during extrusion (Raes et al., 2014). According to Gulati et al., the bioaccessibility of iron was improved after the extrusion of maize and sorghum (Gulati et al., 2020).

Fermentation helps to convert complex food components to simpler or easily digestible forms and eliminate anti-nutritional factors by enzyme activity. It also releases health beneficial bioactive compounds having antioxidant properties. In cereals, starch is the most common carbohydrate, and it is decreased during the fermentation process. Enzymes such as amylase and maltase act on starch during fermentation and cleave them into simple sugars. Microorganisms may quickly digest these simple sugars, resulting in the production of carbon dioxide and ethyl alcohol. Moreover, fermentation improves protein digestibility by breaking down complex proteins partially. Due to starch breakdown, the protein content of pearl millet showed an increase after 24 hours of fermentation (Osman, 2011). The efficiency of fermentation can be determined by the activity of the enzyme phytase, which is directly connected to phytic acid breakdown. Cereals like rice, corn, oats, and millets contain lower enzyme phytase concentrations; hence, it takes a longer fermentation period to destroy phytic acid. Minerals are commonly complexed with undigested dietary components rendering them inaccessible. By degrading phytates and oxalates, fermentation increases the accessibility of magnesium, iron, calcium, phosphorus, and zinc in fermented foods. The tannin content of grains is also reduced by fermentation. Microbial fermentation liberates bound phytochemicals by degrading the grain matrix and enhances cereal antioxidant activity (Nkhata et al., 2018). Fermentation has a dual influence on the glycemic index. The low glycemic index of fermented food is due to short-chain organic acids such as lactic, acetic, and propionic acid (Ostman et al., 2005). Lactic acid inhibits starch hydrolysis in the upper small intestine, whereas propionic and acetic acids reduce enzymatic activity. The rise in glycemic index is related to digestion and absorption of glucose produced by fibre and starch breakdown. The duration of fermentation also influences the rise in post-prandial glycemic response. During microbial fermentation, some of the non-digestible carbohydrates undergo degradation and that reduces abdominal distention and flatulence. The free amino acid level in cereals is increased by proteolysis and metabolic synthesis, especially essential amino acids such as lysine, methionine, and tryptophan. Microbial fermentation also causes the breakdown of starch and enhances starch digestibility. Further, it is fragmented into maltose and glucose.

The malting or germination process improves the nutritional profile of cereals by activating endogenous enzymes. The action of hydrolytic and amylolytic enzymes improves carbohydrate digestibility, particularly starch digestibility. The duration of processing has a significant impact on the extent of enzymatic activity. The reducing sugar content of cereals increased after 12 hours of germination due to enzymatic hydrolysis of starch by α-amylase (Zhang et al., 2015). Germination also increases the fibre content and delays glucose release. Fibre becomes gel in the stomach and provides a satiety feeling. Additionally, intestinal bacteria digest fibre and produce short-chain fatty acids, including butyric, acetic, and propionic acids, which control satiety (McNabney & Henagan, 2017). Butyric and acetic acid suppress genes involved in fatty acid oxidation (Canfora et al., 2015), which is crucial in the early stages of losing weight in obese people. The protein content of cereals varies throughout the germination process because of protein synthesis and breaking down. The increase in protein depends on the type of grain and might be related to amino acid synthesis. The action of enzyme proteases causes a decrease in total protein content, but protein digestibility and biological value are improved by germination. Anti-nutritional factors present in the cereals have a significant impact on mineral bioavailability. Most minerals linked to anti-nutritional factors like tannins and phytic acid are liberated during the malting process. For example, buckwheat reported a decrease in phytic acid level due to phytase hydrolysis when the duration of germination was increased (Zhang et al., 2015). Malting of sorghum and foxtail millet improves the mineral content, especially sodium, potassium, phosphorus, calcium, and magnesium levels. In addition, germination boosts the amount of vitamins in cereals, particularly B vitamins and vitamin E, but a loss of water-soluble vitamins occurs. Malting improves grains nutritional and antioxidant qualities by increasing the ascorbic acid levels (e.g., wheat and ragi). The enzymatic breakdown of starch by amylases and diastases increases the accessibility of glucose. This glucose acts as a precursor for the synthesis of vitamin C. During the steeping stage of the malting process, the fat content of grain increases significantly. However, due to hydrolysis and lipids as an energy source for biochemical processes during germination, it decreases later in the germination phase. Phytochemicals also contribute the enhanced antioxidant activity of germinated grains. The presence of phytochemicals influences the bioavailability of nutrients as well. Extended soaking and fermentation process decrease phytochemical level by leaching. The activity of microorganisms during the fermentation process reduces tannin content. Reduction in tannin and phytic acid concentration in malted grains enhance the bioavailability of minerals.

Along with these major processing treatments, other techniques like popping and flaking also affect the nutritional profile of cereals. Factors such as the quality of the cereal and grain characteristics influence the popping process. Popping improves starch digestibility due to the gelatinization of starch and the breakdown of dietary fibre. Flaking reduces the amount of phytic acid, phosphorus, and dietary fibre while increasing starch digestibility and mineral availability.

2.5 IMPACT OF PROCESSING TREATMENT ON FUNCTIONAL CHARACTERISTICS

In addition to the changes in nutritional characteristics, these processing techniques also affect the functional properties of cereals. Starch is the principal constituent that changes processing. Milling causes the breakdown of starch by the enzyme α-amylase and denaturation of protein by high grinding temperatures. Variations mainly influence the functional characteristics of flour in starch structure. The primary functional qualities such as pasting property, swelling characteristics, solubility, and digestibility are affected during milling. The effect of milling on starch structures varies depending on the source and milling conditions, such as temperature and mechanical force. Non-starch components of the grain, including cell wall matrix, fat, and protein, can protect starch molecules during milling. Changes in starch's molecular, crystalline, and granular structure are caused by the degradation of starch molecules, disruption of double-helical crystallites, and damage to starch granules. The quantity of damaged starch in the flour significantly affects the product quality, particularly bread loaf volume and noodle texture. The degree of starch damage increases as the mechanical force and grinding duration increase. Cereals with high protein content are tougher to mill and cause more damage to starch granules than softer grains. Highly damaged starch molecules have a rougher, more porous surface than undamaged starch molecules. The aggregation of starch granules causes heavily damaged starch to grow. Prolonged grinding also destroys the crystalline structure of starch granules (Li et al., 2014). In addition, high grinding temperature causes the gluten protein in wheat to denaturize, which reduces the water-absorbing capability of flour. Cryogenic milling is a better alternative for conventional milling that eliminates heat generation during milling and grinding processes. Cryogenic milling also reduces the degree of starch degradation and gelatinization.

The quality of the parboiled rice depends on the processing parameters such as soaking time and temperature, steaming time and pressure. The hardness of the kernel can be reduced by extending the soaking period. The greater the steaming time, the greater the hardness of cooked rice. Whereas, longer steaming time decreases the chalkiness and pasting viscosities of cooked parboiled rice. The parboiling technique improves the head rice recovery by lowering sensitivity to breakage during milling. It also reduces the overall pasting profile, swelling, and leaching of amylose. The degree of starch gelatinization is also influenced by the parboiling conditions (Leethanapanich et al., 2016). The gelatinization of starch and the disruption of protein causes the rice kernel to become transparent and glassy during the parboiling process. The swelled starch fills the void spaces in the endosperm and makes the kernel transparent. The parboiling process increases the cooking time of the rice due to strong cohesion between the tightly packed endosperm cells. This causes the starch granules to hydrate slowly. It also reduces the water penetration into the rice kernel and hence results in longer cooking time. Additionally, the cooking quality of the rice is also affected by protein content. The higher the protein content, the higher the gelatinization temperature and cooking time. The hydrophobic property of proteins acts as a barrier to water diffusion into the cooked grain and causes the gelatinization temperature to rise. The water absorption ability of parboiled rice is higher than that of non-parboiled rice. This is due to steaming pressure during parboiling, which further affects the gelatinization of starch (Chukwu & Oseh, 2009).

Malted barley is one of the major ingredients in beer. The beer-making process includes malting of barley, mashing, boiling, and fermentation. Malting allows the seed to germinate and then be kilned to inactivate the enzymes. Moreover, malting decreases the starch level of the kernel while increasing the amylose content. The changes in structure and the alterations of starch during the malting process depend on the grain variety and processing conditions (Wenwen et al., 2019). Mashing also results in the hydrolysis of starch and produces simple sugars. Morever, the barley proteins undergo a sequence of modifications during the malting and brewing processes. The changes include glycation through Maillard reactions during malting, acylation during mashing, and structural unfolding during brewing (Jin et al., 2009). Barley having a protein content of 8–12% is used for brewing. High protein content affects the fermentation process, whereas low protein influences starch degradation rate. Protein also reduces the activity of enzymes by binding with them or by entrapping starch molecules in the protein matrix.

Extrusion cooking is widely used for the manufacture of convenient or ready-to-eat cereals, breakfast cereals, and snacks. It causes alteration in complex food matrices and modifies the functional properties, especially texture. Mainly, amylose/amylopectin ratio is the factor that critically affects the development of textural attributes in products. Amylose is a linear chain molecule and is required for film formation in direct-expanded extruded products. However, high amylose containing products result in softer bites and less crunchiness during processing. Along with amylose, amylopectin is also required to build viscosity and hold expansion during extrusion cooking. Amylopectin is a branched-chain polysaccharide that causes low gel strength due to weaker hydrogen bonding. Due to the high molecular weight of the amylopectin molecule, it builds very high viscosity. In addition, amylopectin is necessary to develop structures that can entrain the micro-steam bubbles produced during direct expansion extrusion processing. An excess level of amylopectin can create a viscosity that causes the starch molecule to be retrograde quickly. Moreover, sugars, salts, and nucleating agents contribute flavor or textural characteristics to the product. The incorporation of sugar can enhance or decrease the overall expansion of the product, cell structure and tenderness of flours. A lower sugar content improves the crispiness of the product. At the same time, higher levels cause the sugar to melt and cause a decline in viscosity and affect the cooking quality of the product. Finally, it reduces the overall expansion of the extrudate (Pichmony et al., 2020). In addition, sugar causes a reduction in the glass transition temperature of the flour, which may result in a broad temperature range during the expansion of flour. Generally, the extrudate undergoes expansion followed by shrinkage during the expansion process. Because of this broad temperature range, the extrudate results in higher shrinkage and lowers final expansion (Fan et al., 1996). This effect may vary with sugar type and flour. Commercial extruded products use chemical nucleating agents (e.g., calcium carbonate, sodium bicarbonate, etc.) to attain desired texture. These ingredients help to increase the expansion of the extrudate and form a finer foam structure.

The popping process improves the texture and sensory characteristics of cereals without affecting the nutritional content. Popping is attained by the gelatinization of starch and the expansion of the grain when exposed to high temperatures for a short duration. During popping, the superheated vapour generated inside the grains cooks them and swells the endosperm quickly, causing the outer skin to burst off. Puffing is almost like popping, but controlled kernel expansion occurs as the vapour pressure escapes through the micropores of the grain structure. This is because of high pressure or temperature gradient. Popping and puffing impart better flavor to the products. The puffing process is affected by the shape and composition of the kernel as well as processing temperature and pressure, whereas expansion volume depends on size, shape, and density of grain. The physicochemical characteristics of flaked cereal products reveal significant starch breakdown and protein-starch interaction during the flaking process. The roller flaking increases protein–starch interaction. The flaked product has a high-water absorption capability because it is manufactured from heat parboiled rice. The flaking process increases the starch damage and surface area of the product by flattening. As a result, the properly thinned flakes become cooked even in cold water.

2.6 CEREAL BIOACTIVES AND THEIR HEALTH BENEFITS

Whole grain cereals and whole grain cereal products are made up of a diverse mix of bioactive ingredients. They are suggested as part of a healthy lifestyle because they contain dietary fibre, antioxidants, and fewer calories than refined grains. These bioactive chemicals are found in the bran or germ component of the grain. Moreover, they are highly effective at lowering the risk of various types of chronic diseases. Some of the beneficial properties of these substances include antioxidant activity, immune system improvement, and toxin removal from the body.

One of the most significant bioactive principles in cereals is phenolic compounds, primarily phenolic acids and flavonoids. They have antioxidant properties and can help to prevent heart disease and cancer. Wheat bran contains ferulic acid, another polyphenol with antioxidant properties. Avenanthramides are polyphenols found only in oats that have anti-inflammatory, anti-atherogenic, and antioxidant effects. Wheat, corn, oats, and rye contain a bioactive compound called lignans. Alkyleresorsinols are phenolic lipids produced from plants that may be found in whole grain cereals like rye and wheat. They have anti-bacterial and anti-fungal and antioxidant properties. Generally, cereals contain lower amounts of flavonoids. Flavonoids are comprised of anthocynins, flavonols, flavones, flavanones, and flavonols. They are mostly present in the pericarp of all cereals. Sorghum contains a wide variety of flavonoids. Epigenin is a flavone present in oats, barley, millet, oat, and sorghum. Barley contains a significant amount of catechin. They have antioxidant, anti-cancer, anti-allergic, anti-inflammatory, anti-carcinogenic, and gastroprotective effects.

Carotenoids are one of the most abundant pigments in some cereal grains. The role of carotenoids as pro-vitamins and antioxidants is important for the normal maintenance of human skin and eyes. The commonly found carotenoids in cereals are lutein, zeaxanthin, beta-cryptoxanthin, α-carotene, and β-carotene. Cereals such as wheat, rice, and corn contain a significant amount of lutein. In the case of rice, colour of the whole grain flour is contributed by lutein and zeaxanthin, which is present in the bran. Phytic acid is a bioactive compound present in the bran fraction of whole grain cereals, particularly within the aleurone layer. Phytic acid shows a negative health effect because it complexes with minerals and makes them unavailable. Plant sterols and stanols are bioactives collectively known as phytosterols with the cholesterol-lowering property. Additionally, cereals such as wheat, rice, barley, oats, and rye contain tocopherols and tocotrienol, which are naturally occurring antioxidants. Rice bran oil contains γ-oryzanol, which reduces cholesterol absorption and platelet aggregation and increases muscle mass. Oats and barley contain a significant amount of soluble fibre, β-glucan polysaccharides. They have cholesterol-lowering properties and control blood sugar (Gani et al., 2012).

Cereals undergo several biochemical changes during long-term storage. The quality of cereals during long-term storage can be maintained by keeping them under optimum storage conditions. Temperature, relative humidity, and moisture content are the major factors determining the quality of the cereals during storage. The quality of the milled grain products is influenced by harvesting and drying methods, storage conditions, and overall handling of grains before milling. The total phenolic content of grains is affected by storage conditions. During storage, ferulic acid experiences damage because of oxidation processes. Cereals contain minerals, vitamins, fibre, and phytochemicals, all of which contribute to their health-promoting qualities. The level of cereal bioactives significantly reduced throughout the refining or milling process. As a result, milling by-products are high in phenolic compounds and may be employed as a natural antioxidant and functional component in value-added food products.

Extrusion, baking, and roasting are high-temperature processing methods that alter the physical and chemical characteristics of grains. Starch gelatinization, protein denaturation, component interactions, and browning reactions are among the most significant changes. Some of the properties such as organoleptic quality, nutritional availability, and antioxidant properties of cereals are improved because of these modifications. During high-temperature processing, heat-sensitive substances such as anti-nutritional factors, toxins, and enzyme inhibitors are inactivated. Thermal processing also aids in the liberation of phenolic acids by the breakage of cellular components. Browning reactions increase the total phenolic content and antioxidant property due to the dissociation of conjugated phenolic moiety followed by polymerization or oxidation and the formation of phenolics. Total phenol concentration is also increased through non-enzymatic browning processes such as the Maillard reaction, caramelization, and chemical oxidation of phenols. The release of phenolic compounds during thermal processing is affected by moisture content, duration, and temperature.

Whole grain cereal-based bakery products are an excellent source of phenolic compounds. Polyphenol concertation increases during baking owing to the Maillard reaction. High-temperature short time extrusion is a hydrothermal process that releases phenolic acids and their derivatives from the cell walls. In cereals, extrusion enhances the bioavailability and concentration of other nutrients. It also increases the stability of food due to the inhibition of endogenous enzymes. During the baking process, Maillard reaction occurs during extrusion as well, resulting in increased antioxidant capacity and phenolic concentration. The phenolic content of grains, including wheat, barley, rye, and oat become altered after extrusion (Zielinski et al., 2001). The greatest amount of free and bound phenolic acids was present in extruded rye and oat. Ferulic acid was found to be the most prevalent component in both raw wholegrain and extruded grain. In addition to these processes, traditional roasting procedures assist in improving the nutritional content, flavour, and shelf life of cereal products. After roasting, barley showed a substantial increase in antioxidant property and total phenols concentration, but sorghum had a considerable decrease. The loss of phenolic chemicals in their free state causes a reduction in antioxidant properties (Sanaa Ragaee, 2014).

Fermentation processes alter the grain matrices by the action of endogenous and microbial enzymes by changing the structure, bioactivity, and nutrient bioavailability. It also helps to improve the flavour and nutritive value of cereals. Moreover, fermentation offers optimum conditions for enzymatic degradation of phytate. The fermentation process reduces anti-nutritional factors such as phytates, tannins, and polyphenols. Soluble minerals such as iron, zinc, calcium, etc., are increased due to decline in phytate content. Lactic acid fermentation reduces the tannin content but increases the of iron. Generally, during fermentation, the cereal grains are soaked in water for a few days. Soaking results in softening of the grain, which makes them easy to grind. The pH of the ground material is reduced by lactic acid and other organic acids generated during fermentation owing to microbial activity. Fermentation also produces some volatile compounds that are responsible for the distinctive flavour of fermented cereal products. This flavour is contributed by diacetyl, acetic acid, butyric acid, etc. In addition, the proteolytic activity of microorganisms and malt enzymes form precursors of flavour compounds such as amino acids (Kohajdov & Karovi, 2007). In the future, more research will be needed to identify and characterize these bioactive compounds and investigate their bioavailability, metabolism, and health benefits in humans.

2.7 IMPACT OF NOVEL PROCESSING METHODS ON CEREALS

Cereal products with improved nutritional, functional, and textural characteristics can develop by implementing novel technologies or combining them with conventional methods. Novel processing techniques produce products having specific functionalities beyond the basic nutritional properties. Innovative post-harvest treatments such as irradiation, ozone technology, and cold plasma can be used instead of disinfection, fumigation, and decontamination of grains. In addition, ozone technology influences the enzyme activity, textural and rheological properties of the product (Tiwari & Pojic, 2021). The conventional milling process results in the loss of nutrients during roller milling. This can be minimized by using some novel milling techniques that use hammer, pin, and turbo mill for fine flour while burr and blade mill to get coarse flour. Jet milling is another high-air-pressure milling used to get ultrafine flour (Lee et al., 2019). But this ultrafine particle size affects the damage starch content and rheological properties of the dough. The dough becomes hard, sticky, and changes relaxation time and rate (Lazaridou et al., 2018). Micronization is also used for ultrafine grinding, which increases the accessibility of the bioactive compounds by processing the whole cereal without any by-products. Novel fractionation methods such as air classification and electrostatic separation can be used to separate flour into various fractions based on size and density.

Innovative primary processing methods such as irradiation, ozone technology, cold plasma, ultrasound, high-pressure processing, and microwave technology alter the techno-functional properties of cereal and its products. Irradiation uses gamma radiations to modify the nutritional, rheological, and textural properties of cereals by affecting starch, protein, and other biological constituents. Irradiation increases the viscosity of the starch by breaking the hydrogen bonds within starch molecules. The degree of starch damage mainly depends on irradiation dose. The changes in physico-chemical properties during irradiation include decreases in pasting properties and an increase in freeze-thaw stability, water solubility, and water absorption capacity (Bashir et al., 2017). Ozonization technology controls the enzymatic activity especially, amylase, protease, lipase, and lipoxygenase activity. This also minimizes the use of a chemical oxidizing agent and change the rheological properties of the dough. Due to ozone treatment, a reduction in pasting temperature, retrogradation, and increase in starch gelatinization was reported in wheat (Çatal & Ibanoğlu, 2014). Cold plasma treatment affects the techno-functional properties of cereals and cereal products by altering the secondary structure of proteins. A study reported that when plasma was applied during the germination process, it reduced germination duration while improving the bioactive compounds (Yodpitak et al., 2019). In the ultrasound method, sound waves with a frequency of 20 kHz are used to induce the cavitation phenomenon. The ultrasound method can cause modification of starch granules in cereals. Earlier studies show that this treatment increased starch's solubility and swelling capacity and increased the level of rapidly digestible and resistant starch (Kaur & Gill, 2019).

High pressure processing (HPP) uses pressure from 400 to 900 MPa, and is mainly used in the cereal processing industry as a preservation technique. Recent investigations on HPP revealed its capabilities to alter the functional properties of starch and protein. It can also modify the physical characteristics and gelatinization properties of starch. In addition, the ability of high-pressure processing to enhance structure formation in gluten-free products has also been explored. The results show that high-pressure processing helps to improve the functional properties of gluten-free bread (Cappa et al., 2016). Moreover, a high hydrostatic pressure process can protect the heat-sensitive constituents and maintain the nutritional and sensory quality of millet-based products (Saleh et al., 2013). The application of microwave technology in cereal processing is a progressing research area. The effect of microwave treatment on wet wheat kernels reported a reduction in gluten level up to 20 ppm (Gianfrani et al., 2017).

Fermentation biotechnology is an innovative secondary processing method of cereals. Generally, the fermentation of cereal by-products increases the bioavailability of minerals and vitamins and improves protein digestibility, while fermentation in combination with the enzymatic treatment results in an extensive breakdown of the cell wall. Preliminary studies reported that the fermentation of wheat with sourdough lactobacilli and fungal proteases could cause complete hydrolysis of gluten. Therefore, this flour can mix with other gluten-free flour to make bread with improved texture (Gobbetti et al., 2019). Recently, enzyme-assisted cereals processing methods captured great attention because they improve the properties of cereal-based foods. For example, instead of mechanical and chemical polishing techniques, exogenous enzymes can polish whole grain cereals without affecting the nutritional value (Singh et al., 2015). Advances in extrusion cooking include adding ingredients into food preparations that can enhance products' nutritional value and overall acceptability. These ingredients may consist of whole grain cereals and high fibre or high protein products (Oliveira et al., 2018). The introduction of barley β-glucan into extruded snack foods improved their total fibre content (Brennan et al., 2013). In addition to conventional extrusion, extrusion-based food printing or 3D printing can be used to design cereal products with desired size, shape, and nutritive value (Severini et al., 2016).

2.8 METHODS TO ENHANCE THE NUTRITIONAL POTENTIAL OF CEREALS

2.8.1 Fortification

Cereals can be used to carry micronutrients such as vitamins and minerals to alleviate micronutrients deficiency as a staple diet. Cereals are good sources of vitamins, especially vitamin A, B (B_1, B_2, B_3, B_5, B_6, B_9) K, and E, but they lack vitamin C, D, and B_{12}. The processing and cooking methods significantly alter the vitamin content of cereals. Therefore, fortification of cereals with essential nutrients or developing vitamin-enriched crops through biofortification is a better solution to restore or maintain the micronutrient profile of cereals. Fortification can be achieved either by direct fortification with vitamins from synthetic sources or by microorganisms such as bacteria and yeast to deliver vitamins (Garg et al., 2021). Fortification of pita bread with whole and pearled barley showed an increase in vitamin E content and antioxidant activity (Garg et al., 2021). The addition of *Lactobacillus plantarum,* isolated from Ogi, to weaning food blends reported an enhancement in the nutritional quality by increasing B vitamins, particularly thiamine, riboflavin, and niacin (Adeyemo & Onilude, 2018). The fortification of bread flour with vitamin B_{12} and folic acid increased vitamin B_{12} and folate in the finished product (Winkels et al., 2008). Iron fortification of wheat, maize, and rice has also been reported to prevent anaemia due to iron deficiency (Uauy et al., 2002).

2.8.2 Nutritionally Enriched Cereals

The introduction of nutritionally enriched cereals becomes a better solution to cater to the population's nutritional needs. In addition to fortification and nutritional enhancement programmes, nutritionally enriched cereals can play an important role in preventing micronutrient malnutrition. To make the cereals nutritionally enhanced, it is necessary to increase their nutritional value and reduce their anti-nutritional factors because the interaction of nutrients with anti-nutritional factors reduces the bioavailability of micronutrients. The nutritional value of the cereals can be improved either by breeding or cultural methods. The plant breeding techniques include the use of species having better quality characteristics to get the desired traits. Cereals with high nutritional profile and disease resistance can also be developed by genetic modification. The development of vitamin E enriched corn and β-carotene enriched rice are examples of genetic modification (Khush, 2021).

Culturing methods such as applying advanced fertilization techniques enhance food quality by changing the mineral content of the plant. Organic farming assists in producing nutritionally enhanced grains without the use of artificial fertilizers and pesticides (Sarwar, 2013).

2.9 ROLE OF CEREALS IN HEALTH

The regular consumption of whole grain cereals improves the nutritional profile by increasing the intake of vitamins, minerals, and fibre while decreasing the intake of total or saturated fat. Furthermore, cereals offer a sensation of fullness due to their low energy density and high fibre content. This aids in the maintenance of a healthy energy balance. The glycemic index (GI) is used to measure the rate of digestion of carbohydrates and their effect on the blood glucose level. A low glycemic index diet lowers LDL cholesterol, which protects against cardiovascular disease. In overweight, middle-aged males, a reduction in postprandial plasma glucose and insulin levels was observed after ingesting a fibre-rich diet (whole grain wheat and rye). (McIntosh et al., 2003). The soluble fibre present in the whole grain improves heart health by reducing cholesterol. Furthermore, the consumption of high-fibre whole grain cereals lowers the incidence of type 2 diabetes, but ingesting refined grain products has the reverse effect (Liu, 2003). The insoluble fibres improve digestive or gut health by absorbing fluid from the gut, promoting the growth of gut bacteria. On the other hand, some people cannot digest the protein, gluten present in wheat, rye, barley, oats, etc. This condition is called gluten intolerance or gluten sensitivity, which may result in celiac disease. Celiac disease is characterized by inflammation of villi in the small intestine that causes malabsorption of nutrients.

2.10 CONCLUSION

Cereals are subjected to various treatments during processing, which may change their physcochemical, nutritional, functional, and organoleptic properties. Whole grain cereals and cereal products have an excellent nutritional profile because most of the bioactive compounds are concentrated on the outer fractions of the cereal grain. Refining cereals through mechanical processing methods like milling, pearling, etc., modifies their nutritional composition and functional characteristics, whereas fermentation, germination, malting, and parboiling process enhance the nutritional composition of cereals by degradation of anti-nutritional factors. Interestingly, the demand for whole grain cereals and their products increase due to their health-promoting properties. Therefore, it is necessary to process them with techniques that have minimal impact on food constituents. Modern or emerging cereal processing techniques such as cryogenic grinding, high-pressure processing, enzyme modifications, modified extrusion techniques, etc., help to develop products with improved nutritional and functional properties and reduced allergenicity.

REFERENCES

Abdelrahim S. M. K and Mudawi, H. (2014). Some Sorghum Milling Techniques versus Flour Quality. *Egyptian Academic Journal of Biological Sciences. C, Physiology and Molecular Biology*, *6*(2), 115–124. https://doi.org/10.21608/eajbsc.2014.16038

Adeyemo, S. M., & Onilude, A. A. (2018). Weaning food fortification and improvement of fermented cereal and legume by metabolic activities of probiotics Lactobacillus plantarum. *African Journal of Food Science*, *12*(October), 254–262. https://doi.org/10.5897/AJFS2017.1586

Arrage, J.M., Barbeau, W.E., Johnson, J. M. (1992). Protein quality of whole wheat as affected by drum-drying and single-screw extrusion. *Journal of Agricultural and Food Chemistry*, *40*, 1943–1947.

Bashir, K., Swer, T.L., Prakash, K.S., and Aggarwal, M. (2017). Physico-chemical and functional properties of gamma irradiated whole wheat flour and starch. *LWT - Food Science and Technology*, *76*, 131–139. https://doi.org/10.1016/j.lwt.2016.10.050

Biel, W., Kazimierska, K., & Bashutska, U. (2020). Nutritional Value of Wheat, Triticale, Barley and Oat Grains. *Acta Scientiarum Polonorum Zootechnica*, *19*(2), 19–28. https://doi.org/10.21005/asp.2020.19.2.03

Brennan, M. A., Derbyshire, E., Tiwari, B. K., andBrennan, C. S. (2013). Integration of β-glucan fibre rich fractions from barley and mushrooms to form healthy extruded snacks. *Plant Foods for Human Nutrition*, *68*, 78–82. https://doi.org/10.1007/s11130-012-0330-0

Canfora, E. E., Jocken, W. J., & Blaak, E. E. (2015). Short chain fatty acids in control of body weight and insulin sensitivity. *Nature Reviews Endocrinology*, *11*, 577–591. https://doi.org/10.1038/nrendo.2015.128

Cappa, C., Barbosa-Cánovas, G.V., Lucisano, M., and Mariotti, M. (2016). Effect of high pressure processing on the baking aptitude of corn starch and rice flour. *LWT - Food Science and Technology*, *73*, 20–27. https://doi.org/10.1016/j.lwt.2016.05.028

Carmen Gianfrani, Gianfranco Mamone, Barbara la Gatta, Alessandra Camarca, Luigia Di Stasio, Francesco Maurano, Stefania Picascia, Vito Capozzi, Giuseppe Perna, Gianluca Picariello, A. D. L. (2017). Microwave-based treatments of wheat kernels do not abolish gluten epitopes implicated in celiac disease. *Food and Chemical Toxicology*, *101*, 105–113. https://doi.org/10.1016/j.fct.2017.01.010

Çatal, H. and Ib̈anoğlu, S. (2014). Effect of aqueous ozonation on the pasting, flow and gelatinization properties of wheat starch. *LWT - Food Science and Technology*, *59*(1), 577–582. https://doi.org/10.1016/j.lwt.2014.04.025

Chukwu, O., & Oseh, F. J. (2009). Response of nutritional contents of rice (Oryza sativa) to parboiling temperatures. *American-Eurasian Journal of Sustainable Agriculture*, *3*(3), 381–387.

Cueto, M., Farroni, A., Schoenlechner, R., Schleining, G., & Buera, P. (2017). Carotenoid and color changes in traditionally flaked and extruded products. *Food Chemistry*, *229*, 640–645. https://doi.org/10.1016/j.foodchem.2017.02.138

Devi, P. B., Vijayabharathi, R., Sathyabama, S., Malleshi, N. G., & Priyadarisini, V. B. (2014). Health benefits of finger millet (Eleusine coracana L.) polyphenols and dietary fiber: A review. *Journal of Food Science and Technology*, *51*(6), 1021–1040. https://doi.org/10.1007/s13197-011-0584-9

Fan, J., Mitchell, J.R., Blanshard, J. M. V. (1996). The effect of sugars on the extrusion of maize grits: I. The role of the glass transition in determining product density and shape. *Int. J. Food Sci. Technol*, *31*(1), 55–65. https://doi.org/10.1111/j.1365-2621.1996.22-317.x

Fapojuwo, O.O., Maga, J.A., Jansen, G.R., 1987. (1987). Effect of extrusion cooking on in vitro protein digestibility of sorghum. *Journal of Food Science*, *52*, 218–219. https://doi.org/10.1111/j.1365-2621.1987.tb14010.x

Gani, A., S M, W., F A, M., & Hameed, G. (2012). Whole-Grain Cereal Bioactive Compounds and Their Health Benefits: A Review. *Journal of Food Processing & Technology Gani*, *3*(3). https://doi.org/10.4172/2157-7110.1000146

Garg, M., Sharma, A., Vats, S., Tiwari, V., Kumari, A., Mishra, V., & Krishania, M. (2021). Vitamins in Cereals: A Critical Review of Content, Health Effects, Processing Losses, Bioaccessibility, Fortification, and Biofortification Strategies for Their Improvement. *Frontiers in Nutrition | Www.Frontiersin.Org*, *1*, 586815. https://doi.org/10.3389/fnut.2021.586815

Gobbetti, M., De Angelis, M., Di Cagno, R. et al. (2019). Novel insights on the functional/nutritional features of the sourdough fermentation. *International Journal of Food Microbiology*, *302*, 103–113. https://doi.org/10.1016/j.ijfoodmicro.2018.05.018

Gulati, P., Brahma, S., & Rose, D. J. (2020). Impacts of extrusion processing on nutritional components in cereals and legumes: Carbohydrates, proteins, lipids, vitamins, and minerals. In *Extrusion Cooking*. Elsevier Inc. https://doi.org/10.1016/B978-0-12-815360-4.00013-4

Gulati, P., Li, A., Holding, D., Santra, D., Zhang, Y., & Rose, D. J. (2017). Heating Reduces Proso Millet Protein Digestibility via Formation of Hydrophobic Aggregates. *Journal of Agricultural and Food Chemistry*, *65*(9), 1952–1959. https://doi.org/10.1021/acs.jafc.6b05574

Guy, R. (2001). Raw materials for extrusion cooking. In R. Guy (Ed.), *Extrusion Cooking: Technologies and Applications*. CRC Press, Boca Raton.

Gwirtz, J. A., & Garcia-casal, M. N. (2014). Processing maize flour and corn meal food products. *ANNALS OF THE NEW YORK ACADEMY OF SCIENCES*. https://doi.org/10.1111/nyas.12299

Ibrahima, O., W. Dhifi, A. R. and B. M. (2004). Study of the variability of lipids in some millet cultivars from Tunisia and Mauritania. *Rivista Italiana Sostanze Grasse*, *81*, 112–116.

Izydorczyk, M. S., Dexter, J. E., & Commission, C. G. (2016). Barley: Milling and Processing. In *Encyclopedia of Food Grains* (2nd ed.). Elsevier Ltd. https://doi.org/10.1016/B978-0-12-394437-5.00154-6

Jin, B., Li, L., Liu, G.-Q., Li, B., Zhu, Y.-K., & Liao, L.-N. (2009). molecules Structural Changes of Malt Proteins During Boiling. *Molecules*, *14*, 1081–1097. https://doi.org/10.3390/molecules14031081

Kanojia, V., Kushwaha, N. L., Reshi, M., & Rouf, A. (2018). Products and by-products of wheat milling process. *International Journal of Chemical Studies*, *6*(4), 990–993.

Kasubuchi, M., Hasegawa, S., Hiramatsu, T., Ichimura, A., Kimura, I. (2015). Dietary gut microbial metabolites, short-chain fatty acids, and host metabolic regulation. *Nutrients*, *7*, 2839–2849. https://doi.org/10.3390/nu7042839

Kaur, H. and Gill, B. S. (2019). Effect of high-intensity ultrasound treatment on nutritional, rheological and structural properties of starches obtained from different cereals. *International Journal of Biological Macromolecules*, *126*, 367–375.

Kent N.L., & A.D., E. (1994). Breakfast cereals and other products of extrusion cooking. In A. D. Kent, N.L., Evers (Ed.), *Technology of Cereals: An Introduction for Students of Food Science and Agriculture*. (pp. 244–256). Pergamon Press, New York, USA.

Khush, G. S. (2021). Plenary Lecture Challenges for meeting the global food and nutrient needs in the new millennium. *Proceedings of the Nutrition Society*, *60*, 15–26. https://doi.org/10.1079/PNS200075

Kim, J.H., Tanhehco, E.J., Ng, P. K. W. (2006). Effect of extrusion conditions on resistant starch formation from pastry wheat flour. *Food Chemistry*, *99*, 718–723. https://doi.org/10.1016/j.foodchem.2005.08.054

Kohajdov, Z., & Karovi, J. (2007). Fermentation of cereals for specific purpose. *Journal of Food and Nutrition Research*, *46*(2), 51–57.

Konik-Rose, Christine Thistleton, J., Chanvrier, H., Tan, I., Halley, P., Gidley, M., Kosar-Hashemi, B., Wang;, H., Larroque, O., Ikea, J., McMaugh, S., Regina, A., Rahman, S., Morell, M., & Li, Z. (2007). Effects of starch synthase IIa gene dosage on grain, protein and starch in endosperm of wheat. *Theoretical & Applied Genetics*, *115*(8), 1053–1065.

Lazaridou, A., Vouris, D.G., Zoumpoulakis, P., and Biliaderis, C. G. (2018). Physicochemical properties of jet milled wheat flours and doughs. *Food Hydrocolloids*, *80*(111–121). https://doi.org/10.1016/j.foodhyd.2018.01.044

Lee, Y.-T., & Min-Jung Shim, Hye-Kyung Goh, Chulkyoon Mok, P. P. (2019). Effect of jet milling on the physicochemical properties, pasting properties, and in vitro starch digestibility of germinated brown rice flour. *Food Chemistry*, *282*, 164–168. https://doi.org/10.1016/j.foodchem.2018.07.179

Leethanapanich, K., Mauromoustakos, A., & Wang, Y. J. (2016). Impacts of parboiling conditions on quality characteristics of parboiled commingled rice. *Journal of Cereal Science*, *69*, 283–289. https://doi.org/10.1016/J.JCS.2016.04.003

Li, E., Dhital, S., & Hasjim, J. (2014). Effects of grain milling on starch structures and flour/starch properties. *Starch/Staerke*, *66*(1–2), 15–27. https://doi.org/10.1002/star.201200224

MacLeod, L., & Evans, E. (2016). Malting. *Reference Module in Food Science*. https://doi.org/10.1016/B978-0-08-100596-5.00153-0

Manay, N. S., & Shadaksharaswamy, M. (2001). Cereals. In *Food Facts and Principles* (Second edi, pp. 197–230). New Age International (P) Ltd., Publishers.

McGoverin, C.M., Snyders, F., Muller, N., Botes, W., Fox, G. Manley, M. (2011). A review of triticale uses and the effect of growth environment on grain quality. *Journal of the Science of Food and Agriculture*, *91*, 1155–1165. https://doi.org/10.1002/jsfa.4338

McIntosh, G. H., Noakes, M., Royle, P. J., & Foster, P. R. (2003). Whole-grain rye and wheat foods and markers of bowel health inoverweight middle-aged men. *The American Journal of Clinical Nutrition*, *77*, 967–974.

Mckevith, B. (2004). Nutritional aspects of cereals. *British Nutrition Foundation Nutrition Bulletin, 29*, 111–142.
McNabney, S. M., & Henagan, T. M. (2017). Short chain fatty acids in the colon and peripheral tissues: A focus on butyrate, colon cancer, obesity and insulin resistance. *Nutrients, 9*, 1348. https://doi.org/10.3390/nu9121348
Mohamed, T. K., Issoufou, A., & Zhou, H. (2012). Antioxidant activity of fractionated foxtail millet protein hydrolysate. *International Food Research Journal, 19*(1), 207–213.
Nkhata, S. G., Ayua, E., Kamau, E. H., & Shingiro, J.-B. (2018). Fermentation and germination improve nutritional value of cereals and legumes through activation of endogenous enzymes. *Food Science & Nutrition, 6*, 2446–245. https://doi.org/10.1002/fsn3.846
Oliveira, L.C., Alencar, N.M.M., and Steel, C. J. (2018). Improvement of sensorial and technological characteristics of extruded breakfast cereals enriched with whole grain wheat flour and jabuticaba (Myrciaria cauliflora) peel. *LWT - Food Science and Technology, 90*, 207–214. https://doi.org/10.1016/j.lwt.2017.12.017
Osman, M. A. (2011). Effect of traditional fermentation process on the nutrient and antinutrient contents of pearl millet during preparation of Lohoh. *Journal of the Saudi Society of Agricultural Sciences, 10*, 1–6. https://doi.org/10.1016/j.jssas.2010.06.001
Ostman, E. M., Granfeldt, Y., Persson, L., & Bjorck, I. M. E. (2005). Vinegar supplementation lowers glucose and insulin responses and increases satiety after a bread meal in healthy subjects. *European Journal of Clinical Nutrition, 59*, 983–988. https://doi.org/10.1038/sj.ejcn.1602197
Papageorgiou, M., Unversity, I. H., & Skendi, A. (2020). Introduction to cereal processing and by-products. In *Sustainable Recovery and Reutilization of Cereal Processing By-Products* (Issue February 2018, pp. 27–30). https://doi.org/10.1016/B978-0-08-102162-0.00001-0
Pastor-Cavada, E., Drago, S.R., Gonza´lez, R.J., Juan, R., Pastor, J.E., Alaiz, M. (2011). The, Effects of the addition of wild legumes (Lathyrus annuus and Lathyrus clymenum) on physical and nutritional properties of extruded products based on whole corn and brown rice. *Food Chem., 128*, 961–967. https://doi.org/10.1016/j.foodchem.2011.03.126
Philippe, S., Saulnie, L., & Fabienne Guillon. (2006). Arabinoxylan and (1fi3),(1fi4)-b-glucan deposition in cell wallsduring wheat endosperm development. *Planta, 224*, 449–461.
Pichmony Ek, Jonathan M. Baner, G. M. G. (2020). Extrusion processing of cereal grains, tubers, and seeds. In *Extrusion Cooking*. Elsevier Inc. https://doi.org/10.1016/b978-0-12-815360-4.00008-0
Raes, K., Knockaert, D., Struijs, K., & Van Camp, J. (2014). Role of processing on bioaccessibility of minerals: Influence of localization of minerals and anti-nutritional factors in the plant. *Trends in Food Science and Technology, 37*(1), 32–41. https://doi.org/10.1016/j.tifs.2014.02.002
Rhaghavendra Rao, S. N., & Juliano, B. O. (1970). Effect of Parboiling on Some Physicochemical Properties of Rice. *Journal of Agriculture and Food Chemistry, 18*, 289-294. https://doi.org/10.1021/jf60168a017
Robin, F., Heindel, C., Pineau, N., Srichuwong, S., & Lehmann, U. (2016). Effect of maize type and extrusion-cooking conditions on starch digestibility profiles. *Internal Journal of Food Science and Technology, 51*, 1319–1326. https://doi.org/10.1111/ijfs.13098
Saleh, A. S. M., Zhang, Q., Chen, J., & Shen, Q. (2013). Millet grains: Nutritional quality, processing, and potential health benefits. *Comprehensive Reviews in Food Science and Food Safety, 12*(3), 281–295. https://doi.org/10.1111/1541-4337.12012
Salmon, R. E. (1984). True Metabolizable Energy and Amino Acid Composition of Wheat and Triticale and Their Comparative Performance in Turkey Starter Diets. *Poultry Science, 63*(8), 1664–1666. https://doi.org/10.3382/ps.0631664
Sanaa Ragaee, K. S. & E.-S. M. A.-A. (2014). The Impact of Milling and Thermal Processing on Phenolic Compounds. *Critical Reviews in FoodScience AndNutrition, 54*(7), 837–849. https://doi.org/10.1080/10408398.2011.610906
Sarwar, H. (2013). The importance of cereals (Poaceae: Gramineae) nutrition in human health: A review. *Journal of Cereals and Oilseeds, 4*(3), 32–35. https://doi.org/10.5897/jco12.023
Severini, C., Derossi, A., and Azzollini, D. (2016). Variables affecting the printability of foods: preliminary tests on cereal-based products. *Innovative Food Science & Emerging Technologies, 38*, 281–291. https://doi.org/10.1016/j.ifset.2016.10.001

Simin Liu. (2003). Whole-grain foods, dietary fiber, and type 2 diabetes: searching fora kernel of truth. *American Journal of Clinical Nutrition*, *77*, 527–529.

Singh, A., Karmakar, S., Jacob, B. S. et al. (2015). Enzymatic polishing of cereal grains for improved nutrient retainment. *Journal of Food Science and Technology*, *52*(6), 3147–3157. https://doi.org/10.1007/s13197-014-1405-8

Slavin, J. L., Jacobs, D., & Marquart, L. (2001). Grain processing and nutrition. *Critical Reviews in Biotechnology*, *21*(1), 49–66. https://doi.org/10.1080/20013891081683

Sramkova, Z., Gregová, E., & Šturdíka, E. (2009). Chemical composition and nutritional quality of wheat grain-Review. *Acta Chimica Slovaca*, *2*(1), 115–138.

Subramanian, V., & Jambunathan, R. (1984). Chemical Composition and Food Quality of Sorghum. *ICRISAT Open Access Repository*.

Taylor, J. R. N., Hugo, L. F., & Yetnerberk, S. (2005). Developments in Sorghum Bread Making. *Using Cereal Science and Technology for the Benefit of Consumers*, 51–56. https://doi.org/10.1533/9781845690632.3.51

Tiwari, U., & Pojic, M. (2021). Introduction to Cereal processing: Innovative Processing aTechniques. In *Innovative Processing Technologies for Healthy Grains* (pp. 9–35).

Uauy, R., Hertrampf, E., & Reddy, M. (2002). Iron Fortification of Foods: Overcoming Technical and Practical Barriers. *Journal of Nutrition*, *132*. https://doi.org/10.1093/jn/132.4.849S

Wenwen, Y., Tao, K., Gidley, M. J., Fox, G. P., & Gilbert, R. G. (2019). Molecular brewing: Molecular structural effects involved in barley malting and mashing. *Carbohydrate Polymers*, *206*(December), 583–592. https://doi.org/10.1016/j.carbpol.2018.11.018

Winkels, R. M., Brouwer, I. A., Clarke, R., Katan, M. B., & Verhoef, P. (2008). Bread cofortified with folic acid and vitamin B-12 improves the folate and vitamin B-12 status of healthy older people: a randomized controlled trial. *The American Journal of Clinical Nutrition*, *88*, 348–355. https://doi.org/10.1093/ajcn/88.2.348

Yodpitak, S., Boonyawan, S. M. D., Sookwong, P., Roytrakul, S., & Norkaew, O. (2019). Cold plasma treatment to improve germination and enhance the bioactive phytochemical content of germinated brown rice. *Food Chemistry*, *289*, 328–339. https://doi.org/10.1016/j.foodchem.2019.03.061

Zhang, G., Xu, Z., Gao, Y., Huang, X., & Yang, T. (2015). Effects of germination on the nutritional properties, phenolic profiles, and antioxidant activities of buckwheat. *Journal of Food Science*, *80*, 1111–1119. https://doi.org/10.1111/1750-3841.12830

Zielinski, H., Kozowska, H. and Lewczuk, B. (2001). Bioactive compounds in the cereal grains before and after hydrothermal processing. *Innovative Food Science and Emerging Technologies*, *2*, 159–169. https://doi.org/10.1016/S1466-8564(01)00040-6

CHAPTER 3

Techno-Functionality of Cereals

Sundaramoorthy Haripriya
Department of Food Science and Technology, Pondicherry University,
Pondicherry, India

CONTENTS

DOI: 10.1201/9781003242192-4

3.1 INTRODUCTION

Cereal grains are eaten as a staple diet throughout the world. Their use in a wide range of food products is essential economically, and they have a lot of industrial applications, especially in food processing and marketing. Cereals are farmed all over the world due to their adaptability to a variety of environmental circumstances. Cereals are essential in India's agricultural economy and human nutritional demands. They account for nearly 80% of the country's food grain production. In terms of production and consumption, wheat and rice are the most important cereals in India. Maize, sorghum, barley, oats, and millets are also eaten as human foods in various parts of the country. Starch is the most abundant carbohydrate in all cereal grains, accounting for 65–85% of total dry matter. The ability to be modified for food applications is one of cereal starch's most essential and valuable characteristics.

The technological properties of starch, which confer several appealing textural qualities to recipe formulations and food product development, account for a large part of its importance in foods. In terms of pasting behavior, gelatinization, retrogradation, and other functionality attribute that influence product quality, starches are generally regarded as the essential constituents of cereals. The textural properties of cereal grains and their processing for food uses are influenced by several physicochemical and functional properties of starches.

Unleavened bread, leavened bread, porridges, pasta, noodles, cookies, soups, boiled meals, and kernels are just a few of the traditional food uses of whole grain cereals. Traditional products can be transformed or reformulated to appeal to consumers based on taste and convenience. For example, modern consumers prefer quick-cooking grains like oatmeal, corn, sorghum, and pearl millet. Cereal milling converts whole grains into forms that are suitable for human consumption. Before food products may be made, cereals must first be transformed into flour. The milling quality of grains is essential for further processing and various food applications. In general, cereals should produce high-quality flour, which is accomplished by separating the endosperm from the bran and germ.

3.2 CEREALS AS FUNCTIONAL INGREDIENTS

The basis of many plant-based foods is cereals. The significant components of grains include carbohydrates, protein, and lipids. Cereal flour has a wide application in the food industry. The contributive role of cereal starch in various food products is widely known. The role of cereal proteins and lipids and their functionality in food products have significance.

3.3 CEREAL PROTEINS

Prolamines or glutenins are the major storage protein cereals. Exceptionally the protein in oats is globulin. Fractionation of cereal protein based on Osborne classification includes salt solution, solubility in water, dilute acid or alkali, and aqueous ethanol methods. Cereal proteins consist of 25–45% prolamines, 3–10% globulins, 5–15% albumins, and 3–40% glutenins based on the source of their origin.

Wheat protein fractions, namely glutenins and gliadins, aid the dough characteristics and the quality of the finished products upon processing. The baking industry's key aspect is the proportional difference between glutenin and gliadin. The molecular mass of cereal protein is an essential parameter in predicting the dough properties. The enhancement of activities of cereal enzymes are utilized from the albumin fraction, and the formation of the starch protein matrix, which influences the preparation of cereal starch, is taken care of by the cereal globulin, and among the cereal protein, wheat protein aid in the prediction of dough properties. The polymeric protein glutenins vary in the range of molecular weight based on the disulfide linkage involved in forming high molecular weight and low molecular weight subunits. Low molecular weight is a more significant proportion of the heterogeneous mixture of the polymer formed. Still, the high molecular weight subunit is crucial in gluten elasticity, thereby critical in bread making (Espinoza-Herrera et al., 2021).

3.4 CEREAL ENZYMES IN THE FOOD INDUSTRY

In most cereal-based formulations, protein enzymes play a significant role. In the germ and bran layers of cereals, lipoxygenases and lipase are available, which play a central role in the product and storage stability. Among the grains, lipase content is highest in oats. Catalases and peroxidases are distributed in cereals. Peroxidases catalyze the oxidative cross-linkage of pentosans in the dough, especially in the rye. Glutathione is oxidized by glutathione dehydrogenase during dough making, which immediately establishes disulfide interchange in flour proteins. The viscosity of the dough decreases if high molecular weight glutenin is depolymerized. Peptidases and proteases present in the cereal grain aid in bread making. The texture of the cereal products is influenced by the proportion of proteases that modify the gluten structure. Proteases play a vital role in producing biscuits, cookies, and crackers. Transglutaminase (TG) has a more significant influence in baked foods where it enhances absorption of water byproducts and later aids in the release utilized by starch during gelatinization during baking. Further, TG also aids in avoiding any negative influence. On dough quality by deep freezing, TG stabilizes gluten structure through cross-linking between polypeptides, strengthening the network and decreasing its sensitivity to freezing conditions.

Among the cereals, barley malt has hydrolytic enzymes (α-β amylases). The proportion of α amylases only increases during sprouting/germination. Excess α-amylases degrade starch during baking, leading to the low-developed and sticky baked products. During malting, α-amylases activity enhances the water-binding capacity of the various starches utilized in producing high-energy food products of cereals.

3.5 SOME APPLICATIONS OF CEREAL PROTEIN

Cereal proteins contribute to various products in the food sector through their unique properties, whether in their native or modified forms. Wheat gluten is used in making films/coatings and adhesions, blending to the suitable barrier properties for water vapor, flavors, and gases. The rheological behavior, mechanical properties, solubility, barrier properties, water solubility, and adhesion to various substitutes can be adjusted by enzymatic and chemical modification of protein functional groups (NH_2, COOH, SH).

3.6 ROLE OF CEREAL PROTEIN IN FOOD PROCESSING

Gliadins' and glutenins' molecular and functional relationship relate significantly to their contribution to the baking process. The contribution of gluten to gas retention and dough elasticity is crucial in bread making. The overall quality and structure of baked products are directly proportional to the role of gluten. The gluten is responsible for the extension of dough life, gas retention, controlled expansion, dough strengthening, and improved water absorption. Gliadins aid in the improvement of dough extensibility and product quality. The characteristic texture of biscuits, cookies, and cracker production primarily depends on gluten present in flour. Extrusion cooking technology is highly versatile and aids in producing a wide range of cereal pasta products. The quality of pasta is highly dependent on the starch protein interaction responsible for the technological properties of flour. Enzymatically hydrolyzed cereal protein is used to produce various flavor hydrolysates.

3.7 CEREAL LIPID

Cereal lipids are essential in the quality of baked cereal products and beer (beverage). Lipid functionality on the physico-chemical properties is through lipid-binding protein, including lipid transfer proteins and puroindolines. These amphiphilic molecules have a favorable or unfavorable impact on the texture and flavor of cereal foods. Lipids contribute to 1–2% of the dry flour in cereals.

3.8 CEREAL LIPID–PROTEIN INTERACTION

The processing and quality output of cereal baked products and beverages are critically dependent on cereal lipids. Lipids in cereal-based food products are contributed internally from the endogenous fats and externally through the ingredients, including fats and oils, and additives like surfactants, including monoglycerides, lecithin, and sucroesters.

Among cereal technology, wheat milling technology has an impact on lipid composition. Increased non-polar lipid content is noted in higher yield in the milling process. The cereal polar lipid fraction includes glycolipids and phospholipids, and non–polar lipid fraction constitutes sterols, free fatty acids, lipophilic pigments, and glycerides (mono-, di-, and triglycerides). The aleurone and embryo layers are concentrated with non-polar lipids, whereas polar lipids are present in starchy endosperm.

Glycolipids are composed of mono-digalactosylacyl glycerides. The amyloplast membrane is mainly provided with galactolipids. In contrast, cell membranes (plasmalemma), membranes of other organelles (protein bodies, vacuoles), and oleosome surface area covered by natural oil droplets form aleurone and embryo. A higher proportion of N-acylphophatidylethanolsamine is present as cereal phospholipids. Monoacylated phospholipids, including lysophosphatidylcholine and free fatty acids, are the main constituents of the lipids that are bound tightly with starch granules.

Mono- and digalactosylacy glycerides are the main constituents of glycolipids. The hydrophilic layer of cutin and waxes also has a significant role in cereals' technological properties. The waxes that are the esters of long-chain fatty acids and alcohols assemble in the hydrophilic polymer, a tridimensional polyester network composed of hydroxy fatty acid derivatives.

3.9 CEREAL LIPID IN FOOD PROCESSING

Numerous lamellar vesicles grow from non-lamellar aggregates during the hydration of cereal flour. In the endosperm of the wheat, the lipids form aggregates from non-layer types creating granular or long tubular structures. Upon hydration, numerous oil droplets are observed: liquid crystalline hexagonal II and cubic phases in wheat dough corresponding to oleosomes (oil bodies) of aleurone and embryo layer upon hydration. These oil bodies during gluten formation are observed as lipid vesicles. The phase diagram can predict the lipid interaction depicting an important textural property in baking bread loaves and sponge cakes. The SEM images show bubbles, gas expansion separated by thin films separated by an aqueous inter-lamellar phase. In bread and cakes, these polar lipids from adsorbed layers of films interact with albumins and globulin forming, acting as surface triglycerides and fatty acids that act as anti-foaming agents. The lipid functionality in baked products is expressed through the foam-forming capacity by transferring lipid vesicles from bulk water to the air–water or oil–water interface.

The rheological property of dough is influenced by the coupled oxidation of polyunsaturated fatty acid and free SH groups of gluten protein by lipooxygenases. The dough mixing tolerance is modified. Non-covalent interaction of the protein with other macromolecular components, especially lipids, is also considered important in processing. It would be interesting to note that membrane protein, apolipoprotein, or lipid-binding protein also compete with the surface protein and lipid for denaturation and adsorption of soluble protein to oil-water interfaces. Lipid transfer protein and puirondolines belonging to the albumin family also play a significant role in dough rheology by monomeric binding lipids and interacting with lipid aggregates. Lipid Transport Protein (LTP 1 and 2) are the low molecular mass protein with a typical pattern of eight cysteine forming disulfide bonds. Among the cereals, the functionality of LTP can be demonstrated in bread making and brewing technology. LTP creates an equilibrium between thin film formed at the oil–water and air interface and lipids in the aqueous dough phase in the bread-making process. They prevent formation of foam in dough, which is generally produced by lipids from destabilization. In brewing technology, LTP 1 is the central portion playing a significant role in the information and stability of head foams.

3.10 TECHNO-FUNCTIONAL ATTRIBUTES AND THEIR SIGNIFICANCE IN FOOD PROCESSING

3.10.1 Hydration Properties

During the processing of cereal grains, the crucial process involves extraction, malting, and cooking. This is critical for industrial processing, where the physico-chemical properties are enhanced – moisture content increases in the hydration process. The primary function of soaking is to reduce the cooking time of the grains. During soaking, the degree of polymerization of rhamnogalacturonan I with the activation of enzymes increases the solubility and shortens the cooking time. Hydration also aids in homogenous gelatinization and denaturation of starch and protein during cooking, which aids in uniform texture in the whole grain. Enhanced water absorption during soaking aids consistent heat transfer, inactivating anti-beta factors including phytates, alkaloids, saponins, lectins, protease inhibitors, and indigestible polysaccharides. Component extraction from grain is also improved by hydration. Wet milling aids in extracting starch from grains by softening the grain. Germination is the natural process after the hydration of grains. This is an essential process for sprout production and the malting process. Cereal grains are germinated for enzyme activation and malt's characteristic aroma, flavor, and color formation to produce alcoholic beverages. Corn kernels, barley, and sorghum kernels are involved in the malting process. The industrialization of grains requires a hydration process; usually, hydration is a batch process.

In germination, seeds absorb water and activate their metabolism; the hydration kinetics can be divided into three stages. The first stage is the absorption of water to start the metabolism. This stage involves the metabolic activity of the seeds and occurs both in live and dead seeds. The water absorption in the seeds is higher than in the living seeds as turgor pressure will counteract hydration. In stage II, the breakdown of the reserve molecule into simpler constituents for germination occurs with a significant increase in water content. In stage III, germination starts, the cell begins to reproduce, and the tissue grows with high moisture content. In food processing, only stage I hydration is essential. The difference in water activity is a driving force in mass transfer unit operation. The process happens through diffusion transportation from a substance of higher concentration (soaking water) to a sense of lower concentration (grain). In the grain, various channels formed in the different tissues and cells of varied composition, structure, sizes, and zones show varied permeability through which water travels through capillary flow and diffusion process. Hydration in grains involves a mass transfer mechanism and fluid flow.

3.10.2 Hydration Behavior

Water passes through the cereal grains' surface area as the cereal grains' bean is permeable to water. The complete hydration process is a long duration as the endosperm structure holds the starch. Decrease in hydration rate from equilibrium moisture value (maximum value) to zero causes the hydration kinetics behavior to have a downward concave shape. "Empty" spaces are also present in some grains with porous structures where water transfer occurs through capillary enhancement. One example is the corn kernel, which has space between germ and endosperm and a porous tip cap structure. The water movement will be through the tip cap and fill the distance between the embryo and the endosperm. As cereal grains have varied layers and structures, Differential scanning calorimetry (DSC) behavior in the initial part of the process has high hydration due to capillary and predominant diffusion in the second stage (Miano & Augusto, 2018).

The most common hydration occurs at higher temperatures, and it has been observed that the higher the temperature, the greater the hydration rate, which increases exponentially. The lag period is reduced when high temperature is employed in grains with sigmoidal behavior. The lag phase is always correlated to the grain seed coat, so when high temperatures are used, the phase time changes due to the seed coat's accelerated hydration; a decrease in the minimum moisture content to change its component state (from glassy to rubbery) and increase its permeability to water. During the hydration process at high temperatures, the equilibrium moisture content will be reduced, supplemented, or maintained without fluctuation. A temperature of 60° can cause significant changes in grains' starch and protein components. As a result, the grain's assessed temperature and thermosensitivity are critical in the hydration process. Because of the expansion of the pores and spaces inside the grain, as well as the solubility of the components, hydration equilibrium and moisture content increase at high temperatures. Reduced equilibrium moisture content is caused by lixiviated water-soluble components breaking cell walls and membranes, as well as a decrease in mass transfer driving force, which leads to rapid saturation of the transparent layer of grains (Miano & Augusto, 2018).

3.11 SURFACE HYDRATION OF WHEAT FLOUR

Cereal flour is used to make the dough and, like cereal, contributes to a few grains' high carbohydrates and protein content. The surface hydration properties of starch in grains help define dough parameters and finished product attributes. The surface wettability features of flours are physical events that affect their functional properties. The physical interactions are determined by the surface energy of a solid, which defines the wettability qualities. Adsorption, adhesion, wettability, and friction are just a few of the surface and contact phenomena that can be predicted using this data. The surface wettability qualities of cereal flour can be estimated by measuring contact angles. The size of contact angles can be utilized to determine the wettability properties of a surface. Contact angles are calculated using the Washburn equation. Capillary rise techniques can also be used to measure the water adsorption capacity of wheat flour. The hydrophilic behavior of compact surfaces in hard wheat is positively related to the presence of water molecules. Compacts' initial values for hard and soft wheat flour are similar, and soft wheat flour has a low apparent absorption rate compared to hard wheat flour, which has a high absorption rate. The transparent rates of water drop absorption are affected by the granular size of the starch particles. The apparent rates of water drop absorption are lower in the more significant granulometric fractions (D50>50m) than in the smaller fraction (D50<50m). The lesser percentage adds to the flour qualities with a high absorption rate.

Low initial contact angle values are characterized by starch at 0% RH. This reveals a "good" surface wettability property exhibited by the compacts based on starch. Starch's apparent high absorption rate is noted, relating to a short duration for water drop absorption seeped into the arrangements (<1S). Starch's high surface wettability property is associated with the hydrophilic structure of starch molecules.

Starch's initial hydration is compact, favoring wettability features with a lower absorption rate. Initial contact angles of gluten soluble and insoluble pentosans range between (38 and 40), similar to starch-based compacts with strong surface wettability. The surface repartition of flour components influences the surface wettability of wheat flour. The structural organization of components on the surface of flour particles is used chiefly to recreate gluten protein. Surface particles can be affected by changes in particle characteristics during compaction.

The surface wettability qualities of flour components are only measured using an indirect method. It will be impossible to directly measure the contact angles of water drops on flour. The qualities of the particles are reflected in the determination of contact angles on the upper surface of compacts. It is also essential to consider potential structural changes in flour components after compaction.

A helium pycnometer is used to measure the density of particles and compacts. Soft and hard wheat has a similar density. The density of the compacts is higher than in particles. Using low pressure in compaction leads to the deformation of particles being elastic and reversible, whereas, during high pressure, the deformation of particles is plastic and irreversible (Shaoxiao, L.W.G.Z.Z., and Baodong, 2013). The wettability of cereal-based protein can be understood by measuring the contact angle of the water drop.

3.12 TRENDING TECHNOLOGY IN THE HYDRATION PROCESS

Ultrasound technology is used to improve the grain hydration process. Acoustic ultrasound waves with frequencies higher than 20 kHz are used in ultrasonic water baths. The water flow in the grain through ultrasound would cause direct and indirect effects. The immediate effect includes acoustic wave, which passes through the grains. This triggers the compression and expansion of the medium in an alternative pattern, causing a pressure difference to enhance hydration with the increased capillary flow. This pattern is also called the inertial flow and sponge effect. Acoustic cavitation is the cause of this ultrasound where micro-cavities are formed due to the implosion of micro water bubbles enhancing the water transfer in grains. Grains with higher water activity tend to significantly affect acoustic cavitation. Increased hydration rates and reduced lag phase in the hydration process are accelerated through ultrasound technology.

In industrialization and direct consumption of grains, hydration plays an important role. The downward concave shape behavior is observed in most cereal grains of the two kinds of hydration kinetic behavior. The behavior is solely dependent on the composition of grain structure.

3.13 RHEOLOGY OF CEREAL FLOURS

Cereal flours are the most crucial crops for humans. Wheat flour can generate three-dimensional structure dough and viscoelasticity when hydrated among the commonly consumed cereals. Rheological properties are the determinable factors in most baking processes like kneading, shaping, and leavening. The complex function of any cereal dough composition and its interaction can be analyzed through the rheological property. The presence of gluten properties mainly determines dough properties. Studies of rheological properties are focused on the gliadins and glutenin subunits (high and low molecular weight). Other flour components and their interaction also contribute to dough rheological properties. Bread-making qualities are affected by the presence of low molecular weight glutenin subunits, whereas high molecular weight glutenin is considered detrimental to dough behavior. During the grain development, varied particle sizes and shapes of starch granules accumulate, and based on the size, they are classified as A-type, B-type, and C-type. Among this classification, B- and C-type granules are considered small granules. These smaller granules enhance the dough's elastic characters. Large particles of A-type granules are associated with poor dough rheological properties. Starch granules generally behave as filling components in the protein starch matrix. During dough development, smaller granules increase the filling degree more than A-type granules, improving the interaction between protein and starch.

The dough rheological properties of baked and extruded products have been positively correlated with adding waxy wheat starch to flour and reconstituted starch flours of various cereals in bread and pasta. Mixolab and texture analyzer aid in determining the dough quality indices and rheological properties. Strong dough strength reflects higher quality and quantity of gluten, which impart the crucial parameters including sough development time and stability. Further, the un-extractable polymeric protein and sulfide bonds, composition, and contents of HMW-GHs determine the wheat dough behavior. The physico-chemical properties of starch influence the development of dough qualities. Dough cohesiveness is the interaction between the dough during compression, and dough adhesiveness is the attraction between the surfaces in contact with the surface of the dough medium. Strong flours are associated with higher values of dough adhesiveness and cohesiveness. The presence of inferior HMW-GSs reflects low dough cohesiveness and adhesiveness, emphasizing that the physical characteristics of starch granules also affect dough rheology.

HMW-GSs strongly contribute to starch granules, flour quality, and dough behavior. Dough springiness is the ability of the dough to respond to the external pressure and its elastic recovery and its elastic recovery. There exists a close relationship between the cereal's starch physico-chemical properties and dough rheological properties. Poor starch properties include thermal, swelling power, and short-range ordered degree, which reflect soft dough rheological properties.

The amylose content of the flour also contributes to the dough's strength. Higher amylose content resists the deformation of dough, high amylopectin decreases dough stability within the granules, and the amylopectin double helix influences the relative crystallinity. Disruption of the helical structure due to heating exposes the hydroxyl groups, which bind with water through a hydrogen bond.

The dough is associated with water through a hydrogen bond – dough associated with higher water mobility results in lower dough stability. In addition, glutamine concentration is higher in gluten. This amino group can aid in the formation of hydrogen bonds with the II and III hydroxyls of the starch glucose molecule. Cereal flours with low crystallinity will have higher interaction of water molecules with gluten resulting in good dough stability.

Gelatinization enthalpy and final viscosity are positively related, and both are related to dough stability, specifying that the dough behavior is affected by the short-range ordered degree of starch. In food industries, the rheology of flour gives a broader scope for utilization in various products. The rheological properties of gluten and dough are often challenging due to their source and other dependence factors. The rheological properties of the cereal flours and their behavior can be understood by inducing stress and strains in various deformation studies.

There is a growing market trend in using composite flour to make baked and extruded products. Rheological analyses of wheat dough are used as predictors of functionality in baking. Cereals devoid of gluten are structurally different and give scope for enhancing their functionality. The structure and functional properties of dough rheology are well studied from the linear viscoelastic testing region. This region helps to understand the functionality of the ingredients. This viscous and elasticity of dough is expressed as Storage (G') and loss moduli (G"), and loss tangent tan δ. Good flour quality dough has a low tan δ value.

The macroscopic homogenous mixture of starch, protein, fat, yeast, salt, and other components (additives) form the dough. Upon optimum mixing, the dough reaches its highest elasticity being fully hydrated. The viscoelastic property of dough is positively related to its water content. With the increase in water content, G' and G" decrease. Water in dough acts as an inert filler, reducing the dynamic properties proportionally, and also aids in relaxation by working as a lubricating agent. These two actions determine the dynamic viscoelastic behavior of the dough.

The cereal starch content, quality, and gluten play a crucial role in dough formation. A continuous network of starch particles and the macromolecule; AR network of the hydrated gluten gives rise to dough's rheological properties. Based on the starch and gluten interaction, the stress levels vary. Starch–starch interaction dominates at low stress, while high-stress protein–protein exchange plays a predatory role in large deformation. The behavior of the dough is based on the nonlinear rheological behavior of starch. This protein content in the dough signifies an increased G'. Constant water content with various starch/gluten blends increases G' with increasing protein content. Similarly, the elasticity of the dough is enhanced with the homogenous dispersal of starch granules in the gluten network. The proportion of gluten varies with the concentration of amylose and amylopectin of a starch source, indicating varied rheological differences depicting the active role of starch.

3.14 PHYSICO-CHEMICAL PROPERTIES WITH ITS GELATINIZATION CAPACITY

Thermal disordering of starch crystal structure causes loss of birefringence. The gelatinization process begins at the hilum as the area around the granule is least organized. Generally, the amylose and amylopectin are orderly aligned in radial order surrounding the hilum. These growth rings around the helium, perpendicular to the granule surface, represent a "maltese cross" under the polarizing light. Disruption of starch granules begins to disrupt the helium upon hydration and heating. This is accompanied by partial swelling, loss of birefringence, and increased viscosity. This disruption pattern varies with the position of the hilum in the cereal starch, either proximal surface or at the central hilum. The temperature and time required for the process are dependent on the cereal source's physicochemical characteristics. Measurement of thermal behavior of starches is done through DSC during heating. The enthalpy (ΔH) is measured from the sample's peak areas of the thermogram in Joules/wt. The thermal transition is defined as onset To, peak Tp, and set Tc transition temperature.

To is the measurement of the beginning of starch gelatinization, which occurs when the starch molecule loses its capacity to bind more water molecules, which is depicted by maximum swelling followed by partial rupture of the granules and loss of birefringence. This leads to the formation of sol coupled with enhanced viscosity due to the leached amylose combining with the amylose and amylopectin dispersed, forming the peak viscosity. The loss of complete birefringence is marked with Tp, brought about by the exclusive breaking of the starch granules enhanced by agitation. The Tc represents the final temperature with the entire gelatinization process by forming a swollen granules network in sol to gel transition. Enthalpy (ΔH) reveals the starch structure damage or degree of crystallinity before gelatinization. Enthalpy values are always inversely related to the damaged starch structure, amylose content, and small-sized granules.

The gelatinization capacity of the starch purely depends on the cereal source, its amylose content, and the granule type. For example, high amylose maize starch takes longer than waxy and normal maize for gelatinization temperature. In contrast, high amylose barley starch has a lower time than the waxy starches. Differences in the proportion of A- and B-type granules alter the gelatinization process. For instance, high amylose barley starch contains a more significant proportion of B-type granules.

In contrast, regular barley has a higher ratio of A-type granules, which affects the gelatinization process. More prominent starch granules take longer to gel and present higher gelatinization temperatures. Smaller starch granules give efficient hydration and swelling capacity. The chain length distribution of amylopectin in small granule fractions influences maximum gelatinization temperature in starch granules. Starches such as maize and waxy barley have a different gelatinization time than the regular starches. The size of the granules is not the only factor determining the gelatinization temperature; starch damaged during the milling process also reduces particle size, increasing and decreasing gelatinization time and capacity.

After gelatinization, the gel structure is affected by the amylose quantity and the proportion of the amylopectin chains. Amylose starch yields soft gels, while hard gels are formed through low amylose starch. The stability of gels in the center of the grain is higher than in the surface area of waxy barley starches and vice versa in the regular barley. Gel stability is minimally affected by the lipid and other minor components. The formation and stability of flexible gels are influenced by the quantity and characteristics of amylopectin and amylose. In food preparation, pure gelatinized starched with ultra-fine particles are achieved through reduced granule size during milling. Paste clarity is a crucial desired attribute of the gel, which is affected by the molecular profile of the amylopectin chain. Starch with high amylose content and high molecular weight granules yield translucent paste as the alignment of the granule yields fine paste that is very complex and different. Waxy starches with a low amount of amylose behave like viscoelastic liquid without intermolecular bonds.

Due to the melting of the crystalline region in thermal processing, disruption of gel structure leads to loss of interaction between networks of disentanglement in the amylopectin molecules.

3.15 RETROGRADATION OF STARCH GRANULES

The gelatinization process follows retrogradation. Recrystallization of the amylopectin occurs in the gelatinized starch granules; a movement observed from the amorphous state to the crystalline state in an ordered form. This results in thickened and softened gels. The two critical factors that limit starch retrogradation are high amylose content. The size of the starch granules also affects retrogradation. Larger starch granules are more stable than smaller ones. A-type granules retrograde easier than B-type, which has a shorter branched amylopectin chain and high lipid content; when starch presents high amylose content with a low proportion of small granules, the retrogradation rate is higher than in starches like in barley with a higher proportion of small granules.

3.16 RHEOLOGICAL PROPERTIES OF CEREAL STARCH

The main macromolecule in cereal grains is starch and protein. (1→3) (1→4) β-glucans and water-soluble pentosans are totally or partially soluble in water at room temperature or when heated. The polymer's solubility and the solvent polymer interaction determine the rheological behavior. Macromolecular characteristics, including molecular weight and hydrodynamic volume, are directly related to its rheological characteristics. Viscosity is a critical parameter used to depict the rheological property of starch, as most starches are used as thickening agents in diverse applications. Viscosity measures the resistance when shear stress is applied to the flowability of fluid or semisolid. At ambient temperature and a high concentration (35–40% w/w), native starch suspension exhibit low viscosity. Starch granules interrupt their structure, captivating water, swelling, and changing viscosity beyond the gelatinization temperature when heated. Starch pasting is the process of the development of viscosity. The thickening power of starch determines its application in the food industry. Cereal starches like maize and wheat form shapes without fluidity at concentration >–6% w/w. These gels exhibit viscoelastic property through immobilization or water in the network formed. Starch concentration, structure, pasting factors (temperature, heating rate, and shear rate), and storage conditions (time and temperature) alter the rheological characteristics of starch gels.

3.17 PASTING PROPERTY OF NATIVE STARCH

During cooling, amylopectin is the primary constituent required for swelling power and viscosity. At the same time, amylose linked with lipids restricts the swelling of starch granules by intertwining with amylopectin. This results in lower peak viscosity and higher pasting temperature. Barley and wheat starches contain a more significant proportion of phospholipids readily complexed with amylose among the cereal starches. Setback viscosity is attributed to the role of amylose in other starches forming networks, and rice and waxy maize exhibit lower setback viscosity than their regular counterparts. Amylograph such as dynamic rheometer in a flow temperature ramp mode, barbender amylograph, and rapid visco analyzer is generally used to measure the pasting properties. The pasting property in cereals starches is dynamically varied with lipids and phosphate monoester derivatives. The viscosity of starch is improved by the addition of fructose, glucose, maltose, lactose, sucrose, and galactose attributed to the water binding ability of sugar. Viscometer-falling ball, capillary flow, orifice, rotational type, or amylograph is used to measure the viscosity of the starch paste. The shear stress of starch paste is expressed as a function of shear rate. This is described in different models, including Bingham, Herschel–Bulkley, and Powel law. The non-Newtonian feature of the viscosity of starch paste intends a nonlinear correlation between shear stress and shear rate. Starch paste exhibits decreased thickness with shear rate and time. Density increases with an increase in starch concentration, limiting its behavior with amylose content and determination temperature.

3.18 GEL FORMATION

Among the various methods used to characterize starch gels' rheological properties, the most frequent methods employed include assessing gel strength by texture analyzer and dynamic modulus analysis using a dynamic rheometer. This determination considers starch gels at various temperatures and shear rates. A dynamic rheometer evaluates the storage modulus (G'), the loss modulus (G"), and the loss tangent (tan δ=G"/G') of starch gel (Narpinder et al., 2003). The gel formation is aided through the interactions between amylose and amylopectin molecules. Amylose gelation contributes to the gel strength or stiffness upon cooking. Amylopectin recrystallizes upon storage and develops a stiff starch gel. Regular maize starch forms a firm gel at 6–8% w/w concentration, whereas waxy maize fails to form a gel at the same concentration. Adding sugars like glucose, maltose, fructose, and sucrose reduces gel strength attributed to local granule swelling due to water binding with sugar molecules. On the contrary, adding salts including Mgcl2, Cacl2, Na2So4, KCl, and NaCl decreases the enthaply (ΔH) and gelatinization temperature by enhancing water into the structure and stabilization of gel at low concentration. Salts including NAI, NASCN, KI, and KSCN decrease gel strength causing depression of starch granules.

3.19 GELATINIZATION

Heating starch in the presence of water or plasticizers (1,4-butanediol, ethylene glycol, and glycerol) or any other alkaline solutions (KOH and NaOH), neutral salt solutions (Licl, Cacl2) solvents like dimethyl sulfoxide (DMSO) brings about gelatinization of starch. The gelatinization temperature of starch is determined by magnetic resonance spectroscopy, thermochemical analysis, polarized light microscopy equipped with a hot stage, differential scanning calorimetry, thermochemical analysis, and other methods including FTIR and X-ray scattering for determining the degree of starch gelatinization. DSC measures gelatinization temperature and enthalpy change (ΔH). With inadequate water, the gelatinization peak broadens towards higher temperatures. The dissociation of crystalline double helices uses energy that reflects as ΔH. Water in the starch sample (at least two times w/w) plays a crucial role in providing consistent gelatinization temperature and ΔH. In the absence of plasticizers and water, gelatinization does not occur, and starch is thermally decomposed above 250°C.

3.20 FACTORS INFLUENCING GELATINIZATION PROPERTY

A positive correlation between the branch chain length (BCL) of amylopectin determines the thermal stability of crystallite formation from the long-chain amylopectin. The presence of long-chain double-helical crystallite of amylose and intermediate compound in maize (amylomaize V [amylose: 52%] and amylomaize VII [amylose: 68 %]) show high conclusion gelatinization temperature compared to other cereal starches. The gelatinization of cereal starches is altered with lipids, sugars, and salts. Gelatinization temperature and ΔH are elevated with the presence of glycerol. It is interesting to understand how incorporating plasticizers like 1, 4 butanediols, glycerol, and ethylene glycol increases gelatinization temperature more than water. This could be attributed to weaker hydrogen bonding capacity with water and plasticizers' molecular weight, which inhibit its penetration into the starch molecules. Similarly, simple sugar, including glucose, sucrose, maltose, and fructose and maltodextrins (maltoheptaose) increase gelatinization temperature and ΔH as sugar binds with water and reduces its accessibility for gelatinization.

3.21 THE FUNCTIONALITY OF MODIFIED CEREAL FLOURS

Cereal flours are modified to improve their physical properties. Cereal flours, their starches, and proteins have wide commercial usage. Maize starches are often used for this purpose and can be changed in various ways, including enzymatic or acid hydrolysis. The starches have been tweaked to improve their functional properties and expand their range of applications in the food business. The main goal of altering starches is to change their physical and chemical properties by physical, chemical, and enzymatic modification. Food businesses produce a variety of products as a result of these starch modification techniques on a vast scale. This form of starch and its modifications have been used commercially in the food processing industry. The sweetness of cereals produced by hydrolyzing their starches is highly varied, ranging from almost little sweetness to extreme sweetness. Hydrolyzed starches are used to make a variety of syrups. The extent of starch hydrolysis employed to manufacture the product determines the syrup quality, which is reported as Dextrose Equivalent (DE), representing the level of total reducing sugars. Maize-derived sweeteners, for example, are made by hydrolyzing the starches produced by wet-milling procedures. The significant techniques of starch hydrolysis in the food industry are acid-catalyzed hydrolysis, acid-enzyme hydrolysis, and enzyme hydrolysis. Furthermore, starch changes can be easily achieved by using cross-linking procedures with the appropriate reagents. The purpose of cross-linking is to reinforce the relatively fragile starch by adding chemical links at random points in the granule.

3.22 MODIFICATION OF CEREAL STARCH

Generally, native cereal starches possess a crystalline and complex granular structure. Various factors alter cereal starches' structural, chemical, physical, and functional properties. Cereal starches are the storage of starch copolymer with uni-physiochemical properties. Cereal starch modification can alter the prime structural properties, functionalizing copolymer. Modification can be carried out by chemical, physical, and enzymatic treatment. The chemical modification includes acetylation, cross-linking, and physical treatment includes microwave heating, heat-sheer extrusion processing, warm water treatment, and gamma irradiation. Starches are sensitive to increased temperature, low and high pH, mechanical stress, radiation, osmotic pressure, ultrasound waves, and light. Gelatinization of starch is obtained by the transition of amorphous to crystalline starch upon heat treatment of starch in an aqueous medium. Microwave radiation affects the crystalline structure and functional properties of starches. The breakdown of amylose–amylopectin linkage with the interaction of starch leads to disruption of crystallinity and swelling of starch.

Chemical factors also alter the structure and function of starches. Several hydroxyl or carboxyl derivatives of hydrocarbons, oxidizing agents, phosphates, carboxylic acids, cationic molecules, and various acid and base cross-linkers modify starches. Starch is very vulnerable to enzymatic and acid hydrolysis, resulting in the degradation of amylose and amylopectin, altering starch's surface properties and morphology, and varying the functional value.

Enzymes are used to improve the protein conformation, which would thereby enhance the functionality. This would be very beneficial in baking products. Compared with the chemical modifications in enzymatic modification, target-specific action, higher reaction rates, and lower toxicity are the advantages (Fan & Picchioni, 2020). Proteases and transglutaminases are the most commonly used enzymes to modify non-gluten proteins. These highly specific and selective enzymes hydrolyze the peptide bond preset in the polypeptide chain. Transglutaminase is a transferase enzyme that catalyzes the formation of isopeptide bonds. The bonds will improve the firmness of the bread. These enzymes change protein structure by hydrolyzing bonds or rearranging the protein network.

3.23 PHYSICAL MODIFICATION OF STARCH

Changes in the three-dimensional structure and morphology of starch result in physical modification. The various factors include ultrasonic waves, pulse–electric field, milling, moisture, pressure, temperature, radiation, and pH. The functional properties modified by physical treatment alter water absorption, swelling capacity, gelation ability, and pasting. The commonly used physical modification methods include microwave treatment, thermal inhibition treatment, osmotic pressure, high pressure, instantaneously controlled pressure, gamma irradiation, UV radiation, treatment by pulse electric field, stirring ball mill, annealing, freeze-thaw, and micronization in vacuum ball mill treatment.

3.24 THERMAL TREATMENT

There are various types of heating treatments. Gentle heating is where starch is heated (45–65°C). This causes trivial changes in the amylose to amylopectin ratio and has no significant differences in the physicochemical characteristics of starch. In superheating treatment, where starch is heated at a high temperature of 180–220°C, the development of spreadable gel particles with a creamy texture on cooking is observed. This change enhances the pasting and gelatinization of starch. In extrusion heating, where a combination of mechanical force is applied along with a low-temperature environment, the starch granules degrade through random chain splitting. Complete loss of crystallinity due to the high degree of granule disruption is also noted. Thermo extrusion is one of the latest and viable advancements in flour production with enhanced techno-functional properties. High shear extrusion is applied over various cereals and pseudocereals flour with specific mechanical energy (SME). Thermo extrusion improved the hydration properties and reduced the final viscosity of flour (Zhang et al., 2014). Significant cold gelling capacity was also noted in the extruded whole grain flour.

When starch undergoes hydrothermal treatment, an aqueous medium is used to heat starch, which causes the physical reorganization of starch granules with improved stability, mobility, and granule size. The gelatinization property of starch is also improved. In Heat Moisture Treatment (HMT) with a limited moisture level of 22–27% and by increasing the temperature higher than the glass transition temperature of 100–120°C for a time period of 1 to 24 hours, alter the shape, size, and crystallinity and granular structure of starches (Neelam et al., 2012). HMT aids the partial or complete alternation of B-type to A-type crystalline starch. The helical structure of starch within the amorphous region is also disrupted, inducing molecular degradation and an increased degree of polarization. HMT decreases peak viscosity, swelling capacity, and amylose leaching and enhances solubility, gelatinization temperature, thermal stability, pasting temperature, pasting time, and interaction properties. HMT also lowers the susceptibility of starch to chemical (acid hydrolysis) and enzymatic (α-amylases) action (Zavareze & Dias, 2011).

3.25 RADIATION TREATMENT

In microwave irradiation of starch subjected to a different range of temperature and moisture alters the dielectric constant: the surface morphology and granular crystallinity change in starch structure. Microwave treatment enhances emulsifying, water, and oil holding capacity, increasing paste viscosity and pasting temperature (Nawaz et al., 2018). Reduced peak viscosity, gelatinization, and a low degree of relative crystallinity are observed. Starch granules exposed to ultraviolet light undergo dextrinization, oxidative photodegradation, depolymerization, and free radical-induced cross-linking. The physical, chemical, and functional properties of starch are all affected by UV exposure.

Starch granules are exposed to high-energy gamma radiation doses. By breaking the amylopectin chain at the amorphous areas, gamma irradiation aids in lowering the amylopectin to amylose ratio (Raffi, J.et al., 1981). Starch undergoes radiolysis and radio depolymerization. The pasting viscosity, enthalpy change of starch and molecular weight, and gyration radius of amylopectin all decrease when exposed to gamma radiation. Solubility, viscosity, swelling power, and gelatinization are all rheological qualities.

3.26 NON-THERMAL MODIFICATION

starch is added as an acid or base to modify the medium's pH. Increased pH causes the partial breakdown of starch granules, accompanied by a decrease in molar size and radius of gyration. Low pH causes starch hydrolysis (especially in the amorphous region), which reduces compression capabilities, swelling power, solubility, and molecular weight. The ability of starch to gel is also enhanced at low pH.

3.27 MECHANICAL TREATMENT

Physical force is applied in pulverizing starch involving simple milling. The ratio of crystallinity/amorphous decreases (Silihe K, Zingue S, et al., 2017). Surface parameters are initially rapidly increased and then gradually decreased. The viscosity of starch decreases and becomes highly susceptible to various physical and chemical factors. Adsorptive capacity, the reactivity of starch, and water binding capacity increase. In micronization of starch using a vacuum ball mill, the starch granules (B- type) are damaged, resulting in loss of double helix content, granular order, and reduction in crystallinity. Starch polymers are also depolymerized. The rheological properties of starch are also influenced. The starch aids in increasing water adsorption, iodine biding capacity, solubility, granule swelling, and starch susceptibility to amylose. The viscosity and elasticity of starch decrease. When starch is subjected to mechanical activation by stirring ball mill causes degradation of crystal structure and leads to the formation of agglomerate to amorphous particles (Huang et al., 2007).

Treating starch by mechanical activation with stirring decreases enthalpy with low gelatinization temperature and shear thinning. The cold water solubility of starch is enhanced by the apparent viscosity (Huang et al., 2007).

3.28 PRESSURE TREATMENT

The microstructure of starch will alter over time due to high-pressure treatment at 400 MPa. The loss of birefringence and melting of amylopectin crystals occurs when pressure is applied. This changes the rheological properties of the starch, enhancing chewiness and hardness while also boosting the freeze-thaw resilience of starch gels. There would be a deformation of the crystalline area and a change from A-type to B-type after ultra-high-pressure treatment (>400 MPa) (Katopo et al., 2002). The rheological properties of the material change as a result of the pressure treatment, with swelling and amylose restriction increasing. The temperature of gelatinization drops as well. In cold water, the median volume diameter of starch increases after instantaneous controlled pressure treatment. Birefringence and gelatinization enthalpy decrease when exposed to polarised light (Loisel et al., 2006). The osmotic pressure treatment (hypertonic sodium sulfate at 100–120 C over the semipermeable membrane) converts B-type crystallinity to A-type crystallinity and alters the morphology of starch granules. This adjustment also raises the temperature of gelatinization. When high pressure (400–900 MPa) is applied, viscosity is reduced along with disintegration and retrogradation of starch.

3.29 ULTRASOUND TREATMENT

When starch is treated with ultrasonic vibrations, the granules get deformed. Temperature and enthalpy of gelatinization affect solubility. The capacity of starch to be pasted and digested is diminished. Viscosity and swelling capacity are improved by ultrasound treatment. Temperature and enthalpy of gelatinization increase and reduce solubility. The starch structure is disrupted in the crystalline area during high-pressure ultrasonication (24 HkZ–360 Hkz) frequency, lowering the enthalpy of gelatinization, crystallinity, starch molecular weight, and consistency coefficient. Increased interaction between amylose–amylopectin and amylose–amylose chain and crystalline area can be achieved by excess annealing water (>65% w/w) or intermediate water content (40–50% w/w) starch at a temperature lower than the onset temperature of gelatinization (Jambrak et al., 2010). Annealing increases the mobility of double-helical chain segments within granules, allowing for subsequent recrystallization, restructuring, or both, which improves molecular order and crystallite uniformity (Zia-ud-Din et al., 2017). With less amylose leaching and granule swelling, the thermal stability of the gel temperature improves (Tester & Debon, 2000). When starch is heated to a high temperature, it dehydrates and becomes anhydrous (1% moisture). The size of the starch granules is reduced. The cohesive texture of the starch improves, and the viscosity stabilizes (Zavareze & Dias, 2011). Cold plasma treatment of starch produces free radicals that cause cross-linking and enhanced amylose leaching because of the low temperature or glow discharge plasma. Depolymerization produced by active plasma species diminishes relative crystallinity – the potential for retrogradation decreases, but pasting and viscosity rise. In an electric field strength of 50Kv/cm, pulsed electric field therapy aids in the modification of starch water suspension. Enthalpy, gelatinization temperature, granule crystallinity, and viscosity are lowered. The diameter of the granules grew as the field strength rose (Han et al., 2009; Han et al., 2012). Freezing of starch is a physical treatment that occurs at extremely low temperatures (below 0 C), resulting in a change in texture and gelatinization qualities and an increase in retrogradation. The freeze-thaw treatment of starch includes heating it to a high temperature (59–79C), then freezing and defrosting it. The number of free-thaw cycles increases starch's complex modulus and phase angle. Starch's rheological properties are influenced. Starch's thermal stability, viscosity, and swelling power all improve. Starch granule surface characteristics are also impacted (Lawal, 2019).

3.30 DUAL MODIFICATION

Heat moisture treatment combined with annealing disrupts the crystalline structure causing an escalation in the enthalpy of starch. The annealing, sonication, and sonication annealing treatment resulted in low relative crystallinity due to the synergic behavior of the treatment. Pasting viscosity increases. Granule disintegration and surface morphology are also noted in the treatment.

3.31 CHEMICAL MODIFICATION

The carboxyl and carbonyl groups are introduced to the native starch using an oxidizing agent in the oxidation process. Adding a carbonyl and a carboxyl group to starch causes depolymerization and a delay in recrystallization (Kuakpetoon & Wang, 2001). Starch's clarity, stability, and binding qualities have all improved, but the viscosity of the dispersion has decreased. When starch is copolymerized with synthetic polymers, structural stability and retrogradation are reduced. Starch-based food items' freeze stability and shelf life have been improved (A. Korma, 2016).

3.32 ETHERIFICATION

3.32.1 Hydroxyethylation

The hydroxyethyl group is incorporated into the starch during this process. This causes a granular structure to change, resulting in increased binding ability (Paleos et al., 2017). Hydroxypropylation, which involves adding a hydroxypropyl group to starch, results in the disruption of inter- and intra-hydrogen bonds and a weakening of the granular structure. It also boosted the amorphous region's motional freedom of starch chains. The water-binding capacity, peak viscosity, swelling power, solubility, and enzymatic digestibility of starch were all improved by hydroxypropylation (Waliszewski et al., 2003). The paste clarity and freeze-thaw stability have also improved. The transition temperature, gelatinization parameters, and gelatinization enthalpy are all reduced. Gels' adhesiveness and hardness are also reduced (A. Korma, 2016).

3.33 ESTERIFICATION

The hydroxyl group of polymeric starch is reacted with an acetyl group in the acetylation process, which prevents crystallization or retrogradation. The suppression of intramolecular hydrogen bonding increases swelling capacity and viscosity. There is a decrease in solubility and a rise in pasting temperature. Fatty acids combine with starch to generate an amylose–fatty acyl complex, which alters starch's optical capacity and thermal behavior. When the hydroxyl group of starch is phosphorylated, monophosphate or diphosphate starch is formed. It increases steric hindrance and prevents molecular chains from being linear. Textural characteristics, viscosity, paste clarity, and freeze-thaw stability of starch are all improved. The temperature at which gelatinization takes place drops. This procedure enhances resilience to low pH, high shear, and high heat. When starch is succinylated using octenyl succinic anhydride, it is then derivatized with alkenyl succinic anhydrides. The peak viscosity, swelling volume, and cool paste viscosity were all enhanced, whereas the gel hardness and gelatinization temperature were decreased. This also boosts the production of resistant and slow-digesting starch. Two techniques of cross-linking in starch include etherification and esterification of granules with cross-linking polymers by reacting with a mixture of sodium trimetaphosphate and sodium tripolyphosphate, or cross-linking in an aqueous alkaline slurry containing sodium sulfate. Intramolecular interactions with small multifunctional molecules with hydroxyl groups on starch are utilized to restrict the mobility of the amorphous chain in starch granules, introducing crosslinking and fortifying granules against various causes. It also improves the internal stability and structure of the system. Cross-linking lowered the solubility of starch in water, reducing its interaction with moisture, lipid, and protein. It also reduces swelling capacity, starch enthalpy, and the rate of retrogradation. It raises the glass transition temperature, melting enthalpy, gel temperature, relaxation enthalpy, and starch stability at high temperatures while increasing the free volume of starch chains.

3.34 ENZYMATIC MODIFICATION

Native starch granules can also be modified to allow an enzymatic breakdown in specific locations. Enzyme assault is more vulnerable to the less well-organized amorphous regions, whereas enzymatic erosion is more resistant to the crystalline lamellae. The small starch granules have unique properties, including storing and releasing sensitive components such as flavors. Wheat starch granules with a diameter of 2 m can resemble lipid micelles and have a fat-like feel. Waxy maize starch nano-crystals have also been made with -amylase. Using starch nanoparticles as a filler in composites in the food industry can increase the composites' mechanical properties and biodegradability.

The molecular size, amylose-to-amylopectin ratio, molecular weight, and branch chain length distribution of starch are all affected by enzyme treatment. As a result, the branch chains produced by enzyme operations cannot be recrystallized and linked to other branch chains to form new hydrogen bonds. Furthermore, specific modifications to amylopectin molecules have been suggested to make their re-association less effective in increasing matrix rigidity during storage. This involves exposing starch suspensions to several enzymes, the most prevalent of which are hydrolyzing enzymes, which produce highly functional derivatives. This method can be traced back to the production of glucose syrup or high fructose corn syrup. Amylomaltases are enzymes that break a -1,4 bond between two glucose units and then produce a new -1,4 bond, resulting in a modified starch that can be employed in foods, cosmetics, pharmaceutics, detergents, adhesives, and drilling fluids. It is also an excellent plant-based gelatin substitute, although it produces a turbid gel, whereas gelatin gels are clear. All modified starches had their amylopectin chain length profiles widened. Cyclomaltodextrinase (CDase; EC 3.2.1.54) was isolated from alkalophilic Bacillus sp. I-5 (CDase I-5). It was utilized to alter rice starch to obtain low-amylose starch products. The amylose content of amylopectin decreased considerably from 28.5 to 9%, while the amylopectin's side-chain length distribution did not change significantly. The retrogradation rate of the modified rice starch was much slower than the control sample while held at 4°C for seven days. Treatment of maize starch with -amylase, -amylase and transglucosidase, maltogenic -amylase, and maltogenic -amylase and transglucosidase resulted in a 14.5%, 29.0%, 19.8%, and 31.0% reduction in digestion rate, respectively, resulting in resistant starch with a lower glycemic index that can be used in diabetes, prediabetes, and obesity. The delayed digestion was thought to be caused by the modified starches' increased starch branch density and crystalline structure. Cyclomaltodextrin glucanosyltransferase (EC 2.4.1.19) generates cyclodextrins (CDs) with a maximum yield of 3.4 and 100% retention inside waxy maize starch granules in the presence of isoamylase. Cyclomaltodextrins are synthesized *in situ* when CDs are retained in the granule, resulting in the development of a unique material with properties comparable to starch granules and cyclomaltodextrins. Creating complexes of organic molecules with cyclomaltodextrins allows for the progressive release of light, heat, and oxygen-sensitive components in starch granules and the delivery of distinct tastes, aromas, and flavors.

3.35 MODIFICATION OF CEREAL PROTEIN FOR DOUGH SYSTEM

Cereals belonging to the triticeace possess a nitrogenous storage protein – gluten. This protein is composed of two proteins: Gliadin and glutenin. In the bread-making process, this protein complex is critical, which plays a vital role in the fermentation process through the retention of carbon dioxide and in the final volume of bread. Other cereals, including rice, maize, and sorghum, do not form this gluten complex due to the absence of gliadin and glutenin. Further, the low protein content of these cereals predominately is prolamines and lysine. These protein types do not entrap gas in bread processing. Modification of protein is done through chemical, enzymatic, and physical processes.

Gluten is the major protein present in cereals, which are composed of two proteins, gliadin and glutenin. These proteins form a gluten matrix that plays a vital role in bread processing. The glutein matrix is responsible for holding carbon dioxide during fermentation and determining the final volume of the product. Other cereals like rice, sorghum, maize, and gluten-free cereals have low prolamines and protein content. Prolamines of gluten-free cereals including zein from maize and kafrin from sorghum cannot be beneficial in processing in their native form because of the encapsulation of prolamines in the matrix inhibiting the interaction among components and thereby failing to limit network formation, retention of gas, and expansion. Further, the predominant forms of non-polar amino acids, as in kaifrins, reflect the stiffness and rigidity of the products.

Like rice, most of the pseudocereals have a predominant globulin. These globulins are present in the endosperm, which is rich in arginine, aspartic, and glutamine. Most of the modifications of cereal protein are only for substituting gluten. Modifying the protein through chemical, thermo-mechanical, and enzymatic methods could address the population of gluten intolerance and Irritable Bowel Syndrome (IBS). Modifying protein content in cereals could aid the market potential of the gluten-free formulation. Modification can alter the functional properties affecting the behavior and texture and the product's thermal and rheological stability behavior.

3.36 CHEMICAL MODIFICATION OF CEREAL PROTEINS

Additive ingredients or chemical agents bring about modification. These induce protein structure changes, whether by conformation or bond modification. Glacial acetic acid and ethanol are the solvents widely used to modify prolamines. Hydrocolloids are also used in chemical modification. These hydrocolloids are water-soluble in polysaccharide, which dominates texture and rheology by forming hydrogen structure of bonds that stabilize the system. Generally, in protein modification of cereals, especially prolamines, ethanol, and glacial acetic acid, are used. An example documented for this modification includes the sorghum γ-kaifrin was studied on the rheology of resin developed and compared with the gluten-containing resins. The removal of gamma fraction, which constitutes cysteine disulfide bonds, was expected to decrease the stiffening of the resin. Still, with larger chain beta kafirin availability, the resin's viscosity surpasses the elasticity's ability. Removal of gamma kafirin reduced the elasticity (stress recovery 36%), changing the resin structure to be more viscous. The force needed for gluten-containing resins is 6.9 N, and the maximum force was increased by 580% with the removal of gamma kafirin. The crucial presence of gamma kaifirn helps to maintain the viscoelasticity behavior.

Dough properties could be improved with the usage of hydrocolloids. Hydrated hydrocolloids aid in forming hydrogen bonds which alter the rheology and texture of the system. A typical example is the addition of inulin and xanthan gum as hydrocolloids to rice flour in the formulation of gluten-free noodles to enhance their physical characteristics. When rice bran was incorporated into the hydrocolloids, a smoother and more consistent hydrocolloid protein matrix was formed, depicting improved tensile strength, elasticity, and firmness.

3.37 PHYSICAL MODIFICATION OF CEREAL PROTEINS

Modification by the mechanical or thermal source is a physical modification. The common physical treatment includes extrusion, freezing, and heating. Most physical treatments modify the functionality and structure of the protein. Physical modification is preferred over the other modification process due to the green process with zero toxicity. The combination process includes extrusion, microwave radiation, ultrasonication, and microfluidization with heat treatments. In mechanical treatments like ultrasonication, microfluidization, and extrusion channels of new interaction and varied particle sizes are observed in the dough system. Ultra sonication and microfludization disintegrate hydrophobicity, aiding in reduced particle size and altering the dough system's character. Thermal treatments result in changes in the structure and stability of the protein. Heat treatments enhance the number of disulfide crosslinking and oxidation of the sulphydryl group in cysteine, enhancing the dough's strength.

Unfolding of protein through denaturation is observed when microwave radiation is done, which enhances the water molecules to agitate, thereby increasing thermal energy. Extrusion improves friction and shear with increased internal temperature. Extrusion aids in improved elasticity through increased disulfide bonding resulting in higher molecular weight protein (Kaur et al., 2018).

Protein modification in cereals can be assessed by studying the thermodynamic properties and dough rheology. Evidence of protein modification can be explored through various techno-functional properties such as electrophoresis, spectroscopy, and microscopy. The rheological modification is studied using a rheometer. Protein denaturation caused by varied thermal behavior is studied through the DSC to assess the endothermic processes. The denaturation and stability of the protein are studied using the DSC, characterized by the melting and glass transition temperature (Tg). Protein stability is studied with the Tg, the temperature above which the relative mobility of the particle is increased. Tg provides the modification in the protein's tertiary structure, indicating the formation of covalent bonds between the structure. DSC aids in understanding the transition enthalpy changes (ΔH).

3.38 ENZYMATIC MODIFICATION OF CEREAL PROTEINS

Enzymes improve protein changes and thereby enhance their functionality in baking systems. This type of modification has several advantages over chemical modifications, including higher reaction rates (up to 1017 times), specificity, and safer (less toxic) reaction conditions, which is why it has been investigated to improve bread performance by increasing protein crosslinking to form a better protein network or hydrolyzing bonds to change protein structure and increase interactions. Proteases, transglutaminases (TGs), oxidases, and amylases have been used to alter gluten-free baked goods in the past; however, the last two cannot be considered as protein modifications because they work with non-protein dough components. Proteases are degradative enzymes that hydrolyze the peptide bond in polypeptide chains and are helpful in protein modification because of their specificity and selectivity. The effects of different proteases on the rheological properties of gluten-free rice bread have been studied. Papain, Protin SD-AY, and Protin SD-NY raised the elastic and viscous moduli by 5.6 and 3.5 times, respectively, indicating more elastic doughs that encouraged volume expansion. Newlase F, the only protease that did not show this behavior, was unable to build microscopic protein aggregates and was necessary to aid in the creation of protein networks, lowering both the specific volume and viscoelastic properties. When rice flour was replaced with quinoa flour, the addition of enzymes resulted in a 15% increase in specific volume (4.3%). This indicated that protease positively affected specific volumes, meaning that protein linkages between hydrolyzed chains can help the dough contain more gas. It is worth mentioning that because of the chemicals used in formulation, the specific volume increase will be more minor, showing that non-protein material can hinder the enzyme's function. Another enzyme investigated in gluten-free cereals is transglutaminase, a transferase that catalyzes the formation of iso-peptide bonds such as "-(-glutamyl) lysine, a bond formed between the -carboxamide group in glutamine and the "-amine of lysine (acyl acceptor). Adding TG to doughs with 20 and 30% sorghum increased bread stiffness by up to 50%; however, the treatment did not affect specific volume. Because zein lacks enough lysine residues to carry out the crosslinking reaction, and kafirin has a low lysine concentration (0.1%), the same inefficiency is anticipated.

3.39 APPLICATION OF CEREAL STARCHES IN THE FOOD INDUSTRY

The starches in cereal grains are the main components. There are numerous culinary options. Cereal starch applications have been studied and described in the literature. Swelling capacity, pasting behavior, viscosity, and other functional characteristics of grain starches have waxy and non-waxy properties, solubility, binding, and bulking capabilities. The food processing industry uses it in a variety of food compositions. Starch's physico-chemical properties and functional qualities in the aqueous environment, as well as its distinctiveness in diverse dietary applications, vary depending on the biological origin. Several metrics can be used to investigate the functional impact of starch on the end-use quality of cereal products. Pasting viscosity is linked to the end-use quality of white salted noodles (udon), and wheat pasting flours with higher viscosity are often favored. The hydrophobic portions of the amylose chain coil up with the helix. This helps them to trap oils and fats, as well as fragrance molecules, inside the helix. This property significantly impacts the ability of starchy grains to absorb oil. The presence of multiple hydroxyl groups on the starch molecules confers the hydrophilic feature. Because of hydrogen bonding, the polymer draws water and is self-attractive, most noticeable in the amylose or straight-chain component. Water-holding capacity, also known as swelling capacity, has a variety of industrial applications. Starches with a high amylose content swell far less than their non-amylose counterparts, showing a high level of intermolecular interaction. The onset of swelling and gelatinization is determined by the swelling characteristic of amylopectin crystallites within the amylopectin molecules. With increasing amylose content, the swelling power decreased linearly. The swelling behavior of cereals has been linked to amylopectin, implying that waxy cereals have the most swelling ability and that high swelling power leads to a less rigid granular structure. The swelling capacity of starches has been employed to screen and correlate with the eating quality of Japanese salt noodles due to its ease of determination. Pasting behaviors, represented in cereal starch viscoamylographic properties, have become more important in many food formulations. Starch exhibits a characteristic viscosity behavior with temperature, concentration, and shear rate changes. The Brabender Viscoamylograph, fast viscoanalyser, and rotational viscometers have been widely utilized for measuring starch paste viscosity and force direct demonstration of starch usage in foods. Starch concentration and starch type have a significant impact on pasting qualities.

The heating treatment is one of the most important aspects of several food processing techniques, and it significantly impacts the end-uses of cereals. The expansion and disruption of starch granules and the gelatinization phenomena occur when starches are heated in water. The fine amylopectin structure (i.e., linear chain distribution, length, and configurations) significantly impacts the starch gelatinization capabilities. Several studies have utilized rheological methods to investigate gelatinization in starch suspensions and to measure the viscoelasticity of starch pastes. For the application of starch, the rheological behavior under various temperature regimes is also of interest.

3.40 MALTS AND DISTILLED LIQUORS FROM CEREAL STARCH

Starch-rich cereal grains, including maize, sorghum, and rice, are promising renewable resources for distilling spirits. Converting cereal starches into alcohol via appropriate fermentation techniques is critical in such manufacturing. Specific enzymes are sometimes added to the solution, primarily when large amounts of starch must be hydrolyzed. In general, two steps are involved: 1) amylolytic enzymes convert starch to soluble sugars, and 2) sugars are fermented to alcohol by enzymes found in yeast primarily utilized for this purpose. The enzymes found in cereal grains, notably barley, hydrolyze carbohydrates during malting. Malting is a term used to describe the process of using the grain's inherent enzymes. This procedure entails controlled germination in which enzymes capable of catalyzing the hydrolysis of starches and other grain components are used. Malting can be done on any cereal, although barley is particularly well suited to the task. In addition, brewing businesses in many countries use grain starches such as rice, corn, sorghum, and millets as adjuncts. Cereals with high starch and low protein content are often favored for malting.

Malt is made from sorghum and millets in many African countries. Wheat malts are also made, but they are only utilized to make wheat-malt beer. Cereal starches make distilled liquors such as whiskies and neutral spirits. Depending on the starches of cereal grains, such as scotch malt whiskey, grain whisky, and so on, there are numerous sorts of whiskies, such as scotch malt whiskey, grain whisky, and so on. The action of enzymes saccharifies cereal starches. For the manufacturing of starch grain whisky, soft wheat with low protein and high starch content has become more popular. Distilleries employ cereals as a source of starch because they are cost-effective and generate liquors (whiskies) with good flavors. Cereals contain considerable amounts of starch or cellulose, making them feasible and renewable ethanol production sources. They may also serve as a sugar platform for a variety of bioproducts. For industrial applications of cereal starches, pretreatment technologies have been developed, including mechanical methods such as decortication and extrusion processes, physical procedures such as steaming and radiation, and chemical methods such as alkaline and acid hydrolysis, and biological methods such as microbial and enzyme degradation.

3.41 INDUSTRIAL NON-NOVEL APPLICATION OF CEREAL FLOUR

In the industrial processing of cereals, starch properties, including high-temperature hydration, swelling, enhanced velocity, and increased gelation capacity, are essential. Cakes, biscuits, other barley products, extrudes, expanded products, gums, candies, etc., depend on the cereal starch for manufacture. Cereal sources are also used in non-food industrial-scale starch production from starch nano-crystal elaboration, dextrin, gums, hydrocolloids, glue, etc. In encapsulation of various numerous food components, including probiotics, cereal starches are utilized. Granule morphology and starch biosynthesis are crucial in manufacturing products using cereal starches. The proportion of small and large granules of starch from barley and wheat have different behavior in brewing and baking, respectively. The highly branched glucans of rice starch are in high solubility in water favoring its utilization in the beverage industry. Waxy, normal, and high amylose barley vary in the structure of amylopectin and, upon physical modification (annealing), show varied responses. The rate of amylose/amylopectin and the agglomeration of starch chains in crystalline and amorphous regions could be the reason.

Cereal starch-based polymers (bioplastics) have recently attracted a lot of attention. Currently, petrochemicals are utilized to make practically all plastics widely used in many industries. Plastics are possibly the most essential of the many different materials on which mankind is currently reliant, given their ubiquitous use in food packaging. The starch-based coating has a number of advantages, including low cost of use, good taste release, compatibility with various processes, and friendly labeling. Bioplastics expand the demand for such products in mass markets.

3.42 CONCLUSION

New cereal industry trends address the ongoing need to develop new food items that are tailored to consumer desires. Using techno-functional qualities to identify flour properties is always an essential criterion for understanding the flour properties. Empirical hydration and rheology procedures, which involve well-known equipment in the cereal industry, are among cereal grains' technical and functional features. Their characteristics influence grain blending and are vital in the technical sheets that define flour usage. The contribution of technical and functional qualities to overcoming the technological constraints of working with grain flours is significant. These features have been highlighted to present a forward-looking picture of the relevant developments in cereal technological functionality. Traditional staple foods can be redesigned and modified to create sustainable food products that promote health and wellness. The capacity of the alteration to build a matrix superior to the native flour in terms of quality characteristics determines the end product's quality. Chemical, enzymatic, and physical modifications are the three primary modifications. There has been a detailed assessment of the most recent research on cereal modification. The documented effects and methodology for examining the changes made with each sort of alteration are outlined, as well as some potential areas for future research in the domain of modification impacts.

REFERENCES

A. Korma, S. (2016). Chemically Modified Starch and Utilization in Food Stuffs. *International Journal of Nutrition and Food Sciences*, *5*(4), 264–272. https://doi.org/10.11648/j.ijnfs.20160504.15

Espinoza-Herrera, J., Martínez, L. M., Serna-Saldívar, S. O. & Chuck-Hernández, C. (2021). Methods for the modification and evaluation of cereal proteins for the substitution of wheat gluten in dough systems. *Foods*, *10*(1). https://doi.org/10.3390/foods10010118

Fan, Y. & Picchioni, F. (2020). Modification of starch: A review on the application of "green" solvents and controlled functionalization. *Carbohydrate Polymers*, *241*(April). 116350.https://doi.org/10.1016/j.carbpol.2020.

Han, Z., Zeng, X. A., Fu, N., Yu, S. J., Chen, X. D. & Kennedy, J. F. (2012). Effects of pulsed electric field treatments on some properties of tapioca starch. *Carbohydrate Polymers*, *89*(4), 1012–1017. https://doi.org/10.1016/j.carbpol.2012.02.053

Han, Z., Zeng, X. an, Zhang, B. shan & Yu, S. juan. (2009). Effects of pulsed electric fields (PEF) treatment on the properties of corn starch. *Journal of Food Engineering*, *93*(3), 318–323. https://doi.org/10.1016/j.jfoodeng.2009.01.040

Huang, Z. Q., Lu, J. P., Li, X. H. & Tong, Z. F. (2007). Effect of mechanical activation on physico-chemical properties and structure of cassava starch. *Carbohydrate Polymers*, *68*(1), 128–135. https://doi.org/10.1016/j.carbpol.2006.07.017

Jambrak, A. R., Herceg, Z., Šubarić, D., Babić, J., Brnčić, M., Brnčić, S. R., Bosiljkov, T., Čvek, D., Tripalo, B. & Gelo, J. (2010). Ultrasound effect on physical properties of corn starch. *Carbohydrate Polymers*, *79*(1), 91–100. https://doi.org/10.1016/j.carbpol.2009.07.051

Katopo, H., Song, Y. & Jane, J. L. (2002). Effect and mechanism of ultrahigh hydrostatic pressure on the structure and properties of starches. *Carbohydrate Polymers*, *47*(3), 233–244. https://doi.org/10.1016/S0144-8617(01)00168-0

Kaur, N., Singh, B. & Sharma, S. (2018). Development of breakfast cereal based on quality protein maize by twin screw extrusion process for improved nutrition. ~ *38* ~ *Journal of Pharmacognosy and Phytochemistry*, *7*(4), 38–48. http://www.phytojournal.com/archives/2018/vol7issue4/PartA/7-3-482-312.pdf

Kuakpetoon, D. & Wang, Y. J. (2001). Characterization of different starches oxidized by hypochlorite. *Starch/Staerke*, *53*(5), 211–218. https://doi.org/10.1002/1521-379X(200105)53:5<211::AID-STAR211>3.0.CO;2-M

Lawal, M. V. (2019). Modified Starches as Direct Compression Excipients – Effect of Physical and Chemical Modifications on Tablet Properties: A Review. *Starch/Staerke*, *71*(1–2), 1–10. https://doi.org/10.1002/star.201800040

Loisel, C., Maache-Rezzoug, Z., Esneault, C. & Doublier, J. L. (2006). Effect of hydrothermal treatment on the physical and rheological properties of maize starches. *Journal of Food Engineering*, *73*(1), 45–54. https://doi.org/10.1016/j.jfoodeng.2005.01.004

Miano, A. C. & Augusto, P. E. D. (2018). The Hydration of Grains: A Critical Review from Description of Phenomena to Process Improvements. *Comprehensive Reviews in Food Science and Food Safety*, *17*(2), 352–370. https://doi.org/10.1111/1541-4337.12328

Narpinder, S., Jaspreet, S., Lovedeep, K., Navdeep, S. S. & Balmeet, S. G. (2003). Morphological, thermal and rheological properties of starches from different botanical sources. *Food Chemistry*, *81*, 1–31.

Nawaz, H., Shad, M. A., Saleem, S., Khan, M. U. A., Nishan, U., Rasheed, T., Bilal, M. & Iqbal, H. M. N. (2018). Characteristics of starch isolated from microwave heat treated lotus (Nelumbo nucifera) seed flour. *International Journal of Biological Macromolecules*, *113*, 219–226. https://doi.org/10.1016/j.ijbiomac.2018.02.125

Neelam, K., Vijay, S. & Lalit, S. (2012). Various Techniques For The Modification of Starch and The Applications of Its Derivatives. *International Research Journal of Pharmacy*, *3*(5), 25–31.

Paleos, C. M., Sideratou, Z. & Tsiourvas, D. (2017). Drug Delivery Systems Based on Hydroxyethyl Starch. *Bioconjugate Chemistry*, *28*(6), 1611–1624. https://doi.org/10.1021/acs.bioconjchem.7b00186

Raffi, J., Frejaville, C., Dauphin, J.F., Dauberte, B., d'Urbal, M. and S. (1981). Gamma Radiolysis of Starches from Different Foodstuffs*Part II. Study of Induced Acidities. *Starch – Stärke*, *33*(7), 235–240.

Shaoxiao, L.W.G.Z.Z. and Baodong, Z. (2013). The influence of ultra high pressure treatment on the physicochemical properties of areca taro starch. *Journal of the Chinese Cereals and Oils Association.*

Silihe K, Zingue S, W. E., Awounfack C, Bishayee A, D. N. & Al, E. (2017). Starch Modification and Applications. *International Journal of Molecular Sciences.*, *18*, 1073.

Tester, R. F. & Debon, S. J. J. (2000). Annealing of starch – A review. *International Journal of Biological Macromolecules*, *27*(1), 1–12. https://doi.org/10.1016/S0141-8130(99)00121-X

Waliszewski, K. N., Aparicio, M. A., Bello, L. A. & Monroy, J. A. (2003). Changes of banana starch by chemical and physical modification. *Carbohydrate Polymers*, *52*(3), 237–242. https://doi.org/10.1016/S0144-8617(02)00270-9

Zavareze, E. D. R. & Dias, A. R. G. (2011). Impact of heat-moisture treatment and annealing in starches: A review. *Carbohydrate Polymers*, *83*(2), 317–328. https://doi.org/10.1016/j.carbpol.2010.08.064

Zhang, C., Zhang, H., Wang, L. & Qian, H. (2014). Physical, functional, and sensory characteristics of cereal extrudates. *International Journal of Food Properties*.17(9). 1921–1933. https://doi.org/10.1080/10942912.2013.767831

Zia-ud-Din, Xiong, H. & Fei, P. (2017). Physical and chemical modification of starches: A review. *Critical Reviews in Food Science and Nutrition*, *57*(12), 2691–2705. https://doi.org/10.1080/10408398.2015.1087379

PART II

Mechanical Processing of Cereals and Its Impact

CHAPTER 4

Milling/Pearling of Cereals

Farhana Mehraj Allai[1], Z.R. Azaz Ahmad Azad[2], Nisar Ahmad Mir[1], and Khalid Gul[3]
[1]Department of Food Technology, Islamic University of Science & Technology, Awantipora, Jammu and Kashmir, India
[2]Department of Post-Harvest Engineering and Technology, Aligarh Muslim University, Aligarh, Uttar Pradesh India
[3]Department of Food Process Engineering, National Institute of Technology, Rourkela, Odisha, India

CONTENTS

4.1 INTRODUCTION

Cereal grains have been considered a keycomponent of a low-fat diet for humans for thousands of years (Seal et al., 2021). The majority of the population relies on cereals due to their energy value and nutritive attributes (Burlando, & Cornara, 2014). Cereals like rice, wheat, corn, barley, millets, rye, pseudocereals, etc., provide essential nutrients (both macro as well as micro) and are also the dominant source of protein, minerals (iron, magnesium, zinc, and phosphorus), and dietary carbohydrate (both soluble and insoluble). They are also rich sources of naturally occurring phytochemicals including carotenoids, phenolics (flavonoids, alkylresorcinols, phenolic acids), vitamin E, and β-glucan, which have a positive impact on human health (Ragaee et al., 2014). The potential health benefit of whole grains increases the demand and supply, thus the annual production of cereal exceeds 2,700 million tons (Saini et al., 2022).

DOI: 10.1201/9781003242192-6

Whole grains can be defined as the intact, cracked, grounded, or flaked caryopsis, consisting of three main parts: endosperm (80–85%), germ (2–3%), and bran (13–17%). The endosperm is embedded in the protein matrix containing cells of starch polymers. Germ is a rich source of fat-soluble vitamins and oils and the bran (outer layer) contains insoluble dietary fiber and minerals (Calinoiu, &Vodnar, 2018). Whole grains are more nutritious and dense as the germ and bran fractions are retained in the same proportion as in the original grain (Allai et al., 2021). Both the layers contain high concentration of bioactive substances including dietary fibers, resistant starch, carotenoids, minerals, anthocyanins, vitamins, and phenolics (Liu & Hou, 2019). As the whole grains and their products are tasteless and the texture is affected negatively, the consumption of whole grains is reduced to lower than the daily recommended allowance (Oliveira, Rosell, & Steel, 2015). Typically, consumers prioritize the taste of food over its nutritional value, even if the food is less nutritious (Kamar, Evans, & Hugh-Jones, 2016). Thus, all the commonly consumed cereal grains are processed before cooking to have good palatability and improve their digestibility (Tosi et al., 2020

Cereals are typically processed before consumption to improve their digestibility, taste, and nutritional value. Processing techniques may vary depending on the type of cereal. The processing of cereals is done by grinding techniques using different types of mills such as hammer, roller, plate, disc, and stone. The size of cereal grains can be reduced through impact and compression; the reduction degree and the resulting particle size can be influenced by method of processing, type of grain, and equipment to be used. Generally, during processing shear force is one of the important mechanisms for reducing the particle size of cereal grains. This force is applied in a parallel direction to the surface of grain, leading to deformation and breakdown. Shear force can be applied through variety of methods such as mixing, grinding, and milling.

Milling is one of the principal pre-processing procedures for grinding grains such as wheat, maize, barley, oats, etc., to break them into smaller sizes. This process decreases cooking time and also improves digestibility. Milling has a significant effect on the nutritional as well as physicochemical properties of the product and can influence the particle size distribution of flour that can in turn affect the textural characteristics and water absorption capacity of the final product. Top of FormMilling also influences the technological, functional, physicochemical, and nutritional aspects of products. Processing of cereals is usuallycomplex and is categorized into two main groups; first primary processing includes milling, malting, or pearling and the secondary processing group involves cooking, baking, extrusion (pasta, breakfast cereals, and snacks), puffing, etc., and the other process is a fractionating group in which the cereal grains are broken into flours of different particle size that have different applications for food and nonfood. Previous literature reported that most of the bioactive substances are concentrated mainly in the bran layer of the grain (Sovarni et al., 2012), which consequently promotes great reduction in the nutritional value of the processed flour (refined flour) (Siebenhandl et al., 2007). The fractionation technique is another alternative strategy that involves the pearling process. The pearling process is done before milling, since it improves the effectiveness of the milling process with the application of friction and abrasion, whichremoves the outer bran layer effectively, allowing nutritious fractions like aleurone layer to be retained in the kernels (Campbell et al., 2012). During milling/pearling, many changes take place that affect the sensory, nutritive, functional, and biological characteristics of cereal products. This chapter aims to provide updated information on the impact of milling techniques on the nutritive, functional, and biological properties of cereal grains.

4.2 PROCESSING TREATMENT

The term milling can be defined as the conversion of whole grains like wheat, oats, maize, rice, and barley into suitable forms such as flour, flakes, or meal that can be utilized as primary ingredients for further processing in the food industries. Whole wheat flour is developed from the entire wheat kernel including all the three parts (i.e., endosperm, germ, and bran); while during milling, two most important layers (i.e., germ and bran) get removed, which contain most of the essential nutrients such as both soluble and insoluble fiber, bioactive compounds, minerals, vitamins, etc., to produice refined flour having smoother texture and longer shelf life. The process of milling can be simple as well as complex, where the grinders are simply used to transform raw material into flour of specific particle size distribution. Moreover, the production of refined flour/white flour is a complex process that involves several different stages of milling, sifting, and purifying in order to remove the germ and bran layer from the starchy endosperm (Liu et al., 2018). After milling, grains can further be processed by soaking, toasting, or cooking to soften and release the starch depending on the desired food product.

The milling process can be applied to an array of cereal grains to develop a wide range of food products ranging from flour for baking to breakfast cereals, beverages, and snacks. The milling of cereal grains such as wheat, rice, oats, and corn can be utilized to develop convienece foods such as breakfast cereals and porridge, while the milling of starchy foods like potatoes can be used to produce potato starch or flour. Also, in the brewing industry, barley milling is used for the production of beer (Owens, 2001). Thus, process of milling can be adapted to meet the specifications and needs of different industries and end products, resulting in a huge range of milling techniques to be used such as hammer mills, roller mills, stone mills, ball mills, and jet mills.

The milling step itself does not involve intentional heating, but in some cases additional processing steps may include drying or heating phases (e.g., after milling, oats are subjected to heating phase also known as toasting or kilning). This process helps to improve aroma, flavor, stability, and shelf-life of oats whereas before milling of maize the moisture content is reduced in drying phase to prevent growth of microorganisms that can detoriate the maize during storage and transportation.

4.3 CLASSIFICATION OF MILLING

Milling processes are classified as wet and dry milling, depending on the degree of size distribution rather than the difference in water, as water is used in both cases for the separation process.

4.3.1 Dry Milling

Dry milling is the simplest technique for the development of a wide variety of food products for human consumption. Grinding of grain as a whole kernel in a roller mill or grindstone is a simple method used across the globe for producing flour (Eckhoff et al., 2003). The roller mill is equipped with roll stands consisting of pair of rolls (back-to-back) placed within the same housing but operated independently. Both the rollers are aligned in a parallel direction but rotate in the opposite to maintain the differential speed. The shelf-life of products made from such flour has limited storability due to the presence of unsaturated fatty acids in the germ of the grain, which make the food products more susceptible to off-flavors and rancidity. The broken cells of the germ can become easily oxidized, resulting in the generation of free radicals and breakdown of the fatty acids into off-flavors. As a result, the food products developed from whole grain flour have shorter shelf-life as compared to refined products that have had the germ removed. To overcome this problem in storage, the removal of germ (10% fat) is the first step in processing to produce products with greater shelf life with much lower fat (Kent, 1994) and also it is important to store the food products away from light and in dry and cool places. Figure 4.1 shows the dry milling process of some common cereals.

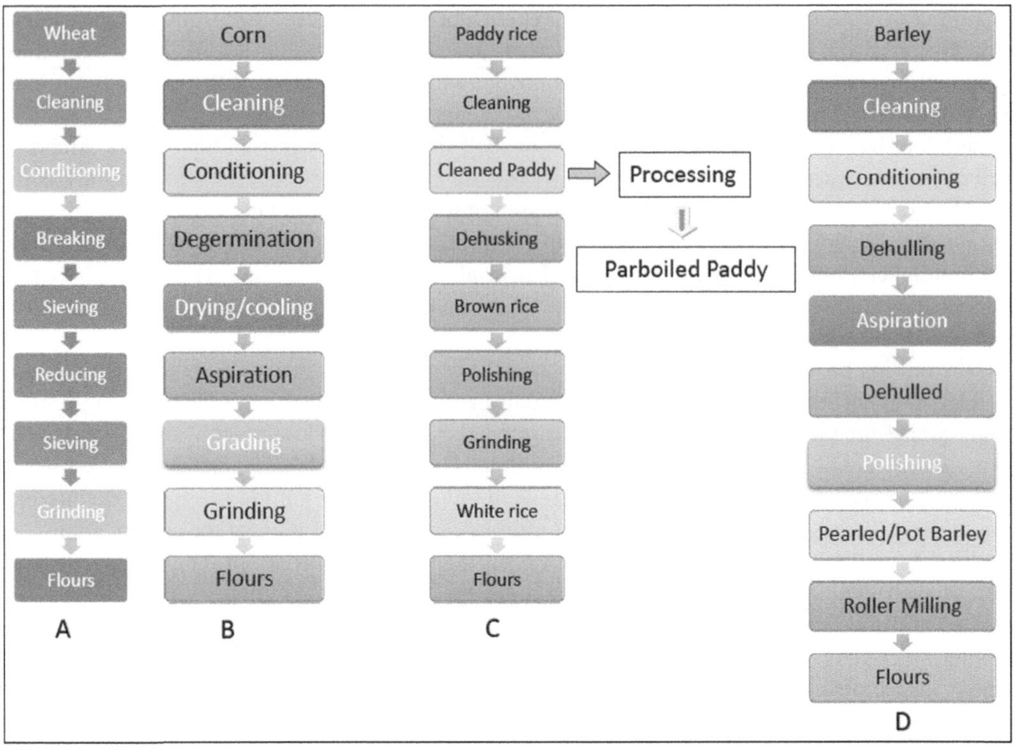

Figure 4.1 Flow chart of dry milling process for (A) wheat, (B) corn, (C) rice, and (D) barley.

In dry milling, damping or tempering is a pre-milling treatment, used as a step in the cleaning process where impurities like dust, dirt, and debris are removed from the cereal grains. It also helps to reduce the amount of broken kernels and thus improves the overall quality and uniformity of grain. Top of FormTempering or conditioning is a preliminary step of processing where water is added under a controlled condition to reach the required moisture level to induce swelling of pericarp, endosperm, and germ, for ease of removal or separation. Conditioning needs much less water than washing, as grains absorb the water, so effluent problems do not arise. In maize, dry milling is done by a degerminator that loses or breaks the pericarp and germ from the endosperm (Dowswell et al., 2019). The material obtained is aspirated, sieved, and milled through a roller to get flour of different particle sizes. The large flaking grits with low fiber and fat content are suitable for the formulation of flaked breakfast cereals while smaller particle sizes are used for the manufacturing of alcoholic beverages, extruded corn snacks, etc. (Eckhoff et al., 2003).

Impact grinder is another system in which machines like turbo crushers or encounters separate the embryos. This method is relatively less expensive and subsequent drying is not needed as it is performed at natural moisture level. Maize germ and maize grits are the products developed from this method. These products have 14.5mm particle size and vary in their fat content (maize germs 18–25% fat and maize grits 1.5% fat). Grits are further processed to reduce the particle size even more such as in semolina and flour. When the particle size of starchy endosperm is reduced to semolina or flour, the fat content of the resulting product is reduced as the fat is concentrated in the germ and bran layer that are being removed during the milling process (e.g., refined wheat flour contains 1–2% fat and semolina, which is prepared from durum wheat containing only 0.8–1.3% fat). In paddy grains, the awns are to be removed before husking as they are not edible; this process is known as deawning (Alizadeh et al., 2012). Conditioning is not necessary for the cleaned paddy rice. Steel rollers or rubber rollers are used to dehusk the cleaned paddy rice. The efficiency of the husking procedure depends on the variety of rice, the humidity of grain, and the tightness of the husk bound to the rice. The husk is then removed by aspiration process in which the coarse particles are trapped and rice passes through the sieve (Bond, 2004). In rice kernel bran is removed by intensive thermal and mechanical stress in a process called pearling. Pearling machines remove bran by friction or abrasive process. The whiteness of rice grain depends on the type of grain, weather, and storage conditions (Singh et al., 2013). After pearling, the loose bran is eliminated during a polishing step. The main focus of the rice milling industry is on the reduction of broken kernels. Mechanical properties like degree of milling, head rice yields, and physical characteristics such as shape, size, and hardness affect the degree of milling (Liu et al., 2009).

4.3.2 Wet Milling

Contrary to dry milling, wet milling involves soaking of grain in water and then separating it into several different components where the chemical and physical changes of basic constituents such as cell wall material, protein, fiber, oil, and starch take place and cause complete damage to the endosperm and release the starch granules that are embedded in a protein network (Wronkowska, 2016). These changes occur due to the presence of water and various enzymes that can break down complex larger molecules of grain into simpler components. Wet milling is commonly used for wheat and corn, but it is also possible to wet mill other cereals successfully such as barley, rice, oats, and sorghum. The wet fractionation method has several steps including handling of grain, steeping, separation, and recovery of basic constituents.

In wet milling, cleaning is the first step where the grains are received in bulk and contain different types of impurities such as shriveled grains, grains with discolored germ, grains damaged by pests, sprouted grains, husks, and dead insects (The European Commission, 2010). After cleaning, steeping is another critical step followed by soaking and shifting. The steeping process is done by different methods like enzymatic, alkali, conventional, and modified wet milling (Eckhoff et al., 1999; Ramirez et al., 2009; Maphosa &Jideani, 2016). The conventional method takes approximately 36 h to complete. It involves the soaking of kernels in sulfuric acid solution to soften the endosperm. In this step, the final moisture content of corn reaches 50% of the weight of kernel and 5–7% of solids present in corn such as lactic acid, minerals, vitamin B, albumin, phytic acid, and globulins are solubilized (Johson and May, 2003). In steeping solution soluble proteins are recovered during centrifugation or filtration process after maintaining the pH value. The conventional method does not change the properties of starch as compared to raw material (Wronkowska and Haros, 2014). The methods employed in wet milling of cereal grains can vary depending on the type of grain being processed and the final product that is intended to be produced.

In corn, the maximum amount of starch is extracted during wet milling. Thus, starch is considered to be the primary product of wet milling. Depending on the primary source, starch is obtained in the form of waxy, regular, and high-amylose starch. The main use of corn starch in the food industry is to prepare syrups, thickeners, soups, bakery, baby foods, confectionery products, brewing adjuncts, and other modified starches. From a cost-effective point of view, the main application of corn wet milling is the conversion of starch to ethanol and sweeteners (Awika, 2011). Apart from the food industry, corn starch finds many nonfood applications such as in the textile industry, pharmaceutical industry, packaging industry for making adhesives, etc. (Papageorgiou & Skendi, 2018).

In rice, wet milling rice flour or rice pulp is produced with fine particle size (10–30 µm) (Chen et al., 1999) to develop a variety of products with a smoother texture like in the case of noodles and baked products (Yoenyongbuddhagal & Noomhorm, 2002). The flour produced from wet milling is superior to that produced by dry milling (Bean et al., 1983). The most essential step in rice wet milling is soaking; this step influences the particle size of rice flour and increases the soaking time from 3 to 168 h. In soaking water diffuses into the cereal grains resulting in the leaching of some compounds and producing fine flour with less starch damage, low pasting properties, high peak viscosity, and high retrogradation (Chiang & Yeh, 2002). In dry milled rice, more damaged starch is observed, thus giving a lower peak with better solubility and fewer final viscosities.

The wheat wet milling process involves the separation of flour into its basic ingredients, primarily gluten and starch. Dry milling of wheat flour leads to the loss of some protein and starch and thus minimizes the recovery of minor components such as pentosan and bran (Velicogna & Shea, 2016). This is because heat is generated during the milling process, resulting in the denaturation of protein and causing some starch to disintegrate into simple sugars. After cleaning of wheat, steeping is done in which double the weight of water is taken and is kept for 10 days where gluten is separated from starch and produces gluten protein network. The water is then drained off and the soaked wheat slurry is enclosed in a cloth and the starch released out. The starch is dissociated from the gluten network without damaging it and disintegrated into smaller particles. Soluble compounds pass with water through the sieves while wet gluten (insoluble) is retained. The other fractions – germ, bran, and gluten – are removed and used as fodder. Yuan et al. (1998) reported that the steeping of hard red winter wheat is done in different acid media like HCl, SO_2, and lactic acid at different concentrations and at different time intervals. It was observed that after 16, 20, and 24 h steeping acid concentration and time affected the protein value, yield, color, and consistency of the obtained starch.

Wet milling of other cereal grains such as rye, triticale, barley, and oat starches is very limited. The extraction of starch from these cereals is difficult as rye has poor gluten forming ability and higher pentosan content; in the case of oat wet milling hydrated protein and bran layers are formed. Barley starch contains a high level of β-glucan that produces solutions with higher viscosity, thus making the separation of starch very difficult by centrifugation or screening method (Serna Saldivar, 2022. In wet milling, barley can be utilized for the production of malt, which is an essential ingredient for the preparation of fermented beverages such as beer. Initially the barley grains are steeped in water to initiate the germination process where the enzymes get activated and can disintegrate the protein and starch of cereal grains. Then the grains are dried and further processed to eliminate husk, leaving behing malted barley.

Whole buckwheat milling is similar to wheat milling but here first dehulling is done by abrasion or impact milling with grinding plates or emery stones. In wet milling starch extraction rate is higher in hulled grains as compared to dehulled buckwheat and the characteristics of isolated starch are insignificantly changed as compared to raw material (Sharma et al., 2021).

4.4 IMPACT OF MILLING ON NUTRITIONAL CHARACTERISTICS

Cereal grains are dehulled, milled, refined, polished, etc., after cleaning and grading to yield products for industrial usage. These pre-processing treatments change the nutritional value of a resultant product to varying degrees. It also alters the matrices, cells in which nutrients are ingrained, removal of bran and germ fractions, etc. Although milling is a prime step in the post-production of some cereal grains, the main aim of milling is to remove the outer shell (husk) to produce an edible fraction in the form of flour with varying particle sizes (Oghbaei, & Prakash, 2016). In addition to producing powder of varying protein composition and size, modern milling techniques are used to separate the flour particles of varying sizes to develop flour with exact protein values used for the preparation of different bakery items like cookies, doughnuts, cakes, and bread. Similarly, whole brown rice takes 1 h for cooking to soften while removal of the hull in white rice requires only 20 minutes of cooking. Dried rice or fully cooked rice can be cooked in a minute. Thus, milling saves cooking time energy and provides convenience (Ruth & Cheryll, 2017). Moreover, rice milling is done to improve sensory and physical attributes and also enhance the shelf-life stability of rice grains. The physical characteristics of rice grains play an important role in calculating the textural properties and cooking time; the thickness of rice grain affects the cooking time and texture as the grain thickness increases the duration of cooking (Bhat et al., 2019). Parboiling is another method that involves partial boiling of paddy before de-husking. The objective of parboiling is to decrease the breakage in milling and enhance the nutritive value (Kalita et al., 2021).

Structurally, whole cereal grains contain all the three fractions intact:endosperm, bran, and germ. Nutrients (both micro and macro) and phytonutrients are distributed throughout the grain but the highest concentration of nutrients is found in the outer layer of the grain, so refining or differential milling decreases the nutrient value except starch (Slavin et al., 1999). In the milling of wheat, only the starchy endosperm is retained containing carbohydrates. Many nutritionally valued biochemical substances like minerals, dietary fiber, vitamins, and bioactive compounds are lost during conventional milling, which plays a vital role in decreasing chronic diseases (Mellen et al., 2008). When bran and germ are removed, white flour is produced with very low nutritive value, resulting in nutritional disorders and digestive disturbances (Iuliana et al., 2012). Table 4.1 depicts the impact of the milling process on the nutritional characteristics of cereal grains as reported in different studies. The milling extraction rate of white flour is only 68%, which means the rest of 32% is lost during milling while whole grain flour has a 100% milling extraction rate, which means all the parts of the seed are there in the flour (Heshe et al., 2016). Germ and bran fractions of grains contain 44.45% of minerals and vitamins. About 70–80% of vitamins are lost during the milling of grains such as vitamin B_6, thiamine, riboflavin, biotin, folic acid, pantothenic acid, and niacin. There are also great losses of minerals (25–90%) (Fardet, 2010; Truswell, 2002) including 85% zinc loss, 80% potassium, copper, and magnesium loss and 90% loss of manganese loss and 25% protein loss (Ramberg & McAnalley, 2002; Redy & Love, 1999).

Rice is a rich source of dietary fiber, but during processing, several layers of rice grain are removed affecting the quality of rice samples (Fernando, 2013). Brown rice and white rice contain 0.6–1 g/100g and 0.2–0.5 g/100g of crude fiber, respectively. Milled unparboiled rice shows lower fiber content as compared to milled parboiled rice as the parboiling method retains most of the nutrients present in the bran and transfers them into the starchy endosperm (Lovegrove et al., 2019). The insoluble fiber present in parboiled rice has several health benefits like giving the feeling of fullness, reducing weight, decreasing the risk of diabetes, slow release of insulin, etc. The flaking and puffing process causes changes in dietary fiber, phytin phosphorus, and phosphorus content. Flaked rice has a direct relation with the thickness; the lower the thickness the lesser the constituents, whereas calcium and iron contents remained unaffected (Suma et al., 2007).

The process of milling, though having its limitations, has many positive aspects as well such as improved digestibility and starch content of grain (Oghbaei & Prakash, 2012, 2013). The milling method and particle size are directly proportional to the starch content of flour. Reduced mesh size increases the starch content (Kerr et al., 2000) as more fiber is separated and fine flour with higher starch content is produced. Anti-nutritional factors are found in the bran layer of grains like phytate and tannin, which can bind with protein and enzymes and thus decrease their activity (Saleh et al., 2013).

4.5 IMPACT OF MILLING ON THE FUNCTIONAL CHARACTERISTICS

Functional modification of cereal grains using milling has gained attention as it has the ability to alter the functionality of cereal products. The altered functionality can be due to the change in structural conformation like in the case of ball-milled starch and food products enriched with starch. As the milling time increases enthalpy value gradually reduces, regardless of the stress applied.

Table 4.1 Impact of Milling Processes on the Chemical Composition of Common Cereal Grains (per 100g)

	Differential milling		Dry and wet milling		Sieving		Dry and wet milling		Commercial milling	
	Wheat flour		Rice		Finger millet flour		Sorghum		Pearl millet	
Components	**Whole**	**Refined**	**Whole (PR-113)**	**Milled**	**Whole**	**Refined**	**Whole**	**Flour**	**Whole**	**Refined**
Fat (g)	2.50	1.54	1.65	1.24	1.78	1.29	1.9	2.80	1.5-6.8	5.8
Protein (g)	14.28	11.6	7.63	9.16	7.15	6.33	10.04	6.20	19.4	12.4
Ash (g)	1.89	0.78	0.52	4.43	2.21	1.80	1.6	1.6		
Soluble dietary fiber (g)	0.51	0.27	0.61	-	1.55	1.79	1.73	1.59	9.0	7.5
Insoluble dietary fiber (g)	12.45	3.39	2.44	1.04	20.23	12.15	8.49	8.10		-
Zinc (mg)	1.62	0.70	1.35	0.2–6.3	2.50	1.98	1.60	1.3	2.4	-
Iron (mg)	7.54	3.24	0.43	0.2–0.8	6.52	3.29	4.10	8.4	10.8	-
Calcium (mg)	50.75	34.20	7.33	10–30	404.3	294.8	25.0	10.03	41	188
Thiamine (mg)	0.64	0.33	0.13	0.02–0.11	0.55	0.34	0.35	2.31	0.3	0.25
Riboflavin (mg)	0.21	0.13	0.037	0.02–0.06	0.24	0.19	0.14	0.38	0.2	0.06
Niacin (mg)	-	-	4.3	1.3–2.4	-	-	2.10	3.1	2.9	1.32
Tannins (mg)	385.7	127.8	-	-	851.2	563.9	-	-	-	-
Phytates (mg)	604.0	396.5	1.85	-	628.2	432.0	-	-	-	-
Phytin (g)	-	-	-	0.02–0.07			-	-	-	-
References	Oghbaei, & Prakash (2013)		Priya et al., 2019		Oghbaei, & Prakash (2012)		Dayakar et al., 2017		Embashu, &Nantanga (2019)	

Change in enthalpy can be measured by the reduction levels of amylose-lipid complexes, which further lose the crystallinity that can be interpreted by FTIR spectroscopy. A similar result was found in rice flour or rice starch processed in a planetary ball mill (Han et al., 2007; Loubes et al., 2012). The ball mill native cassava starch lost its crystallinity and was highly amorphous. Thus, milling also damages the helical structure of amylose and hence fragments the amylose-lipid complex (Roa et al., 2014). The milling process disintegrates a part of starch polymer, which causes higher susceptibility to hydrolysis and maximum levels of hydration. These properties in the dough can contribute to greater water absorption capacity and enhanced gassing power, and influences the final texture and quality of other developed products (Price et al., 2021). The highest water absorption capacity was observed in hulled barley followed by hulled oat, triticale, rye, and wheat. This sequence can be explained by the distribution, composition, and amount of fiber present in the grain. Cellulose is found in the outer layer of grain while arabinoxylans location varies according to the grain type.In the case of rye, wheat, and triticale it is found in the aleurone and endosperm layer and barley and oat arabinoxylans are present in the aleurone and bran layer. In barley, β-glucans are distributed in the endosperm whereas in wheat the main role of water absorption is mainly due to the gluten protein. Rye, triticale, hulled oat, and barley has the maximum amount of fiber that compete with the protein for water (Aprodu, & Banu, 2017).

In a planetary ball mill degree of modification depends on the nature of starch, the intensity of the process, amorphous phase enhancement, and deformation in ordered structure (Tan et al., 2015). Previous studies also reported that the change by planetary mill also produces changes in less ordered areas of double helix structure and crystalline regions (Liu et al., 2011). Ramadhan, & Foster (2018) observed that in oat bran protein flour, the structure of starch and protein was altered by the ball milling process and also decreased the enthalpy changes of lipid-starch complexes melting characteristics. In ball milling, higher speed decreased the particle size distributions from 116.45 to 27.67mm resulting in minimum utilization of particle volume packing. Moreover, the heating-cooling cycle allowed the starch to make a stronger network structure. Ball milled samples showed thermal behavior that remained unchanged in peak transition temperature upon the second heating cycle.

Fractionation is a process in which the particle size of fiber-rich plant constituents is reduced, which modifies the surface area, structure, and functional characteristics of the particle. Wu et al. (2007) and Chau et al. (2007) assumed that insoluble fiber present in the plant material is subjected to different milling treatments like jet milling, high-pressure micronization, and ball milling resulting in a change in the structure of bran, and thus improving physical, chemical, and functional characteristics, which have a positive effect on human health. These treatments rearrange the insoluble fiber to more soluble form, which improves the nutrient availability of the fiber and thus digestibility. Additionally, to improve the digestibility and solubility of the fiber, these milling techniques can also increase functional properties of fiber such as texture, viscosity, and water-holding capacity as the compounds present in the outer layer (i.e., bran) interact with water through hydrogen bonding (Tulse et al., 2014). This can make the fiber have many food applications such as preparation of snacks, baked goods, and other processed foods. Bottom of FormMilling time is another factor that affects the viscosity of starch. Less milling time results in high viscosity as most of the granule integrity is conserved. Viscosity is decreased with an increase in moisture content and milling time, attributed to the disintegration of starch polymers by the milling process (Martínez-Bustos et al., 2007). Rao et al. (2015) also reported that apparent viscosity reduced with an escalation in shear rate, which means all the samples showed shear thinning properties. The maximum viscosity values were shown at the minimum shear rate and vice versa. A similar result was shown by Ibanoglu and Ibanoglu (1999) for semi-liquid breakfast cereal.

4.6 IMPACT OF MILLING ON BIOLOGICAL CHARACTERISTICS

The bran layer of whole grain contains essential vitamins, minerals, bioactive compounds, insoluble dietary fibers as well as anti-nutrients that bind with proteins and enzymes and inhibits their activity. The milling process separates the bran layer from the grains (i.e., removal of anti-nutrients such as lectins, phytic acid, and tannins takes place), but at the same time, it also removes important nutrients (micro as well as a macro) (Gupta et al., 2015). Recent research has reported the beneficial effect of anti-nutrients on health as shown in Figure 4.2. Heating and milling significantly reduce polyphenols and phytic acid, and improves the digestibility of proteins and starch (Chowdhury and Punia 1997). Anti-nutrients are powerful chelating agents that decrease mineral bioavailability by producing insoluble complexes (Weaver and Kanna, 2001). In cereals, almost 3–6% of the weight of seed contains phytic acid that is present in high levels while tubers contain less phytic acid (Alabaster et al., 1996). Before milling, other processing methods like soaking significantly increase mineral content for soaked grains but reduce mineral values for malted cereal grains. An escalation in minerals in the soaked grains is due to the adsorption of minerals from water (Claver et al., 2011). Moreover, during soaking more minerals can be leached out due to the release of minerals bounded with phytate as phytate is water-soluble; thus sorghum is reported to increase calcium, zinc, and sodium when soaked for 6 to 24 h. Similar results were shown by oats and wheat for calcium, sodium, and potassium (El-Safy & Salem, 2013). Furthermore, during malting, the mineral content can be reduced as the germination process uses the minerals for embryo growing (Udeh et al., 2018) (e.g., reduction in zinc could be due to the utilization of Zn in tissue growth and cell reproduction). Condensed tannins form stable complexes with metal ions, proteins, and other polysaccharides, resulting in the decrease of protein digestibility and limiting the availability of essential nutrients in the gut (Keyata et al., 2021). Another important anti-nutritional factor, oxalic acid, can harm human

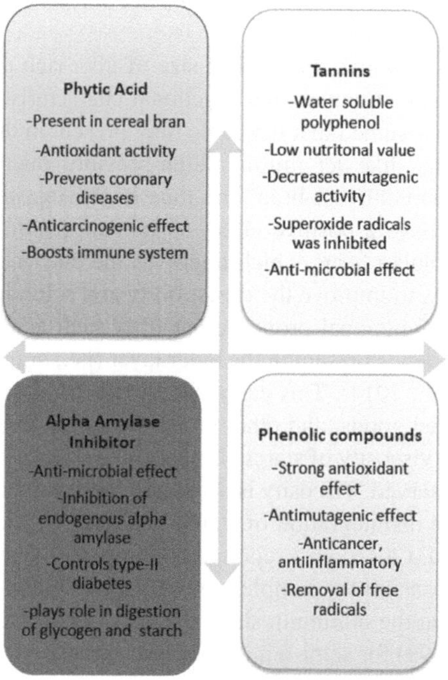

Figure 4.2 Beneficial effects of anti-nutrients present in cereal grains.

health by reducing the absorption of calcium and aiding in the formation of kidney stones (Gemede, 2020). Thus, different pre-milling methods (washing, soaking, and malting) can be used to decrease the anti-nutritional content and improve the bioavailability of minerals.

Antioxidant activity is contributed by phytochemical compounds. These compounds have a diverse range of health-promoting benefits. Whole grains are a rich source of bioactive compounds and are released during the digestion process from the fiber complex due to enzyme action (Siddiq & Prakash, 2015). The refining and milling process can increase the antioxidant activity as milling breaks the grain matrix and cell wall, which improves the availability of digestion enzymes to substances that are bound with the food matrix (Prom et al., 2006). Phenolics are found in the germ and bran layer of cereal grains that are removed by different mechanical methods (abrasion, compression, impact, and shear) and then re-combined to develop a variety of products with a range of functional and nutritional characteristics (whole vs. refined grain flour). Mechanical methods break the protective layer of whole grain and consequently enhance the phenolic exposure to different processing conditions and environmental factors that could change the stability of final products. Furthermore, refinement of whole grains, bioprocessing, and recombination methods fine-tune phenolic contents and also alterother factors that affect the bioaccessibility of phenolics present in grain (e.g., in wheat bran ultrafine grinding in association with electrostatic separation enriches the phenolic content) (Debelo & Ferruzzi, 2020).

In cereal grains, ferulic acid is abundantly found andcontributes approximately 8–20mg/100g in wheat and is considered as a main dietary source (Lempereur et al., 1998). Dietary fiber is associated with ferulic acid and is linked through ester bonds to hemicelluloses. It also can inhibit the formation of superoxide, preventing the clumping of blood platelets and lowering cholesterol (Nystrom et al., 2007; Kayahara, 2004). In rice biofunctional compounds are mainly found in bran and germ fractions,most of which are removed during milling and polishing (e.g., in the case of brown rice, the essential nutrients and other macro and micronutrients such as vitamin E, phytic acids, vitamin B, GABA, dietary fiber are present in higher amounts in comparison to the ordinary milled rice). At the same time, brown rice takes a longer time to get cooked and is harder to chew, and is not as tasty as white rice (Champagne et al., 2004). Lee et al. (2019) investigated the impact of hammer and jet milling on the digestibility of starch of germinated brown rice. During jet milling, coarse and fine particle fractions are produced with higher starch damage and can be digested easily in comparison with samples milled with hammer milling. Beta et al. (2005) studied the effect of pearling on the antioxidant and phenolic content in wheat fractions. Wheat fractions from the first and second pearling contain more than 4000mg/kg of phenolics and after the third and fourth pearling wheat still has more than 3000mg/kg of phenolic content as the inner layer of the crease are intact and contains a significant amount of phenolics. In rolling mill wheat fractions showed decreasing trend in phenolics (shorts>bran>flour>refined endosperm fraction) as pentosans are present in the aleurone layer of wheat that is being removed during milling and the degree of refinement reduces (Symon and Dexter, 1993). It was also reported that pearled fractions had higher or similar levels of phenolic content compared with shorts and bran obtained from roller milling.

4.7 CONCLUSION

A significant amount of dietary energy and nutrients are provided by cereals and cereal products, which are staple foods in the majority of human diets in both developed and developing nations. Today's grain processing sector is as varied as the spectrum of goods it produces. Practically every meal produced contains grains in some form and the variety of non-food applications is expanding daily. As a result, food manufacturers face tremendous processing challenges. New methods of food processing have been developed to enhance nutritional value, physicochemical properties, microbial safety, production efficiency, and process effectiveness. More thorough research is needed to explore and comprehend how pearling affects the performance of flour. To evaluate the impact of pearling on flour performance, future studies may entail rheological and baking experiments on doughs made from the two types of flour (pearled and unpearled), as well as comparisons of bread loaves in terms of appearance, volume, texture, and flavour, among other factors.

REFERENCES

Alabaster, O., Tang, Z., &Shivapurkar, N. (1996). Dietary fiber and the chemopreventive modulation of colon carcinogenesis. *Mutation Research/Fundamental and Molecular Mechanisms of Mutagenesis*, *350*(1), 185–197.

Alizadeh, M. R., Minaei, S., Rahimi-Ajdadi, F., Tavakoli, T., Khoshtaghaza, M. H., &Zareiforoush, H. (2012). Flow properties of awned and de-awned paddy grains through a horizontal hopper orifice. *Particulate Science and Technology*, *30*(4), 343–353.

Allai, F. M., Azad, Z. R. A. A., Gul, K., & Dar, B. N. (2021). Wholegrains: a review on the amino acid profile, mineral content, physicochemical, bioactive composition, and health benefits. *International Journal of Food Science & Technology*.

Aprodu, I., &Banu, I. (2017). Milling, functional and thermo-mechanical properties of wheat, rye, triticale, barley and oat. *Journal of Cereal Science*, *77*, 42–48.

Awika, J. M. (2011). Major cereal grains production and use around the world. In *Advances in cereal science: implications to food processing and health promotion* (pp. 1–13). American Chemical Society.

Bean, M. M., EA, E. H., &Nishita, K. D. (1983). Rice flour treatment for cake-baking applications [with water just before use]. *Cereal Chemistry*, 60, 445–449.

Beta, T., Nam, S., Dexter, J. E., &Sapirstein, H. D. (2005). Phenolic content and antioxidant activity of pearled wheat and roller-milled fractions. *Cereal Chemistry*, *82*(4), 390–393.

Bhat, F. M., Riar, C. S., &Sood, S. (2019). Effects of milling on the bran removal, nutritional and cooking characteristics of traditional rice cultivars. *Food Science and Nutrition Technology, 4*(4), 1–9.

Bond, N., 2004. Rice milling. In: Champagne, E.T. (Ed.), Rice: Chemistry and Technology. AACC International, Inc., St. Paul, MN, USA

Burlando, B., &Cornara, L. (2014). Therapeutic properties of rice constituents and derivatives (*Oryza sativa* L.): A review update. *Trends in Food Science & Technology,40,* 82–98.

Calinoiu, L. F., &Vodnar, D. C. (2018). Whole grains and phenolic acids: A review on bioactivity, functionality, health benefits and bioavailability. *Nutrients*, *10*(11), 1615.

Campbell, G. M., Webb, C., Owens, G. W., &Scanlon, M. G. (2012). Milling and flour quality. In S. P. Cauvain (Ed.), Breadmaking – Improving quality (2nd ed., pp. 188–215). Cambridge: Woodhead Publishing.

Champagne, E. T., Wood, D. F., Juliano, B. O., &Bechtel, D. B. (2004). The rice grain and its gross composition. *Rice Chemistry and Technology*, *3*, 77–107.

ChauCF, WuSC, LeeMH (2007) Physicochemical changes upon micronization process positively improve the intestinal health enhancement ability of carrot insoluble fibre. *Food Chem, 104*, 1569–1574.

Chen, J. J., Lu, S., &Lii, C. Y. (1999). Effects of milling on the physicochemical characteristics of waxy rice in Taiwan. *Cereal Chemistry*, *76*(5), 796–799.

Chiang, P. Y., &Yeh, A. I. (2002). Effect of soaking on wet-milling of rice. *Journal of Cereal Science*, *35*(1), 85–94.

Chowdhury, S., &Punia, D. (1997). Nutrient and antinutrient composition of pearl millet grains as affected by milling and baking. *Food/Nahrung, 41*(2), 105–107.

Claver, I. P., ZHOU, H. M., ZHANG, H. H., ZHU, K. X., Qin, L. I., &Murekatete, N. (2011). The effect of soaking with wooden ash and malting upon some nutritional properties of sorghum flour used for impeke, a traditional Burundian malt-based sorghum beverage. *Agricultural Sciences in China*, *10*(11), 1801–1811.

Dayakar Rao, B., Bhaskarachary, K., Arlene Christina, G. D., Sudha Devi, G., Vilas, A. T., &Tonapi, A. (2017). Nutritional and health benefits of millets. *ICAR_Indian Institute of Millets Research (IIMR): Hyderabad, Indian*, 112.

Debelo, H., Li, M., &Ferruzzi, M. G. (2020). Processing influences on food polyphenol profiles and biological activity. *Current Opinion in Food Science*, *32*, 90–102.
Dowswell, C. R., Paliwal, R. L., &Cantrell, R. P. (2019). *Maize in the third world*. CRC Press.
Eckhoff, S. R., Du, L., Yang, P., Rausch, K. D., Wang, D. L., Li, B. H., &Tumbleson, M. E. (1999). Comparison between alkali and conventional corn wet-milling: 100-g procedures. *Cereal Chemistry*, *76*(1), 96–99.
Eckhoff, S.R., PaulsenM.R., YangS.C. (2003). Maize. *Encyclopedia of food sciences and nutrition* (Second Edition). Editor(s): Benjamin Caballero, Academic Press, 3647–3653.
El-Safy, F., &Salem, R. (2013). The impact of soaking and germination on chemical composition, carbohydrate fractions, digestibility, antinutritional factors and minerals content of some legumes and cereals grain seeds. *Alexandria Science Exchange Journal*, *34*(October-December), 499–513.
Embashu, W., &Nantanga, K. K. M. (2019). Pearl millet grain: A mini-review of the milling, fermentation and brewing of ontaku, a non-alcoholic traditional beverage in Namibia. *Transactions of the Royal Society of South Africa*, *74*(3), 276–282.
European Commission, 2010. Commission Regulation (EU) No. 165/2010 of 27 February 2010 amending Regulation (EC). No 1881/2006 setting maximum levels for certain contaminants in foodstuffs as regards aflatoxins. *Official Journal of the European Union*, *50*, 8e12.
Fardet, A. (2010). New hypotheses for the health-protective mechanisms of whole-grain cereals: What is beyond fibre? *Nutrition Research Reviews*, *23*, 65–134.
Fernando, B. (2013). Rice as a Source of Fibre. *Journal of Rice Res*earch, *1*(2), 2–4
Gemede, H. F. (2020). Nutritional and antinutritional evaluation of complementary foods formulated from maize, pea, and anchote flours. *Food Science & Nutrition*, *8*(4), 2156–2164.
Gupta, R. K., Gangoliya, S. S., &Singh, N. K. (2015). Reduction of phytic acid and enhancement of bioavailable micronutrients in food grains. *Journal of Food Science and Technology*, *52*(2), 676–684.
Han, M. R., Chang, M. J., &Kim, M. H. (2007). Changes in physicochemical properties of rice starch processed by ultra-fine pulverization. *Journal of Applied Biological Chemistry*, *50*(4), 234–238.
Heshe, G. G., Haki, G. D., Woldegiorgis, A. Z., &Gemede, H. F. (2016). Effect of conventional milling on the nutritional value and antioxidant capacity of wheat types common in Ethiopia and a recovery attempt with bran supplementation in bread. *Food Science & Nutrition*, *4*(4), 534–543.
İbanoğlu, Ş., &İbanoğlu, E. (1999). Rheological properties of cooked tarhana, a cereal-based soup. *Food Research International*, *32*(1), 29–33.
Iuliana, B., Georgeta, S., Sorina, I. V., &Iuliana, A. (2012). Effect of the addition of wheat bran stream on dough rheology and bread quality. *The Annals of the University Dunarea de Jos of Galati. Fascicle VI-Food Technology*, *36*(1), 39–52.
Johnson, L. A., &May, J. B. (2003). Wet milling: the basis for corn biorefineries. *Corn: Chemistry and Technology*, (Ed. 2), 449–494.
Kalita, T., Gohain, U. P., &Hazarika, J. (2021). Effect of Different Processing Methods on the Nutritional Value of Rice. *Current Research in Nutrition and Food Science Journal*, *9*(2).
Kamar, M., Evans, C., &Hugh-Jones, S. (2016). Factors influencing adolescent whole grain intake: A theory-based qualitative study. *Appetite*, *101*, 125–133.
Kayahara, H. (2004). *Germinated brown rice*. Nagano: Department of Sciences of Functional Foods, Shinshu University.
Kent, N. L. (1994). *Kent's Technology of Cereals: An introduction for students of food science and agriculture*. Elsevier.
Kerr, W., Ward, C., McWatters, K., &Resurreccion, A. (2000). Effect of milling and particle size on functionality and physicochemical properties of cowpea flour. *Cereal Chemistry*, *77*, 213–219.
Keyata, E. O., Tola, Y. B., Bultosa, G., &Forsido, S. F. (2021). Premilling treatments effects on nutritional composition, antinutritional factors, and in vitro mineral bioavailability of the improved Assosa I sorghum variety (Sorghum bicolor L.). *Food Science & Nutrition*, *9*(4), 1929–1938.
Lee, Y. T., Shim, M. J., Goh, H. K., Mok, C., &Puligundla, P. (2019). Effect of jet milling on the physicochemical properties, pasting properties, and in vitro starch digestibility of germinated brown rice flour. *Food Chemistry*, *282*, 164–168.
Lempereur, I., Surget, A., &Rouau, X. (1998). Variability in dehydrodiferulic acid composition of durum wheat (*Triticum durum Desf.*) and distribution in milling fractions. *Journal of Cereal Science*, *28*, 251–258.

Liu, F., He, C., Wang, L., &Wang, M. (2018). Effect of milling method on the chemical composition and antioxidant capacity of Tartary buckwheat flour. *International Journal of Food Science & Technology, 53*(11), 2457–2464

Liu, T. Y., Ma, Y., Yu, S. F., Shi, J., &Xue, S. (2011). The effect of ball milling treatment on structure and porosity of maize starch granule. *Innovative Food Science & Emerging Technologies, 12*(4), 586–593.

Liu, T., &Hou, G. G. (2019). Trends in whole grain processing technology and product development. In *Whole Grains* (pp. 257–279). CRC Press.

Liu, T., Mao, D., Zhang, S., Xu, C., &Xing, Y. (2009). Fine mapping SPP1, a QTL controlling the number of spikelets per panicle, to a BAC clone in rice (Oryza sativa). *Theoretical and Applied Genetics, 118*(8), 1509–1517.

Loubes, M. A., Resio, A. C., Tolaba, M. P., &Suarez, C. (2012). Mechanical and thermal characteristics of amaranth starch isolated by acid wet-milling procedure. *LWT-Food Science and Technology, 46*(2), 519–524.

Lovegrove, A., Kosik, O., Bandonill, E., Abilgos-Ramos, R., Romero, M., Sreenivasulu, N., &Shewry, P. (2019). Improving Rice dietary fibre content and composition for human health. *Journal of Nutritional Science and Vitaminology, 65*(Supplement), S48–S50.

Maphosa, Y., &Jideani, V. (2016). Physicochemical characteristics of Bambara groundnut dietary fibres extracted using wet milling. *South African Journal of Science, 112*(1–2), 01–08.

Martínez-Bustos, F., López-Soto, M., San Martin-Martinez, E., Zazueta-Morales, J. J., &Velez-Medina, J. J. (2007). Effects of high energy milling on some functional properties of jicama starch (Pachyrrhizuserosus L. Urban) and cassava starch (Manihot esculenta Crantz). *Journal of Food Engineering, 78*(4), 1212–1220.

Mellen, P. B., Walsh, T. F., &Herrington, D. M. (2008). Whole grain intake and cardiovascular disease: a meta-analysis. *Nutrition, Metabolism and Cardiovascular Diseases, 18*(4), 283–290.

Nystrom, L., Achrenius, T., Lampi, A. M., Moreau, R. A., &Piironen, V. (2007). A comparison of the antioxidant properties of sterylferulates with tocopherol at high temperatures. *Food Chemistry, 101*(3), 947–954.

Oghbaei, M., &Prakash, J. (2012). Bioaccessible nutrients and bioactive components from fortified products prepared using finger millet (*Eleusine coracana*). *Journal of the Science of Food and Agriculture,92*, 2281–2290.

Oghbaei, M., &Prakash, J. (2013). Effect of fractional milling of wheat on nutritional quality of milled fractions. *Trends in Carbohydrate Research,5*, 53–58.

Oghbaei, M., &Prakash, J. (2016). Effect of primary processing of cereals and legumes on its nutritional quality: A comprehensive review. *Cogent Food & Agriculture, 2*(1), 1136015.

Oliveira, L. C., Rosell, C. M., &Steel, C. J. (2015). Effect of the addition of whole-grain wheat flour and of extrusion process parameters on dietary fibre content, starch transformation and mechanical properties of a ready-to-eat breakfast cereal. *International Journal of Food Science & Technology, 50*(6), 1504–1514.

Owens, G. (Ed.). (2001). *Cereals processing technology* (Vol. 53). CRC Press.

Papageorgiou, M., &Skendi, A. (2018). Introduction to cereal processing and by-products. In *Sustainable Recovery and Reutilization of Cereal Processing By-Products* (pp. 1–25). Woodhead Publishing.

Price, C., Kiszonas, A. M., Smith, B., &Morris, C. F. (2021). Roller milling performance of dry yellow split peas: Mill stream composition and functional characteristics. *Cereal Chemistry, 98*(3), 462–473.

Priya, T. R., Nelson, A. R. L. E., Ravichandran, K., &Antony, U. (2019). Nutritional and functional properties of coloured rice varieties of South India: a review. *Journal of Ethnic Foods, 6*(1), 1–11.

Prom-u-thai, C., Huang, L., Glahn, R. P., Welch, R. M., Fukai, S., &Rerkasem, B. (2006). Iron (Fe) bioavailability and the distribution of anti-Fe nutrition biochemicals in the unpolished, polished grain and bran fraction of five rice genotypes. *Journal of the Science of Food and Agriculture, 86*(8), 1209–1215.

Ragaee, S., Seetharaman, K., &Abdel-Aal, E. S. M. (2014). The impact of milling and thermal processing on phenolic compounds in cereal grains. *Critical Reviews in Food Science and Nutrition, 54*(7), 837–849.

Ramadhan, K., &Foster, T. J. (2018). Effects of ball milling on the structural, thermal, and rheological properties of oat bran protein flour. *Journal of Food Engineering, 229*, 50–56.

Ramberg, J., &McAnalley, B. (2002). From the farm to the kitchen table: A review of the nutrient losses in foods. *GlycoScience& Nutrition*, 3, 1–12.

Ramírez, E. C., Johnston, D. B., McAloon, A. J., &Singh, V. (2009). Enzymatic corn wet milling: engineering process and cost model. *Biotechnology for Biofuels*, *2*(1), 1–9.

Redy, M., &Love, M. (1999). The impact of food processing on the nutritional quality of vitamins and minerals. In L. S. Jackson, M. G. Knize, &J. N. Morgan (Eds.), *Impact of processing on food safety* (pp. 99–106). New York, NY: Plenum.

Roa, D. F., Baeza, R. I., &Tolaba, M. P. (2015). Effect of ball milling energy on rheological and thermal properties of amaranth flour. *Journal of Food Science and Technology*, *52*(12), 8389–8394.

Roa, D. F., Santagapita, P. R., Buera, M. P., &Tolaba, M. P. (2014). Ball milling of Amaranth starch-enriched fraction. Changes on particle size, starch crystallinity, and functionality as a function of milling energy. *Food and Bioprocess Technology*, *7*(9), 2723–2731.

Ruth MacDonald, Cheryll Reitmeier. (2017). Chapter 6–*Food processing. Understanding food systems*. Editor(s): Ruth MacDonald, Cheryll Reitmeier, Cambridge, MA: Academic Press, 179–225.

Saini, P., Islam, M., Das, R., Shekhar, S., Sinha, A. S. K., & Prasad, K. (2022). Wheat bran as potential source of dietary fiber: prospects and challenges. *Journal of Food Composition and Analysis*, 105030.

Saleh, A. S., Zhang, Q., Chen, J., &Shen, Q. (2013). Millet grains: nutritional quality, processing, and potential health benefits. *Comprehensive Reviews in Food Science and Food Safety*, *12*(3), 281–295.

Seal, C. J., Courtin, C. M., Venema, K., &de Vries, J. (2021). Health benefits of whole grain: effects on dietary carbohydrate quality, the gut microbiome, and consequences of processing. *Comprehensive Reviews in Food Science and Food Safety*, *20*(3), 2742–2768.

Serna-Saldivar, S. O. (2022). Production of cereal-based raw materials for the snack industry. In *Snack foods* (pp. 131–154). CRC Press.

Sharma, K. D., Sharma, B., &Saini, H. K. (2021). Processing, value addition and health benefits. In *Millets and Pseudo Cereals* (pp. 169–184). Woodhead Publishing.

Siddiq A, A., & Prakash, J. (2015). Antioxidant properties of digestive enzyme-treated fibre-rich fractions from wheat, finger millet, pearl millet and sorghum: A comparative evaluation. *Cogent Food & Agriculture*, *1*(1), 1073875.

Siebenhandl, S., Grausgruber, H., Pellegrini, N., Del Rio, D., Fogliano, V., Pernice, R., &Berghofer, E. (2007). Phytochemical profile of main antioxidants in different fractions of purple and blue wheat, and black barley. *Journal of Agricultural and Food Chemistry*, *55*, 8541–8547.

Singh, A., Das, M., Bal, S., Banerjee, R., 2013. Rice processing. In: de Pinho Ferreira Guiné, R., dos Reis Correia, P.M. (Eds.), *Engineering aspects of cereal and cereal-based products*. CRC Press, Boca Raton, FL, USA.

Slavin, J. L., Martini, M. C., Jacobs, D. R., &Marquart, L. (1999). Plausible mechanisms for the protectiveness of whole grains. *The American Journal of Clinical Nutrition*, *70*, 459S–463S.

Sovrani, V., Blandino, M., Scarpino, V., Reyneri, A., Coïsson, J. D., Travaglia, F., …Arlorio, M. (2012). Bioactive compound content, antioxidant activity, deoxynivalenol and heavy metal contamination of pearled wheat fractions. *Food Chemistry*, *135*, 39–46.

Suma, R. C., Sheetal, G., Jyothi, L. A., &Prakash, J. (2007). Influence of phytin phosphorous and dietary fibre on *in vitro* iron and calcium bioavailability from rice flakes. *International Journal of Food Sciences and Nutrition*, *58*, 637–643.

Symons, S. J., &Dexter, J. E. (1993). Relationship of flour aleurone fluorescence to flour refinement for some Canadian hard common wheat classes. *Cereal Chemistry*, *70*, 90–90.

Tan, X., Zhang, B., Chen, L., Li, X., Li, L., &Xie, F. (2015). Effect of planetary ball-milling on 619 multi-scale structures and pasting properties of waxy and high-amylose cornstarches. *620 Innovative Food Science & Emerging Technologies*, *30*, 198–207.

Tosi, P., Hidalgo, A., &Lullien-Pellerin, V. (2020). The impact of processing on potentially beneficial wheat grain components for human health. In G. Igrejas, T. M. Ikeda, &C. Guzman (Eds.), *Wheat quality for improving processing and human health* (pp. 387–420). Switzerland: Springer International Publishing.

Truswell, A. S. (2002). Cereal grains and coronary heart disease. *European Journal of Clinical Nutrition*, *56*, 1–14.

Tulse, S. B., Reshma, V., Inamdar, A. A., & Sakhare, S. D. (2014). Studies on multigrain milling and its effects on physical, chemical and rheology characteristics of milled streams. *Journal of Cereal Science*, *60*(2), 361–367.

Udeh, H. O., Duodu, K. G., &Jideani, A. I. (2018). Effect of malting period on physicochemical properties, minerals, and phytic acid of finger millet (Eleusine coracana) flour varieties. *Food Science & Nutrition*, *6*(7), 1858–1869.

Velicogna, R., Shea Miller, S. (2016). Wet milling of wheat. Reference module in food science. Elsevier.

Weaver, C. M., &Kannan, S. (2001). Phytate and mineral bioavailability. In N. Rukma Reddy & Shridhar K. Sathe (Eds.), *Food phytates* (pp. 227–240). CRC Press.

Wronkowska, M. (2016). Wet-milling of cereals. *Journal of Food Processing and Preservation*, *40*(3), 572–580.

Wronkowska, M., &Haros, M. (2014). Wet-milling of buckwheat with hull and dehulled–the properties of the obtained starch fraction. *Journal of Cereal Science*, *60*(3), 477–483.

WuSC, ChienPJ, LeeMH, ChauCF (2007) Particle size reduction effectively enhances the intestinal health-promotion ability of an orange insoluble fibres in hamsters. *Journal of Food Science*, *72*, S618–S621.

Yoenyongbuddhagal, S., &Noomhorm, A. (2002). Effect of physicochemical properties of high-amylose Thai rice flours on vermicelli quality. *Cereal Chemistry*, *79*(4), 481–485.

Yuan, J., Chung, D. S., Seib, P. A., &Wang, Y. (1998). Effect of steeping conditions on wet-milling characteristics of hard red winter wheat. *Cereal Chemistry*, *75*(1), 145–148.

CHAPTER 5

Flaking of Cereals

Gunjana Deka and Himjyoti Dutta
Department of Food Technology, Mizoram University, Aizawl, Mizoram, India

CONTENTS

5.1 INTRODUCTION

Cereals can be defined as grains or edible seeds of plants from the grass (Poaceae) family (Bender, 2006). Cereals are important contributors to the nutrient intake of human beings and are considered as staple foods by most populations. They also indirectly contribute to human

DOI: 10.1201/9781003242192-7

nutrition by being the primary livestock feed. Products made from cereal grains can be considered as complex multicomponent systems that contain mixtures of vitamins, minerals, oils, proteins, polysaccharides, and phytochemicals or bioactive compounds (Moreno et al., 2018). Cereals are traditionally processed to provide a variety of ready to eat (RTE) expanded products like puffed and popped cereals and flaked products. Some undergo dehusking as an essential step while others undergo milling to generate their desired flour. Certain processing treatments are applied on the grains to impart special characteristics and improve organoleptic properties (Oghbaei & Prakash, 2016). Processing of grains can be categorized into non-thermal and thermal processes. Non-thermal processes include hammer and roller milling, and thermal processes include dry processing (popping, roasting, and micronizing) and wet processing (autoclaving, steam pelleting, expanding, steam flaking, extruding, and toasting) often involving pre-moistening of the grains followed by the hydrothermal processes (Safaei & Yang, 2017).

Flaked cereals are traditionally and commonly available RTE products. Flaking involves thermal as well as mechanical treatments. Cereals are subjected to traditional pounding, pressing, or modern mechanical flaking methods involving the use of industrial machineries. The process initiates with hydrating the grains up to their equilibrium moisture content levels, gelatinizing the starch by steaming, and drying to about 16 to 20% moisture followed by flaking/flattening using mechanical devices (Itagi et al., 2012). Flaking of cereals grains is believed to have originated with traditional rice flaking in the rice growing regions of India. Flaked rice is ethnically consumed as a popular breakfast cereal, snack, or savory (S. N. Kumar & Prasad, 2013). The flaking process necessitates the conversion of raw starch to gelatinized starch. In the process, some resistant starch can also form that functions as a prebiotic source and can be used in enriching breakfast cereal as a source of dietary fibre (Kumar et al., 2018).

Various methods to produce rice flakes have been developed using the basic edge runner machine (Ananthachar et al., 1982). This was followed by a developed continuous process (Narasimha et al., 1982), by the Central Food Technological Research Institute, India for making rice flakes where a heavy-duty roller flaker was used to flake soaked, roasted, shelled, polished (gently), and tempered paddy. This process resulted in 6–7% higher yield of flakes than the edge-runner process. Compared to edge runner flakes, roller flakes were found to be moist, tender, and with a greater tendency for lumping (Ekanayake & Narasimha, 1997). A method of flaking RTE cereal-based foods was patented (Robie & Hilgendorf, 2001), which involves steeping the grains to about 20% moisture in water, followed by tempering, flaking, and toasting. A method for steam flaking involving passing of the grain through a steam chest at predetermined time and pressure followed by compression (Brown, 2003). These processes are being modified in terms of time, temperature, and pressure and tempering to produce flakes of various cereals.

The processing treatments involved in flaking changes the structure and composition of the grain. They can either alter the nutritional quality by reducing the nutrients, phytochemical and antinutritional factors, or might improve the digestibility or availability of nutrients (Oghbaei & Prakash, 2016). Rolling of grains results in cracks in the hull and pericarp and helps in rumen digestibility. Incorporation of moisture in the flaking process can result positively in maximizing utilization. Steam is supplied at either high, low, or atmospheric pressures to the grains before rolling in the steam flaking process. This process uses longer conditioning time and produces thinner rolled flakes as compared to steam rolling. Gelatinization is induced in the starch granules due to the combined effect of heat moisture and pressure, which disrupts the granules and increases the digestibility (Safaei & Yang, 2017). The structure of the starch granule with its structural components, namely amylose and amylopectin, are responsible for the functional characteristics attributed by the flaked cereals. Generally, the functional properties of cereals and cereal products are based on its water absorption capacity. Flaked cereals undergo gelatinization in the starch granules and prevent or retard retrogradation in them due to the drying treatment applied, which in turn increases the hydration properties of the flaked product (Shinde & Durgadevi, 2018). This also increases the slurry viscosity of the starch owing to its higher water absorption at room temperature (Mujoo & Ali, 2000). Gelatinization temperature increases with decrease in moisture content (about less than 30%) (Holm et al., 1988). The bran layer, which is rich in bioactive compounds, is removed before flaking. This results in reduced bioactive content in the flaked product. Antinutrients, on the other hand, combine or react with the nutrients or their hydrolytic enzymes and reduce their bioavailability (Samtiya et al., 2020). Thus, reduction in their content can be considered as a positive effect. The phytochemicals and antinutrients also help to reduce oxidative stress in the body due to their antioxidant properties. Therefore, an optimum degree of flaking can also help in retention of these nutrients as well as regulate starch digestion (Safaei & Yang, 2017).

5.2 PROCESSING TREATMENT

The common primary processing steps involved in processing of any cereal are cleaning, sorting, and removal of extraneous matters from the grain surfaces. The general processing outline includes all or a few of the steps including mixing, cooking, de-lumping, drying, cooling, tempering, flaking, and drying/toasting and are often coupled with some preprocessing treatments of the cereals (Thielecke et al., 2021). The main preprocessing and processing steps differ for different types of cereals and the intended products. Accordingly, flaked cereal products can be broadly classified as formed-flaked, extrusion-flaked, and whole-flaked.

Wheat kernels are broken open by a process known as bumping where steaming of the grain followed by passing between coarse rollers to slightly crush the kernels takes place. Crushing should be gentle to avoid the formation of soft and gluey flour after cooking due to abundance and overexposure of broken starch granules. This preprocessing step is followed by formulation, cooking in a cooker (steam 15 psi, 30 minutes), lump breaking, drying until 16–18%, cooling (below 43°C) and tempering, flaking, and toasting (1–3% moisture content). The desired grit size for flaking is about 0.375 in. (1 cm) in diameter. Some small grits of 0.125 in. (0.3 cm) and some large up to 0.50 in. (1.25) are also seen (Fast, 2000). Corn is subjected to dry break milling, resulting in release of the germ and bran and collection of the coarse endosperm grits to be further refined. The corn germ carries most of its oil. Milling increases the shelf life of the corn flour and flaked product by preventing oxidative and hydrolytic rancidity development. The obtained flours are subjected to formed flaking or extrusion flaking. The process of producing extruded flakes involves mixing, hydrothermal treatment, extrusion, drying, cooling, and tempering, flaking, drying/toasting. For formed flaking, the mixed cooked or uncooked dough is sheeted, cut to size uniformity, flaked by compressive rolling (often coupled with cooking), and cooled and dried. Extrusion cooking omits the cooking and de-lumping process involved in the traditional process. Extrusion, apart from cooking the grains, helps in forming uniform pellets resulting in better flakes from single as well as multicomponent formulations (Tribelhorn et al., 1991). Rice flakes are often whole flaked. Popular flaked rice products like Indian *chira/poha* are made from whole milled grains or larger broken pieces of whole kernels (Thielecke et al., 2021; Fast, 2000).

5.2.1 Fundamentals of Treatment

The process of flaking involves thermo-mechanical treatments (Itagi et al., 2012). According to the flaking approaches, the methods for flaked cereals can be dry rolling, steam rolling, or steam flaking. In dry rolling, the dry grain is broken into several pieces by feeding through large rollers. Steam rolling uses the process of steaming the grains (up to 22% moisture content) followed by crushing to produce a thick flake (Theurer et al., 1999).

Grain quality, water, temperature (heat), time, and flake density are the main determinants that effect starch gelatinization. The steam flaking method uses moisture and heat to enhance gelatinization of the starch granules (Dehghan-banadaky et al., 2007). It is a more extensive process where the grain is fed to a closed steam chest and steam is applied at different pressures and times for about 30 to 60 minutes to facilitate moisture absorption of 18 to 20% (Theurer et al., 1999). Decrease in flake density leads to increase in flaking pressure (Xiong et al., 1991). The combined and increased effect of moisture, heat, and pressure helps in adequate flaking and desired gelatinization. This also increases the digestibility of starch by disorganizing the starch granules and disrupting the protein matrix, which surrounds the starch granules in the endosperm (Rooney & Pflugfelder, 1986; Svihus et al., 2005). Studies have shown that processing of the grains including steam flaking have improved the starch digestibility of ruminants by increasing the surface area exposure compared to whole grains (Rastgoo et al., 2020). Therefore, this method is generally preferred. The increase in temperature will result in water penetration into the starch granule and loss of birefringence resulting in the characteristic appearance. Proper gelatinization helps in desired extent of fracturing and flattening of the grain with mechanical pressure. This is critical, as its failure leads to rapid retrogradation of starch and re-attainment of kernel-like shape and texture upon cooling.

5.2.2 Process and Machinery Involved

Flakes are prepared by soaking the grains to their equilibrium moisture content followed by hydrothermal treatment to gelatinize the starch, drying (16–20% moisture) and tempering, and flaking by either traditional or mechanical means, drying and sizing (Itagi et al., 2012). Conventionally used flaking units are roller flakers and edge runners. Roller flaking uses a single step pressing. The paddy to be flaked is previously soaked and roasted (175 °C) with sand in a continuous mechanical roaster until attainment of 20–22% moisture content, followed by dehusking in a centrifugal sheller, debranning in a cone type rice polisher, and tempering for moisture equilibration. The tempered grain is then passed through the heavy-duty roller flaker that bears two parallel rolls rotating in opposite directions. Thick (1.0 mm), medium (0.4 mm), and thin (0.3 mm) flakes can be obtained by adjusting the clearance between the rolls. This continuous process was developed at Central Food Technological Research, Mysore, and yielded about 70% rice flakes (6–7% more than traditional process) (Narasimha et al., 1982).

The edge runner unit consists of a perforated, round, flat, thick metallic sieve with raised edge, mounted with a roller at the end of the sieve. This process results in thick (1.3 mm), medium (0.9 mm), and thin (0.7 mm) flakes after being flaked for 35, 45, and 60 s, respectively. The process of flaking for edge runner flakes starts with the soaking of paddy in hot water in a barrel overnight (12 h), resulting in moisture content of about 30% (wet basis). This is followed by manually roasting at a temperature of 250°C in fine hot sand to about 18–20% moisture content (wet basis). The roasted paddy is then flaked in this unit at a speed of 300 rpm. Simultaneously the dried husk and the part of the bran layers are removed with the aid of the perforations in the sieve (Ananthachar et al., 1982). Flakes produced by the edge runner unit were found to be drier, tougher, and with a lower tendency for lumping compared to the roller flaked flakes (Ekanayake & Narasimha, 1997).

A process for the preparation of Ready-to-Eat (RTE) flakes exhibiting noticeable grain fragments of approximately 1 mm thickness was developed (Robie & Hilgendorf, 2001). The method involved steeping of the grains with warm water until 20% moisture content, cooking in a twin-screw extruder with a short residence time to form cooked dough, and forming the cooked dough into pellet followed by drying, tempering, flaking, and toasting.

A steam flaking method was invented with the objective of optimizing gelatinization during the process. The process starts with feeding the grain into a steam chest for a predetermined time, where steam enters through steam delivery pipes from the boiler at predetermined temperature and pressure. The grain then falls through two rotating corrugated steel rollers. The method was used to lessen the variability in gelatinization by determining the density of the flakes by adjusting the pressure, temperature, holding time, and gap between the rollers. Therefore, the resulting flakes can have a predetermined density and extent of gelatinization (Brown, 2003).

Currently, modifications are being made in the mentioned processes of cereal flaking (rice, wheat, corn, sorghum, oats) using an edge runner and/or roller flaker by altering the processing conditions like soaking time, roasting temperature, flaking time, flaking pressure, retention time, and rolling temperature generation to study their impacts on various product properties. However, the general basic processes outlined remain unaltered. Table 5.1 shows some of the process treatments undertaken by some researchers for producing cereal flakes.

Table 5.1 Processing Treatments Applied on Different Grains for the Preparation of Cereal Flakes

Grain	Processing conditions and Method	Key Findings	References
Sorghum hybrid (NC+ 363y)	Tempering: 17, 20, and 23%; 25–30°C, 24 hours Steam Flaker	More moisture: stronger flakes Tempered Flakes: Starch extensively gelatinized SEM: Good Quality flakes-translucent, strong little chalky; Poor Quality flakes – opaque, chalky	(McDonough et al., 1997)
Foxtail millet	Heavy duty RF (single pass)	Successful application of processing might be achieved to prepare RTE/use products	(Ushakumari et al., 2004)
Rice (IR-64, BPT-5204, MTU-1001	Soaking (overnight); Roasting and Tempering ER, HDRF, ER+HDRF	EMC: High in HDRF flakes, lower in ER flakes; Protein: High in ER flakes, lower in HDRF flakes; Solubility: High in ER flakes, lower in ER+HDRF flakes Nutrient loss: High in ER+HDRF flakes	(Deepa & Singh, 2011)
Sorghum (*Sorghum bicolor* (L.) Moench) varieties	Hydrating, Pearling, Hydrothermal Treatment Roller Press Roaster: drying and blistering	Flaking caused Decrease in moisture, protein, crude fat & total minerals Increase in carbohydrates and crude fibre Decrease in phytic acid, total phenols and flavanols.	(Sreeramaiah, 2012)
Rice (Gurjari Variety)	Soaking: 7-8 hour, RT till 30±2%; Roasting: HTST, 170-180°C, 28s, till 17-20%m.c Flaking machine: 900rpm, 15HP electric motor (thick flakes) + roller (145±5kg/cm²); 75±5°C (ETFR)	WSI and WAI: more in ETFR than raw, roasted, and thick flakes Viscosity: More in brown rice than roasted, thick flaked and ETFR after cooling at 50°C. SEM: disintegration of starch granules within ETFR; caused by high mechanical force and temperature.	(S. Kumar et al., 2018)
Rice (Gurjari Variety)	Soaking: 7-8 hour, RT till 30±2%; Roasting: HTST, 170-180°C, 28s, till 17-20%m.c Flaker: 900rpm, 15HP electric motor (thick flakes) + roller (145±5kg/cm²); 75±5°C (ETFR). Roasting: 260°C to 340°C; 20to 60s (RFR)	Combination of temperature and time of 302°C and 41 s: more yield of highly acceptable Roasted Flaked rice.	(S. Kumar & Prasad, 2017)

ER: edgerunner, HDRF: heavy-duty roller flaker, HTST: High temperature short time, ETFR: Extra thin flaked rice, RFR: Roasted flaked rice, SEM: scanning electron microscopy, WSI: Water solubility index, WAI: Water absorption index, RT: Room Temperature

5.3 IMPACT OF PROCESSING TREATMENT ON NUTRITIONAL CHARACTERISTICS

Micronutrients (carbohydrates, proteins, and lipids) and micronutrients (vitamins and minerals) contribute to the nutritional properties of cereal grains. Grains are subjected to processing treatments like rolling, pelleting, steamrolling, or steam-flaking, which improve their palatability and digestibility. The combined application of heat, moisture, time, and mechanical pressure involved in flaking considerably affect the nutritional quality of the processed grains (Kokić et al., 2013). These processing treatments break down the barriers, namely the husk, pericarp, and protein matrix. This mainly enhances the release of starch granules from the amyloplasts and protein matrix. This is depicted in Figure 5.1. The molecular order is disrupted during the gelatinization process, which in turn helps the host and microbial amylolytic enzymes in easily accessing the starch to hydrolyze (Safaei & Yang, 2017). Also, the thermo-mechanical processing can easily lead to inter-component interactions, forming new micro- or nanostructured composites. In order to achieve best sensory acceptance, maximize starch digestion, and avoid metabolic disturbances, the processing should be optimal (Koenig & Beauchemin, 2013). Overprocessing might lead to excessive hydrothermal degradation and rapid release of oligosaccharide, which in turn leads to digestive disorders because of increased load of simple structures in the colon. Underprocessing, on the other hand, renders the starch unavailable for rumen fermentation (Koenig & Beauchemin, 2013; Ahmad et al., 2010). Flaking treatment decreases the phosphorous, phytin phosphorous, and dietary fiber contents in rice, which have been found to be inversely proportional to the flake thickness. Iron and calcium contents have been found to be unaffected during the flaking process (Suma et al., 2007). This is due to rice being devoid of oxalates and tannins (inhibitors of iron absorption), which rules out the possibility of inhibiting the bioavalability by these factors.

Controlled hydrothermal processes reportedly enhances resistant starch (RS) in grains. Retrogradation causing the formation of complex structure inaccessible by human amylolytic enzymes have been mostly considered responsible for this (Roy et al., 2019).

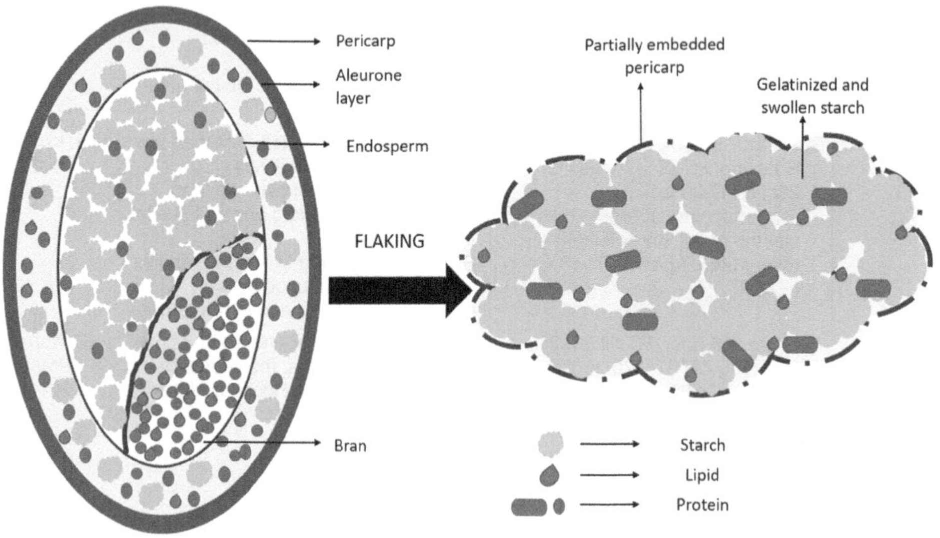

Figure 5.1 Effect of flaking on the changes in structure and compositional distribution in the grain kernel.

However, extrusion cooking method in modern cereal flaking results in reduced RS formation compared to other methods. This may be related to complete gelatinization of the starch (Vaidya & Sheth, 2011). Extrusion conditions affect the product quality. Higher protein and starch digestibility, higher retention of amino acids and vitamins, higher absorption of minerals, decreased lipid oxidation, and increased soluble dietary fiber can be achieved after mild extrusion at high moisture content, low temperature, and low residence time (Singh et al., 2007). Nutritional degradation can be caused by use of extreme conditions (≥200°C and ≤ 15% moisture content).

5.3.1 Starch

Cereal grains are mostly composed of carbohydrates, mostly in the form of starch. Starch content ranges from 40% of dry matter in oats up to 80% in rice. Endosperm, the central part of the grain, accumulates the starch granules deposited in concentric layers (Safaei & Yang, 2017; Thielecke et al., 2021). The quality of any processed grain depends upon the starch chemistry. The ratio of the two starch fractions, namely amylose and amylopectin and their chain lengths, vary with source and behave differently on hydrothermal processing. The process of flaking using steam can alter the granule structure and chemistry of starch. Flaking process requires moisture, heat, and pressure, which when applied, induce gelatinization. Starch gelatinization is a process occurring in the presence of heat and sufficient or excess moisture. Swelling of granules takes place coupled with water absorption, leading to loosening and opening of the bonds between glycosidic starch chains, finally resulting in granular disruption, complete loss of inherent crystallinity, and dissolution of the chains in the surrounding water. If processed at high temperature under insufficient water, dextrinization occurs leading to formation of small glycosidic units called dextrins (Rooney & Pflugfelder, 1986). It also generates the characteristic "roasted" flavor and aroma in the product. Dextrinization can be enhanced by lowering of pH and addition of salts. Dextrin does not retrograde. Retrogradation is a partially opposite phenomenon to gelatinization occurring while gelatinized starch is cooled. It involves re-association of the disorganized amylose and amylopectin chains to a more ordered structure having a tendency for recrystallization. However, the native type or extent of crystallinity is never regained, and some newer glycosidic structures may get formed during retrogradation. Severe processing like extrusion often exerts higher degree gelatinization compared to milder processing methods like pelleting and expanding. As compared to pelleting, extrusion treatment results in increased ruminal degradation of starch in sorghum and maize. This is attributed to higher water content, temperature, and longer residence time, which lead to the formation of completely gelatinized starch upon extrusion (Svihus et al., 2005). Gelatinization significantly increases starch availability by increasing the susceptibility for amylolytic degradation due to loss of crystalline structure (Murray et al., 2001; Peres & Oliva-Teles, 2002; Ljøkjel et al., 2004). Resistant starch can be fractions of retrograded starch (Eerlingen et al., 1994). One major factor responsible for reduced or no RS formation may be the unavailability of free water in the product. The phenomenon of reorganization of disorganized glycosidic chains during retrogradation takes place under sufficient availability of water. Gelatnization carried out at low water availability in flaked rice fails to undergo the phase reversion and the starch remains in the gelatinized state itself (Dutta et al., 2016). Compared to other processing methods, extrusion cooking, which results in fully gelatinized starch and decrease in molecular entanglement, helps in reducing the formation of resistant starch (Thielecke et al., 2021). Flaking induces stress in the entire kernel body leading to plastic deformation, and hence mechanical damage is observed in starch granules. This raises the hydration power of the flaked rice. During flaking increase in equilibrium moisture content (EMC) has been reported due to starch damage (Zakiuddin Ali & Bhattacharya, 1976). Edge runner flaking method for roasted parboiled paddy does not change the amylose and amylopectin content, while roller flaking decreases the amylopectin content, which indicates molecular breakdown of starch (Mujoo & Ali, 1999). The gelatinized starch granules are flattened during flaking, which results in disintergration of the granules and are distributed as fragments (Kumar et al., 2018).

5.3.2 Protein

Proteins in cereal grains range from 7 to 11% and are considered as the second most important macronutrient after starch (Moreno et al., 2018). Starch granules in the endosperm of cereal grains are surrounded by protein matrix for structural integrity. They also act as a barrier to enzymatic hydrolysis of whole grains. The protein matrix blocks the site of absorption and/or binds with enzyme, which might reduce the surface availability of starch for ruminal bacterial or the host enzymes (Safaei & Yang, 2017). The flaking process in cereals results in physical structural changes and starch gelatinization. These result in the denaturation of proteins in the starch–protein matrix (Xu et al., 2019). Protein denaturation opens the structures for tannin–protein interactions and forms complexes, which helps in retention of antioxidant in flaked grains (Moreno et al., 2018). During the flaking of wheat, treatments (i.e., soaking, steaming, and tempering) induce change in properties of proteins by causing structural changes in them. Heating results in denaturation of proteins involving breakdown of intra-molecular disulfide bonds and formation of new intermolecular disulfide bonds and peptide bonds. Such protein–protein interaction often causes sufficient cross-linking (Rao et al., 2002). In this reported study, proteins in wheat undergo considerable changes in flaking. Total extractability of protein decreases by half. Amount of glutenin decreases while proportions of low molecular weight protein fractions and sulfhydryl content in extractable protein increase. Significant Maillard reaction during the process of hot extrusion results due to the presence of reducing sugars and amino acids in the products. This often reduces the nutritional value of proteins. The essential amino acid lysine is said to be potentially unavailable on heating the dough (Slavin et al., 2001). Studies on the changes in protein and release of α-zeins due to processing treatments (conventionally pressed or extrusion flaking) in cornflakes was investigated by Batterman-Azcona & Hamaker (1998). They revealed that zein proteins formed large disulfide bonds upon cooking. Furthermore, extrusion process was found to be harsher than the conventional flaking process since protein bodies were destroyed and dispersion of α-zeins took place upon extrusion. Conventional flaking resulted in flattening or misshaping the protein bodies, partially releasing the α-zeins. Therefore, the release of zein and disintegration of protein bodies were thought to result in the formation of visco-elastic dough, which can be the factors influencing the final product texture.

5.3.3 Lipids

The germ in the cereal grain is the main hub of lipids. The aleurone layer of bran contains a smaller portion of it. The compositional ranges of linoleic, oleic, and palmitic acid in cereals are 39–69%, 11–36%, and 18–28%, respectively (Moreno et al., 2018). The processing treatments used in flaking of cereals use thermal energy, which affects the nutritional properties. Along with gelatinization and dextrinization, which positively affects starch hydrolysis, the formation of starch–lipid complexes might reduce the availability of starch for digestion. Lipid oxidation, which is the main cause of food deterioration with formation of rancid odor, can be minimized by extrusion cooking for flaking. Extrusion significantly denatures lipase enzyme, helps in the forming amylose-lipid complexes, promotes the release of endogenous antioxidants, and creates Maillard reaction products with antioxidant activity (Viscidi et al., 2004; Thachil et al., 2014; Camire et al., 2005). These factors contribute to reduction of lipid oxidation phenomena and help in increasing oxidative stability of the product. Some reviews and investigations show that extrusion leads to substantial reduction in fatty acids (linoleic, linolenic, oleic, and palmitic) and α-, β-, and γ-tocopherols, which may be attributed to the involvement of these fatty acids in the formation of amylose-lipid complexes (Moreno et al., 2018).

5.3.4 Micronutrients

Micronutrients in cereal grains are found in the form of minerals and vitamins (mostly vitamin B). These are mainly concentrated in the bran layer. The primary processing steps in cereal flakes processing results in the removal of bran layers and can therefore result in major reduction of micronutrients (Thielecke et al., 2021). Bran layer removal along with soaking decreases the mineral content of rice kernel due to leaching effect (de Lumen & Chow, 1991). Minerals are not significantly affected by heat treatment and remains almost stable. Absorption of minerals can be improved by reducing other factors that inhibit their absorption. Extrusion cooking can sufficiently deactivate or suppress such antinutritional factors. Fiber components can also reduce the absorption of minerals. Use of high temperature can modify their chelating properties by reorganizing the fiber components, which can significantly tackle this problem (Alonso et al., 2001). Higher content of the major minerals such as calcium (Ca), potassium (K), and sodium (Na) have been found in brown rice compared to roasted, thick flaked, and extra thin flaked rice (ETFR) (Kumar et al., 2018).

Vitamins vary in stability during thermal treatments, but the stability of hydro-soluble vitamins can be affected by flaking. Degradation of ascorbic acid can occur due to higher barrel temperature and low feed moisture during extrusion flaking, while short barrel extruders can result in higher retention of vitamin B types (Moreno et al., 2018). Moisture content and temperature plays important roles in vitamin retention. The effect of temperature on corn extrudates was studied and it was found that riboflavin content did not differ significantly at barrel temperature of 80–110°C. Also, at 20% feed moisture and 130°C process temperature, riboflavin content was higher than the feed sample with 25% moisture processed under the same processing temperature (Bilgi Boyaci et al., 2012). Significant decrease in fat-soluble vitamins A and E, tocopherols and tocotrienols on extrusion cooking have been observed in cereal grain (oat, barley, wheat, rye, and buckwheat). α-tocopherol and α-tocotrienol (isomers of vitamin E) have been found to be the least resistant to high temperature (Zielinski et al., 2001).

5.4 IMPACT OF PROCESSING TREATMENTS ON FUNCTIONAL CHARACTERISTICS

Functional properties of cereal grains and their products are generally determined by the water absorption capacity and its related attributes. The structure and chemistry play vital roles in deciding the functional properties of any starchy food. Adding to these, processing types, conditions, and severity are substantially important for designing processed products with desired functional characteristics. Changes in the functional properties of the products are mainly dependent upon the structural changes in the two primary components of starch, namely amylose and amylopectin (Jane & Chen, 1992). Flaking involving thermal treatment and mechanical pressure significantly alters the state of starch and eventually impart functional changes in the product (Mujoo & Ali, 1999). Generally, starch solubility of flaked cereal products is notably higher than the raw grain, suggesting formation of newer leachable fractions in those. Gelatinized starch is simpler and has a tendency to entrap water molecules or itself get released in surrounding water, forming a colloidal solution. Contrary to this, retrogradation generally involves reordering of amylopectin and amylose, thereby enhancing structural confinement, rigidity, and inhibiting dissolution (Mujoo & Ali, 2000; Biliaderis & Perez, 1993; Ong & Blanshard, 1995).

5.4.1 Hydration Properties

Starch gelatinization increases the hydration capacity or water absorption capacity of flaked rice (Shinde & Durgadevi, 2018), while retrogradation decreases the same. Degree of gelatinization is a factor that determines the equilibrium moisture content (EMC) at room temperature of flakes. Flaking involves mechanical damage of starch granules with simultaneous gelatinization. As discussed before, gelatinized starch in flaked rice and other dry heat processed starchy grains is a form of "dehydrated gelatinized starch" with tremendous affinity for water. Flaked grain remains in its gelatinized form with its characteristic high tendency to absorb water molecules.

Structures of amylose and amylopectin structure has direct effects on gelatinization characteristics. Bonds between amylose, lipids phosphorylated lipids, and long lateral chain amylopectin (starch-lipid) restrict water uptake, while high amylopectin with short lateral chains promotes hydration to form gels and retrograde. The major component of starch is the amylopectin with its multiple branched chain of (1-4) α-D glucose residues interlinked by (1-6)-α-D glucosidic linkages while amylose has long linear chains of (1-4)-linked α-D-glucopyranose residues, some with a few (>10) branches. The branch chain length of amylopectin is related to the crystalline structure in starch, and it also affects gelatinization and retrogradation as well as pasting properties of the starch (Jane et al., 1999).

Gelatinization is also found to be affected by granule size. Greater crystallinity of large starch granules can be confirmed by higher gelatinization enthalpy of large starch granules as compared to the small starch granules. Also, gelatinization temperature has been reported to be higher in smaller starch granules, despite having lower crystallinity (Svihus et al., 2005). The hydration of rice flakes has been reported (Ali & Bhattacharya, 1976). Along with retarded retrogradation, mechanical damage of rice starch led to its increased hydration. Decrease in swelling and solubility behavior of parboiled rice was observed (Ali & Bhattacharya, 1972) (Unnikrishnan & Bhattacharya, 2007). This decrease can be attributed to the degree of retrogradation occurred in parboiled rice during cooling of the steamed (gelatinized) paddy and storage. Edge runner flakes prepared from varieties of rice that underwent lesser degree of gelatinization showed lower values of EMC (Deepa & Singh, 2011). Furthermore, the same flakes when passed through roller exhibited increased EMC values. This was attributed to removal of bran layers in the latter and exposure of higher starchy portions to the water used for absorption. Also, the gelatinized starch was not allowed to retrograde and was flaked immediately after the roasting step, delaying the amylose precipitation.

5.4.2 Viscosity and Rheology

As processed grain products are primarily consumed after sufficient hydration, paste viscosity is the key functional quality attribute for them. Most commonly, Rapid Visco Analyser (RVA), which is a rotational viscometer that uses variable heating (for gelatinization), cooling (for retrogradation), and shear (for controlled measurement) capabilities, is used to determine the pasting characteristics of starch or any starchy grain (Kumar et al., 2018). For raw starch, in the initial heating stage, the starch granules take up water, swell, and a high viscosity (peak viscosity, PV) value is developed until the gelatinization temperature (GT) is reached. At and above GT, the swollen granules rupture, releasing variable amount of absorbed water, causing a drop (breakdown, BD) in viscosity to a level termed as hot paste viscosity (HPV). This is gelatinized starch paste. During the cooling stage, this paste retrogrades, resulting in recrystallization and sedimentation, which again increases (setback, SB) the viscosity value to a higher cold paste viscosity (CPV) value. This pattern of change in paste viscosity varies with starch source and primarily due to differences in the proportion of amylose and amylopectin in them. Mostly linear amylose molecules and highly branched amylopectin show different behaviors to these hydrothermal changes. Pasting characteristics are a result of these changes shown by the starch during application of heat and shear.

Amylose and amylopectin content determines the viscosity of brown rice (Chen, 1995). Higher amylose content lowers its viscosity while higher content of amylopectin leads to its increased viscosity (Miles et al., 1985). Flaking incorporates thermal and mechanical treatments that are supposed to result in substantial gelatinization of starch with minimal retrogradation. Gelatinization-induced swelling and simultaneous starch leaching, especially amylose, is a characteristic of the flaked grain matrix. This results in the formation of a 3D network structure. So, unlike raw starch or grain, flaked rice is already in the gelatinized state and would show a distinctly different pattern of the RVA curve. Hence, with dynamic temperature viscometry as in RVA, including a static viscosity measurement at room temperature, can be considered as more appropriate. Value of PV in flaked rice (Gurjari variety) was found to be 57.83 cP higher than its raw variety (Kumar et al., 2018). The PV would indicate the water absorption and holding capacity of the starch in the processed product upon heating. However, as the sample systems are structurally different, running the same heating-cooling cycle does not give the precise indication of hygroscopicity. In practice, slowing down the heating rate can result in much higher water absorption in the flaked sample, despite keeping lower PV retention due to pre-occurence of rupture granules in them. More appropriately, gelatinized and heat damaged starch shows increased slurry viscosity resulting in faster water absorption and swelling at room temperature. Apparent viscosity decreases with increased shear rate. Rheological behavior of edge runner (ER) flakes, roller flakes (RF), and roasted parboiled rice show the highest to lowest apparent viscosities, respectively, over a range of shear stress. Additionally, thin flakes of ER show the highest apparent viscosity as compared to RF due to increased molecular starch damage in RF compared to only granular starch damage in ER flakes (Mujoo & Ali, 2000). Higher extent of gelatinization, molecular degradation, and structural modification of starch granules during processing treatments like flaking leads to higher initial viscosity (Chang & Yang, 1992 Whalen & Walker, 1997). Amylose–lipid complex and low amylopectin content in high amylose starches results in low swelling power and viscosity even at high temperatures, whereas high amylopectin content imparts higher swelling power and viscosity at low temperatures (Song & Jane, 2000).

Hardness is one of the important textural attributes. First crack of the product at one point indicated by maximum peak value can be recorded and the obtained value can be used for determination of hardness. For the breakdown of samples, the applied compression force (N) can vary significantly on different levels of processing. On processing of raw rice (Gurjari variety) to flaked rice, and further to roasted flaked rice, decrease in degree of harness was observed from 197.03N to 176.60N and to 107.23 N, respectively. This may be due to dry heat roasting combined with mechanical compression to flatten the grain, which has lots of surface cracks or damaged starch that appears on the surface of the flaked rice. Contrastingly, roasting leads to further gelatinization of the partially gelatinized starch and conversion to the expanded form of roasted flaked rice leads to reduction in its hardness (Kumar & Prasad, 2017).

5.4.3 Color Profile

Color is a significant factor for any food product. The color of a flaked product depends on factors like color of the raw grain, changes or degradation of native color due to processing, maillard browning, and translucency development due to gelatinization. Extent of gelatinization and the presence of water in the grain determines the translucency and opacity of the flakes. Reduction in CIE lightness (L*) value resulted from change of starch color from dirty white to translucent due to onset of gelatinization (Roy et al., 2019). Tempering before processing can result in better availability and distribution of water required for gelatinization. The resultant product is more translucent flakes than non-tempered less gelatinized starch giving more opaque flakes. The increased translucency results from partially gelatinized starch granules surrounded by a continuous phase dehydrated gelatinized starch causing reduction of air spaces within the flake [30]. Removal of bran and husk during production of extra thin flaked rice (ETFR) increased the L* value significantly indicating a brighter shade of the product, while a* value significantly decreased during processing of brown rice to ETFR. Increasing b* value (17.63) was also observed upon roasting of rice indicating development of a yellowish tinge (Kumar & Prasad, 2017). High heat applied during soaking and steaming can result in maillard browning of the flaked grains (Dutta & Mahanta, 2014).

5.4.4 Thermal Properties

Like any other processed grain product, gelatinization and retrogradation are the key thermal transitions related to processing of flaked rice and the product itself. Protein matrix covering the starch granules within the grain endosperm can delay starch gelatinization (Eliasson, 1983). Flaking disrupts the protein matrix, enhances gelatinization tendency, and hence lowers gelatinization temperature (Holm et al., 1988). The major part of flaked cereal is already gelatinized starch. Therefore, a continuous temperature rise in the Differential Scanning Calorimeter (DSC) to the gelatinization temperature of the specific rice starch would not necessarily give the characteristic thermal transition peak for gelatinization. However, in actuality, the peak is obtained with variable intensities due to general occurrence of ungelatinized and partially gelatinized swollen granules within the flaked product matrices. Also, as there is limited level of retrograded starch, the characteristic peak for melting of retrograded starch is also absent or miniature. Conditions of flaking also determine the gelatinization temperature (GT). Slight decrease in GT was observed upon steam flaking under mild conditions, whereas slight increase was observed upon steam flaking under sever conditions compared to that of raw wheat (Holm et al., 1988). In the gelatinization process, T_o (onset temperature) is defined as the initial temperature, T_p (peak temperature) is the begin of gelatinization or crystal melting, T_e (endset temperature) is the final temperature, and ΔH is gelatinization enthalpy. Increase in time, shear, temperature, and pressure during extrusion of cereal flours increases the T_o, $T_{p,}$ and T_e (Kaur et al., 2016). Gelatinization temperature (GT) of starch increases steeply immediately after water level reduces below 30%. In general, the general practice of adding water to the DSC sample of rice can alter the state of gelatinization and retrogradation, resulting in formation of major resistant starch melting peak and often gives inconclusive or erroneous results about the status of thermal transition within the material. Flaking involves starch gelatinization, which is inducted by a combination of heat, moisture, and pressure. Molecular behavior of the starch described by its mean thermal transition is related to heat and moisture. Occurrence of starch–lipid interactive complexes in processed grain systems has been widely reported and ensured using DSC analyses giving its characteristic peak at near 100°C (Dutta & Mahanta, 2014). Formation of starch–lipid complexes affects textural and nutritional properties of the processed products and is often enhanced upon addition of extraneous oil. Traditional flaked-puffed products like *hurum* preparation involves significant inclusion of oil into the product matrix. This creates major direction for research in such hydrothermally processed flaked cereal products.

5.4.5 Structural Properties

Higher water uptake is observed in smaller starch granules, due to their larger superficial area. The presence of surface pores and channels in the granules can further enhance water uptake by starches (Cornejo-Ramírez et al., 2018). However, in pregelatinized flaked products, these basic criteria do not play the key role. Flaking involves high mechanical force, which results in gelatinization leading to disintegration and distribution of the starch granules as large fragments (Kumar et al., 2018). Tempering and steaming results in swollen starch granule, which exudes amylose upon flaking. This breaks the pericarp that remains adherent to the surface of the flake (McDonough et al., 1997). Flaked and roller dried products results in complete disorganized granular and cellular structures of starch. Prominently visible protein and cell wall matrix in milled millet was found to be absent in processed (decorticated, roller dried, flaked, popped, and extruded) millet (Ushakumari et al., 2004). SEM studies of raw (brown), soaked paddy, roasted paddy, thick flaked rice, and ETFR was carried out by Kumar et al. (2018). Raw rice exhibited its typical irregular hexagonal small starch granules (2.5–3.0 μm). Increase in size of the granules was observed upon soaking (7.0–7.5 μm) that further increased after roasting (8.0–8.5 μm), resulting from water absorption and increase in internal pressure of moist granules during gelatinization, respectively. Furthermore, fragmentation and changes in the starchy matrix of ETFR developed from thick flaked rice was attributed by the high temperature and mechanical pressure exerted by the flaking machine rolls. Expansion of grain takes place both longitudinally and axially during flaking (Levine, 1993). Steamed sorghum kernels when forced through flaking rollers expanded in the same manner. SEM studies showed that developed translucency resulting from dehydrated gelatinized starch was attributed to good quality sorghum flakes (McDonough et al., 1997).

X-ray diffraction (XRD) is used for phase identification of crystalline materials. It can provide information about unit cell dimensions. The diffraction peaks are attributed to the amorphous and semi-crystalline lamellas of starch granules, based on the patterns of which starches are classified as A, B, and C types. Cereal starches commonly have A-type pattern, tubers and amylose-rich starches show B-type pattern, whereas C type is a mixture of A and B type, commonly found in leguminous starches. Formation of complex between amylose and a complexing functional group such as iodine, alcohol, cyclohexane, protein, fatty acids, etc., gives rise to V-type pattern. Relative crystallinity by X-ray diffractogram of raw, roasted, flaked, and roasted flaked rice was studied (Kumar & Prasad, 2017). It showed distinct peak angle of 2θ at 15°, 17°, 18°, and 23° for raw rice, indicating A-type starch crystallinity pattern. Conversion of crystalline structure of rice to amorphous form can be confirmed by the right shift with peak broadening during roasting. Further conversion with increase in resistant starch content was observed during flaking by edge runner.

Fourier Transform Infrared (FTIR) spectra of raw, roasted, flaked, and roasted flaked rice were observed (Kumar & Prasad, 2017). Stretching bands in all the samples at wave number 1018 $cm^{-1}$and 1658 cm^{-1}confirms presence of functional groups -C-O and C=O. The presence of free fatty acids in all the samples was observed from band range 1744.14–1743.85 cm^{-1}. Roasted flaked rice showed -O-H bond stretching corresponding to band at 3621.23 cm^{-1} to 3803.84 cm^{-1} wave number and symmetrical and asymmetrical CH2- stretching at wave number 2854.65 cm^{-1} to 2926.01 cm^{-1}. In the finger point region of all the samples, the bands at 759.95–707.88 cm^{-1} indicated α-linkage in starch. Processing of raw rice to roasted and flaked rice showed shift of band to lower wave number. Some degree of modification or stretching was observed in all the samples as compared to raw rice. No major difference in bands was observed during processing of roasted flaked rice. The transmittance value increased from raw rice to roasted flaked rice. All the samples represented crystalline character corresponding to the band ratio of the intensities of two bands (1082.07 and 1022.27 cm^{-1}), but raw rice showed highest crystallinity.

5.5 IMPACT OF PROCESSING TREATMENT ON BIOLOGICAL CHARACTERISTICS

Biological characteristics are defined by the bioactive compounds, antinutritional factors, phytochemical profile, and digestibility of cereal grains. The phytochemical compounds are generally higher in concentration in the outer layers of food grain. The whole grain would therefore comprise starch, protein, lipids, dietary fibre, and associated compounds, which are phytic acid, tannin, polyphenol, and some enzyme inhibitors like trypsin inhibitor along with minerals and vitamin (Oghbaei

& Prakash, 2016). Preprocessing operations like milling and refining reduce their content (Slavin et al., 1999). Antinutrients and phytochemicals are generally affected by the dehulling process and its extent. These compounds exert both positive as well as negative effects on the grain. While reduction of these compounds (phytates, tannins, and phenolic elements) lead to higher digestibility of protein and carbohydrates as well as improved availability of minerals, they also help to reduce oxidative stress in the human body by exhibiting strong antioxidant properties and restricting free radical activity (Harland & Morris, 1995). Processing treatments like rolling, steam rolling, or steam flaking break the hull, pericarp, and protein matrix and expose the starch in the endosperm to the microbial and internal enzymes (Safaei & Yang, 2017).

5.5.1 Antinutritional Factors

The naturally occurring antinutritional factors in cereals include phytate, tannins, phenolic elements, and enzyme inhibitors, which affect the nutritional profile of the cereal grains by negatively altering absorption. The antinutrients combine with nutrients and reduce their bioavailability of various compounds by reducing the digestibility of proteins, carbohydrates, and mineral absorption (Samtiya et al., 2020). But they can also reduce oxidative stress in the human body because they exhibit antioxidant properties (Harland & Morris, 1995). The antinutrients are mostly present in the outer layers of the grain and therefore get reduced in the hulling, milling, and refining processes, which are generally the preprocessing treatments for flaking. Flaking therefore reduces their content. The hydrothermal treatments of processing used in flaking can help change the nutrient content. The storage form of phosphorous in cereals, phytic acid, acts as chelating agent for minerals and prevents intestinal absorption by reducing the bioavailability of divalent cations by forming insoluble complexes (Lestienne et al., 2005; Weaver & Kannan, 2001). The phytic acid content can be reduced by preprocessing and processing treatments (Oghbaei & Prakash, 2016). The total and bioavailable minerals and related constituents in rice flakes of different thickness were estimated (Suma et al., 2007). They found that phosphorous, phytin phosphorous, and dietary fibre content decreased in proportion with decreasing thickness of flaked rice, but the iron and calcium contents remained unaffected. This is because rice does not contain oxalates and tannins, which are the inhibitors of iron absorption, and these factors therefore have no effect on their bioavailability. Reduction in phytic acid content of wheat bran was reported due to particle size reduction, hydrothermal treatment, and fermentation treatment (Majzoobi et al., 2014).

5.5.2 Phytochemical Profile

Polyphenols are phytochemicals that are considered as the non-nutritive compounds that are contained in the phenolic acids, flavonoids, and lignans of cereals. They mainly exist in three forms: free (ferulic acid, protocatechuic acid, gallic acid, parabiosanoic acid, caffeic acid, and erucic acid), soluble binding, and insoluble binding form. The majority of polyphenols are found as the binding form, which generally are made up of ferulic acid, caffeic acid, vanillic acid, and syringic acid. They are considered to have strong antioxidant and antimutagenic effects (Deng et al., 2012; Tian et al., 2019). The phytochemical contents in cereal grains are found to get reduced after flaking treatments. The total phenolic content (TPC) and total flavonoid content (TFC) decreased from grains to their respective flakes in most of the conducted research. This can be attributed to the outer layers being removed during processing, which contain most of the phytochemicals. The cortex, aleurone, and germ contain most of the antioxidant components that are affected due to removal of these layers during pre-processing treatment prior to flaking (Tian et al., 2019). TPC in sorghum varieties decreased from the range of 637.00 to 2476.00 GAE μg/g in grains to 453.33 to 1746.33 GAE μg/g in flakes, while TFC decreased from the range of 907.00 to 2128.00 (RE) μg/g in grains to 380.00 to 1237.00 (RE) μg/g in flakes (Sreeramaiah, 2012). Moderate loss of tocotrienols, caffeic acid, and avenanthramide Bp (N-(4′-hydroxy)-(E)-cinnamoyl-5-hydroxy-anthranilic acid) was observed after steaming and flaking of dehulled oat groats while there was increase in ferulic acid and vanillin. Avenanthramides predominantly exist in oat grains as oat phytoalexins (antibiotics of plants). Three predominant in oat grains are Bc (also called avenanthramide C), Bf (also called avenanthramide B), and Bp (also called avenanthramide A). Steaming did not affect the tocopherols and the avenanthramides Bc (N-(3′,4′-dihydroxy-(E)-cinnamoyl-5-hydroxy-anthranilic acid) and Bf (N-(4′-hydroxy-3′-methoxy)-(E)-cinnamoyl-5-hydroxy-anthranilic acid) (Bryngelsson et al., 2002). Soluble polyphenol content reduced by 50% due to flaking and increased in blistered flakes, while bound polyphenols decreased with flaking (except maize) and decreased in blistered flakes (except pearl millet and maize) (Itagi et al., 2012). Decrease due to flaking may be because of the removal of bran layers while increase can be credited to the liberation of bound polyphenols as well as higher extractability due to changes in starch structure on dry heat treatment (Finocchiaro et al., 2007).

5.5.3 Bioactivities

Cereals are good sources of biologically active compounds such as dietary fibre, tocopherols, tocotrienols, phenolic compounds, vitamins, and microelements (Sofi, 2019). Phenolic compounds, carotenoids, vitamin E, dietary fibre, and β-glucan are considered the most important groups of bioactive compounds (Okarter & Liu, 2010). These compounds exhibit positive impacts on human health. The compounds like phenolics, tocopherols, carotenoids anthocyanins flavonoids, tannin, and other bioactive compounds possess antioxidant properties. Therefore, the presence of these compounds can help protect against cardiovascular diseases, diabetes, hypertension, cancer, and asthma. Infections may also be caused if these are consumed in abundance in cereals (Okarter & Liu, 2010; Dasgupta & Klein, 2014). Use of whole grains for processing would retain most of the biologically active compounds since most of them are concentrated in the external layers, but dehusking and milling would bring about reduction in these compounds, as discussed earlier. Also, the use of thermal treatments if used in flaking can cause some physical and chemical changes due to starch gelatinization, protein denaturation, interaction of compounds, and browning reactions (Ragaee et al., 2013). Maillard browning may induce complex formation with its reaction byproducts along with high moisture content causing thermal degradation of phenolic compounds and its polymerization. This in turn can affect their antioxidant activity. Thermal processing can also break down cellular constituents and cell walls and help in releasing the bound polyphenols (Dewanto et al., 2002). Reactions such as maillard browning, chemical oxidation, and caramelization can help in increasing the total phenol content and free radical scavenging capacity during thermal processing, which can be attributed to the dissociation of conjugated phenolic moiety followed by polymerization and/or oxidation and formation of phenolics. These phenolics are not endogenous to the grains (Ragaee et al., 2013). Flaking treatments also affect the phosphorus, phytin phosphorus, and dietary fiber content of flaked rice. It was found that the lesser the thickness, the lower the constituent (Suma et al., 2007).

5.5.4 Starch and Protein Digestibility

Flaking methods involve roller flaking, steam flaking, temper flaking, and extrusion flaking. Processing treatments used in flaking increases microbial rumen digestibility of starch. The endosperm in the cereal grain is surrounded by a protein matrix that interrupts enzymatic hydrolysis and reduces surface availability of starch for host enzyme and rumen bacteria. Processing by mechanical roller flaker (dry) cracks the hull and the pericarp granting rumen microorganisms and enzymes access to the endosperm, increasing the surface area for attachment. However, incorporating moisture along with thermal treatments as in steam flaking and temper flaking can be more beneficial due to combined effect resulting in extensive rolling, which in turn helps in maximum utilization (Safaei & Yang, 2017; McAllister et al., 1994). During flaking processing of grains, starch gelatinization takes place due to heat and moisture and pressure used as aid in processing treatments, which increases the microbial enzymatic digestion by changing the protein matrix and structure of the starch McAllister et al., 1994; Alvarado G et al., 2009). Studies show that degree of flaking is one of the major determinants for the availability of grain starch. From a study conducted on barley and sorghum it was stated that, with increased degree of flaking, *in vitro* digestibility of starch improved (Osman et al., 1970). Studies show that steam flaking can improve starch digestibility of sorghum grain-based diets in cattle as compared to whole, ground, or dry rolled products. Also, *in vitro* amylolytic attack of starch in cereals by both ruminal microbial and pancreatic enzyme sources can be improved by optimum conditions of moist heat and pressure (Theurer et al., 1999).

Protein digestibility can be modified by processing treatments that include temperature and pressure, which helps in denaturation. Processing treatments used prior to flaking such as milling and grinding enhance the protein digestibility of the grain, due to size reduction phenomena, which obviously opens up the cellular structures and exposes the matrix to hydrolytic enzymes (Joye, 2019). Extrusion process used for flaking can also improve protein digestibility due to high shear effect on its structure and conformation, which leads to its denaturation (Moreno et al., 2018). External (cell structure and antinutritional factors) as well as internal factors (amino acid sequence and folding or cross linking) affect protein digestibility. Even though processing treatments are aimed at increasing protein digestibility, it may also lead otherwise. Protein digestibility is affected by the grain structure and composition, the presence of disulfide bonds, surface functional groups, and protein hydrophobicity and conformation (Duodu et al., 2003). Along with these properties, intensity of treatment determines that the proteins might lead to high accessibility to hydrolytic enzymes by losing their tightly folded structure, or they might unfold. High-intensity thermal treatments might lead to degradation of amino acid residues (Swaisgood & Catignani, 1991). According to research, *in vitro* loss of digestibility is affected by hydrophobic interactions between proteins (Annor et al., 2017). Aggregate formation will take place upon unfolding of the proteins due to denaturation and conformational changes and exposure of the hydrophobic protein patches along with other interactions, which will eventually lead to impaired protein digestibility (Joye, 2019; Gulati et al., 2017). Heat treatment on sorghum leads to formation of disulfide bonds between the proteins and reduced protein digestibility (Annor et al., 2017). Thermal treatments may have positive and negative effects on protein digestibility. This can be confirmed from research where trypsin inhibitors inactivation by thermal treatment positively affects the digestibility of proteins, but inversely, the denaturation and aggregation caused by the thermal treatment decreases the protein digestibility (Sarwar Gilani et al., 2012).

5.6 CONCLUSION

Flaking is a thermo-mechanical process. The traditional cereal flaking methods have been modified and developed for better yield and product versatility (Figure 5.2). Edge runner and developed heavy duty roller flaker are among the earliest methods. The flaking treatments lead to changes in the structure of the cereal grains and their components, particularly starch. The product characteristics are related to the physico-chemical alteration in the starch. These also relate to the general physical, cooking, and digestibility characteristics of the flaked products. Flaked cereals have a unique pre-gelatinized water deficit state, which is the key flaked product characteristic. It enhances the water absorption capacity of the grains. Use of roller flaker breaks the hull and the pericarp and increases enzyme accessibility, thereby improving digestion. Furthermore, treatments

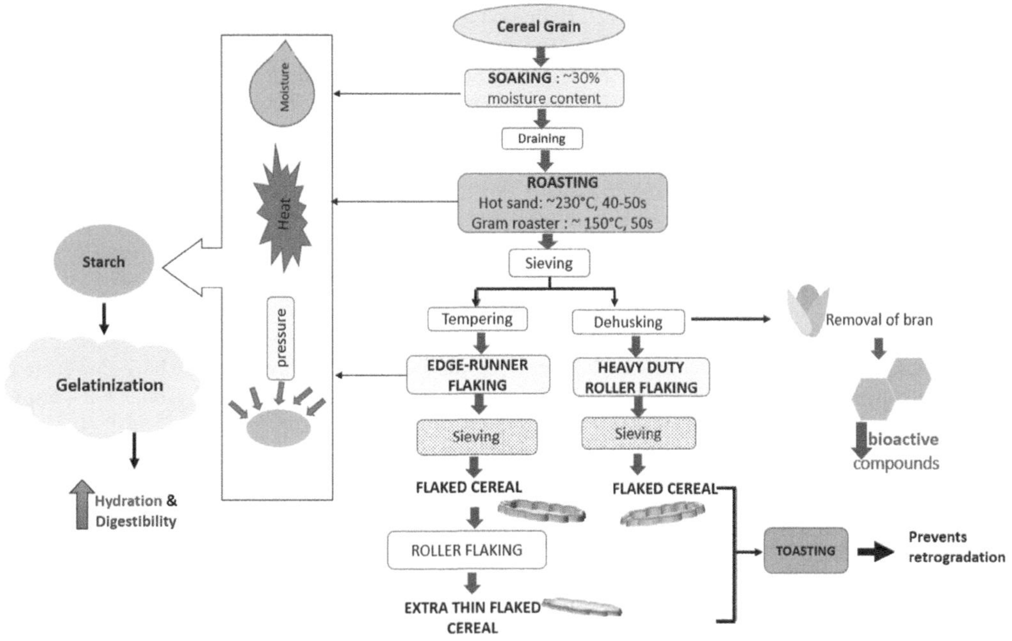

Figure 5.2 Process flow chart of cereal flake preparation and its impact on thegrain.

like steam flaking and temper flaking methods incorporate moisture which can induce combined beneficial effect due to extensive rolling, maximizing product variants and their utilization. The optical properties of the flaked cereal grain are attributed to the colour of the raw cereal grain, change or degradation of colour due to processing, maillard browning, and translucency development due to gelatinization. Gelatinization temperature is an important thermal property that directly relates to moisture availability in any processed grain matrix. Proteinaceous substances within the matrices can delay gelatinization. Protein disruption during flaking leads to enhances the gelatinization tendency of the otherwise embedded starch granules. The bioactive compounds in cereal bran can help reduce oxidative stress in the human body and can act as potential antioxidants. These mostly get reduced during dehusking and abrasive milling. Therefore, optimum degree and method of processing is necessary for the retention of the essential nutrients and bioactive compounds without compromising digestibility. The traditional and modernized flaking methods and the product characteristics have been scientifically explored. Findings have also created newer scope for investigations on nutrition holding, release, and absorption from the flaked RTE grain-based products. Bearing ethnic significance to numerous societies and being palate-friendly for large proportion of the global population, flaked grains remain as a basic processed food with technological importance.

REFERENCES

1. Ahmad, M., Gibb, D. J., McAllister, T. A., Yang, W. Z., Helm, J., Zijlstra, R. T., & Oba, M. (2010). Adjusting roller settings based on kernel size increased ruminal starch digestibility of dry-rolled barley grain in cattle. *Canadian Journal of Animal Science, 90*(2), 275–278. https://doi.org/10.4141/cjas09062

2. Ali, S. Z., & Bhattacharya, K. (1972). Hydration and amylose solubility behaviour of parboiled rice. *Lebensm. Wiss. u. Technol, 5*, 207–212.
3. Ali, S. Zakiuddin, & Bhattacharya, K. R. (1976). Starch retrogradation and starch damage in parboiled rice and flaked rice. *Die Starke, 28*(7), 233–240. https://doi.org/10.1002/star.19760280706
4. Alonso, R., Rubio, L. A., Muzquiz, M., & Marzo, F. (2001). The effect of extrusion cooking on mineral bioavailability in pea and kidney bean seed meals. *Animal Feed Science and Technology, 94*(1–2), 1–13. https://doi.org/10.1016/s0377-8401(01)00302-9
5. Alvarado G, C., Anrique G, R., & Navarrete Q, S. (2009). Effect of including extruded, rolled or ground corn in dairy cow diets based on direct cut grass silage. *Chilean Journal of Agricultural Research, 69*(3). https://doi.org/10.4067/s0718-58392009000300008
6. Ananthachar, T. K., Narasimha, R., & Gopal, H. S. (1982). Improvement of the traditional process for rice flakes. *J Food Sci Technol, 19*, 47–50.
7. Annor, G. A., Tyl, C., Marcone, M., Ragaee, S., & Marti, A. (2017). Why do millets have slower starch and protein digestibility than other cereals? *Trends in Food Science & Technology, 66*, 73–83. https://doi.org/10.1016/j.tifs.2017.05.012
8. Batterman-Azcona, S. J., & Hamaker, B. R. (1998). Changes occurring in protein body structure and α-Zein during cornflake processing. *Cereal Chemistry, 75*(2), 217–221. https://doi.org/10.1094/cchem.1998.75.2.217
9. Bender, D. A. (2006). *Benders' dictionary of nutrition and food technology* (7th ed.). Woodhead Publishing. ISBN: 9781845690519
10. Bilgi Boyaci, B., Han, J.-Y., Masatcioglu, M. T., Yalcin, E., Celik, S., Ryu, G.-H., & Koksel, H. (2012). Effects of cold extrusion process on thiamine and riboflavin contents of fortified corn extrudates. *Food Chemistry, 132*(4), 2165–2170. https://doi.org/10.1016/j.foodchem.2011.12.013
11. Biliaderis, C. G., & Perez, B. (1993). Thermophysical properties of milled rice starch as influenced by variety and parboiling method. *Cereal Chemistry, 70*, 512–516.
12. Brown, D. R. (2003). Method for steam flaking grain (Patent No. 6586028). In *US Patent* (No. 6586028). https://patents.google.com/patent/US6586028B1/en
13. Bryngelsson, S., Dimberg, L. H., & Kamal-Eldin, A. (2002). Effects of commercial processing on levels of antioxidants in oats (Avena sativa L.). *Journal of Agricultural and Food Chemistry, 50*(7), 1890–1896. https://doi.org/10.1021/jf011222z
14. Camire, M. E., Dougherty, M. P., & Briggs, J. L. (2005). Antioxidant-rich foods retard lipid oxidation in extruded corn. *Cereal Chemistry, 82*(6), 666–670. https://doi.org/10.1094/cc-82-0666
15. Chang, S. M., & Yang, H. (1992). Thermal processing effects on rice characteristics. *Food Structure, 11*, 373–382.
16. Chen, J. J. (1995). Effect of Milling Methods on the Physicochemical Properties of Waxy Rice Flour. *Cereal Chem, 76*, 796–799.
17. Cornejo-Ramírez, Y. I., Martínez-Cruz, O., Del Toro-Sánchez, C. L., Wong-Corral, F. J., Borboa-Flores, J., & Cinco-Moroyoqui, F. J. (2018). The structural characteristics of starches and their functional properties. *CyTA – Journal of Food, 16*(1), 1003–1017. https://doi.org/10.1080/19476337.2018.1518343
18. Dasgupta, A., & Klein, K. (2014). *Antioxidants in Food, Vitamins and Supplements: Prevention and Treatment of Disease.*
19. de Lumen, B. O., & Chow, H. (1991). Nutritional quality of rice endosperm. In *Rice* (pp. 782–814). Springer US.
20. Deepa, C., & Singh, V. (2011). Nutrient changes and functional properties of rice flakes prepared in a small scale industry. *ORYZA-An International Journal on Rice, 48*, 56–63.
21. Dehghan-banadaky, M., Corbett, R., & Oba, M. (2007). Effects of barley grain processing on productivity of cattle. *Animal Feed Science and Technology, 137*(1–2), 1–24. https://doi.org/10.1016/j.anifeedsci.2006.11.021
22. Deng, G.-F., Xu, X.-R., Guo, Y.-J., Xia, E.-Q., Li, S., Wu, S., Chen, F., Ling, W.-H., & Li, H.-B. (2012). Determination of antioxidant property and their lipophilic and hydrophilic phenolic contents in cereal grains. *Journal of Functional Foods, 4*(4), 906–914. https://doi.org/10.1016/j.jff.2012.06.008
23. Dewanto, V., Wu, X., & Liu, R. H. (2002). Processed sweet corn has higher antioxidant activity. *Journal of Agricultural and Food Chemistry, 50*(17), 4959–4964. https://doi.org/10.1021/jf0255937

24. Duodu, K. G., Taylor, J. R. N., Belton, P. S., & Hamaker, B. R. (2003). Factors affecting sorghum protein digestibility. *Journal of Cereal Science, 38*(2), 117–131. https://doi.org/10.1016/s0733-5210(03)00016-x
25. Dutta, H., & Mahanta, C. L. (2014). Laboratory process development and physicochemical characterization of a low amylose and hydrothermally treated ready-to-eat rice product requiring no cooking. *Food and Bioprocess Technology, 7*(1), 212–223. https://doi.org/10.1007/s11947-012-1037-9
26. Dutta, H., Mahanta, C. L., Singh, V., Das, B. B., & Rahman, N. (2016). Physical, physicochemical and nutritional characteristics of Bhoja chaul, a traditional ready-to-eat dry heat parboiled rice product processed by an improvised soaking technique. *Food Chemistry, 191*, 152–162. https://doi.org/10.1016/j.foodchem.2014.10.144
27. Eerlingen, R. C., Jacobs, H., & &,;. J. (1994). Enzyme-resistant starch V. Effect of retrogradation of waxy maize starch on enzyme susceptibility. *Cereal Chemistry, 71*(4), 351–355.
28. Ekanayake, S., & Narasimha, H. V. (1997). Comparative properties of rice flakes prepared using edge runner and roller flaker. *Journal of Food Science and Technology, 34*, 291–295.
29. Eliasson, A.-C. (1983). Differential scanning calorimetry studies on wheat starch—gluten mixtures. *Journal of Cereal Science, 1*(3), 199–205. https://doi.org/10.1016/s0733-5210(83)80021-6
30. Fast, R. B. (2000). Chapter 2: Manufacturing technology of ready-to-eat cereals. In *Breakfast Cereals and How They Are Made* (pp. 17–54). AACC International, Inc. https://doi.org/10.1094/1891127152.002
31. Finocchiaro, F., Ferrari, B., Gianinetti, A., Dall'Asta, C., Galaverna, G., Scazzina, F., & Pellegrini, N. (2007). Characterization of antioxidant compounds of red and white rice and changes in total antioxidant capacity during processing. *Molecular Nutrition & Food Research, 51*(8), 1006–1019. https://doi.org/10.1002/mnfr.200700011
32. Gulati, P., Li, A., Holding, D., Santra, D., Zhang, Y., & Rose, D. J. (2017). Heating reduces proso millet protein digestibility via formation of hydrophobic aggregates. *Journal of Agricultural and Food Chemistry, 65*(9), 1952–1959. https://doi.org/10.1021/acs.jafc.6b05574
33. Harland, B. F., & Morris, E. R. (1995). Phytate: A good or a bad food component? *Nutrition Research (New York, N.Y.), 15*(5), 733–754. https://doi.org/10.1016/0271-5317(95)00040-p
34. Holm, J., Björck, I., & Eliasson, A.-C. (1988). Effects of thermal processing of wheat on starch: I. Physico-chemical and functional properties. *Journal of Cereal Science, 8*(3), 249–260. https://doi.org/10.1016/s0733-5210(88)80036-5
35. Itagi, H. N., Rao, B. V. R. S., Jayadeep, P. A., & Singh, V. (2012). Functional and antioxidant properties of ready-to-eat flakes from various cereals including sorghum and millets: Properties of RTE flakes. *Quality Assurance and Safety of Crops & Foods, 4*(3), 126–133. https://doi.org/10.1111/j.1757-837x.2012.00136.x
36. Jane, J., Chen, Y. Y., Lee, L. F., McPherson, A. E., Wong, K. S., Radosavljevic, M., & Kasemsuwan, T. (1999). Effects of amylopectin branch chain length and amylose content on the gelatinization and pasting properties of starch. *Cereal Chemistry, 76*(5), 629–637. https://doi.org/10.1094/cchem.1999.76.5.629
37. Jane, J. L., & Chen, J. (1992). Effect of amylose molecular size and amylopectin branch chain length on paste properties of starch. *Cereal Chemistry, 69*, 60–65.
38. Joye, I. (2019). Protein digestibility of cereal products. *Foods (Basel, Switzerland), 8*(6), 199. https://doi.org/10.3390/foods8060199
39. Kaur, G., Sharma, S., Singh, B., & Dar, B. N. (2016). Comparative study on functional, rheological, thermal, and morphological properties of native and modified cereal flours. *International Journal of Food Properties, 19*(9), 1949–1961. https://doi.org/10.1080/10942912.2015.1089892
40. Koenig, K. M., & Beauchemin, W. (2013). Processing feed grains: factors affecting the effectiveness of grain processing for beef and dairy cattle production. &quot. In *34th Western Nutrition Conference – Processing* (pp. 62–73).
41. Kokić, B., Mária, C., Zuzana, F., Mária, P., Matúš, R., & Rade, D. (2013). Influence of thermal treatments on starch gelatinization and in vitro organic matter digestibility of corn. *Food and Feed Research, 40*, 93–99.
42. Kumar, S., Haq, R.-U., & Prasad, K. (2018). Studies on physico-chemical, functional, pasting and morphological characteristics of developed extra thin flaked rice. *Journal of the Saudi Society of Agricultural Sciences, 17*(3), 259–267. https://doi.org/10.1016/j.jssas.2016.05.004

43. Kumar, S. N., & Prasad, K. (2013). Effect of Paddy Parboiling and Rice Puffing on Physical, Optical and Aerodynamic Characteristics. *International Journal of Agriculture and Food Science Technology, 4*(8), 765–770.
44. Kumar, S., & Prasad, K. (2017). Optimization of flaked rice dry roasting in common salt and studies on associated changes in chemical, nutritional, optical, physical, rheological and textural attributes. *Asian Journal of Chemistry, 29*(6), 1380–1392. https://doi.org/10.14233/ajchem.2017.20563
45. Lestienne, I., Icard-Vernière, C., Mouquet, C., Picq, C., & Trèche, S. (2005). Effects of soaking whole cereal and legume seeds on iron, zinc and phytate contents. *Food Chemistry, 89*(3), 421–425. https://doi.org/10.1016/j.foodchem.2004.03.040
46. Levine, L. (1993). Musing on the mechanics of flaking rolls. *Cereal Foods World, 38*, 873–874.
47. Ljøkjel, K., Sørensen, M., Storebakken, T., & Skrede, A. (2004). Digestibility of protein, amino acids and starch in mink (Mustela vison) fed diets processed by different extrusion conditions. *Canadian Journal of Animal Science, 84*(4), 673–680. https://doi.org/10.4141/a01-089
48. Majzoobi, M., Pashangeh, S., Farahnaky, A., Eskandari, M. H., & Jamalian, J. (2014). Effect of particle size reduction, hydrothermal and fermentation treatments on phytic acid content and some physicochemical properties of wheat bran. *Journal of Food Science and Technology, 51*(10), 2755–2761. https://doi.org/10.1007/s13197-012-0802-0
49. McAllister, T. A., Bae, H. D., Jones, G. A., & Cheng, K. J. (1994). Microbial attachment and feed digestion in the rumen. *Journal of Animal Science, 72*(11), 3004–3018. https://doi.org/10.2527/1994.72113004x
50. McDonough, C. M., Anderson, B. J., & Rooney, L. W. (1997). Structural characteristics of steam-flaked sorghum. *Cereal Chemistry, 74*(5), 542–547. https://doi.org/10.1094/cchem.1997.74.5.542
51. Miles, M. J., Morris, V. J., Orford, P. D., & Ring, S. G. (1985). The roles of amylose and amylopectin in the gelation and retrogradation of starch. *Carbohydrate Research, 135*(2), 271–281. https://doi.org/10.1016/s0008-6215(00)90778-x
52. Moreno, C. R., Fernández, P. C. R., Rodríguez, E. O. C., Carrillo, J. M., & Rochín, S. M. (2018). Changes in nutritional properties and bioactive compounds in cereals during extrusion cooking. In *Extrusion of Metals, Polymers and Food Products*. InTech. http://dx.doi.org/10.5772/intechopen.68753.
53. Mujoo, R., & Ali, S. Z. (1999). Molecular degradation of rice starch during processing to flakes. *Journal of the Science of Food and Agriculture, 79*(7), 941–949. https://doi.org/10.1002/(sici)1097-0010(19990515)79:7<941::aid-jsfa309>3.0.co;2-l
54. Mujoo, R., & Ali, S. Z. (2000). Changes in physico-chemical and rheological properties of rice during flaking. *International Journal of Food Properties, 3*(1), 117–135. https://doi.org/10.1080/10942910009524620
55. Murray, S. M., Flickinger, E. A., Patil, A. R., Merchen, N. R., Brent, J. L., & Fahey, G. C. (2001). In vitro fermentation characteristics of native and processed cereal grains and potato starch using ileal chyme from dogs. *Journal of Animal Science, 79*(2), 435. https://doi.org/10.2527/2001.792435x
56. Narasimha, H. V., Ananthachar, R., & Gopal, H. S. (1982). Development of a continuous process for making rice flakes. *Journal of Food Science and Technology, 19*, 233–236.
57. Oghbaei, M., & Prakash, J. (2016). Effect of primary processing of cereals and legumes on its nutritional quality: A comprehensive review. *Cogent Food & Agriculture, 2*(1). https://doi.org/10.1080/23311932.2015.1136015
58. Okarter, N., & Liu, R. H. (2010). Health benefits of whole grain phytochemicals. *Critical Reviews in Food Science and Nutrition, 50*(3), 193–208. https://doi.org/10.1080/10408390802248734
59. Ong, M. H., & Blanshard, J. M. V. (1995). The significance of starch polymorphism in commercially produced parboiled rice. *Die Starke, 47*(1), 7–13. https://doi.org/10.1002/star.19950470104
60. Osman, H. F., Theurer, B., Hale, W. H., & Mehen, S. M. (1970). Influence of grain processing on in vitro enzymatic starch digestion of barley and sorghum grain. *The Journal of Nutrition, 100*(10), 1133–1139. https://doi.org/10.1093/jn/100.10.1133
61. Peres, H., & Oliva-Teles, A. (2002). Utilization of raw and gelatinized starch by European sea bass (Dicentrarchus labrax) juveniles. *Aquaculture (Amsterdam, Netherlands), 205*(3–4), 287–299. https://doi.org/10.1016/s0044-8486(01)00682-2
62. Ragaee, S., Seethraman, K., & Abdel-Aal, E.-S. (2013). Effects of processing on nutritional and functional properties of cereal products. In *Engineering Aspects of Cereal and Cereal-Based Products*. CRC Press. https://doi.org/10.1201/b15246-15

63. Rao, U. P., Vatsala, & Rao, P. H. (2002). Changes in protein characteristics during the processing of wheat into flakes. *European Food Research and Technology*, *215*(4), 322–326. https://doi.org/10.1007/s00217-002-0553-7
64. Rastgoo, M., Kazemi-Bonchenari, M., HosseinYazdi, M., & Mirzaei, M. (2020). Effects of corn grain processing method (ground versus steam-flaked) with rumen undegradable to degradable protein ratio on growth performance, ruminal fermentation, and microbial protein yield in Holstein dairy calves. *Animal Feed Science and Technology*, *269*(114646), 114646. https://doi.org/10.1016/j.anifeedsci.2020.114646
65. Robie, S. C., & Hilgendorf, D. J. (2001). R-T-E cereal and method of preparation (Patent No. 6291008). In *US Patent* (No. 6291008).
66. Rooney, L. W., & Pflugfelder, R. L. (1986). Factors affecting starch digestibility with special emphasis on sorghum and corn. *Journal of Animal Science*, *63*(5), 1607–1623. https://doi.org/10.2527/jas1986.6351607x
67. Roy, M., Dutta, H., Jaganmohan, R., Choudhury, M., Kumar, N., & Kumar, A. (2019). Effect of steam parboiling and hot soaking treatments on milling yield, physical, physicochemical, bioactive and digestibility properties of buckwheat (Fagopyrum esculentum L.). *Journal of Food Science and Technology*, *56*(7), 3524–3533. https://doi.org/10.1007/s13197-019-03849-9
68. Safaei, K., & Yang, W. (2017). Effects of grain processing with focus on grinding and steam- flaking on dairy cow performance. In *Herbivores*. InTech. https://doi.org/10.5772/67344
69. Samtiya, M., Aluko, R. E., & Dhewa, T. (2020). Plant food anti-nutritional factors and their reduction strategies: an overview. *Food Production, Processing and Nutrition*, *2*(1). https://doi.org/10.1186/s43014-020-0020-5
70. Sarwar Gilani, G., Wu Xiao, C., & Cockell, K. A. (2012). Impact of antinutritional factors in food proteins on the digestibility of protein and the bioavailability of amino acids and on protein quality. *The British Journal of Nutrition*, *108 Suppl 2*(S2), S315–32. https://doi.org/10.1017/S0007114512002371
71. Shinde, G., & Durgadevi, M. (2018). Development of iron fortified chocolate flavoured rice flakes. *Journal of Nutrition & Food Sciences*, *08*(01). https://doi.org/10.4172/2155-9600.1000649
72. Singh, S., Gamlath, S., & Wakeling, L. (2007). Nutritional aspects of food extrusion: a review. *International Journal of Food Science & Technology*, *42*(8), 916–929. https://doi.org/10.1111/j.1365-2621.2006.01309.x
73. Slavin, J. L., Jacobs, D., & Marquart, L. (2001). Grain processing and nutrition. *Critical Reviews in Biotechnology*, *21*(1), 49–66. https://doi.org/10.1080/20013891081683
74. Slavin, Joanne L., Martini, M. C., Jacobs, D. R., Jr, & Marquart, L. (1999). Plausible mechanisms for the protectiveness of whole grains. *The American Journal of Clinical Nutrition*, *70*(3), 459s–463s. https://doi.org/10.1093/ajcn/70.3.459s
75. Sofi, S. A. (2019). Cereal Bioactive Compounds: A Review. *International Journal of Agriculture Environment and Biotechnology*, *12*(2). https://doi.org/10.30954/0974-1712.06.2019.5
76. Song, Y., & Jane, J. (2000). Characterization of barley starches of waxy, normal, and high amylose varieties. *Carbohydrate Polymers*, *41*(4), 365–377. https://doi.org/10.1016/s0144-8617(99)00098-3
77. Sreeramaiah, H., & Goudar, G. (2012). Effect of flaking on nutrient and phytochemical retention in sorghum[*Sorghum bicolor (L.). Moench*] varieties. *Biotechnology, Bioinformatics & Bioengineering*, *2*(2), 653–658.
78. Suma, R. C., Sheetal, G., Jyothi, L. A., & Prakash, J. (2007). Influence of phytin phosphorous and dietary fibre on in vitro iron and calcium bioavailability from rice flakes. *International Journal of Food Sciences and Nutrition*, *58*(8), 637–643. https://doi.org/10.1080/09637480701395515
79. Svihus, B., Uhlen, A. K., & Harstad, O. M. (2005). Effect of starch granule structure, associated components and processing on nutritive value of cereal starch: A review. *Animal Feed Science and Technology*, *122*(3–4), 303–320. https://doi.org/10.1016/j.anifeedsci.2005.02.025
80. Swaisgood, H. E., & Catignani, G. L. (1991). Protein digestibility: in vitro methods of assessment. *Advances in Food and Nutrition Research*, *35*, 185–236. https://doi.org/10.1016/s1043-4526(08)60065-0
81. Thachil, M. T., Chouksey, M. K., & Gudipati, V. (2014). Amylose-lipid complex formation during extrusion cooking: effect of added lipid type and amylose level on corn-based puffed snacks. *International Journal of Food Science & Technology*, *49*(2), 309–316. https://doi.org/10.1111/ijfs.12333

82. Theurer, C. B., Huber, J. T., Delgado-Elorduy, A., & Wanderley, R. (1999). Invited review: summary of steam-flaking corn or sorghum grain for lactating dairy cows. *Journal of Dairy Science*, *82*(9), 1950–1959. https://doi.org/10.3168/jds.S0022-0302(99)75431-7
83. Thielecke, F., Lecerf, J.-M., & Nugent, A. P. (2021). Processing in the food chain: do cereals have to be processed to add value to the human diet? *Nutrition Research Reviews*, *34*(2), 159–173. https://doi.org/10.1017/S0954422420000207
84. Tian, S., Sun, Y., Chen, Z., Yang, Y., & Wang, Y. (2019). Functional properties of polyphenols in grains and effects of physicochemical processing on polyphenols. *Journal of Food Quality*, *2019*, 1–8. https://doi.org/10.1155/2019/2793973
85. Tribelhorn, R. E., Lorenz, K. J., & Kulp, K. (1991). Breakfast Cereals In: *Handbook of Cereal Science and Technology*, edited by Klaus J. Lorenz and Karel Kulp, Routlegde. 741–762.
86. Unnikrishnan, K. R., & Bhattacharya, K. R. (2007). Swelling and solubility behaviour of parboiled rice flour. *International Journal of Food Science & Technology*, *16*(4), 403–408. https://doi.org/10.1111/j.1365-2621.1981.tb01831.x
87. Ushakumari, S. R., Latha, S., & Malleshi, N. G. (2004). The functional properties of popped, flaked, extruded and roller-dried foxtail millet (Setaria italica). *International Journal of Food Science & Technology*, *39*(9), 907–915. https://doi.org/10.1111/j.1365-2621.2004.00850.x
88. Vaidya, R. H., & Sheth, M. K. (2011). Processing and storage of Indian cereal and cereal products alters its resistant starch content. *Journal of Food Science and Technology*, *48*(5), 622–627. https://doi.org/10.1007/s13197-010-0151-9
89. Viscidi, K. A., Dougherty, M. P., Briggs, J., & Camire, M. E. (2004). Complex phenolic compounds reduce lipid oxidation in extruded oat cereals. *Lebensmittel-Wissenschaft Und Technologie [Food Science and Technology]*, *37*(7), 789–796. https://doi.org/10.1016/j.lwt.2004.03.005
90. Weaver, C., & Kannan, S. (2001). Phytate and Mineral Bioavailability. In *Food Phytates*. CRC Press. https://doi.org/10.1201/9781420014419.ch13.
91. Whalen, P. J., & Walker, P. J. (1997). Measurement of extrusion effects by viscosity profile using the Rapid ViscoAnalyser. *Cereal Foods World*, *42*, 469–475.
92. Xiong, Y., Bartle, S. J., & Preston, R. L. (1991). Density of steam-flaked sorghum grain, roughage level, and feeding regimen for feedlot steers. *Journal of Animal Science*, *69*(4), 1707–1718. https://doi.org/10.2527/1991.6941707x
93. Xu, N., Wang, D., & Liu, J. (2019). Variance of Zein protein and starch granule morphology between corn and steam flaked products determined starch ruminal degradability through altering starch hydrolyzing bacteria attachment. *Animals: An Open Access Journal from MDPI*, *9*(9), 626. https://doi.org/10.3390/ani9090626
94. Zielinski, H., Kozlowska, H., & Lewczuk, B. (2001). Bioactive compounds in the cereal grains before and after hydrothermal processing. *Innovative Food Science & Emerging Technologies: IFSET: The Official Scientific Journal of the European Federation of Food Science and Technology*, *2*(3), 159–169. https://doi.org/10.1016/s1466-8564(01)00040-6

PART III

Biological Processing of Cereals and Its Impact

CHAPTER 6

Germination of Cereals
Effect on Nutritive, Functional, and Biological Properties

Pratik Nayi[1], Mamta Thakur[2], and Vikas Nanda[3]
[1]Department of Tropical Agriculture and International Cooperation, National Pingtung University of Science and Technology, Neipu, Pingtung, Taiwan
[2]Department of Food Technology, School of Sciences, ITM University, Gwalior, Madhya Pradesh, India
[3]Department of Food Engineering and Technology, Sant Longowal Institute of Engineering and Technology, Longowal, Punjab, India

CONTENTS

DOI: 10.1201/9781003242192-9

6.1 INTRODUCTION

Globally, cereals are the primary component of nearly every human diet. Rice, wheat, maize, barley, and sorghum are by far the most commonly cultivated crops in the world. Rye and millets as well as species such as oats and pseudocereals are regionally important. Plants of the Poaceae family are all monocotyledonic and create grains, which are dry and one-seeded fruits, made up of a pericarp and a seed. The seed or grain contains germ, nucellar epidermis, endosperm, and seed coat as different parts. A variety of active compounds are found in cereal grains that are organized into compartments and kept separate by cell walls or other barriers (Lemmens et al., 2019).

Cereals are widely regarded as excellent sources of protein, fibre, minerals and other essential nutrients. They also contain other bioactive compounds, such as tocopherols, phytosterols, lignans, phytochemicals, and folates, which are becoming more popular because of the growing demand for healthier nutrition (Tacer-Caba et al., 2015; Wrigley 2017; Divte et al., 2022). Few constituents in cereal grains scavenge the free radicals when others behaves like a enzyme cofactor or indirect antioxidants which generally delivers the positive impact on health in addition to fundamental nutrients for the minimizing the risk of severe diseases because of synergistic effect and their additive (Singh and Sharma, 2017).

Anti-vitamins, enzyme inhibitors, haemaglutinins, and other anti-nutritional factors lessen the nutritive benefits and functionality of cereal grains. Dehulling, germination, fermentation, soaking and heat processes like boiling, canning, and infrared heating are just a few of the processing technologies that can be used to minimize or remove such anti-nutritional components. Grain germination increases micronutrient content while also improving nutritional value, functionality, and sensory perception (Nkhata et al., 2018; Gabriele and Pucci, 2022). There are increase in essential amino acids like lysine, vitamin bioavailability, and protein digestibility whereas for anti-nutrients like phytic acid, trypsin inhibitors are reduced (Benincasa et al., 2019; Kumar and Anand, 2021). Soaking of grains in water is usually the initial step of germination process. There are several biochemical changes that take place during germination to prepare the grain for seedling development. Sprouted grains and shoots and also malt can be made from germination only, using which a wide range of ingredients can be made from cereal grains. The germination process also alter the grain's aesthetic, aroma, structure and nutritional value (Nkhata et al., 2018; Li et al., 2021; Thakur et al., 2021).

Numerous recent publications (Gan et al., 2017; Benincasa et al., 2019; Lemmens et al., 2019; Li et al., 2021) have reviewed the cereal germination considering the bioactive compounds after germination, but none have highlighted the significance of germination for improving the nutrition, functionality, and biological properties of grains together. This chapter mainly focused on the nutritive and functional quality of germinated cereals and their biological effect. In this chapter, an overview of germination process as well as safety aspects of germinated grains (sprouts) are given.

6.2 GERMINATION OF CEREALS: AN OVERVIEW

Almost everyone in the world relies on cereal grains as a main source of nutrition. There are a wide range of species and varieties of cereal grains like w heat, maize, rice, sorghum, barley, oats, rye, and numerous millets. In addition to providing significant amounts of energy and nutrients such as carbohydrates, protein, and a small amount of anti-nutrients, cereal grains contain bioactive components (Tacer-Caba et al., 2015; Copeland, 2020). In addition to reduce blood cholesterol, whole grain cereal products have been shown to reduce the cancer risk, as well as the risk of heart attack (Tomé-Sánchez et al., 2022). The functionality of cereal grains can be further improved by the process of germination.

A non-thermal biological process known as germination enhances the nutritional profile of cereals by boosting the digestibility of nutrients, decreasing anti-nutritional compounds, increasing the levels of free amino acids and available carbohydrates, and providing better functionality of the cereals (Nkhata et al., 2018; Lemmens et al., 2019). Amylase, protease, and lipase are three enzymes involved in the hydrolysis of starch, protein, and fat during the germination process. Both endogenous and newly synthesized enzymes initiate to alter the grain components when grains are steeped and held under room conditions for germination. It is thus possible to decompose complex macromolecules into lower molecular weight molecules that are easier for the body to digest and absorb (Hübner and Arendt, 2013; Nkhata et al., 2018).

Normally, germination is triggered due to the absorption of water by grains. Several simple methods can be used to germinate seeds, including sterilization, soaking, and sprouting (Gan et al., 2017; Hussein et al., 2019). To encourage metabolic activities, germination necessitates a combination of favourable humidity, oxygen availability, and temperature (Buriro et al., 2011, Springer and Mornhinweg, 2019, Gómez-lvarez and Pucciariello, 2022). Initially, the seeds are sterilized prior to soaking in order to prevent the microbial growth. The most frequently used sterilization reagents for germination is sodium hypochlorite (NaClO) solution, which can be employed at room temperature for 5–30 minutes in different concentrations (Limón et al., 2014). Grains can be sterilized for little as three minutes using ethanol or 70% ethanol (Pajak et al., 2014). While some studies (Hefni et al., 2012; Jahan et al., 2021) did not sterilize seeds prior to soaking, this may have been due to concerns about the sterilization reagents' potential harmful effects on the grains and the resulting food safety risks for consumption (Jahan et al., 2021). Consequently, grain germination does not necessitate sterilization.

Following this is the need to soak the grains in water to rehydrate. The grain weight (g): water volume ratio should be taken into account when soaking the grains in addition to consider the soaking temperature and time. At room temperature (between 20 to 30°C), the grains can be soaked for several hours to up to 24 hours. As the grains are hydrated, the metabolic processes of the dry grains increase significantly. After soaking, it is possible to germinate/sprout grains by keeping them in special germinators or incubators. Grains are typically sprouted in the dark at a temperature between 20 and 30 °C. In this period, water should be sprayed over grains daily in order to maintain a relatively high humidity level, and water should be replaced regularly, at least twice a day. The removal of water assists in the elimination of the metabolic products of germinated grains and reduce the microbiological proliferation (Kandil et al., 2015).

There are many factors affecting the nutritional, biological, and functional quality of germinated seeds/grains which play a significant role during the germination process. The different variables like salt stress, high and low temperature, light modulation, genotype and seed source, hypoxia stress, and germination conditions like humidity, time, temperature, etc. affect the germination process, as shown in Figure 6.1. One of the most significant abiotic stressors on plants is salinity especially during the pre-seedling growth which is phase of salt sensitivity. Many researchers study on high and low temperature for the investigating the effects of extreme temperatures which applied during germination process. This also affects the nutritional quality of germinated seeds. Light is one of the main variable which mainly affects the plant growth and its development (Flores et al., 2016). It improves the nutritional quality of seedlings also. Genotype play a most important role for the determination of the nutritional value of germinated seeds and biochemical changes composition too. Hypoxia stress is essential for γ-aminobutyric acid (GABA) accumulation in seedlings in response to hypoxia stress throughout the germination process. Germination conditions improve germination and post-germination seedling growth. They are beneficial for the biochemical changes that occur during the germination process.

Thus, germination, in general, is a low-cost, simple and eco-friendly method of developing sprouts and germinated grains in a brief span of time. Functional germinated grains and sprouts can be produced using germination as a bioprocessing technique in the food industry. A number of biochemical changes change take place in grains during germination, including the release from embryo of a number of hormones (gibberellin), which reach the aleuronic layer and stimulate enzymes like amylases and proteases release from endosperm. Further, there is reduction of enzyme inhibitor activity. Such biochemical changes function on components found in germ and endosperm and, as a result, turn into active metabolism in the grain from dormant metabolism (Iordan et al., 2013; Shu et al., 2016; Nkhata et al., 2018). Stored carbohydrates and proteins are broken down into smaller molecules by such enzymes. Polyphenolic compounds, glucosinolates, and compounds rich in selenium are all found in germinated grains. Sprouts are also rich in enzymes, minerals, vitamins, and essential amino acids, as well as a variety of phytochemicals that have been shown the physiological advantages (Gan et al., 2017). The germination process resulted in a significant boost in nutrients and other valuable substances.

6.3 EFFECT OF GERMINATION

Generally, most people consumed cereals or grains in the form of sprouts, dried, roasted, and as malt. For the improvement of nutritional properties and some aspects related to technologies, the grains need to be soaked in water and germinated before consumption. Many changes have been observed during germination like appearance, taste, flavor, and nutritional value of the grain. When it comes to metabolic processes, conditions of germination and water content after steeping play a large role. Malt quality has been extensively studied as a result of changes in germination conditions (Hübner and Arendt 2013).

The metabolic action of dry seed multiplies during the soaking process. There are no additional nutrients added to the seed during germination, and the germinated seed only consumes water and oxygen as it undergoes complex biochemical changes. As a result of the breakdown of nutritionally undesirable components and the disintegration of complex substances into more simple forms, favourable nutritional changes occur during germination. Biochemical processes that take place during activation generally produce bioactive compounds like ascorbic acid and phenolic compounds which possess physiological benefits. Some components synthesized during grain germination include arginoxylns, ferulic acid, prolyl endopeptide inhibitor, γ-aminobutyric acid, inoxolnols, sterols and tocopherols, minerals and tocotrienols (Chung et al., 2012). Different studies with diverse findings have been observed and the differences are mainly due to changes in the soaking process, germination temperature, biological activation rates and drying techniques of sprouts (Singh and Sharma 2017). Table 6.1 illustrates the different recent studies showing the effect of germination on nutritional, functional and biological properties of different cereal grains.

6.3.1 Nutritional Properties

Germination boosts the enzyme reactions of grains, resulting in the conversion of complex carbohydrates, proteins, and lipids into simpler compounds, as well as the activation of proteases that degrade proteins, resulting in increased nutritional bioavailability. In barley, wheat, oat, and

Table 6.1 Effect of Germination on Nutritional, Functional, and Biological Properties of Different Cereal Grains

	Germination conditions						
	Soaking		Sprouting				
Cereal(s)	Time (h)	Temperature (°C)	Time (h)	Temperature (°C)	RH (%)	Findings	References
Nutritional properties							
Amaranth, quinoa, and buckwheat	12 and 24	Room temperature	24, 48, and 72	25	-	A significant rise of 7.01, 74.67, 126.62, and 87.47% was observed in crude protein, crude fiber, phenolic content, and antioxidant activity of amaranth with a decrease of 32.30% and 29.57% in tannin and phytic acid, respectively; increase of crude proteins, crude fiber, phenolic content, and antioxidant activity noticed in buckwheat and quinoa in which tannin and phytic acid decreased by 59.91 and 17.42%, in buckwheat and 27.08% and 47.57%, in quinoa, respectively.	Thakur et al., (2021)
Biofortified wheat genotypes (MB-64-1-1, 48-41-23-6-4-4, 75-1-4-16-3-5, 77-46-6-8-2-1, 1-1-7-18-5-5-18, 49-1-73-8-5, & 49-1-11-9-7-1)	12	Room temperature	12, 24, 48, 72, and 96	Room temperature	-	Increase in soluble proteins and starch contents and higher total phenolic and anthocyanin contents of all wheat types	Verma et al., (2021)
Red rice	8	25 ± 2	8, 16, 24, 32, and 40	25 ± 1	90	A decrease in ash, moisture, and pH but no change observed in other basic constituents with increase in germination time; a rise in caffeic, coumaric, and ferulic acids, and myricetin after 16 h of germination; and increase in GABA (gamma-aminobutyric acid) amount in the germinated red rice flour	Müller et al. (2021)

(continued)

Table 6.1 (Continued)

	Germination conditions						
	Soaking		Sprouting				
Cereal(s)	Time (h)	Temperature (°C)	Time (h)	Temperature (°C)	RH (%)	Findings	References
Maize	24	Room temperature	72	28	85	Rise in the levels of calcium (47 to 58%), magnesium (90 to 108%), iron (0.50 to 1.25%), potassium (2.10 to 4.30), and sodium (0.50 to 0.70%) and a significant increase in the contents of vitamins (A, B1, C, D, and E) of germinated maize	Gnanwa et al. (2021)
Barley (hull-less H13 cultivar)	4	Room temperature	1.6–6.19 days	12.1–19.9	>90	Rise in the levels of vitamin B1, B2 and C, GABA, total phenolic content, and proteins and decrease in fat, carbohydrates, fibre, and β-glucan after germination of barley grains	Rico et al., (2020)
Wheat (PBW-550), paddy (PR-113) and triticale (TL-2908)	7-8 (wheat & triticale) and 9-10 (paddy)	40±2	94 (wheat & triticale) and 96 (paddy)	28±2 (wheat and triticale) and 35±2 (paddy)	-	Increased protein content by 25.06% in wheat, 24.26% in triticale and 20.36% in brown rice; an increment of 43.28%, 58.01%, and 44.09% in total sugars of wheat, triticale, and brown rice, respectively; non-significant change in fat and ash content of all grains except fat content rise in rice by 21.12% and improved essential amino acids and lysine content in all grains	Sibian et al. (2017)
Barley, oats, rye, and wheat	24 ± 1	22 ± 2	12, 24, 36, and 48	35 ± 2	95 ± 2	A significant decrease of starch by 26.40% in oats but in other grains, decrease was non-significant; reduction in b-glucan content of barley grains by 20.50% and non-significant variations observed in sucrose content	Senhofa et al. (2016)

Rice and wheat	24	Room temperature	72	25	-	An increase of two-fold in iron content of rice; raised availability of zinc from rice and wheat to 80% and 87%, respectively; and increased manganese and calcium availability in rice as a result of germination	Luo et al. (2014)
Functional properties							
Maize	24	Room temperature	72	28	85	Increase in water absorption capacity (from 205.36 to 304.90%), foaming capacity (from 22.08 to 26.74%), emulsification activity (from 29.01 to 37.08%) and emulsification stability (19.23 to 21.75%) of germinated maize flour; decrease in bulk density and sedimentation value (from 60.46 to 52.39 mL) of germinated maize flour; increase in dispersibility and oil absorption capacity but decrease in foaming stability of germinated samples	Gnanwa et al. (2021)
Maize, wheat and white sorghum	24	Room temperature	42	35	-	Increase in bulk and tapped density of germinated maize but reduction in case of wheat and sorghum; rise in foaming capacity, foaming stability and swelling potential of wheat and sorghum flour but decrease in maize flour; increase in water-holding capacity (WHC), oil absorption capacity (OAC), acidity, emulsion activity, emulsion stability, emulsion activity, and emulsion stability of germinated wheat, maize, and sorghum samples than ungerminated ones	Siddiqua et al. (2019)
Brown rice, oat, sorghum, and millet	20 (brown rice), 12 (sorghum & millet) & 8 (oat)	25	60	25 (oats) and 30 (brown rice, sorghum & millets)	95	Reduction in pasting temperature of germinated grains flour with increased germination time and significant decrease in swelling factor in germinated flour than raw flour	Li et al., (2020)

(*continued*)

Table 6.1 (Continued)

	Germination conditions						
	Soaking		Sprouting				
Cereal(s)	Time (h)	Temperature (°C)	Time (h)	Temperature (°C)	RH (%)	Findings	References
Barley	37	24	5 days	21	-	Reduction in hectoliter and a thousand kernel weight with rise in increasing the germination temperature and time; increase in water absorption capacity, swelling power, water solubility index and dispersibilty of germinated barley flour	Tura and Abera (2020)
Wheat (PBW-550), paddy (PR-113) and triticale (TL-2908)	7-8 (wheat & triticale) and 9-10 (paddy)	40±2	94 (wheat & triticale) and 96 (paddy)	28±2 (wheat and triticale) and 35±2 (paddy)	-	An increase in the water and oil absorption capacity, emulsification activity, and stability and a decrease in bulk density and sedimentation value of all grain samples	Sibian et al. (2017)
Sorghum (variety SL 44)	10	25	12, 24, 36,, and 48	25, 30, and 35	-	A decrease in water absorption and swelling power and increase in oil absorption capacity, gel consistency, foaming and emulsifying properties of germinated sorghum flour	Singh et al. (2017)
Sorghum	12	Room temperature	24, 48 & 72	Room temperature	-	Reduction in bulk density but rise in water and oil absorption capacity, swelling power, foaming capacity, emulsion capacity, and foaming stability of germinated flour	Ocheme et al. (2015)
Biological properties							
Barley	3	Room temperature	24, 48, 72 & 96	Room temperature	-	Anti-wrinkle activity was observed in the test group treated with germinated barley extract, which was linked to higher amount of oligomeric components of procyanidin and prodelphinidin	Park et al. (2021)

Rice sprouts	-	-	-	-	-	Activation of macrophages by rice sprouts through TLR2/4-mediated activation of p38 and JNK signalling pathway that led to higher generation of immunomodulators such as NO, iNOS, IL-1β, IL-6 and TNF-α and phagocytosis in RAW264.7 cells	Geum et al. (2021)
Sorghum	24	Room temperature	72	Room temperature	-	Better ACE inhibition in non-germinated sorghum than germinated one and improved erythrocytes protection in germinated sample	Arouna et al., (2020)
Wheat	4	Room temperature	1-7 days	12–21	>90	Longer the germination time, there was increase in antioxidant activity of germinated wheat; Increased concentration of individual polyphenols, better radical scavenging potential and decreased tumor necrosis factor α and interleukin 6 in macrophages was observed after germination	Tomé-Sánchez et al., (2020)
Rice	3 days	Room temperature	Till sprout length was 15–25 mm	37	80	A significant increase in α-amylase and α-glucosidase inhibition and glucose uptake by 3T3-L1 adipocytes and administering germinated rice extract orally for eight weeks improved the insulin levels and reduced the blood glucose levels in a C57BLKS/J-db/db mice model.	Lee et al. (2019)
Corn	6	Room temperature	24	Room temperature	-	Higher levels of carotenoids, GABA, total phenolic content and total anthocyanin content in sprouts of corn genotypes but very high levels of these in seedlings	Chalorcharoenying et al. (2017)

rice, the germination process activates the hydrolytic enzymes which deconstruct starch, non-starch carbohydrates and proteins leading to increased oligosaccharides and amino acids (Singh et al., 2015).

As a result of the lack of essential amino acids like lysine, threonine, and tryptophan in grains, protein quality in cereals is poorer compared to those present in animal meat and milk. The digestibility of cereal protein is less than animal food (meat or milk) because of anti-nutritional factors like phytic acid, tannins, and polphenols that bind to the proteins (Berrazaga et al., 2019). Grain germination reduces the level of tannin and phytic acid, and improves the protein availability in cereals (Table 6.1). Riboflavin, thiamin, niacin and lysine concentrations were found to be higher in germinated grains.

6.3.2 Carbohydrates and Sugars

Carbohydrates undergo biological reactions during germination that lower their energy content (Zhao et al., 2018). The breakdown of carbohydrates into simpler molecules occurs as a result of germination, which activates enzymatic activity in sprouting grains. There is a rise in reducing sugars and a release of insoluble polyphenols which are attached using covalent bonds to cell wall carbohydrates due to the degradation of starch and non-starch carbohydrates (Hung et al., 2012). Starch is broken down by a combination of the enzymes α- and β – amylases – debranching enzyme, and α-glucosidase. After 24–144 hours of germination, the starch content of oat seeds dropped to 21% from 60% whereas the free sugar content rose from 5% to 28% (Tian et al., 2010).

The germination process also improves the fibre content in food (Jan et al., 2017). Several research findings have shown that an increase in fibre is associated with decreased dry matter due to the enzymatic hydrolysis of starch molecules in addition to microbiological breakdown of cell components like lipids, proteins, and carbohydrates (Laxmi et al., 2015; Megat et al., 2016). When grains germinate, the amount of cellulose, lignin, and hemicelluloses in the crude fibre increases dramatically. This is because the plant cells are constantly synthesising new cell components. Since glucose release is slowed down by dietary fibre, it may be therefore important for diabetic people to increase their fibre intake. Because fibre helps to slow down the digestion of starch and gastric emptying in the stomach, it also contributes to feelings of fullness as well (Barber et al., 2020). Short-chain fatty acids like acetic acid, butyric acid, and propionic acid are synthesized by colonic bacteria after fermenting the fibre because salivary and pancreatic amylases cannot break down the fibre.

6.3.3 Proteins and Amino Acids

The anti-nutrients and protease inhibitors, that establish complexes with proteins and protease enzymes, decrease the bioavailability and digestibility of dietary protein, which may restrict the health benefits of protein. Anti-nutritional factors can be reduced by germination, allowing the full nutritional benefits of cereal proteins to be achieved (Samtiya et al., 2020). After germination, cereals' in-vitro protein digestibility improved. As per Hejazi et al., (2016), after 48 hours of germination, amaranth grain's in-vitro protein digestibility increased to 83% from 76%. Grain germination has been linked with rise in protein digestibility because of enzymatic hydrolysis and alterations in the grain composition following the degradation of constituents like phytic acid and polyphenols. As a result, during germination, the essential amino acids and crude protein become significantly more readily available (Kim et al., 2018).

Depending on the grain type, germination has been shown to increase the digestibility of protein (Laxmi et al., 2015). After quinoa germination, some researchers have found that the amount of total protein in the plant has decreased despite the rise in the levels of essential amino acids like methionine and tryptophan (Bhathal and Kaur, 2015). During germination, the carbohydrates and lipids are employed in respiration leading to loss of dry weight, but some amino acids are synthesized during germination, therefore, there is an increase in proteins (Jan et al., 2017). Furthermore, proteases have been implicated in protein losses during germination. Protein breakdown for growth-related nucleic acid production appears to be outweighed net protein synthesis that can lead to an overall increase in proteins. Buckwheat protein content increased significantly after 72 hours of germination, most likely as a result of a greater rate of synthesis of proteins than proteolysis (Zhang et al., 2015).

6.3.4 Lipids and Fatty Acids

Since lipids are required for respiration, their level in cereals rise mildly during soaking phase but decreases later in germination (El-Safy et al., 2013; Laxmi et al., 2015). During germination, hydrolytic activities and lipids' utilization as energy source for biochemical processes have been reported in some studies to reduce fat content (Jan et al., 2017; Thakur et al., 2021). Crude lipids, oleic acid and linoleic acid increased in germinated barley, as reported by Beaulieu et al., (2022) and Ortiz et al. (2021). Distinct germination times and grains from different sources may be responsible for the inconsistencies between such studies.

6.3.5 Vitamins

Vitamin B1, B2, B3, and B5, biotin, vitamins C and E, tocopherols and phenolic compounds can be produced and made more readily available in the biochemical processes that occur during germination. Malted ragi and wheat were found to have increased levels of vitamin C (ascorbic acid) (Desai et al., 2010; Guo et al., 2012; Laxmi et al., 2015). Ascorbic acid can be derived from glucose, mannose, and galactose found in plants and animals, and these sugars can be converted into vitamin C. Because amylases and diastases hydrolyze starch to increase glucose availability for vitamin C biosynthesis, malting/germination results in an increase in vitamin C (Guo et al., 2012). Grains that germinate at 20–25°C have twice α-tocopherol and minerals after 102 h germination, according to Ozturk et al., (2012).

6.3.6 Minerals

The main anti-nutritional factor found in cereals is phytic acid which tightly holds the minerals and reduces their bioavailability. The phytic acid can be hydrolyzed by phytase enzyme that converts it into phosphoric acid and myoinositol. Thus the action of phytase makes minerals more bioavailable, decreases buckwheat's phytic acid content as germination time increased, according to a study on the buckwheat (Zhang et al., 2015). The availability of minerals varies with grain type, for instance, iron was maximum in wheat and zinc was highest in rice and wheat while calcium and manganese were the most widely available in rice (Luo et al., 2014). A lot of trace elements are bound in cereals that are high in anti-nutritional factors, like tannins, phytates, etc. Malting causes grains to release more of the bound minerals (Gupta et al., 2015). Anti-nutritional factors that bind the minerals, are leached thereby increasing the mineral's bioavailability. Phytate content, phytase activation, the degree to which minerals are bound or their interaction thereof, may explain the differences in mineral availability after germination from diverse cereal grains which are germinated for same duration. In pearl millet, the hydrochloric acid extraction of calcium, iron and zinc were triggered by 2-16%, 15% and 12-25%, respectively, when the germination time was increased (Badau et al., 2005).

6.3.7 Anti-nutrients

Trypsin inhibitor, phenolic compounds, phytic acid, oligosaccharides, and other anti-nutritional factors are all reduced during germination of grains (Nkhata et al., 2018; Samtiya et al., 2020). Many nutrients, including protein and minerals, are unavailable or inactive for digestion because of the presence of anti-nutrients. Anti-nutrient reduction mechanisms during germination are different for each nutrient. Tannins are thought to be released into the germination medium or to be degraded by enzymes once it is placed in the seed (Patel and Dutta, 2018). Polyphenol oxidase and other catabolic enzymes have become more active, which has led to reduction in the total amount of free phenolic acids. Inorganic phosphate is released by the endogenous phytase enzyme, which breaks down phytates (Vashishth et al., 2017). The decline in phytic acid during germination would increase the available mineral content in cereals because it is one of the contributing factors for decreasing the bioavailability of minerals (Gupta et al., 2015; Kumar and Anand 2021). Trypsin inhibitors are also found in a number of cereals. The body's ability to digest protein is greatly hampered by the inhibition of trypsin's activity. A nutritional challenge is that trypsin inhibitor is highly resistant to heat and stable at higher temperature. It is therefore impossible to reduce its activity using many cooking methods. After germination, trypsin inhibitor activity decreases gradually, which results in higher trypsin activity (Zhang et al., 2015).

6.3.7.1 Functional Properties

The functional attributes mainly depend on the quality properties of end food products and their physicochemical characteristics which alter the processing as well as protein behavior in the food systems. The protein size and structure, its functionality and interaction with the other components of food like fats and carbohydrates are modified by several treatments. For the particular aspect, the modified proteins have diverse functionality, which is linked to parent protein, so they may be incorporated to a minor amount of product. Based on mechanism of action, the functional characteristics are generally categorized as mentioned: (i) hydration properties: thickening, wettability, water, solubility, and oil; (ii) protein structure and rheological properties: aggregation, gelation, consistency, elasticity, and adhesiveness; and (iii) protein surface attributes: whippability, emulsification and foaming properties, and protein formation (Shevkani et al., 2015). Nevertheless, as predicted by Murekatete (2012), the function of functional properties contributes to the amino acids (energy levels from the molecular orbitals and its calculation, hydrophilicity, electronic properties, molecular size, and heat formations). Details on the functional properties of germinated cereals are given in Table 6.1.

6.3.8 Protein Solubility

The protein and solubility as a part of functional properties are most critical because they change the other properties mainly gelation, emulsification, and foaming. Factors affecting protein solubility include source, pH, treatment variations, ionic strength, and the addition of some substances. In sorghum, the protein solubility profile increases with the increases in germination. In germinated samples, the maximum solubility of protein was observed at pH 6 and their values raised further to a maximum in the germinated sorghum flour on fifth day. Elkhalifa and Bernhardt (2010) mentioned the minimum protein solubility at pH 4 in all germinated samples, which is mainly the isoelectric pH. For the lowest protein solubility of raw samples, the pH level was found to be 2 and 4, but after germinating the samples, the protein had better solubility at pH level 4 compared to control. This happened mainly because of the maximum proteolytic activity during germination, which increases the protein solubility of the storage proteins resulting from hydrolysis. Therefore, the germinated flour can be employed in food formulations where protein solubility is required (Singh and Sharma, 2017).

6.3.9 Water Absorption Capacity

When grains are germinated, their water-absorption capacity (WAC) is greatly increased. Sorghum flour's WAC was increased by germination for the first three days of germination but decreased as germination time increased further (Elkhalifa and Bernhardt, 2010). Siddiqua et al. (2019) noticed similar rises in WAC of germinated maize, wheat, and sorghum flour than non-germinated samples. The amount of water linked with proteins is in complex relation with amino acid composition and increases as the quantity of charged residues, conformation, hydrophobicity, pH, temperature, and ionic strength of proteins. The breakdown of carbohydrate molecules, change in protein quality, and rise in protein level during germination is responsible for increase in WAC, as more water absorbing sites are accessible (Elkhalifa and Bernhardt, 2010).

6.3.10 Oil Absorption Capacity

The physical confinement of oil is thought to be the cause of its high absorption capacity. In order to preserve the flavour of food and enhance the mouth feel, fat is essential. A rise in amino acid availability due to removing the non-polar residues from the protein molecules could explain the improved oil absorption capacities (OAC) after germination. This is because oil binding is dependent on hydrophobic amino acid's surface availability (Karami and Akbari-Adergani, 2019). The OAC of sorghum, wheat, and maize flour improved as the grains germinated (Siddiqua et al., 2019). There was increase in the OAC of sorghum flour up to the third day of germination in another study, while further germination reduced OAC (Elkhalifa and Bernhardt 2010). Germinated flour has a higher oil-binding potential, suggesting that it could be useful in food formulations that require a high oil-holding capacity. OAC has a positive effect on the food products' taste, flavour, and lipophilicity. Due to rise in lipophilic protein's level, the fat globule-holding capability of grains increases as well, which is an indicator of protein's ability to absorb and retain oil. The increased OAC might be attributed to such factors. Germinated flour samples may have been able to absorb more oil because of their lower fat content (Elkhalifa and Bernhardt, 2010).

6.3.11 Gelation Capacity

A gel is somewhere in between solid and liquid. There are molecular networks in food systems obtained from – proteins and polysaccharides, or a combination of both. Due to the fact that big molecules may create crosslinks in three dimensions, proteins are more effective gelling agents than carbohydrates (Moure et al., 2006). Gelation is aided by protein flexibility and denatureability, as well as by the big molecule's ability to crosslink in three dimensions. Molecules that have been denatured clump together into gel. Initial two days of sprouting and past three days of germination sorghum had gelation concentrations of 14 and 12%, respectively, but the control sorghum flour had 18% gelation (Elkhalifa and Bernhardt, 2010).

6.3.12 Emulsification Properties

For cakes, coffee whiteners, and frozen desserts, the protein's ability to promote the emulsion synthesis and stability is critical. A variety of emulsifying and stabilising qualities are needed in such items due to the wide range of formulations and conditions to which they have been exposed. Protein type, quantity, and solubility affect the emulsification's effectiveness. Emulsifying capacity of the flours of sorghum (Elkhalifa and Bernhardt, 2010), wheat and maize (Siddiqua et al., 2019), and amaranth (Adeniyi and Obatolu, 2014) greatly increased after germination. The rise in emulsion capacity can be understood as how dietary components such as low fat, high protein, and high protein-to-fat ratios interact with each other in the germinated flour samples. Further, the rise in the amount of stabilized oil droplets may be found at the interface of the emulsion causing increase in emulsification capacity. The oil and protein content of a substance determines its emulsification properties (Siddiqua et al., 2019). Dissociation, limited unfolding, and denaturation of protein polypeptides led to the release of amino acid hydrophobic sites that facilitate the hydrophobic association of peptide chains having fat droplets. This resulted in an increase in protein volume and surface area available as well as enhanced emulsification capacity (Jiménez-Martínez et al., 2012).

Germinating wheat and maize (Siddiqua et al., 2019), sorghum (Elkhalifa and Bernhardt, 2010), and amaranth (Adeniyi and Obatolu, 2014) had considerably higher emulsion stability than native grains with longer germination times (Siddiqua et al., 2019). When grains are germinated, there is improvement in emulsifying capability, but some emulsions are not particularly stable and change with the temperature of germination (Adeniyi and Obatolu, 2014).

6.3.13 Swelling Power

The tendency of starch grains to swell on heating in the presence of water determines swelling power. Amorphous and crystallized starch granules have different degrees of crystallinity, which affects the amount of water molecules that may interact with the starch chains, which in turn affects the amount of swelling (Cornejo-Ramírez et al., 2018). Hydrogen bonding determines the water-holding capacity of glucose molecules, which affects starch's ability to swell. The double helices in crystallites are stabilized by hydrogen bonds that are broken during gelatinization process and substituted using water molecules. As a result, the crystalline nature is disrupted, which weakens the bonding between the granules and allows them to swell. Further, parameters such as amylose:amylopectin ratio, molecular weight and distribution of individual part, the extent of branching and the dimension of outer branches in amylopectin influence starch's ability to swell at the molecular level (Sasaki and Matsuki, 1998; Elkhalifa and Bernhardt, 2013; Cornejo-Ramírez et al., 2018).

During grain germination, amylases and proteases likely disrupted the hydrogen atoms of sugars and amino acids, respectively, decreasing the swelling potential of maize and sorghum flours (Elkhalifa and Bernhardt 2013; Adedeji et al., 2014). At this point, an enzyme called α-amylase is activated and begins to break down starch molecules at α-1,4 glucosidic linkages, producing oligosaccharides with reduced molecular weights like dextrins, maltose, and glucose, which usually do not swell.

Xu et al. (2012) and Elkhalifa and Bernhardt (2013), on the other hand, found that the swelling potential of germinated brown rice grains as well as sorghum increased as the temperature was raised, despite previous studies showing that the swelling power of these grains decreased. The gelatinization of starch occurs as a result of increased granule water uptake, which is thought to be the cause of rise in swelling power with increasing temperature. Grain flour needs considerable increased temperatures to accomplish complete granule swelling because the starch vibrates more strongly, disrupting the inter-molecular bonds thereby permitting the hydrogen-bonding sites for attracting and binding further water molecules (Elkhalifa and Bernhardt, 2013).

6.3.14 Foaming Properties

For leavening food products, like bread, cakes, and biscuits, the ability to generate stable foams with gas through the formation of impermeable protein films is a critical quality for flours that are utilized in these applications. A food material's ability to foam is determined by its protein's surface-active characteristics (Wang et al., 2019). During whipping, the foam is generated by the increased surface area of air/liquid interphase, denaturing proteins. In the absence of germination, amaranth flour (Adeniyi and Obatolu, 2014), wheat flour (Siddiqua et al., 2019), and sorghum flour (Elkhalifa and Bernhardt, 2010) showed no foaming capacity, but as germination progressed, the foaming capacity increased and continued to increase with increasing germination time. The protein surface denaturation is triggered by germination, which lowers the air-water interface's surface tension, thereby allowing for the absorption of solubilized protein molecules, leading to an increase in foaming capacity (Elkhalifa and Bernhardt, 2010; Siddiqua et al., 2019).

Since the efficacy of whipping substances hinges on their capacity to keep the whip for longer periods, the foaming stability is critical. The foaming stability of sorghum, wheat, and amaranth flours were increased by germination, and the effect was more pronounced the longer the germination period was (Elkhalifa and Bernhardt, 2010; Adeniyi and Obatolu, 2014; Siddiqua et al., 2019). Cereal flour foam stability may be affected by conformational changes during germination (Elkhalifa and Bernhardt, 2010). For germinated grains, increased foaming stability may have been due to the bioavailability of intrinsic proteins that were bound by anti-nutritional substances.

6.3.14.1 Biological Properties

Grains develop enzymes, new proteins, vitamins, and phytochemicals throughout the germination process in order to deliver the bioactives and health advantages (Figure 6.1). Biological activities such as antioxidant, anti-inflammatory, and anti-aging properties have been found in

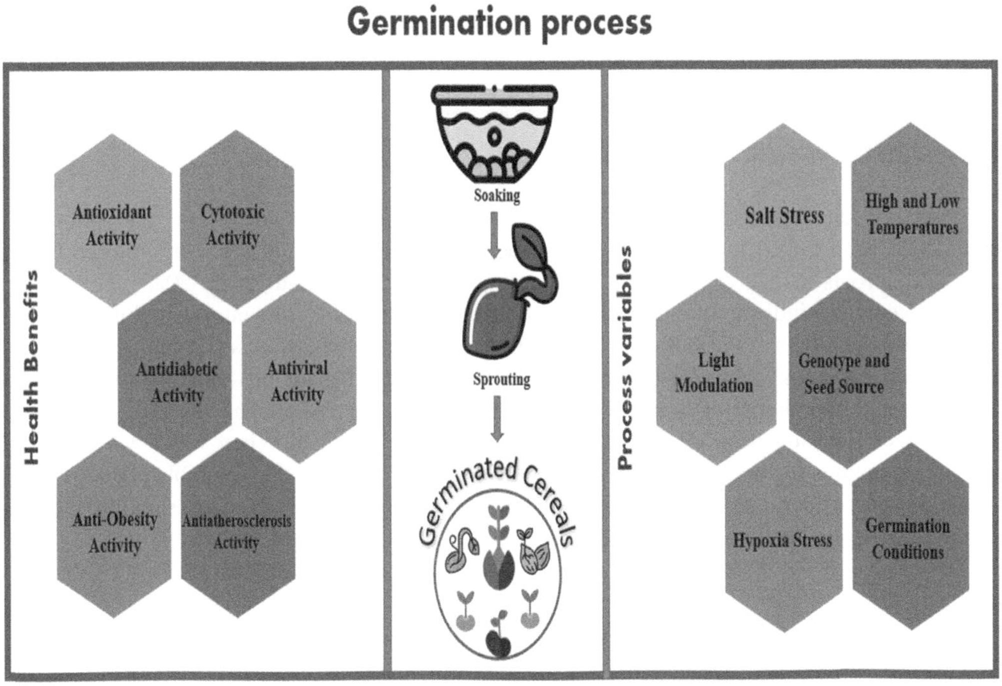

Figure 6.1 Germination process, related health benefits, and process variables.

the metabolomes created in the process of germination (Table 6.1). Because of this, germinated grains, particularly eatable germinated grains, are attracting considerable interest as potential sources of the active chemicals utilized in healthy functional foods and nutraceuticals (Donkor et al., 2012).

6.3.15 Antioxidant Properties

Oxidative stress is created by oxygen radicals, and antioxidants are required to protect the cells from this stress. Numerous substances with antioxidant capabilities have been found in grains, such as polyphenols, vitamins, and sterols (Gani et al., 2012). Bran and germ, either alone or in combination, contain a high concentration of most of the listed bioactive compounds. To varying degrees, they contribute to the antioxidant capabilities and were impacted by the germination process. There have been indirect antioxidant effects linked to several minerals and trace elements that can work along with enzyme co-factors. It is still not known exactly how antioxidants affect the human body in vivo despite numerous in vitro studies showing their beneficial effects (Tang and Tsao 2017). Antioxidant substances present in a variety of grain tissues may provide a robust defence against oxidative stress-related diseases (Donker et al., 2012). Hence, natural sources of antioxidant compounds like germinated grains must be emphasized because they do not only provide a range of bioactives, but also do so in naturally occurring ratios. Strengthening the body's antioxidant defences is made possible in part by the high concentration of bioactive chemicals in grain sprouts (Singh and Sharma, 2017).

Antioxidants and bioactive molecules including phenolic compounds and γ-aminobutyric acid (GABA) were increased by germination, and the antioxidant activity was enhanced (AL-Ansi et al., 2020). Bread made from germinated barley provides the European Food Safety Authority's recommended daily GABA intake, according to AL-Ansi et al. (2022). When grains are germinated, their overall phenolic content and antioxidant activity rise by several orders of magnitude, according to various studies (Garzón and Drago, 2018; Singh et al., 2019). However, some investigations have found a considerable drop in the total phenolic content following germination (Khoddami et al., 2017; Arouna et al., 2020).

6.3.16 Other Properties

Biological functions including anti-bacterial, anti-inflammatory, anti-diabetic, and anti-cancer effects have been linked to the buildup of bioactive substances like polyphenols in germinated grains and sprouts (Singh et al., 2015; Lee et al., 2019; Gabriele and Pucci, 2022). Germinated grains' bioactivities indicate that they may have potential therapeutic advantages and that they can be included in our diets to avoid the development of certain chronic diseases (Table 6.1). Activation of signalling cascades, contact with cell receptors, and reduction of pro-inflammatory cytokine transcription in endothelial cells, monocytes, and macrophages are all known anti-inflammatory mechanisms of wheat phenolic compounds (Calabriso et al., 2020). When wheat germinates at higher temperatures (up to 21°C) and for longer periods of time (up to 7 d), soluble phenolic acids, flavonoid C-glycosides, and lignans are more abundant, which increases the plant's ability to reduce oxidative stress and inflammation (Tomé-Sánchez et al., 2020).

6.4 SAFETY AND STABILITY

It is commonly accepted that consuming germinated cereal grains has nutritional and health advantages. Germinated grains, however, are not well-documented in terms of their safety. Owing to absence of a post-germination kill phase, sprout ingestion has been linked to various foodborne illnesses (Gensheimer and Gubernot, 2016; Miyahira and Antunes, 2021; Bazaco et al., 2021). However, despite several investigations, little is known about cereal and pseudo-cereal sprouts (Benincasa et al., 2019). Numerous probable pre- and post-harvest causes of contaminants, including as seed, germination media, and soaking water besides transportation, handling, and storage conditions of seedlings, can result in microbial contamination of sprouts (Warriner and Smal, 2014; Miyahira and Antunes, 2021). Due to microbial multiplication that occurs throughout the sprouting phase, this is the most essential stage. Bacteria, yeasts, and mould mycelium proliferate rapidly, while dormant spores are triggered simultaneously. While sprouting, the amount of potential bacteria and yeasts reaches its peak. Steeping and sprouting results in an increase of 7- to 15-fold in the microbial load, which typically rises from 10^6 to 10^9 CFU/g dm for bacteria and from 10^3 to 10^6 CFU/g dm for yeast in sprouts (Nagar and Bandekar, 2009; Juste et al., 2011). Mycotoxins, which are hazardous to humans in levels greater than 4 to 20 g/kg food, are produced when mould, particularly *Fusarium* species, develop in large quantities (Juste et al., 2011). To suppress bacteria growth during sprouting, the techniques have been explored in previous attempts where microbial loads were reduced from 10^5 to 10^7 CFU/g dm for bacteria and to 10^3 to 10^4 CFU/g dm in case of fungi. The food products with low levels of microorganisms are generally considered safe for intake by humans.

Therefore, there is a need for grain treatments that from the beginning should be optimized during the production process. Physico-chemical and biological applications are used in all methods to mitigate health risks (Yang et al., 2013). For example, irradiation and supercritical carbon dioxide are examples of physical treatment approaches, while antagonistic microbes and anti-bacterial metabolites are examples of biological intervention programs. Ozone and chlorine disinfectants and sanitizers, as well as electrolyzed water, are some examples of chemical interventions. Buckwheat sprouts stored at 4°C for 8 days were found to have a reduced count of *E. coli* O157:H7 and *S. typhimurium* inoculation counts, without influencing colour properties, as a result of the combination of a sanitizer mixture (ClO_2 and 0.3% fumaric acid), UV-C (2 kJ/m^2), and modified atmospheric packing (air and CO_2 gas) (Chun and Song, 2013). For the elimination of microbiological risks involved with wheat grains infected with *Listeria monocytogenes* and *Botrytis cinerea*, the photoactivated chlorophyllin-chitosan complex (Chl-KCHS) has been an ecofriendly and economic technique (Buchovec and Luksiene, 2015). Additionally, ionising radiations treatment (X-rays, electron beam, or γ-rays) and their conjunction with sodium hypochlorite could be used to protect the safety of sprouts. This irradiation technology has limits and is not approved in all countries. Further, the negative effects on several seed growth metrics constitute considerable downsides. There are, however, concerns about the safety of consuming germinated grains, which must be studied by researchers and scientists.

6.5 FUTURE ASPECTS

Cereal grains continue to be an essential staple food around the world, contributing significantly to the intake of many nutrients, like carbs, proteins, vitamins, minerals, and fibre in addition to several phytochemical compounds. The bioactive chemicals included in cereal grains are responsible for a variety of health advantages. The germination process releases the minerals from phytate salts and makes them available to the intestine for absorption, while vitamin synthesizes and builds up during sprouting. In comparison to raw grains, germination improves both the nutritional profile and the functionality of grains, resulting in higher nutrient absorption in the body. As a result of germination, protein digestibility and bioaccessibility are enhanced because protease inhibitors are either eliminated or suppressed. However, when it comes to maximizing the nutritional value of grain sprouts, the process conditions and kind of cereal must be optimized.

After germination, more research is needed to isolate and characterize these bioactive chemicals that have a positive impact on health. Because grains contain a wide spectrum of chemicals with antioxidant potential, research is also required for the determination of metabolism, bioavailability, and physiological effects of these compounds in humans. Furthermore, sprouting's metabolic alterations have been associated with certain health advantages, but there is a lack of clinical evidence to back claims about specific health benefits. Sprouting grains, according to preliminary research, may improve levels of cholesterol and glucose in the blood, as well as blood pressure and mineral absorption. Many studies must be carried out using in vitro and animal models in this direction.

For food, biopharmaceutical, and nutraceutical applications, germination could be used like a powerful bioprocessing technique to improve the quality and composition of grains. The ultimate sustainability and effectiveness of these germinated cereal grains rests on the buyer's recognition, which must be persuaded. As a result, care must be taken to ensure that the biotechnological advancements do not undermine the conventional positive attitude toward germinated foods.

6.6 CONCLUSION

Considering the intensification of bioactive compounds – minerals, vitamins, and polyphenolic compounds particularly through germination – it can be concluded that germination is the useful technique in cereal grains. Enzymatic activities during the germination process boost the antioxidant capacity, which has health-promoting properties besides changing the nutritional content and biochemical activities. Grains' functional properties can be enhanced by germination-driven biological activation, and this could be used to generate new healthy functional foods. As of now, more research is needed to determine whether or not such bioactive compounds have any health benefits for humans.

REFERENCES

Adedeji, O. E., O. D. Oyinloye, and O. B. Ocheme. 2014. Effects of germination time on the functional properties of maize flour and the degree of gelatinization of its cookies. African Journal of Food Science 8:42–47 DOI: 10.5897/AJFS2013.1106

Adeniyi, P. O. and V. A. Obatolu. 2014. Effect of germination temperature on the functional properties of grain amaranthus. American Journal of Food Science and Technology 2:76–79. DOI: 10.12691/ajfst-2-2-5

AL-Ansi W., A. A. Mahdi, Q. A. Al-Maqtari, et al., 2020. The potential improvements of naked barley pretreatments on GABA, β-glucan, and antioxidant properties. *LWT* Food Science and Technology 130:109698. DOI: 10.1016/j.lwt.2020.109698

Al-Ansi, W., Y. Zhang, T. A. A. Alkawry, et al., 2022. Influence of germination on bread-making behaviors, functional and shelf-life properties, and overall quality of highland barley bread. LWT Food Science and Technology 159:113200. DOI: 10.1016/j.lwt.2022.113200

Arouna, N., M. Gabriele, and L. Pucci. 2020. The impact of germination on sorghum nutraceutical properties. Foods 9:1218. DOI: 10.3390/foods9091218

Badau, M. H., I. Nkama, and I. A. Jideani. 2005. Phytic acid content and hydrochloric acid extractability of minerals in pearl millet as affected by germination time and cultivar. Food Chemistry 92:425–435. DOI: 10.1016/j.foodchem.2004.08.006

Barber, T. M., S. Kabisch, A. F. H. Pfeiffer, and M. O. Weickert. 2020. The health benefits of dietary fibre. Nutrients 12:3209. DOI: 10.3390/nu12103209

Bazaco, M. C., S. Viazis, D. C. Obenhuber, P. Homola, F. Shakir, and A. Fields. 2021. An overview of historic foodborne illness outbreak investigations linked to the consumption of sprouts: 2012–2020. Food Safety Magazine 2021 https://www.food-safety.com/articles/7202-an-overview-of-historic-foodborne-illness-outbreak-investigations-linked-to-the-consumption-of-sprouts-20122020 (accessed January 10, 2022)

Beaulieu, J. C., R. A. Moreau, M. J. Powell, and J. M. Obando-Ulloa. 2022. Lipid profiles in preliminary germinated brown rice beverages compared to non-germinated brown and white rice beverages. Foods 11:220. DOI: 10.3390/foods11020220.

Benincasa, P., B. Falcinelli, S. Lutts, F. Stagnari, and A. Galieni. 2019. Sprouted grains: A comprehensive review. Nutrients 11:421. DOI: 10.3390/nu11020421

Berrazaga, I., V. Micard, M. Gueugneau, and S. Walrand. 2019. The role of the anabolic properties of plant- versus animal-based protein sources in supporting muscle mass maintenance: A critical review. Nutrients 11:1825. DOI: 10.3390/nu11081825

Bhathal, S. and N. Kaur. 2015. Effect of germination on nutrient composition of gluten free Quinoa (*Chenopodium Quinoa*). International Journal of Scientific Research 4:423–425

Buchovec, I. and Z. Luksiene. 2015. Novel approach to control microbial contamination of germinated wheat sprouts: Photoactivated chlorophillin-chitosan complex. International Journal of Food Processing Technology 1:1–5. DOI: 10.15379/2408-9826.2015.02.01.4

Buriro, M., F. C. Oad, M. I. Keerio, et al., 2011. Wheat seed germination under the influence of temperature regimes. Sarhad Journal of Agriculture 27:539–543.

Calabriso, N., M. Massaro, E. Scoditti, A. Pasqualone, B. Laddomada, and M. A. Carluccio. 2020. Phenolic extracts from whole wheat biofortified bread dampen overwhelming inflammatory response in human endothelial cells and monocytes: Major role of VCAM-1 and CXCL-10. European Journal of Nutrition 59:2603–2615.

Chalorcharoenying, W., K. Lomthaisong, B. Suriharn, and K. Lertrat. 2017. Germination process increases phytochemicals in corn. International Food Research Journal 24:552–558.

Chun, H. H. and K. B. Song. 2013. The combined effects of aqueous chlorine dioxide, fumaric acid, and ultraviolet-C with modified atmosphere packaging enriched in CO2 for inactivating preexisting microorganisms and *Escherichia coli* O157:H7 and *Salmonella typhimurium* inoculated on buckwheat sprouts. Postharvest Biology and Technology 86:118–124 DOI: 10.1016/j.postharvbio.2013.06.031

Chung, H. J., A. Cho, and S. T. Lim. 2012. Effect of heat-moisture treatment for utilization of germinated brown rice in wheat noodle. LWT Food Science and Technology 47(2):342–347.

Copeland, L. 2020. Cereal grains for nutrition and health. Cereal Chemistry 97:7–8. DOI: 10.1002/cche.10244

Cornejo-Ramírez, Y. I., O. Martínez-Cruz, C. L. D. Toro-Sánchez, F. J. Wong-Corral, J. Borboa-Flores, and F. J. Cinco-Moroyoqui. 2018. The structural characteristics of starches and their functional properties. CyTA-Journal of Food 161:003–1017. DOI: 10.1080/19476337.2018.1518343

Desai, A. D., S. S. Kulkarni, A. K. Sahoo, R. C. Ranveer, and P. B. Dandge. 2010. Effect of supplementation of malted ragi flour on the nutritional and sensorial quality characteristics of cake. Advance Journal of Food Science and Technology 2:67–71.

Divte, P. R., N. Sharma, S. Parveen, S. Devika, and A. Anand. 2022. Cereal grain composition under changing climate. In: Climate Change and Crop Stress. ed. A. K. Shanker, C. Shanker, A. Anand, M. Maheswari, 329–360. Academic Press.

Donkor, O. N., L. Stojanovska, P. Ginn, J. Ashton, and T. Vasiljevic. 2012. Germinated grains–sources of bio-active compounds. Food Chemistry 135:950–959. DOI: 10.1016/j.foodchem.2012.05.058

Elkhalifa, A. E. O. and R. Bernhardt. 2010. Influence of grain germination on functional properties of sorghum flour. Food Chemistry 121:387–392. DOI: 10.1016/j.foodchem.2009.12.041

Elkhalifa, A. E. O. and R. Bernhardt. 2013. Some physicochemical properties of flour from germinated sorghum grain. Journal of Food Science and Technology 50:186–190.

El-Safy, F. and R. Salem. 2013. The impact of soaking and germination on chemical composition, carbohydrate fractions, digestibility, antinutritional factors and minerals content of some legumes and cereals grain seeds. Alexandria Science Exchange Journal 34:499–513.

Flores, J., C. González-Salvatierra, and E. Jurado. 2016. Effect of light on seed germination and seedling shape of succulent species from Mexico. Journal of Plant Ecology 9:174–179. DOI: 10.1093/jpe/rtv046

Gabriele, M. and L. Pucci. 2022. Fermentation and germination as a way to improve cereals antioxidant and antiinflammatory properties. In: Current Advances for Development of Functional Foods Modulating Inflammation and Oxidative Stress. ed. Hernandez-Ledesma and C. Martinez-Villaluenga, 477–497. Academic Press.

Gan, R. Y., W. Y. Lui, K. Wu, et al., 2017. Bioactive compounds and bioactivities of germinated edible seeds and sprouts: An updated review. Trends in Food Science and Technology 59: 1–14. DOI: 10.1016/j.tifs.2016.11.010

Gani, A., S. M. Wani, F. A. Masoodi, and G. Hameed. 2012. Whole-grain cereal bioactive compounds and their health benefits: A review. Journal of Food Processing and Technology 3:146–56. DOI: 10.4172/2157-7110.1000146

Garzón, A. G. and S. R. Drago. 2018. Free α-amino acids, γ-Aminobutyric acid (GABA), phenolic compounds and their relationships with antioxidant properties of sorghum malted in different conditions. Journal of Food Science and Technology 55:3188–3198. DOI: 10.1007/s13197-018-3249-0

Gensheimer, K. and D. Gubernot. 2016. 20 years of sprout-related outbreaks: FDA's investigative efforts. Open Forum Infectious Diseases 3:1438. DOI:10.1093/ofid/ofw172.1140

Geum, N. G., J. H. Yeo, J. H. Yu, et al., 2021. Rice sprouts exert immunostimulatory activity in mouse macrophages, RAW264. 7 cells. Food and Agricultural Immunology 32:349–359.

Gnanwa, M. J., J. B. Fagbohoun, K. C. Ya, S. H. Blei, and L. P. Kouame. 2021. Assessment of minerals, vitamins and functional properties of flours from germinated yellow maize (*Zea mays* L.) seeds from Daloa (Côte D'Ivoire). International Journal of Food Science and Nutrition Engineering 11:35–42 DOI:10.5923/j.food.20211102.01

Gómez-Álvarez, E. M. and C. Pucciariello. 2022. Cereal germination under low oxygen: Molecular processes. Plants 11:460. DOI: 10.3390/plants11030460.

Guo, X., T. Li, K. Tang, and R. H. Liu. 2012. Effect of germination on phytochemicals profiles and antioxidant activity of mung beans sprouts (*Vigna radiata*). Journal of Agricultural and Food Chemistry 60:11050–11055. DOI: 10.1021/jf304443u

Gupta, R. K, S. S. Gangoliya, and N. K. Singh. 2015. Reduction of phytic acid and enhancement of bioavailable micronutrients in food grains. Journal of Food Science and Technology 52: 676–684. DOI:10.1007/s13197-013-0978-y

Hefni, M. and C. M. Witthöft. 2012. Effect of germination and subsequent oven-drying on folate content in different wheat and rye cultivars. Journal of Cereal Science 56:374–378. DOI: 10.1016/j.jcs.2012.03.009

Hejazi, S. N., V. Orsat, B. Azadi, and S. Kubow. 2016. Improvement of the in vitro protein digestibility of amaranth grain through optimization of the malting process. Journal of Cereal Science 68:59–65. DOI: 10.1016/j.jcs.2015.11.007

Hübner, F. and E. K. Arendt. 2013. Germination of cereal grains as a way to improve the nutritional value: A review. Critical Reviews in Food Science and Nutrition 53:853–861. DOI: 1080/10408398.2011.562060.

Hung, P. V., T. Maeda, S. Yamamoto, and N. Morita. 2012. Effects of germination on nutritional composition of waxy wheat. Journal of the Science of Food and Agriculture 92:667–672. DOI: 10.1002/jsfa.4628

Hussein, T. H. A., A. El-Shafea, U. A. A. El-Behairy, and M. M. F. Abdallah. 2019. Effect of soaking and sprouting using saline water on chemical composition of wheat grains. Arab Universities Journal of Agricultural Sciences 27:707–715. DOI: 10.21608/ajs.2019.43688

Iordan, M., Stoica, A., and Popescu, E. C. 2013. Changes in quality indices of wheat bread enriched with biologically active preparations. Annals Food Science and Technology 14(2):165–170.

Jahan, S., F. Bisrat, M. O. Faruque, M. J. Ferdaus, S. S. Khan, and T. Farzana. 2021. Formulation of nutrient enriched germinated wheat and mung-bean based weaning food compare to locally available similar products in Bangladesh. Heliyon 7:e06974.

Jan, R., D. C. Saxena, and S. Singh. 2017. Physico-chemical, textural, sensory and antioxidant characteristics of gluten-free cookies made from raw and germinated Chenopodium (*Chenopodium album*) flour. LWT – Food Science and Technology 71:281–287. DOI: 10.1016/j.lwt.2016.04.001

Jiménez-Martínez, C., A. C. Martínez, A. L. M. Ayala, M. Muzquiz, M. M. Pedrosa, and G. Dávila-Ortiz. 2012. Changes in protein, nonnutritional factors, and antioxidant capacity during germination of *L. campestris* seeds. International Journal of Agronomy 2012: 387407 DOI: 10.1155/2012/387407.

Juste, A., S. Malfliet, M. Lenaerts, et al., 2011. Microflora during malting of barley: Overview and impact on malt quality. Brewing Science 64:22–31.

Kandil, A. A., A. E. Sharief, S. E. Seadh, and J. I. K. Alhamery. 2015. Germination parameters enhancement of maize grain with soaking in some natural and artificial substances. Journal of Crop Science 6:142–149

Karami, Z. and B. Akbari-Adergani. 2019. Bioactive food derived peptides: A review on correlation between structure of bioactive peptides and their functional properties. Journal of Food Science and Technology 56:535–547. DOI: 10.1007/s13197-018-3549-4

Khoddami, A., M. Mohammadrezaei, and T. H. Roberts. 2017. Effects of sorghum malting on colour, major classes of phenolics and individual anthocyanins. Molecules 22:1713. DOI: 10.3390/molecules22101713

Kim, M. J., H. S. Kwak, and S. S. Kim. 2018. Effects of germination on protein, γ-aminobutyric acid, phenolic acids, and antioxidant capacity in wheat. Molecules 23:2244. DOI: 10.3390/molecules23092244

Kumar, S. and R. Anand. 2021. Effect of germination and temperature on phytic acid content of cereals. International Journal of Research in Agricultural Sciences 8:2348–3997.

Laxmi, G., N. Chaturvedi, and S. Richa. 2015. The impact of malting on nutritional composition of foxtail millet, wheat and chickpea. Journal of Nutrition and Food Sciences 5:1000407. DOI: 10.4172/2155-9600.1000407

Lee, Y. R., S. H. Lee, G. Y. Jang, et al., 2019. Antioxidative and antidiabetic effects of germinated rough rice extract in 3T3-L1 adipocytes and C57BLKS/J-db/db mice. Food and Nutrition Research 63. DOI: 10.29219/fnr.v63.3603

Lemmens, E., A. V. Moroni, J. Pagand, et al., 2019. Impact of cereal seed sprouting on its nutritional and technological properties: A critical review. Comprehensive Reviews in Food Science and Food Safety 18:305–328.

Li, C., D. Jeong, J. H. Lee, and H. J. Chung. 2020. Influence of germination on physicochemical properties of flours from brown rice, oat, sorghum, and millet. Food Science and Biotechnology 29:1223–1231.

Li, R., Z. J. Li, N. N. Wu, and B. Tan. 2021. Effect of pre-treatment on the functional properties of germinated whole grains: A review. Cereal Chemistry DOI: 10.1002/cche.10500

Limón, R. I., E. Peñas, C. Martínez-Villaluenga, and J. Frias. 2014. Role of elicitation on the healthpromoting properties of kidney bean sprouts. LWT Food Science and Technology 56(2):328–334

Luo, Y. W., W. H. Xie, X. X. Jin, Q. Wang, and Y. J. He. 2014. Effects of germination on iron, zinc, calcium, manganese and copper availability from cereals and legumes. CyTA – Journal of Food 12:22–26. DOI: 10.1080/19476337.2013.782071

Megat, R. M. R., A. Azrina, and M. E. Norhaizan. 2016. Effect of germination on total dietary fibre and total sugar in selected legumes. International Food Research Journal 23:257–61.

Miyahira, R. F., and A. E. C. Antunes. 2021. Bacteriological safety of sprouts: A brief review. International Journal of Food Microbiology 352:109266. DOI: 10.1016/j.ijfoodmicro.2021.109266

Moure, A., Sineiro, J., Domínguez, H., and Parajó, J. C. 2006. Functionality of oilseed protein products: A review. Food Research International 39(9):945–963.

Müller, C. P., J. F. Hoffmann, C. D. Ferreira, G. W. Diehl, R. C. Rossi, and V. Ziegler. 2021. Effect of germination on nutritional and bioactive properties of red rice grains and its application in cupcake production. International Journal of Gastronomy and Food Science 25:100379. DOI: 10.1016/j.ijgfs.2021.100379

Murekatete, N., Y. Hua, Kong, X., & Zhang, C. (2012). Effects of fermentation on nutritional and functional properties of soybean, maize, and germinated sorghum composite flour. International Journal of Food Engineering 8:1–15.

Nagar, V. and J. R. Bandekar. 2009. Microbiological quality of packaged sprouts from supermarkets in Mumbai, India. International Journal of Food Safety, Nutrition and Public Health 2:165–175.

Nkhata, S. G., E. Ayua, E. H. Kamau, and J. B. Shingiro. 2018. Fermentation and germination improve nutritional value of cereals and legumes through activation of endogenous enzymes. Food Science and Nutrition 6:2446–2458. DOI: 10.1002/fsn3.846

Ocheme, O. B., O. E. Adedeji, G. Lawal, and U. M. Zakari. 2015. Effect of germination on functional properties and degree of starch gelatinization of sorghum flour. Journal of Food Research 4:159. DOI: 10.5539/jfr.v4n2p159

Ortiz, L. T., S. Velasco, J. Treviño, B. Jiménez, and A. Rebolé. 2021. Changes in the nutrient composition of barley grain (*Hordeum vulgare* L.) and of morphological fractions of sprouts. Scientifica 2021:9968864. DOI: 10.1155/2021/9968864

Ozturk, I., O. Sagdic, M. Hayta, and H. Yetim. 2012. Alteration in α-tocopherol, some minerals, and fatty acid contents of wheat through sprouting. Chemistry of Natural Compounds 47(6):876–879

Pająk, P., R. Socha, D. Gałkowska, J. Rożnowski, and T. Fortuna. 2014. Phenolic profile and antioxidant activity in selected seeds and sprouts. Food Chemistry 143:300–306

Park, S. C., Q. Wu, E. Ko, et al., 2021. Secondary metabolites changes in germinated barley and its relationship to anti-wrinkle activity. Scientific Reports 11:1–9. DOI: 10.1038/s41598-020-80322-0

Patel, S. and S. Dutta. 2018. Effect of soaking and germination on anti-nutritional factors of garden cress, wheat and finger millet. International Journal of Pure and Applied Bioscience 6:1076–1081.

Rico, D., E. Peñas, M. del Carmen García, et al., 2020. Sprouted barley flour as a nutritious and functional ingredient. Foods 9:296. DOI: 10.3390/foods9030296

Samtiya, M., R. E. Aluko, and T. Dhewa. 2020. Plant food anti-nutritional factors and their reduction strategies: An overview. Food Production, Processing and Nutrition 2:1–14. DOI: 10.1186/s43014-020-0020-5

Sasaki, T. and J. Matsuki. 1998. Effect of wheat starch structure on swelling power. Cereal Chemistry 75:525–529.

Senhofa, S., T. Ķince, R. Galoburda, I. Cinkmanis, and I. M. Sabovics. 2016. Effects of germination on chemical composition of hull-less spring cereals. Research for Rural Development 1:91–97

Shevkani, K., N. Singh, A. Kaur, and J. C. Rana. 2015. Structural and functional characterization of kidney bean and field pea protein isolates: A comparative study. Food Hydrocolloids 43 679–689. DOI:10.1016/j.foodhyd.2014.07.024

Shu, K., X. Liu, Q. Xie, and Z. He. 2016. Two faces of one seed: Hormonal regulation of dormancy and germination. Molecular Plant 9:34–45. DOI: 10.1016/j.molp.2015.08.010

Sibian, M. S., D. C. Saxena, and C. S. Riar. 2017. Effect of germination on chemical, functional and nutritional characteristics of wheat, brown rice and triticale: A comparative study. Journal of the Science of Food and Agriculture 97:4643–4651. DOI: 10.1002/jsfa.8336.

Siddiqua, A., M. S. Ali, and S. Ahmed. 2019. Functional properties of germinated and non-germinated cereals: A comparative study. Bangladesh Journal of Scientific and Industrial Research 54:383–390.

Singh, A. K., J. Rehal, A. Kaur, and G. Jyot. 2015. Enhancement of attributes of cereals by germination and fermentation: A review. Critical Reviews in Food Science and Nutrition 55:1575–1589. DOI: 10.1080/10408398.2012.706661.

Singh, A., S. Sharma, and B. Singh. 2017. Effect of germination time and temperature on the functionality and protein solubility of sorghum flour. Journal of Cereal Science 76:131–139. DOI: 10.1016/j.jcs.2017.06.003

Singh, A., S. Sharma, B. Singh, and G. Kaur. 2019. In vitro nutrient digestibility and antioxidative properties of flour prepared from sorghum germinated at different conditions. Journal of Food Science and Technology 56:3077–3089.

Springer, T. L., and D. W. Mornhinweg. 2019. Seed germination and early seedling growth of barley at negative water potentials. Agronomy 9:671. DOI: 10.3390/agronomy9110671

Tacer-Caba, Z., D. Nilufer-Erdil, and Y. Ai. 2015. Chemical composition of cereals and their products. In: Handbook of Food Chemistry. ed. P. C. K. Cheung and B. M. Mehta. 301–329. Switzerland: Springer Nature

Tang, Y. and R. Tsao. 2017. Phytochemicals in quinoa and amaranth grains and their antioxidant, anti-inflammatory, and potential health beneficial effects: A review. Molecular Nutrition and Food Research, 61:1600767. DOI: 10.1002/mnfr.201600767

Thakur, P., K. Kumar, N. Ahmed, et al., 2021. Effect of soaking and germination treatments on nutritional, anti-nutritional, and bioactive properties of amaranth (*Amaranthus hypochondriacus* L.), quinoa (*Chenopodium quinoa* L.), and buckwheat (*Fagopyrum esculentum* L.). Current Research in Food Science 4:917–925. DOI: 10.1016/j.crfs.2021.11.019

Tian, B., B. Xie, J. Shi, et al., 2010. Physico-chemical changes of oat seeds during germination. Food Chemistry 119(3):1195–1200. DOI: 10.1016/j.foodchem.2009.08.035

Tomé-Sánchez, I., A. B. Martín-Diana, E. Peñas, et al., 2020. Soluble phenolic composition tailored by germination conditions accompany antioxidant and anti-inflammatory properties of wheat. Antioxidants 9:426. DOI: 10.3390/antiox9050426

Tomé-Sánchez, I., E. Peñas, B. Hernández-Ledesma, and C. Martínez-Villaluenga. 2022. Role of cereal bioactive compounds in the prevention of age-related diseases. In: Current Advances for Development of Functional Foods Modulating Inflammation and Oxidative Stress, ed. Hernandez-Ledesma and C. Martinez-Villaluenga, 247–286. Academic Press.

Tura, A. G. and S. Abera. 2020. Study of some physical and functional properties of the malted Temash Barley (*Hordeum Vulgare* L.) grains and its flour. Cogent Food & Agriculture 6:1855841. DOI: 10.1080/23311932.2020.1855841

Vashishth, A., S. Ram, and V. Beniwal. 2017. Cereal phytases and their importance in improvement of micronutrients bioavailability. 3 Biotech 7:42. DOI: 10.1007/s13205-017-0698-5

Verma, M., V. Kumar, I. Sheikh, et al., 2021. Beneficial effects of soaking and germination on nutritional quality and bioactive compounds of biofortified wheat derivatives. Journal of Applied Biology and Biotechnology 9:2–5. DOI: 10.7324/JABB.2021.9503

Wang, Y. H., Y. Lin, and X. Q. Yang. 2019. Foaming properties and air–water interfacial behavior of corn protein hydrolyzate–tannic acid complexes. Journal of Food Science and Technology 56:905–913. DOI: 10.1007/s13197-018-03553-0

Warriner, K., and B. Smal. 2014. Microbiological safety of sprouted seeds: Interventions and regulations. In: The produce contamination problem. ed. K. R. Matthews, G. M. Sapers, and C. P. Gerba, 237–268. New Jersey: Academic Press.

Wrigley, C. 2017. Chapter 4 Cereal-grain morphology and composition. In: Cereal grains: Assessing and Managing Quality. ed. C. Wrigley, I. Batey, and D. Miskelly, 55–87. Sydney: Woodhead Publishing.

Xu, J., H. Zhang, X. Guo, and H. Qian. 2012. The impact of germination on the characteristics of brown rice flour and starch. Journal of the Science of Food and Agriculture 92:380–387. DOI: 10.1002/jsfa.4588.

Yang, Y., F. Meier, J. Lo Ann, et al., 2013. Overview of recent events in the microbiological safety of sprouts and new intervention technologies. Comprehensive Reviews in Food Science and Food Safety 12:265–280. DOI: 10.1111/1541-4337.12010.

Zhang, G., Z. Xu, Y. Gao, X. Huang, and T. Yang. 2015. Effects of germination on the nutritional properties, phenolic profiles, and antioxidant activities of buckwheat. Journal of Food Science 80:H1111–H1119. DOI:10.1111/1750-3841.12830

Zhao, M., H. Zhang, H. Yan, L. Qiu, and C. C. Baskin. 2018. Mobilization and role of starch, protein, and fat reserves during seed germination of six wild grassland species. Frontiers in Plant Science 9: 234. DOI: 10.3389/fpls.2018.00234

CHAPTER 7

Fermentation of Cereals

Tapasya Kumari, Arup Jyoti Das, and Sankar Chandra Deka
Department of Food Engineering and Technology, School of Engineering, Tezpur University, Assam, India

CONTENTS

DOI: 10.1201/9781003242192-10

7.1 INTRODUCTION: CEREAL INDUSTRY OVERVIEW AND ITS BY-PRODUCTS

Cereals belong to the grass family called *poaceae,* also commonly recognised as *gramineae,* which is cultivated in order to obtain the edible components of its cereals, and their cultivation had been recorded for a long decades. In this group wheat, rice, maize, barley, millet, sorghum, rye, and oat are the cereals that are grown widely on a global scale (Hill and Li, 2016). Cereal crops are cultivated in substantial quantities and dispense more food energy worldwide as compare to others. In year 2000, cereals comprised 60% of the world's food production in 73% of the total world's harvested area (Charalampopoulos et al., 2002). In comparison, wheat and rice are the most preeminent crops in Asian as well as in western countries, respectively (FAO, 2012; Samota et al., 2017). For human daily consumption cereals are the most important food sources, and the production is over 2 billion tons every year, but unfortunately around 30% of this production is wasted due to various factors. In developing nations, food losses are seen during agricultural production to a great extent, whereas in industrialized countries the losses are within the distribution and consumption stages. Thus, these unused foods can not be incorporated within the food chain for reutilizing. Therefore, in order to recapture the functional components, the waste products are used as substrate in terms of by-products, which leads to the enhancement of fortified products having an immense market value (Galanakis et al., 2012). In the processing of cereals, bran and germ are the major by-products produced during the milling procedures. The initial step of conventional milling was mostly to grind and the starchy endosperm was separated from the outermost layers (dry milling) which become more dominant. The rationality behind the removal of germ and bran is that they unfavourably influence the processing conditions of flour, which is one of the main reasons for the major consumption of cereal foods made up of refined flour (Patel, 2012; Poutanen et al., 2014). However, due to this process refined flour is deprived of numerous compounds needed for nutrition as it is processed through various steps of refining (Poutanen et al., 2014). The most common and frequent processes is to discard and reuse them as compost or as feed. To relieve the economic as well as environmental burden, different approaches have been explored such as to it to produce biofuels such as ethanol (Ravindran and Jaiswal, 2016), which are frequently generated from the fraction of cereal bran cellulose, such as maize. it is calculated that the production of ethanol is based on 4% of the global grain utilization (Kalscheuret al., 2012). In the human diet cereals are considered to be one of the most essential food resources but give rise to a huge amount of waste generation throughout processing. The by-products of the cereal industry are fortified with an abundant amount of nutrients, but still wind up as fuel, feed, and biorefinery substrate, or waste products. This is the reason the development of innovative biotechnologies is essential to enhance the utility of generated by-products, potentially making them novel and commercially competitive functional foods.

Lactic acid is isolated from carbon sources from a wide range of waste generated by wheat bran, maize cob, and brewer's spent grains (BSG) and is also used in food as well as pharmaceutical industries (Koutinas et al., 2014), whereas rice, maize, and wheat bran are frequently utilized to formulate phytic acid. Cereal waste may be used to manufacture commercially valuable enzymes, because of their high nutrient content, availability, and low cost as cultivation substrates. Utilization of these cereal waste-product is regarded as one of the most constructive in terms of valorization of different food products (Ravindran and Jaiswal, 2016). In order to alter the cereal matrices, fermentation is combined with various technological as well as biological techniques. The oldest form of biotechnology, fermentation of barley, was done to produce beer about 5000 years ago. During fermentation, both endogenous as well as the enzymes isolated from bacteria are capable of modifying the constituents of grain by affecting the composition, bioactivity, as well as the bioavailability of nutrients (Hole et al., 2012). Fermentation is commonly used in various sectors in order to enhance the functional properties, nutritional value, shelf life, texture, flavour, taste as well as the digestibility of foods (Sanlier et a., 2019; Angelescu et al., 2019). These processes are widely used in the field of beverages. A popular section of beverages that are fermented from cereals include: maize (*zea mays*), barley (*hordeum vulgare*), millet (*panicum miliaceuml.*), sorghum (*sorghum bicolor*), oats (*avena sativa*), rye (*secale cereale*), wheat (*triticum aestivum* L.) and rice (*oryza glaberrima/ oryza sativa*) (Blandino et al., 2003). All these cereals are used as good sources of fermentation substrate that act as functional properties due to the existence of various nutrients which might be later on correlated by probiotic activities (Tolun et al., 2019). Despite the beneficial effects cereal fermentation is affected by numerous factors such as temperature and pH, length of the fermentation, nutrients present, growth factor requirements, as well as the moisture content of the grains, which are controlled by technological processes to standardize these attributes. Fermented beverages are common, and production of these products includes traditional methods to sustain product quality, as well as reliability (Tolun et al., 2019; Blandino et al., 2003). Lactic acid bacteria (LAB) influence fermentation by lowering pH, which is antagonistic to the growth of bacteria having pathogenic behavior, so only enhancing the shelf durability which ultimately leads to product security safety (Phiri et al., 2019). Traditional practices show homogenous cultures of various useful favorable microorganisms, turned as probiotic (Achi et al., 2019). Fermented beverages are indispensable to humans as they act as carriers for crucial microbes that play a significant role in health and hygiene due to their nutritional and pharmaceutical properties (Anal et al., 2019; Phiri et al., 2019; Achi et al., 2019). Furthermore, fermentation enhances the digestibility of protein as well as the bioavailability of minerals and micronutrients. Among the leading advantages of cereal-based beverages include the popularity among vegetarians, vegans, and lactose-intolerant consumers. Fermentation has been used to enhance the technological, sensory, nutritional, and functional properties of the by-products of the cereal manufacturing industry. It has several impacts on the nutritional profile of food and improves preservation and safety of food, along with reducing the anti-nutritional compounds as well enhancing flavor. In several cases, more compelling benefits have been seen such as in baked food manufacturing, where fermented cereal by-products have increased nutritional properties. To a wider extent, the application of cereal by-products improves the eco-sustainability of food systems, and provides a different way to reduce malnutrition and hunger. There are two types of fermentation: submerged fermentation (SmF), which combines liquid, and solid-state fermentation (SSF), which involves growth of microbes on solid particles. SSF has gained much interest due to the utility of low-cost agro-industrial as well as agricultural residue. A variety of microorganisms are seen in the process of fermentation due to their advancement of growth in situations with low moisture content. The potential use of SSF has been calculated on the basis of total phenolic content (TPC) and antioxidant potential. Around 22-fold increment of was seen in *A. Oryzae* fermented wheat grain (Karlund et al., 2020).

7.1.1 Probiotics

Depending on the cereal substrate and production procedure, fermented non-alcoholic cereal-based beverages have a vast array of diversified probiotics. The processing treatment should confirm the sustainability on the composition of bacteria during formulation the final product having probiotic functionality (Kort et al., 2012). The microorganisms used predominantly during the process of fermentation of the African *mahewu*, a sour beverage with non-alcoholic properties manufactured by cornmeal in Africa subcontinents and other Arabian gulf countries, belong to the *lactococcuslactis* subsp. In a study of Turkish *boza*, the application of lab and yeast isolates used as starter cultures that permit controlled fermentation was carried out. The choice of right starter culture with probiotic impacts and antimicrobial properties builds the functional properties of *boza* (Altay et al., 2013). In African countries, the main challenges for the expansion and enlargement of fermented cereal-based beverages with probiotics are the lack of proper knowledge on the ground of well-being and nutritional properties of food and beverages. Scientifically, there is vital importance to sustain proper potential for starter cultures having probiotic properties, shown that through continuous fermentation, the functional and organoleptic attributes of the final products are variable (Pasqualone et al., 2021).

The main factor in the case of proper utilization of cereal by-products in food product fortification is the use of bioprocessing engineering, together with fermentation of chosen starter cultures. In the way along a more prominent and advanced food chain, biotechnology can be used towards the value addition of agricultural side streams which can be treated as very valuable help.

7.2 CURRENT TRENDS

Cereals have been considered a staple food for centuries. In most countries, diets incorporate cereal as a fundamental staple food. Cereals such as wheat, rice, and maize provide more than 90% of the total calories obtained from cereals globally. These cereals are considered as a main staple food in Asia, United States, and Europe. In human nutrition the contribution of wheat is much more important due to its versatility and its use in products like baked foods and pasta that are consumed worldwide. According to the Food and Agricultural Organization (FAO), the total cereal production worldwide (mainly wheat) in 2018 and 754.1 million tons respectively. Wheat is flanked by rice; indeed, the FAO has augmented its predictions in the area of production of paddy in 2017 by 2.9 million tons to 759.6 million tons worldwide. Due to their low yield generation the cultivation of ancient as well as use of minor cereals decreased progressively where the low technological properties took major role. For this reason, these minor cereals were replaced by modern cultivars of high yielding variety such as wheat, rice, and maize having less genetic diversity. But in recent times, increased demand for healthy and nutritious products brought the attention to the minor and ancient cereals by selective breeding. This process is much more essential towards the economic growth of developing nations.

Especially they are gathering attention in the western market as they are treated to be healthier than the grains of modern era, due to the evaluation of high mount of dietary fiber (DF), essential vitamins, proteins, resistant starch (RS), minerals and phenols. In take such food incorporated with earliest and minor cereal gave us a highly balanced constitution and possess moderate amount of glycemic index (GI) having large particle diameter, huge ratios of bran and germ to endosperm, existence of fiber having soluble properties along with huge Resistant starch. For making of highly nutritious, innovative and functional baked food, the utility of ancient and minor cereals has been considered in terms of valorization, palatability (inflated sensory value), and products satisfaction (extended shelf life), throughout the processing. Consequently the newly valorized food had no wheat grain blend. Such a way perfectly gathers the consumer's interest in novel, natural as well as inventive model having huge nutritional content and functional characteristics (Angioloni et al., 2011).

7.3 NUTRITIONAL CHARACTERISTICS

Out of the minor species discussed, the pseudo-cereals buckwheat, amaranth, and quinoa are largely accepted due to their gluten-free properties, where the cereal species (minor) are correlated with wheat such as emmer, spelt, einkorn and kamut having gluten-containing behaviour. The Mediterranean region are the well example of hulled wheat associated species (spelt, emmer and einkorn) which are more prominent among the cereal crops of ancient times. Einkorn has a huge amount of protein and compounds with bioactive properties such as carotenoids and significantly a lower level of α-amylase, beta-amylase, as well as lipoxygenase properties. In addition, einkorn shows less T-cell stimulatory gluten peptides (Abdel et al., 2002; Hidalgo et al., 2006). The earliest tetraploid cereal emmer (farro) was one of the very first cereals to be shown in the Fertile Crescent, and it was the optimum day to day ration of the roman legions. But from a long decades, emmer was continuously rejected because of the intense use of durum wheat, which is much easier to hull. Till this time, the growth had been traditionally confined to minimal hilly places of Turkey, Italy, and the Balkan countries, where it is utilized for human nutrition as well as for animal feed. An additional ancient wheat-related species is spelt which was broadly grown until the expansion of fertilizers. The harvesting process with mechanical tool left it behind by the wayside in context of wheat which was efficiently adaptable with the growing industrialization advancement. It is among the husked hexaploid cereals, that shows a genome alike to soft wheat (T. Aestivum L.) (Yan et al., 2003). Nutriment profile of spelt was calculated to be relatively huge, so based on this protein content and composition, as well as its lipids, and crude fiber content, where the vitamin and mineral also play a significant role (Kohajdova et al., 2008.) The cereal consisting modern durum wheat, the size is 2-3 times larger than the common wheat with high amount of other biological nutrients such as protein, amino acid, minerals, lipids, and fatty acids comparing to other modern wheat. The most distinct supremacy of kamut wheat is noticed to be its protein content. Due to the huge percentage of lipids profile, generate excess energy other than carbohydrates. Kamut may be well explained as a "high-energy grain" (Tomoskozi et al., 2016). Until the 16th century barley was main source of bread flour and has remained a main food in European countries of northern side throughout the twentieth century. Despite of its high nutrients value, barley is known to be the "poor man's bread," because of less gluten content present in it (Newman et al., 2006). Millets and sorghum possess main source proteins, energy, vitamins, and minerals for the people of semiarid tropics of Asia where these are considered to be their staple food. It is abounded with minerals where the bioavailability ranges from lower than 1% for some conformations of iron, to a significantly a higher value of 90% for potassium and sodium. The protein content of sorghum is generally moderate in essential amino acids (EAAs) such as threonine and lysine (c. 2/100 g protein) where in case of oil and legume seeds, they have some restricting effects due to the presence of anti-nutritional factors (ANFs) like amylase inhibitors and trypsin, tannins and phytic acids (Noha et al., 2011). Millets are small-sized seed of cereals inclusive of species such as finger millet (eleusinecoracana), pearl millet (pennisetumglaucum), kodo millet (paspalumsetaceum), foxtail millet (setaria italic), little millet (panicumsumatrense), proso millet (penicummiliaceum), and barnyard millet (echinochloautilis). In terms of its productivity, India found to be the top producer in the past with an annual incremental production of 334,500 tons (43.85% of the world production) of millet. Finger millet is also known for its potential health benefits due to its polyphenol content. Teff is a well-known source of fiber and carbohydrate and also contain high amount of zinc, calcium, and iron other than sorghum, wheat, and barley (Abebe et al., 2014). In association with production of healthy cereal food products teff is a valuable assets sue to its nutritional background. As comparing to other cereals, oats be a member of the Poaceae family and are also identify on the Indian subcontinent as jai or javi. Common oats (Avenasativa L.) is the most prominent variety of the oats which was cultivated having huge amounts of extravagant nutrients such as fibers (soluble), vitamins, minerals, EAAs, and fatty acids (unsaturated), along with important phytochemicals (Flander et al., 2007). Health benefits associated with oats was ascribed mainly to the fraction of β-glucan with high density, which has witnessed to decrease blood cholesterol level and the intestinal absorption of glucose. Rye is considered to be very much important grains in Europe continent used for breads, whereas the consumption was found to low over worldwide. In northern side of Europe it is cultivated in huge quantity where the major producers are Russia, Germany, Poland, and Belarus along with Ukraine. In the side of Denmark and Finland, it is taken as whole grain bread of rye, due to the presence of DF. The prominent acid in the form of hydroxycinnamic acid in rye is ferulic acid, which is accompanied by para-coumaric acid and sinapic acids. For the incorporation into the gluten-free diet, amaranth, quinoa, and buckwheat are widely suggested by enhancing its nutritional attributes. Quinoa is also defined as a well-known source of essential minerals, carbohydrates, maltose, and D-xylose. Especially, quinoa showed high protein source, and an amino acid structural significant which is well maintained, other than the important cereals groups. The oil along with water absorptions is better, which influences its potentiality in making human food and drink formulations. A comparative study with other grains, amaranth contains maximum amount of protein source, double the amount of lysine, excessive DF, and 5–20 times the content of iron and calcium (Venskutonis et al., 2013). The amaranth is an important crop for developing nations which is rich in DF, high source of protein, squalene, tocols, and compounds showing cholesterol-lowering functionality (Johns et al., 2007). The annual plant buckwheat, mainly farmed in china, Ukraine, Russian federation, and Kazakhstan. It is having many nutraceutical elements and packed with huge amount of vitamins, dominantly the vitamin-b group. Composition of the amino acids profile of protein content in buckwheat is well equalized having high biological value in spite of the digestibility of protein is comparatively less. Buckwheat flour has an essential group of microelements such as copper (Cu), manganese (Mn), selenium (Se) and zinc (Zn), where the macroelements are sodium (Na), potassium (K), magnesium (Mg) and calcium (Ca) (Li et al., 2001).

7.4 FERMENTATION OF CEREALS AND ITS PRODUCTS

A huge loss of biodiversity was observed in food diversity caused by the focus on a very few specific crops also on the disappearance of traditional meals and customs in food preparation. The rising demand of nutritional cereals has brought the attention towards the ancient variety of grains which has been grown in marginal area just to preserve the genetic diversity. The fermentation process was continued by LAB (lactic acid bacteria), that got popularity with the dawn of civilization because most of the studies had been focused on the preservation of food to enhance their functional, nutritional and organoleptic properties (Prajapati et al., 2008). Fermented food and drinks have become popular due to their therapeutic and nutritional value. The vast range of fermented food based on cereal is a testament to diversity having traditional benefits and the human abilities to discover alternative foods to a varying context (Table 7.2). The energy uptake in the developing nation is fulfilled by the cereal-based fermented food (Guyot et al., 2012).

Sourdough fermentation is based on the dough that is fermented by LAB and yeast found to be the oldest biotechnological method (Gobbetti et al., 1998). It has been suggested that fermentation by LAB significantly improves the nutritional, functional, technological, and sensory properties and lowers ANFs (Coda et al., 2014; Moroniet et al., 2012). Sourdough main affects the GI, and has a great impact on the human health (Table 7.1). Fermentation of sourdough showed immense effect on the macromolecules present in the dough which promotes gluten and starch interaction and thus reduced the starch bioavailability (Katina et al., 2005; Coda et al., 2014). The literature also suggests that the proper use of LAB can improve the texture as well as the flavor of minor and ancient cereal foods (Gobbetti et al., 2014; Pontonio et al., 2019). Wheat bran has 37–53% DF where 95% is insoluble, mainly cellulose, arabinoxylans, and lignin, and oat bran consists of about 15–20% DF where 61% is soluble associated β-glucan (5–20%) (Sahin 2021).

7.4.1 Wheat-Related Products

The utility of einkorn and spelt has been rising behind an absolute desertion because of their huge content of nutritional value where these cereals have been incorporated in the fermentation of sourdough and bread formulation. Although the application of sourdough fermentation with

Table 7.1 Sourdough Biotechnology of Bran and Bran-Fortified Cereals

Bran source	Fermentation type	Effects	References
Wheat	*Lactobacillus brevis E95612 and Kazachstania exigua C81116 with enzymes*	Increase of peptides and free amino acids concentration; increase of protein digestibility, soluble fiber concentration. Decrease of pungent flavor and bitter taste of fortified bread	Coda et al. (2014b)
Wheat	*Commercial baker's yeast and Lactobacillus brevis L62*	Improvement of the texture and sensory properties of fortified bread; increase of starch gelatinizaton	Salmenkallio-Marttila, Katina, and Autio (2001)
Wheat	*Lactobacillus sanfranciscensis DE9 and Lactobacillus plantarum 3DM*	Increase o textural and sensory properties of fortified bread. Increase of antioxidant and phytase activities; increase of free amino acids concentration and protein digestibility. Decrease of starch digestibility and glycemic response.	Rizzello et al. (2012)
Wheat	*Spontaneous fermentation*	Increase of folates and phenols bioavailability; pentosan solubilization	Katina et al. (2012)
Wheat	*Lactobacillus plantarum DSM 32248 and Lactobacillus rossiae DSM 32249*	Increase of fiber content, protein Digestibility, and nutritional indexes of fortified bread. Decrease of glycemic index.	Pontonio et al. (2017)
Oat	*Streptococcus thermophilus, Lactobacillus rhamnosus, Saccharomyces cerevisiae,* and *Candida milleri*	Folic acid fortification	Gobbetti et al. (2020)
Oat	*Candida milleri*	Increase of fiber solubility	Degutyte-Fomins et al. (2002)
Rye	*Weissella confusa*	Exopolysaccarides synthesis	Kajala et al. (2016)
sorghum (24 hr)	*Lactic acid bacteria (LAB) sp.*	Reduced phytic acid, trypsin inhibitors, and tannins Increased in vitro protein digestibility	Nkhata et al. (2018)
Sorghum (36 hr)	*Lactic acid bacteria (LAB) sp.*	Increased titratable acidity, crude protein, protein digestibility, and total solids	Nkhata et al. (2018)
Pearl millet (4 hr)	*Saccharomyces cerevisiae*	Increase in glucose Decrease in total carbohydrates Decrease in AIA No change in fructose	Nkhata et al. (2018)
Pearl millet (24 hr)	*Saccharomyces cerevisiae*	Increase in total protein Decrease in specific amino acids such as lysine, glycine, and arginine	Nkhata et al. (2018)
Wheat, barley, rice, and maize	*Lactic acid bacteria (LAB) sp.*	Increase in lysine content	Gobbetti et al. (2020)
Wheat, barley, rice, millet, and maize (22–25°C) and 37°C	*Lactic acid bacteria (LAB) sp.*	Increase in available lysine	Nkhata et al. (2018)

(*continued*)

Table 7.1 (Continued)

Bran source	Fermentation type	Effects	References
Pearl millet	*Saccharomyces cerevisiae*	Reduction in trypsin inhibitors Increased protein digestibility	Gobbetti et al. (2020)
Finger millet	*Lactic acid bacteria (LAB) sp.*	Increased bioavailability of calcium, phosphorus, and iron	Gobbetti et al. (2020)
High-carotenoid biofortified maize	*Lactic acid bacteria (LAB) sp.*	Loss of carotenoids with modest losses after 24 and 72 hr but bigger losses after 120 hr. Reduced bioavailability	Nkhata et al. (2018)
Soybeans (48 hr)	*Aspergillus species*	Decrease in phytosterols, glycosylated saponins, and tocopherols	Hubert et al. (2008)
Wheat flour made into sourdough bread	*Lactic acid bacteria (LAB) sp.*	Decreased glycemic index	Scazzina et al. (2008)

starters (i.e., *Lactobacillus sanfranciscensis* BB12, *Lactobacillus plantarum* 98a, and *Lactobacillus brevis* 3BHI) require more time for processing, einkorn is a better substrate to use in the bakery industry due to a number of reasons: (1) enhancement of phenolic acid configuration of breads, (2) increased emulsification of fiber, and (3) preservation of carotenoids to increase their bioavailability. The Bulgarian version of a traditional and popular beverage of the Balkan region (Boza) is also prepared by einkorn, which has a huge amount of health benefits such as healthy blood pressure, greater colonic health, enhanced milk production in the case of lactating women, lower plasma cholesterol, assisted digestion by secretion of gastric juice, and energizing of hepatic and pancreatic cells (Table 7.1). The valuable impacts of Boza on the human body are because of two main characterises of this drink: the prebiotic properties of cereal source associated with quickest intake of probiotic lactic acid bacteria. Huge biodiversity was seen on selected starters during laboratory fermentation on the grounds of their continuous extension and acidification and the proportion to lose free amino acids where spelt sourdoughs. *L. brevis* 20S, *Weissellaconfusa* 24S, and *L. plantarum* 6E were selected. The dietary bioavailability of these cultures is reduced by phytic acid, which when acting as an ANF makes a complicated compounds with cations. For upgrading their nutritional quality the elevated phytase activity was seen in spelt and emmer sourdoughs which perform an important role. A similar proportion was utilized for emmer flour, and similar results were obtained after fermentation with the starters L. plantarum 6E, L. plantarum 10E, and W. confusa 12E. Spelt and emmer sourdoughs were also used to consolidate wheat bread, which was clearly desired worldwide for its appetite as compared to that of wheat bread (Petrova et al., 2020).

7.4.2 Rice-Related Products

Using traditional solid-state starter, beer has been prepared in many Asian countries (Tsuyoshi et al., 2005). Due to its wide range of proteins, vitamins, sugars, bioactive compounds, and other organic acids these beverages are very much nutritious and shows functional properties (Kim et al., 2009). They also possess a gluten free alternative to the conventional barley beer for the intake which is not suitable for gluten-sensitive celiac patient (Hager et a., 2014). The production of this beer are quite different from the conventional beers made from barley or wine made from grapes (Table 7.1). The pH is acid due to the presence of lactic acid, tartaric acid, citric acid, and acetoin,

Table 7.2 Studies on Fermentation of Cereal-Based Media Using Probiotic LAB

Fermented extract	Lactic acid bacteria	Viable cell concentration after fermentation (Log10 cfu mL_1)	Fermentation temperature (C)	Fermentation time (h)	References
Barley malt, barley grains, and barley grains and malt mixture	*L. plantarum NCIMB 8826 and L. acidophilus NCIMB 8821*	7.9–8.5*	30	28	Rathore et al. (2012)
Barley malt, barley grains, oat grains, and wheat grains	*L. plantarum NCIMB 8826*	NA	37	24	Salmeron et al. (2009)
Barley malt, barley, and wheat grains	*L. fermentum, L. reuteri, L. acidophilus, L. plantarum, and*	7.2–10.1*	37	48	Charalampopoulos et al. (2002a)
Barley malt or grains, each mixed with whey powder and tomato pulp	*L. acidophilus NCDC16*	7.7–8.8*	37	12	Arora et al. (2010)
Oat grains	*Bifidobacterium adolescentis (2C, 5, 9J, 15, 19, 34, 56, and 111), B. breve 134, B. longum (3A, 12B, 46 and 51) and B. lactis Bb12*	5.5–7.8*	37	24	Laine et al. (2003)
Oat grains	*L. plantarum B28*	10.9	37	8	Angelov et al. (2006)
Oat grains	*L. plantarum B28, L. casei spp paracasei B29, Candida rugosa Y28 and C. lambica Y30*	9.4–11*	37	8–10	Angelov et al. (2006)
Millet and rice grains fortified with pumpkin and sesame seed milk	*S. thermophiles, L. acidophilus and Bifidobacterium BB-12*	9.6	37	16	Hassan et al. (2012)
Malt extract	*L. reuteri 11951, mixed culture with yeast*	8.1–8.9*	37		Kedia et al. (2007)
Malt extract	*B. breve NCIMB 702257*	9.3–9.4	37	16–23	Kedia et al. (2007)
Malt extract enriched with yeast extract or peptone	*B. adolescentis NCIMB 702204, B. infantis NCIMB 702205, B. breve NCIMB 702257 and B. longum NCIMB 702259*	8.7–9*	37	24	Hassani et al. (2016)

which give a sour taste and pleasant flavor. Some selective parts are utilized in the preparation of these rice beer starters and various category of endophytic entity of plant source serve as the active microbes for mixed culture and multistage rice fermentation (Chen et al., 2010). Earlier presence of amylolytic fungus in beer starters have been recorded, such as *Mucor*, *Rhizopus* and *Aspergillus* in takju of Korea (Kim et al., 2010), *Mucorcircinelloides*, *Rhizopus*chinensis and *Saccharomycops isfibuligera*inmarcha of Nepal and *Mucorcircinelloides*, *Rhizopuschinensis*and *Rhizopus stolonifer* in marcha of Darjeeling, Sikkim and other parts of 8 states of India in Northeast area. Also in 39 samples of nuruk (Korea), 174 filamentous fungal strains were isolated among which 6 genera (*Lichtheimia, Aspergillus, Rhizopus, Rhizomucor, Mucor,* and *Syncephalastrum*) and 17 fungal species were reconized (Tamang et al., 2006).

7.4.2.1 Rice Beer

In Asian countries the preparation of beer from rice is very common and is called names like *sake* in Japan, *shaosingiju* and *lao-chao* in China, *chongju* and *takju* in Korea, *tapuy* in the Philippines, *brembali* and *tape-ketan* in Indonesia, *khaomak* in Thailand, *rou nep than* in Vietnam, and *tapai-pulul* in Malaysia (Aidoo et al., 2006). Using local food crops, fermented food and beverages are prepared in the Indian subcontinent and these practices have gone on for centuries (Roy et al., 2004). The north-eastern cluster of the India subcontinent is categorized by a vast array of tribal association. Out of 450 tribes 250 tribes are resides in this area only. They have a high stock of traditional grip due to their occupation in this steep region (Chatterjee et al., 2006). Knowledge of consumable plants and their benefits along with conservation of foods for intake and medicinal intent is common. They predominantly cultivate rice and consume it in the form of fermented alcoholic beverages. The products are sweet and alcoholic (Thapa et al., 2004). This type of drink has both socio-economical events as well as a great importance in making medicinal products and treated as a drug (Singh et al., 2006). It has been used to treat headache, insomnia, body ache, and inflammation of different body parts, diarrhoea and urinary problems, for expelling worms, and as treatment of cholera (Deka et al., 2010).

7.4.3 Other Cereals

Due to its enhanced nutritional value, barley sourdough is a promising substance that upgrades the bread based on barley (Harth et al., 2016) but it is not commonly followed because of its negative impacts on bread dough loaf capacity as well as rheology. Barley sourdoughs are dominated by LAB bacteria species of *Lactobacillus fermentum, L. plantarum, L. brevis, Leuconostoc mesenteroides, W. confuse* in the laboratory and bakery industries (Table 7.1).

Fermentation is a superior alternative for expanding the digestibility of sorghum proteins (Belton et al., 2004) and swapping the properties and microstructure of starch by reducing ANFs (e.g., tannins and phytic acid) (Osman et al., 2004), which leads to improvement of amino acid balance and enhances vitamin content (Pranato et al., 2013). Sorghum has low nutritional value because of its low starch and protein digestibility and poor organoleptic base (Wong et al., 2009). Similarly, in Ben-saalga, a thin porridge made by cooking the fermented part of pearl millet (Pennisetumglaucum), *Lactobacillus fermentum, L. plantarum,* and Pediococcuspentosaceus, which typically influence the spontaneous fermentation, are manage for the improvement of its nutritional profile (Nout et al., 2009). The impacts of spontaneous fermentation on inhibitors of enzyme, tannin content, phytic acid, and in vitro protein digestibility (IVPD) of sorghum flour are enhanced. The enhancement of IVPD has been linked to separation of protein and tannin combined with protein, making it valuable in the digestion of pepsin and tannin in sorghum; indeed, *L. plantarum* has tannase activity (Pranato et al., 2013) that breaks down the tannin compound with protein, which makes it valuable in pepsin digestion. Sourdough fermentation allows the utility of the flour that has been used in bread making which bring health benefits.

7.4.4 Pseudo-cereals

There has been great interest in gluten-free bakery products made of gluten-free flours like buckwheat, quinoa, and amaranth. Buckwheat is found to be a great origin of starch and may products such as dietary fibers, proteins, antioxidants along with some trace elements. High-quality proteins, lipids, and minerals like calcium, potassium, and phosphorus are also found in amaranth (de la et al., 2010). Pseudo-cereals are also used to make food as well as feed. Amaranth, quinoa (both *Amaranthaceae*), buckwheat (*Polygonaceae*), and chia (Lamiaceae) are the most common pseudo-cereals. Quinoa along with amaranth are essential food crops in Latin America where buckwheat was cultured around 6000BC in Asia, Yamuna, and Tibet. Today buckwheat is common in Europe and widely eaten used in porridge in Ukraine, Estonia, and Russia and in pancakes in France and Belgium (Wallonia). Buckwheat has health-promoting effects such as high amount of polyphenols and vitamins that help in the prevention of cancer, cardiovascular diseases, and diabetes (Gandhi et al., 2013).

7.4.4.1 Quinoa

Quinoa has a better nutritional profile than common cereals as it has essential amino acids, up to 23% non-gluten proteins in dry matter, and other phytochemicals such as phenols, flavonoids, and steroids (Kuljanabhagavad et al., 2008). It also has high water binding capacity, high solubility, and comes to gelatinization in less time than other cereals (Bolıvar-Monsalve et al., 2018). However, it has some disadvantages. For example, due to the presence of triterpenic saponins it has a bitter taste (Gomez-Caravaca et al., 2014). Quinoa flour is used to prepare bread or other bakery goods. Quinoa fortification improves protein content, enhances flavonoid and total phenolic contents, and also improves antioxidant activity, but at the same time increased hardness and reduces specific volume of the product. Thus, the sensory attributes of bread and its dough are also affected (Wolter et al., 2014; Stikic et al., 2012). To promote technological advantages and nutritional value, fermentation of quinoa flour or blends of flours have been found to be acceptable alternatives (Dallagnol et al., 2013). Fermentation of sourdough eventually lowers the saponin content and thus improves rheology and sensory attributes.

7.4.4.2 Amarnanth

The growth of amaranth as a niche product to the popular natural resources was changed with due course of time which was originated from Mexico. The flavour impressions, outstanding nutritional content as well as the gluten-free content are the main reasons for its popularity (Jekle et al., 2010). Amaranth possesses up to 16.5% protein, with optimum levels of amino acid content and vitamins and minerals (Olusegun, 1983). Many health benefit studies have been reported on amaranth as it lowers blood glucose and cholesterol levels, improves anemia and hypertension, and causes antitumor activity (Caselato-Sousa and Amaya-Farfan, 2012). Amaranth seeds have bioactive compounds including protease inhibitors, antimicrobial peptides, anticancer compounds, lectins, antioxidants, and some phytonutrients that inhibit both oxidative chain reactions and free radicals within human cell membranes. Amaranth flour may be used as a full or partial substitute in many food products such as cookies, bread, pastries, pasta, cereal flakes, tortillas, and many more (Silva-Sanchez et al., 2004). Amaranth flour (10–40%) is also used in recipes either containing gluten or gluten-free (Onyango et al., 2013; Schoenlechner et al., 2010). Lactic acidification has been found to be the best method to process amaranth as it improves nutritional, technological, and sensory attributes. The starters used in sourdough fermentation of amaranth flour are LABs (i.e., *Lactobacillus plantarum, Lactobacillus sakei,* and *Pediococcuspentosaceus*) (Sterr et al., 2009). Consumers also appreciate gluten-free fortified bread with amaranth sourdough (10%, w/w) (Rozyło et al., 2015). Even baked products made with wheat flours and amaranth promote the development of probiotic strains (*Lactobacillus rhamnosus*) through processing (Matejcekova et al., 2016).

7.4.4.3 Buckwheat

Buckwheat is comprised of dietary fiber, lysine, proteins, and antioxidants, mainly rutin and quercetin and micronutrients, which have prebiotic properties and cholesterol-lowering activity (Alvarez- Jubete et al., 2009; Li et al., 2010). Buckwheat fortification is a challenging process with some technological parameters and due to tannins and phytate, ultimately influences the digestibility negatively and deliver bitterness (Li and Zhang, 2001). Buckwheat sourdough baking performance depends on its manufacturing, digestibility, and palatability. Fermented buckwheat flour used to make muffins showed a high amount of potassium and magnesium as macronutrients and zinc and manganese as microelements (Ciesarova et al., 2016). Buckwheat flour after fermentation (i.e., buckwheat sourdough) contains a high amount of gluten-free nutrients, enhanced protein, and better sensory acceptability (Saturni et al., 2010; Wronkowska et al., 2013). Fermentation performed with selected LAB strains enhanced the bioactive attributes of buckwheat by releasing GABA (Coda et al., 2010).

7.4.4.4 Hemp

Hemp seed possesses proteins (20–25%), oils (25–35%), fibers (10–15%), and carbohydrates (20–30%), along with some trace amount of minerals (Deferne and Pate, 1996). Its seed is rich in polyunsaturated fatty acids that are 80% containing α-linolenic acid and linoleic acids with a 3:1 ratio (Da Porto et al., 2015). Hemp protein contains a fair amount of sulphur-containing amino acids, glutamic acid and arginine, which show greater affinity towards digestibility (House et al., 2010). Depending on the variety, the seed contains D-9 tetrahydrocannabinol. Hemp is an underutilized crop with significant nutritional value. Lack of knowledge about its functional and technological properties and anti-nutritional factors (ANF) such as cyanogenic glycosides, trypsin inhibitors, phytic acid, condensed tannins, and saponins make them undesirable and responsible for its limited use (Russo and Reggiani, 2015). Hemp flour fermented with LAB (Pediococcusacidilactici, Lactobacillus plantarum, and Leuconostocmesenteroides) combined with mixed starter has been used to develop hemp sourdough. The fermentation prevents the development of cyanogenic glycosides, phytic acid, total saponins, and tannins. Sourdough of hemp fortified with wheat breads (5–15%, w/w) enhanced rheology properties without affecting its sensory attributes. Fementation with fortification proportionally increased the bread's protein digestibility but decreased the predicted glycemic index (Nionelli et al., 2018).

7.5 FERMENTATION IMPACT ON NUTRIENT PROFILE OF CEREALS

Cereal fermentation is mainly focused on enhancement of protein content as well as digestibility. Many studies have been done on fermentation (submerged or solid-state) and its macronutrients. Studies have shown that for oats, the single and co-inoculums analyzed for solid-state fermentation. Comparing *R. oryzae* and *L. plantarum*, *R. oryzae* (+104.7%) fungus strain showed higher enhancement of soluble proteins (Tables 7.1 and 7.3). It was found that with synergetic action by both the microorganisms, *R. oryzae* produced greater high-soluble protein with fungal activity (Wu et al., 2018). For this reason, the use of LABs is rare in SSF as it requires optimum moisture and nutrition (nitrogen). To convert complex polymers into simpler forms LABs are co-cultured with fungus (e.g., *R. oryzae*) as an energy and nutrient source (Wu et al., 2018; Wang et al., 2019). Studies have also reported that fermentation carried out with fungus showed higher protein in quinoa. The highest was in *Agaricus bisporus* (+133.6%), than *Helvella lacunose* (+90%) followed by *Fomitiporiayanbeinsis* (+58.8%). Stoffel et al. (2019) also reported that there is protein increment showed in SSF of wheat, rice, and corn by fungi *Agaricusblazei*, *Auriculariafuscosuccinea,* and *Pleurotusalbidus* (+30%, +19%, and + 46%, respectively). Yeasts such as *S. cerevisiae* found in rice-black gram mixed flour have been used to increase the protein concentration in the fermentation. Studies reported that around 10% protein (dry basis) increment due to the presence of yeast cells (Rani et al., 2018). When protein comparison was done between cereals and fermented legumes, protein was much higher in the fermented legumes as compared to cereals with the highest rate in legumes being tepary bean (+35%) compared to +133.6% in quinoa (Mora-Uzeta et al., 2020). Moreover, it was found that fungus strains contribute larger extent comparing to yeasts or bacteria. In sorghum, there was 12.39% of protein increment by SmF with *L. plantarum* and after 36 h of incubation there was slight reduction in the amino acids conversion into flavors compounds (Pranoto et al., 2013). In fermented sorghum, IVPD has been studied using *plantarum* species. Apparently, these bacteria have the proteolytic and tannase activities that can enhance IVPD. Protein fragments were hydrolysed into smaller amino acids and peptide molecules whereas tannins-proteins came out resulting in high IVPD (Pranoto et al., 2013).

Physiochemically, titratable acidity enhancement and pH reduction were also seen in fermentation where in fermented rice-black gram flour titratable acidity was increased by fungus, which converted carbohydrates into fermentable sugars and organic acids containing acetic acid, citric acid, and lactic acid (Rani et al., 2018). *Rhizopus* also plays an important role in saccharification and liquefaction because of its amylolytic ability (Wu et al., 2018). In sorghum, starch has been significantly reduced using SmF (*L. plantarum*as) where microorganisms were treated as proteolytic bacterium and hydrolyzed the starch granules into simple sugars and thus enhanced *in vitro* starch digestibility (IVSD) (Pranoto et al., 2013). Studies also show that after fermentation quinoa contains high quantities of vitamins and minerals and a wide range of antioxidants, making the value of TPC highly significant (Xu et al., 2019). Compared to other cereals, rice, corn, and oat have significantly less TPC and also no β-carotene or tocopherols after removal of the husk. It has been seen that by enzymatic activity (amylases, xylanases, and glucosidases), the bioavailability of the compound increases significantly. Literature reported that after solid-state fermentation of 6 h in rice-black gram mixed flours by yeasts, the TPC content increased by 0.44 mg GAE/g. In barley after co-fermentation with LAB and fungus, FPC (1.29 mg GAE/g) release was seen along with phenolic acids (Wang et al., 2019). Even in wheat and rice, TPC was well evaluated after fermentation by fungus (*Aspergillus oryzae)* (Saharan et al., 2017). The increment in wheat was about 6 times (+ 460%) and 9 times (+758.8%) in rice. Thus, the antioxidant property and bioavailability strongly depends upon grain variety, species, cultivation, and processing conditions (S´anchezMaga˜na et al., 2019). *Bifidobacterium* bacteria strains have also been found to synthesize new phenolic compounds and thus increase TPC. Studies have shown that when using *B. longus,* there is an increase in TPC for wheat (~35 mg GAE/g) and quinoa (~41 mg GAE/g) (Ayyash et al., 2018; Stoffel et al., 2019). Even the luminosity (L*) increased after fermentation with *P. albidus* in rice and wheat flours. However, in fermentation of corn by *A. fuscosuccinea,* there was a decrease in b* parameter leading to a dull yellowish sample. Incubation time is also one of the important factors after fermentation that can significantly impact change. Thus, 6 h was found to be the optimal fermentation time with *S. cerevisiae* (Rani et al., 2018), and for filamentous fungus (*F. yanbeiensis, H. lacunose,*

A. bisporus, A. cosuccinea, P. albidus or A. blazei, 35 days of fermentation were required (Stoffel et al., 2019; Xu et al., 2019). It was also revealed that longer fermentation time (108 h) in corn grains fermented by fungal leads to higher undesirable off-odours (S´anchezMaga˜na et al.,2019). The phenolic compounds were also enhanced in oats after fermentation and those phenolics present in the grains neutralized the free radicals by donating protons and electrons (Xiao et al., 2015; Ayyash et al., 2018). It was found that with fermentation with the fungus *L. rhamnosus* in cereals (i.e., wheat, buckwheat, rye, and barley), there is an enhancement in the scavenging activity of α-diphenyl-β-picrylhydrazyl (DPPH) where the highest amount was seen in buckwheat (Đorđevi´c et al., 2010). From a health perspective, obesity was found to be controlled by using wheat-fermented product. *P. albidus* and *A. fuscosuccinea* inhibited lipase activity by +413% and +40%, respectively (Stoffel et al., 2019). Thus, fermentation could be used as an efficient tool for controlling diabetes with cereals (Garrido-Galand et al., 2021).

7.6 NUTRITIONAL VALUE OF FERMENTED ALCOHOLIC AND NON-ALCOHOLIC CEREAL BEVERAGE

The transportation of nutrients and bioactive compounds to the body can be carried out by beverages to facilitate their bioavailability. Barley foods have been recognized as fermented food products like beer along with distilled beverages that possess various health stimulating impacts (Zarnkow et al., 2008; Celus et al., 2006; Nsogning Dongmoet al., 2016). Those bioactive compounds were ferulic acids, proanthocyanidins, chalcones, benzoic acid, cinnamic acid, flavones, and flavanones (Kim et al., 2007). Traditionally barley grains are used to formulate malt and those malted products as a raw ingredient are classified as alcoholic or non-alcoholic beverages. Malt beverages have been classified into beer-brewing industries and non-alcoholic malt beverages industries which has the similar process for malt raw materials production (Mahmoudi et al., 2015). The main difference between beer-brewing and non-alcoholic fermented cereal beverages is the nutritional point. In commercial malt beverages the reducing sugars amounts were 604–944 mg/ dL, and admissible amount of vitamins A and C, whereas the beverage of malt had less amount of, protein, iron, oxalate, phytate, hydrogen cyanide, and zinc.

7.7 LACTIC ACID FERMENTATION OF CEREALS AND PSEUDOCEREALS

Usually LABs are auxotrophic in nature and have the ability to synthesize water-soluble vitamins, which mainly consist of B vitamins like B2, B7, B9, and B12 (Patel et al., 2013). Therefore, it is necessary to study the metabolic behavior of LABs to formulate fortified functional products with other vitamins (Table 7.3). Studies have shown that an Indonesian fermented soybean product called tempeh has a great amount of riboflavin (B2) concentration where *Streptococcus* and *Enterococcus* species are mainly utilized for the biosynthesis of vitamin (Keuth&Bisping, 1993). It has been shown that among the LABs *Lactococci spp.* and *Lactobacillus spp.* (i.e., *Lactococcilactis, Lactobacillus bulgaricus, Lactobacillus plantarum, Enterococcus spp.,* and *Streptococcus thermophilus*) have the ability to generate folate (B9) (LeBlanc et al., 2007). Various circumstances affect B9 production such as high external pH, pamino benzoic acid concentration, which stimulates folate biosynthesis, and huge tyrosine amounts reduced folate concentration (O'Connor et al., 2005). Certain members of Lactobacillus groups have the potential to produce cobalamin (B12) and also the probiotic strain of Lactobacillus reuteri. On the other hand, it was observed that in fermented carrot juices, the auxotrophic practice towards certain vitamins depends on usages of Bifidobacterim which get degraded by 15–45% in its carotenoid content. However, in non-dairy products probiotic lactobacilli strains or

Table 7.3 Non-alcoholic Fermented Cereal Beverages and Their Microorganisms and Nutritional Composition

Beverage/Place of Origin	Cereals	Microorganisms	Functional Compounds	References
Amazake–Japan	rice *koji*	*Lactobacillus sakei*, *Aspergillus oryzae;*	amino acids; vitamins B1, B2, B6; pantothenic acid, vitamin E, flavonoids, dietary fiber, polysaccharides, sterols;	Pasqualone, et al. (2018)
Bors/Borsht–Central and Eastern Europe, Romania	wheat bran, corn flour	*Lactobacillus delbrueckii* ssp. *Delbrueckii;*	lipophilic and hydrophilic antioxidants (tocopherols, tocotrienols), phenolic compounds, *group B vitamins*, vitamin E, alkylresorcinols; lignans;	Pasqualone, et al. (2018)
Boza–Turkey, Greece, Bulgaria, Albania, Romania, Bosnia Herzegovina; South Afr	barley, oats, rye, millet, maize, wheat, rice	*Lactobacillus plantarum*, *Lactobacillus rhamnosus*, *Lactobacillus pentosus*, *Lactobacillus paracasei*, *Lactobacillus fermentum*, *Lactobacillus brevis*, *C. inconspicua*, *C. pararugosa*;	vitamin A, vitamins B1, B2, B6, nicotinamide; Ca, Fe, P, Zn, Na, β-glucan, dietary fibers;	Pasqualone et al. (2021)
Busa–Syria, Egypt, Kenya, Turkistan;	rice or millet	*Lactobacillus* sp. *Saccharomyces* spp.	dietary fiber, amino acids, fatty acids, vitamins B1, B2;	Pasqualone et al. (2021)
Gowé–West Africa, Benin	malted and non-malted sorghum, maize	*Lb. fermentum, Weissella confusa, Weissella kimchii, Lactobacillus mucosae, Pediococcus acidilactici, Pediococcus pentosaceus*;	amino acids (glutamic acid and leucine), minerals (Fe, Ca, Zn, P);	Pasqualone et al. (2021)
Kunun-zaki–Nigeria	wheat and sorghum/ millet, wheat, malted rice	*Lb. plantarum, Lb. fermentum, Lactococcus lactis; Saccharomysces cerevisiae;*	minerals (Fe, Ca, Mg, K);	Pasqualone et al. (2021)
Kvass–Lithuania, Russia, Eastern Poland	extruded rye, malted barley	*Lactobacillus casei, Lb. sakei, P. pentosaceus, S. cerevisiae;*	vitamins B1, B3, B2, B6; dietary fibers, Zn, Cu, maltose, maltotriose, glucose, fructose;	Pasqualone et al. (2021)
Mahewu/Amahewu–Africa (Botswana, South Africa, Zimbabwe)	maize, sorghum, millet malt or wheat flour	*Lb. brevis, L. casei, L. lactis, Lb. plantarum, S. cerevisiae, S. pombe;*	Na, K, Ca, Fe, Zn, Mn; dietary fiber, carbohydrates, *group B vitamins*;	Fadahunsi et al. (2017)
Munkoyo–Zambia, Democratic Republic of Congo	maize	*Lb. plantarum, Weissella confusa, L. lactis Enterococcus italicus;*	fiber, vitamins B1, B2, B3, B6, B12, Ca, Fe, Zn, proteins, crude fat;	Byakika et al. (2019)

(continued)

Table 7.3 (Continued)

Beverage/Place of Origin	Cereals	Microorganisms	Functional Compounds	References
Obushera–Uganda	sorghum flour or millet, maize	*L. Lactis*, *Lb. plantarum*, *Lb. fermentum*, *Lb. delbrueckii*, *Weissella confusa*;	proteins, minerals, fiber;	Misihairabgwi et al. (2018)
Oshikundu/ *Ontaku*–Namibia	pearl millet meal, sorghum, or pearl millet malt	*Lb. plantarum, L. lactis, Lb. delbrueckii ssp. delbrueckii, Lb. fermentum, Lb. pentosus, Lactobacillus curvatus;*	shikimic acid, maleic acid, phytic acid, succinic acid; vitamins B1, B2; Ca, Cu, Fe, K, Mg, Mn, Na, S, Zn, P;	Waters et al. (2015)
Pozol–South Eastern Mexico, Central America	maize	*Lb. plantarum, Lb. fermentum, Lb. casei, Lb. delbrueckii, Leuconostoc sp., Bifidobacterium sp., Streptococcus sp., Saccharomyces sp.;*	group B vitamins; dietary fiber;	Coskun et al. (2017)
Shalgam–Turkey	bulgur flour (wheat)	*Lb. plantarum, Lb. paracasei, Lb. brevis Lb. fermentum, S. cerevisiae;*	β-carotene, group B vitamins, Ca, Na, Fe;	Blandino et al. (2003); Oi et al. (2003); Waters et al. (2015)

some starter cultures utilize more folate than they produce. The literature also revealed that by fermentation there is an improvement in novel functional foods with an increment of vitamin content which eliminates the fortification of chemically synthesised vitamins, thus enhancing the nutritional and commercial value in the market (Stanton et al., 2005). There should be a balance between the autotrophic and auxotrophic properties for the production food products with high nutritional content. The two main advantages of fermentation with probiotic LABs are the health bolstering service that prompted by the probiotic's synergy with the human and enhancing raw materials nutritional value by producing bioactive microbial metabolites during fermentation of lactic acid. Vitamins, fatty acids, or organic acids are included by these metabolites (Stanton et al., 2005). During fermentation with LABs, there is huge degradation of myo-Inositol hexaphosphate (IP6) and thus used as a prominent method to discardanti-nutritional element and increase the mediumnutritional value. Most research has been carried out on phytatedegradation during food fermentation using LAB. The bacteria synthesis phytases took the charges for the devaluing of IP6 lowering its functional and bioactive properties (Hussin et al., 2010). However, there was a decrease in the pH by the lactic acid fermentation results in the huge bacterial count of lactic and acetic acid which promotes the phytase properties lessen the amount of IP6 in fermented medium. It has been also shown the LAB results are not fixed as few strains activate low phytase properties and many have no significant degradation activity. In sourdough fermentation, LABs significantly showed the degradation of phytate. Among all the LABs, *Lactobacillus sanfranciscens* has been found to be the best phytase producer (De Angelis et al., 2003).

7.8 CEREAL-BASED PROBIOTIC PRODUCTS FROM LACTIC ACID FERMENTATION

Fermentation using strains of probiotic bacteria have been used mainly for the probiotication of food materials. Milk-based products are the most common fermented probiotic supplements on the market (Rivera Espinoza & Gallardo-Navarro, 2010). However, cereal grains are a plant-based option that could be used as a promising alternative to milk-based probiotic formulas without the drawbacks relating to fermented dairy products such as high cholesterol levels (Prado et al., 2008). Moreover, cereal-based fermented probiotication shows combined function of probiotics as well as prebiotics, namely dietary fibers, arabinoxylans, b-glucans, and many more. Developing a novel functional food made of both cereals and probiotics is challenging. Thus, several technological directions have been introduced in the composition and processing of cereal grains, organoleptic acceptance, productivity of the starter culture, stability of the probiotic strain during storage, and the nutritional value of the final product (Charalampopoulos et al., 2002). Cereal-based media can be used as a transportation vehicle for probiotic LAB, which is widely used and accepted these days. The gap of bacterial growth patterns between malted and non-malted grains-based media has also been successfully demonstrated. When fermentation is carried out in malt-based products, there is an increase in the selected LAB strains (i.e., *Lactobacillus plantarum, L. reuteri, L. fermentum,* and *L. acidophilus*) compared to non-malted grain media because of differences in glucose, fructose, sucrose, maltose, and free amino nitrogen content. For malt-based fermentation, the use of microbial culture mixture has been studied. Addition of yeast with varying ratios to LAB inoculum (*L. reuteri*) emphasized the growth of the bacteria. In mixed yeast/bacteria culture, the pH is lower and lactic acid production and ethanol found are higher (Kedia et al., 2007). To ensure fermented probiotic product efficiency, it is recommended to contain a satisfactory number of bacterial viable cells while consumption. Thus, during fermentation the bacterial strain's ability to attain huge cell population is very important. It is suggested to have the minimum therapeutic daily dose to be between 8 and 9 log10 cfu mL^{-1} viable cells that can be fulfilled by the 100 mL or gram of daily intake of a product. Mainly the research is based on fermentation parameters such as initial pH, inoculation dosage, and temperature, which significantly impact the concentration of viable cells) (Hassani et al., 2015). In one of study, *L. acidophilus* NCIMB 8821, *L. plantarum* NCIMB 8826, and *L. reuteri* NCIMB 1195 were exposed to acidic conditions for 4 h to simulate gastric tract acidic parameters. By exposure to acidic conditions, all strains of the cell population were found to be significantly reduced. With the incorporation of cereal extracts such as barley malt, barley, and wheat, survivability of all the strains significantly increased. This was seen only when sugar was present in the cereals (Charalampopoulos et al., 2003). Moreover, the same strains were subjected to salty environment acting bile conditions (2% bile for 4 h in the phosphate-saline buffer at pH 7). *L. reuteri* NCIMB 1195 among all the tested strains was found to have the best resistance to bile without any additional components and *L. acidophilus* NCIMB 8821 was the most sensitive strain. On bile tolerance of probiotic LAB strains, media composition plays a vital role. Cereal extracts show greater positive influence on the survivability of strains mainly by the malt. Among all the cereal extracts wheat, barley, and barley malt showed the most impact on resistance of strains and later had greater positive influence. The effectiveness of strain resistance was found to be based on soluble sugar and free amino nitrogen concentration present in the medium (Patel et al., 2004). In addition, immobilization by cereal fiber on *L. plantarum* vitality in the gastrointestinal tract has also been studied. There is a synergetic protective effect on viability by the cereal fiber as an immobilization of bacterial cells. Although the cell survivability along with gastric and bile conditions there is a significant difference where the protective effect in gastric acidic conditions is slight and pronounced in salty bile conditions (Michida et al., 2006)

7.9 PROBIOTIC AND NON-ALCOHOLIC FERMENTED CEREAL-BASED PRODUCTS

Probiotics can be found in a diverse variety in non-alcoholic beverages, depending on the cereal substrate and production methods. The stability is provided by the processing method of the bacterial composition to get the probiotic functionality as the final output. While the beverage showing the probiotic property should have better than 107 CFU/ml, not all types of LAB show probiotic characteristics. Studies have shown that *L. Rhamnosus* in non- alcoholic fermented cereal beverages (NFCBs) is can prevent and treat gastrointestinal issues (Kort et al., 2012). Since ancient times, *bors* has been consumed by the traditional Romanian NFCB as a intestinal remedy. There are several probiotic bacteria isolated from bors such as *L. casei, L. brevis, L. plantarum,* and *L. Fermentum* for curative purposes (Grosu-Tudor et al., 2019). Whereas, Bulgarian bozamainly consists in the microbiota identification that is lactic acid producing bacteria and yeasts such as *Lactobacillus acidophilus, L.Plantarum,L.coprophilus, L.fermentum, L.euconostocraffinolactis, Ln. brevis, Ln. Mesenteroides, Saccharomyces cerevisiae, Geotrichumpenicillatum, G. candidum, Candida tropicalis,* and *C. Glabrata,* respectively (Blandino et al., 2003). In the case of *boza* found in Turkey, the utility of yeast as well as LAB isolates as beginning cultures were studied using controlled fermentation. The selection and identification of proper strains showing probiotic as well as antimicrobial properties intensify *boza's* functional properties (Altay et al., 2013). In African subcontinents, the principal difficulty for the growth of fermented cereal fortified probiotic drinks is insufficient understanding of the nutritional and health benefits of such fermented beverages along with consumer skepticism. Moreover, there is a need for proper facilities for developing probiotic starter cultures through spontaneous fermentation (Setta et al., 2020; Pasqualone et al., 2021).

7.10 CONCLUSION AND FUTURE PERSPECTIVES

The process of fermentation gets rid of toxic chemicals along with harmful microorganism in foods and adds beneficial bacteria. When considering the impacts of cereal fermentation on human health, some factors should be evaluated: (i) the advantageous probiotic properties of LAB strains incorporated in the process, (ii) the grain's beneficial content, (iii) biochemical changes occurs during fermentation, and (iv) enzymatic and genetic capacity of LAB strains, which give them the power to utilize carbohydrate. The relationship between cereals and LABs shows synergetic effect and thus is a connective bridge between its preferable substrate and bacterial strain by its enzymatic ability to digest various carbohydrates or prebiotic dietary fibers (resistant starch, fructans, beta-glucans, and xyloglucans). Fermentation of cereals offers the ability to develop new functional fortified foods with maximum health beneficial properties. Fermented cereals and pseudo-cereals can be used to treat various health issues such as cardiovascular disease, cancers, metabolic and allergic disorders, etc. However, different crop variants fermentation or their metabolic pathways have not been investigated and there is need to be incorporated in the future value on the joint efforts of nutrition science, biotechnology, plant selection and medicine. Therefore, more research on their bioavailability, bioactivity, and bioaccessibility is needed to develop new bioactive natural compounds.

REFERENCES

Abdel-Aal ESM, Young JC, Wood PJ, Rabalski I, Hucl P, Falk D, Fre'geau-Reid J. Einkorn: a potential candidate for developing high lutein wheat. Cereal Chem 2002;79:455–7.

Abebe W, Ronda F. Rheological and textural properties of Teff [Eragrostis Tef (Zucc.) Trotter] grain flour gels. J Cereal Sci 2014;60:122–30.

Achi, O.K.; Asamudo, N.U. Cereal-Based Fermented Foods of Africa as Functional Foods. *Int. J. Microbiol. Appl.* **2019**, 1527–1558.

Aidoo, K.E., Nout, M.R., and Sarkar, P.K. Occurrence and function of yeasts in Asian indigenous fermented foods. *FEMS Yeast Research*, 6: 30–39, 2006.

Altay, F.; Karbancioglu-Guler, F.; Daskaya-Dikmen, C.; Heperkan, D. A review on traditional Turkish fermented non-alcoholic beverages: Microbiota, fermentation process and quality characteristics. *Int. J. Food Microbiol.* **2013**, *167*, 44–56.

Alvarez-Jubete, L., E. K. Arendt, and E. Gallagher. 2009. Nutritive value and chemical composition of pseudocereals as gluten-free ingredients. International Journal of Food Sciences and Nutrition 60 (sup4):240–57.

Anal, A.K. Quality ingredients and safety concerns for traditional fermented foods and beverages from Asia: A review. *Fermentation* **2019**, *5*, 8

Angelescu, I.R.; Zamfir, M.; Stancu, M.M.; Grosu-Tudor, S.S. Identification and probiotic properties of lactobacilli isolated from two different fermented beverages. *Ann. Microbiol.* **2019**, *69*, 1557–1565.

Angelov, A., Gotcheva, V., Kuncheva, R., & Hristozova, T. (2006). Development of a new oat-based probiotic drink. *International Journal of Food Microbiology*, *112*(1), 75–80.

Angioloni A, Collar C. Nutritional and functional added value of oat, Kamut, spelt, rye and buckwheat versus common wheat in breadmaking. J Sci Food Agric 2011;91:1283–92.

Arora, S., Jood, S., & Khetarpaul, N. (2010). Effect of germination and probiotic fermentation on nutrient composition of barley based food mixtures. *Food Chemistry*, *119*(2), 779–784.

Ayyash, M., Johnson, S. K., Liu, S., Al-Mheiri, A., & Abushelaibi, A. (2018). Cytotoxicity, antihypertensive, antidiabetic and antioxidant activities of solid-state fermented lupin, quinoa and wheat by Bifidobacterium species: In-vitro investigations. *Food Science & Technology, 95*, 295–302.

Belton PS, Talor JRN. Sorghum and millets: protein sources for Africa. Trends Food Sci Technol 2004;15:94–8.

Blandino, A., Al-Aseeri, M.E., Pandiella, S.S., Cantero, D., and Webb, C. Cereal-based fermented foods and beverages. *Food Res. Int.* **2003**, *36*, 527–543.
Bolıvar-Monsalve, J., C. Ceballos-Gonz_alez, C. Ram_ırez-Toro, and G. A. Bol_ıvar. 2018. Reduction in saponin content and production of glutenfree cream soup base using quinoa fermented with Lactobacillus plantarum. Journal of Food Processing and Preservation 42 (2):13495.
Caselato-Sousa, V. M., and J. Amaya-Farf_an. 2012. State of knowledge on amaranth grain: A comprehensive review. Journal of Food Science 77:93–104.
Celus, I., Brijs, K., Delcour, J.A., 2006. The effects of malting and mashing on barley protein extractability. Cereal Sci. 44, 203–211.
Charalampopoulos, D., Pandiella, S.S. & Webb, C. (2003). Evaluation of the effect of malt, wheat and barley extracts on the viability of potentially probiotic lactic acid bacteria under acidic conditions. International Journal of Food Microbiology, 82, 133–141.
Charalampopoulos, D., Wang, R., Pandiella, S. S., & Webb, C. (2002). Application of cereals and cereal components in functional foods: a review. *International journal of food microbiology*, *79*(1–2), 131–141.
Chatterjee, S., Saikia, A., Dutta, P., Ghosh, D., Pangging, G., and Goswami, A.K. Technical report on Biodiversity Significance of North East India for the study on Natural Resources, Water and Environment Nexus for Development and Growth in North Eastern India, WWF-India, New Delhi, 2006.
Chen, S., and Xu, Y. The influence of yeast strains on the volatile flavour compounds of Chinese rice wine. *Journal of the Institute of Brewing*, 116(2): 190–196, 2010.
Ciesarova, Z., E. Basil, K. Kukurov_a, L. Markov_a, H. Zieli_nski, and M. Wronkowska. 2016. Gluten-free muffins based on fermented and unfermented buckwheat flour-content of selected elements. Journal of Food and Nutrition Research 55 (2):108–13.
Coda R, Rizzello CG, Gobbetti M. Use of sourdough fermentation and pseudo-cereals and leguminous flours for the making of a functional bread enriched of γ-aminobutyric acid (GABA). Int J Food Microbiol 2010;137:236–45.
Coda, R., Rizzello, C. G., Curiel, J. A., Poutanen, K., & Katina, K. (2014). Effect of bioprocessing and particle size on the nutritional properties of wheat bran fractions. *Innovative Food Science & Emerging Technologies*, *25*, 19–27.
Da Porto, C., D. Decorti, and A. Natolino. 2015. Potential oil yield, fatty acid composition, and oxidation stability of the hempseed oil from four Cannabis sativa L. cultivars. Journal of Dietary Supplements 12 (1):1–10.
Dallagnol, A. M., M. Pescuma, G. F. De Valdez, and G. Roll_an. 2013. Fermentation of quinoa and wheat slurries by Lactobacillus plantarum CRL 778: Proteolytic activity. Applied Microbiology and Biotechnology 97 (7):3129–40.
De Angelis, M., Gallo, G., Corbo, M.R. et al. (2003). Phytase activity in sourdough lactic acid bacteria: purification and characterization of a phytase from Lactobacillus sanfranciscensis CB1. International Journal of Food Microbiology, 87, 259–270.
de la Barca AMC, Rojas-Martínez ME, Islas-Rubio AR, Cabrera-Chávez F. Gluten-free breads and cookies of raw and popped amaranth flours with attractive technological and nutritional qualities. Plant Foods Hum Nutr 2010;65:241–6.
Deferne, J. L., and D. W. Pate. 1996. Hemp seed oil: A source of valuable essential fatty acids. Journal of the International Hemp Association 3:4–7.
Deka, D., andSarma, G.C. Traditionally used herbs in the preparation of rice-beer by the rabha tribe of Goalpara district, Assam.*Indian Journal of Traditional Knowledge*, 9(3): 459–462, 2010.
Đorđevi´c, T. M., ˇSiler-Marinkovi´c, S. S., & Dimitrijevi´c-Brankovi´c, S. I. (2010). Effect of fermentation on antioxidant properties of some cereals and pseudo cereals. *Food Chemistry, 119*(3), 957–963.
FAO (2012). Cereal Supply and Demand Brief.
Flander L, Salmenkallio-Marttila M, Suortti T, Autio K. Optimization of ingredients and baking process for improved wholemeal oat bread quality. Food Sci Technol 2007;40:860–70.
Galanakis, C.M. (2012). Recovery of high added-value components from food wastes: conventional, emerging technologies and commercialized applications. *Trends Food Sci. Technol.* 26:68–87.
Gandhi A, Dey G. Fermentation responses and in vitro radical scavenging activities of Fagopyrum esculentum. Int J Food Sci Nutr 2013;64:53–7.

Garrido-Galand, S., Asensio-Grau, A., Calvo-Lerma, J., Heredia, A., & Andrés, A. (2021). The potential of fermentation on nutritional and technological improvement of cereal and legume flours: a review. *Food Research International*, 110398.

Gobbetti M. The sourdough microflora: Interactions of lactic acid bacteria and yeasts. Trends Food Sci Technol 1998;9:267–74.

Gobbetti, M., De Angelis, M., Di Cagno, R., Polo, A., & Rizzello, C. G. (2020). The sourdough fermentation is the powerful process to exploit the potential of legumes, pseudo-cereals and milling by-products in baking industry. *Critical reviews in food science and nutrition*, *60*(13), 2158–2173.

Gobbetti M, Rizzello CG, Di Cagno R, De Angelis M. How the sourdough may affect the functional features of leavened baked goods. Food Microbiol 2014;37:30–40.

Gomez-Caravaca, A. M., G. Iafelice, V. Verardo, E. Marconi, and M. F. Caboni. 2014. Influence of pearling process on phenolic and saponin content in quinoa (Chenopodium quinoa Willd). Food Chemistry 157:174–8.

Grosu-Tudor, S.-S.; Stefan, I.-R.; Stancu, M.-M.; Cornea, C.-P.; DeVuyst, L.; Zamfir,M. Microbial and nutritional characteristics of fermented wheat bran in traditional Romanian bor, s production. *Rom. Biotechnol. Lett.* **2019**, *24*, 440–447.

Guyot JP. Cereal-based fermented foods in developing countries: ancient foods for modern research. Int J Food Sci Technol 2012;47:1109–14.

Hager, A.S., Taylor, J.P., Waters, D.M., and Arendt, E.K. Gluten free beer–A review. *Trends in Food Science & Technology*, 36(1): 44–54, 2014.

Harth H, Van Kerrebroeck S, De Vuyst L. Community dynamics and metabolite target analysis of spontaneous, backslopped barley sourdough fermentations under laboratory and bakery conditions. Int J Food Microbiol 2016;228:22–32.

Hassani, A., Procopio, S., & Becker, T. (2016). Influence of malting and lactic acid fermentation on functional bioactive components in cereal-based raw materials: a review paper. *International Journal of Food Science & Technology*, *51*(1), 14–22.

Hassani, A., Zarnkow, M. & Becker, T. (2015). Optimisation of fermentation conditions for probiotication of sorghum wort by Lactobacillu acidophilus LA5. International Journal of Food Science & Technology, 50, 2271–2279.

Hidalgo A, Brandolini A, Pompei C, Piscozzi R. Carotenoids and tocols of einkorn wheat (Triticum monococcum ssp. monococcum L.). J Cereal Sci 2006;44:182–93.

Hill, C.B., and Li, C. (2016). Genetic architecture of flowering phenology in cereals and opportunities for crop improvement. *Front. Plant Sci.* 7:1906.

Hole, A.S., Rud, I., Grimmer, S., Sigl, S., Narvhus, J., and Sahlstrøm, S. (2012). Improved bioavailability of dietary phenolic acids in whole grain barley and oat groat following fermentation with probiotic *Lactobacillus acidophilus, Lactobacillus johnsonii,* and *Lactobacillus reuteri. J. Agric. Food Chem.* 60:6369–6375.

House, J. D., J. Neufeld, and G. Leson. 2010. Evaluating the quality of protein from hempseed (Cannabis sativa L.) products through the use of the protein digestibility corrected aminoacid score method.Journal of Agricultural and Food Chemistry 58 (22):11801–7.

Hussin, A.S.M., Farouk, A.-E. & Salleh, H.M. (2010). Phytate-degrading enzyme and its potential biotechnological aplication: a review. Journal of Agrobiotechnology, 1, 1–16.

Jekle, M., A. Houben, M. Mitzscherling, and T. Becker. 2010. Effects of selected lactic acid bacteria on the characteristics of amaranth sourdough. Journal of the Science of Food and Agriculture 90 (13): 2326–32.

Johns T, Eyzaguirre PB. Biofortification, biodiversity and diet: a search for complementary applications against poverty and malnutrition. Food Policy 2007;32:1–24.

Kalscheur, K. F., Garcia, A. D., Schingoethe, D. J., Royón, F. D., and Hippen, A. R. (2012). "Feeding biofuel co-products to dairy cattle" in Biofuel Co-products as Livestock Feed Ed. H. P. S. Makkar, 115–154.

Karlund, A.; Gomez-Gallego, C.; Korhonen, J.; Palo-Oja, O.M.; El-Nezami, H.; Kolehmainen, M. Harnessing microbes for sustainable development: Food fermentation as a tool for improving the nutritional quality of alternative protein sources. *Nutrients* **2020**, *12*, 1020

Katina K, Arendt EK, Liukkonen H, Autio K, Flander L, Poutanen K. Potential of sourdough for healthier cereal products. Trends Food Sci Technol 2005;16:104–12.

Katina, K., Juvonen, R., Laitila, A., Flander, L., Nordlund, E., Kariluoto, S., ... & Poutanen, K. (2012). Fermented wheat bran as a functional ingredient in baking. *Cereal chemistry*, *89*(2), 126–134.

Kedia, G., Wang, R., Patel, H. & Pandiella, S.S. (2007). Use ofmixed cultures for the fermentation of cereal-based substrates with potential probiotic properties. Process Biochemistry, 42, 65–70.

Keuth, S. & Bisping, B. (1993). Formation of vitamins by pure cultures of tempe moulds and bacteria during the tempe solid substrate fermentation. Journal of Applied Bacteriology, 75, 427–434.

Kim, H.R., Kim, J.H., Bae, D.H., and Ahn, B.H. Characterization of Yakju brewed from glutinous rice and wild-type yeast strains isolated from Nuruks. *Journal of Microbiology and Biotechnology*, 20(20): 1702–1710, 2010.

Kim, J.H., Shoemaker, S.P., and Mills, D.A. Relaxed control of sugar utilization in *Lactobacillus brevis*. *Microbiology*, 155(4): 1351–1359, 2009.

Kim, M.-J., Hyun, J.-N., Park, J.-C., Kim, J.G., Lee, S.-J., Chun, S.-C., Chung, I.-M., 2007. Relation between phenolic compounds, anthocyanins content and antioxidant activity in colored barley germplasm. J. Agric. Food Chem. 55, 4802–4809.

Kohajdová Z, Karovičová J. Nutritional value and baking applications of spelt wheat. Acta Sci Pol Technol Aliment 2008;7:5–14.

Kort, R.; Sybesma,W. Probiotics for every body. *Trends Biotechnol.* **2012**, *30*, 613–615.

Koutinas, A.A., Vlysidis, A., Pleissner, D., Kopsahelis, N., Garcia, I.L., Kookos, I.K., et al. (2014). Valorization of industrial waste and by-product streams via fermentation for the production of chemicals and biopolymers. *Chemi. Soc. Rev.* 43:2587–2627.

Kuljanabhagavad, T., P. Thongphasuk, W. Chamulitrat, and M. Wink. 2008. Triterpene saponins from Chenopodium quinoa Willd. Phytochemistry 69 (9):1919–26.

Laine, R., Salminen, S., Benno, Y., & Ouwehand, A. C. (2003). Performance of bifidobacteria in oat-based media. *International Journal of Food Microbiology*, *83*(1), 105–109.

LeBlanc, J.G., de Giori, G.S., Smid, E.J., Hugenholtz, J. & Sesma, F. (2007). Folate production by lactic acid bacteria and other food-grade microorganisms. Communicating Current Research and Educational Topics and Trends in Applied Microbiology, 1, 329–339.

Li, D., X. Li, and X. Ding. 2010. Composition and antioxidative properties of the flavonoid-rich fractions from tartary buckwheat grains. Food Science and Biotechnology 19 (3):711–6.

Li, S, Zhang, QH. Advances in the development of functional foods from buckwheat. Crit Rev Food Sci Nutr 2001;41:451–64.

Mahmoudi, T., Oveisi, M.R., Jannat, B., Behzad, M., Hajimahmoodi, M., Sadeghi, N., 2015. Antioxidant activity of Iranian barley grain cultivars and their malts. Afr. J. Food Sci. 9, 534–539.

Matejcekova, Z., D. Lipt_akov_a, and L'. Valık. 2016. Evaluation of the potential of amaranth flour for lactic acid fermentation. Journal of Pharmacy and Nutrition Sciences 6 (1):1–6.

Michida, H., Tamalampudi, S., Pandiella, S.S., Webb, C., Fukuda, H. & Kondo, A. (2006). Effect of cereal extracts and cereal fiber on viability of Lactobacillus plantarum under gastrointestinal tract conditions. Biochemical Engineering Journal, 28, 73–78.

Mora-Uzeta, C., Cuevas-Rodriguez, E., Lopez-Cervantes, J., Mil´an-Carrillo, J., Guti´errez Dorado, R., & Reyes Moreno, C. (2020). Improvement nutritional/antioxidant properties of underutilized legume tepary bean (Phaseolus Acutifolius) by solid state fermentation. *Agrociencia, 53*, 987–1003.

Moroni AV, Zannini E, Sensidoni G, Arendt EK. Exploitation of buckwheat sourdough for the production of wheat bread. Eur Food Res Technol 2012;235:659–68.

Newman CW, Newman RK. A brief history of barley foods. Cereal Foods World 2006;51:4–7.

Nionelli, L., M. Montemurro, E. Pontonio, M. Verni, M. Gobbetti, and C. G. Rizzello. 2018. Pro-technological and functional characterization of lactic acid bacteria to be used as starters for hemp (Cannabis sativa L.) sourdough fermentation and wheat bread fortification. International Journal of Food Microbiology 279:14–25.

Nkhata, S. G., Ayua, E., Kamau, E. H., & Shingiro, J. B. (2018). Fermentation and germination improve nutritional value of cereals and legumes through activation of endogenous enzymes. *Food science & nutrition*, *6*(8), 2446–2458.

Noha, A, Mohammed IA, Mohamed A, Elfadil EB. Nutritional evaluation of sorghum flour (Sorghum bicolor L. Moench) during processing of Injera. World Acad Sci Int J Nutr Food Eng 2011;51:99–103.

Nout, MJR. Rich nutrition from the poorest – Cereal fermentations in Africa and Asia. Food Microbiol 2009;26:685–92.

Nsogning Dongmo, S., Procopio, S., Sacher, B., Becker, T., 2016. Flavor of lactic acid fermented malt based beverages: current status and perspectives. Trends Food Sci. and Technol. 54, 37–51.

O'Connor, E., Barrett, E., Fitzgerald, G., Hill, C., Stanton, C. & Ross, R. (2005). Production of vitamins, exopolysaccharides and bacteriocins by probiotic bacteria. Probiotic Dairy Products, 167–194.

Olusegun, O. L. 1983. Handbook of tropical foods. In Food science and technology, ed. H. T. Chan, Jr. 1st ed., 1–28. New York, NY: Marcel Dekker Inc.

Onyango, C., E. A. Mewa, A. W. Mutahi, and M. W. Okoth. 2013. Effect of heat-moisture-treated cassava starch and amaranth malt on the quality of sorghum-cassava-amaranth bread.

Osman MA. Changes in sorghum enzyme inhibitors phytic acid, tannins and in vitro protein digestibility occurring during Khamir (local bread) fermentation. Food Chem 2004;88:129–34.

Pasqualone, A., & Summo, C. (2021). Qualitative and Nutritional Improvement of Cereal-Based Foods and Beverages.

Patel, A., Shah, N. & Prajapati, J. (2013). Biosynthesis of vitamins and enzymes in fermented foods by lactic acid bacteria and related genera-A promising approach. Croatian Journal of Food Science and Technology, 5, 85–91.

Patel, H.M., Pandiella, S.S., Wang, R.H. & Webb, C. (2004). Influence of malt, wheat, and barley extracts on the bile tolerance of selected strains of lactobacilli. Food Microbiology, 21, 83–89.

Patel, S. (2012). Cereal bran: the next super food with significant antioxidant and anticancer potential. *Med. J. Nutrition Metab.* 5:91–104.

Petrova, P., & Petrov, K. (2020). Lactic acid fermentation of cereals and pseudocereals: Ancient nutritional biotechnologies with modern applications. *Nutrients*, *12*(4), 1118.

Phiri, S.; Schoustra, S.E.; van den Heuvel, J.; Smid, E.J.; Shindano, J.; Linnemann, A. Fermented cereal-based Munkoyo beverage: Processing practices, microbial diversity and aroma compounds. *PLoS ONE* **2019**, *14*, e0223501.

Pontonio, E., & Rizzello, C. G. (2019). Minor and ancient cereals: Exploitation of the nutritional potential through the use of selected starters and sourdough fermentation. In *Flour and Breads and Their Fortification in Health and Disease Prevention* (pp. 443–452). Academic Press.

Poutanen, K., Sozer, N., and Della Valle, G. (2014). How can technology help to deliver more of 777 grain in cereal foods for a healthy diet?. *J. Cer. Sci.* 59:327–336.

Prado, F.C., Parada, J.L., Pandey, A. & Soccol, C.R. (2008). Trends in non-dairy probiotic beverages. Food Research International, 41,111–123.

Prajapati JB, Nair BM. The history of fermented foods. In: Farnworth ER, editor. Handbook of fermented functional foods. Boca Raton: CRC Press; 2008. p. 2–22.

Pranoto, Y., Anggrahini, S., & Efendi, Z. (2013). Effect of natural and Lactobacillus plantarum fermentation on in-vitro protein and starch digestibilities of sorghum flour. *Food Bioscience, 2*, 46–52

Rani, P., Kumar, A., Purohit, S. R., & Rao, P. S. (2018). Impact of fermentation and extrusion processing on physicochemical, sensory and bioactive properties of rice-black gram mixed flour. *LWT – Food Science and Technology, 89*, 155–163.

Rathore, S., Salmerón, I., & Pandiella, S. S. (2012). Production of potentially probiotic beverages using single and mixed cereal substrates fermented with lactic acid bacteria cultures. *Food Microbiology*, *30*(1), 239–244.

Ravindran, R., and Jaiswal, A.K. (2016). Exploitation of food industry waste for high-value products. *Trends Biotechnol.* 34:58–69.

Rivera-Espinoza, Y. & Gallardo-Navarro, Y. (2010). Non-dairy probiotic products. Food Microbiology, 27, 1–11.

Rizzello, C. G., Coda, R., Mazzacane, F., Minervini, D., & Gobbetti, M. (2012). Micronized by-products from debranned durum wheat and sourdough fermentation enhanced the nutritional, textural and sensory features of bread. *Food Research International*, *46*(1), 304–313.

Roy, B., Kala, C.P., Farooquee, N.A., and Majila, B.S. Indigenous Fermented food and beverages: a potential for economic development of the high altitude societies in Uttaranchal. *Journal of Human Ecology*, 15(1): 45–49, 2004.

Rozyło, R., S. Rudy, A. Krzykowski, and D. Dziki. 2015. Novel application of freeze-dried amaranth sourdough in gluten-free bread production. Journal of Food Process Engineering 38 (2):135–43.

Russo, R., and R. Reggiani. 2015. Evaluation of protein concentration, amino acid profile and antinutritional compounds in hempseed meal from dioecious and monoecious varieties. American Journal of Plant Sciences 06 (01):14–22.

S´anchez Maga˜na, L., Reyes Moreno, C., Mil´an-Carrillo, J., Mora Rochin, S., Le´on- L´opez, L., Guti´errez Dorado, R., & Cuevas-Rodriguez, E. (2019). Influence of solid-state bioconversion by rhizopusoligosporus on antioxidant activity and phenolic compounds Of Maize (Zea Mays L.). *Agrociencia, 53*, 45–57.

Saharan, P., Sadh, P. K., & Singh Duhan, J. (2017). Comparative assessment of effect of fermentation on phenolics, flavanoids and free radical scavenging activity of commonly used cereals. *Biocatalysis and Agricultural Biotechnology, 12*, 236–240.

Sahin, A. W., Coffey, A., & Zannini, E. (2021). Functionalisation of wheat and oat bran using single-strain fermentation and its impact on techno-functional and nutritional properties of biscuits. *European Food Research and Technology*, 1–13.

Salmenkallio-Marttila, M., Katina, K., & Autio, K. (2001). Effects of bran fermentation on quality and microstructure of high-fiber wheat bread. *Cereal Chemistry*, *78*(4), 429–435.

Salmeron, I., Fuciños, P., Charalampopoulos, D., & Pandiella, S. S. (2009). Volatile compounds produced by the probiotic strain Lactobacillus plantarum NCIMB 8826 in cereal-based substrates. *Food Chemistry*, *117*(2), 265–271.

Samota, M.K., Sasi, M., Awana, M., Yadav, O.P., Amitha Mithra, S.V., Tyagi, A., et al. (2017). Elicitor-induced biochemical and molecular manifestations to improve drought tolerance in rice (*Oryza sativa* L.) through seed-priming. *Front. Plant Sci.* 8:934.

Sanlier, N.; Gokcen, B.B.; Sezgin, A.C. Health benefits of fermented foods. *Crit. Rev. Food Sci. Nutr.* **2019**, *59*, 506–527.

Saturni, L., G. Ferretti, and T. Bacchetti. 2010. The gluten-free diet: Safety and nutritional quality. Nutrients 2 (1):16–34.

Schoenlechner, R., I. Mandala, A. Kiskini, A. Kostaropoulos, and E. Berghofer. 2010. Effect of water, albumen and fat on the quality of gluten-free bread containing amaranth. International Journal of Food Science & Technology 45 (4):661–9

Setta, M.C.; Matemu, A.; Mbega, E.R. Potential of probiotics from fermented cereal-based beverages in improving health of poor people in Africa. *J. Food Sci. Technol.* **2020**.

Silva-S_anchez, C., J. Gonz_alez-Casta~neda, A. De Le_on-Rodr_ıguez, and A. P. Barba de la Rosa. 2004. Functional and rheological properties of amaranth albumins extracted from two Mexican varieties. Plant Foods for Human Nutrition 59:169–74.

Singh, P.K., and Singh, K.I. Traditional alcoholic beverage, Yu of Meitei communities of Manipur, *Indian J. Traditional Knowledge*, 5(2): 184–190, 2006.

Stanton, C., Ross, R.P., Fitzgerald, G.F. & Sinderen, D.V. (2005). Fermented functional foods based on probiotics and their biogenic metabolites. Current Opinion in Biotechnology, 16, 198–203.

Sterr, Y., A. Weiss, and H. Schmidt. 2009. Evaluation of lactic acid bacteria for sourdough fermentation of amaranth. International Journal of Food Microbiology 136 (1):75–82

Stikic, R., D. Glamoclija, M. Demin, B. Vucelic-Radovic, Z. Jovanovic, D. Milojkovic-Opsenica, S. E. Jacobsen, and M. Milovanovic. 2012. Agronomical and nutritional evaluation of quinoa seeds (Chenopodium quinoa Willd.) as an ingredient in bread formulations. Journal of Cereal Science 55 (2):132–8.

Stoffel, F., Santana, W. D. O., Fontana, R. C., Gregolon, J. G. N., Kist, T. B. L., De Siqueira, F. G., Mendonça, S., & Camassola, M. (2019). Chemical features and bioactivity of grain flours colonized by macrofungi as a strategy for nutritional enrichment. *Food Chemistry, 297*, 124988

Tamang, J.P., and Thapa, S. Fermentation dynamics during production of *bhaati jaanr*, a traditional fermented rice beverage of the Eastern Himalayas. *Food Biotechnology*, 20(3): 251–261, 2006.

Thapa, S., and Tamang, J.P. Product characterization of *kodo ko jaanr*: fermented finger millet beverage of the Himalayas. *Food Microbiology*, 21: 617–622, 2004.

Tolun, A.; Altintas, Z. Medicinal Properties and Functional Components of Beverages. In *Functional and Medicinal Beverages*; Elsevier Inc.: Amsterdam, The Netherlands, 2019

Tomoskozi S. Ancient wheats and pseudocereals for possible use in cereal-grain dietary intolerances. In: B_ ek_es F, Schoenlechner R, Wrigley C, Batey I, Miskelly D, editors. Cereal grains. 2nd ed. Woodhead Publishing; 2016. p. 353–78.

Tsuyoshi, N., Fudou, R., Yamanaka, S., Kozaki, M., Tamang, N., Thapa, S., and Tamang, J.P. Identification of yeast strains isolated from marcha in Sikkim, a microbial starter for amylolytic fermentation. *International Journal of Food Microbiology*, 99(2): 135–146, 2005.

Varnam, A.H., and Sutherland, J.P. *Beverages: Technology, Chemistry and Microbiology, Vol 2*, Chapman and Hall, London, 1994.

Venskutonis PR, Kraujalis P. Nutritional components of amaranth seeds and vegetables: a review on composition, properties, and uses. Compr Rev Food Sci Food Saf 2013;12:381–412.

Wang, K., Niu, M., Song, D., Liu, Y., Wu, Y., Zhao, J., ... Lu, B. (2019). Evaluation of biochemical and antioxidant dynamics during the co-fermentation of dehusked barley with Rhizopus oryzae and Lactobacillus plantarum. *Journal of Food Biochemistry, 44*(2)

Wolter, A., A. Hager, E. Zannini, M. Czerny, and E. K. Arendt. 2014. Influence of dextran-producing Weissella cibaria on baking properties and sensory profile of gluten-free and wheat breads. International Journal of Food Microbiology 172:83–9.

Wong, J.H., Lau, T., Cai, N., Singh, J., Vensel, P.J.F., Vensel, W.H., Hurkman, W.J., Wilson, J.D., Lemauxa, P.G., and Buchanan, B.B. 2009. Digestibility of protein and starch from sorgum (Sorghum bicolor) is linked to biochemical and structural features of grain endosperm. Journal of Cereal Science 49, 73–82.

Wronkowska, M., M. Haros, and M. Soral-_Smietana. 2013. Effect of starch substitution by buckwheat flour on gluten-free bread quality. Food and Bioprocess Technology 6 (7), 1820–7

Wu, H., Rui, X., Li, W., Xiao, Y., Zhou, J., & Dong, M. (2018). Whole-grain oats (Avena sativa L.) as a carrier of lactic acid bacteria and a supplement rich in angiotensin I-converting enzyme inhibitory peptides through solid-state fermentation. *Food & Function, 9*(4), 2270–2281.

Xiao, Y., Rui, X., Xing, G., Wu, H., Li, W., Chen, X., ... Dong, M. (2015). Solid state fermentation with Cordyceps militaris SN-18 enhanced antioxidant capacity and DNA damage protective effect of oats (Avena sativa L.). *Journal of Functional Foods, 16*, 58–73.

Xu, L. N., Guo, S., & Zhang, S. W. (2019). Effects of solid-state fermentation on the nutritional components and antioxidant properties from quinoa. *Emirates Journal of Food and Agriculture, 31*(1), 39–45.

Yan Y, Hsam SLK, Yu JZ, Jiang Y, Ohtsuka I, Zeller FJ.HMWandLMWglutenin alleles among putative tetraploid and hexaploid European spelt wheat (Triticum spelta L.) progenitors. Theor Appl Genet 2003;107:1321–30.

Zarnkow, M., Almaguer, C., Burberg, F., Back, W., Arendt, E.K., Gastl, M., 2008. The use of response surface methodology to optimise malting conditions of Tef (Zucc.) Trotter as a raw material for gluten-free foods and beverages. Brew. Sci. Monatsschr. Brauwiss 61, 94–104.

CHAPTER 8

Soaking of Cereals

Vandana Yalakki and Arya S.S.
Food Engineering and Technology Department, Institute of Chemical Technology, NM Parikh Marg, Matunga, Mumbai, Maharashtra, India

CONTENTS

DOI: 10.1201/9781003242192-11

8.1 INTRODUCTION

Cereals are grains or edible seeds of the *Gramineae*/grass family. Cereals are cultivated worldwide and include wheat, rice, maize, barley, millets, triticale, sorghum, etc. Among them, rice and wheat are the most cultivated ones (50% of cereal production in the world). In general, the majority of cereals contain a germ/embryo with the genetic materials required for a new plant to grow and an endosperm, which is mostly filled with starch. Cereals are the main staple foods in the human diet and nourish the body with macro- and micronutrients. They are enriched with macronutrients such as carbohydrate, energy, fibre, and protein and minor nutrients such as B-complex vitamins, vitamin E, zinc, and magnesium. They can even be fortified with vitamins and minerals such as iron and calcium. Apart from these, cereals are also good sources of bioactive constituents, which provide additional health benefits and that are gaining in popularity. The consumption of cereals on a regular basis, especially consumption of whole grain cereals, will help the body to prevent lifestyle-related disorders such as diabetes mellitus, high blood pressure, high cholesterol, and some diseases like cancer and coronary heart disease (McKevith, 2004).

Cereals (including millets) are good sources of not only major and minor nutrients but also anti-nutrients such as tannins, saponins, phytic acids, lectins, amylase inhibitors and protease inhibitors, and goitrogens. The main problem with them is they bind with nutrients and hinder their bioavailability. Thus, in order to reduce the anti-nutrients or reduce their effects, traditional processing techniques such as soaking, germination, autoclaving, fermentation, debranning, etc., are used (Jaybhaye & Srivastav, 2015; Handa et al., 2017; Samtiya et al., 2020). Soaking is the most common process among these.

Soaking is a very important process that involves keeping the grains in water with or without additives. Soaking is also a prior step to germination, fermentation, parboiling, wet milling of grains, etc., which causes hydration and water uptake through the grains, facilitates germination, affects starch content, causes changes in protein, lipids, micronutrients, bioactives, phytochemicals, anti-nutrients, starch and protein digestibility, and also affects rheological, thermal, surface, and structural properties of the grains. Hence, this chapter thoroughly explains the soaking process of cereals and the changes in the nutritional, functional, and biological properties of the cereals after soaking.

8.2 PROCESSING TREATMENT

8.2.1 Fundamentals of the Treatments

Soaking or hydration is a process of immersing grains in water to increase their water content or to hydrate the grains. The process of soaking activates gibberellic acid synthesis, which is crucial for germination. Soaking is a critical step in food industries before any processing and has a great impact on nutritional and physicochemical parameters of the grains. Hydration or soaking is a required process before extraction, cooking, germination, fermentation, and malting (Miano & Augusto, 2018).

Usually for the soaking of the cereals and millet grains, the grains should be thoroughly washed and cleaned of foreign matter. After that, they should be immersed in distilled water or normal water or any other soaking medium with or without additives according to requirements (Wu et al., 2013). The soaking medium can be at room temperature or higher temperature and at atmospheric pressure or high pressure or it can also be subjected to vacuum. The water used for soaking should have suitable chemical composition and pH in order to serve the purpose of the further process for value addition of grains to suitable food products (Serna-Saldivar, 2016). After soaking of the grains for required period of time and required moisture content, the water is drained and the grains can be further subjected to different processes (Saleh et al., 2018). When the grains are soaked, the part called "hilum" present in the germ layer takes up the water first and the water absorption is initially very fast and slows down as it reaches saturation (Serna-Saldivar, 2016).

8.2.2 Process and Machinery Involved

Soaking is a process that requires simple equipment. Usually for the soaking of cereals and millet grains, the grains should be thoroughly washed and cleaned of foreign matter. After that, they should be immersed in distilled water or normal water or any other soaking medium or sprayed by water with or without additives for about 24 to 72 hours according to the requirements (Wu et al., 2013). The moisture content of the grains increase to about 42–48%.

The main equipment involved are the vertical cylinders or tanks with cone bases. The air incorporation takes place through the perforated pipelines in the containers. Temperature of the water is very crucial as it affects the microbiological aspects and rate of hydration. The water temperature can be between 10 and 20°C but preferably 15°C. If the temperature is a bit higher, the microbial growth will be higher and the availability of oxygen for the developing embryo lower. Along with temperature, regular supply of air is also crucial as the air will carry the heat created through respiration and CO_2 as well as supplies oxygen. The soaking water is drained after 12–24 hours, and may contain soluble solids liberated by grains that are broken and new fresh water will be incorporated (Serna-Saldivar, 2016).

8.3 IMPACT OF SOAKING TREATMENT ON NUTRITIONAL CHARACTERISTICS

8.3.1 Starch

Starch is the storage polysaccharide in plants that is the reservoir of food. It accounts for about two thirds of caloric intake of carbohydrates in the majority of humans. The majority of starches are obtained commercially from different cereals such as rice, wheat, corn (waxy corn & high amylose corn), as well as roots or tubers like tapioca, taro, sweet potato, potato, etc. (Whistler & Daniel, 2000). But starch is rarely consumed in raw form because it is difficult for the enzymes involved in digestion in humans to digest the crystalline structure of starch (Alcázar-Alay & Meireles, 2015; Alvarez-Ramírez et al., 2019). Therefore, cooking has to be done to make the starch more digestible by facilitating enzymatic digestion in the gastrointestinal tract so that more glucose is released into bloodstream (Alvarez-Ramírez et al., 2019).

Cooking of starch is an energy demanding process and is usually assisted by the presence of a lot of water to partially or fully rupture the crystalline starch. This process leads to the leached amylose entering the intergranular space (Conde-Petit et al., 2001; Alvarez-Ramírez et al., 2019). Thus, to minimize the energy consumption, soaking is commonly used to cause hydration of the starch by adding moisture content into the cereal grains, so that it can help to solubilize the chains of starch. Soaking is a part of wet milling of cereal grains.

Rocha-Villarreal et al. (2018) reviewed the changes in physical, nutritional, and functional characteristics of rice as well as other cereals after parboiling and other hydrothermal treatments. Parboiling involves three major steps: soaking/hydration, steaming, and drying. In the step of soaking, the grains are soaked in water at room temperature for a certain time or preheated water is used to increase absorption/hydration rate. Therefore, soaking in water is very important for gelatinization (Luh & Mickus, 1991; Rocha-Villarreal et al., 2018). Hot water is very useful in breaking hydrogen bonds between amylose and amylopectin, so that there is an increase in solubility and surface absorption, which is very difficult with cold water (Rocha-Villarreal et al., 2018). It has been found that soaking at higher temperature contributes to an increase in hardness value of rice kernels (Mir & Bosco, 2013; Rocha-Villarreal et al., 2018). Due to soaking in preheated water followed by steaming, the axial dimensions of the grains will increase as a result of swelling in starch/gelatinization of starch caused by hydrothermal treatment. But it was found for parboiled sorghum that the grain dimensions were decreased (Young et al., 1993). Changes in sensory properties like aroma and flavours were also connected with soaking parameters like soaking time and temperature of water (Bello et al., 2015; Rocha-Villarreal et al., 2018).

Soaking plays an important role in wet milling of cereals. For producing cereal flours, such as rice flour, wet milling, dry milling, and semi-dry milling have been used out of which wet milling is known for producing flour with higher quality for baking applications (Bean et al., 1983; Chiang & Yeh, 2002). It was observed by Preston et al. (1987) and Chiang & Yeh (2002) that the water uptake by the wheat kernel during soaking resulted in softening of the same and decreased damaged starch content in it, which in turn depend on the temperature. However, soaking did not significantly affect the gelatinization temperature according to the results obtained by Chiang & Yeh (2002).

Kale et al. (2015) studied changes in starch properties, chemical composition, and glycaemic index of basmati rice after soaking treatment. When rice is soaked, its constituents are leached into water (Otegbayo et al., 2001; Ibukun, 2008; Sareepuang et al., 2008). The starch decreases in amount (amylose content) after soaking in hot water as a result of leaching. Also, the crystallinity of starch, which is determined by amylose, can be reduced as a result of hot water soaking (Kale et al., 2015). Similarly, the amylose-to-amylopectin ratio, which is an indicator of starch digestibility, is reduced. SEM revealed the increased swelling of rice after soaking in hot water due to partial gelatinization of the rice (Kale et al., 2015). When the rice grains (both waxy and normal) were subjected to high hydrostatic pressure (HHP) soaking, the crystallinity of starch was reduced (Hu et al., 2011; Tian et al., 2014). The retrogradation was lower in the normal rice subjected to HHP soaking compared to that of the rice soaked in atmospheric pressure because of the lower amylose leakage in HHP soaked rice, resulting in formation of a much thinner film on the surface of the grains reducing the quick retrogradation (Stolt et al., 2001; Tian et al., 2014).

8.3.2 Protein

Proteins are one of the most important macronutrients and are essential for the human body. These are the polymers of building blocks called amino acids, which are joined through peptide bonds. Proteins present in food are crucial for sensory, textural, nutritional as well as some functional properties of food such as emulsification, gelling, foaming ability, fat binding capacity, etc. (Zayas, 2012). Protein-rich foods include meat and poultry products, fish, egg, dairy products as well as cereal grains, millets, and pulses. Cereals are considerable sources of proteins, and millets, which are small-seeded cereals, are are rich in protein content as well as other nutrients. But the main problem with nutrition of millets is the presence of anti-nutrients that hinder the bioaccessibility and bioavailability of protein and minerals such as iron, zinc, etc. (Yousaf et al., 2021). Thus, in order to reduce the adverse effects of anti-nutrients, processing techniques such as soaking, germination, malting, fermentation, boiling, etc., are used (Sheela et al., 2018; Yousaf et al., 2021).

Soaking is an important processing technique used to hydrate grains. Soaking has an important role in reducing anti-nutritional factors when combined with germination, which results in activation of hydrolytic enzymes, exo- and endopeptidases along with increase in protein, minerals, vitamins, and unsaturated fatty acids (Dicko et al., 2006; Inyang and Zakari, 2008; Yousaf et al., 2021). Little and barnyard millets when soaked have an increase in protein content because the longer duration of soaking increases the non-protein nitrogen content. Also, pearl millet when subjected to soaking followed by germination has increased protein content as well protein digestibility compared to non-treated grains (Hassan et al., 2006). In the case of parboiling of millets (which involves soaking as the first step), the protein extractability is decreased (especially for prolamin and globulin), but the digestibility of proteins is increased after parboiling (Dharmaraj and Malleshi, 2011; Rocha-Villarreal et al., 2018). El-Safy & Salem (2013) studied the impact of soaking on different physico-chemical properties and functional properties of cereals and legumes, and found that soaking and sprouting significantly increased the protein content.

8.3.3 Lipids

Lipids are very important macromolecules that have concentrated sources of energy. They include diverse organic compounds such as fats, oils, wax, hormones, steroids, etc. Lipids have various functions in the body such as storage of energy, thermal insulation, membrane functions, message transmission, etc. (Thompson, 2020). Lipids are made up of three fatty acids and a glycerol moiety (Ahmed, 2020). Lipids are present in large quantities in foods like meat and poultry products, milk & dairy products, and oils present in various oilseeds, such as safflower, sunflower, corn, other cereals, etc. Cereals contain lipids (mostly having unsaturated fatty acids) mainly in their germ part, but also some of them are spread throughout the endosperm. Lipids in cereals are responsible for important properties like perfect bread production by wheat flour, the colour of pasta products, storage stability of cereals as well as oatmeal flavour (Barnes, 1984).

During cereal processing, the composition of cereal grains also changes including changes in lipid content. During soaking treatment of cereals, followed by germination there might be changes in lipid content, especially decreases in fat content due to liberation of lipolytic enzymes/disruption of fat due to β-oxidation (El-Safy & Salem, 2013). Soaking can cause diffusion of globules fat from layers of bran and husk into the endosperm inside during parboiling of basmati rice (Otegbayo et al., 2001; Dutta & Mahanta, 2012; Kale et al., 2015). At higher soaking temperatures, the crude fat is reduced due to diffusion and rupture of fat droplets (Otegbayo et al., 2001; Kale et al., 2015), which is in line with the observation by Sareepuang et al. (2008) in the case of aromatic rice. However, in the case of polished soaked rice, the crude fat content was higher than the non-polished one, which was as a result of diffusion of droplets of fat into the endosperm (Kale et al., 2015). Ocheme & Chinma (2008) studied the changes in physicochemical properties of millet flour subjected to soaking and germination. They found that the fat content was decreased considerably after soaking and germination.

8.3.4 Micronutrients

Vitamins and minerals are commonly called micronutrients, which are essential in small quantities for body maintenance. They perform various functions in the body such as enzyme production and production of hormones and other substances, which are crucial for growth and development of the body. Although they are required in minimal quantities, deficiency of micronutrients cause diseases and some serious health conditions as stated by the WHO. Zinc, vitamin A, and iron deficiency are the most commonly found serious illnesses throughout the world (Gupta et al., 2015).

Cereal grains contain a variety of micronutrients such as iron, zinc, vitamins, etc. When subjected to different processing conditions, there will be changes in the micronutrients. During soaking of cereals, minerals commonly get leached into the soaking medium, resulting in reduction of these minerals (Saharan et al., 2001; Lestienne et al., 2005). When the cereals and legumes are soaked, the iron content of the grains has been found to be considerably lower when compared with untreated ones. The highest reduction of iron content has been found in rice whereas maize showed the lowest reduction. In the case of millets, rice and maize, zinc quantity has also been shown to decrease but lower than that of iron content, which might be due to the difference in locations of zinc and iron or the difference in bound molecules (Lestienne et al., 2005). It was also observed that the zinc leaching was higher in the case of cereals whose hull part contained protein layers (aleurone part), while in the case of legumes zinc leaching was not found. The possible reason for this is because zinc is present in many proteins and enzymes where it plays important structural roles (Pernollet, 1978; Lestienne et al., 2005).

It is obvious that the parboiling (which involves soaking too) considerably changes the minerals and water-soluble vitamin content of cereal grains. The soaking of cereal grains followed by parboiling results in diffusion of micronutrients into the endosperm from the aleurone layer and bran, which include increases in thiamine, calcium, riboflavin, and phosphorus (Bello et al., 2015; Rocha-Villarreal et al., 2018). However, soaking followed by steaming of oat grain does not alter the calcium, phosphorus, and iron (Rocha-Villarreal et al., 2018), and in the case of finger millet, hydrothermal treatment results in decreased mineral content, especially phosphorus and copper, without affecting zinc, iron, and calcium (Dharmaraj & Malleshi, 2011; Rocha-Villarreal et al., 2018). Soaking can change the mineral and ash composition of rice during parboiling where the minerals are leached into the soaking water when paddy is soaked in water. A study by Kale et al. (2015) showed that the soaking temperature affects the mineral composition where the mineral content (including P, K, Mn, Fe, S, Mg, etc.) is decreased with increase in temperature of soaking medium of PB1121 variety rice. However, brown rice when polished after soaking has greater amounts of minerals compared to directly polished ones. The ash content is reduced with soaking and germination of some cereals and legumes (El-Safy & Salem, 2013). Table 8.1 shows the changes in different properties of cereal grains when subjected to soaking.

Table 8.1 Changes in Different Properties of the Cereal Grains When Subjected to Soaking

Properties/ constituents	Changes after soaking	References
Starch	Hydration and solubilization of starch chains, enhanced gelatinization/swelling, mellowing of endosperm, reduced crystallinity & retrogradation (HHP soaking)	Chiang & Yeh, (2002), Manful et al. (2008), Hu et al. (2011), Tian et al. (2014), Kale et al. (2015), Rocha-Villarreal et al. (2018)
Protein	Increase in protein content, increased non-protein nitrogen	Adnan & Shafie, Dicko et al. (2006), Hassan et al. (2006), Inyang and Zakari, (2008), El-Safy & Salem, (2013), Yousaf et al. (2021)
Lipids	Decrease in fat content, diffusion of fat from bran to husk (parboiling)	Otegbayo *et al*, (2001), Ocheme & Chinma (2008), Dutta & Mahanta, (2012), El-Safy & Salem, (2013), Kale et al. (2015)
Micronutrients	Leaching of minerals, reduced mineral content such as iron & zinc, increase in thiamine, calcium, riboflavin, and phosphorus (after parboiling)	Saharan et al. (2001), Lestienne et al. (2005), Dharmaraj & Malleshi, (2011), El-Safy & Salem, (2013), Bello et al. (2015), Kale et al. (2015), Rocha-Villarreal et al. (2018)
Hydration properties	Activation of enzymes, homogeneous protein denaturation, uniform starch gelatinization, improved starch extraction, reduced energy consumption, increased moisture	Sefa-Dedeh and Stanley (1979), Hu et al. (2011), Tian et al. (2014), Wood, (2016), Yu et al. (2017), Miano & Augusto, (2018), Alvarez-Ramírez et al. (2019)
Surface properties	Increase in surface area and volume, loss in integrity of the aleurone layer	Bolaji et al. (2017), Yu et al. (2017)
Rheological properties	More viscoelastic nature of cooked rice, reduction in hardness after high-temperature soaking, reduction in the peak viscosity, setback viscosity, and final viscosity	Hu et al. (2011), Tian et al. (2014), Kale et al. (2015), Zhu et al. (2019)

Table 8.1 (Continued)

Properties/ constituents	Changes after soaking	References
Colour profile	Lower L^* values and higher a^* and b^* values (parboiling), increase in yellowness of endosperm due to diffusion from bran, increase in lightness (HHP & vacuum soaking)	Pillaiyar & Mohandas (1991), Kimura et al. (1993), Bhattacharya, (1996), Lamberts et al. (2006), Tian et al. (2014), Yu et al. (2017), Rocha-Villarreal et al. (2018), Saleh et al. (2018)
Thermal properties	The higher the degree of hydration/moisture content in the grain, the lower is the ΔH, T_0 and T_c, greater enthalpy as soaking temperature decreased	Han et al. (2009), Loubes et al. (2012)
Structural properties	Prevention of damage in microstructure of the grain with higher structure of network with greater cohesiveness and springiness, reduction in hole/cavity size (HP soaking), production of uniform channels between connection parts and the molecules of starch, grains became translucent, darker, and glassy state (parboiling), increase in crystallinity of starch	Young et al. (1990), Ushakumari, (2009), Tian et al. (2014), Rocha-Villarreal et al. (2018)
Anti-nutrients	Reduction in anti-nutritional factors such as saponins, polyphenols, oxalates and phytates, trypsin inhibitors, etc., increase in nutrient accessibility and bioavailability and in-vitro solubility	Lestienne et al. (2005), Hotz and Gibson, (2007), Gupta et al. (2015), Kumari (2018), Rathore et al. (2019), Samtiya et al. (2020)
Phytochemical profile and bioactives	Decrease in phenolic compounds, leaching of phenolics into water, decrease in proanthocyanins and anthocyanins, but, antioxidants such as tocopherols, tocotrienols, and γ-oryzanol were increased, which are lipophilic (parboiling)	Hurst et al. (2009), Fasahat et al. (2013), Min et al. 2014; Rocha-Villarreal et al. (2018), Owolabi et al. (2019)
Starch and protein digestibility	Increase in starch digestibility due to hydration, decreased retrogradation, improved protein digestibility	Mwikya et al. (2000), Alonso et al. (2000), Albarracín et al. (2013) Mbithi – Osman & Gassem, (2013), Meng et al. (2018), Rocha-Villarreal et al. (2018), Alvarez-Ramírez et al. (2019), Hasan et al. (2019), Samuel & Peerkhan, (2020); Yousaf et al. (2021)

8.4 IMPACT OF PROCESSING TREATMENT ON FUNCTIONAL CHARACTERISTICS

8.4.1 Hydration Properties

Hydration can be described as a process of increasing the moisture content of grains by soaking in water. This is a crucial step before any processing and has a great impact on nutritional and physicochemical parameters of grains. Hydration is a mandatory process before extraction, cooking, germination, fermentation, and malting (Miano & Augusto, 2018).

Hydration enhances the cell wall enzyme activation, homogeneous protein denaturation while cooking, and uniform starch gelatinization resulting in uniform texture throughout the grain (Miano & Augusto, 2018). The cooking time can be reduced by enhanced heat transfer through absorbed water and improved reduction of anti-nutrients (Sefa-Dedeh and Stanley 1979; Miano & Augusto, 2018).

The hydration/steeping can enhance the extraction of substances from the grains. For example, starch can be extracted from cereals through wet-milling process. Hydration helps in mellowing the grains, enhancing wet-grinding process, and purification of starch (Singh and Eckhoff 1996; Miano & Augusto, 2018). Hydration is essential for fermentation as the microbes require higher water activity for their growth.

Miano & Augusto (2018) explained the hydration pathway of cereal grains. In cereals, usually the hydration pathway follows two steps with a Downward Concave Shape (DCS). In the first step, capillary flow is predominant with greater rate of hydration due to the transfer of water through capillary into the empty spaces of the grains. In the second stage, the hydration rate is lower due to the diffusion mechanism through the bran layer throughout the surface, and the hydration rate falls to zero when it reaches the equilibrium moisture content. When Japonica rice was subjected to vacuum soaking, the water absorption was increased due to vacuum and the density as well as the mobility of the protons of the water internal to the rice kernel were affected while soaking (Li et al., 2021). Soaking/hydration can result in reduced energy consumption of starch containing food matrices (Alvarez-Ramírez et al., 2019). It was found that the soaking of the rice when combined with high hydrostatic pressure resulted in more increase in moisture content compared to rice soaked at atmospheric pressure as the high pressure forces more water molecules into grains from the surface/periphery of the rice (Hu et al., 2011; Tian et al., 2014; Yu et al., 2017). Also, the temperature of the soaking water affects the hydration rate into the grains, which was clearly indicated in some studies where the higher the temperature of the soaking water the better the absorption of moisture into the grains (Thakur & Gupta, 2006; Han & Lim, 2009; Yu et al., 2017).

8.4.2 Surface Properties

The soaking process of cereals also considerably affects the surface properties of the grains. When maize grains were soaked before the preparation of ogi, it was found that the surface area as well as the volume of the grains were increased significantly, compared to the non-soaked ones (Bolaji et al., 2017). Usually, brown rice at its surface consists of an episperm, a pericarp, and an aleurone layer (Bhatnagar, 2014; Yu et al., 2017). But after soaking and high-pressure treatments, it was evident from SEM (Scanning Electron Microscopy) that there was a loss in integrity of the aleurone layer exposing the gaps between the different aleurone layers. The gap/cavities were enlarged with increase in pressure concealing the boundaries in between the aleurone layers. Hence, this surface damage caused resulted in elevated uptake of water, which in turn facilitated the cooking process thereafter (Yu et al., 2017).

8.4.3 Rheological Properties

Rheology can be called the science that describes materials' flow and deformation behaviour. The relationship among time, strain, and stress is measured in the rheology. It studies the meshing between food interactions between macromolecules such as gliadin-glutenin interactions and the sensory feel or perception, which is objectively measured and then correlated (AACC international). In rheology, food can be categorized into elastic, viscoelastic, and viscous materials (Zheng, 2019).

When rice is soaked at high temperature, more viscoelastic nature is observed for the cooked rice (Zhu et al., 2019). Liu et al. (2020) observed that the flours of mild-parboiled germinated rice, germinated brown rice, and white rice showed non-Newtonian, viscoelastic (gel-like), thixotropic, and shear thinning behaviour after the mild-parboiling treatment. Parboiling involves a hydrothermal treatment with three major steps: soaking, steaming, and drying (Oli et al., 2014; Liu et al., 2020). It is performed for a variety of grains, especially different kinds of rice such as paddy, brown rice, white rice, germinated brown rice, etc., to impact their sensory, cooking, nutritional, and physicochemical properties (Koh & Surh, 2016; Tian et al., 2018; Cheng et al., 2019; Hu et al., 2019; Liu et al., 2020). When rice is soaked with heated water, as the temperature of the water is increased, the hardness is reduced due to the weakening in macrostructure of rice with increase in temperature. At lower temperature of soaking of rice, the water molecules are concentrated at the periphery of the grain and the grain may still be glass-like, but as the temperature increases, the water penetrates into a much wider area than before. The higher temperature and the high water content lead to better movement of the amorphous region and transition of the grain into a rubber-like loose state (Zhu et al., 2019). As the temperature of the soaking water increases in parboiling, the peak viscosity, setback viscosity, and final viscosity are considerably reduced (Kale et al., 2015). When the rheological properties were studied with respect to soaking and sprouting of dough prepared from wheat grains, it was found that the G" (viscous modulus) and G' (elastic modulus) increased with increase in oscillation. However, the dough of the sprouted wheat showed decreased G" and G' compared to sound wheat, which indicates reduction of dough strength or elasticity after sprouting. The tan δ values were also enhanced with soaking and sprouting (Singh et al., 2001). The hardness of brown rice was reduced when subjected to high-pressure soaking and the reduction was higher when there was increase in pressure of the same soaking treatment (Yu et al., 2017). It was found that when rice grains were subjected to high hydrostatic pressure soaking, there was reduction in hardness value of the cooked waxy grain and normal rice grain. This might be due to the prevention of leaching of amylopectin and amylose. Also, the springiness is reduced and cohesiveness is increased (Hu et al., 2011; Tian et al., 2014).

8.4.4 Colour Profile

Colour is a major attribute in the food industries because the first thing that attracts the consumer is the appearance of food, especially the colour. Colour is sometimes the indication of flavour. Colour can be measured visually or instrumentally. However, the instrumental methods are more reliable and specific than visual methods. There are three main types of colour instruments: tristimulus and monochromatic colorimeters and colorimetric spectrophotometers. The CIE L^*, a^*, b^* scale, and Hunter L, a, b are the 3D colour scales most commonly used in the food industry to measure colour.

The processing conditions applied to grains such as soaking, parboiling, cooking, extrusion, etc., might affect the colour. It was found that soaking followed by parboiling resulted in changes in rice colour compared to non-treated ones. The L^* values were lower and a^* and b^* values were higher for the parboiled samples compared to non-parboiled ones, which is due to the Maillard or non-enzymatic browning reaction (Kimura et al., 1993; Bhattacharya, 1996; Saleh et al., 2018). Also, the pigment particles move from bran layers of the cereal into the endosperm while soaking, which results in change in colour. For instance, yellow pigments are diffused into the endosperm and enzymatic action, which results in elevated yellow colour of the endosperm in parboiled rice (Lamberts et al., 2006; Saleh et al., 2018). The yellow and red pigments from the bran are migrated into endosperm, which has been proven by colour analysis (Lamberts et al., 2006; Rocha-Villarreal et al., 2018). The degree of change in colour is related to parameters such as temperature of soaking water, time of soaking and heating, etc. (Lamberts et al., 2006; Saleh et al., 2018). When brown rice is subjected to high-pressure soaking, there is an increase in lightness compared to nonhigh-pressure-soaked rice. This might be due to water diffusion to the rice caused by high-pressure treatment. Also, brown rice subjected to two cycles of high-pressure treatment has lower b^* values than that of a single cycle with overall reduction in yellowness and increase in lightness of the brown rice (Yu et al., 2017). Sometimes, soaking at high temperature results in elevated colour of rice (Yamakura et al., 2005; Tian et al., 2014). It was found that the high hydrostatic pressure soaking and the vacuum soaking of normal and waxy rice grains results in much higher lightness value and reduced colour intensities, which is very important for consumer acceptance (Tian et al., 2014). This is because of the reduced oxygen availability to the enzymes causing browning in the presence of vacuum and the inactivation of the polyphenol oxidase and catalase in the rice grains by the application of high hydrostatic pressure (Knorr et al., 2006; Mabashi et al., 2009; Tian et al., 2014).

8.4.5 Thermal Properties

The properties of the solid/liquid/gases change when they are heated or cooled because the potential as well as kinetic energy of the molecules change with addition or removal of heat from the body. Hence, the thermal properties of materials are related to how the body (solid/gas/liquid) reacts or responds when the heat is applied or removed. Also, the response of the material body might be in terms of phase change, increase in temperature, changes in length/volume, chemical reaction initiation, etc. (Buck & Rudtsch, 2011). The thermal properties can usually be measured by Differential Scanning Calorimetry (DSC) where the samples can be heated with specific rate in a temperature range and the thermal transitions in the samples measured in terms of onset (T_0) and conclusion temperature (T_c) as well as melting enthalpy (ΔH). Han et al. (2009) found that the extent to which the hydration takes place in the grains while soaking affects the gelatinization temperature. The higher the degree of hydration/moisture content in the grain, the lower the ΔH, T_0 and T_c. Also, the rice when soaked in lower temperature showis greater enthalpy than when soaked at higher temperature. In another study, it was found that the soaking temperature and also SO2 concentration had significant effect on the thermal properties of the amaranth grain (Loubes et al., 2012).

8.4.6 Structural Properties

Soaking can also cause changes in the structural properties of cereals. It was found that the rice when soaked under high pressure and under vacuum, the damage in microstructure of the grain was prevented, due to formation of higher structure of network with greater cohesiveness and springiness. The high-pressure soaking reduced the hole/cavity size. The difference in pressure between inside and outside of the grains helped in maintaining the structural integrity of the starch and for the production of uniform channels between connection parts and the molecules of starch (Tian et al., 2014). It was found that when the rice was parboiled (where soaking is an important step), the grains became more translucent, darker, and glassy (Ushakumari, 2009; Rocha-Villarreal et al., 2018). The formation of the crystalline pattern, which is due to the complex between amylose and lipid, was determined by the parboiling conditions as well as the gelatinization temperature (Maache-Rezzoug et al., 2008; Rocha-Villarreal et al., 2018). The crystallinity of starch in parboiled rice increases while soaking and steaming. The gelatinized starch is recrystallized into crystalline structures of A or B type during cooling of the parboiled rice (Young et al., 1990; Rocha-Villarreal et al., 2018).

8.5 IMPACT OF PROCESSING TREATMENT ON BIOLOGICAL CHARACTERISTICS

8.5.1 Anti-nutritional Factors

The anti-nutritional factors are defined as the compounds that affect the nutritional profile of grains and that are formed in the grains as a result of their metabolism and also from some different processes such as nutrient inactivation, hindrance to the digestion process and feed utilization. Cereals are good sources of not only major and minor nutrients but also some anti-nutrients such as tannins, saponins, phytic acids, lectins, amylase inhibitors and protease inhibitors, and goitrogens. The main problem with them is they bind with nutrients and hinder their bioavailability. In serious cases, they cause malnutrition and nutrient deficiency. Thus, in order to reduce anti-nutrients or reduce their effects, traditional processing techniques are used such as soaking, germination, autoclaving, fermentation, debranning, etc. (Jaybhaye & Srivastav 2015; Handa et al., 2017; Samtiya et al., 2020).

Soaking is a simple yet very useful method for reducing anti-nutrient content, and it can also decrease cooking time. The soaking process can aid in liberating the enzymes (such as phytases) in cereal and millet grains. Soaking hydrates the grains and facilitates the germination process, which is associated with reduction in anti-nutritional factors and some enzyme inhibitors so that they can improve the digestibility of nutrients and nutrient accessibility and availability (Samtiya et al., 2020). Fermentation also needs soaking as a prerequisite, which helps to lower the level of anti-nutrients (Gupta et al., 2015; Samtiya et al., 2020). Anti-nutritional factors are usually water soluble in nature so that their removal is facilitated by soaking as they leach into soaking water. Soaking can increase the phytase activity, which in turn facilitates phytate reduction in grains. Soaking followed by fermentation can reduce the number of phytochemicals as a result of diffusion of water-soluble minerals and vitamins of cereals (Ogbonna et al., 2012; Kruger et al., 2014). Soaking is helpful for lowering the level of tannins, phytates, etc. Even cereals like barley and wheat are recommended to be eaten after soaking (Gupta et al., 2015; Samtiya et al., 2020). It was found that if ragi/finger millet is soaked in water for 24–48 hours at ambient temperature, anti-nutrients present such as saponins, polyphenols, oxalates and phytates, trypsin inhibitors, etc., can be reduced to some extent (Hotz and Gibson, 2007; Rathore et al., 2019) resulting in improvement of bioaccessibility as well as bioavailability of nutrients and minerals. The tannin and phytic acid content can be reduced by soaking finger millet in lye solution or distilled water for about 8 hours (Rathore et al., 2019). It was also found that the in vitro solubility of zinc and iron can be increased by soaking pearl millet with phytase enzyme (exogeneous or endogenous) (Lestienne et al., 2005; Gupta et al., 2015).

8.5.2 Phytochemical Profile

Cereals are a major part of the human diet and provide macronutrients such as carbohydrates, proteins and fats, energy, minerals, B vitamins, and also phytochemicals, which provide additional health benefits (Xiong et al., 2020). The phytochemical profile of cereals complements those of vegetables and fruits (Liu 2007; Xiong et al., 2020). Phytochemicals include phenolic compounds, γ-oryzanol, phytic acid, phytosterols, carotenoids as well as dietary fibres, which are present in minute quantities and act as accessory constituents (Kris-Etherton et al., 2002; Xiong et al., 2020), which perform specific functions in our body such as anti-inflammatory, antioxidants, hormonal regulation, and immunity improvement (Johnson 2013; Xiong et al., 2020). The phenolic phytochemicals also include derivatives of cinnamic acids consisting of ferulic acid, p-coumaric acid, caffeic acid, and sinapic acid, and the derivatives of benzoic acid include salicylic acid, syringic acid, *p*-hydroxybenzoic acid, and vanillic acid (Bach Knudsen et al., 2017).

According to Wu et al. (2013), soaking caused a decrease in phenolic compounds as a part of preparation of sorghum tea, which might be because of the leaching of phenolic conjugates in soaking medium (Igbedioh, 1995). It was found that the parboiling of rice, which involves steeping/soaking as the first step, reduced the phenolic compounds and also the antioxidant activity due to the leaching process in soaking water (Rocha-Villarreal et al., 2018), but in the case of polished rice, parboiling before milling partially preserved the phenolics (Rocha-Villarreal et al., 2018). In the case of pearl millet and proso millet, the quantity of free phenolic compounds was substantially increased by parboiling in the preparation of porridge as well as couscous where gallic acid and protocatechuic acid were detected as major phenolics and p-coumaric acid was found in every parboiled product except for that of couscous prepared with proso millet (parboiled) (Bora, 2013; Rocha-Villarreal et al., 2018).

8.5.3 Bioactives

Bioactive components can be defined as "the biomolecules which are non-essential components of the food and which have modulatory effects on one or more of the metabolic processes in the body resulting in improvement of health" (Solomon & William, 2003; Guaadaoui et al., 2014). The cereal grains and their products are nice sources of biologically active substances like sterols, tocotrienols, dietary fibres such as lignin, β-glucans, lignan, cellulose and arabinoxylan, tocopherol, phenolic acids, polyphenols, alkylresorcinol, microelements, and vitamins. The bioactive components are usually located in the bran layer of cereals (Kulawinek et al., 2007; Bartłomiej et al., 2012). Although these are present in minute quantities, they have a good impact on human health such as prevention of cancer, diabetes, cardiovascular diseases as well as obesity (Anderson et al., 2000; Bartłomiej et al., 2012). In Thailand, pigmented rice varieties are cultivated that are rich in minerals, fibre, lignan, B-vitamins, γ-oryzanol, polyphenols, phytic acid, selenium, etc. (Irakli & Katsantonis, 2017; Reddy et al., 2017; Owolabi et al., 2019).

Soaking as well as germination processes can improve the health benefits as well as quality of cereals. When soaking cereals such as rice, the grains will be mellowed by the water resulting in better sensory attributes (Kushwaha 2016; Owolabi et al., 2019). It was found that soaking followed by germination slightly lowered the total phenolic compounds. Conversion to cinnamic acid from L-phenylpropanoid by PAL or phenylalanine ammonialyase can be observed during soaking and germination, which in turn results in production of phenolic acids like caffeic acid, ferulic acid as well as *p*-coumaric acid. Otherwise, the phenolics could also be transformed to flavonoids or lignin. The process of soaking might cause leaching of the phenolics, which are water soluble, resulting in lowering their number (Owolabi et al., 2019).

In the case of parboiling of (which involves soaking) of red and purple bran rice, the amount of proanthocyanins and anthocyanins with high DP (degree of polymerization) was decreased (Rocha-Villarreal et al., 2018), but the number of antioxidants such as tocopherols, tocotrienols, and γ-oryzanol were increased, which are lipophilic (Min et al., 2014; Rocha-Villarreal et al., 2018).

8.5.4 Starch and Protein Digestibility

Starch plays a key role in human nutrition as a major carbohydrate. It gives food importnt technical properties such as texturizing capacity. The ability of starch to provide nutrition highly depends on the form in which the starch is present and the processing method used. The physiological properties of starch are mainly the liberation of glucose into blood and the pace of digestion in the body (Lehmann & Robin, 2007). Starch can be digested easily as in case of products of starch hydrolysis or very slowly digested as in the case of resistant starch (Englyst et al.,1992; Lehmann & Robin, 2007). The amylose-to-amylopectin ratio is responsible for the degree of digestion of starch and the physiological reactions of the same (Behall et al.,1988; Lehmann & Robin, 2007). Starches with high amylose content commonly serve as resistant starches (RS) and those with high or fully amylopectin content serve as rapidly digestible starch (RDS) (e.g., waxy and fully gelatinized starch) (Tester et al., 2004; Lehmann & Robin, 2007).

When the starch molecules are soaked, water penetrates into the crystalline as well as amorphous regions of the starch causing solubilization into water. Due to the hydration of starch, the cooking process can be more efficient, which in turn helps to digest the starch more easily in the body by enzymes (Alvarez-Ramírez et al., 2019). Meng et al. (2018) studied the effect of pressure soaking on black rice by high hydrostatic pressure technology and found that the pressure soaking resulted in lowering of the amylose leaching in water with decreased retrogradation, which can be correlated with increase in digestibility of the starch. In the case of parboiling, the soaking of the rice followed by steaming resulted in change in amylose-to-amylopectin ratio and the amylose composition, which might be due to retrogradation from the leached amylose (Dharmaraj & Malleshi, 2011; Rocha-Villarreal et al., 2018). The soaking of millets can improve the starch digestibility according to some sources, which might be due to the removal of the anti-nutrients thta hinder the activity of enzyme amylase (starch digesting enzyme) through diffusion to soaking water (Samuel & Peerkhan, 2020; Yousaf et al., 2021).

Protein is a very important part of our diet. Proteins are polymers and have to be hydrolyzed by enzymes (peptidases) into simpler forms such as peptides and amino acids in the body before they get absorbed. The peptidases are present in pancreas, small intestine, and stomach. After conversion into smaller molecules, they will be deliberately absorbed by small intestine (Lundquist & Artursson, 2016; Joye, 2019). When they are in native state, proteins have a specific state of conformation because of the unique amino acid alignments. Because of the tightly folded structure, the digestibility of the protein is hampered. Proteins occur in structures called protein bodies (Duodu et al., 2003; Joye, 2019) or may be enclosed physically in cell structure (Bhattarai et al., 2017; Joye, 2019), which hinders the accessibility of the enzyme to its substrate. Hence, food processing methods must be used in order to improve the digestibility of some proteins that involve disruption of the cell wall and inactivation of anti-nutritional compounds thereby improving the availability of biopolymers, etc. According to Dharmaraj & Malleshi (2011) and Rocha-Villarreal et al. (2018), parboiling, which includes soaking of the rice, resulted in improvement of protein digestibility. The soaking of brown rice resulted in increase in protein digestibility (Albarracín et al., 2013). It was evident that soaking followed by germination improved the digestibility of the protein (Mbithi–Mwikya et al., 2000; Alonso et al., 2000; Osman & Gassem, 2013). However, when sorghum grains were soaked, there was not much improvement in digestibility of protein. The protein digestibility of millets can be improved after soaking according to some studies (Sharma & Sharma 2021). This was because of the reduction in the anti-nutrients after soaking, which would otherwise bind with the protein and the involved enzymes resulting in hindrance of the digestion process (Hasan et al., 2019; Yousaf et al., 2021).

8.6 CONCLUSION

Cereals are main staple foods in the human diet that nourish the body with macro- and micronutrients. They are enriched with macronutrients such as carbohydrate, energy, fibre, and protein and minor nutrients such as B-complex vitamins, vitamin E, zinc, and magnesium. Cereals (including millets) are good sources of not only major and minor nutrients but also some anti-nutrients such as tannins, saponins, phytic acids, lectins, amylase inhibitors and protease inhibitors, and goitrogens, which may hinder the absorption of nutrients in the body. In order to reduce the anti-nutrients and also to facilitate different processes like germination, malting, fermentation, parboiling, etc., soaking is a preliminary process that is commonly followed. The process of soaking can result in changes in nutritional, functional as well as biological characteristics of cereals. When cereal grains are soaked, the starch gets hydrated, mellowed, and solubilized to some extent and also improves gelatinization in the case of parboiling. The crystallinity of the starch and the ratio of amylose-to-amylopectin is reduced. Similarly, the protein content of the grains will also increase after soaking, which may be due to the increase in non-protein nitrogen content and also reduction of some anti-nutritional factors. However, the fat/lipid content will reduce after soaking due to the release of lipolytic enzymes. During soaking of cereals, the minerals are commonly leached into soaking medium, resulting in reduction of these minerals. Along with these, there will also be significant changes in hydration, surface properties, thermal and rheological properties, colour as well as structural properties of the grains after soaking. The soaking process results in reduction of anti-nutritional factors present in cereals and millets such as phytates, tannins, saponins, etc., which in turn increases the bioavailability of minerals and vitamins. But phytochemicals such as phenolics and bioactive constituents are reduced after soaking due to their leaching in soaking water. The starch digestibility is elevated due to the increased hydration and softening of the endosperm after soaking, resulting in better absorption in the body. Similarly, there will be an increase in protein digestibility due to the release of protein from the anti-nutrients.

Hence, there are numerous advantages and some drawbacks of the process of soaking. However, soaking as a cheap and low-energy process plays an important role as a pre-treatment/preliminary process in many processing methods. Soaking can be included in industrial processes for reducing energy consumption and to reduce costs.

REFERENCES

Ahmed, S., Shah, P., & Ahmed, O. (2020). Biochemistry, Lipids. *StatPearls [Internet].*

Albarracín, M., González, R. J., & Drago, S. R. (2013). Effect of soaking process on nutrient bio-accessibility and phytic acid content of brown rice cultivar. *LWT-Food Science and Technology*, *53*(1), 76–80.

Alcázar-Alay, S. C., & Meireles, M. A. A. (2015). Physicochemical properties, modifications and applications of starches from different botanical sources. *Food Science and Technology*, *35*, 215–236.

Alonso, R., Aguirre, A., & Marzo, F. (2000). Effects of extrusion and traditional processing methods on antinutrients and in vitro digestibility of protein and starch in faba and kidney beans. *Food chemistry*, *68*(2), 159–165.

Alvarez-Ramírez, J., Vernon-Carter, E. J., Carrillo-Navas, H., & Meraz, M. (2019). Impact of soaking time at room temperature on the physicochemical properties of maize and potato starch granules. *Starch-Stärke*, *71*(3-4), 1800126.

Anderson, J. W., Hanna, T. J., Peng, X., & Kryscio, R. J. (2000). Whole grain foods and heart disease risk. *Journal of the American College of Nutrition*, *19*(sup3), 291S-299S.

Bach Knudsen, K. E., Nørskov, N. P., Bolvig, A. K., Hedemann, M. S., & Laerke, H. N. (2017). Dietary fibers and associated phytochemicals in cereals. *Molecular nutrition & food research*, *61*(7), 1600518.

Barnes, P. (1984). Cereal lipids. *Nutrition & Food Science*.

Bartłomiej, S., Justyna, R. K., & Ewa, N. (2012). Bioactive compounds in cereal grains–occurrence, structure, technological significance and nutritional benefits–a review. *Food Science and Technology International*, *18*(6), 559–568.

Bean, M. M., MM, B., & EA, E. H. (1983). Rice flour treatment for cake-baking applications.

Behall, K. M., Scholfield, D. J., & Canary, J. (1988). Effect of starch structure on glucose and insulin responses in adults. *American Journal of Clinical Nutrition, 47,* 428–432.

Bello, M. O., Loubes, M. A., Aguerre, R. J., & Tolaba, M. P. (2015). Hydrothermal treatment of rough rice: Effect of processing conditions on product attributes. *Journal of Food Science and Technology, 52*, 5156–5163. https://doi.org/10.1007/s13197-014-1534-0

Bhatnagar, A. S., Prabhakar, D. S., Kumar, P. P., Rajan, R. R., & Krishna, A. G. (2014). Processing of commercial rice bran for the production of fat and nutraceutical rich rice brokens, rice germ and pure bran. *LWT-Food Science and Technology*, *58*(1), 306–311.

Bhattacharya, S., (1996). Kinetics on color changes in rice due to parboiling. *Journal of Food Engineering 29*, 99–106.

Bhattarai, R. R., Dhital, S., Wu, P., Chen, X. D., & Gidley, M. J. (2017). Digestion of isolated legume cells in a stomach-duodenum model: Three mechanisms limit starch and protein hydrolysis. *Food & function*, *8*(7), 2573–2582.

Bolaji, O. T., Awonorin, S. O., Shittu, T. A., & Sanni, L. O. (2017). Changes induced by soaking period on the physical properties of maize in the production of Ogi. *Cogent Food & Agriculture*, *3*(1), 1323571.

Bora, P. (2013). *Nutritional properties of different millet types and their selected products* (Doctoral dissertation).

Buck, W., & Rudtsch, S. (2011). Thermal properties. In *Springer handbook of metrology and testing* (pp. 453–483). Springer, Berlin, Heidelberg.

Cheng, K. C., Chen, S. H., & Yeh, A. I. (2019). Physicochemical properties and in vitro digestibility of rice after parboiling with heat moisture treatment. *Journal of Cereal Science*, *85*, 98–104.
Chiang, P. Y., & Yeh, A. I. (2002). Effect of soaking on wet-milling of rice. *Journal of Cereal Science*, *35*(1), 85–94.
Conde-Petit, B., Nuessli, J., Arrigoni, E., Escher, F., & Amado, R. (2001). Perspectives of starch in food science. *CHIMIA International Journal for Chemistry*, *55*(3), 201–205.
Dharmaraj, U., & Malleshi, N. G. (2011). Changes in carbohydrates, proteins and lipids of finger millet after hydrothermal processing. *LWT-food Science and Technology*, *44*(7), 1636–1642.
Dicko, M. H.; Gruppen, H.; Zouzouho, O. C.; Traoré, A. S.; Van Berkel, W. J. & Voragen, A. G. (2006). Effects of germination on the activities of amylases and phenolic enzymes in sorghum varieties grouped according to food end use properties. *Journal of the Science of Food and Agriculture, 86*(6):953–963.
Duodu, K. G., Taylor, J. R. N., Belton, P. S., & Hamaker, B. R. (2003). Factors affecting sorghum protein digestibility. *Journal of cereal science*, *38*(2), 117–131.
Dutta, H., & Mahanta, C. L. (2012). Effect of hydrothermal treatment varying in time and pressure on the properties of parboiled rices with different amylose content. *Food Research International*, *49*(2), 655–663.
El-Safy, F., & Salem, R. (2013). The impact of soaking and germination on chemical composition, carbohydrate fractions, digestibility, antinutritional factors and minerals content of some legumes and cereals grain seeds. *Alexandria Science Exchange Journal*, *34*(October-December), 499–513.
Englyst, H. N., Kingman, S. M., & Cummings, J. H. (1992). Classification and measurement of nutritionally important starch fractions. *European Journal of Clinical Nutrition, 46*, S33-S50.
Guaadaoui, A., Benaicha, S., Elmajdoub, N., Bellaoui, M., & Hamal, A. (2014). What is a bioactive compound? A combined definition for a preliminary consensus. *International Journal of Nutrition and Food Sciences*, *3*(3), 174–179.
Gupta, R. K., Gangoliya, S. S., & Singh, N. K. (2015). Reduction of phytic acid and enhancement of bioavailable micronutrients in food grains. *Journal of food science and technology*, *52*(2), 676–684.
Han, J. A., & Lim, S. T. (2009). Effect of presoaking on textural, thermal, and digestive properties of cooked brown rice. *Cereal chemistry*, *86*(1), 100–105.
Handa, V., Kumar, V., Panghal, A., Suri, S., & Kaur, J. (2017). Effect of soaking and germination on physicochemical and functional attributes of horse gram flour. *Journal of Food Science and Technology, 54*(13), 4229–4239.
Hasan, M., Maheshwari, C., Garg, N. K., & Kumar, M. (2019). Millets: Nutri-Cereals.
Hassan, A. B., Ahmed, I. A. M., Osman, N. M., Eltayeb, M. M., Osman, G. A., & Babiker, E. E. (2006). Effect of processing treatments followed by fermentation on protein content and digestibility of pearl millet (Pennisetum typhoideum) cultivars. *Pakistan Journal of Nutrition*, *5*(1), 86–89.
Hotz C & Gibson RS. (2007). Traditional food-processing and preparation practices to enhance the bioavailability of micronutrients in plant-based diets. *Journal of Nutrition, 37*, 1097–100.
Hu, X., Xu, X., Jin, Z., Tian, Y., Bai, Y., & Xie, Z. (2011). Retrogradation properties of rice starch gelatinized by heat and high hydrostatic pressure (HHP). *Journal of Food Engineering*, *106*(3), 262–266.
Hu, Z., Shao, Y., Lu, L., Fang, C., Hu, X., & Zhu, Z. (2019). Effect of germination and parboiling treatment on distribution of water molecular, physicochemical profiles and microstructure of rice. *Journal of Food Measurement and Characterization*, *13*(3), 1898–1906.
Ibukun, E. O. (2008). Effect of prolonged parboiling duration on proximate composition of rice. *Scientific Research and Essays*, *3*(7), 323–325.
Igbedioh, S. O., Shaire, S., & Aderiye, B. J. I. (1995). Effects of processing on total phenols and proximate composition of pigeon pea (*Cajanus cajan*) and climbing bean (*Vigna umbellate*). *Journal of Food Science and Technology, 32*, 497–500.
Inyang, C. U & Zakari, U. M. (2008). Effect of germination and fermentation of pearl millet on proximate, chemical and sensory properties of instant "Fura"-a Nigerian cereal food. *Pakistan Journal of Nutrition, 7*(1):9–12.
Irakli, M. N., Katsantonis, D. N., & Ward, S. (2017). Rice bran: a promising natural antioxidant component in breadmaking. *J Nutraceuticals Food Sci*, *2*(3), 14.
Jaybhaye, R. V., & Srivastav, P. P. (2015). Development of barnyard millet ready-to eat snack food: Part II. *Food Science Research Journal, 6*(2), 285–291.

Johnson, I. T. (2013). Phytochemicals and health. *Handbook of plant food phytochemicals: Sources, stability and extraction*, 49–67.

Joye I. (2019). Protein Digestibility of Cereal Products. *Foods (Basel, Switzerland)*, *8*(6), 199. https://doi.org/10.3390/foods8060199

Kale, S. J., Jha, S. K., Jha, G. K., Sinha, J. P., & Lal, S. B. (2015). Soaking induced changes in chemical composition, glycemic index and starch characteristics of basmati rice. *Rice Science*, *22*(5), 227–236.

Kimura, T., Bhattacharya, K. R., & Ali, S.Z. (1993). Discoloration characteristics of rice during parboiling (I): effect of processing conditions on the color intensity of parboiled rice. *Journal of the Agricultural Chemical Society of Japan 24*, 23–30.

Knorr, D., Heinz, V., & Buckow, R. (2006). High pressure application for food biopolymers. *Biochimica et Biophysica Acta (BBA)-Proteins and Proteomics*, *1764*(3), 619–631.

Koh, E., & Surh, J. (2016). Parboiling improved oxidative stability of milled white rice during one-year storage. *Food science and biotechnology*, *25*(4), 1043–1046.

Kris-Etherton, P. M., Hecker, K. D., Bonanome, A., Coval, S. M., Binkoski, A. E., Hilpert, K. F., & Etherton, T. D. (2002). Bioactive compounds in foods: their role in the prevention of cardiovascular disease and cancer. *The American journal of medicine*, *113*(9), 71–88.

Kruger, J., Oelofse, A., & Taylor, J. R. (2014). Effects of aqueous soaking on the phytate and mineral contents and phytate: Mineral ratios of wholegrain normal sorghum and maize and low phytate sorghum. *International Journal of Food Sciences and Nutrition, 65*(5), 539–546.

Kulawinek, M., Jaromin, A., & Kozubek, A. (2007). Mechanizmy działania składników pełnych ziaren zbóż. *Lek w Polsce*, *17*(2), 85–96.

Kushwaha, U.K.S. (2016). Black rice: Research, History and Development, 1st edn. Switzerland: Springer International Publishing.

Lamberts, L., De Bie, E., Derycke, V., Veraverbeke, W. S., De Man, W., & Delcour, J. A. (2006). Effect of processing conditions on color change of brown and milled parboiled rice. *Cereal Chemistry, 83*, 80–85. https://doi.org/10.1094/CC-83-0080

Lehmann, U., & Robin, F. (2007). Slowly digestible starch–its structure and health implications: a review. *Trends in Food Science & Technology*, *18*(7), 346–355.

Lestienne, I., Caporiccio, B., Besançon, P., Rochette, I., & Trèche, S. (2005). Relative contribution of phytates, fibers, and tannins to low iron and zinc in vitro solubility in pearl millet (Pennisetum glaucum) flour and grain fractions. *Journal of agricultural and food chemistry*, *53*(21), 8342–8348.

Li, Q., Li, S., Guan, X., Huang, K., & Zhu, F. (2021). Effects of vacuum soaking on the hydration, steaming, and physiochemical properties of japonica rice. *Bioscience, Biotechnology, and Biochemistry*, *85*(3), 634–642.

Liu, Q., Kong, Q., Li, X., Lin, J., Chen, H., Bao, Q., & Yuan, Y. (2020). Effect of mild-parboiling treatment on the structure, colour, pasting properties and rheology properties of germinated brown rice. *LWT*, *130*, 109623.

Liu, R. H. (2007). Whole grain phytochemicals and health. *Journal of cereal science*, *46*(3), 207–219.

Loubes, M. A., Resio, A. C., Tolaba, M. P., & Suarez, C. (2012). Mechanical and thermal characteristics of amaranth starch isolated by acid wet-milling procedure. *LWT-Food Science and Technology*, *46*(2), 519–524.

Luh, B. S., & Mickus, R. R. (1991). Parboiled rice. In *Rice* (pp. 470–507). Springer, Boston, MA.

Lundquist, P., & Artursson, P. (2016). Oral absorption of peptides and nanoparticles across the human intestine: Opportunities, limitations and studies in human tissues. *Advanced drug delivery reviews*, *106*, 256–276.

Maache-Rezzoug, Z., Zarguili, I., Loisel, C., Queveau, D., & Bul_eon, A. (2008). Structural modifications and thermal transitions of standard maize starch after DIC hydrothermal treatment. *Carbohydrate Polymers, 74,* 802–812. https://doi.org/10.1016/j.carbpol.2008.04.047.

Mabashi, Y., Ookura, T., Tominaga, N., & Kasai, M. (2009). Characterization of endogenous enzymes of milled rice and its application to rice cooking. *Food research international*, *42*(1), 157–164.

Mbithi-Mwikya, S., Van Camp, J., Yiru, Y., & Huyghebaert, A. (2000). Nutrient and antinutrient changes in finger millet (Eleusine coracan) during sprouting. *LWT-Food Science and Technology*, *33*(1), 9–14.

McKevith, B. (2004). Nutritional aspects of cereals. *Nutrition Bulletin*, *29*(2), 111–142.

Meng, L., Zhang, W., Wu, Z., Hui, A., Gao, H., Chen, P., & He, Y. (2018). Effect of pressure-soaking treatments on texture and retrogradation properties of black rice. *LWT, 93*, 485–490.

Miano, A. C., & Augusto, P. E. D. (2018). The hydration of grains: A critical review from description of phenomena to process improvements. *Comprehensive Reviews in Food Science and Food Safety, 17*(2), 352–370.

Min, B., McClung, A., & Chen, M. H. (2014). Effects of hydrothermal processes on antioxidants in brown, purple and red bran whole grain rice (Oryza sativa L.). *Food Chemistry, 159*, 106–115.

Mir, S. A., & Bosco, S. J. D. (2013). Evaluation of physical properties of rice cultivars grown in the temperate region of India. *International Food Research, 20,* 1521–1527

Ocheme, O. B., & Chinma, C. E. (2008). Effects of soaking and germination on some physicochemical properties of millet flour for porridge production. *Journal of Food Technology, 6*(5), 185–188.

Ogbonna, A. C., Abuajah, C. I., Ide, E. O., & Udofia, U. S. (2012). Effect of malting conditions on the nutritional and anti-nutritional factors of sorghum grist. Annals of the University Dunarea de Jos of Galati Fascicle VI--*Food Technology, 36*(2), 64–72 https://doaj.org/article/b14b46c570b74603b13c5beb9421e0b5

Oli, P., Ward, R., Adhikari, B., & Torley, P. (2014). Parboiled rice: understanding from a materials science approach. *Journal of Food Engineering, 124*, 173–183.

Osman, M. A., & Gassem, M. (2013). Effects of domestic processing on trypsin inhibitor, phytic, acid, tannins and in vitro protein digestibility of three Sorghum varieties. *Inter J Agric Tech, 9*, 1187—1198.

Otegbayo, B. O., Osamuel, F., & Fashakin, J. B. (2001). Effect of parboiling on physico-chemical qualities of two local rice varieties in Nigeria. *Journal of Food Technology in Africa, 6*(4), 130–132.

Owolabi, I. O., Chakree, K., & Takahashi Yupanqui, C. (2019). Bioactive components, antioxidative and anti-inflammatory properties (on RAW 264.7 macrophage cells) of soaked and germinated purple rice extracts. *International Journal of Food Science & Technology, 54*(7), 2374–2386

Pernollet, J. C. (1978). Protein bodies of seeds: Ultrastructure, biochemistry, biosynthesis and degradation. *Phytochemistry, 17*(9), 1473–1480.

Preston, K. R., Kilborn, R. H., & Dexter, J. E. (1987). Effects of starch damage and water absorption on the alveograph properties of Canadian hard red spring wheats. *Canadian Institute of Food Science and Technology Journal, 20*(2), 75–80.

Rathore, T., Singh, R., Kamble, D. B., Upadhyay, A., & Thangalakshmi, S. (2019). Review on finger millet: Processing and value addition. *The Pharma Innovation Journal, 8*(4), 283–291.

Reddy, C. K., Kimi, L., Haripriya, S., & Kang, N. (2017). Effects of polishing on proximate composition, physico-chemical characteristics, mineral composition and antioxidant properties of pigmented rice. *Rice science, 24*(5), 241–252.

Rocha-Villarreal, V., Serna-Saldivar, S. O., & García-Lara, S. (2018). Effects of parboiling and other hydrothermal treatments on the physical, functional, and nutritional properties of rice and other cereals. *Cereal Chemistry, 95*(1), 79–91.

Saharan, K., Khetarpaul, N., & Bishnoi, S. (2001). HCl-extractability of minerals from ricebean and fababean: Influence of domestic processing methods. *Innovative Food Science and Emerging Technologies, 2*(4), 323–325.

Saleh, M., Akash, M., & Ondier, G. (2018). Effects of temperature and soaking durations on the hydration kinetics of hybrid and pureline parboiled brown rice cultivars. *Journal of Food Measurement and Characterization, 12*(2), 1369–1377.

Samtiya, M., Aluko, R. E., & Dhewa, T. (2020). Plant food anti-nutritional factors and their reduction strategies: An overview. *Food Production, Processing and Nutrition, 2*(1), 1–14.

Samuel, K. S., & Peerkhan, N. (2020). Pearl millet protein bar: Nutritional, organoleptic, textural characterization, and in-vitro protein and starch digestibility. *Journal of Food Science and Technology, 57*(9), 3467–3473. https://doi.org/10.1007/s13197- 020-04381-x.

Sareepuang, K., Siriamornpun, S., Wiset, L., & Meeso, N. (2008). Effect of soaking temperature on physical, chemical and cooking properties of parboiled fragrant rice. *World Journal of Agricultural Sciences, 4*(4), 409–415.

SEFA, D., & DW, S. (1979). The relationship of microstructure of cowpeas to water absorption and dehulling properties.

Serna-Saldivar, S. O. (2016). *Cereal grains: properties, processing, and nutritional attributes.* CRC press.

Sharma, R., & Sharma, S. (2021). Anti-nutrient & bioactive profile, in vitro nutrient digestibility, techno-functionality, molecular and structural interactions of foxtail millet (Setaria italica L.) as influenced by biological processing techniques. *Food Chemistry*, 130815.

Sheela, P.; UmaMaheswari, T.; Kanchana, S.; Kamalasundari, S. & Hemalatha, G. (2018). Development and evaluation of fermented millet milk-based curd. *Journal of Pharmacology and Phytochemistry, 7*, 714–717.

Singh, H., Singh, N., Kaur, L., & Saxena, S. K. (2001). Effect of sprouting conditions on functional and dynamic rheological properties of wheat. *Journal of Food Engineering*, *47*(1), 23–29.

Singh, N., & Eckhoff, S. R. (1996). Wet milling of corn-A review of laboratory-scale and pilot plant-scale procedures. *Cereal chemistry*, *73*(6), 659–667.

Solomon, H. K., & William, W. W. (2003). Bioactive Food Components, Encyclopedia of Food & Culture. *Acceptance to Food Politics, B Letter, Charles Scribner's Sons*, *1*, 201.

Stolt, M., Oinonen, S., & Autio, K. (2001). Effect of high pressure on the physical properties of barley starch. *Innovative Food Science and Emerging Technologies, 1,* 167–175.

Tester, R. F., Karkalas, J., & Qi, X. (2004). Starch structure and digestibility enzyme -substrate relationship. *World's Poultry Science Journal, 60,* 186–195.

Thakur, A. K., & Gupta, A. K. (2006). Water absorption characteristics of paddy, brown rice and husk during soaking. *Journal of Food Engineering*, *75*(2), 252–257.

Thompson, T. E. (2020, February 21). *Lipid. Encyclopedia Britannica.* https://www.britannica.com/science/lipid

Tian, J., Cai, Y., Qin, W., Matsushita, Y., Ye, X., & Ogawa, Y. (2018). Parboiling reduced the crystallinity and in vitro digestibility of non-waxy short grain rice. *Food chemistry*, *257*, 23–28.

Tian, Y., Zhao, J., Xie, Z., Wang, J., Xu, X., & Jin, Z. (2014). Effect of different pressure-soaking treatments on color, texture, morphology and retrogradation properties of cooked rice. *LWT-Food Science and Technology*, *55*(1), 368–373.

Ushakumari, S. R. (2009). *Technological and physico-chemical characteristics of hydrothermally treated finger millet* (Doctoral dissertation, University of Mysore).

Whistler, R. L., & Daniel, J. R. (2000). Starch. *Kirk-othmer encyclopedia of chemical technology.*

Wu, L., Huang, Z., Qin, P., & Ren, G. (2013). Effects of processing on phytochemical profiles and biological activities for production of sorghum tea. *Food Research International*, *53*(2), 678–685.

Xiong, Y., Zhang, P., Warner, R. D., Shen, S., & Fang, Z. (2020). Cereal grain-based functional beverages: from cereal grain bioactive phytochemicals to beverage processing technologies, health benefits and product features. *Critical Reviews in Food Science and Nutrition*, 1–25.

Yamakura, M., Haraguchi, K., Okadome, H., Suzuki, K., Tran, U. T., Horigane, A. K., & Ohtsubo, K. I. (2005). Effects of soaking and high-pressure treatment on the qualities of cooked rice. *Journal of Applied Glycoscience*, *52*(2), 85–93.

Young, R., Gomez, M., McDonough, C., Waniska, R., & Rooney, L. (1993). Changes in sorghum starch during parboiling. *Cereal Chemistry, 70*, 179–183.

Young, R., Haidara, M., Rooney, L. W., & Waniska, R. D. (1990). Parboiled sorghum: development of a novel decorticated product. *Journal of Cereal Science*, *11*(3), 277–289.

Yousaf, L., Hou, D., Liaqat, H., & Shen, Q. (2021). Millet: A review of its nutritional and functional changes during processing. *Food Res earch International*, *142*, 110197.

Yu, Y., Pan, F., Ramaswamy, H. S., Zhu, S., Yu, L., & Zhang, Q. (2017). Effect of soaking and single/two cycle high pressure treatment on water absorption, color, morphology and cooked texture of brown rice. *Journal of food science and technology*, *54*(6), 1655–1664.

Zayas, J. F. (2012). *Functionality of proteins in food.* Springer science & business media.

Zheng, H. (2019). Introduction: measuring rheological properties of Foods. In *Rheology of Semisolid Foods* (pp. 3–30). Springer, Cham.

Zhu, L., Cheng, L., Zhang, H., Wang, L., Qian, H., Qi, X., & Wu, G. (2019). Research on migration path and structuring role of water in rice grain during soaking. *Food Hydrocolloids*, *92*, 41–50.

CHAPTER 9

Enzymatic Processing of Cereals

Anju Boora Khatkar, Sunil Kumar Khatkar, and Narender Kumar Chandla
College of Dairy Science and Technology,
Guru Angad Dev Veterinary and Animal Sciences University (GADVASU),
Ludhiana, Punjab, India

CONTENTS

DOI: 10.1201/9781003242192-12

9.1 INTRODUCTION

Cereals (also known as grains) are members of the *Gramineae* family. They are the growing food and feed for humans and livestock because they provide energy, essential nutrients, including phytonutrients or phytochemicals, and dietary fibers in staple diets. The most commonly used and produced cereals are corn (Zea), rice/paddy (Oryza), wheat (Triticum), barley (Hordeum), oat (Avena), rye (Secale), Sorghum (Sorghum), millet (Pennisetum), and triticale (a hybrid of wheat and rye). They are a rich source of starch and are mostly cultivated for their fruit/edible seeds (i.e., caryopsis), for human beings, for fruits and straw for animal feed, and several industrial applications (Papageorgiou and Skendi, 2018). On the other hand, they also have some technological and nutritional complications. For example, most cereals are (1) deficient in many essential amino acids like methionine, lysine, and threonine at varying levels; (2) contain antinutritional factors; (3) exhibit impaired digestibility of protein and carbohydrate; (4) their germ is very sensitive to oxidation; (5) exhibit cultivar variations; and (6) require different environmental conditions for harvesting and storage and many more. Thus, cereals need to undergo several processing steps or stages that start from prerequisite (harvesting and storage) to primary processing (sorting, grading, cleaning, sieving, removal of inedible parts by dehusking/dehulling, reduction of size/grinding, soaking, tempering, parboiling, dehydration, drying, milling, refining, polishing, etc.), which is further followed by secondary processing (enzymatic processing or fermentation, puffing, baking, flaking, frying, extrusion, etc.) for their value addition, for adding vanity of products in the food chain, and finally for making them a significant and nutritious component of diets for humans and animals.

A food product's components' nutritional and functional quality plays a big role in consumer satisfaction. Despite the fact that whole cereal grains are recognized for their bioactive components, there is a need for new methods in the food industry to improve the nutritional status, sensory profile, and overall quality of final products. Besides the common processing techniques of cereals, the enzymatic processing or the uses of enzymes (both exogenous and endogenous) as an alternative to chemical additives, as a reaction catalyst to start and speed up the biochemical reactions, and as a process regulator, modifier, activator, or enhancer and nutritional profile improver like enhancement of prebiotic oligosaccharides or phenolic compounds (Moller and Svensson, 2021) for cereals and cereals-based foods have also shown their potential. While enzymes are highly efficient and frequently catalyze specific reactions with a high yield and minimal side effects, they perform under fairly strict conditions, requiring a specific range of pH, ionic strength, and temperature, as well as the addition of cofactors and coenzymes in many cases (Wong, 1995). Enzymes also impart key roles in facilitating the processing behavior of cereals and cereals-based food products like mixing, baking, brewing, etc., and enhancing the quality attributes like nutritional, functional, biological, microstructure, and organoleptic properties. Thus, enzymes have also shown numerous applications in cereal processing, and today cereals' enzymatic processing is increasing steadily. Enzymatic processing eventually replaced the addition of chemical additives from various food products due to their efficacy and environmental friendliness (Miguel et al., 2013, Hassan et al., 2014). This chapter aims to highlight and review the significance of the enzymatic processing of cereals for enhancing the nutritive, functional, biological, microstructure, and organoleptic quality of the final product.

9.2 NATURE AND OCCURRENCE OF ENZYMES

All enzymes are natural proteins and are available in mainly two forms: (1) exogenous or industrial form of enzymes, which are produced employing microbial (bacteria, yeast, and molds) fermentation, and (2) endogenous, which is secreted by the animal inside the body during digestion process or also naturally available in fruit portion of plants or cereals. They are being used as a supplemented or substituted process, natural or raw ingredients/components, in combination as a designed mixture, and worked at the molecular level in the processing chain of cereals for their value addition. The most common food for enzymes in cereals is starch, polysaccharides (cell walls), dietary fibers, lignins, and proteinaceous materials. Generally, microbial enzymes, if added externally, are most commonly used in the cereal industry due to their designed or controlled ability to enhance functionality, retain nutrition, alter the nature of the components in a positive direction or their bioavailability, and provide uniform characteristics to the product throughout the year. Several types of enzymes like amylolytic enzymes (α-amylase, β-amylase, pullulanase, and glucoamylase), hemicellulases, and cellulases (α-L-arabinosidase, cellulase, endo 1-4-β-D-xylanase, ferulic acid esterase, laminarinase, and lichenase), protease, esterases and lipases (lipase and lysophospholipase), phytase, oxidase (glucose-oxidase, lipoxygenase), pectinases (polygalacturonase), β-glucanases, α-glucosidase, limit dextrinase, etc., are active in cereal processing (Poutanen, 1997).

Enzymes are mostly present in the outer portion of the kernel (aleurone and bran layers) as well as in the germ. Enzyme activity varies between fractions of cereals, and the germ's endosperm has the lowest enzyme activity. Conversely, bran and dietary fiber-rich fractions may also be high in endogenous enzymes. Several enzymes (α-L-arabinofuranosidases, arabinoxylan α-L-arabinofuranohydrolases, and β-D-xylopyranosidases) have been found in the bran portion of the wheat that may be responsible for the variance in liquid loss across wheat dough from various cultivars. Endogenous enzymes can affect dietary fiber components' solubilization and viscosity, including β-glucans and arabinoxylans. Adverse effects of endogenous enzymes are widely established in preharvest sprouting.

9.3 TYPES OF ENZYMES AND THEIR MODE OF ACTION

The various type of enzymes with their potential application in food industry is presented in Table 9.1.

9.3.1 Amylases

Amylases (α-, β-, glucoamylase), pullulanase, α-glucosidase, and cyclodextrin glycosyltransferase are examples of amylolytic enzymes that degrade starch. Amylases are categorized into three types: α-amylase, β-amylase, and glucoamylase. α-amylase randomly cleaves α-1,4 bonds while β-amylase systematically cleaves them at the polysaccharide's ends. Glucoamylase produces glucose by cleaving the α-1, 6 bonds. The amylolytic enzymes are used extensively in the brewing, distilling, and baking sectors (Wong, 1995) and in the production of starch sweeteners and high-maltose syrups.

9.3.1.1 α-Amylases

The α-amylase is a starch-degrading enzyme that hydrolyzes the α-1,4-glycosidic linkages of starch and produces short-chain dextrins (Sindhu et al., 2017). These enzymes are present in all living organisms in one or another way. The availability of calcium ion is essential for the stability, activity, and integrity of the majority of metalloenzymes such as α-amylases (Raveendran et al., 2018). The α-amylases are widely used in baking, brewing, starch liquefaction, and as a digestive aid in the food industry (Couto and Sanromán, 2006). They are widely used in the bread-making industry as a flavour enhancer and antistaling agent. The α-amylases, which help break down starch into smaller dextrins that yeast can ferment, are added to the dough before baking. It enhances the bread's flavour, crust colour, and toastability (Van Der Maarel et al., 2002). Branched dextrins with a large molecular mass may also be made using α-amylases, which are used as a glazing additive

Table 9.1 Different Types of Enzymes and Their Potential Application in Food Industry

Enzymes	Application	References
Amylases	Brewing, distilling, and baking sectors. High-maltose syrups, starch sweeteners, as a digestive aid, glazing additive in rice cakes and other powdered foods, ethanol production, etc.	Wong, 1995, Van Der Maarel et al., 2002, Aiyer, 2005, Shiv, 2015, Garg et al., 2016, Fernandes and Carvalho, 2017, Miguel et al., 2013, Else et al., 2013
Glucoamylases	Development of high-glucose and high-fructose syrups, baking sector, sake, soy sauce, and light beer manufacturing	James et al., 1996, Raveendran et al., 2018
Proteases	Bread and baked food, improve digestibility of potein	Miguel et al, 2013, Hassan et al., 2014 Khatun et al., 2019
Transglutaminase	In baking products, noodles, pasta, edible film production, cheese and dairy products, and processing of meat	Kuraishi et al., 2000, 2001, Fernandes and Carvalho., 2017, Kieliszek and Misiewicz, 2014, Kieliszek and Błażejak, 2017.
Lipases	Baking, dairy industry, vegetable oils industry, to improve the emulsification capabilities of egg yolk lipids, in development of cocoa butter equivalents and human milk fat substitutes, fruit juice, beer and wine industries	Guerrand, 2017
Esterase	Aroma and flavour development, and waste management	Gallage et al., 2014, Faulds, 2010
Lipoxygenase	Bleaching agent, improves dough rheology, and enhances mixing tolerance	Faubion and Hoseney, 1981
Glucose oxidase	Bakery products	Bonet et al., 2006
Pectinase	Extraction, clarification, and stabilisation of fruit juices, coffee, cocoa, and tea, as well as the making of jams and jellies, pickling, oil extraction, retting and degumming processes, and bio-scouring of cotton fibre	Kubra et al., 2018
Xylose isomerase	High-fructose corn syrup, maize sweeteners	Wong, 1995
Xylanases	Baking, increased starch digestibility, bioethanol production	Bender et al., 2017, Korompokis et al., 2019 Juodeikiene et al., 2014
Catalase	Baking industry and in food wrappers to prevent oxidation of food and control food perishability	Raveendran et al., 2018

in rice cakes and other powdered foods (Aiyer, 2005). In the starch industry, the three stages of enzyme conversion of starch are gelatinization, liquefaction, and saccharification. Starch granules are dissolved to form a viscous solution by the process of gelatinization. During starch liquefaction α-amylases act on gelatinised starch and produce regular sized chains (i.e., dextrin, maltose, malt-triose as well as maltpentose) (Van Der Maarel et al., 2002), which are further hydrolyzed to glucose by glucoamylase during saccharification. Liquefaction results in partial hydrolysis of starch and reduction in viscosity. More glucose and maltose are produced as a result of saccharification. The thermostable α-amylases from *Bacillus amyloliquefaciens, Stearothermophilus,* or *Bacillus licheniformis* are the most common enzymes used for starch saccharification in the food industry (Van Der Maarel et al., 2002). The α-amylases are used in ethanol production, where starch is broken down into sugars with the help of enzyme, and sugar is fermented into alcohol by *Saccharomyces cerevisiae*. α-amylases, along with cellulases and pectinases, are also used to clarify fruit beverages to prevent post-process haze formation and cloudiness (in apple and kiwi juices) and increase the juice yield (Shiv, 2015 and Garg et al., 2016).

9.3.1.2 β-Amylases

The β-amylases (4-α-D-glucan maltohydrolase) hydrolyzes the second α-1,4 glycosidic bond present on the non-reducing end of starch and break off into maltose (two glucose units). The β-amylase may be found in yeasts, moulds, bacteria, and plants (particularly in the seeds). Heat-stable β-amylase is mainly obtained from bacteria and cereals. The β-amylase converts starch into maltose during fruit ripening, and imparts sweet flavour to the ripe fruits. The β-amylases are the main component of the diastase mixture that removes starchy sizing factors from textiles and turns cereal grains into fermentable sugar. The β-amylase prefers a pH range of 4.0–5.0. Starch is made up of two components: amylose and amylopectin. The β-amylase (maltogenic) reacts on straight-chain amylose but can not act on most branch-chain amylopectins. When just β-amylase is present, mainly maltose is formed along with a fraction of high molecular weight dextrin, or a residue of the amylopectin component. When α-amylase (dextrinogenic) reacts to starch, dextrins with a low molecular weight are produced. These gummy dextrins may induce a sticky crumb in bread. Because enzymatic reactions on starch in bread manufacturing are time-limited, only "attackable" or "damaged" starch granules can create the fermentable sugar required to raise the dough. The β-amylase has negligible influence on viscosity. However, α-amylase significantly reduces the viscosity of gelatinized starch, making it useful in the manufacturing of syrup and dextrose (Anonymous, 2021).

The synergistic effect of α- and β-amylases is used in the bread making process for hydrolysis of starch to dextrins and further to maltose, respectively. As a result, whereas the α-amylase releases dextrins from starch, the β-amylase hydrolyzes these dextrins to maltose, which yeast may use for further fermentation. Overall the amylases activity imparts many benefits, including increased bread volume and crumb texture and reduced dough viscosity. Additionally, the resultant reducing sugars promote maillard reactions, result in the crust' browning and enhancement of the pleasant flavour (Fernandes and Carvalho, 2017, Miguel et al., 2013). The combined activity of amylases, namely maltogenic amylases, and glucoamylase, may be employed to reduce bread staling. Staling is a chemical and physical process that diminishes the palatability of baked foods, and it is sometimes confused with a normal drying-out process caused by water evaporation. Staling is characterized by a rise in crumb stiffness, a loss in crumb elasticity, and a tough, leathery crust aspect. Staling is caused in part by starch retrogradation. This occurs when moisture from the starch granules migrates into the interstitial spaces, causing amylose and amylopectin molecules to realign. Retrogradation begins shortly after baking, when the starch gelatinizes and a substantial quantity of water is absorbed. Amylose retrogrades more quickly than amylopectin. The partial hydrolysis caused by enzyme activity, which is best done after gelatinization, considerably changes the structure of starch since the hydrolysis fragments are too tiny to retrograde (Else et al., 2013).

9.3.1.3 Glucoamylases

Glucoamylases are the exo-acting enzymes that catalyze the hydrolysis of starch polysaccharides from the non-reducing end, releasing β-glucose. These are also referred to as saccharifying enzymes, and they are found in large quantities in all living creatures. Glucoamylases are mostly produced by *Aspergillus niger* and *Aspergillus awamori*, but this enzyme obtained from *Rhizopus oryzae* is frequently used in industry (Coutinho and Reilly, 1997). At low temperatures, most of the glucoamylases are stable, and lose activity due to conformational change at elevated temperatures. Glucoamylases have a wide range of uses in the food sector, including the development of high-glucose and high-fructose syrups. They are also used in the baking sector to improve flour quality, minimize dough staling, and enhance the colour and quality of high-fiber baked goods (James et al., 1996). Glucoamylases are enzymes that break down the starch in the flour to produce maltose and fermentable sugars, and then further fermentation by yeast leads to the dough's rise. These enzymes are also utilized to make glucose, which is then fermented by *Saccharomyces cerevisiae* to produce ethanol. Glucoamylases are essential in the manufacturing of sake, soy sauce, and light beer. They break down dextrins into fermentable sugars with a lower calorific value and alcohol concentration in beer (Raveendran et al., 2018).

9.3.2 Xylose Isomerase (D-Xylose Ketol-Isomerase)/Glucose Isomerase

In the presence of divalent metal ions, xylose isomerase, also known as glucose isomerase, catalyzes the reversible isomerization of D-xylose and D-glucose to their respective ketoses, D-xylulose and D-fructose. The enzyme is primarily a "xylose" isomerase. In physiological systems, the isomerization of glucose is not the principal activity. Hemicellulose, which makes up about 40% of plant biomass, is home to various microbes. In bacteria, xylose from hemicellulose breakdown is transformed to xylulose by xylose isomerase, which is subsequently phosphorylated to xylulose-5-phosphate and channeled to the pentose phosphate pathway. The maize sweetener sector is acknowledged as the most significant use of xylose isomerase. The enzyme is immobilized in many methods for the commercial manufacture of high-fructose corn syrup, including whole-cell entrapment, covalent crosslinking, and adsorption on cellular materials (Wong, 1995).

9.3.3 Proteases

Proteases are enzymes that break down peptide bonds. Another name for the same enzyme is peptidase or peptide hydrolase. This group of enzymes may be divided into exopeptidases and endopeptidases based on whether they work outside (exo-acting) or within (endo-acting). Both proteinase and endopeptidase refer to the same enzyme (Wong, 1995). Proteases have been used to manufacture bread and baked foods for centuries. The proteolytic action of proteases is beneficial for both gluten and dough. As a result of proteolytic action, mixing times are shortened, dough consistency is lowered, and the dough becomes more uniform. Additionally, precise hydrolysis assists in regulating gluten strength and facilitates the pulling and kneading of the dough. Endopeptidases support the most of these effects since their action is more visible in the gluten network and dough rheology. The exopeptidase effects are more visible in the development of flavour and colour as a result of Maillard reactions between amino acids released and sugars present. Proteases have eventually replaced sodium metabisulfite in dough conditioning due to their efficacy and environmental friendliness (Miguel et al., 2013; Hassan et al., 2014).

9.3.4 Transglutaminases

The enzyme transglutaminase catalyzes the creation of isopeptide bonds between proteins. The acyl-transfer process between the γ-carboxyamide group of glutamine residues in peptide bonds and primary amines is catalyzed by transglutaminase (TG). Furthermore, transglutaminase catalyzes the deamination process in the absence of free amine groups. Water is an acyl acceptor in this scenario (Kuraishi et al., 2001). TG may alter proteins by amine insertion, cross-linking, and deamination. Transglutaminase's catalytic activity requires a pH of 5.5 and a temperature of 40°C. The only exception is transglutaminase obtained from *Streptomyces* sp., which works best at a temperature of 45°C (Kieliszek and Misiewicz, 2014). Its cross-linking characteristic is widely employed in different food applications, including the production of cheese and dairy products, the processing of meat, edible film production, noodles, pasta, and bakery products (Kuraishi et al., 2000, 2001; Fernandes and Carvalho, 2017; Kieliszek and Misiewicz, 2014). Transglutaminases are utilized in baking because their cross-linking effect on gluten proteins enhances dough stability, volume, elasticity, and robustness. TG leads in the protein modification by intra- or intermolecular cross-linking, increasing the protein's end-use utility (Kieliszek and Błażejak, 2017). By catalyzing the cross-linking of various proteins, such as soy proteins, whey proteins, gluten, meat, and fish protein, transglutaminases can be employed to change the functional characteristics of dietary proteins (Fernandes and Carvalho, 2017). The TG modification of dietary proteins has demonstrated several benefits. It contributes to the protection of lysine in food proteins against various chemical reactions, thereby extending the shelf life of food. Cross-linking of various proteins having complementary limiting essential amino acids can create food proteins with a better nutritional value. Transglutaminases can encapsulate lipids and lipid-soluble compounds, generate water- and heat-resistant films, eliminate the necessity of heat for gelation, increase water-holding capacity and elasticity, and affect solubility and functional characteristics. Additionally, it has been observed that transglutaminases increase the wool quality during felting, whitening, handling, and shrinking (Zhu et al., 2011).

9.3.5 Lipases

Lipases are triacylglycerol acylhydrolases (EC 3.1.1.3) and phospholipases A1 (3.1.1.32) and A2 (3.1.1.4), which are phosphoglyceride acyl hydrolases. These are two major types of lipolytic enzymes. On the other hand, phospholipases C (3.1.4.3) and D (3.1.4.4) are commonly referred to as lipolytic enzymes, despite the fact that they are not acylhydrolases. Animals, plants, and microbes all have triacylglycerol lipases. Animal lipases include pancreatic, gastric, and intestinal lipases and lipases present in milk (Wong, 1995). Lipases are enzymes that catalyze the hydrolysis of long-chain triglycerides. They are naturally present in the stomach and pancreas of humans and other animal species to digest fats and lipids. Industrial lipases have a wide range of uses in a variety of sectors, including food processing, oleo-chemicals, polymers, detergents, pharmaceuticals, waste management, cosmetics, and biodiesel. Lipases have a wide range of uses and benefits in the food and agricultural industries, where they can have quantitative and/or qualitative effects. Lipases, for example, provide a significant increase in oil yield while also improving the aesthetic of the finished product in the processing of vegetable oils. Lipases are used in the baking and dairy industry to enhance and speed up the formation of aromatic elements. Lipases are also used to improve the emulsification capabilities of egg yolk lipids in the egg processing industry. Lipases are also produced to develop novel functional components and functional food ingredients like cocoa butter equivalents and human milk fat substitutes (Guerrand, 2017). In the food and beverage industry, lipases find major applications in dairy, baking, fruit juice, beer, and wine industries (Guerrand, 2017). Lipase hydrolyzes triacylglycerols into mono- and diacylglycerols, glycerol, and free fatty acids. Lipase activity on the flour lipids (naturally present in flour or added fat) are utilized to improve handling, strength, and machinability of dough, and improve bread oven spring along with volume and crumb structure in white bread.

9.3.6 Esterases

Esterases can facilitate the separation of esters into acid and alcohol in an aqueous medium. Additionally, esterases hydrolyze short-chain acylglycerols instead of long-chain acylglycerols, distinguishing them from lipases. Esterases are widely employed in the food and alcoholic beverage sectors, primarily utilized to modify oil and fat and create aromas and flavours (Raveendran et al., 2018, Panda and Gowrishankar, 2005). The fruity flavours found in cheese result from various methyl or ethyl esters of short-chain fatty acids. It has been reported that bacteria produce ethyl esters and thioesters. Alvarez-Macarie and Baratti (2000) described the development of a new thermostable esterase from *Bacillus licheniformis* (highly thermotolerant) heterologously produced in *E. coli* for the synthesis of short-chain flavour esters. Feruloyl esterase is a critical enzyme in producing ferulic acid, a precursor to vanillin, an aroma agent used in different food products and beverages (Gallage et al., 2014). Feruloyl esterases, a significant group of the esterase family, hydrolyze the ester link between ferulic acid and various polysaccharides in the plant cell wall. Since feruloyl esterases hydrolyze lignocellulosic biomass, they are a necessary component of waste management (Faulds, 2010).

9.3.7 Lipoxygenase

Lipoxygenase is a dioxygenase that catalyzes the oxygenation of polyunsaturated fatty acids comprising a cis, cis-1,4-pentadiene system to hydroperoxides. Lipoxygenases may be extracted from both animal and plant sources. Lipoxygenase activity is found in a broad variety of plants like soybeans, navy beans, mung beans, peas, green beans, and peanuts, and cereals, like wheat, rye, barley, oat, corn, and rice bran. They are generally responsible for off-flavour and off-odour in foods (Wong, 1995). Lipoxygenase affects wheat flour dough in a variety of ways. It works well as a bleaching agent, improves dough rheology, and enhances mixing tolerance. The oxidation of pigments and unsaturated fatty acids by ambient oxygen is assumed to cause the bleaching effect. Lipoxygenase improves mixing tolerance. Lipoxygenase has no impact on defatted flour, and when lipid removed from lipoxygenase-treated dough is reintroduced into flour, it has been shown to reduce mixing tolerance. The oxidation of sulfhydryl groups by the lipid peroxides created by lipoxygenase's action on lipids has been proposed as a mechanism for lipoxygenase's improved dough rheology (Faubion and Hoseney, 1981). As a result of oxidation, hydroxyl radicals are generated, which react with the yellow carotenoid in wheat flour and the peptides/proteins in the dough, forming hydroxyacids. As a result, the yellow tint is reduced, resulting in a whiter crumb. Apart from the bleaching effect, the oxidation of thiol groups of gluten leads to the rearrangement of disulfide linkages and cross-linking of tyrosine residues, resulting in increased loaf volume (Raveendran et al., 2018).

9.3.8 Glucose Oxidase

Glucose oxidase (β-D-glucose: oxygen 1-oxidoreductase) commercially extracted from *Aspergillus niger* and *Penicillium amagasakiense* catalyzes the oxidation of β-D-glucose to δ-D-glucono-1,5-lactone along with the reduction of oxygen to water. In addition, it can also be extracted from various fungi like *Aspergillus oryzae, Penicillium notatum, Penicillium glaucum, Hanerochaete chrysosporium,* and *Talaromyces flavins* (Wong, 1995). Various oxidases have been employed in bread manufacturing in place of chemical oxidants like potassium bromate or potassium iodate to increase dough strength and handling attributes, as well as to improve the texture and aesthetic value of the baked product. Glucose oxidase and hexose oxidase are frequently used to refer to this group of enzymes. Excessive oxidase can result in excessive cross-linking, interfering with gas retention and dough handling, and resulting in a low-quality product (Bonet et al., 2006).

9.3.9 Polyphenol Oxidase

Polyphenol oxidase, scientifically designated as 1, 2-benzenediol: oxygen oxidoreductase, was also initially found in mushrooms in 1856 by Schoenbein. It is also known as tyrosinase, phenolase, catechol oxidase, monophenol oxidase, cresolase, and catecholase (Whitaker, 1995). This enzyme protects higher plants from insects and microorganisms. When damaged, it develops an impermeable scab of melanin that protects them from other microbe assault and desiccation (Szent-Györgyi and Vietorisz, 1931). Polyphenol oxidase is a copper-containing metalloprotein that catalyzes the oxidation of phenolic components to quinones, forming brown pigments in injured tissues. This enzymatic process results in post-harvest losses.

9.3.10 Pectinases

Pectinase is a collection of enzymes that break down pectin compounds. Microorganisms and plants are the primary producers. Pectinases are widely used in the food, agriculture, and environmental industries. Their most recent applications is in the production of bioenergy. These enzymes are employed in the food industry for extraction, clarification, and stabilization of fruit juices. They also aid in the fermentation of coffee, cocoa, and tea, as well as in making jams and jellies. They are employed in the softening process in pickling. They are widely used in the agricultural industry for the purification of plant viruses, oil extraction, retting and degumming processes, and bio-scouring of cotton fibre. Pectinases are also utilized to extract pure DNA from plants, maceration of plant tissue, protoplast separation from plant cells, and biomass liquification and saccharification. They are a kind of animal feed enzyme that aids in the absorption of nutrients. They play a critical function in wastewater treatment by promoting pectin breakdown ... Pectinases are utilized to increase bioethanol levels to produce bioenergy. Pectinases are mostly used in the wine and paper industries in the industrial sector. The enzymes pectinases, cellulases, and hemicellulases break down and dissolve the cell wall matrix, liquefying material and the liberation of intracellular carbohydrates (Kubra et al., 2018).

9.3.11 Catalase

In aerobic species, the enzyme catalase (hydrogen peroxide: hydrogen-peroxide oxidoreductase) catalyzes the hydrogen peroxide (toxic oxidizing agent) into water and oxygen [$2H_2O_2 \rightarrow 2H_2O + O_2$]. Thus, the enzyme protects cells against the harmful effects of hydrogen peroxide. The enzyme is a tetramer, with each component comprising a single polypeptide linked to a high-spin ferric protoporphyrin IX prosthetic group. Between pH 3 and 10, catalase is reasonably stable. In the presence of detergents, denaturation via dissociation occurs at high pH (Wong, 1995). It has applications in food to prevent them from oxidizing and is generally used in food wrappers and in the cheese industry to remove the H_2O_2 of raw milk. Catalase is mainly found in aerobic organisms and may be synthesized from microorganisms such as *Aspergillus niger* and *Micrococcus luteus*, and bovine liver. Microorganisms are frequently favoured as sources of enzymes due to their benefits in terms of rapid growth, ease of handling, and genetic modification to get the desired result (Raveendran et al., 2018). Catalase is used in the textile industry to remove excess hydrogen peroxide. This enzyme is commonly employed in the food processing sector in conjunction with other enzymes. Catalase is frequently used with glucose oxidases to preserve food. The development of acetaldehydes after using a glucose oxidase/catalase cocktail to eliminate oxygen from wine before bottling improves colour with stability (Röcker et al., 2016). Catalase is used in milk processing to remove peroxide from milk, in the baking industry to remove glucose from egg white, and in food wrappers to prevent oxidation of food and control food perishability. Catalase is not used very often when making cheese (Raveendran et al., 2018).

9.4 ENZYMES IN NUTRITION, DIGESTIBILITY, AND SENSORY PROFILING

A food product's components' nutritional and functional quality plays a big role in consumer satisfaction. Even though whole cereal grains are recognized for their bioactive components, scientists desperately need new methods to reduce nutritional losses caused by traditional food processing. Different kinds of enzymes are used in the food industry to improve the nutritional status, sensory profile, and overall quality of final products. Many industries use various types of enzymes for the enzymatic processing of food products to satisfy the needs of the food industry. The use of enzymes to depolymerize carbohydrates in grain's bran layer is proving to be an effective strategy for phenolic mobilization and dietary fiber solubilization (Singh et al., 2015).

Enzymes in sensory profiling: Enzymatic processing of cereals or composite foods improves the sensory profiling of products, including texture, flavour, mothfeel, etc.

Aroma is sometimes improved by adding exogenous enzymes. Composite bread made with rice flour taste better with fungal α-amylase added (Noomhorm et al., 1994). The ability of hemicellulases to change intestinal viscosity and nutritional absorption has been demonstrated in both animals and *in vitro* investigations. These enzymes degraded cell walls, allowing more wheat protein and glucose to be extracted. Soluble and insoluble dietary fibre ratios can shift when hemicellulases are employed in baking to improve texture. As a result, the physiological consequences of these components would be influenced. Enzymes can enhance oligosaccharides in cereal raw materials. Prebiotic meals for specific activation of health-beneficial bacteria in the colon are currently being developed. The use of xylanolytic enzymes to produce arabino-xylo-oligosaccharides (Yamada et al., 1993) has been suggested. The viscosity of an oat flummery was controlled by α-amylases in a new yoghurt-like cereal-based snack (Jaskari et al., 1995). In addition to increasing β-glucan concentration and providing lactic acid bacteria substrates, starch hydrolysis and subsequent decrease in viscosity enabled the production of a product rich in dietary fibre. Enzyme preparations should not depolymerize the physiologically active soluble β-glucan. The α-amylase alone or in conjunction with protease has been used to create wheat bran components to boost dietary fibre content (Rasco et al., 1991). Aspects of flavour and hue in bread occurs due to the Maillard reaction response between dextrins (low molecular weight) and proteinaceous substances. The α-amylases activation also enhances the darker crust of the resultant bread. A need for browning processes in microwavable items has not been particularly addressed by enzymes, even though this may be a desirable attribute. When it comes to bread making, lipoxygenases have long been used as flour whiteners. Martinez-Anaya (Martinez-Anaya, 1996) has studied the role of enzymes in creating bread taste. Endogenous and added enzymes, yeast, and starter cultures, offer precursors for fragrance chemicals generated in fermentative and thermal processes. Proteins can also be turned into bitter peptides when carbohydrate degradation products react with protein degradation product (Martinez-Anaya, 1996). To ensure optimal baking performance, barley and wheat malts are used to standardize the enzyme activity in flour (Mäkinen and Arendt, 2015, Guardado-Félix et al. 2020). Malt adds flavour and aroma to different food products.

Enzymes in digestibility of bio-molecules: Enzyme processing has the potential to change and increase the digestibility of starch by promoting the synergy of starchy with non-starchy components, but hampered enzyme accessibility, resulting in slower starch digestion rates (Khatun et al., 2019). Deproteinization increased *in vitro* rice flour starch digestibility (Ye et al., 2018). Utilizing xylanases help in the isolation of arabinoxylans from rye bran and offers a component that improves the rheology and gelation of dough (Bender et al., 2017). The glycoside hydrolase (GH) 10 and 11 based commercial endo-β-(1,4)-xylanases have been used in baking for their sensitivity and high activity for unsubstituted regions of the xylan and opposed the activity of GH 10 thus, which can work on the xylose linkages nearer to the side chain decorations. On the other hand, proteinaceous xylanase inhibitors from wheat and other cereals can be solved by the structure-guided engineering method on the production of restructured GH 10 xylanases from *Trichoderma reesei* and *Penicillium canescens* where a loop of 5-amino-acid residue was introduced, thus reducing their susceptibility to inhibitory proteins (Denisenko et al., 2019).

Enzymes in nutrition: Enzymatic processing of cereals results in improved nutritional status of final products not only by increasing the digestibility and nutritional profile but also by degrading the anti-nutritional factors and by hindering the unfavourable bio-chemical reactions responsible for nutrient degradation. To prevent starch retrogradation in starch polymers, microbial amylases (maltogenic) are added in flour (Woo et al., 2020). As a result, the bran, containing fibres and bio-active molecules, is removed, thus reducing the nutritional value of the grain. The bio-polishing of cereal grains using amylases, cellulases, esterases, β-glucanases, proteases, pectinases, and xylanases exposes and mobilizes soluble fibres and phenols of bran and also results in healthier and more palatable and acceptable products (Singh et al., 2016). Enzymes are also used to hydrolyze starch, proteins, polysaccharides, and phenolics of cell walls, and seed germination is used to activate their activities.

Additionally, the germination of cereals also increases the activities of β-glucanase, xylanase, proteases, and lipases, which are cell wall-erecting enzymes (Guzmán-Ortiz et al., 2019). Mäkinen and Arendt (2015) recommend malted/sprouted rye, millets, or sorghum as flour substitutes. Enzymes activity also helps to enhance nutrient digestion (Singh et al., 2019). A low abundant allele encoding a variation of the starch debranching enzyme LD (Limit Dextrinase) was shown to be responsible for significantly enhanced sorghum starch digestibility. When this is compared to the LD expressed by the predominant allele, this LD's N-terminal putative starch-binding region had a few residues of substituted amino acid (Gilding et al., 2013; Moller et al., 2015). The function of these sites in activity toward starch was validated by mutational investigation of the corresponding domain of barley LD, although it did not match the increased activity seen for uncommon allele in sorghum (Andersen et al., 2020).

Celiac disease is caused by gluten proteins found in wheat, rye, and barley (Cela et al., 2020). Consumption of gluten-containing meals and beverages should be limited so that no celiac disease autoimmune responses are triggered; proteases that degrade proline and immunoreactive sequences that are rich in glutamine can be used. It was recently shown that combined expression of a glutamine-specific endoproteinase from barley and a prolyl endopeptidase from *Flavobacterium meningosepticum* in wheat endosperm might detoxify immunogenic gluten. Disadvantageous sequences in barley C hordeins (storage protein) is predicted by endoproteinase from barley (Kerpes et al., 2017). Wheat genotypes with barley endo-proteinase and *Flavobacterium meningosepticum* prolyl endopeptidase have been developed by transformation, but their thermostability has been improved by sequence-guided site-saturation mutagenesis (Osorio et al., 2020). The use of *Sphingomonas capsulate*- and *Aspergillus niger*-based prolyl endo-proteases in conjunction with endoproteinase from barley is another method for obtaining gluten-free grain products (Cavaletti et al., 2019). Enzymatic processing is also utilized to make gluten-free beer (Kerpes et al. 2017). Detoxifying hordein (prolamin gluco-protein) using endoproteinase from barley enabled the production of beer (gluten-free). However, the lack of thermostability may pose a problem when extracting proteases from malt.

Phenol acids extracted enzymatically are also used to scavenge free radicals from wheat bran and are a low-cost and useful biotech product (Katileviciute et al., 2019). The esterases release ferulic acid, whereas xylanase treatment may create feruloyl oligosaccharides from insoluble dietary fibers (Wu et al., 2017, Dupoiron et al., 2018). Dietary fibers were isolated from oat and barley fractions utilizing β-glucanase jointly with other enzymes that degrade cell wall to produce varied-sized soluble β-glucan appropriate for use in drinks (Aktas-Akyildiz et al., 2018). The treatment of various wheat-milling fractions with xylanase increases starch digestibility and reduces physicochemical barriers to nutrients (Korompokis et al., 2019). Resistant starch is created when cell barriers slow down pancreatic amylase. *In vitro* starch digestion can be slowed by keeping cell walls intact, but in wheat, cell wall degradation by xylanase speeds up digestion (Korompokis et al., 2019). Starch and protein digestion are also improved using protease treatment (Khatun et al., 2019). To improve starch digestibility and to produce resistant starch, which is absorbed by the microbiota of human gut further to host valuable short-chain fatty acids, enzyme management is a possibility. Also, starch breakdown using salivary α-amylase resulted in increased postprandial blood glucose levels, indicating the need of chewing. *In vitro* studies investigating starch digestibility should thus include a chewing phase (Tamura et al., 2017).

9.5 ENZYMES FOR FUNCTIONALITY AND QUALITY IMPROVEMENT

The use of bacterial and fungal α-amylases to enhance flour is an old practice. The capacity of endogenous α-amylase to lower the viscosity of heat-treated flour slurries is extensively utilized to determine flour baking quality. The permitted range depends on the kind of cereal product and the α-amylase activity of the flour. For example, the activation of α-amylase results in a sticky and gummy bread crumb in low-dropping number sprouted wheat. Recent research on wheat and germinated wheat xylanolytic enzymes has yielded promising results. It is possible to improve dough rheology and gel formation using xylanases to extract arabinoxylans from rye bran (Bender et al., 2017). External α-amylase has been obtained from wheat and barley malt. The detailed mechanism for the firm texture in bread is unclear; however, the role of retrogradation of starch is clear. The enzyme creates low molecular weight branched-chain starch polymers as hydrolysis products, which may interfere with amylopectin recrystallization. In addition, dextrins may interfere with connections between inflated starch granules and the continuous protein network in bread (Akers and Hoseney, 1994), and amylopectin-lipid complexes may be formed via amylopectin-bond breakage (Kweon et al., 1994). Because hemicellulases can extend bread shelf life, the processes that decrease staling must be clarified. Enzyme activity and temperature stability in the oven determine the degree of α-amylolysis and other enzymatic processes that occur during baking. Bacterial α-amylases are thermostable and can cause dextrin overproduction if not dosed properly. This would produce a stickiness in the crumb portion of bread, comparable to the impact of high endogenous α-amylase levels (Every and Rose, 1996).

Yeast ferments some of α-amylase's hydrolysis products. Changes in dough rheology caused by enzymes also contribute to increased bread volume. Softening allows the dough to expand more, increasing oven spring (Cauvain and Chamberlain, 1988). The dough's gas retention properties may also be improved due to different compositions and a more stable liquid film at the gas-liquid interface. Gas cell stabilization is critical in the manufacturing of frozen dough and is a key target for novel enzyme applications. Degradation and alteration of pentosans are known to significantly alter their functional characteristics. Unextractable cell wall material can release high molecular weight, water-soluble arabinoxylans (Gruppen et al., 1993, Rouau, 1993). This is because the extracted arabinoxylans have a high apparent viscosity (Rouau, 1993). They can either solubilize insoluble substrates or hydrolyze them further (Gruppen et al., 1993; Rouau, 1993). Achieving selective solubilization of polysaccharides requires a close study of the tissue surrounding cereal cell walls (Poutanen et al., 1995). Inaccessible substrates require many enzymes to complete reactions. In pentosans, ferulic acid plays an important function. In addition to its involvement in cell wall design, ferulic acid is believed to be involved in crosslinking of arabinoxylan molecules. The β-glucans and pentosans from barley cell walls were recently found to be affected by ferulic acid esterase (Moore et al., 1996). These modifications occur in wholemeal rye baking, where arabinoxylans are critical for bread quality (Autio et al., 1996; Harkonen et al., 1995). Softer dough resulted from cell wall breakage and depolymerization (Autio et al., 1996). The α-amylase and pre-harvest sprouting are both responsible for the lower falling number of wheat flour due to the mechanism of biopolymer degradation. However, over-expression of wheat' α-amylase in the grain to mimic α-amylase caused improvement in the volume of loaf and Maillard browning (Ral et al., 2016). Pre-harvest sprouting reduces loaf volume in general (Olaerts et al., 2018). Volume, texture, and stability are all said to be improved by enzymes. Extensibility and ovenspring are improved by using xylanase (Maat et al., 1991). Amylase's dextrins may also compete with gelatinized starch, gluten, and other insoluble components of bread crumb for water (Every and Rose, 1996). In addition to improving bread volume (Cauvain and Chamberlain, 1988; Gruppen et al., 1993; Rouau et al., 1994), α-amylases and xylanolytic enzymes have also been shown to have additive effects (Haseborg and Hlmmelstem, 1998; Rouau et al., 1994; Laurikainen, et al., 1998). For example, poor flour baking quality (Cauvain and Chamberlain, 1988; Rouau et al., 1994) or the use of non-wheat raw material (Cauvain and Chamberlain, 1988) reduces bread volume (Krishnarau and Hoseney, 1994; Noomhorm et al., 1994). Further, yeast strains have been shown to improve bread quality by increasing volume and decreasing staling when both α-amylase and xylanase genes are cloned into baker's yeast (Randez-Gil et al., 1995; Monfort et al., 1996). There is evidence that dietary fibre (Haseborg and Hlmmelstem, 1998) and insoluble pentosans (Laurikainen, et al., 1998) can ameliorate some of the issues produced by hemicellulase-containing enzyme preparations (Krishnarau and Hoseney, 1994). Gradual solubilization of non-starch polysaccharides enhances bread quality. Degradation of water-binding pentosans and partial hydrolysis of proteins enhance product texture in biscuit manufacturing. To keep crackers soft, non-brittle, and shelf-stable, xylanase is used in a proprietary manner (Craig et al., 1992). This reduces the swellable water-soluble hemicellulose content in low-water-soluble goods, boosting moisture tolerance during storage (Slade et al., 1993). To reduce dough viscosity by 25–85%, just one or two cuts per water-soluble pentosan molecule were claimed. Also, when chemical additives are replaced, enzymes keep bread and biscuits fresh. To replace the bleaching agent bromate, hemicellulases and ascorbic acid are appropriate substitutes (Sprössler, 1996). These include α-amylase-derived bromate replacers. Protease-based enzyme preparations can replace sodium metasulphite in biscuits (Sprössler, 1996).

9.6 ENZYMES AS PROCESS MODIFIERS AND IMPROVERS

The primary source of sugars, which are fermentable, in beer manufacturing processes are malted barley. In contrast, raw barley or/and other grains, the adjunct grains, can also be used to provide enzymes for the fermentation process (Kok et al., 2019). In Europe, barley and corn, sorghum in Africa, rice in Asia, and corn in the United States are some of the grains that can be used as an auxiliary. The high adjuncts containing cocktails of several enzymes like amylase, β-glucanase, cellulase, and protease to make 100% raw grain wort (Kok et al., 2019; Zhuang et al., 2017) can also be used. As a result of the low proteolytic activity of barley, the sensory qualities of beers that are prepared solely using barley are often light and lacking mouthfeel and body (Steiner et al., 2012). To understand this, a significant deal of research has been done on the role of enzymes in the germination process of barley. A semi-quantitative microarray method of carbohydrate for identifying cell wall polysaccharide epitopes using antibodies (monoclonal) has been recently used to monitor specific phases in the manufacturing of beer (Fangel et al., 2018) where the enzymatically catalyzed molecular conversions of arabinoxylans, 1,3;1,4-β-glucans, galactomannans, mannans, xyloglucans, and pectin components were studied in detail. Additionally, this research rated 60 distinct beers in a way that might be useful for deciding the most cost-effective enzyme additives to utilize during the mashing process.

Some of the residues of distilling or brewing as a source of carbohydrates, proteins, lipids, and minerals may be utilized as animal feed, but also as a source of food with the help of enzymes (Connolly et al., 2019). Testing on probiotics and a model of human faecal fermentation have shown that xylanase hydrolysis, which produces prebiotic oligosaccharides, is an effective method for valuing brewers' leftover grains (Sajib et al., 2018). The protein included in brewers' leftover grains may be used for both food and feed (Wen et al., 2019). Enzyme pretreatment was one method when extracting the phenolics from barley malt rootlets in brewing (Budaraju et al., 2018). To boost bioethanol production from fusarium-infected barley, commercial xylanase and amylolytic enzymes may be used, and distillers' dried grains containing solubles can be inoculated with antimicrobial lactic acid bacteria to liberate the residue and make it appropriate for animal feed (Juodeikiene et al., 2014). Preprocessing methods like soaking, germination, fermentation, phytase treatment, etc., lower the phytic acid level in grains (Gupta, Gangoliya, and Singh, 2015; Li et al., 2014; Rasane et al., 2015a; Rasane et al., 2015b). With the use of *Aspergillus sp.* and *Trichoderma sp.*-based xylanase and cellulase enzymes, Das et al. (2008) developed a system to polish rice where the brown rice was soaked for 24 hours at 35°C in water to achieve moisture content of 35.5 g/100. The enzyme treatment lasted 2 hours at 50°C. Enzymes operated on bran layers' non-starch polysaccharides, releasing monomeric sugar components. Enzymes also lowered the brown rice's crude oil (16 g/100 g) and crude fiber (20 g/100 g). Dukare et al. (2021) used an enzymatic technique to extract nano-starch from maize, tuber of potato, and cassava and observed that the lowest size was attained by acid hydrolysis followed by enzyme hydrolysis for starch of maize. The starches from maize, potato, and cassava yielded nano-starch 18, 29, and 41 wt%, respectively, with significant reduction in the crystalline area, melting enthalpy, and amorphous zone. Enzyme hydrolyzed nano-starch has the potential to be used as a filler in biocomposites, not only because of its reinforcing capabilities but also because of its renewability and biodegradability. Gabaza et al. (2018) used exogenous enzymes, for African cereals, such as phytase and cocktails of enzymes (phytase, laccase, and tannase - P + L + T) for degrading phytic acid and phenolic and to explore iron and zinc bioaccessibility and observed that phytase therapy increased total soluble zinc content whereas the cocktails had a favorable impact on total soluble iron. On the other hand, both the treatments lowered the bioaccessibility of iron and zinc due to the interactions of minerals with enzymes. The potential of the high molecular weight soluble fiber, β-glucan, to reduce serum cholesterol and post-prandial blood glucose levels has been studied (Gamel et al., 2014) using the effects of amylases, proteases, and lipases on solubility. By accelerating the release of β-glucan from the food matrix, enzymes improved the ultimate viscosity whereas inclusion of digestive enzymes reduced it by the partial depolymerization of β-glucan and also by hindering starch and protein's role in viscosity. The viscosity and solubility of β-glucan isolated from oat crackers were not affected by the addition of lipase. When lichenase was added to the solution, it was discovered that β-glucan was the predominant source of viscosity, with little interference from other components. Gruppi et al. (2017) devised an enzymatic wheat conditioning procedure to partially hydrolyze the fiber fraction in order to improve whole flour milling yield and antioxidant capacity. Viscozyme® L (complex with various carbohydrase activities) and a 50:50 combinations of Celluclast® BG and Fungamyl® Super AX (purified 1,4-xylanase and a fungal α-amylase mixture) were employed along with other processing settings and noticed that enzymes had an impact on the cell wall components, thus enhancing the bran friability substantially, free glucose level (350% greater), free xylose, and antioxidant capacity.

ACKNOWLEDGMENT

The authors are thankful to the book's editor for allowing us to compile the information on the trending topic of enzymatic processing of cereals.

REFERENCES

Aiyer, P. V. (2005). Amylases and their applications. *African Journal of Biotechnology*, 4(13).

Akers, A. A., and Hoseney, R. C. (1994). Water-soluble dextrins from α-amylase-treated bread and their relationship to bread firming. *Cereal Chemistry,* 71(3), 223–226.

Aktas-Akyildiz, E., Sibakov, J., Nappa, M., Hytönen, E., Koksel, H. A. M. İ. T., and Poutanen, K. (2018). Extraction of soluble β-glucan from oat and barley fractions: Process efficiency and dispersion stability. *Journal of Cereal Science*, *81*, 60–68.

Alvarez-Macarie, E., and Baratti, J. (2000). Short chain flavour ester synthesis by a new esterase from Bacillus licheniformis. *Journal of Molecular Catalysis B: Enzymatic,* 10(4), 377–383.

Andersen, S., Svensson, B., and Møller, M. S. (2020). Roles of the N-terminal domain and remote substrate binding subsites in activity of the debranching barley limit dextrinase. *Biochimica et Biophysica Acta (BBA)-Proteins and Proteomics*, *1868*(1), 140294.

Anonymous, (2021). https://www.britannica.com/science/β-amylase. Retrieved on 16th December 2021.

Autio, K., Härkönen, H., Parkkonen, T., Frigård, T., Poutanen, K., Siika-aho, M., and Åman, P. (1996). Effects of purified endo-β-xylanase and endo-β-glucanase on the structural and baking characteristics of rye doughs. *LWT-Food Science and Technology*, *29*(1-2), 18–27.

Bender, D., Nemeth, R., Wimmer, M., Götschhofer, S., Biolchi, M., Török, K., and Schoenlechner, R. (2017). Optimization of arabinoxylan isolation from rye bran by adapting extraction solvent and use of enzymes. *Journal of Food Science*, *82*(11), 2562–2568.

Bonet, A., Rosell, C. M., Caballero, P. A., Gómez, M., Pérez-Munuera, I., & Lluch, M. A. (2006). Glucose oxidase effect on dough rheology and bread quality: a study from macroscopic to molecular level. *Food Chemistry,* 99(2), 408–415.

Budaraju, S., Mallikarjunan, K., Annor, G., Schoenfuss, T., and Raun, R. (2018). Effect of pre-treatments on the antioxidant potential of phenolic extracts from barley malt rootlets. *Food Chemistry*, *266*, 31–37.

Cauvain, S. P., and Chamberlain, N. (1988). The bread improving effect of fungal α-amylase. *Journal of Cereal Science*, *8*(3), 239–248.

Cavaletti, L., Taravella, A., Carrano, L., Carenzi, G., Sigurtà, A., Solinas, N., and Mamone, G. (2019). E40, a novel microbial protease efficiently detoxifying gluten proteins, for the dietary management of gluten intolerance. *Scientific Reports*, *9*(1), 1–11.

Cela, N., Condelli, N., Caruso, M. C., Perretti, G., Di Cairano, M., Tolve, R., and Galgano, F. (2020). Gluten-free brewing: issues and perspectives. *Fermentation*, *6*(2), 53.

Connolly, A., Cermeño, M., Crowley, D., O'Callaghan, Y., O'Brien, N. M., and FitzGerald, R. J. (2019). Characterisation of the in vitro bioactive properties of alkaline and enzyme extracted brewers' spent grain protein hydrolysates. *Food Research International*, *121*, 524–532.

Coutinho, P. M., and Reilly, P. J. (1997). Glucoamylase structural, functional, and evolutionary relationships. *Proteins: Structure, Function, and Bioinformatics,* 29(3), 334–347.

Couto, S. R., and Sanromán, M. A. (2006). Application of solid-state fermentation to food industry—a review. *Journal of Food Engineering,* 76(3), 291–302.

Craig, S. A., Mathewson, P. R., Otterburn, M. S., Slade, L., Levine, H., Deihl, R. T., and Magliacano, A. M. (1992). *U.S. Patent No. 5,108,764*. Washington, DC: U.S. Patent and Trademark Office.

Das, M., Gupta, S., Kapoor, V., Banerjee, R., and Bal, S. (2008). Enzymatic polishing of rice–A new processing technology. *LWT-Food Science and Technology*, *41*(10), 2079–2084.

Denisenko, Y. A., Gusakov, A. V., Rozhkova, A. M., Zorov, I. N., Bashirova, A. V., Matys, V. Y., and Sinitsyn, A. P. (2019). Protein engineering of GH10 family xylanases for gaining a resistance to cereal proteinaceous inhibitors. *Biocatalysis and Agricultural Biotechnology*, *17*, 690–695.

Dukare, A. S., Arputharaj, A., Bharimalla, A. K., Saxena, S., and Vigneshwaran, N. (2021). Nanostarch production by enzymatic hydrolysis of cereal and tuber starches. *Carbohydrate Polymer Technologies and Applications*, *2*, 100121.

Dupoiron, S., Lameloise, M. L., Bedu, M., Lewandowski, R., Fargues, C., Allais, F., and Rémond, C. (2018). Recovering ferulic acid from wheat bran enzymatic hydrolysate by a novel and non-thermal process associating weak anion-exchange and electrodialysis. *Separation and Purification Technology*, *200*, 75–83.

Else, A.J., Tronsmo, K.M., Niemann, L.A., Moonen, J.H.E. (2013). Use of an anti-staling enzyme mixture in the preparation of baked bread. Patent Application US 20130059031 A1.

Every, D., and Ross, M. (1996). The Role of Dextrins in the Stickiness of Bread Crumb made from Pre-Harvest Sprouted Wheat or Flour Containing Exogenous α-Amylase. *Journal of Cereal Science*, *23*(3), 247–256.

Fangel, J. U., Eiken, J., Sierksma, A., Schols, H. A., Willats, W. G., and Harholt, J. (2018). Tracking polysaccharides through the brewing process. *Carbohydrate Polymers*, *196*, 465–473.

Faubion J M and Hoseney. R. C. (1981). Lipoxygenase: Its Biochemistry and Role in Breadmaking. *Cereal Chemistry,* 58:175–180.

Faulds, C. B. (2010). What can feruloyl esterases do for us? *Phytochemistry Reviews,* 9(1), 121–132.

Fernandes, P., and Carvalho, F. (2017). Microbial enzymes for the food industry. In Biotechnology of microbial enzymes (pp. 513–544). Academic Press.

Gabaza, M., Shumoy, H., Muchuweti, M., Vandamme, P., and Raes, K. (2018). Enzymatic degradation of mineral binders in cereals: Impact on iron and zinc bioaccessibility. *Journal of Cereal Science*, *82*, 223–229.

Gallage, N. J., Hansen, E. H., Kannangara, R., Olsen, C. E., Motawia, M. S., Jørgensen, K. and Møller, B. L. (2014). Vanillin formation from ferulic acid in Vanilla planifolia is catalysed by a single enzyme. *Nature Communications,* 5(1), 1–14.

Gamel, T. H., Abdel-Aal, E. S. M., Ames, N. P., Duss, R., and Tosh, S. M. (2014). Enzymatic extraction of β-glucan from oat bran cereals and oat crackers and optimization of viscosity measurement. *Journal of Cereal Science*, *59*(1), 33–40.

Garg, G., Singh, A., Kaur, A., Singh, R., Kaur, J., and Mahajan, R. (2016). Microbial pectinases: an ecofriendly tool of nature for industries. *3 Biotechnology*, 6(1), 1–13.

Gilding, E. K., Frere, C. H., Cruickshank, A., Rada, A. K., Prentis, P. J., Mudge, A. M., and Godwin, I. D. (2013). Allelic variation at a single gene increases food value in a drought-tolerant staple cereal. *Nature Communications*, *4*(1), 1–6.

Gruppen, H., Kormelink, F. J. M., and Voragen, A. G. J. (1993). Enzymic degradation of water-unextractable cell wall material and arabinoxylans from wheat flour. *Journal of Cereal Science*, *18*(2), 129–143.

Gruppi, A., Duserm Garrido, G., Dordoni, R., De Faveri, D. M., and Spigno, G. (2017). Enzymatic wheat conditioning. *Chemical Engineering Transactions,* 1777–1782.

Guardado-Félix, D., Lazo-Vélez, M. A., Pérez-Carrillo, E., Panata-Saquicili, D. E., and Serna-Saldívar, S. O. (2020). Effect of partial replacement of wheat flour with sprouted chickpea flours with or without selenium on physicochemical, sensory, antioxidant and protein quality of yeast-leavened breads. *LWT-Food Science & Technology*, *129*, 109517.

Guerrand, D. (2017). Lipases industrial applications: focus on food and agroindustries. OCL *Oilseeds and Fats Crops and Lipids,* 24(4), D403.

Gupta, R. K., Gangoliya, S. S., and Singh, N. K. (2015). Reduction of phytic acid and enhancement of bioavailable micronutrients in food grains. *Journal of Food Science and Technology, 52*, 676–684.

Guzmán-Ortiz, F. A., Castro-Rosas, J., Gómez-Aldapa, C. A., Mora-Escobedo, R., Rojas-León, A., Rodríguez-Marín, M. L., and Román-Gutiérrez, A. D. (2019). Enzyme activity during germination of different cereals: A review. *Food Reviews International*, *35*(3), 177–200.

Härkönen, H., Lehtinen, P., Suortti, T., Parkkonen, T., Siika-aho, M., and Poutanen, K. (1995). The effects of a xylanase and α β-glucanase from Trichoderma reesei on the non-starch polysaccharides of whole meal rye slurry. *Journal of Cereal Science*, *21*(2), 173–183.

Haseborg, E. T., and Himmelstein, A. (1988). Quality problems with high-fiber breads solved by use of hemicellulase enzymes. *Cereal Foods World. 38*(5), 419–20.

Hassan, A. A., Mansour, E. H., El Bedawey, A. E. A., & Zaki, M. S. (2014). Improving dough rheology and cookie quality by protease enzyme. *American Journal of Food Science and Nutrition Research, 1*(1), 1–7.

James, J., Simpson, B. K., and Marshall, M. R. (1996). Application of enzymes in food processing. *Critical Reviews in Food Science and Nutrition,* 36(5), 437–463.

Jaskari, J., Henriksson, K., Nieminen, A., Suortti, T., Salovaara, H., and Poutanen, K. (1995). Effect of hydrothermal and enzymic treatments on the viscous behavior of dry- and wet-milled oat brans. *Cereal Chemistry*, *72*(6), 625–631.

Juodeikiene, G., Cernauskas, D., Vidmantiene, D., Basinskiene, L., Bartkiene, E., Bakutis, B., and Baliukoniene, V. (2014). Combined fermentation for increasing efficiency of bioethanol production from Fusarium sp. contaminated barley biomass. *Catalysis Today*, *223*, 108–114.

Katileviciute, A., Plakys, G., Budreviciute, A., Onder, K., Damiati, S., and Kodzius, R. (2019). A sight to wheat bran: high value-added products. *Biomolecules*, *9*(12), 887.
Kerpes, R., Fischer, S., and Becker, T. (2017). The production of gluten-free beer: Degradation of hordeins during malting and brewing and the application of modern process technology focusing on endogenous malt peptidases. *Trends in Food Science and Technology*, *67*, 129–138.
Khatun, A., Waters, D. L., and Liu, L. (2019). A review of rice starch digestibility: effect of composition and heat-moisture processing. *Starch-Stärke*, *71*(9–10), 1900090.
Kieliszek, M., and Błażejak, S. (2017). Microbial transglutaminase and applications in food industry. *Microbial Enzyme Technology in Food Applications*, 180.
Kieliszek, M., and Misiewicz, A. (2014). Microbial transglutaminase and its application in the food industry. A review. *Folia Microbiologica*, 59(3), 241–250.
Kok, Y. J., Ye, L., Muller, J., Ow, D. S. W., and Bi, X. (2019). Brewing with malted barley or raw barley: what makes the difference in the processes? *Applied Microbiology and Biotechnology*, *103*(3), 1059–1067.
Korompokis, K., De Brier, N., and Delcour, J. A. (2019). Differences in endosperm cell wall integrity in wheat (Triticum aestivum L.) milling fractions impact on the way starch responds to gelatinization and pasting treatments and its subsequent enzymatic in vitro digestibility. *Food and Function*, *10*(8), 4674–4684.
Krishnarau, L. and Hoseney, R.C. (1994). Enzymes increase loaf volume of bread supplemented with starch things and soluble Pentosans. *Journal of Food Science,* 59, 1251- 1254.
Kubra, K.T., Ali, S., Walait, M and Sundus, H. (2018). Potential Applications of Pectinases in Food, Agricultural and Environmental Sectors. *Journal of Pharmaceutical, Chemical and Biological Sciences,* 6(2):23–34.
Kuraishi, C., Nakagoshi, H., Tanno, H., and Tanaka, H. (2000). Application of transglutaminase for food processing. In: Hydrocolloids (pp. 281–285). Elsevier Science.
Kuraishi, C., Yamazaki, K., and Susa, Y. (2001). Transglutaminase: its utilization in the food industry. *Food Reviews International*, 17(2), 221–246.
Kweon, M. R., Park, C. S., Auh, J. H., Cho, B. M., Yang, N. S., and Park, K. H. (1994). Phospholipid hydrolysate and antistaling amylase effects on retrogradation of starch in bread. *Journal of Food Science*, *59*(5), 1072–1076.
Laurikainen, T., Härkönen, H., Autio, K., and Poutanen, K. (1998). Effects of enzymes in fibre enriched baking. *Journal of the Science of Food and Agriculture*, *76*(2), 239–249.
Li, J., Vasanthan, T., Gao, J., Naguleswaran, S., Zijlstra, R. T., and Bressler, D. C. (2014). Resistant starch escaped from ethanol production: Evidence from confocal laser scanning microscopy of distiller's dried grains with solubles (DDGS). *Cereal Chemistry*, *91*(2), 130–138.
Maat, J., Roza, M., Verbakel, J., Stam, H., Santos de Silva, M. J., Bosse, M., and Hessing, J. G. M. (1991). Xylanases and their application in bakery. *Food Technology and Biotechnology*, *46*(1) 22–31.
Mäkinen, O. E., and Arendt, E. K. (2015). Nonbrewing applications of malted cereals, pseudocereals, and legumes: A review. *Journal of the American Society of Brewing Chemists*, *73*(3), 223–227.
Martinez-Anaya, M., Gómez-Cabellos, S., Giménez, M. J., Barro, F., Diaz, I., and Diaz-Mendoza, M. (2019). Plant proteases: from key enzymes in germination to allies for fighting human gluten-related disorders. *Frontiers in Plant Science*, *10*, 721.
Miguel, A. M., Martins-Meyer, T. S., Figueiredo, E. V. D. C., Lobo, B. W. P., and Dellamora-Ortiz, G. M. (2013). Enzymes in bakery: current and future trends. *Food Industry,* 278–321.
[#]Møller, M. S., and Svensson, B. (2021). Enzymes in grain processing. *Current Opinion in Food Science,* 37, 153–159.
Møller, M. S., Windahl, M. S., Sim, L., Bøjstrup, M., Abou Hachem, M., Hindsgaul, O. and Henriksen, A. (2015). Oligosaccharide and substrate binding in the starch debranching enzyme barley limit dextrinase. *Journal of Molecular Biology*, *427*(6), 1263–1277.
Monfort, A., Blasco, A., Prleto, A. and Sanz, P. (1996) Combined expression of *Aspergillus nidulans* Endoxylanase X 24 and *Aspergillus oryzae* α-Amylase in industrial Baker's Yeasts and their use in Bread Making. *Applied Environmental Microbiology, 62*, 3712–3715.
Moore, J., Bamforth, C. W., Kroon, P. A., Bartolome, B., and Williamson, G. (1996). Ferulic acid esterase catalyses the solubilization of β-glucans and pentosans from the starchy endosperm cell walls of barley. *Biotechnology Letters*, *18*(12), 1423–1426.

Noomhorm, A., Bandola, D. C., and Kongseree, N. (1994). Effect of rice variety, rice flour concentration and enzyme levels on composite bread quality. *Journal of the Science of Food and Agriculture,* 64(4), 433–440.

Olaerts, H., and Courtin, C. M. (2018). Impact of preharvest sprouting on endogenous hydrolases and technological quality of wheat and bread: a review. *Comprehensive Reviews in Food Science and Food Safety, 17*(3), 698–713.

Osorio, C. E., Wen, N., Mejías, J. H., Mitchell, S., von Wettstein, D., and Rustgi, S. (2020). Directed-mutagenesis of Flavobacterium meningosepticum prolyl-oligopeptidase and a glutamine-specific endopeptidase from barley. *Frontiers in Nutrition, 7,* 11.

Panda, T., and Gowrishankar, B. S. (2005). Production and applications of esterases. *Applied Microbiology and Biotechnology,* 67(2), 160–169.

Papageorgiou, M., and Skendi, A. (2018). Introduction to cereal processing and by-products. *In Sustainable Recovery and Reutilization of Cereal Processing By-Products* (pp. 1–25). Woodhead Publishing.

#Poutanen, K. (1997). Enzymes: An important tool in the improvement of the quality of cereal foods. *Trends in Food Science and Technology, 8*(9), 300–306.

Poutanen, K., Härkönen, H., Laurikainen, T., Suortti, T., Autio, K., Luonteri, E., and Tenkanen, M. (1995). Enzymatic accessibility of rye cell wall polysaccharides. In *International rye symposium: Technology and Products* (pp. 121–130). VTT Technical Research Centre of Finland.

Ral, J. P., Whan, A., Larroque, O., Leyne, E., Pritchard, J., Dielen, A. S., and Newberry, M. (2016). Engineering high α-amylase levels in wheat grain lowers F alling N umber but improves baking properties. *Plant Biotechnology Journal, 14*(1), 364–376.

Randez-Gil, F., Prieto, J. A., Murcia, A., and Sanz, P. (1995). Construction of baker's yeast strains that secrete Aspergillus oryzae α-amylase and their use in bread making. *Journal of Cereal Science,* 21(2), 185–193.

Rasane, P., Jha, A., Kumar, A., and Sharma, N. (2015b). Reduction in phytic acid content and enhancement of antioxidant properties of nutricereals by processing for developing a fermented baby food. *Journal of Food Science and Technology, 52,* 3219–3234.

Rasane, P., Jha, A., Sabikhi, L., Kumar, A., and Unnikrishnan, V. S. (2015a). Nutritional advantages of oats and opportunities for its processing as value added foods – a review. *Journal of Food Science and Technology, 52,* 662–675.

Rasco, B. A., Borhan, M., Yegge, J. M., Lee, M. H., and Siffring, K. (1991). Evaluation of enzyme and chemically treated wheat bran ingredeitns in yeast-raised breads. *Cereal Chemistry, 68*(3), 295–299.

Raveendran, S., Parameswaran, B., Ummalyma, S. B., Abraham, A., Mathew, A. K., Madhavan, A., Rebello, S., and Pandey, A. (2018). Applications of Microbial Enzymes in Food Industry. *Food Technology and Biotechnology,* 56(1), 16–30.

Röcker J, Schmitt M, Pasch L, Ebert K, Grossmann M. (2016). The use of glucose oxidase and catalase for the enzymatic reduction of the potential ethanol content in wine. *Food Chemistry,* 210:660–70.

Rouau, X. (1993). Investigations into the effects of an enzyme preparation for baking on wheat flour dough pentosans. *Journal of Cereal Science, 18*(2), 145–157.

Rouau, X., El-Hayek, M. L., and Moreau, D. (1994). Effect of an enzyme preparation containing pentosanases on the bread-making quality of flours in relation to changes in pentosan properties. *Journal of Cereal Science, 19*(3), 259–272.

Sajib, M., Falck, P., Sardari, R. R., Mathew, S., Grey, C., Karlsson, E. N., and Adlercreutz, P. (2018). Valorization of Brewer's spent grain to prebiotic oligosaccharide: Production, xylanase catalyzed hydrolysis, in-vitro evaluation with probiotic strains and in a batch human fecal fermentation model. *Journal of Biotechnology, 268,* 61–70.

Shiv, K. (2015). Role of enzymes in fruit juice processing and its quality enhancement. *Advances in Applied Science Research,* 6(6), 114–124.

Sindhu, R., Binod, P., Madhavan, A., Beevi, U. S., Mathew, A. K., Abraham, A., ... and Kumar, V. (2017). Molecular improvements in microbial α-amylases for enhanced stability and catalytic efficiency. *Bioresource Technology,* 245, 1740–1748.

Singh, A., Karmakar, S., Jacob, B. S., Bhattacharya, P., Kumar, S. P., and Banerjee, R. (2015). Enzymatic polishing of cereal grains for improved nutrient retainment. *Journal of Food Science and Technology, 52*(6), 3147–3157.

Singh, A., Sharma, S., Singh, B., and Kaur, G. (2019). In vitro nutrient digestibility and antioxidative properties of flour prepared from sorghum germinated at different conditions. *Journal of Food Science and Technology*, *56*(6), 3077–3089.

Singh, A., Sharma, V., Banerjee, R., Sharma, S., and Kuila, A. (2016). Perspectives of cell-wall degrading enzymes in cereal polishing. *Food Bioscience*, *15*, 81–86.

Slade, L., Levine, H., Craig, S., Arciszewski, H., and Saunders, S. (1993). U.S. Patent No. 5,200,215. Washington, DC: U.S. Patent and Trademark Office.

Sprössler, B. (1996). Entwicklung und Einsatzmöglichkeiten von Enzymen bei der Backwarenherstellung. *Getreide, Mehl und Brot (1972)*, *50*(5), 281–283.

Steiner, E., Auer, A., Becker, T., and Gastl, M. (2012). Comparison of beer quality attributes between beers brewed with 100% barley malt and 100% barley raw material. *Journal of the Science of Food and Agriculture*, *92*(4), 803–813.

Szent-Gyrgyi, A., & Vietorisz, K. (1931). Function and significance of polyphenol oxidase from potatoes. *Biochemistry*, Z, 233, 236–239.

Tamura, M., Okazaki, Y., Kumagai, C., and Ogawa, Y. (2017). The importance of an oral digestion step in evaluating simulated in vitro digestibility of starch from cooked rice grain. *Food Research International*, *94*, 6–12.

Van Der Maarel, M. J., Van der Veen, B., Uitdehaag, J. C., Leemhuis, H., and Dijkhuizen, L. (2002). Properties and applications of starch-converting enzymes of the α-amylase family. *Journal of Biotechnology*, 94(2), 137–155.

Wen, C., Zhang, J., Duan, Y., Zhang, H., and Ma, H. (2019). A Mini Review on Brewer's Spent Grain Protein: Isolation, Physicochemical Properties, Application of Protein, and Functional Properties of Hydrolysates. *Journal of Food Science*, *84*(12), 3330–3340.

Whitaker, J.R. (1995). Polyphenol Oxidase. In: Food Enzymes. Springer, Boston, MA.

#Wong, D.W.S. (1995). In: Food Enzymes. Springer, Boston, MA.

Woo, S. H., Shin, Y. J., Jeong, H. M., Kim, J. S., Ko, D. S., Hong, J. S., and Shim, J. H. (2020). Effects of maltogenic amylase from Lactobacillus plantarum on retrogradation of bread. *Journal of Cereal Science*, *93*, 102976.

Wu, H., Li, H., Xue, Y., Luo, G., Gan, L., Liu, J., and Long, M. (2017). High efficiency co-production of ferulic acid and xylooligosaccharides from wheat bran by recombinant xylanase and feruloyl esterase. *Biochemical Engineering Journal*, *120*, 41–48.

Yamada, H., Itoh, K., Morishita, Y., and Taniguchi, H. (1993). Advances in cereal chemistry and technology in Japan Structure and Properties of Oligomers From Wheat Bran'. *Cereal Foods World, 38,* 490–492.

Ye, J., Hu, X., Luo, S., McClements, D. J., Liang, L., and Liu, C. (2018). Effect of endogenous proteins and lipids on starch digestibility in rice flour. *Food Research International*, *106*, 404–409.

Zhu, D., Wu, Q., & Hua, L. (2011). Industrial enzymes. In: Comprehensive Biotechnology (Second Edition), Murray Moo-Young (Ed.). Academic Press, 3–13.

Zhuang, S., Shetty, R., Hansen, M., Fromberg, A., Hansen, P. B., and Hobley, T. J. (2017). Brewing with 100% unmalted grains: barley, wheat, oat and rye. *European Food Research and Technology*, *243*(3), 447–454.

#Annotated references – Provide core information about the enzymatic processing of cereals and referred as the prime source to compile the information for this book chapter.

PART IV

Thermal Processing of Cereals and Its Impact

CHAPTER 10

Conventional Heating (Dry and Wet Heating)

Mohamad Mazen Hamoud-Agha[a] and Arashdeep Singh[b]
[a]Tech Vegetal, Le moulin de Cadillac Noyal-Muzillac, France
[b]Department of Food Science and Technology,
Punjab Agricultural University, Ludhiana, Punjab, India

CONTENTS

DOI: 10.1201/9781003242192-14

10.1 INTRODUCTION

Cereals are defined as a grain or edible seed of the grass family, *Gramineae* (Bender & Bender, 1999), and include wheat, barley, buckwheat, maize, rye, oats, rice, sorghum, and millet (Singh and Sharma, 2017). Commonly, the cereal grain comprises the embryo plant and starchy endosperm, which are surrounded by the aleurone (protein and lipids) layer as a protective seed coat. Cereals and cereals-based foods have been one of the most important sources of energy and nutrients for humans since ancient times (FAO, 2002). Cereal products are rich in protein, vitamins, and minerals and also contain a wide range of bioactive and health benefits substances such as dietary fiber, phenolic compounds, and antioxidants. Several studies reported that regular consumption of cereals, particularly whole grains, have a key role in the prevention of diabetes, colorectal cancer, and chronic diseases (Singh and Sharma, 2017; Irondi et al., 2019). The health benefits of whole grain cereal consumption were reviewed by Liu (2007). The cereals need to be processed to maximize the digestibility and bioavailability of the nutrients and to improve the organoleptic properties. After milling, the thermal process, such as backing, steaming, cooking, frying, extrusion, drying, and toasting, is applied in the production of cereal products.

Starch is the major component of the cereal grain. It has an important role in the human diet and is also a functional constituent of foods (Wang et al., 2021) such as a thickening, gelling, and stabilizer agent. Starch granule structure is characterized by its shape, size, and composition, according to its botanical origin (Wang et al., 2021). Nevertheless, starch applications are limited by its retrogradation tendency and naturally poor solubility (Zhu et al., 2020). Chemical, physical, and enzymatic treatments are often applied to overcome these natural limitations of starch and to improve its functional properties.

Among the different processes applied to cereals, the thermal processes, within the physical methods, are particularly of interest, in recent years, as no chemical additives are involved and these processes are generally recognized as cheap, safe, and simple (Maniglia et al., 2021a). The hydration level plays an important role in defining the final functionality of starch, as the presence of water controls the gelatinization degree (Maniglia et al., 2021b). The thermal treatments give either pregelatinized or nonpregelatinized starches depending on water content conditions. However, the negative consequences of heating on the bioactive compounds and nutritional values of cereals have already been reported. Generally, vitamins are thermally unstable. During backing, for example, 25% of thiamine, 15% of riboflavin, and 50% of folate values are lost (Bhattacharya, 2014).

In this chapter, the conventional dry and wet heating methods of cereal are reviewed and their effects on functional, biological, and nutritive properties of cereal are discussed. Conventional cereal toasting, as a particular dry heat treatment, is also studied in detail.

10.2 DRY HEATING OF CEREAL

In dry heat treatments, cereals are exposed to heat at 120–250°C under moister content conditions of 7–13% (Zou et al., 2020) for a certain time (a few minutes to several hours, usually less than 1 hr) (Zhou et al., 2021). Dry heat treatment (DHT) is considered as green technology that is used particularly to increase swelling properties and water absorption capacity of starch (Maniglia et al 2021a; Maniglia et al., 2021b), with improved oil-binding ability, as an alternative solution to flour chlorination, which is forbidden in numerous countries. DHT consists of two steps: a drying process, followed by a thermal treatment. The first step allows the flour to reach a suitable moisture content that prevents starch gelatinization and glutenin denaturation (Maniglia et al., 2021a). The second step is thermal treatment at different times and temperatures depending on the desirable results and the nature of the raw materials. The final product is cooled by a moisture adjustment step. Agglomerates are usually formed in this step (Chandrapala et al., 2012).

Conventional heating methods such as ovens, heated rotary drums, screw conveyors, and agitated beds are still widely used. The selection of emerging technologies such as microwave and radio frequency heating has also been applied (Pei-Ling et al., 2010; Vittadini et al., 2008). The development of innovative methods for DHT of cereal is still challenging (Maniglia et al., 2021a; Maniglia et al., 2021b).

DHT affects the different properties of starch granules such as their shape, surface, and their physicochemical behavior like pasting, thermal properties, solubility, and swelling capacity. Buscella et al. (2016) reported that DHT treatment significantly increases the viscosity of wheat flour suspension (Bucsella et al., 2016). In addition, DHT can affect protein functionality, cereal digestibility, and bioactive compounds' bioavailability. Several studies, for example, reported that DHT improves the structural characteristics of glutinous rice, waxy corn, and maize starches (Zhou et al., 2021).

In addition to starch modification, the improvement of DHT-treated flours are also related to the modification of proteins and lipids (Mann et al., 2014). Protein and lipids are generally located at the surface of starch granules and act as barriers against water absorption and starch swelling during dough preparation. The protein partial denaturation, oxidation, and structure disruption by heat liberate the entrapped lipids to contribute to the flour functionality. Under these conditions, water moves easily towards starch granules during gelatinization. Furthermore, the hydrophobic surfaces improve the air incorporation by the Pickering method in backed products to reduce the collapse (Falsafi et al., 2019; Kaur et al., 2019). DHT-treated flour in bread and cake making as a partial substitute for wheat flour (Miano et al., 2017; Neill et al., 2012).

10.2.1 Impact of Dry Heating

10.2.1.1 Effects of Dry Heating on Morphological Properties

The morphological characteristics of starch granules were studied after DHT. Native wheat starch granules of A (large oval or disk-like particles) or B (small spherical or polygonal particles) types are generally smooth and intact. It is reported that starch granule surfaces were distinctly eroded after repeated and continuous DHT (Zhang et al., 2021a). Similarly, Zou et al. (2019) also reported that waxy corn starch had distortion and notches on the granules because of the amylose leaching after DHT (Zou et al., 2019).

10.2.1.2 Effects of Dry Heating on Crystalline Structure

Similarly, the effects of DHT on the crystalline structure of starch have been studied. The crystalline structure of starch is categorized as A, B, and C type according to amylopectin chain arrangement. Zou et al. (2019) reported that DHT did not change the crystal type of waxy corn starch, although the crystallinity was reduced, in accordance with numerous researches on proso millet flour, proso millet starch, quinoa starch, high amylose rice starch, rice flour, normal rice starch, and waxy rice starch (Y. Li et al., 2013; Oh et al., 2018; Qiu et al., 2015; Sun et al., 2014; Y. Zhou et al., 2021). These studies confirmed that starch property modifications after DHT are mainly due to the the amorphous region changes (Zhou et al., 2021). After dry heat treatments, the relative crystallinity of quinoa starch increased. This increase was higher for repeated dry heat treatment (44.83%) than continuous one (42.96%) compared to the native starch (38.6%). These results indicate that DHT makes the starch molecule arrangement more ordered and consistent. In contrast, some studies reported lower relative crystallinity after DHT, as in the case of high amylose rice starch (Oh et al., 2018), waxy rice starch (Li et al., 2013), and millet starch (Sun et al., 2014), by the destruction of the molecular structure and the changes in crystallite orientation (Zhou et al., 2019).

10.2.1.3 Effects of Dry Heating on Thermal Properties

Thermal properties (or gelatinization parameters) of cereal starches under DHT were studied. The initial temperature, peak temperature, termination temperature, and temperature range reflect the gelatinization properties of starch. The enthalpy (ΔH) indicates the energy of dissociating the double-helix structure and describes the crystal structure of the crystalline region and the double-helix structure of amylose in the amorphous region (Cooke & Gidley, 1992). Low thermal parameter values indicate starch is easy to gelatinize (Y. Zhou et al., 2021). A decrease in the enthalpy values reflects a partial gelatinization of the starch granules and a decrease in double-helix structure content. An increase of enthalpy value describes a stable structure and is due to the recrystallization of starch after treatment. Compared to native quinoa starch, the DHT-treated samples had higher gelatinization transition temperature (Zhou et al., 2021). Similarly, the thermal properties of maize starch were decreased after repeated and continuous DHT (at 140°C) (Zou et al., 2020) compared to native samples. Rice starch also shows a gradually decreased tendency for the thermal properties with the extended treatment time (1, 2, 4 h) at 110, 130, and 150°C, respectively (Oh et al., 2018).

10.2.1.4 Effects of Dry Heating on Water Solubility and Swelling Power Properties

The water-solubility and swelling power properties describe the interaction between the amorphous and crystalline regions of starch granules (Zhang et al., 2021b). High contents of amylose result in poor solubility and swelling power properties. The study of Zhang et al. (2021b) showed that the solubility of normal and waxy wheat starch increased by DHT (Zhang et al., 2021b). However, the swelling power of waxy wheat starch decreases as the healing time is prolonged. Zou et al. (2020) reported an improvement of solubility and swelling power of maize starch after DHT (Zhou et al., 2020). Similarly, Zhou et al. (2021) reported that water solubility increased under repeated and continuous DHT compared to the native quinoa starch due to leakage of amylose outside the starch granules (Zhou et al., 2021). The values were higher for the continuous DHT (176.87% compared to native) in accordance with the results of the Zou et al. (2019) study of waxy corn starch under continuous and repeated DHT (Zou et al., 2019). The damaged granule surface by DHT absorbs more water, which increases, as a result, the swelling power (Gong et al., 2017). However, prolonged DHT resulted in a decrease in swelling power and water solubility of quinoa starch (Zhou et al., 2021).

10.2.1.5 Effects of Dry Heating on Pasting Properties

Several factors can influence the pasting properties and the viscosity of starch, such as starch source, particle size, amylose and lipid contents, amylopectin side chain length, and starch crystal structure (Miao et al., 2009). DHT could enhance the pasting stability of starch and the resistance to heat and shear. Zou et al. (2019) reported improved peak viscosity after 4 h of heating at 140°C of waxy corn starch. However, upon prolonged heating cycles and time, the peack viscosity was decreased because of the destruction of starch crystal structure during heating. The authors also reported that the final viscosity and setback decreased with the increase of heating time, which improve the stability of cold starch paste after repeated or continuous DHT. After DHT (at 130°C for heating times up to 18 h), the contents of peak, trough, and final viscosities, setback, and the peak time of normal and waxy wheat starches were reduced progressively with prolonged cycles and time (Zhang et al., 2021b). Similar results were also reported by (Zhou et al., 2021) for quinoa starch. DHT-treated starch showed increased, initially, and then decreased of pasting properties as the number of heating cycling or treatment time increased.

10.2.1.6 Digestibility of Dry-Heated Cereal

The effects of DHT on the digestibility of treated cereal were also studied as it is a very important subject. Depending on the digestion rate, starch can be categorized as rapidly and slowly digested starches and resistant starch (Englyst & Hudson, 1996) according to the particle size, the crystallite type, and the amylose content of starch. Oh, et al. (2018) reported that DHT decreased the predicted glycemic index and the rapidly digestible starch content of the high amylose rice starch under DHT (temperatures of 110, 130, and 150°C and times of 0, 1, 2, and 4 h) (Oh et al., 2018). The effect of DHT on in vitro digestibility of maize starch was also studied (Chi et al., 2019). DHT did not affect significantly the rapidly digestible starch content. Nonetheless, the DHT-treated samples had higher slowly digestible starch content compared to native starch. DHT-treated maize starch presented increased contents of slowly digestible starch (Zou et al., 2020). Another study showed that normal wheat starch exhibited greater rapidly digested starch thanks to DHT compared to the native sample (Zhang et al., 2021b). Slowly digested starch content increased at first and then decreased with prolonged heating times. A reduction of resistant starch was reported; however, waxy cereals had a reversed behavior. Slow digested and resistant starches increased after DHT in this case. The improved digestibility following DHT could be explained by the disruption of starch granules and the formation of a porous structure with promotes the attack of enzymes.

10.3 TOASTING AND ROASTING OF CEREAL

Traditionally, roasting (or toasting) is an intense dry thermal treatment, where food is heated at high temperatures (150–350°C) (Murthy et al., 2008; Sruthi et al., 2021) to meet some required sensorial and functional properties. It is a complex process involving various physicochemical and structural modifications that result in the development of desirable color, aroma, flavor, and texture characteristics to improve its digestibility and overall palatability. Some studies reported that roasting may improve the nutritional quality by increasing the total available starches and in vitro hydrolysis rate (Carrera et al., 2015), reducing the anti-nutritional factors, such as the case of Samh flour, for example (Alderaywsh et al., 2019), and improving, as well, the bioavailability of minerals because of the greater loss of phytic acid by heat (Khan et al., 1991). Furthermore, it causes the inactivation of microorganisms and enzymes that extend the shelf-life (Zhu, 2016), and the destruction of toxins, such as ochratoxins in oats (Lee, 2020). Roasting also leads to puffing of cereals and food grains.

There is no precise definition for toasting and roasting in the literature. In general, considering the time-temperature combination, toasting is applied at a higher temperature for a short time (< 1 min), while roasting is conducted at a lower temperature but for longer processing periods (several minutes) (Bhattacharya, 2014). Furthermore, some authors use the term roasting for grains, nuts, and coffee heating while toasting is associated, rather, with cereal flake heating.

The roasting of cereals is seeing increased interest in the food industry. For example, roasting is an important process to produce healthy and crunchy cereal-based snacks, breakfast cereals, and high-energy cereal bars (Mrad et al., 2014; Oboh et al., 2010). Toasting cereals (maize and wheat mainly) is an essential process to prepare the traditional food "gofio" in the Canary Islands (Hernández et al., 2014). Flour of roasted maize is widely used in Mexico and the United States (Carrera et al., 2015).

Numerous publications studied the roasting of cereals like barley (Chen et al., 2019; Sharma et al., 2011), wheat (Murthy et al., 2008), chia seeds (Hatamian et al., 2020, oat (Gujral et al., 2011; Schlörmann et al., 2020), samh seeds (Mohamed Ahmed et al., 2020), maize (Schoeman et al., 2017), and buckwheat (Ma et al., 2020; Małgorzata et al., 2016; Tanwar et al., 2019). Buckwheat is a gluten-free grain and has numerous nutritional benefits, but its particular bitter flavor reduces consumer acceptability. Furthermore, tartary buckwheat contains potential allergens like protein inhibitors and toxins such as fagopyrin that harm human health. For these reasons, roasting is commonly applied for buckwheat processing to reduce the bitterness (Małgorzata et al., 2016; Zhu, 2016) and to reduce the protein inhibitors (Bhinder et al., 2019). Roasting applied before milling also reduces energy consumption (Schoeman et al., 2017). In addition to cereals, several food ingredients and products may be roasted such as pulses, nuts, spices, coffee, and meats (Rizki et al., 2015; Sruthi et al., 2021).

Roasting is a complicated but sensible process. The roasted cereal quality and process efficiency depend on several parameters such as product composition, heating technology, time/temperature combination, and several other operating conditions. The choice of each parameter influences the quality of the final product characteristics. Several research reports have shown that roasting results in the formation of acrylamide and hydroxymethylfurfural, which are toxic compounds (Bertuzzi et al., 2020). Severe roasting may also cause damage to bioactive components (i.e., anthocyanins and polyphenols) thereby reducing the antioxidative activities (Mrad et al., 2014). In this section, the roasting of cereals will be briefly reviewed. The different conventional roasting techniques and their effect on cereal properties will be detailed.

10.3.1 Conventional Methods of Roasting

Roasting and toasting are simultaneous heat and mass transfer processes. Three distinct phases can be defined in the roasting process: drying, typical roasting, and flavor/color development (Fadai et al., 2017). In addition to temperature/time combinations, the final characteristics of roasted products may also be affected by the roasting method employed. The high demand for roasted cereals of good quality led to the development of different innovative roasting methods; however, the conventional roasting processes, despite some inconvenience, are still largely applied. Some of the conventional roastings are summarized here.

10.3.1.1 Pan Roasting

In this method, stainless steel, cast iron, coated enamelware, or clay-heated pot may be used. The roasting surface is in direct contact with flames or another heat source. Stirring or mixing may improve the roasting homogeneity. In this method, the surface temperature is higher than the core and it is often over-roasted (Ozel et al., 2014). Kanagaraj et al. (2019) reported charring in rice and barnyard millet roasted by this method at 190°C (Kanagaraj et al., 2019).

10.3.1.2 Sand Roasting

This method is a traditional cereal and grain processing method, particularly in India (Sharma & Gujral, 2011). Using hot sand as a heating medium, the roasting is performed in a pan at a temperature varying from 250 to 350°C for a few seconds to produce roasted ready-to-eat grains, like rice, corn, barley, oat, black gram, and chickpeas (Sharma et al., 2011; Sharma & Gujral, 2011). The high temperature converts cereals' moisture into superheated vapor rapidly, which increases pressure within the grain causing puffing and structural and chemical modifications (Jogihalli et al., 2017). This roasting method has shortcomings including uneven temperature distribution, sand contamination and high silica content in products, low productivity, and high energic consumption. Furthermore, the workers have to work in direct contact with the smoke and flame (Murthy et al., 2008; Sruthi et al., 2021).

10.3.1.3 Oven Roasting

This method is commonly applied in household applications and small plants where the product is loaded into a roasting cell, mostly, on plats. Roasting ovens can be heated with hot air or electrical resistance. The traditional ovens have a stationary heat source (natural convection), while the later models are equipped with a fan to circulate the heat inside the roasting cell, to obtain, thus, more uniform products and to accelerate the process than the conventional ones. Air current velocity needs to be adjusted. Fast air current may dry out products and negatively impact their shape, texture, and overall quality. On the other hand, slow hot air currents reduce the efficiency of heat transfer and increase the roasting time (Sruthi et al., 2021). Lack of uniformity (as the product reset in a static position), high labor costs, low production rate, limitations of continuous ovens, and prolonged manually loading and unloading periods are the major drawbacks of the batch roasting oven (Yang et al., 2010).

10.3.1.4 Drum-Type Rotary Roasters

In this method, the roasting chamber is a cylindrical rotating drum at the desired angle for uniform heating and good mixing. The drum walls are heated directly by gas flames or by circulated hot air or hot gas of burning fuel resources (Fast et al., 1990). A permeant product movement guarantees a uniform roasting. This method is used for cereal flakes in the breakfast industry.

10.3.1.5 Fluidized-Bed Roaster

In this method, hot air is blown upwards through the cereal bed at adjusted flow to suspend the grains in a gentle boiling motion over a perforated plate. In a continuous fluidized bed roaster, the cereal particles move on a vibratory porous belt through which heated air passes. The inclination and the vibration of the belt control the residence time in this process. Some ameliorations were reported in the literature to improve the roasting homogeneity and reduce the residence time (Murthy et al., 2008).

In conventional roasting methods, the products are, often, of poor quality as a consequence of severe roasting conditions (high temperatures and long times). Non-uniform heating within the same batch, the contamination with hazardous pollutants, and the formation of toxic substances are the major drawbacks for these methods. The effect of roasting on different cereal properties and components is reviewed in the next section.

10.3.2 Impact of Roasting and Toasting

10.3.2.1 Effects of Roasting on Functional Characteristics of Cereal

The structural, mechanical, and functional properties of toasted cereal, including size, shape, density, porosity, milling behavior, etc., have been reported in the literature (Sruthi et al., 2021).

Rapid evaporation of moisture during the roasting process increases the porosity, volume, and surface area of cereal grains and decreases, and as a result, its bulk density and hardness (Bhattacharya, 2014; Gujral et al., 2011). Schoeman et al. (2017) studied the effect of conventional convective oven roasting at 180°C for 140 sec on the microstructure of white maize using X-ray micro-computed tomography. The three-dimensional images showed the presence of air cavities and internal cracks and confirmed an increase in total kernel volume by approximately 11% and a significant decrease in relative density by 6%. However, these modifications had no significant effects on dry milling yield. The authors also reported that the conventional oven roasting resulted in more important microstructure changes compared to the innovative forced convection continuous tumble roasting method under the same roasting conditions. Roasting lowered the bulk density of barley by 47.8 to 59.1% and the thousand kernel weight by 2.1–13.7% (Sharma et al., 2011). Roasting also resulted in lowering the bulk density of oat groats by 31–44% (Gujral et al., 2011).

The viscosity and pasting properties of roasted cereal were also studied. These properties indicate the textural characteristics have a great influence on the further utilization of roasted cereal products in the food industry. Viscosities of raw oat kernels and thin and thick oat flakes considerably reduced with increasing roasting temperature due to protein denaturation (Schlörmann et al., 2020). Denatured proteins are known to have higher hydrophobicity than native ones (Ozawa et al., 2009). The denatured protein, which is absorbed in the surface of starch granules, could reduce starch swelling, in addition to the gelatinization and the thermal damage of starch, which therefore reduces the overall viscosity compared to native oat products. Similar findings of oat flour were previously reported (Cenkowski et al., 2006; D. Zhang et al., 1997). Conventional hot sand roasting resulted in a significant increase in the peak, final, breakdown and setback viscosities whereas an increase was measured in pasting temperature and time to peak of barley flour (Sharma et al., 2011). Using microwave roasting, Sharanagat et al. (2019) reported a decrease in viscosity and pasting properties of sorghum due to starch deterioration and lipide-amylose structure creation.

Water absorption capacity (WAC) measures the amount of absorbed water, and it represents the gelatinization index of starch (Singh et al., 2017a). Water solubility index (WSI) describes the number of soluble solids and it is used to define the degradation of starch molecules and the released polysaccharide from starch granules (Jogihalli et al., 2017). Due to starch gelatinization and the formation of porous structure, roasting may increase the WAC of roasted cereals like barley, for example (Gujral et al., 2011). Roasting of chia seeds at 160°C had no significant influence on WAC; however, an increase in roasting time and temperature (180°C) resulted in increasing in WAC (Hatamian et al., 2020). Another study reported an increase of WAC of roasted buckwheat (120°C for 10 min) by 23.31% (Tanwar et al., 2019). However, the water solubility index decreases upon roasting due to the formation of lipid-amylose structures during the heating process (Shamekh et al., 1994). Conversely, Chung et al. (2011) reported an increase in the soluble solid contents of roasted corn kernels due to texture softening and the decomposition of the insoluble polymer during roasting at high temperatures.

Oil absorption capacity (OAC) measures the amount of absorbed oil. This index involves organoleptic aspects like taste and mouthfeel, and it is important for food formulations (Singh et al., 2017b; Sruthi et al., 2021). The study of Hatamian et al. (2020) on chia seeds roasting indicated that roasting at 160°C did not affect OAC while an increase of roasting time at 180°C increased significantly the OAC index. These results may be due to protein denaturation and decomposition. Tanwar et al. (2019) also reported an increase of OAC of roasted buckwheat flour by 5.54% in agreement with the results of Abulude et al. (2005) who reported an increase in OAC in roasted finger millet (Abulude et al., 2005).

10.3.2.2 Effects of Roasting on Nutritional Characteristics of Cereal

Several studies showed that roasting alters the structures of the starch granules and gelatinizes starch. Ma et al. (2000) reported that roasting raw buckwheat grain flour at 200°C for 5 seconds completely gelatinized starch and damaged its structure. Therefore, no peak (gelatinization temperature) was detected in the DSC curve of roasted flour. Similar results were reported by Sharma et al. (2011) who studied the roasting of barley grains at 280°C. A high degree of gelatinization, from 86.2 to 97.6%, was obtained, which agrees with the decrease of DHgel value from 86 to 65% as compared to control samples. The authors also reported that the starch became more susceptible to enzymatic breakdown after roasting, and it ranged from 28.8 to 43.1% due to the gelatinization during roasting. Furthermore, Ma et al. (2020) reported that roasting treatment reduces remarkably the starch contents by thermal decomposition of five varieties of buckwheat flours. Gujral et al. (2011) studied the effect of sand roasting (at 280°C for 15 sec) on the damaged starch of oat. The damaged starch content significantly increased after roasting that varied from 72 to 82 g/100 g of oat flour compared to 0.5 to 4.9 g/100 g of control oat flour samples. This increase may be associated with the breaking of the starch granules due to heat during toasting. These results agree with the previous results reported by Mariotti et al. (2006) for toasted oat flour. Both the soluble and insoluble fractions of dietary fiber of cereals have an important functional and nutritional role (Santos et al., 2018). Roasting modulates fibers with important loss of insoluble dietary fiber and a significant increase in high molecular weight soluble dietary fiber (Sruthi et al., 2021) due to the disruption of chemical bonds of carbohydrates resulting in smaller and more soluble fractions. Schlörmann et al. (2020) reported that the amounts of insoluble dietary fibers of roasted oat kernels were lower (5.9–7.9%) compared to control untreated samples (9.7%). Insoluble dietary fibers contents were also lower in thin and thick oat toasted flakes compared to raw samples. Among the soluble dietary fibers of cereals, β-glucan has particularly high health benefits as it is effective in reducing serum cholesterol concentration and postprandial blood glucose level. Regular consumption of β-glucan is associated with lower risks of cardiovascular diseases and type 2 diabetes (Schlörmann et al., 2020; Whitehead et al., 2014). Furthermore, β-glucan has interesting functional properties as an emulsion stabilizer agent, and it is used to improve food products' texture. Compared to other cereals, barley and oat are known for their high content of β- glucan (Charalampopoulos et al., 2002). The effect of roasting on β-glucan in these cereals, in particular, has been studied. For example, Schlörmann et al. (2020) studied the impact of roasting (140–180°C for 20 min in universal drum roaster) on β-glucan of oat kernels and flakes. In general, no impact of roasting oat products on the content of β-glucan were reported in this study, which agrees with the study of Hu et al. (2010) who reported no significant loss of β-glucan after steaming and roasting of oat. Similarly, Gujral et al. (2011) studied the effect of sand roasting on the extractability of β-glucan of oat. The authors reported that roasting significantly increased the extractable β-glucan content by 9.8–61.1 g/100 g oat flour due to heating effects on the cell wall structure of grains. Sharma et al. (2011) also studied the influence of sand roasting on the β-glucan of barley. The roasting process did not have any significant effect on the total β-glucan contents or its extractability. However, the roasting lowered the ratio of soluble to insoluble β-glucan contents. The results show a decrease in soluble β-glucan content, and this decrease ranged from 4.9 to 25.3%.

The effect of roasting on the proteins of cereals was also studied. Several studies reported that the roasting process does not affect the protein profile and essential amino acids of the same seeds

(Alderaywsh et al., 2019; Mohamed Ahmed et al., 2020). Similar results were reported for roasted flaxseeds (160°C, 8 min) by Waszkowiak et al. (2018). However, the authors reported that increasing the roasting time and temperature significantly altered the proteins of flaxseeds. Ma et al. (2020) investigated the effect of roasting on the proteins of two varieties of buckwheat. After roasting (200°C for 50 sec), the protein content of Tartary buckwheat significantly decreased by 12.40–43.42%, whereas that of common buckwheat increased by 43.18–75.55%. The authors did not explain these results. Małgorzata et al. (2016) also studied the consequence of roasting on the protein profile of buckwheat. Increasing roasting time resulted in a significant decrease in tryptophan fluorescence and total protein content, which indicates deterioration of protein nutritional quality during the processing. Similarly, Bhinder et al. (2019) reported a decrease in protein content of buckwheat after infrared roasting.

The roasting effect on the fat contents of cereal was studied. Several studies reported that roasting does not affect fat contents and fatty acid profiles of cereal such as oat products (Schlörmann et al., 2020) and chia seed flours (Hatamian et al., 2020). In contrast, some authors reported that the roasting of seeds leads to improved fat content due to heat-induced microstructural modifications of cell walls enabling the release of oil (Yodkaew et al., 2017). Another study established that roasting enhances the concentration of stearic, linoleic, and behenic acids. However, in the same study, the authors reported that roasting decreases the quantities of palmitic, arachidic, and linolenic acids due to the oxidative and non-oxidative destruction of saturated and unsaturated fatty acids following a heat treatment (Mohamed Ahmed et al., 2020). Similarly, Tanwar et al. (2009) reported that roasting buckwheat flour reduced lightly the fat content due to the formation of lipid starch complex, which is resistant to lipid extraction. The effect of roasting on fat contents depends on the cereal variety. Ma et al. (2020) reported that roasting increases the fat contents of common buckwheat up to 18.6%, whereas roasting decreases the fat contents of tartary buckwheat by 30–43%.

Roasting cereals showed variable effects on minerals. It was reported that roasting of the same seeds resulted in an increase of K, Na, Ca, and Cd contents; however, it reduced the levels of Mg, P, Fe, Mn, and Zn. Roasting did not affect the other trace elements such as B, Co, Cr, Cu, Mo, Ni, and Pb in the same study (Mohamed Ahmed et al., 2020). Obadina et al. (2016) reported that roasting of pearl millet grains reduced K, Ca, Mg, and P compared to unroasted raw samples (Obadina et al., 2016). Similarly, a reduction in K, Mg, Ca, P, S, Mn, Cu, and Zn levels in roasted chia seeds were reported (Ghafoor et al., 2018).

10.3.2.3 Effects of Roasting on Biological Characteristics of Cereal

The roasting process involves multiple reactions, including Maillard reaction, caramelization, and lipide thermal oxidation that produce distinct flavors such as N- and O-heterocyclic compounds, such as pyrazines, pyrroles, furans and pyranone, and multiples aldehydes, alcohols, phenols, and ketones, which lead to the desirable aroma of roasted cereals (Schlörmann et al., 2019). For example, the study of Majcher et al. (2013) demonstrated that the roasting of barley brew (at 180–200°C) resulted in the formation of 3-butanedione, 3-(methylthio)propanal, and 2-furfurylthiol that created the special coffee and nutty flavor of roasted products. Other composites such as methyl pyrazine, 2-furaldehyde, and maltol were detected in roasted barley malt (230°C, 130 min) (Sruthi et al., 2021).

Principally, the Maillard reaction products are responsible for the natural flavor of roasted cereals. The Maillard reaction depends on the concentration of amino acids and reducing sugars, moisture content, temperature, and time of treatment. The quantification of Millard reaction changes is defined by the available lysine, FAST index (fluorescence of advanced Maillard reaction products and soluble tryptophan), loss in ophthaldialdehyde reactivity, and browning index (Bhinder et al.,

2019). The FAST index is calculated as the ratio of the fluorescence exhibited by free fluorescent intermediate compounds to soluble tryptophan. The FAST index is a reliable and sensitive indicator of the nutritional damage induced by thermal treatments. Małgorzata et al. (2016) reported increased FAST index for roasted buckwheat grains, which is connected to the observed decrease in fluorescence of tryptophan and decrease in protein and available lysine contents and the nutritional value of buckwheat, as a result, along with increasing times of roasting. Additionally, the Maillard reaction during roasting also leads to the formation of acrylamide, a probable human carcinogen, which may reduce the nutritional value of roasted cereals. The study of Schlörmann et al. (2020) indicated that the acrylamide contents in roasted oat products increase with increasing toasting temperature. The acrylamide was below the detection limit of 10 μg/kg in raw oat kernels, but it increased to 11.5 μg/kg and 449 μg/kg for kernels roasted at 140°C and 180°C, respectively. Higher acrylamide levels were measured on roasted oat flakes than in kernels. The acrylamide contents in thin and thick flakes, roasted at 180°C, were 651 μg/kg and 875 μg/kg, respectively.

Color is an important quality control indicator of cereal products (Singh et al., 2021). In addition to flavor development, heat-induced reactions, particularly caramelization and browning reactions, are behind the typical colors of roasted cereal products (Sruthi et al., 2021). In the case of roasted buckwheat flours Ma et al. (2020) reported that roasting decreases lightness value (L*), showing the development of darker color increases the redness value (a*) and the yellowness value (b*), indicating the formation of brown pigments via Maillard reaction. Browning index measures the amount of color change to brown. This index increased at higher roasting temperatures and time. Schoeman et al. (2017) reported that traditional oven toasting led to a darker yellow-brown color compared to innovative forced convection continuous tumble (FCCT) roasting under similar conditions (180°C for 140 s), due to using superheated steam in the FCCT method, which results in less color deterioration in comparison to hot air. Similar results were also reported for oat flour (Gujral et al., 2011). After roasting the lightness (L*) decreased by 2–4% approximately and the redness (a*) and yellowness (b*) increased significantly up to 46.7% and 15.6%, respectively. These results are in perfect agreement with those of Sharma et al. (2011) who studied barley flour roasting. Upon hot sand roasting, the lightness (L*) lowered by 2.7–7,0%, and the redness (a*) and yellowness (b*) values were increased by 102–255% and 12.0-44.8%, respectively (Sharma et al., 2011).

Roasting affects cereal bioactive components such as phenolic and flavonoids compounds and their antioxidant properties have also been studied. These components are important in terms of food quality and human health. There have been conflicting results on the effect of roasting on the antioxidant activity of roasted cereal, which is basically determined by amounts of the phenolic and flavonoid compounds. Roasting affects the phenolic and flavonoid compounds in different manners. The different reactivities to the roasting process are due to the differences in chemical structures and binding statutes to cells tissues of these compounds. Mohamed Ahmed et al. (2020) studied the effect of roasting on the antioxidant activity, total phenolic contents, and total flavonoid contents of some seeds. For example, among the phenolic acids, roasting reduced the amounts of protocatechuic, syringic, caffeic, and trans-ferulic acids, whereas it improved the levels of gallic, p-coumaric, and trans-cinnamic acids. Concerning the flavonoids, roasting decreases the contents of flavan-3-ol (catechin) and flavanones (naringenin), whereas it enhanced the levels of stilbene, flavones, and flavonols. The authors resumed that roasting improved the bioactive properties and antioxidant activity of samh seeds due to the global increase of total phenolic and total flavonoid concentrations in treated samples (Mohamed Ahmed et al., 2020). This increase is likely due to the disruption of cellar structure and the release of free phenolic and flavonoids compounds. Furthermore, the production of new phenolic compounds during roasting, due to Maillard reaction products, can also be the cause of this increase (Sacchetti et al., 2016). The increase in antioxidant activity of the roasted cereals as the levels of phenolic compounds decreased with increasing heating temperature suggests that the antioxidant activity is not exclusively based on the phenolics. Instead, non-enzymatic browning reactions products that may have been formed during roasting probably contributed to the antioxidant activity (Irondi et al., 2019). Tanwar et al. (2019) measured an increase of 21.71% in the antioxidant activity of roasted buckwheat flour thanks to the formation of non-enzymatic browning products, especially melanonids, which contribute to the antioxidant activity. Another study reported comparable modifications in phenolic compounds, such as trans-ferulic, p-hydroxybenzoic, p-coumaric, vanillic acids, kaempferol, and quercetin for roasted quinoa seeds (Carciochi et al., 2016). The effect of the different roasting processes on the antioxidant activities, in vitro and cells, of barley malts, were studied. The authors reported a positive impact on the biological functions of barley malts due to roasting. This observation may be related to the liberation of bound phenolics under thermal treatment. However, a slight decrease of soluble ester and glycoside phenolics and flavonoids was recorded with increasing roasting temperature (Chen et al., 2019). Similarly, the roasting effects on the antioxidant quality of chia seeds flour was studied (Hatamian et al., 2020). Increasing roasting time at 160°C had no significant effect on total phenolic contents, while an increase in roasting time from 15 to 25 min at 180°C decreased the total phenolic content. However, the antioxidant activities increased at this stage due to the formation of melanoidins resulting from the non-enzymatic browning reaction. Further increase of roasting time to 35 min significantly increased the total phenolic but decreased the antioxidant activities of roasted seeds. In contrast, another study reported that roasting buckwheat flour results in a decrease in total phenolics and flavonoids and antioxidative activities (Bhattacharya, 2014). This reduction is due to heat-induced polymerization, which augment the molecular weights of phenolic compounds and decrease their solubility (Carciochi et al., 2016). The loss may also be attributed to thermal decomposition and heat-induced oxidation of the phenolic compounds (Rawson et al., 2013). Similarly, roasting of buckwheat reduced the total flavonoid content by 30.02%–39.03% in common buckwheat and by 4.76%–8.95% in tartary buckwheat, respectively (Ma et al., 2020). The total phenol content of common buckwheat increased by 30.98%–49.55% after roasting compared to raw samples, whereas the total phenols of tartary buckwheat decreased by 18.37%–36.35% after roasting. As a result, the total antioxidant capacity of common buckwheat increased upon roasting up to 49.45% but decreased for the tartary buckwheat by 13.24 to 24.33%. In summary, the global effect of roasting on the cereal's total antioxidant capacity depends on the balance between the thermal degradation of naturally existing antioxidant compounds and the formation of new compounds having antioxidant activity (Rizki et al., 2015).

10.4 WET HEATING OF CEREAL

10.4.1 Methods and Machinery of Wet Heating

Wet heating treatments include cooking, steaming, autoclaving, annealing, boiling, parboiling, and heat-moisture treatment. Wet heating treatments are applied in cereals for improvement in palatability, stability, color, nutritional properties, physicochemical properties, and shelf life of cereals (Kadiri, 2017; Matin et al., 2021). Cooking is a technological process of grain treatments. Raw grains are cooked to improve their nutritional properties and digestibility mainly through the gelatinization of starch granules that are embedded in the endosperm (Matin et al., 2021). It is generally carried out in a hermetically sealed high-pressure vessel. Matin et al. (2021) treated wheat and triticale grains in a similar vessel with another perforated vessel inside it to prevent direct contact of grains with high temperature and water for 10 and 15 min at 0.5 bar pressure. Thermometers and manometers are installed with the device to control temperature and pressure along with safety valves and steam releasing valves. For treatment of grains through steaming, grains are directly exposed to steam at atmospheric pressure as opposed to autoclaving. Karanam et al. (2020) steamed barley grains by spreading them on steel trays of thickness 2.5 cm and exposed to live steam at 98°C at atmospheric pressure in an autoclave for a time ranging from 5 to 30 min. In autoclaving, samples are exposed to steam at temperatures above 100°C and pressure for about 15–30 minutes (Karanam et al., 2020). In the hydrothermal treatment of maize grains, after acquiring moisture content in samples of 15%, 25%, and 35%, grains were steamed in an autoclave at 110°C, 5 psi for 15–30 minutes (Rocha-Villarreal et al., 2018).

Annealing is a hydrothermal method for the physical modification of starch as it modifies starch properties without rupturing the starch granule (Mathobo et al., 2021). Chung et al. (2012) carried out annealing of germinated brown rice grains. They placed 150 g of grains in 1000 ml and incubated these at 50°C for 24 hrs and subsequently dried at 40°C (Chung et al., 2012). Parboiling is a hydrothermal treatment, commonly applied in paddy, to improve nutritional quality and milling yield. It is a three-step process involving soaking, steaming, and drying. Partial gelatinization of starch during the process causes physical, chemical, and organoleptic modification in cereal grains (Rocha-Villarreal et al., 2018).

10.4.2 Impact of Wet Heating

10.4.2.1 Effects of Wet Heating on Macronutrients and Micronutrients of Cereal

Wet heating or hydrothermal treatments leads to various changes in the nutritional properties of cereal grains. It is responsible for the modification of starches and protein and also affects its digestibility. Hydrothermal treatments can be employed in the preparation of low GI cereal foods as sequential gelatinization and retrogradation of starch leads to the formation of resistant starches (Kaur et al., 2021). This is also observed in the parboiling process. Also, during the steaming process, sugars such as glucose, fructose, maltose, and sucrose leach out and are consumed in Maillard's reaction. Protein extractability (globulins, glutelins, and prolamines) is generally reduced but their digestibility is improved. This is due to the association of proteins with starch granules under excessive conditions of steaming (Rocha-Villarreal et al., 2018; Rocha-Villarreal et al., 2018). Maize grains were attained with a moisture content of 15, 25, and 35%, which are further autoclaved for 15 and 30 min. Maize grains with 25% moisture content, steamed for 15 and 30 mins, were found to have a maximum fat content of 1.76 and 1.96%. This is due to the disruption of fat globules located in the scutellum and aleurone layer, which migrate to the endosperm. Thiamine content in flour obtained with the whole caryopsis through each treatment indicates that soaking causes migration of thiamine from pericarp and germ to inner layers. Thus, preventing its losses. Similar phenomena have been observed during parboiling in paddy increases riboflavin, thiamine, calcium, and phosphorus concentration due to the inward diffusion from bran and aleurone layer to endosperm. But during steaming, pressure cooking, and atmospheric cooking, destruction or complexation of thiamine take place (Rocha-Villarreal et al., 2018; Rocha-Villarreal et al., 2018). Soaking and steaming in barley and oats does not significantly affect the concentration of minerals such as calcium, iron, phosphorus, zinc, and copper (Rocha-Villarreal et al., 2018). Steaming of pearl millet for 20 and 15 minutes before pearling and after pearling, respectively, was able to inactivate the lipase enzyme completely. Due to the absence of the outer layer on the grain, the time taken to inactivate lipase was lower in steam after pearling treatment (Yadav et al., 2012).

10.4.2.2 Effects of Wet Heating on Rheological and Pasting Properties

Modification of starch granules upon wet heating led to an alteration in pasting properties of treated flours. Upon heat-moisture treatment of wheat flour and wheat starch with moisture at 25 and

30% and temperature at 100 and 120°C, pasting temperature increases with an increase in moisture content and temperature in heat-moisture treatment (Li et al., 2019). Also, wheat flour has a higher pasting temperature as compared to the wheat starch samples due to the presence of wheat proteins. The increase in pasting temperature is due to the rearrangement of the starch granules that leads to the formation of double-helical amylopectin clusters. Thus, increasing the region of crystallinity. This limits the starch swelling. The formation of starch-lipid complexes thus contributes to reduction of swelling of starch granules and results in increase in pasting temperature. The viscosities of wheat flour decreased upon treatment due to the denaturation of proteins, influencing starch-chain and starch-protein interactions (Li et al., 2019).

Germinated brown rice grains after annealing and heat-moisture treatment were found to have increased peak, breakdown, setback, and final viscosity. The increase in viscosities was higher in heat-moisture treatment as compared to annealing. Annealing restricts amylose leaching and intragranular binding leads to an increase in viscosity whereas under heat-moisture treatment granular rigidity increases due to an increase in starch chain interactions in amorphous regions owing to the volume occupied by swollen grains. Also, under heat-moisture treatment, reordering of starch molecules takes place, and increased mobility of starch leads to an increase in its tendency to aggregate to form a new crystalline structure. An increase in pasting viscosity is majorly attributed to the improved gelatinization and improved thermal stability of germinated grains after heat-moisture treatment (Chung et al., 2012).

10.4.2.3 Effects of Wet Heating on Functional and Physical Properties

Water-holding capacity is the ability of the flour to hold water physically. Water-holding capacity and swelling index of amaranth grains increase with increase in steaming time up to 3 and 6 min and further decreases up to 9 min. The decrease in water-holding capacity and swelling index was observed due to the breakdown of the starch-protein complex that leads to a loss in amylose content and solubilization of starch in water. Excessive heating promotes amylose leaching and its solubilization, which renders it unavailable for interaction with amylopectin. Hence, reducing the water-holding capacity and swelling index at high temperatures for a longer duration (Malik et al., 2021). In the hydrothermal treatment of corn, swelling power decreases at steaming of grains with high moisture content (25 and 35%) when steaming time is increased up to 30 min. The reduction in swelling power is due to reduced hydration and increased amylose and amylopectin interaction along with arrangement in the crystalline region of starch. This lowers the ability of double helices of amylopectin to retain and absorb water, whereas water solubility in corn increased with an increase in steaming and was maximum at 30 minutes. This could be due to leaching out of amylose and partial disruption of amylopectin helices in heat-damaged granules. Steaming could also facilitate the solubility of proteins (Rocha-Villarreal et al., 2018).

Barley grains steamed for 5, 10, 15, 20, 25, and 30 min reduced L* values and increased a* and b* values with an increase in time of the steaming process. The color change is attributed to non-enzymatic browning reactions and caramelization reactions (Karanam et al., 2020). Similar changes were observed after the steaming of Amaranth grains. The a* value increased with an increased in steaming time due to an increase in the production of melanoidin pigment (Malik et al., 2021). Steaming in pearl millet before and after pearling also led to a decrease in L* value due to the polymerization of anthocyanins and polyphenolic compounds. Conversion of flavonol pigment compounds into intermediate compounds aids in a change in color (Yadav et al., 2012). Bulk density in pearl millet was found to be increased upon steaming before pearling. However, in steaming after pearling operation, bulk density increased initially but lowered after 15 minutes (Yadav et al., 2012).

10.4.2.4 Effects of Wet Heating on Structural Properties

The hardness of grains increases upon hydrothermal treatment. With the increase in steaming time from 5 to 30 min, barley grains are shown to have a higher hardness value. The force required to break the grains increases from 152.21 N to 240.35 N. Hardness in grains is due to strong adhesion between starch granules and protein matrix in the endosperm. SEM analysis of the barley grains upon hydrothermal treatment showed more compactly packed starch granules in the protein matrix as opposed to native barley grains in which starch granules were loose packed and less intact (Karanam et al., 2020). Heat-moisture treatment of wheat flour and wheat starches at 25 and 30% moisture content and 100 and 120°C causes some changes in morphological characteristics in starch granules as observed under SEM analysis. In comparison to round- and oval-shaped starch granules, starch granules in samples have dents and were broken down during treatment indicating partial gelatinization. In wheat flour samples, protein structure was degraded with an increase in temperature and moisture. Aggregation and clump formation were more evident with an increase in moisture (Li et al., 2019).

Dough prepared from steamed amaranth grains for 3, 6, and 9 min was assessed for its microstructure. It was observed that under the effect of hydrothermal processing, starch particles became closely bound to the protein, and the surface of dough became comparatively smooth and flat. Starch granules are bound tightly with protein, and it is also suggested that gelatinized starch and coagulated protein acts as cementing material in steamed samples. However, upon increasing the duration of steaming to 9 min, the protein network appears to be disrupted and clusters and protein and starch granules give it a coarser appearance (Malik et al., 2021).

10.4.2.5 Effects of Wet Heating on Bioactive Profiles and Antinutritional Factors

In barley, corn, red sorghum, and finger millet, total phenolic content and antioxidant activity can be improved by cooking under pressure or boiling. These treatments contribute to increasing free phenolics such as vanillin, free ferulic acid, and p-coumaric acid in corn (Ragaee et al., 2014). Some of the wet heating treatments may have detrimental effects on the bioactive compounds in cereals while others may improve the extractability of these compounds. In rice cultivars, free phenolic compounds reduce by 16–91% upon hydrothermal processing. Cooking in brown and polished rice decreases the concentration of flavonoids, free and bound phenolics, and subsequently decreases the antioxidant capacity due to degradation of phenolic compounds upon thermal treatment leading to decarboxylation and polymerization of free phenolic compounds. During autoclaving of wheat bran, total phenolic compounds increased by 50% in water-soluble extract as, under the effect of high temperature and pressure, partial hydrolysis of fiber polysaccharides takes place facilitating the breakdown of a cell wall structure and releasing bound phenolic compounds (Kadiri, 2017).

Steaming in corn has been shown to reduce the carotenoids content due to oxidation or degradation of carotenoid compounds (Rocha-Villarreal et al., 2018). Also, free phenolic content decreases by 15% and 25% when steamed for 15 and 30 minutes after soaking. However, samples with a moisture content of 35% (attained by soaking) had an increase in phenolic content concentration as compared to non-steamed counterparts as high temperature for longer duration leads to increase the extractability and release of bound phenolics due to detrimental effect on cell-wall structure and its binding properties.

The research studied the comparative effects of dry heating and wet heating methods on bioactive compounds of sorghum flour. It was concluded that cooked sorghum grains had total retention of 3-deoxy anthocyanidins, total phenolic content, and vitamin E of 87.6%, 73.9%, and 86.3%, respectively. The content of luteolinidin, apigeninidin, 7-methoxy-apigeninidin, and

5-methoxy-luteolinidin was significantly reduced upon cooking whereas total carotenoid retention in cooking was 98.8%, which was higher than in dry heating treatments. Loss of phenolic compounds in the cooking water is responsible for reduced total phenolic content. Lower retention of vitamin E is due to its susceptibility towards the action of heat. The reduced TPC upon cooking is attributed to the loss on phenolic compounds in the cooking water. However, wet heat treatment improves the accessibility of carotenoid compounds and compensates for the minor loss due to leaching (Cardoso et al., 2014). Hydrothermal treatment (steaming) in pearl millet enables significant reduction in total phenolic content and tannin content, which are responsible for bitterness. The reduction in phenolic compounds is due to the effect of high temperature (Yadav et al., 2012). Antioxidant activity of pearl millet cultivars decreased by 9.6 to 37.4% upon cooking in boiling water due to leaching out of antioxidants in water (Siroha & Sandhu, 2017). Antinutrients such as phytic acid, tannins, saponins, lectins, and trypsin inhibitors are significantly reduced by autoclaving and cooking. Application of heat activates phytase enzyme, which is responsible for the reduction of phytic acid content, and leads to increased availability of free mineral content. Reduction in tannin content improves protein digestibility as tannins can interact with protein and reduce its absorption in the human body (Samtiya et al., 2020).

10.4.2.6 Effects of Wet Heating on Digestibility of Nutrients

Assessment of in vitro digestibility in heat-moisture treated wheat flour and wheat starch samples was carried out by Li et al. (2019). With the increase in heat and moisture, in both flour and starch samples, rapidly digestible starch (RDS) and slowly digestible starch (SDS) content decrease whereas resistant starch (RS) content increases. The presence of high temperature and moisture content ease information of resistant starch due to interaction between starch, lipids, and proteins that restrict enzyme-mediated hydrolysis of starch chains (Li et al., 2019). Upon annealing and heat-moisture treatment of germinated brown rice grains, rapidly digestible starch decreased whereas slowly digestible and resistant starches increased significantly. Decrease in starch digestibility is more prominent in heat-moisture treatment than annealing, which is due to an increase in structural rigidity associated with chain association and rearrangement (Chung et al., 2012).

10.5 CONCLUSION

The heating of cereal, dry or wet heating, results in quality enhancement in terms of macro- and micronutrients, improve its digestibility, palatability, and functional properties. Thermal treatments are considered green, safe, and efficient as they significantly reduce the level of antinutritional factors and enhance the nutritional quality of cereals. Furthermore, cereal flours and starches as a result of dry and wet heating modulate their functionality, thus making them useful as functional ingredients in gluten-free products. Also, the products prepared from modified cereals have better sensory acceptability and nutritional value.

REFERENCES

Abulude, F. O., Lawal, L. O., & Kayode, A. O. (2005). Effect of processing on some functional properties of millet (Eleusine coracana) flour. *Journal of Food Technology*, *3*(30), 460–463.

Alderaywsh, F., Osman, M. A., Al-Juhaimi, F. Y., Gassem, M. A., Al- Maiman, S. A., Adiamo, O. Q., Özcan, M. M., & Ahmed, I. A. M. (2019). Effect of Traditional Processing on the Nutritional Quality and *in vivo* Biological Value of Samh (*Mesembryanthemum forsskalei* Hochst) Flour. *Journal of Oleo Science*, *68*(10), 1033–1040. https://doi.org/10.5650/jos.ess19095

Bender, D., & Bender, A. (1999). *Benders' dictionary of nutrition and food technology* (7th éd.). Woodhead Publishing.

Bertuzzi, T., Martinelli, E., Mulazzi, A., & Rastelli, S. (2020). Acrylamide determination during an industrial roasting process of coffee and the influence of asparagine and low molecular weight sugars. *Food Chemistry*, *303*, 125372. https://doi.org/10.1016/j.foodchem.2019.125372

Bhattacharya, S. (2014). Roasting and toasting operations in food: Process engineering and applications. In S. Bhattacharya (Éd.), *Conventional and Advanced Food Processing Technologies* (p. 221–248). John Wiley & Sons, Ltd. https://doi.org/10.1002/9781118406281.ch10

Bhinder, S., Singh, B., Kaur, A., Singh, N., Kaur, M., Kumari, S., & Yadav, M. P. (2019). Effect of infrared roasting on antioxidant activity, phenolic composition, and Maillard reaction products of Tartary buckwheat varieties. *Food Chemistry*, *285*, 240–251. https://doi.org/10.1016/j.foodchem.2019.01.141

Bucsella, B., Takács, Á., Vizer, V., Schwendener, U., & Tömösközi, S. (2016). Comparison of the effects of different heat treatment processes on rheological properties of cake and bread wheat flours. *Food Chemistry*, *190*, 990–996. https://doi.org/10.1016/j.foodchem.2015.06.073

Carciochi, R. A., Galván D'Alessandro, L., & Manrique, G. D. (2016). Effect of roasting conditions on the antioxidant compounds of quinoa seeds. *International Journal of Food Science & Technology*, *51*(4), 1018–1025. https://doi.org/10.1111/ijfs.13061

Cardoso, L. de M., Montini, T. A., Pinheiro, S. S., Pinheiro-Sant'Ana, H. M., Martino, H. S. D., & Moreira, A. V. B. (2014). Effects of processing with dry heat and wet heat on the antioxidant profile of sorghum. *Food Chemistry*, *152*, 210–217. https://doi.org/10.1016/j.foodchem.2013.11.106

Carrera, Y., Utrilla-Coello, R., Bello-Pérez, A., Alvarez-Ramirez, J., & Vernon-Carter, E. J. (2015). In vitro digestibility, crystallinity, rheological, thermal, particle size and morphological characteristics of pinole, a traditional energy food obtained from toasted ground maize. *Carbohydrate Polymers*, *123*, 246–255. https://doi.org/10.1016/j.carbpol.2015.01.044

Cenkowski, S., Ames, N., & Muir, W. E. (2006). Infrared processing of oat grits in a laboratory scale electric micronizer. *Canadian Biosystems Engineering*, *48*, 17–25.

Chandrapala, J., Oliver, C., Kentish, S., & Ashokkumar, M. (2012). Ultrasonics in food processing. *Ultrasonics Sonochemistry*, *19*(5), 975–983. https://doi.org/10.1016/j.ultsonch.2012.01.010

Charalampopoulos, D., Wang, R., Pandiella, S. S., & Webb, C. (2002). Application of cereals and cereal components in functional foods: A review. *International Journal of Food Microbiology*, *79*(1–2), 131–141. https://doi.org/10.1016/S0168-1605(02)00187-3

Chen, Y., Huang, J., Hu, J., Yan, R., & Ma, X. (2019). Comparative study on the phytochemical profiles and cellular antioxidant activity of phenolics extracted from barley malts processed under different roasting temperatures. *Food & Function*, *10*(4), 2176–2185. https://doi.org/10.1039/C9FO00168A

Chi, C., Li, X., Lu, P., Miao, S., Zhang, Y., & Chen, L. (2019). Dry heating and annealing treatment synergistically modulate starch structure and digestibility. *International Journal of Biological Macromolecules*, *137*, 554–561. https://doi.org/10.1016/j.ijbiomac.2019.06.137

Chung, H.-J., Cho, A., & Lim, S.-T. (2012). Effect of heat-moisture treatment for utilization of germinated brown rice in wheat noodle. *LWT*, *47*(2), 342–347. https://doi.org/10.1016/j.lwt.2012.01.029

Chung, H.-J., Cho, D.-W., Park, J.-D., Kweon, D.-K., & Lim, S.-T. (2012). In vitro starch digestibility and pasting properties of germinated brown rice after hydrothermal treatments. *Journal of Cereal Science*, *56*(2), 451–456. https://doi.org/10.1016/j.jcs.2012.03.010

Chung, H.-S., Chung, S.-K., & Youn, K.-S. (2011). Effects of roasting temperature and time on bulk density, soluble solids, browning index and phenolic components of corn kernel. *Journal of Food Processing and Preservation*, *35*(6), 832–839. https://doi.org/10.1111/j.1745-4549.2011.00536.x

Cooke, D., & Gidley, M. J. (1992). Loss of crystalline and molecular order during starch gelatinisation: Origin of the enthalpic transition. *Carbohydrate Research*, *227*, 103–112. https://doi.org/10.1016/0008-6215(92)85063-6

Englyst, H. N., & Hudson, G. J. (1996). The classification and measurement of dietary carbohydrates. *Food Chemistry*, *57*(1), 15–21. https://doi.org/10.1016/0308-8146(96)00056-8

Fadai, N. T., Melrose, J., Please, C. P., Schulman, A., & Van Gorder, R. A. (2017). A heat and mass transfer study of coffee bean roasting. *International Journal of Heat and Mass Transfer*, *104*, 787–799. https://doi.org/10.1016/j.ijheatmasstransfer.2016.08.083

Falsafi, S. R., Maghsoudlou, Y., Rostamabadi, H., Rostamabadi, M. M., Hamedi, H., & Hosseini, S. M. H. (2019). Preparation of physically modified oat starch with different sonication treatments. *Food Hydrocolloids*, *89*, 311–320. https://doi.org/10.1016/j.foodhyd.2018.10.046

FAO. (2002). *World Agriculture: Towards 2015/2030. Summary Report.* FAO. https://www.fao.org/3/y4252e/y4252e00.htm

Fast, R. B., Fred, J. S., William, J. T., Donald, D. T., & Suson, J. G. (1990). Toasting and toasting ovens for breakfast cereals. *Cereal Food World*, *35*(3), 299.

Ghafoor, K., Aljuhaimi, F., Özcan, M. M., Uslu, N., Hussain, S., Babiker, E. E., & Fadimu, G. (2018). Effects of roasting on bioactive compounds, fatty acid, and mineral composition of chia seed and oil. *Journal of Food Processing and Preservation*, *42*(10), jfpp.13710. https://doi.org/10.1111/jfpp.13710

Gong, B., Xu, M., Li, B., Wu, H., Liu, Y., Zhang, G., Ouyang, S., & Li, W. (2017). Repeated heat-moisture treatment exhibits superiorities in modification of structural, physicochemical and digestibility properties of red adzuki bean starch compared to continuous heat-moisture way. *Food Research International*, *102*, 776–784. https://doi.org/10.1016/j.foodres.2017.09.078

Gujral, H. S., Sharma, P., & Rachna, S. (2011). Effect of sand roasting on beta glucan extractability, physicochemical and antioxidant properties of oats. *LWT – Food Science and Technology*, *44*(10), 2223–2230. https://doi.org/10.1016/j.lwt.2011.06.001

Hatamian, M., Noshad, M., Abdanan-Mehdizadeh, S., & Barzegar, H. (2020). Effect of roasting treatment on functional and antioxidant properties of chia seed flours. *NFS Journal*, *21*, 1–8. https://doi.org/10.1016/j.nfs.2020.07.004

Hernández, O. M., Fraga, J. M. G., Jiménez, A. I., Jiménez, F., & Arias, J. J. (2014). Characterization of toasted cereal flours from the Canary Islands (gofios). *Food Chemistry*, *151*, 133–140. https://doi.org/10.1016/j.foodchem.2013.11.039

Hu, X., Xing, X., & Ren, C. (2010). The effects of steaming and roasting treatments on β-glucan, lipid and starch in the kernels of naked oat (*Avena nuda*): Effects of steaming and roasting in kernels of *Avena nuda*. *Journal of the Science of Food and Agriculture*, *90*(4), 690–695. https://doi.org/10.1002/jsfa.3870

Irondi, E. A., Adegoke, B. M., Effion, E. S., Oyewo, S. O., Alamu, E. O., & Boligon, A. A. (2019). Enzymes inhibitory property, antioxidant activity and phenolics profile of raw and roasted red sorghum grains in vitro. *Food Science and Human Wellness*, *8*(2), 142–148. https://doi.org/10.1016/j.fshw.2019.03.012

Jogihalli, P., Singh, L., Kumar, K., & Sharanagat, V. S. (2017). Novel continuous roasting of chickpea (Cicer arietinum): Study on physico-functional, antioxidant and roasting characteristics. *LWT*, *86*, 456–464. https://doi.org/10.1016/j.lwt.2017.08.029

Kadiri, O. (2017). A review on the status of the phenolic compounds and antioxidant capacity of the flour: Effects of cereal processing. *International Journal of Food Properties*, *20*(sup1), S798–S809. https://doi.org/10.1080/10942912.2017.1315130

Kanagaraj, S. P., Ponnambalam, D., & Antony, U. (2019). Effect of dry heat treatment on the development of resistant starch in rice (Oryza sativa) and barnyard millet (Echinochloa furmantacea). *Journal of Food Processing and Preservation*, *43*(7). https://doi.org/10.1111/jfpp.13965

Karanam, M., Theertha, D. P., Kumar, A., Inamdar, A. A., & Sakhare, S. D. (2020). Effect of hydrothermal treatment on physical and semolina milling properties of barley. *Journal of Food Engineering*, *287*, 110142. https://doi.org/10.1016/j.jfoodeng.2020.110142

Kaur, J., Kaur, K., Singh, B. Singh, A., & Sharma, S. (2021). Insights into the latest advances in low glycemic foods, their mechanism of action and health benefits. *Journal of Food Measurement and Characterization*. https://doi.org/10.1007/s11694-021-01179-z

Kaur, M., Punia, S., Sandhu, K. S., & Ahmed, J. (2019). Impact of high pressure processing on the rheological, thermal and morphological characteristics of mango kernel starch. *International Journal of Biological Macromolecules*, *140*, 149–155. https://doi.org/10.1016/j.ijbiomac.2019.08.132

Khan, N., Zaman, R., & Elahi, M. (1991). Effect of heat treatments on the phytic acid content of maize products. *Journal of the Science of Food and Agriculture*, *54*(1), 153–156. https://doi.org/10.1002/jsfa.2740540117

Lee, H. J. (2020). Stability of ochratoxin A in oats during roasting with reducing sugars. *Food Control*, *118*, 107382. https://doi.org/10.1016/j.foodcont.2020.107382

Li, M.-N., Zhang, B., Xie, Y., & Chen, H.-Q. (2019). Effects of debranching and repeated heat-moisture treatments on structure, physicochemical properties and in vitro digestibility of wheat starch. *Food Chemistry*, *294*, 440–447. https://doi.org/10.1016/j.foodchem.2019.05.040

Li, Y., Zhang, H., Shoemaker, C. F., Xu, Z., Zhu, S., & Zhong, F. (2013). Effect of dry heat treatment with xanthan on waxy rice starch. *Carbohydrate Polymers*, *92*(2), 1647–1652. https://doi.org/10.1016/j.carbpol.2012.11.002

Liu, R. H. (2007). Whole grain phytochemicals and health. *Journal of Cereal Science*, *46*(3), 207–219. https://doi.org/10.1016/j.jcs.2007.06.010

Ma, Q., Zhao, Y., Wang, H.-L., Li, J., Yang, Q.-H., Gao, L.-C., Murat, T., & Feng, B.-L. (2020). Comparative study on the effects of buckwheat by roasting: Antioxidant properties, nutrients, pasting, and thermal properties. *Journal of Cereal Science*, *95*, 103041. https://doi.org/10.1016/j.jcs.2020.103041

Majcher, M. A., Klensporf-Pawlik, D., Dziadas, M., & Jeleń, H. H. (2013). Identification of aroma active compounds of cereal coffee brew and its roasted ingredients. *Journal of Agricultural and Food Chemistry*, *61*(11), 2648–2654. https://doi.org/10.1021/jf304651b

Małgorzata, W., Konrad, P. M., & Zieliński, H. (2016). Effect of roasting time of buckwheat groats on the formation of Maillard reaction products and antioxidant capacity. *Food Chemistry*, *196*, 355–358. https://doi.org/10.1016/j.foodchem.2015.09.064

Malik, A., Khamrui, K., & Prasad, W. (2021). Effect of hydrothermal treatment on physical properties of amaranth, an underutilized pseudocereal. *Future Foods*, *3*, 100027. https://doi.org/10.1016/j.fufo.2021.100027

Maniglia, B. C., Castanha, N., Le-Bail, P., Le-Bail, A., & Augusto, P. E. D. (2021a). Starch modification through environmentally friendly alternatives: A review. *Critical Reviews in Food Science and Nutrition*, *61*(15), 2482–2505. https://doi.org/10.1080/10408398.2020.1778633

Maniglia, B. C., Castanha, N., Rojas, M. L., & Augusto, P. E. (2021b). Emerging technologies to enhance starch performance. *Current Opinion in Food Science*, *37*, 26–36. https://doi.org/10.1016/j.cofs.2020.09.003

Mann, J., Schiedt, B., Baumann, A., Conde-Petit, B., & Vilgis, T. A. (2014). Effect of heat treatment on wheat dough rheology and wheat protein solubility. *Food Science and Technology International*, *20*(5), 341–351. https://doi.org/10.1177/1082013213488381

Mariotti, M., Alamprese, C., Pagani, M. A., & Lucisano, M. (2006). Effect of puffing on ultrastructure and physical characteristics of cereal grains and flours. *Journal of Cereal Science*, *43*(1), 47–56. https://doi.org/10.1016/j.jcs.2005.06.007

Mathobo, V. M., Silungwe, H., Ramashia, S. E., & Anyasi, T. A. (2021). Effects of heat-moisture treatment on the thermal, functional properties and composition of cereal, legume and tuber starches—A review. *Journal of Food Science and Technology*, *58*(2), 412–426. https://doi.org/10.1007/s13197-020-04520-4

Matin, A., Grubor, M., Ostroski, N., Jurisic, V., Bilandzija, N., Kontek, M., Zdunic, Z., & Kricka, T. (2021). Effect of hydrothermal treatment on the improvement of wheat and tritcale grain properties. *Agriculturae Conspectus Scientificus*, *86*(3), 243–250.

Miano, A. C., Rojas, M. L., & Augusto, P. E. D. (2017). Other Mass Transfer Unit Operations Enhanced by Ultrasound. In *Ultrasound: Advances for Food Processing and Preservation* (p. 369–389). Elsevier. https://doi.org/10.1016/B978-0-12-804581-7.00015-4

Miao, M., Zhang, T., & Jiang, B. (2009). Characterisations of kabuli and desi chickpea starches cultivated in China. *Food Chemistry*, *113*(4), 1025–1032. https://doi.org/10.1016/j.foodchem.2008.08.056

Mohamed Ahmed, I. A., Al Juhaimi, F. Y., Osman, M. A., Al Maiman, S. A., Hassan, A. B., Alqah, H. A. S., Babiker, E. E., & Ghafoor, K. (2020). Effect of oven roasting treatment on the antioxidant activity, phenolic compounds, fatty acids, minerals, and protein profile of Samh (Mesembryanthemum forsskalei Hochst) seeds. *LWT*, *131*, 109825. https://doi.org/10.1016/j.lwt.2020.109825

Mrad, R., Debs, E., Saliba, R., Maroun, R. G., & Louka, N. (2014). Multiple optimization of chemical and textural properties of roasted expanded purple maize using response surface methodology. *Journal of Cereal Science*, *60*(2), 397–405. https://doi.org/10.1016/j.jcs.2014.05.005

Murthy, K. V., Ravi, R., Bhat, K. K., & Raghavarao, K. S. M. S. (2008). Studies on roasting of wheat using fluidized bed roaster. *Journal of Food Engineering*, *89*(3), 336–342. https://doi.org/10.1016/j.jfoodeng.2008.05.014

Neill, G., Al-Muhtaseb, A. H., & Magee, T. R. A. (2012). Optimisation of time/temperature treatment, for heat treated soft wheat flour. *Journal of Food Engineering, 113*(3), 422–426. https://doi.org/10.1016/j.jfoodeng.2012.06.019

Obadina, A., Ishola, I. O., Adekoya, I. O., Soares, A. G., de Carvalho, C. W. P., & Barboza, H. T. (2016). Nutritional and physico-chemical properties of flour from native and roasted whole grain pearl millet (Pennisetum glaucum [L.]R. Br.). *Journal of Cereal Science, 70*, 247–252. https://doi.org/10.1016/j.jcs.2016.06.005

Oboh, G., Ademiluyi, A. O., & Akindahunsi, A. A. (2010). The effect of roasting on the nutritional and antioxidant properties of yellow and white maize varieties: Roasting changes maize quality & antioxidants. *International Journal of Food Science & Technology, 45*(6), 1236–1242. https://doi.org/10.1111/j.1365-2621.2010.02263.x

Oh, I. K., Bae, I. Y., & Lee, H. G. (2018). Effect of dry heat treatment on physical property and in vitro starch digestibility of high amylose rice starch. *International Journal of Biological Macromolecules, 108*, 568–575. https://doi.org/10.1016/j.ijbiomac.2017.11.180

Ozawa, M., Kato, Y., & Seguchi, M. (2009). Investigation of Dry-Heated Hard and Soft Wheat Flour. *Starch – Stärke, 61*(7), 398–406. https://doi.org/10.1002/star.200800142

Ozel, M. Z., Yanık, D. K., Gogus, F., Hamilton, J. F., & Lewis, A. C. (2014). Effect of roasting method and oil reduction on volatiles of roasted Pistacia terebinthus using direct thermal desorption-GCxGC-TOF/MS. *LWT – Food Science and Technology, 59*(1), 283–288. https://doi.org/10.1016/j.lwt.2014.05.004

Pei-Ling, L., Xiao-Song, H., & Qun, S. (2010). Effect of high hydrostatic pressure on starches: A review. *Starch – Stärke, 62*(12), 615–628. https://doi.org/10.1002/star.201000001

Qiu, C., Cao, J., Xiong, L., & Sun, Q. (2015). Differences in physicochemical, morphological, and structural properties between rice starch and rice flour modified by dry heat treatment. *Starch – Stärke, 67*(9–10), 756–764. https://doi.org/10.1002/star.201500016

Ragaee, S., Seetharaman, K., & Abdel-Aal, E.-S. M. (2014). The Impact of Milling and Thermal Processing on Phenolic Compounds in Cereal Grains. *Critical Reviews in Food Science and Nutrition, 54*(7), 837–849. https://doi.org/10.1080/10408398.2011.610906

Rawson, A., Hossain, M. B., Patras, A., Tuohy, M., & Brunton, N. (2013). Effect of boiling and roasting on the polyacetylene and polyphenol content of fennel (Foeniculum vulgare) bulb. *Food Research International, 50*(2), 513–518. https://doi.org/10.1016/j.foodres.2011.01.009

Rizki, H., Kzaiber, F., Elharfi, M., Ennahli, S., & Hanine, H. (2015). *Effects of roasting temperature and time on the physicochemical properties of sesame (Sesamum indicum .L) seeds. 11*(1), 9.

Rocha-Villarreal, V., Hoffmann, J. F., Vanier, N. L., Serna-Saldivar, S. O., & García-Lara, S. (2018). Hydrothermal treatment of maize: Changes in physical, chemical, and functional properties. *Food Chemistry, 263*, 225–231. https://doi.org/10.1016/j.foodchem.2018.05.003

Rocha-Villarreal, V., Serna-Saldivar, S. O., & García-Lara, S. (2018). Effects of parboiling and other hydrothermal treatments on the physical, functional, and nutritional properties of rice and other cereals. *Cereal Chemistry, 95*(1), 79–91. https://doi.org/10.1002/cche.10010

Sacchetti, G., Ioannone, F., De Gregorio, M., Di Mattia, C., Serafini, M., & Mastrocola, D. (2016). Non enzymatic browning during cocoa roasting as affected by processing time and temperature. *Journal of Food Engineering, 169*, 44–52. https://doi.org/10.1016/j.jfoodeng.2015.08.018

Samtiya, M., Aluko, R. E., & Dhewa, T. (2020). Plant food anti-nutritional factors and their reduction strategies: An overview. *Food Production, Processing and Nutrition, 2*(1), 6. https://doi.org/10.1186/s43014-020-0020-5

Santos, J., Alvarez-Ortí, M., Sena-Moreno, E., Rabadán, A., Pardo, J. E., & Beatriz PP Oliveira, M. (2018). Effect of roasting conditions on the composition and antioxidant properties of defatted walnut flour: Effect of roasting on the chemical properties of defatted walnut flour. *Journal of the Science of Food and Agriculture, 98*(5), 1813–1820. https://doi.org/10.1002/jsfa.8657

Schlörmann, W., Zetzmann, S., Wiege, B., Haase, N. U., Greiling, A., Lorkowski, S., Dawczynski, C., & Glei, M. (2019). Impact of different roasting conditions on chemical composition, sensory quality and physicochemical properties of waxy-barley products. *Food & Function, 10*(9), 5436–5445. https://doi.org/10.1039/C9FO01429B

Schlörmann, W., Zetzmann, S., Wiege, B., Haase, N. U., Greiling, A., Lorkowski, S., Dawczynski, C., & Glei, M. (2020). Impact of different roasting conditions on sensory properties and health-related compounds of oat products. *Food Chemistry, 307*, 125548. https://doi.org/10.1016/j.foodchem.2019.125548

Schoeman, L., du Plessis, A., Verboven, P., Nicolaï, B. M., Cantre, D., & Manley, M. (2017). Effect of oven and forced convection continuous tumble (FCCT) roasting on the microstructure and dry milling properties of white maize. *Innovative Food Science & Emerging Technologies*, *44*, 54–66. https://doi.org/10.1016/j.ifset.2017.07.021

Shamekh, S., Forssell, P., & Poutanen, K. (1994). Solubility Pattern and Recrystallization Behavior of Oat Starch. *Starch – Stärke*, *46*(4), 129–133. https://doi.org/10.1002/star.19940460403

Sharanagat, V. S., Suhag, R., Anand, P., Deswal, G., Kumar, R., Chaudhary, A., Singh, L., Singh Kushwah, O., Mani, S., Kumar, Y., & Nema, P. K. (2019). Physico-functional, thermo-pasting and antioxidant properties of microwave roasted sorghum [Sorghum bicolor (L.) Moench]. *Journal of Cereal Science*, *85*, 111–119. https://doi.org/10.1016/j.jcs.2018.11.013

Sharma, P., & Gujral, H. S. (2011). Effect of sand roasting and microwave cooking on antioxidant activity of barley. *Food Research International*, *44*(1), 235–240. https://doi.org/10.1016/j.foodres.2010.10.030

Sharma, P., Gujral, H. S., & Rosell, C. M. (2011). Effects of roasting on barley β-glucan, thermal, textural and pasting properties. *Journal of Cereal Science*, *53*(1), 25–30. https://doi.org/10.1016/j.jcs.2010.08.005

Singh, A., & Sharma, S. (2017). Bioactive components and functional properties of biologically activated cereal grains: A bibliographic review. *Critical Reviews in Food Science and Nutrition*, 57(14), 3051–3071.

Singh, A., Bobade, H., Sharma, S., Singh, B., & Gupta, A. (2021). Enhancement of digestibility of nutrients (in vitro), antioxidant potential and functional attributes of wheat flour through grain germination. *Plant Foods for Human Nutrition*, 76(1), 118–124.

Singh, A., Gupta, A., Surasani, V. K. R., & Sharma, S. (2021). Influence of supplementation with pangas protein isolates on textural attributes and sensory acceptability of semolina pasta. *Journal of Food Measurement and Characterization,* 15(2), 1317–1326.

Singh, A., Sharma, S., & Singh, B. (2017a). Effect of germination time and temperature on the functionality and protein solubility of sorghum flour. *Journal of Cereal Science*, 76, 131–139.

Singh, A., Sharma, S., & Singh, B. (2017b). Influence of grain activation conditions on functional characteristics of brown rice flour. *Food Science and Technology International*, 23(6), 500–512.

Siroha, A. K., & Sandhu, K. S. (2017). Effect of heat processing on the antioxidant properties of pearl millet (Pennisetum glaucum L.) cultivars. *Journal of Food Measurement and Characterization*, *11*(2), 872–878. https://doi.org/10.1007/s11694-016-9458-1

Sruthi, N. U., Premjit, Y., Pandiselvam, R., Kothakota, A., & Ramesh, S. V. (2021). An overview of conventional and emerging techniques of roasting: Effect on food bioactive signatures. *Food Chemistry*, *348*, 129088. https://doi.org/10.1016/j.foodchem.2021.129088

Sun, Q., Gong, M., Li, Y., & Xiong, L. (2014). Effect of dry heat treatment on the physicochemical properties and structure of proso millet flour and starch. *Carbohydrate Polymers*, *110*, 128–134. https://doi.org/10.1016/j.carbpol.2014.03.090

Tanwar, B., Lamsal, N., Goyal, A., & Kumar, V. (2019). Functional and physicochemical characteristics of raw, roasted and germinated buckwheat flour. *Asian Journal of Dairy and Food Research*, *of*. https://doi.org/10.18805/ajdfr.DR-1452

Vittadini, E., Carini, E., Chiavaro, E., Rovere, P., & Barbanti, D. (2008). High pressure-induced tapioca starch gels: Physico-chemical characterization and stability. *European Food Research and Technology*, *226*(4), 889–896. https://doi.org/10.1007/s00217-007-0611-2

Wang, Y., Chen, L., Yang, T., Ma, Y., McClements, D. J., Ren, F., Tian, Y., & Jin, Z. (2021). A review of structural transformations and properties changes in starch during thermal processing of foods. *Food Hydrocolloids*, *113*, 106543. https://doi.org/10.1016/j.foodhyd.2020.106543

Waszkowiak, K., Mikołajczak, B., & Kmiecik, D. (2018). Changes in oxidative stability and protein profile of flaxseeds resulting from thermal pre-treatment: Changes in stability of flaxseeds resulting from thermal pre-treatment. *Journal of the Science of Food and Agriculture*, *98*(14), 5459–5469. https://doi.org/10.1002/jsfa.9090

Whitehead, A., Beck, E. J., Tosh, S., & Wolever, T. M. (2014). Cholesterol-lowering effects of oat β-glucan: A meta-analysis of randomized controlled trials. *The American Journal of Clinical Nutrition*, *100*(6), 1413–1421. https://doi.org/10.3945/ajcn.114.086108

Yadav, D. N., Kaur, J., Anand, T., & Singh, A. K. (2012). Storage stability and pasting properties of hydrothermally treated pearl millet flour. *International Journal of Food Science & Technology*, *47*(12), 2532–2537. https://doi.org/10.1111/j.1365-2621.2012.03131.x

Yang, J., Bingol, G., Pan, Z., Brandl, M. T., McHugh, T. H., & Wang, H. (2010). Infrared heating for dry-roasting and pasteurization of almonds. *Journal of Food Engineering, 101*(3), 273–280. https://doi.org/10.1016/j.jfoodeng.2010.07.007

Yodkaew, P., Chindapan, N., & Devahastin, S. (2017). Influences of superheated steam soasting and sater sctivity sontrol as sxidation mitigation methods on physicochemical properties, lipid oxidation, and free fatty acids compositions of roasted rice. *Journal of Food Science, 82*(1), 69–79. https://doi.org/10.1111/1750-3841.13557

Zhang, B., Zhang, Q., Wu, H., Su, C., Ge, X., Shen, H., Han, L., Yu, X., & Li, W. (2021a). The influence of repeated versus continuous dry-heating on the performance of wheat starch with different amylose content. *LWT, 136*, 110380. https://doi.org/10.1016/j.lwt.2020.110380

Zhang, B., Zhang, Q., Wu, H., Su, C., Ge, X., Shen, H., Han, L., Yu, X., & Li, W. (2021b). The influence of repeated versus continuous dry-heating on the performance of wheat starch with different amylose content. *LWT, 136*, 110380. https://doi.org/10.1016/j.lwt.2020.110380

Zhang, D., Doehlert, D. C., & Moore, W. R. (1997). Factors affecting viscosity of slurries of oat groat flours. *Cereal Chemistry Journal, 74*(6), 722–726. https://doi.org/10.1094/CCHEM.1997.74.6.722

Zhou, L., Fang, D., Wang, M., Li, M., Li, Y., Ji, N., Dai, L., Lu, H., Xiong, L., & Sun, Q. (2020). Preparation and characterization of waxy maize starch nanocrystals with a high yield via dry-heated oxalic acid hydrolysis. *Food Chemistry, 318*, 126479. https://doi.org/10.1016/j.foodchem.2020.126479

Zhou, W., Song, J., Zhang, B., Zhao, L., Hu, Z., & Wang, K. (2019). The impacts of particle size on starch structural characteristics and oil-binding ability of rice flour subjected to dry heating treatment. *Carbohydrate Polymers, 223*, 115053. https://doi.org/10.1016/j.carbpol.2019.115053

Zhou, Y., Cui, L., You, X., Jiang, Z., Qu, W., Liu, P., Ma, D., & Cui, Y. (2021). Effects of repeated and continuous dry heat treatments on the physicochemical and structural properties of quinoa starch. *Food Hydrocolloids, 113*, 106532. https://doi.org/10.1016/j.foodhyd.2020.106532

Zhu, F. (2016). Chemical composition and health effects of Tartary buckwheat. *Food Chemistry, 203*, 231–245. https://doi.org/10.1016/j.foodchem.2016.02.050

Zhu, P., Wang, M., Du, X., Chen, Z., Liu, C., & Zhao, H. (2020). Morphological and physicochemical properties of rice starch dry heated with whey protein isolate. *Food Hydrocolloids, 109*, 106091. https://doi.org/10.1016/j.foodhyd.2020.106091

Zou, J., Xu, M., Tang, W., Wen, L., & Yang, B. (2020). Modification of structural, physicochemical and digestive properties of normal maize starch by thermal treatment. *Food Chemistry, 309*, 125733. https://doi.org/10.1016/j.foodchem.2019.125733

Zou, J., Xu, M., Tian, J., & Li, B. (2019). Impact of continuous and repeated dry heating treatments on the physicochemical and structural properties of waxy corn starch. *International Journal of Biological Macromolecules, 135*, 379–385. https://doi.org/10.1016/j.ijbiomac.2019.05.147

CHAPTER 11

Extrusion of Cereals

Navnidhi Chhikara[a], Anil Panghal[b], and D.N. Yadav[c]
[a]Guru Jambheshwar University of Science & Technology, Hisar, Haryana, India
[b]Department of Processing & Food Engineering, AICRP-PHET, Chaudhary Charan Singh Haryana Agricultural University, Hisar, Haryana, India
[c]Central Institute of Post-Harvest Engineering & Technology, Punjab, Ludhiana, India

CONTENTS

DOI: 10.1201/9781003242192-15

11.1 INTRODUCTION

Extrusion is a high-temperature, high-pressure, and short-time process where the conditioned starch and proteinaceous food materials are plasticized and cooked leading to various structural modifications, molecular transformations, and thus different finished products (Castells, Marin, Sanchis, & Ramos, 2005). Extrusion results in melting of ingredients and melted form is forced to flow through a die, which is designed to form and/or puff-dry the ingredients. The extrusion process is also known as a high-temperature short-time (HTST) process due to high cooking temperature (180–190C) and low residence time (20–40 seconds). The process also offers energy savings, flexibility, controllability, product shaping, continuous production, and very little effluent during processing. Extrusion technology provides the opportunity to process a variety of food products by just changing a minor ingredient and processing condition on the same machine. Several different shapes, textures, colors, and appearances can be processed by making minor changes in hardware and processing conditions. Extrusion is energy efficient and low cost compared to other processes. Presently, most extruders are available with automation, which can increase productivity. The raw material undergoes various chemical, molecular, and structural transformations, like protein denaturation, starch gelatinization, cross-linking molecules to form expandable matrices, and degradation reactions of vitamins, pigments, etc.

Major changes in living standards, eating habits, and health awareness have motivated consumers towards healthy food to provide nutrition as well as health benefits (Panghal et al., 2018). The food industries are facing challenges in developing new products with special health-enhancing characteristics. Extrusion is advanced technology for food formulation with minimal impact on nutritional attributes of food in comparison to conventional thermal processing technologies. Extrudates are microbiologically safe, can be stored for long periods because of low moisture without the need for refrigeration, and require less labor for handling and less packaging materials and storage space. Wide consumer acceptance of extruded products is due to convenience, value, attractiveness, texture, and appearance of the products. Cereals are most commonly used for formulation of extruded products like breakfast cereals, pasta, noodles, and puffed products. Wheat, rice, and corn are commonly used for extrusion (however, rye, barley, pseudocereals, millets, and legumes have been utilized in the last ten years for product and nutrient diversification). Cereals are unique blends of starch and proteins and provide puffing and glossiness to the developed product. Grains mainly contain 66–76% carbohydrates, which includes starch (55–70%), sugars (~3%), cellulose (~2.5%), arabinoxylans (1.5–8%), b-glucans (0.5–7%), and gluco-fructans (~1%) and various phytoconstituents like flavonoids, carotenoids, phenolics, tannins, etc. Some anti-nutritional components like phytates, saponins, trypsin inhibitors, etc., are also present. Whole grain consumption has health-promoting effects like prevention of insulin resistance, coronary disease, diabetes, obesity, breast cancer, ischemic stroke, asthma, and premature death. The composition of cereals strongly affects the nutritional, functional, and sensory attributes of developed products and thus affects different process parameters and conditions for extrusion. Extrusion has been well explored for fabrication of various products such as snack food, baby food, breakfast cereals, pasta, pet foods, and ready-to-eat snacks. The cooking process affects the polysaccharides, physicochemical properties, and bioavailability characteristics of food constituents due to structural modification of nutrients as well as reduction of anti-nutritional factors. Nutritional concerns about extrusion cooking may increased when extrusion is used specifically to produce nutritionally balanced or enriched foods, like weaning foods, dietetic foods, and meat replacers. Thus, here we discuss the effect of extrusion on nutrients and phytochemical components of cereals along with process technology.

In the last 25 years, snack food consumption has increased due to its low cost, attractive shapes, and color and convenience. Snack foods have become an integral component of the diet due to hectic lifestyles and major changes in living standards and eating habits of the world's population. Snacks fulfil 15–20% of mineral, 13–17% of vitamins, and 15–25% of our daily energy intake (Chaplin & Smith, 2011). Extrusion is involved in preparation of ready-to-eat products from cereals; co-extruded snacks; sweet and salty snacks; croutons for soups and salads; expanded products; dry pet and fish foods; texturized vegetable proteins; weaning foods, breakfast cereal, pasta, and confectionery products. This versatility offers a huge scope for enrichment with bioactive compounds to cope with nutritional problems. Extruded snacks have universal appeal and can be good carriers for supplying nutrition to large population segments.

11.2 EXTRUSION PROCESS

Extrusion is a thermo-mechanical process that converts preconditioned cereal grits/flour into valuable convenience foods. Other ingredients might include sugar, fats, proteins, water, fruits, and vegetables to develop products with improved health benefits. Food material is forced to flow, under one or more varieties of conditions of mixing, heating, and shear, through a die designed to give attractive shapes to products. The cooking is completed in the barrel where raw material is compressed and sheared at elevated temperature and pressure to a molten state. The cooking temperature can be high (100–170°C) with residence time of usually 20–40 seconds, suggesting extrusion as HTST. During this, different changes occurs at a molecular level such as breaking of covalent bonds, protein denaturation, starch gelatinization, polymeriziation of proteins, and cross-linking of molecules from expandable matrices.

11.2.1 Types of Extruders

There are two types of extruders: single-screw and twin-screw extruders. The choice depends on the product requirements. In both machinery systems, different unit operations like conveying, mixing, homogenization, heating/cooling, cooking, sterilization, forming/shaping, expansion, texturisation, flash drying, and center filling take place.

11.2.1.1 Single-Screw Extruders

Single-screw extruder is the most common cooking extruder with three zones (feed zone, compression zone, melting and plasticizing zone) and die are the important parts of a single-screw extruder. A single-screw extruder consists of feeder, extruder barrel with single-screw, pressure and temperature controlling system along with a motor. The feed section function is for conveying the raw material in proper amounts and speed to the screw section. The feeder can be in the form of a vibratory feeder, variable speed auger, and weigh belts depending upon the nature and dimensions of the raw material. The section in between feed zone and metering zone is the compression zone, where the raw material is heated up to a coherent mass during movement. Heat is generated in the barrel due to mechanical movement of the screw as well as provided by the heating system in the extruder. Thus, rise in temperature results in starch geletatinization leading to rise in viscosity and thus the material becomes more cohesive. The barrel of the extruder is made up of diffrenet segments to alter the interior conformation as well as for easier replacement of the exit portion due to more wear and tear. The barrel has grooves on surface for efficient and faster mixing and conveying of material. The screw speed, its configuration, and material quality strongly control the mass flow rate of material in the barrel. The molten compressed material now passes through a die. The die is basically to provide different shapes and designs to make attractive products. The die dimension is crucial for determining the product attributes. The power to the extruder is maintained through a power supply electric motor. The power of the motor used is dependent on the capacity and performance of the extruder. The co-figurational variation accounts for different types of extruders: Solid single-screw extruders, interrupted-flight extruder expander, single, segmented-screw extruders. The barrel is jacketed for steam flow to provide additional heat in the extruder. A single-screw extruder is used for the formulation of direct expanded corn snacks, texturized vegetable protein, ready-to-eat breakfast cereal, production of full fat soy, pet foods, floating and sinking aquatic feed, production of baby foods, rice bran stabilization, precooked or thermally modified starches, flours, and grain breading.

11.2.1.2 Twin-Screw Extruders

The use of twin-screw extruders in the food processing industry started in the 1970s, with an expanding number of applications in the 1980s. This extruder has better mixing and pumping performance producing a more uniform flow of the product through the barrel. Twin-screw extruders are generally more expensive in terms of capital and operational cost and are more mechanically complex than a single-screw machine for the same capacity. Twin-screw has two parallel screws in the barrel either rotating in same direction (co-rotating) or in opposite direction (counter-rotating). The counter-rotating extruders are preferably used for materials requiring lower screw speed and higher residence time and relatively no viscous behavior (e.g., confectioneries, gums, jellies, etc.). Commonly, industries are using co-rotating screw extruders due to process uniformity, efficiency, easy cleaning, and better control process parameters as well as product quality. In twin-screw extruders, the preconditioned raw material is fed through hopper with a controlled feed rate. Preconditioning can be done by adding water or any other liquid ingredients like milk, whey, fruits and vegetable juices or purees, emulsifiers, fats, etc. The feeder supplies raw material into the inlet of extruder at preoptimized feed rate and speed. The raw material melts, geltainizes, and is forced to move through an electrically heated barrel. The viscous mass is passed through die allowing the material to expand and a cutter is used to control the length of the extruded sample. Different shaped dies enable huge variations in products (e.g., balls, rings, stars, alphabet letters, rod shaped, etc.). After this, the products are dried in tray driers and then packaged.

11.3 IMPACT OF PROCESSING TREATMENT ON NUTRITIONAL CHARACTERISTICS (MACRO- AND MICRONUTRIENTS)

Extrusion is a themomechanical process for development of healthy snacks, breakfast cereals, baby food, etc. Thus, it is important to study the effect of extrusion processing on nutrient status.

11.3.1 Starches

Starches are reserve carbohydrates of plants and occur as amylose and amylopectin granules in the cell in plastids, separated from the cytoplasm. Amylose is a linear chain polysaccharide linked by α-1, 4 glycosidic bonds and amylopectin is a branched polymer joined by α-1, 6 glycosidic bonds. The amylose content of most cereal starches lies between 20 and 35%, and the remaining is amylopectin (Khatkar, Panghal, & Singh, 2009). The arrangements and ratio of these granules regulate starch behavior and properties like cooking, gel formation, pasting behavior, gelatinization, retrogradation, and other functionality attributes and thus affect the finished product quality. Starch granule amorphous regions absorb water and become hydrated on heating and get swelled. The unwinding of double helices, breakdown of hydrogen bonds, and thus loss of optical birefringence is known as starch gelatinization. Starch is instantly digested in digestive systems resulting in rapid release of glucose in the blood stream, which has been associated with high risk for cardiovascular disease, diabetes, and cancer. On the basis of digestibility, starch exists in three forms: rapidly digestible starch, slowly digestible starch, and resistant starch.

Slowly digested and resistant starches are not digested in small intestine and are helpful in maintaining blood glucose level and favorable for beneficial colonic microflora growth. Complete gelatinization of starch requires at least 30–40% moisture at atmospheric pressure. As compared to other thermal processing methods, extrusion gelatinizes starch at much lower moisture content (12–22%). Extrusion results in partial gelatinization and starch fragmentation due to the mechanical shearing effect on starch granules (i.e., starch resistant to α-amylase in the gastrointestinal tract is passed into the large intestine undigested). High-temperature and pressure treatment of starch results in granular and molecular changes and thus degrades starch to dextrin. High shear force physically tears apart starch granules and allows rapid water penetration into the interior of starch molecules, leading to formation of dextrinized starch. Amylopectin being branched is easily sheared off in the barrel and thus affects the textural properties of extruded products. Reduction in weight of both amylopectin and amylose molecules has been reported. Faraj, Vasanthan, & Hoover (2004) studied the effect of extrusion process on the development of resistant starch retrograded at different temperature, moisture content, and screw speed in hull-less barley flour from CDC-Candle (waxy) and Phoenix (regular) variety. Retrograded starch content in the isolates was lower than extruded samples before enzymatic isolation. The study indicated that the retrograded starch did not concentrate, and showed that hydrolysis of the other components such as beta-glucan and proteins may have exposed resistant starch that initially escaped hydrolysis to alpha-amylase.

Brennan et al. (2012) reported that addition of dietary fibers in flour base significantly reduced the rate and extent of carbohydrate hydrolysis of the extruded products. As such the addition of dietary fibers to extruded products reduced the amount of readily digestible starch components of breakfast products, and increased the amount of slowly digestible carbohydrates. López-Barrios, Gutiérrez-Uribe, & Serna-Saldívar (2014) fabricated a variety of snacks and ready-to-eat foods using lentil, chickpea, and dry pea flours. Total available carbohydrate (TAC) ranged from 625 g/kg to 657 g/kg dry matter. Extrusion did not have a significant effect on TAC, whereas a significant decline in oligosaccharides (raffinose and stachyose) and flatulence factors was observed. The extruded flour is a good option for fabrication of value-added, nutritious snacks with good dietary fiber, thus leading to commercialization of pulse-based snacks.

11.3.2 Dietary Fibers

Dietary fibers are analogous carbohydrates (lignin, oligosaccharides, polysaccharides, and associated plant substances), or the edible parts of plants that are resistant to absorption in the small intestine with partial or complete fermentation in the large intestine of humans. Studies have shown that high consumption of these fibers significantly lowers the risk of cardiovascular disease, diabetes, obesity, hypertension, and gastrointestinal diseases. Dietary fibers are categorized on the basis of their water solubility: SDF, soluble dietary fiber (pectins, pentosans, gums and mucilage) and IDF, insoluble dietary fiber (lignin, cellulose, and parts of hemicelluloses). The soluble fraction of dietary fiber present in fruits and gums increases the intestinal viscosity and thus forms gel in the small intestine and accounts for reduced glucose absorption, affecting the postprandial serum glucose levels in diabetics. SDF has stronger antioxidant activity than IDF. Extrusion can be a possible technique for fiber modification and thus enhanced functional properties. Altan et al. (2009) revealed that extrusion increased digestibility of barley extrudates when prepared with any tomato/grape pomace but both tomato and grape pomace together decreased the digestibility of barley-based extrudates.

Wheat bran, a major by-product of flour milling, is comprised of about 53% dietary fiber (xylans, lignin, cellulose, galactan, and fructans) and comparatively higher vitamins and minerals than endosperm. Bran is around 13–19% of total wheat grain weight, and about 90% is used for animal feed and only 10% in different food preparations in food. Wheat bran dietary fiber and phytochemicals (phenolic acids, sterols, alkyl-resorcinol, vitamin E, and minerals) can be possibly used to reduce diverticulosis, gallstone formation, and prevent diabetes, and has a positive impact on colon cancer. Extrusion cooking solubilizes some fiber components. Homogenization and disruption of bran particles by severe mechanical processing during extrusion cooking renders more availability of dietary fiber for fermentation. During extrusion process, the mechanical shearing, attrition, and expansion alter the bulk density of the product. High fiber ingredients are normally low in moisture and take on moisture slowly, which requires some type of preconditioning before extrusion. Extrusion significantly increases soluble fiber content but decreases insoluble fiber content.

Yan et al. (2015) extruded bran to significantly increase SDF content from 9.82% to 16.72% with improved swelling capacity and water retention capacity. Water-soluble polysaccharide isolated from extruded SDF was found to possess good antioxidant properties and have applications in functional food, pharmaceutical, and cosmetic industries.

Rashid et al. (2015) also reported that extrusion has a positive impact on total SDF and decreased IDF. This might be due to breakdown of covalent and non-covalent bonds in the carbohydrate and protein matrix, thus producing smaller and more soluble fragments. Therefore, extrusion cooking can be applied to modify wheat bran beneficially for further value addition.

Gluten allergy and celiac disease are making gluten-free diets and fiber deficiency more common. Stojceska et al. (2009) used extrusion for increasing total dietary fiber in gluten-free cereal flour along with different fruits and vegetables. The materials were added up to 30% into balanced formulations made of gluten-free products from rice flour, cornstarch, potato starch, soya flour, and milk powder. Different process parameters, such as solid feed rate 15–25 kg/h, water feed rate 12%, screw speed 200–350 rpm, and barrel temperatures (80°C at feed entry and 80–150°C at die exit) were used. The results showed that extrusion has the potential to enhance the availability of total dietary fiber in gluten-free products made from gluten-free cereals, fruits, and vegetables.

Dietary fiber affects digestibility when added to extruded starchy foods. Long cellulosic fibers have low solubility, because of trans-glycosidation. Extrusion is responsible for mechanical stress and breakdown of glycosidic bonds of polysaccharides, which leads to liberation of oligosaccharides. This process increased the SDF amount and rate of Maillard reaction producing more melanoids and thus increased antioxidant activity. During extrusion, insoluble dietary fiber changes into soluble fiber and formation of resistant starch occurs. Extrusion processing of fibrous raw material has added potential to improving the nutritional quality of fiber. Branched dietary fiber molecules are more vulnerable to shear during extrusion. Smaller molecules could amalgamate to form large insoluble compounds or Mallard compounds that may be considered as lignin.

11.3.3 Proteins

Protein energy malnutrition is a major threat among children and other vulnerable groups due to inadequate dietary habits and high cost of protein food in developing countries. Food products should be formulated with advanced techniques so that protein quantity and quality are minimally affected. Proteins are the complex compounds made up of essential and non-essential amino acids. Quantity, quality, digestibility, and availability of essential amino acids describe the nutritional value of protein. Proteins undergo conformational and molecular changes during high-temperature processing. Shear, pressure, and temperature during extrusion process results in conformational and molecular changes leading to protein insolubilization due to non-covalent interactions accompanied with disulphide bond formation. Due to heat transfer from heated barrel along with increase in pressure, shear, and intensive mixing, protein undergoes denaturation, and non-polar amino acids, the reactive free sulfhydryl (SH) groups, and peptides are exposed. In general, protein denaturation along with aggregation leads to decreased hydrophobicity, as a result three-dimensional network formation with higher water-holding capacity, less oil absorption capacity, and protein solubility. During past extrusion, protein solubility of wheat decreased even at lower temperatures. However, the impact of extrusion on protein also varies with process parameters and ingredient characteristics. All processing parameters have variable effects on digestibility of protein (Table 11.1). Protein digestibility value is comparatively higher for extruded products than non-extruded products. This might be due to denaturation of proteins and inactivation of anti-nutrient factors/enzymes that hinders digestion/absorption of food. Increase in digestibility of vegetables by mild extrusion process increases the nutritional value of vegetable protein.

Increased speed of screw may enhance the protein digestibility of extruded corn gluten, due to protein denaturation by shear forces, thus facilitating enzyme hydrolysis. Singh et al. (2000) suggested that inactivation of anti-trypsin factors takes place during extrusion resulting in improved enzymatic hydrolysis of proteins. Protein denaturation due to extrusion opens different reactive sites for pepsin hydrolysis and thus enhances digestibility.

During extrusion free amino acids are much more susceptible to damage. Millard reaction takes place leading to a decrease in lysine availability, which may be used as indicator of protein damage during extrusion. Shear results in starch and sucrose to reducing sugars.

Reducing sugars present and formed by shear of starch react with lysine and high barrel temperature and low moisture during extrusion promotes Milliard reaction and thus lowers overall nutritional quality of protein. High temperature breaks down the secondary structure resulting in protein denaturation, only not the destruction of amino acids. Protein quality of cereals is diminished by their amino acid composition as they are deficient in lysine, tryptophan, and threonine. Depletion of these essential amino acids during extrusion processing can result in food products that may lead to growth retardation in children and young animals. Paes and Maga (2004) studied the impact of extrusion on essential amino acid profile in maize-based extrudates prepared from normal maize and Quality Protein Maize (QPM). Extrusion was carried out by tempering flour to 150 g/kg of moisture, then passing through single-screw extruder at screw speed of 80 rpm, compression ratio of 3:1, and die head temperature of 130°C, using different die nozzle diameters (3 and 5 mm). Decline in essential amino acids (leucine, isoleucine, threonine, valine, and lysine) was observed in extruded flour, whereas methionine, histidine, tryptophan, and phenylalanine did not show any significant change in both maize samples. While QPM have high amounts of lysine, methionine, and tryptophan compared to normal maize samples ($P<0.05$), despite adverse effect of extrusion on amino acid retention, QPM flours may be extruded to provide maize extrudates with superior protein quality.

11.3.4 Lipids

Fat is a high-energy food associated with major health issues like high cholesterol, heart disease, cancer, and obesity. Food samples with more than 10% lipids can not be extruded due to slip issues in the extruder. Sometimes, extruded snack foods are fried after extrusion to evaporate moisture and to modify texture and flavor.

Extrusion cooking causes a decrease in extractable fat, which may be due to formation of monoglycerides-amino acid complex. Fat or oil is encapsulated within the cells of oil seeds. When a raw oil seed is ground, the resulting product will be dry. During extrusion, oil is released from cells

Table 11.1 Effect of Extrusion on Nutritional Components and Phytochemicals of Products

Product Name	Ingredients/Process	Effect on nutrients	References
Chenopodium extrudates	18-20%, m.c. barrel temperature – 115-135°C, screw speed – 225-275 rpm	Increase in antioxidant activity, Total phenolics and decline in total dietary fiber	Jan et al., 2017
Fibre fortified corn extrudates	Defatted soybean, germinated brown rice meal, mango peel fiber, Corn grits	Fibres increased from 4.82% to 5.92-17.80%, Protein 5.03% to 5.46- 13.34% Phenolics increased upto 22.33-33.53% Antioxidants increased to 5.30-11.53%	Korkerd et al., 2016
Pigeon pea & unripe plantain blends	Pigeon pea, unripe plantain flour (25%), Process conditions: 25% moisture content, Screw speed – 80 rpm, Barrel diameter – 20 mm	Increase in Fat and energy, Decrease in Protein, ash, crude fiber, vitamin-A, B_1, B_2, and C	Anuonye et al., 2012
Purple potato and yellow pea flour exrudates		Significant losses in total anthocyanins (60-70%), 15-20% loss in phenolics	Nayak et al., 2011
Achha-soyabean blend		6% decrease in Riboflavin (B_2), 86.36%, decrease in Pyridoxine (B_6), No significant change on Vit C	Anuonye et al., 2010
Red bean cornstarch extrudates		Reduction of 70% of total phenol content	Anton et al., 2009
Acha- soyabean extrudates	Acha, soyabean	Significant rise in Amino acids and Minerals Serine- 17.09%, Glutamine- 16.55%, Proline-4.23%, Glycine-12.02%, Cystenine-5.85%, Leucine-3.87%, Iron-69.44%, Nickel-28.57%, Selenium-6.25%, Sodium-38.89% Some amino acids were reduced like Lysine- 15.67%, Histidine- 5.88%, Arginine–14.80%,Aspartame–7.40%, Threonine – 10.26%, Alanine-4.85%, Valine – 20.72%, Methionine – 40%, Isoleucine – 24.04%, Tryptophan – 4.79%, Phenylalanine – 2.96%	Anuonye et al., 2010
Extruded red beans		Increase by 14% total phenolics, 84% quercetin, Ferulic acid 40%, losses in chologenic acid by 33%, caffeic acid 9%	Korus et al., 2007
Maize Extrudates	Short-barrel and Long-barrel	Short-barrel have higher retention of Vitamin B group – 44-62% as compared to long-barrel- 20%	Athar et al., 2006
Buckwheat		63-94% decrease in tocopherols and tocotrienls	Zieeinsics et al., 2006
Maize	Maize grits, Screw speed – 80 rpm, Die head temperature – 30°C, Barrel diameter – 19.05 mm, 15% m.c.	% loss in Isoleucine – 8.9%, Leucine – 8.0-9.4%, Lysine – 10.1%, Threonine-8-8.7%, Valine – 10.2-9.6%	Paes & Maga, 2004

making products oily. Starch granules present in raw material may form a hydrophobic layer on fat globules, thus preventing moisture absorption and restricting starch gelatinization. Starch-fat and starch-protein complexes have significant effect on the rheological characteristics of cereal melts and thus their expansion behavior.

Rice bran, a major byproduct of the rice milling industry, can be used as health food as it contains about 34.0–62.0% carbohydrates, 11.3–14.9% protein, and 15.0–19.7% oil. Bran separation during milling releases lipases resulting in bran rancidity, thus making it less stable and unfit for human consumption. Thus, lipase inactivation is crucial for beneficial bran utilization, which can be done by extrusion.

Moisio et al. (2015) extruded rye bran of variable particle size (633 and 15 μm) under different processing conditions, moisture (13–30%), temperature (80–140 °C), and studied lipid stability as measured by volatile compounds and losses in tocopherols and tocotrienols. Highly expanded, porous, and stable lipid during extrusion was produced from low water content (13–16%). Enhanced lipid stability was due to more formation of Milliard reaction products acting as antioxidants. Fine rye extrudates at 13% moisture content resulted in extensive Milliard reaction accounting for undesirable flavor. The study revealed that coarse rye bran extruded at lowest moisture content was better in terms of expansion, porosity, and lipid stability and thus extended shelf life.

Fish contains abundant amount of high value protein and lipids having omega-3 fatty acids, especially DHA (docosa-hexaenoic acids) and EPA (eicosa- pentaenoic acid). These omega-3 fatty acids are essential for physiological developments and also prevent different like conditons hypertension, coronary artery disease, arthritis, diabetes, cancer, and inflammatory and autoimmune disorders.

11.3.5 Vitamins

Fat-soluble and water-soluble vitamins both are essential organic compounds that are required in small amounts by humans and are requisite for normal growth, health, metabolism, and reproduction. These are heat sensitive and are lost or reduced during thermal processing of food. Vitamin A, folic acid, and biotin are heat labile, whereas vitamins A, C, and D are sensitive to oxidation. Therefore, processed foods can not meet these nutrient demands. Snack foods are widely consumed by school-age children and other specific groups with high nutrition requirements, so the nutrient losses in such foods are quite crucial. Vitamins A, B_1, C, E, and folic acid are the least stable, whereas the vitamin B complex group (except B_1) is the most stable depending on sensitivity during extrusion.

The shorter residence time and lower expander temperature results in only 20% losses in vitamins; on the contrary, losses increased up to 65% on increasing resident time and expander temperature.

Cereal products are considered important sources of vitamin B complex group and so there is tremendous need to retain all these health benefits during processing. Extrusion is involved in preparation of breakfast cereals, muesli bars, and snack food, which contribute a major share of cereal products. Athar et al. (2006) compared the sensitivity of vitamin B complex group in crisp extruded snack food using single-screw short-barrel snack extruder. The study revealed that short-barrel extruders retain 44–62% of vitamins B in snacks, while long extruders retain only 20% in maize extrudates. Niacin has high stability whereas similar values for riboflavin were observed in both extruders. Pyridoxine was more stable in maize extrudates as compared to maize and pea blend and oat extrudates. Thiamin had minimum stability during extrusion. The study revealed that heat-labile vitamins B groups are better retained in short-time, high-temperature extrusion as compared to lower temperature and longer-time extrusion cooking processes.

Cheftel (1986) studied vitamin stability in extrusion of crisp bread products at mechanical energy of 0.09 kWh/kg to 0.13kWh/kg, and retention times from 0.5 min to 1 min with temperature at 178°C. Thiamine and pyridoxine being most heat labile decreased linearly with temperature. Thiamine losses varied from 5–100% during extrusion; higher losses were observed with either no water usage in formulation or high barrel temperature in extrudates of wheat flour. Similarly, thiamine is also reduced with high temperature of barrel and high screw speed. Thiamine is found to be more heat sensitive than riboflavin, whereas riboflavin is more shear sensitive.

Vitamin C is highly sensitive and prone to oxidation in the presence of heat and oxygen, the destruction becomes faster in the presence of alkali, heat, light, metal ions, and oxidative enzymes. Charlton and Ewing (2007) reported complete destruction of vitamin C with extrusion.

Extruded products are not rich in fat-soluble vitamins; still the vitamin stability is important especially in case of fortified foods. Tocopherols and carotenoids are sensitive to oxidation and thermal degradation. Beta carotene, an antioxidant, coloring agent, and precursor of vitamin A, can be added to meet the vitamin A requirement. Though effort can be made to reduce the oxidative losses of carotene either using inert atmosphere or with addition of antioxidant BHT (butylated hydroxyl toluene), Suknark et al. (2001) observed 52 and 73% retention of retinyl palmitate in tapioca starch mixed with either fish or protein flour, respectively.

11.3.6 Minerals

Minerals are heat-stable micronutrients essential for different physiological functions in the body. Thus, minerals are retained during extrusion processing and sometimes increased due to abrasion of interior barrel and screws by bran/fiber present in food samples.

Gbenyi et al. (2016) studied the impact of different composition feed, moisture of feed, and temperature of extrusion on the proximate composition and essential amino acids of sorghum and cowpea extrudates. Sorghum and cowpea flour (90:10, 80:20, and 70:30 ratios) was conditioned to 20, 22.5, and 25% moisture and extruded at 120, 140, and 160°C in a single-screw extruder. Addition of cow pea in extrudates significantly increased the carbohydrate, energy, protein, fat, moisture, crude fiber content, and ash. Lysine and methionine content were also increased. Extrudates have sufficient amount of protein and all essential amino acids to meet nutritional requirements of children.

11.4 IMPACT OF PROCESSING PARAMETERS ON FUNCTIONAL CHARACTERISTICS (RHEOLOGICAL, THERMAL, STRUCTURAL, AND SENSORY)

Extrusion is a versatile process used to improve the nutritional and organoleptic qualities of food products. Raw materials are put through a shaped die with a predetermined rate under variable conditions and produce products like pasta, baby foods, snack foods, breakfast cereals, breadcrumbs, biscuits, crackers, croutons, confectionery items, granules, chewing gum, modified starch, texturized vegetable proteins, pet foods, dried soups, and dry beverage mixes. Processing parameters such as material composition, material feed rate, process temperature (barrel and die) and pressure, dough moisture, screw speed, take-up unit speed, and die diameter determine the size, shape, texture, crispness, hardness, taste, and other quality attributes of these extruded products. The raw materials (RM) components, the granularity of RM, pretreatment conditions, food additives along with the screw configuration, screw length-to-diameter ratio, and die design also affect the quality of products. Feed composition, moisture content, and particle size have the greatest effect on extrusion. Raw materials such as corn, wheat, rice, and soybean used in the extrusion process are starch- and protein-based materials. The typical composition of any blend consists of starch, protein, lipid/fat, and fiber, which contribute to the product quality. The raw material in extrusion covers various combinations of ingredients including cereals, grains, starches, tubers, legumes, oilseeds, cereals as well as animal fats and proteins.

11.4.1 Feed Moisture

The moisture is a critical variable that has multiple fractions in starch gelatinization, protein denaturation, barrel lubrication, and the final product quality. A dry extruder can process materials with 8–22% moisture with no additional drying of extrudates. Most extrudate snacks have a moisture content between 8–12% and require additional drying to impart desired texture and mouthfeel. Higher moisture leads to harder extrudates, less starch depolymerization, and more protein aggregation through disulfide bonds. All these attributes change the technological, sensory, and microstructure parameters of the products. Polypeptides of legume proteins degrade during extrusion, and this degradation increases with a decrease in feed moisture (Ghumman et al., 2016). The low moisture of the raw material is responsible for increased extrudate expansion during extrusion. Low feed moisture (11%) results in a higher expansion ratio in cassava extrudates and higher feed moisture (16%) results in poor expansion. The reason is that at higher feed moisture, the extruded material becomes moist and soft. The high feed moisture content is responsible for low expansion, higher density, lower WAI (water absorption index), higher WSI (water solubility index), lower puncture energy, and high hardness in extrudates. The starch degradation usually reduces product expansion, and high moisture content helps in gelatinization and degradation of starch. Lower feed moisture helps in the production of extrudates with a higher expansion ratio, WAI, WSI, and lower density with softer products. Due to increase moisture of feed, screw speed and barrel temperature reduce the pressure on die pressure. The high feed moisture increases the melt plasticity while the high temperature decreases the viscosity. Further, Sharma et al. (2015) reported the effect of higher (24%) and lower (20%) feed moisture on the structural, morphological, and functional characteristics as well as digestibility of corn, field pea, and kidney bean extrudates. They found that compared to raw form, the extrudates had reduced WAI, retrogradation (syneresis), and paste viscosities, but greater WSI. Further, they observed that WAI, WSI, and retrogradation were increased at lesser feed moisture than those at higher feed moisture. In addition, the digestibility of those extrudates at both moisture levels was observed. They found that compared to those extruded at lower feed moisture, all extrudates extruded at higher feed moisture had a substantial reduction in rapidly digestible starch (RDS) content. A similar trend was also observed in the case of resistant starch (RS) content. Another study conducted by Thymi et al. (2005) also reported the effects of feed moisture content on the structural properties of maize starch extrudates. They found that the higher input moisture reduces the extrudate radial expansion ratio, leading to higher apparent density and lower porosity values. Further, Liu et al. (2000) reported the effect of processing conditions on the physical and sensory properties of extruded oat-corn puff. The feed moisture content of 18%, 19.5%, and 21% was used. The influence of feed moisture content on the expansion index was distinct in different formulations. The expansion index of the standard formula (0% corn flour) increased as feed moisture increased, but the expansion ratio of the high cornflour (45%) formula was reduced. The influence of feed moisture on the expansion of the medium cornflour (15 and 30%) formulation was not noteworthy. As a result, they concluded that at higher feed moisture content, the expansion index of the extruded product was reduced except for 100% oat flour extruded puff.

11.4.2 Screw Speed

Increase in screw speed in twin-screw hot extruders increase the transport rate and convection with mixing and thus decrease the residence time of the material in the extrusion barrel; therefore, it is necessary to optimize the screw speed. Lower screw speed causes food component degradation due to longer residence time and longer exposure to the extrusion temperature (Figure 11.1). When increasing the screw speed, the melt viscosity of the food was found to decrease until the food melt fracture is reached. These conditions are ideal for processing food products. However, when the

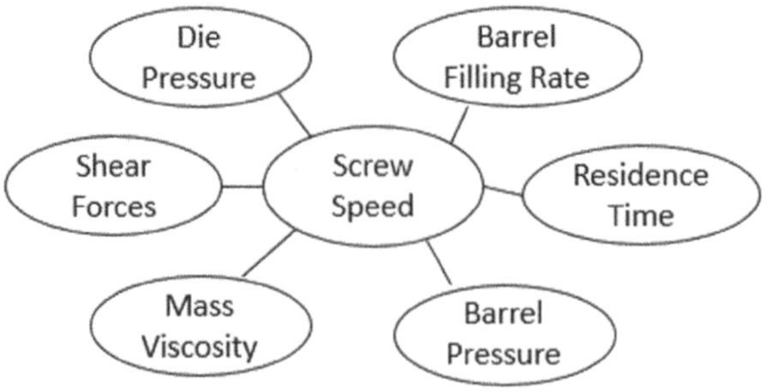

Figure 11.1 Factors influenced by screw speed.

screw speed is further increased to much higher speeds, the food components degrade due to the heat produced by the mechanical energy of the screws. Bouasla et al. (2016) found that screw speed had no significant effect on the preparation time of pasta. They also reported water absorption capacity and cooking loss of pasta is more if extruded at slow screw speed. Screw speed slightly reduced the density and hardness of wheat extrudate, but screw speed did not affect the hardness of dry pasta. Higher screw speed affects the shear rate and influences the heat exchange inside the barrel, which reduces the viscosity of the material. Geetha et al. (2012) conducted a study on the twin-screw extrusion of Kodo millet-chickpea blend and determined the process parameter optimization and physicochemical properties. The physical characteristics of the extruded product investigated were bulk density, hardness, expansion index, and crispness. The extrudates showed bulk density values ranged from 0.19–0.88 g/cm^3. According to the regression coefficients, screw speed had a substantial adverse linear influence on bulk density, whereas the response surface data implied that bulk density and screw speed have an inverse correlation. The value of bulk density was maximum at a screw speed of 250 rpm and subsequently decreased as the screw speed increased, reaching 0.19 g/cm^3 at 300 rpm. This is consistent with the findings of a snack extruded from chickpea flour. Further, the influence of screw speed on the expansion index was reported to be quadratic. Higher screw speed causes an increase in shear, which causes a reduction in melt viscosity. With a screw speed of 80–200 rpm, both total and radial expansion increase. The effect of screw speed on the hardness values of extrudates was found to have an adverse linear effect. That is, by increasing the speed of the screw, there is a reduction in the hardness values of the extrudate.). There was a positive linear impact of screw speed on the crispness of the extrudates. That is, by increasing the screw speed, the crispiness of the extrudates was also improved. Increased screw speed may have resulted in better puffing of the starchy feed material into hot melts, which can expand as they exit from the die, producing bubbles that grow within the melt to generate continuous foam (Guy, 2001). In the extrusion process, higher screw speed increases shear stress in the active cooking section, resulting in greater product temperatures and material overheating. Browning events, and consequently color changes, would be favored under these circumstances.

11.4.3 Feed Rate

Variable feed rate at constant screw speed affects the fill level of the extrusion barrel, since increasing the feed rate will increase the fill level and barrel and thus the barrel pressure. Thus, a balance needs to be found between feed rate and screw speed to maintain a constant raw material and product flow. A high feed rate results in a higher degree of fill and changes the melt rheological characteristics that further affect starch degradation and elastic effect with changes in density.

11.4.4 Barrel Temperature

Barrel temperature is one of the major factors that control the quality of the extrusion process. The food raw material has to be processed above its glass transition temperature but below its degradation temperature. The different components of the raw material recipe can be processed below or above its cooking temperature depending on the final product. It is well known that the temperature influences the viscosity of the batch. Therefore, an equilibrium has to be found between, on the one hand, a low temperature where the batch shows high viscosity, and thus a high torque, and on the other hand, an elevated temperature where the torque is reduced due to the low viscosity of the batch but where all the ingredients could be degraded. High temperature and large shearing forces of extrusion disintegrate the quaternary and tertiary structure of food components and increase the interactions between food components and moisture, which further influences the rheological behavior of the food material in the extruder by reducing the shear load (Alam et al., 2016). The product temperature can consequently be a major determinant factor in the quality of the final product. It is important to note that the product temperature will be different from the barrel temperature. Indeed, mechanical energy is often transferred from the screws into the molten material. In single-screw hot extruders, temperature influences flow rate most, as the flow properties of raw materials are proportional to their viscosity. An increase in the barrel temperature increased the feeding rate but decreased the residence time of the product within the barrel. Increased barrel temperature reduces the density and hardness; however, WAI, WSI, and puncture energy were increased. The barrel temperature influences residence time for the twin-screw extrusion process also. According to previous studies, increasing feed moisture during extrusion decreased WSI but did not influence WAI (Bryant et al., 2001). Rice flour constituents may exhibit biochemical modifications during incubation at elevated temperature (99°C). For instance, starch can caramelize and proteins can denaturize, and both effects have an impact on fat absorption.

Another study conducted by Ali et al. (2020) reported the effect of extruder variables (barrel temperature) on the functional, morphological, sensory, and digestible properties of maize and potato starch extrudates. They observed that the maximum L* value was found in potato starch extruded at 100°C, while the lowest was found in maize starch extruded at 160°C. The L* values for all extrudates were decreased as the barrel temperature increased. It is widely recognized that high extrusion temperature induces the synthesis of melanoidins (brown pigments) via the Maillard reaction, leading to non-enzymatic browning and consequently darkening of the end product (Bhat et al., 2019). Furthermore, using a high temperature during the extrusion process boosts starch disintegration, resulting in the liberation of free sugars, which provide ideal circumstances for caramelization processes (Sharma et al., 2015). On the other hand, all extrudates a* and b* values are negatively correlated with extrusion barrel temperature. In the same study, the maximum a* and b* values were found in maize starch extruded at 100°C, whereas the minimum was found in potato starch extruded at 160°C. The reduction in a* and b* values by increasing the barrel temperature may be attributed to the decomposition of pigments during extrusion cooking (Nayak et al., 2011). As a result, they concluded that the L*, a*, and b* values of extruded starches reduced as the extrusion temperature increased. The functional properties such as WSI and WAI were also investigated in this study as a function of barrel temperature. The minimum WAI value was found in potato starch extruded at 160°C, whereas the maximum WAI value was found in cornstarch extruded at 100°C. Extrusion temperature has a substantial impact on extruded starch WAI. Extruded corn and potato starches showed reduced WAI values when processing at higher temperatures than at lower temperatures. It is worth mentioning that extrusion cooking can disrupt starch molecules, leading to a reduction in the water absorption capacity. Singh et al. (2015) also reported that the high-temperature extrusion process enhances starch breakdown and dextrinization, lowering extrudate WAI values. Further, it was also reported that the extrusion temperature and starch source had a substantial effect on the WSI of extruded starches, based on F values. The minimum WSI value was found in cornstarch extruded at 100°C, while the maximum was found in potato starch extruded at 160°C. The WSI of extrudates increasing behavior when increasing the barrel temperature. The degree of gelatinization was maximum with increased barrel temperature, resulting in enhanced breakdown and dextrinization of starch molecules, leading to an increased value of WSI. The variation in WSI between maize and potato starch extrudates could be due to the greater amylose concentration of potato starch compared to maize starch (Singh et al., 2003). On the other hand, the impacts of barrel temperature on WSI of corn and potato starch extrudates were identical, exhibiting a similar pattern. As a result, they concluded that both potato and maize starch extrudates showed maximum WSI values extruded at higher barrel temperature whereas the trend was opposite in the case of WAI values of both the starch extrudates. The digestibility of maize and potato starch extrudates was also investigated in the same study. In the presence of moisture, high temperatures and high shearing pressures employed during the extrusion cooking induced significant starch gelatinization, increasing the sensitivity of starch to enzymatic breakdown. Gelatinization of starch during extrusion has a significant impact on starch digestibility because crystalline structures are changed, glucosidic linkages are broken, and molecules become more easily available to digestive enzymes, resulting in enhanced digestibility. The digestibility of starch was maximum when increasing the barrel temperature. The minimum starch digestibility value was found in potato starch extruded at 100°C, whereas the maximum was found in maize starch extruded at 160°C. The reason for the maximum digestibility of starch after extrusion at high barrel temperature was attributed to the maximum degree of gelatinization. The minimum starch digestibility of potato starch extrudates compared to maize starch extrudates could be because potato starch is a resistant starch type II (Ali et al., 2019). Hence, they concluded that the digestibility of starch was maximum for both extrudates when extruding at higher barrel temperatures.

11.4.5 Barrel Pressure

The expansion of extruded food material depends on the difference in pressure inside the barrel and the outside atmosphere. When the raw materials enter the barrel the starch granules are compressed, deformed, and start to gelatinize due to shear heat and heat supplied from the heater. Starch loses its structure due to the breakdown of hydrogen bonds between molecules. The recipe water diffuses into starch molecules, which builds the pressure in the barrel along with the shear energy supplied by reverse pitch screw elements. Barrel pressure along with other critical parameters determine the hardness, crispness, and other characteristics of the product, which influence the acceptance and further success of the product in the market.

11.4.6 Die Temperature

The expansion ratio of extruded products increased with a decrease in die temperature. Degradation of legume polypeptides increased with an increase in extrusion temperature (Ghumman et al., 2016). At the die, the molten mass is compressed and shaped by the different shapes of the die before the pressure and temperature drop result in the material expanding as it exits the extruder die. Higher die temperature leads to a high expansion ratio, WAI, WSI, and lower density and hardness.

11.4.7 Mass Temperature

Mass temperature is the product temperature measured at the exit or die. Mass temperature is significantly dependent on the barrel temperature and pressure; it is usually higher than barrel temperature. The friction between the barrel surface and the feed pellets as they flow through the screw flight is the most likely reason for the increase in product temperature. Furthermore, the beneficial impact of feed rate induces denser packing of the material in the barrel and at the die, which leads to significant shear and heat development, as well as a high mass temperature. The melt temperature was increased by 7–9°C because of the shearing process, while moisture was decreased by 7–9%, resulting in a temperature increase of 18–20°C (Onwulata et al., 2001). The process of extrusion of hulled rice was conducted by Xu et al. (2015). The study reported that addition of thermostable α-amylase resulted in break down of starch and dextrinization leading to decline in viscosity and reduction in mechanical energy requirement.

11.4.8 Die Pressure and Torque

Die pressure is a measurement of the resistance to material flow through a die aperture, which is influenced greatly by its design. The amount of feed material accumulated at the die aperture causes it to increase. The torque of the extruder is an estimate of work needed to move the feed material out of the die aperture during the process and is an indication of the machine's safe handling. Torque serves two purposes: (i) operating the product and (ii) turning the screw under full and no-load conditions. An increase in feed moisture, temperature, and screw speed minimizes die pressure and torque, as an increase in feed moisture improves melt plasticity whereas an increase in temperature minimizes viscosity (De Pilli et al., 2012; Singh et al., 2007). This could be linked to starch molecules colliding during the extrusion cooking, causing gelatinization, swelling, and disruption of delicate molecules. Increased screw speed increases the shear rate and heat transfer in the barrel, causing the material's viscosity to drop. As a result, the extruder's torque also decreases. Singh Gujral et al. (2002) found a comparable impact of fat on-die pressure reduction during extrusion of brown rice grits. Furthermore, rice with a reduced amylose content produces high torque as it gets sticky during the extrusion process, whereas rice a high amylose concentration produces reduced torque as the extruder temperature rises (Guha & Ali, 2006). In another investigation, a decrease in die pressure was reported in the extrusion of rice-pea flour mix owing to an increase in amylose concentration (Singh et al., 2007). Conversely, high shear extrusion cooking of whey protein concentrate (WPC) and sweet whey solids (SWS) incorporated with rice flour increased die pressure (Onwulata et al., 2001).

11.4.9 Specific Mechanical Energy (SME)

The ratio of total mechanical energy gained from material transferring due to screw spinning within the extruder to the mass flow rate of feed material to the extruder is known as specific mechanical energy (SME) (Kantrong et al., 2018). The transformation of starch during extrusion cooking was influenced by the amount of mechanical energy applied to the feed material, which promotes the breaking of intermolecular hydrogen bonds. The more starch gelatinization and product expansion happen at a greater rate of SMEs (Pardhi et al., 2019. SME is also a measure of expansion index and is proportional to the melt temperature. When starch gelatinizes and the solid mass flow transforms into a viscoelastic flow at elevated temperatures, the viscosity of the melt viscosity is reduced. SME reduced for high amylose rice flour extrusion at high temperatures but improved somewhat for low amylose rice flour extrusion (Guha & Ali, 2006). The SME was raised when the green gram percentage in the feed mixture was high (Chakraborty & Banerjee, 2009). This might be due to the higher protein concentration in the green gram. Conversely, Singh et al. (2007) reported that the extrusion of rice grit with pea grit resulted in a reduction in SME efficiency. Moreover, adding apple pomace to rice powder resulted in an increase in SME (Mehraj et al., 2018). The lubricating effect has been shown to improve as input moisture increases, resulting in reduced SME. SME, on the other hand, tends to grow at low moisture levels, resulting in starch gelatinization and considerable product expansion (Onwulata et al., 2001). Further investigation showed that the inclusion of thermostable α-amylase resulted in decreased SME input because of the enzymatic hydrolysis and breakdown of starch (Xu et al., 2015). Increased screw speed and shear of the feed material screw causes significant friction between the material, the screw, and the barrel surface, resulting in a higher SME (Kantrong et al., 2018). The same trend was observed during the extrusion cooking of oat and rice flour (Sandrin et al., 2018).

11.5 IMPACT OF PROCESSING TREATMENT ON BIOLOGICAL CHARACTERISTICS

11.5.1 Anti-nutritional Factors

Anti-nutirents are the secondary metabolites secreted in plant sources for self-defense. These compounds inhibit the digestion/absorption pathway of particular nutrients (e.g., cyanogens, glucosinolates, mycotoxins, phytic acid, lectins, α-amylase inhibitors, tannins, etc.). Traditional processes like soaking, roasting, and blanching also reduce these anti-nutirents but on a small level. Extrusion destroys various anti-nutrients and natural toxins and improves digestibility and safety of foods (Table 11.2). Extrusion can be used to enhance the protein digestibility and reduce anti-nutrients especially phytates, trypsininhibitors, and haemaglutinin tannins. Extrusion significantly improves the protein digestibility of pea seeds, faba, and kidney beans, as compared with soaking, dehulling, and germination.

On increasing the level of moisture content and extrusion temperature, there is increased destruction of trypsin inhibitors. Inactivation of anti-nutrients increases with more residence time and moisture content at constant barrel temperature. Singh et al. (2000) extruded blends of 20% wheat bran and broken rice and reported total destruction of trypsin inhibitor activity (TIA). However, when bran concentration is increased above 20%, inactivation of trypsin inhibitors is decreased from 92 to 60%. High bran concentration in blend form lowers the degree of expansion and thus lowers the impact of heat and results in less enzyme inactivation.

Lectin (haemaglutinating activity) is heat resistant and interacts with di-peptides, disaccharides, and other enzymes involved in nutrient digestion and uptake. Lectin combines with the cell lining of the intestinal walls in almost the same way as it combines with RBCs, thus causing impairment with the absorption of amino acid. Lectins are found in kidney beans, soya bean field beans, white beans, and horse gram with the highest amount in kidney bean. Aqueous heat treatment can reduce the amount of anti-nutirents but is not able to completely destroy these.

Kelkar et al. (2012) prepared extruded products from Navy and Pinto beans after soaking for 8 hr in water with a feed moisture content of 36% and a temperature of 85°C. Good quality extruded products were prepared with reduction of 85–95% in phyto-hemaglutinin.

Extrusion has been found to be efficient in removing lectin in legume flour (Alonso et al., 2000). Delgado et al. (2012) extruded bean (30%) and corn (70%) with 20% moisture content at 150°C and 25 rpm and observed the total destruction of TIA (initial level – 10,857 TIU/g) and lectin activity (initial level – 640 HU/mg proteins). This study by Delgado et al. (2012) supported the fact that low moisture, high temperature, and severe mechanical stress is sufficient to completely destroy the anti-nutritional factors. These studies clearly show that extrusion process is effective in eliminating or reducing lectin as compared with conventional aqueous heat treatment.

Cereals are commonly infested by fungus in fields or storage producing secondary metabolites mycotoxins (e.g., fumonisins, aflatoxins, ochratoxins, patulin, moniliformin, HT-2 toxin and T-2 toxin, zearalenone and derivatives, deoxynivalenol [vomitoxin], and derivatives) that are quite harmful for both human and animal consumption. Castells et al. (2004) reported that extrusion can

Table 11.2 Impact of Extrusion on Bioactive Compounds in Food Systems

Food	Process Parameter	Impact	References
Nutrient Dense Porridge	Amaranth, groundnut, iron-rich beans, pumpkin, orange-fleshed sweet potato, carrot, and maize. Process parameters were: Barrel temperature (130–170°C) and feed moisture content (14%–20%)	Increasing barrel temperature and feed moisture content leads to a decrease in polyphenol, phytic acid contents, and iron extractability, but increases zinc extractability.	Akande et al., 2017
Lentil (*Lens culinaris* Medik)	Twin-screw extruder, Feed moisture content (14, 18, 22%), Feed rate (20.4 kg/h), Compression zone temperature (140,160, and 180ºC), mixing zone (110, 130, and 165ºC), conveying zone (95,115, and 135ºC), Screw speed: 150, 200, and 250 rpm,	Decline in phytate (99.30%), trypsin inhibitors (99.54%) and tannin (98.83%) with optimized 160ºC compression zone temperature, 18% feed moisture content, and 200 rpm screw speed.	Rathod & Annapure, 2016
Bran of wheat, rice, barley and oat	Twin-screw extruder, moisture content: 14, 17 and 20%, Die temperature: 115ºC, 140ºC, and 165ºC, Barrel Temperature: 25°C, 75ºC, and 100ºC, Screw speed: 400 rpm, Feed rate: 20 kg/h	Polyphenols, phytates, trypsin inhibitors, and oxalates were reduced by 73.38, 54.51, 72.39, and 36.84%, respectively. Highest reduction observed at 140°C and 20% feed moisture content.	Kaur, Sharma, Singh & Dar (2015)
African Bread fruit-soybean-corn snack	Single-screw extruder, Feed moisture content 21%, Barrel temperature: 140 °C, Screw speed: 140 rpm	Oxalates, trypsin inhibitors phytate, and tannins declined by 50, 69.77, 66.67, and 34.78%, respectively.	James & Nwabueze (2013
Beans (*Phaseolus vulgaris L.*)	Pre-soaked beans (in water for 8h) were oven-dried (65 ± 2 °C for 12 h), Twin-screw extruder, Feed moisture (36%), Feed rate (120 g/min), Temperature (85, 100, and 120°C)	Combined treatments decreased the lectin by 85-95%.	Kelkar et al. (2012)
Bean (*Phaseolus vulgaris)* & Corn (*Zea maize*)	Single screw extruder, Feed moisture content (20%), Compression ratio: 1:6, Temperature (150°C), Screw speed (25 rpm)	Complete inactivation of lectins and trypsin inhibitors	Delgado et al. (2012)
Beans (*Phaseolus vulgaris)*	Single-screw extruder, Feed moisture content (12.3, 14, 18, 22 and 23.7%), Temperature (150, 154, 164, 174 and 178 °C), Screw speed 414 rpm	The activities of the α-amylase inhibitors and trypsin and haemagglutinins were completely eliminated	Lopes et al., (2012)
Pigeon Pea and 25% unripe plantain flour		% decrease in Phytate- 69-76%, Tannins 61.22%, Hydrogen cyanide- 78.18%, Saponon 84.48%, Lectin- 58.82%	Anuonye et al., 2012

(*continued*)

Table 11.2 (Continued)

Food	Process Parameter	Impact	References
Kidney bean	Twin-screw extruder, Feed moisture (25%), feed rate (350 g/min), Temperature (150°C), Screw speed (100 rpm)	The contents of tannins, phytates and lectins, trypsin, α-amylase inhibitory and chymotrypsin activities were reduced.	Marzo et al., (2011)
Pea (*Pisum Sativum)* Chickpea (*Cicer Arietinum*) Faba pea (*Vicia Faba*)	Twin-screw extruder, Feed rate (107-116 g/min), Barrel temperature (70 °C (middle), 95 °C (front) and 110 °C (Outlet die), Screw speed (380 rpm)	Total tannin and phytates were greatly reduced.	Adamidou, Nengas, Grigorakis, Nikolopoulou, and Jauncey (2011)
Blend of acha and soybean	Single screw extruder	Significant reduction in saponin, phytate and oxalate	Anuonye et al., (2010)
Common beans (*Phaseolus vulgaris* L.)	Single-screw extruder, Feed moisture content (20%), Central temperature (150°C), Screw speed: 150 rpm	Inactivation of α-amylase inhibitory and haemagglutination activity	Batista, Prudêncio, & Fernandes (2010a)
Cowpea	Single-screw extruder, Moisture content: 25%, Temperature: 150° C, Screw speed: 150 rpm.	Inactivation of α-amylase and lectin, reduction of phytate content (33.2%) and trypsin inhibitor activity (38.2%)	Batista, Prudêncio, & Fernandes (2010b)
Lin seed (*Linum usitatissimum*)	Single-screw extruder, Feed moisture content (30-50%), Barrel temperature (60-100 °C), Screw speed (60-100 rpm)	Highest decline in tannin content (61.27%) at barrel temperature of 80°C, screw speed of 96.8 rpm and feed moisture content of 40%.	Mukhopadhyay et al., (2007)
Maize and finger millet	Single-screw extruder, Screw speed (200 rpm), Barrel temperatures (150°C (feed section), 180°C (compression section) and 180°C (metering section).	Phytate content was unaffected. Tannin content decreased to 697 mg/100 g after extrusion of the unfermented blend and to 551 mg/100 g after fermentation and extrusion.	Onyango et al. (2005)
Corn-bean-sardine meal Bean meal Sorghum-beansardine meal Rice-bean sardine meal	Twin-screw extruder, Feed moisture content (12-14.4%), Screw speed (300 rpm), Feed rate (3.61 kg/h), Temperature zone (70 °C, 100 °C, 127°C, 141°C, 131°C).	Significant destruction of all ANFs investigated.	Mosha, Bennink, & Ng (2005)
Sesame (Sesamum indicum) meal	Single screw extruder, Barrel temperature (X1): 63.18-96.80°C, Feed moisture content (X3): 31.59-48.41%, Screw speed (X2): 63.18-96.80 rpm.	Significant reduction in tannin content (61.25%)	Mukhopadhyay & Bandyopadhyay (2003)

Faba beans, Pea seeds, chickpea, kidney beans	Moisture content- 18.22%, Barrel temperature- 140°C-180°C, Inlet temperature – 100°C, Screw speed – 250 rpm, Screw compression – 4:1, Feeding screw speed – 160 rpm, Die diameter – 3 mm	Significant reduction of phytic acid, tannins, phenols, α-amylase, trypsin inhibitors, and improved protein digestibility	Hady and Habiba, 2003
Rice, wheat maize	Twin-screw extruder, Extraction rate 60 and 80%, Maximum temperature and pressure 160 °C and 10 MPa, respectively.	Reduction effect on phytate content was minimal.	Hurrell, Reddy, Burri, and Cook (2002)
Lathyrus sativus seeds	Twin-screw extruder, Feed moisture content: 14,18, 22, 26, or 30%, Temperature: zone 1: 90-140°C, zone 2: 100-180°C, zone 3: 120-220°C, and zone 4: 100-200°C.	The content of β-ODAP and tannins were reduced while the activity of trypsin inhibitors was completely inhibited. Reduction/inactivation effect on ANFs was enhanced with an increase of moisture content in seeds (from 14 to 30%) before extrusion and with an increase of barrel temperature from 90/100/120/100°C to 140/180/220/200°C.	Grela, Studzinski, & Matras (2001)
Kidney beans and Faba bean	Twin-screw extruder, Screw speed 100 rpm, Feed moisture content 25%, Die temperature 152 and 156 °C, Feed rate 383 and 385 g/min.	Inhibition of haemagglutinins activity was possible without modifying the protein content.	Alonso, Aguirre, & Marzo, (2000)
Peas (*Pisum sativum* L.)	Twin-screw extruder, Screw speed 100 rpm, Feed rate 21.5 kg/h, Die temperature 145°C, Feed moisture content 25%.	Trypsin inhibitory and lectin activities were completely inhibited. Chymotrypsin inhibitory activity was reduced. A minor reduction in the phytate content of peas was found.	Alonso et al., (2000)

be applied to reduce mycotoxins and thus help to produce safe food. Destruction rate is dependent on type of extruder, screw type, die configuration, screw speed, barrel temperature, moisture content, additives used, and initial mycotoxin concentration of the raw material in extrusion. Fumonisins, aflatoxins, and zearalenone can be reduced up to 100, 95, and 83% whereas deoxynivalenol, ochratoxins A, and moniliformin can be reduced to 55, 40, and 30%, respectively.

Frias et al. (2011) extruded pea (*Pisumsativum*) to form pea-derived extruded products with improved nutrition value at °C. Peas are good source of carbohydrates (53 g/100 g), dietary fiber (18 g/100 g), protein (24 g/100 g), riboflavin, and thiamine (0.1–0.2 mg/100 g) and energy (330 kcal/100 g). Pea being rich in lysine can be used with cereals conforming with FAO reference pattern. Simultaneously, pea contains anti-nutrients like phytic acid (0.4 g/100 g), a-galactosides (4 g/100 g), and TIA (2 TIU/mg). Extrusion slightly increases protein and fat content, whereas dietary fiber, thiamine and a-galactosides, and TIA levels were reduced significantly. Pea sample extruded at 135°C possess had the highest chemical score and protein digestibility corrected amino acid score.

Legumes contain abundant amounts of carbohydrates, dietary fibers, vitamins, and some minerals including trace elements and are inexpensive. However, the presence of anti-nutritional factors (e.g., phytates, polyphenols, enzyme inhibitors [trypsin, chymotrypsin, and a-amylase] and hemag-glutinins remarkably decreases their food utilization. Extrusion is more energy efficient and has better process control, and high production capacities can be employed to reduce anti-nutrients as compared to other thermal treatments. Hady and Habiba (2003) soaked legumes (pea, chickpea, faba, and kidney beans) for 16 h in water and extruded at barrel temperature of 140°C and 180°C and feed moisture of 18% and 22% and studied the impact on anti-nutrients, total and phytate phosphorus, and protein digestibility. Both soaking and extrusion drastically reduced anti-nutirents; however, in extruded samples, phytic acid phosphorus availability and in vitro digestibility was improved. Aflatoxins level (59%) decline was achieved in the naturally contaminated peanut meal at moisture of 35 g/100 g. In vitro protein digestibility and fluoro-dinitro-benzene available lysine of the extrudates were not significantly different from non-extruded peanut meal. Extrusion conditions for aflatoxin reduction did not adversely affect protein nutritional quality.

Aflatoxins cause severe health complications and hazards in human beings. Saalia and Phillips (2011) evaluated the effect of extrusion on aflatoxin level in peanut meal in the presence of calcium chloride with lysine and methylamine. Extrusion cooking reduced aflatoxins from an initial 417.72 mg/kg to 66.87 mg/kg (i.e., 84% reduction) in the peanut meal. Methylamine and lysine favored aflatoxin reduction whereas calcium chloride hampered aflatoxin degradation.

Berrios et al. (2010) studied the value addition of extruded product by reducing the flatulence factors and increasing dietary fiber made from lentil, chickpea, and dry pea. The decline in the raffinose family of oligosaccharides (raffinose and stachyose) concentration was observed in pulse extrudates. Formulated pulse flours showed remarkable increase in dietary fiber. Flatulence-causing oligosachraides such as reffinose and stachyose also decreased significantly in extruded high starch fractions of pinto beans, which may improve the consumer acceptability of legume-based extruded products.

11.5.2 Phytochemicals

Phenolic compounds present in plants can protect food against oxidation and humans against disease and predation. These compounds, including the large number in the flavonoid family, are the focus of numerous studies to illuminate their role in human health. In potato peels total free phenolics such as chlorogenic acid predominates, which decreased significantly due to extrusion.

Higher barrel temperature and feed moisture protect free phenolics. Red and blue anthocyanin pigments provide attractive colors and are believed to serve as antioxidant that protect vision and cardiovascular health.

11.5.2.1 Antioxidants

Phytochemicals in whole grain consist of phenolic compounds, different carotenoids, and vitamins. Phenolics include phenolic acids (p-coumaric, caffeic, ferulic, vanillic, and syringic acids) as well as flavonoids (flavonols, flavonones, catechins, and anthocyanins). Whole grain phenolics possess potent antioxidant activities and are able to scavenge free radicals that may increase oxidative stress and potentially damage large biological molecules such as lipids, proteins, and DNA. Extrusion impact on phytochemicals has been studied by various researchers.

Black rice phytochemicals and antioxidant activity was compared for unprocessed and extruded milled fractions (Ti et al., 2015). Extrusion increased the free phenolics, anthocyanins, and oxygen radical absorbance capacity (ORAC) and decreased the bound forms. In extruded bran, total phenolics, anthocyanins, and ORAC was increased by 12.6%, 5.4%, and 19.7%, respectively. On the contrary, these parameters decreased by 46.5%, 88.4%, and 33.1%, respectively, in polished rice and by 71.2%, 87.9%, and 14.7%, respectively, in brown rice. One study suggested that development of different milled fractions of extruded black rice can be suitable base with balanced nutritional characteristics for today's functional food markets. Sharma et al. (2012) evaluated the extrusion impact on antioxidant activity of barley. TPC, TFC, and reducing power was declined, whereas non-enzymatic browning index, metal chelating activity, and DPPH radical scavenging activity was increased after extrusion. The feed moisture and barrel temperature significantly affected the antioxidant properties of barley.

11.5.2.2 Pigments

Color is an important quality parameter responsible for aesthetic appeal as well as nutritional value. Due to increased consumer awareness and safety concerns, natural colors are preferred over synthetic colors. Color usage in foods is under the control of regulatory bodies of individual countries. Natural color besides providing color also imparts some health benefits so their use has increased in health foods. Natural colors are easily soluble in food dispersions making their incorporation into food systems easier, and there are no numerical limits or side effects. Anthocyanin (E 163) is a flavonoid phenolic compound responsible for red, purple, or blue color and is present in berries, fruits, and flowers. Anthocyanin has an important role in neuronal, cardiovascular illness, cancer, diabetes, and important from a nutritional and biological standpoint. Anthocyanin color stability depends on the presence of complexing agents such as phenols and metals, temperature, pH, and oxygen. Significant reduction of anthocyanin was observed in maize extrudates when blueberry (Khanal et al., 2009 or cranberry pomace was added to cornstarch (White et al., 2009) during extrusion.

Durge et al. (2013) studied anthocyanin suitability for healthy rice flour with neutracuetical properties during extrusion. Anthocyanin retention was improved with high moisture content of feed and screw speed, whereas it decreased with increase in temperature of die. Sodium carbonate and citric acid was also added to extrudates and stored at ambient room conditions (30±2 °C). Citric acid added at the level of 1% can enhance anthocyanin retention up to 18.2%. Metallized polyethylene has superior packaging material for extrudate storage than LDPE by reducing oxidative losses due to light.

Gat and Ananthanarayan. (2016) used paprika oil extract for coloring the extrudates of rice and also optimized parameters for extrusion such as moisture content, die temperature, and screw speed.

Samples with 3% paprika oil content and 25% feed moisture were passed through the barrel at 120°C and 100 rpm screw speeds were most acceptable. The study revealed that red color retention was increased with increase in screw speed and feed moisture and decreased with rise in temperature of barrel. Kosinska-Cagnazzo et al. (2017) incorporated goji berries in rice-based extrudates using a twin-screw extruder with constant screw speed and feed moisture and analyzed them in order to determine antioxidant activity and contents of rutin, zeaxanthindipalmitate, and 2-O- β-D-glucopyranosyl-L-ascorbic acid. About a 20 time increase in antioxidant activity and a decrease in rutin content of goji extrudates was observed. Irrespective of the extrusion condition, rutin retention in extrudates was not lower than 60% as compared to raw mixtures; for zeaxanthindipalmitate the retention level was around 40%. 2-O-β-dglucopyranosyl- L-ascorbic acid was stable at the conditions applied.

Hegazy et al. (2017) formulated snack foods using germinated dehulled chickpea (10, 20, and 30%), tomato pomace (5%), and corn. Extrusion led to remarkable reduction in protein and fat percentage of the extrudates, while fiber, ash, and carbohydrates contents were not affected. Also, phytic acid and tannic acid were reduced by 40.64–46.07% and 40.46–44.88%, respectively, while in vitro protein digestibility was improved by 6.37–7.71% in extrudates. Total phenolic content of germinated chickpea and tomato pomace enriched snacks increased from 1.07 to 5.55%, and antioxidant activity observed by DPPH radical scavenging activity increased by 1.92–7.94% after extrusion. Addition of germinated chickpea up to 30% and tomato pomace to corn extrudates caused an increase in bulk density, apparent density, and WAI, while expansion ratio, porosity, and WSI values followed an opposite trend. Germinated chickpea up to 20% as a protein source as well as tomato pomace as an antioxidant can successfully be added to develop new healthy corn snacks.

Cueto et al. (2017) studied the carotenoid losses in cornflakes prepared by traditional and extrusion cooking. In traditional toasting, reduction in lutein and zeaxanthin was 60% and 40%, respectively, due to higher susceptibility of lutein to isomerization and decomposition. In contrast, only 35% reduction of lutein and zeaxanthin was observed in extruded product. The study showed that carotenoid content is reduced during processing and extrusion can be a better approach for nutritional enrichment.

11.6 CONCLUSION

The idea of nutritionally balanced extruded products is a wide research area. Extrusion is suitable for nutrient retention, antinutritional destruction, and nutrient availability. Extrusion is recommended and used for health foods, weaning foods, and convenient snacks. Thus, better understanding of all nutritional alteration aspects during extrusion might be helpful in appropriate processing with better nutrient retention at an industrial level. This knowledge of nutrient stability can be beneficial in the manufacturing of specifically designed foods like meat replacers, dietetic foods, weaning foods, etc., and also in formulation of nutrient-dense foods.

REFERENCES

Adamidou S, Nengas I, Grigorakis K, Nikolopoulou, D, & Jauncey, K (2011) Chemical composition and antinutritional factors of field peas (Pisumsativum), chickpeas (Cicerarietinum), and faba beans (Viciafaba) as affected by extrusion preconditioning and drying temperatures. Cereal Chemistry, 88(1), 80–86.

Akande OA, Nakimbugwe D, Mukisa IM (2017) Optimization of extrusion conditions for the production of instant grain amaranth-based porridge flour. Food Science & Nutrition,. 2017;5:1205–1214.

Alam, M. S., Kaur, J., Khaira, H. & Gupta, K. (2016) Extrusion and extruded products: changes in quality attributes as affected by extrusion process parameters: a review. *Critical reviews in food science and nutrition*, **56**(3), 445–473.

Ali, S., Singh, B. & Sharma, S. (2019) Impact of feed moisture on microstructure, crystallinity, pasting, physico-functional properties and in vitro digestibility of twin-screw extruded corn and potato starches. *Plant Foods for Human Nutrition*, **74**(4), 474–480.

Ali, S., Singh, B. & Sharma, S. (2020) Effect of processing temperature on morphology, crystallinity, functional properties, and in vitro digestibility of extruded corn and potato starches. *Journal of Food Processing and Preservation*, **44**(7), 14531.

Alonso R, Aguirre A, & Marzo, F. (2000). Effects of extrusion and traditional processing methods on antinutrients and in vitro digestibility of protein and starch in faba and kidney beans. Food Chemistry, 68(2), 159–165.

Alonso R, Grant G, Dewey P, Marzo F (2000) Nutritional assessment in vitro and in vivo of raw and extruded peas (Pisum s ativum L.). Journal of Agricultural and Food Chemistry, 48(6), 2286–2290.

Altan A, McCarthy KL, & Maskan, M (2009) Effect of extrusion cooking on functional properties and in vitro starch digestibility of barley-based extrudates from fruit and vegetable by-products. *Journal of Food Science*, *74*(2).

Anton AA, Fulcher RG, Arntfield SD (2009) Physical and nutritional impact of fortification of corn starch-based extruded snacks with common bean (Phaseolus vulgaris L.) flour: Effects of bean addition and extrusion cooking. Food Chemistry, 113(4), 989–996.

Anuonye JC, Jigam AA, & Ndaceko GM (2012) Effects of extrusion-cooking on the nutrient and anti-nutrient composition of pigeon pea and unripe plantain blends.

Anuonye JC, Onuh JO, Egwim, E, Adeyemo, S. O. (2010). Nutrient and antinutrient composition of extruded acha/soybean blends. Journal of Food Processing and Preservation 34(S2), 680–691.

Batista KA, Prudêncio SH, Fernandes KF (2010a) Changes in the functional properties and antinutritional factors of extruded hard-to-cook common beans (Phaseolus vulgaris, L.). Journal of Food Science 75(3), C286–C290.

Batista KA, Prudêncio SH, Fernandes KF (2010b) Changes in the biochemical and functional properties of the extruded hard-to-cook cowpea (Vignaunguiculata L. Walp). International Journal of Food Science & Technology 45(4), 794–799.

Berrios JDJ, Morales P, Cámara M, Sanchez-Mata MC (2010) Carbohydrate composition of raw and extruded pulse flours. Food research international *43*(2), 531–536.

Bhat, N. A., Wani, I. A., Hamdani, A. M. & Gani, A. (2019) Effect of extrusion on the physicochemical and antioxidant properties of value added snacks from whole wheat (*Triticum aestivum* L.) flour. *Food Chemistry*, **276**, 22–32.

Bouasla, A., Wojtowicz, A., Zidoune, M. N., Olech, M., Nowak, R., Mitrus, M. & Oniszczuk, A. (2016) Gluten-free precooked rice-yellow pea pasta: Effect of extrusion-cooking conditions on phenolic acids composition, selected properties and microstructure. *Journal of Food Science*, **81**(5), 1070–1079.

Brennan MA, Derbyshire EJ, Brennan CS, Tiwari BK (2012) Impact of dietary fibre-enriched ready-to-eat extruded snacks on the postprandial glycaemic response of non-diabetic patients. Molecular Nutrition & Food Research 56(5), 834–837.

Bryant, R.J., Kadan, R.S., Chamapagne, E.T., Vinyard, B.T. & Boykin, D. (2001) Functional and digestive characteristics of extruded rice flour. *Cereal Chemistry*, **78**(2), 131–7

Castells M, Marin S, Sanchis V, Ramos AJ (2005) Fate of mycotoxins in cereals during extrusion cooking: a review. Food Addit Contam *22*(2), 150–157.

Chakraborty, P. & Banerjee, S. (2009) Optimization of extrusion process for production of expanded product from green gram and rice by response surface methodology. *Journal of Scientific and Industrial Research*, **68**(2), 140–148.

Chaplin K, Smith AP (2011) Definitions and perceptions of snacking. Current Topics in Nutraceuticals Research 9(1/2), 53.

Charlton SJ, Ewing WN (2007) The vitamin directory.Context Products Ltd. England.

Cheftel JC (1986) Nutritional effects of extrusion cooking. Food Chemistry 20, 263–283.

Cueto M, Farroni A, Schoenlechner R, Schleining G, Buera P (2017) Carotenoid and color changes in traditionally flaked and extruded products. Food chemistry 229, 640–645.

De Pilli, T., Derossi, A., Talja, R. A., Jouppila, K. & Severini, C. (2012) Starch-lipid complex formation during extrusion-cooking of model system (rice starch and oleic acid) and real food (rice starch and pistachio nut flour). *European Food Research and Technology*, **234**(3), 517–525.

Delgado E, Vences-Montaño M, Rodríguez JH, Rocha-Guzmán N, Rodríguez-Vidal A, Herrera-Gonzalez S, Medrano-Roldan H, Solís-Soto A, Ibarra-Perez F (2012) Inhibition of the growth of rats by extruded snacks from bean (*Phaseolus vulgaris*) and corn (*Zea mays*). Emirates Journal of Food and Agriculture 24(3), 255.

Durge AV, Sarkar S, Singhal RS (2013) Stability of anthocyanins as pre-extrusion coloring of rice extrudates. Food research international 50(2), 641–646.

El-Hady EA, Habiba RA (2003) Effect of soaking and extrusion conditions on antinutrients and protein digestibility of legume seeds. LWT-Food Science and Technology, 36(3), 285–293.

Faraj A, Vasanthan T, Hoover R (2004) The effect of extrusion cooking on resistant starch formation in waxy and regular barley flours. Food Research International 37(5), 517–525.

Frias J, Giacomino S, Peñas E, Pellegrino N, Ferreyra V, Apro N, Vidal-Valverde C (2011) Assessment of the nutritional quality of raw and extruded Pisumsativum L. var. laguna seeds. LWT-Food Science and Technology, 44(5), 1303–1308.

Gat Y, Ananthanarayan L (2016) Use of paprika oily extract as pre-extrusion colouring of rice extrudates: impact of processing and storage on colour stability. Journal of food science and technology 53(6), 2887–2894.

Gbenyi DI, Nkama I, Badau MH (2016) Optimization of Physical and Functional Properties of Sorghum-Bambara Groundnut Extrudates. Journal of Food Research 5(2), 81.

Geetha, R., Mishra, H. N. & Srivastav, P. P. (2012) Twin screw extrusion of kodo millet-chickpea blend: process parameter optimization, physico-chemical and functional properties. *Journal of Food Science and Technology*, **51**(11), 3144–3153.

Ghumman, A., Kaur, A., Singh, N. & Singh, B. (2016) Effect of feed moisture and extrusion temperature on protein digestibility and extrusion behaviour of lentil and horsegram. *LWT*, **70**, 349–357.

Grela E, Studzinski T, Matras J (2001) Antinutritional factors in seeds of Lathyrussativus cultivated in Poland. Lathyrus Lathyrism Newsletter 2(2), 101–104

Guha M Ali SZ (2006) Extrusion cooking of rice: Effect of amylose content and barrel temperature on product profile. *Journal of Food Processing and Preservation*, **30**(6), 706–716.

Guy R (Ed.) (2001) Extrusion cooking: Technologies and applications. Woodhead publishing.

Hegazy HS, El-Bedawey AE, Rahma ES, Gaafar AM (2017) Effect of Extrusion Processs on Nutritional, Functional Properties and Antioxidant Activity of Germinated Chickpea Incorporated Corn Extrudates. American Journal of Food Science and Nutrition Research 4(1), 59–66.

Hurrell RF, Reddy MB, Burri J, Cook JD (2002) Phytate degradation determines the effect of industrial processing and home cooking on iron absorption from cereal-based foods. British Journal of Nutrition 88(02), 117–123.

James S, Nwabueze T (2013) Quality evaluation of extruded full fat blend of African breadfruit-soybean-corn snack. International Journal of Scientific and Technology Research 2(9), 212–216.

Jan R, Saxena DC, Singh S (2017) Effect of storage conditions and packaging materials on the quality attributes of gluten-free extrudates and cookies made from germinated chenopodium (Chenopodium album) flour. Journal of Food Measurement and Characterization, 11, 1071–1080.

Kantrong, H., Charunuch, C., Limsangouan, N. & Pengpinit, W. (2018) Influence of process parameters on physical properties and specific mechanical energy of healthy mushroom-rice snacks and optimization of extrusion process parameters using response surface methodology. *Journal of Food Science and Technology*, **55**(9), 3462–3472.

Kaur S, Sharma S, Singh B, Dar B (2015) Effect of extrusion variables (temperature, moisture) on the antinutrient components of cereal brans. Journal of Food Science and Technology 52(3), 1670–1676.

Kelkar S, Siddiq M, Harte J, Dolan K, Nyombaire G, Suniaga H (2012) Use of low temperature extrusion for reducing phytohemagglutinin activity (PHA) and oligosaccharides in beans (*Phaseolus vulgaris* L.) cv. Navy and Pinto. Food Chemistry 133(4), 1636–1639.

Khanal RC, Howard LR, Prior RL (2009) Procyanidin content of grape seed and pomace, and total anthocyanin content of grape pomace as affected by extrusion processing. Journal of food science 74(6).

Khatkar B, Panghal A, Singh U (2009) Applications of Cereal Starches in Food Processing. Indian food industry 28(2), 37–44.
Korkerd S, Wanlapa S, Puttanlek C, Uttapap D, Rungsardthong V (2016) Expansion and functional properties of extruded snacks enriched with nutrition sources from food processing by-products. Journal of food science and technology 53(1), 561–570.
Korus J, Gumul D, Czechowska K (2007) Effect of extrusion on the phenolic composition and antioxidant activity of dry beans of Phaseolus vulgaris L. Food Technology and Biotechnology 45(2), 139.
Kosińska-Cagnazzo A, Bocquel D, Marmillod I, & Andlauer W (2017). Stability of gojibioactives during extrusion cooking process. Food Chemistry 230, 250–256.
Liu, Y., Hsieh, F., Heymann, H. & Huff, H. E. (2000) Effect of process conditions on the physical and sensory properties of extruded oat-corn puff. *Journal of Food Science,* **65**(7), 1253–1259.
Lopes LC, de Aleluia Batista K, Fernandes KF, de Andrade Cardoso Santiago R (2012) Functional, biochemical and pasting properties of extruded bean (Phaseolus vulgaris) cotyledons. International Journal of Food Science and Technology 47(9), 1859–1865.
López-Barrios L, Gutiérrez-Uribe JA, Serna-Saldívar SO (2014) Bioactive peptides and hydrolysates from pulses and their potential use as functional ingredients. Journal of food science 79(3).
Marzo F, Milagro FI, Urdaneta E, Barrenetxe J, Ibanez FC (2011) Extrusion decreases the negative effects of kidney bean on enzyme and transport activities of the rat small intestine. Journal of Animal Physiology and Animal Nutrition, 95(5), 591–598.
Mehraj, M., Naik, H., Reshi, M., Mir, S. & Rouf, A. (2018) Development and evaluation of extruded product of rice flour and apple pomace. *The Bioscan*, **13**(1), 21–26.
Moisio T, Damerau A, Lampi AM, Partanen R, Forssell P, Piironen V (2015) Effect of extrusion processing on lipid stability of rye bran. European Food Research and Technology 241(1), 49–60.
Mosha TC, Bennink MR, Ng PK (2005) Nutritional quality of drum-processed and extruded composite supplementary foods. Journal of Food Science 70(2), C138–C144.
Mukhopadhyay N, Bandyopadhyay S (2003) Extrusion cooking technology employed to reduce the antinutritional factor tannin in sesame (Sesamum indicum) meal. Journal of Food Engineering, 56(2–3), 201–202.
Mukhopadhyay N, Sarkar S, Bandyopadhyay S (2007) Effect of extrusion cooking on anti-nutritional factor tannin in linseed (Linumusitatissimum) meal. International Journal of Food Sciences and Nutrition 58(8), 588–594
Nayak B, Berrios JD, Powers JR, Tang J (2011) Effect of extrusion on the antioxidant capacity and color attributes of expanded extrudates prepared from purple potato and yellow pea flour mixes. Journal of food science 76(6), 874–883.
Onwulata, C. I., Smith, P. W., Konstance, R. P. & Holsinger, V. H. (2001) Incorporation of whey products in extruded corn, potato or rice snacks. *Food Research International*, **34**(8), 679–687.
Onyango C, Noetzold H, Ziems A, Hofmann T, Bley T, Henle T (2005) Digestibility and antinutrient properties of acidified and extruded maize-finger millet blend in the production of uji. LWT-Food Science and Technology 38(7), 697–707
Paes MCD, Maga J (2004) Effect of extrusion on essential amino acids profile and color of whole-grain flours of quality protein maize (qpm) and normal maize cultivars. Revistabrasileira de milho e sorgo 3(01).
Panghal A, Janghu S, Virkar K, Gat Y, Kumar V, Chhikara N (2018) Potential Non-Dairy Probiotic Products–A Healthy Approach. Food Bioscience 21, 80–89.
Pardhi, S. D., Singh, B., Nayik, G. A. & Dar, B. N. (2019) Evaluation of functional properties of extruded snacks developed from brown rice grits by using response surface methodology. *Journal of the Saudi Society of Agricultural Sciences*, **18**(1), 7–16.
Rashid S, Rakha A, Anjum FM, Ahmed W, Sohail M (2015) Effects of extrusion cooking on the dietary fibre content and Water Solubility Index of wheat bran extrudates. International Journal of Food Science & Technology, 50(7), 1533–1537.
Rathod RP, Annapure US (2016) Effect of extrusion process on anti-nutritional factors and protein and starch digestibility of lentil splits. LWT-Food Science and Technology 66, 114–123.
Saalia FK, Phillips RD (2011) Degradation of aflatoxins by extrusion cooking: effects on nutritional quality of extrudates. LWT-Food Science and Technology 44(6), 1496–1501.

Sandrin R, Caon T, Zibetti AW, De Francisco A (2018) Effect of extrusion temperature and screw speed on properties of oat and rice flour extrudates. *Journal of the Science of Food and Agriculture*, **98**(9), 3427–3436.

Sharma P, Gujral HS, Singh B (2012) Antioxidant activity of barley as affected by extrusion cooking. Food Chemistry 131(4), 1406–1413.

Sharma, S., Singh, N. & Singh, B. (2015) Effect of extrusion on morphology, structural, functional properties and in vitro digestibility of corn, field pea and kidney bean starches. *Starch* 67(9–10), 721–728.

Singh, D., Chauhan GS, Suresh I, Tyagi SM (2000) Nutritional quality of extruded snacks developed from composite of rice brokens and wheat bran. International Journal of Food Properties 3(3), 421–431.

Singh Gujral, H. & Singh, N. (2002) Extrusion behaviour and product characteristics of Brown and milled rice grits. *International Journal of Food Properties*, **5**(2), 307–316.

Singh, B., Rachna, Hussain, S. Z. & Sharma, S. (2015). Response surface analysis and process optimization of twin screw extrusion cooking of potato-based snacks. *Journal of Food Processing and Preservation*, **39**(3), 270–281.

Singh, J., Singh, N., Sharma, T. R. & Saxena, S. K. (2003) Physicochemical, rheological and cookie making properties of corn and potato flours. *Food Chemistry*, **83**(3), 387–393.

Singh, S., Gamlath S, Wakeling L (2007) Nutritional aspects of food extrusion: a review. International Journal of Food Science & Technology, 42(8), 916–929.

Stojceska V, Ainsworth P, Plunkett A, İbanoğlu Ş (2009) The effect of extrusion cooking using different water feed rates on the quality of ready-to-eat snacks made from food by-products. Food Chemistry, 114(1), 226–232.

Suknark K, Lee J, Eitenmiller RR, Phillips RD (2001) Stability of tocopherols and retinyl palmitate in snack extrudates. Journal of Food Science, 66(6), 897–902.

Thymi, S., Krokida, M. K., Pappa, A. & Maroulis, Z. B. (2005) Structural properties of extruded corn starch. *Journal of food engineering*, **68**(4), 519–526.

Ti H, Zhang R, Zhang M, Wei Z, Chi J, Deng Y, Zhang Y (2015) Effect of extrusion on phytochemical profiles in milled fractions of black rice. Food chemistry, 178, 186–194.

White BL, Howard LR, Prior RL (2009) Polyphenolic composition and antioxidant capacity of extruded cranberry pomace. Journal of Agricultural and Food Chemistry, 58(7), 4037–4042.

Xu, E., Wu, Z., Long, J., Wang, F., Pan, X., Xu, X. & Jiao, A. (2015) Effect of thermostable α-amylase addition on the physicochemical properties, free/bound phenolics and antioxidant capacities of extruded hulled and whole rice. *Food and Bioprocess Technology*, 8(9), 1958–1973.

Yan X, Ye R, Chen Y (2015) Blasting extrusion processing: the increase of soluble dietary fiber content and extraction of soluble-fiber polysaccharides from wheat bran. *Food Chemistry*, 180, 106–115.

CHAPTER 12

Frying of Cereals

Nalla Bhanu Prakash Reddy[a], P.S. Gaikwad[b], Monica Ostwal[c], and B.K. Yadav[d]
[a]Department of Food Processing Business Incubation Centre
[b]Department of Food Packaging and System Development
[c]Department of Food Product Development
[d]Liaison Office
National Institute of Food Technology, Entrepreneurship and Management – Thanjavur,
Formerly Indian Institute of Food Processing Technology,
Ministry of Food Processing Industries, Govt. of India,
Thanjavur, Tamil Nadu, India

CONTENTS

DOI: 10.1201/9781003242192-16

12.1 INTRODUCTION

Frying is one of the popular ancient techniques followed for the preparation of food products through contact with heated oil. Frying is a simple and rapid process that is performed at a high temperature and results in high heat transfer that provides a pleasant and attractive characteristic including color, texture, and taste to finished food products (Zeb, 2019a). Commercial frying is carried out using two methods: deep-fat frying and shallow frying. Deep-fat frying is also known as submerge or immerse frying, in which food products are covered with hot oil and heat transfer takes place to fry the food products. In shallow frying, food products are kept on a frying pan that contains a thin layer of hot oil and frying takes place with the help of heat transfer from the hot oil to the food products (Oreopoulou et al., 2006, Berk, 2018).

Cereals are essential for the development of human health since they comprise a high amount of energy, dietary fiber, minerals, protein, phytochemicals, and vitamins (Oghbaei and Prakash, 2016). The principal cereal crops include rice (paddy), wheat, maize (corn), sorghum (jowar), barley, millets, oats, and rye (Manay and Shadaksharaswamy, 2008; Wrigley, 2018). Generally, direct frying of cereals encompasses dry heat and a small quantity of hot oil or fat for cooking that prevents sticking to pan or vessel. While frying cereal-based food products there will be high heat transfer between food components that causes moisture evaporation and results in volume expansion leaving behind pore space filled with oil. Also, the frying of cereals is responsible for causing denaturation of protein, gelatinization of starch formation of color and crust, and increase in oil content and oil uptake in fried cereal-based food products (Onipe et al. 2015). Therefore, this chapter mainly discusses changes during frying on functional, compositional, and biological properties of fried cereal-based food products.

12.2 MECHANISM OF FRYING AND OIL ABSORPTION

Frying is a process of transfer of heat from a high heat reservoir of oil to the food product. Initially, the medium of frying is preheated at the desired temperature range between 150–200°C and later food products are immersed in hot oil (Zeb, 2019a). While frying, there will be high heat transfer that takes place from heated oil to immersed food products by external surface (gas or hot air), and uniform heat distribution supplied throughout the process and results in the formation of

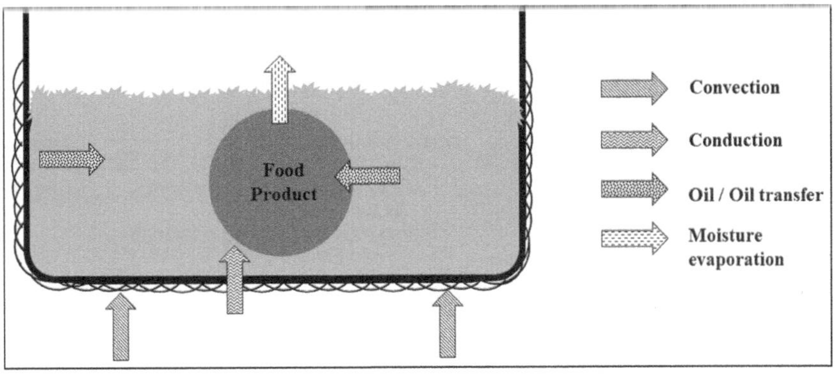

Figure 12.1 Heat transfer and oil absorption during frying of food product.

fried food products (Achir et al., 2008). Figure 12.1 shows the transfer of heat and absorption of fat or oil during frying at a high temperature. At the time of frying, the immersed food products absorb a small amount of oil in the replacement of vaporized moisture or water loss, which is mainly dependent on the removal of moisture and how the moisture is lost from the food products (Gaikwad et al., 2021).

As shown in Figure 12.1, the transfer of heat occurs through convection from heated fat or oil to the fried food product surface and later by conduction to the center of a fried food product. While frying food products at such a high temperature, the water molecules present inside the food product get heated to their boiling point and escape in the form of vapor, and leave behind pore space. Post frying, during the cooling stage, the available surface oil tries to penetrate through pore space and is absorbed inside the food product (Gaikwad et al., 2021; Oreopoulou et al., 2006). According to Zeb (2019a) at the time of frying, the moisture vaporization is faster compared to surrounding oil that eliminates the steam by convection.

12.3 EFFECT OF FRYING ON CEREALS

Frying is a popular, conventional, and relatively quick process of cooking extensively used at both domestic and industrial level, and there is a wide range of products developed by treatment in edible oil ("frying medium" is used sometimes synonymously from here on) keeping the characteristics like taste and texture in view for consumer acceptance (Paesani and Gómez, 2021). The heat transfer during the frying of cereals affects the physicochemical properties of starch and its structure, thereby improving the overall product quality of starch-based foods.

12.3.1 Effect of Frying on Functional Properties

Changes in the food products during frying take place due to multiple factors including characteristics of food, type of the oil employed in frying, the surface-to-volume ratio of the oil, the temperature of the frying medium, rate of air incorporation into the oil, type of heating process, duration of immersion in frying medium, and the frying container material as well (Bordin et al., 2013). The changes during frying may change the functional properties of cereal-based food products (Table 12.1). The changes in functional properties of fried cereal-based food products are discussed below.

Table 12.1 Impact of Frying on Functional and Compositional Properties of Cereal-Based Food Products

S. No.	Cereal or cereal-based food products	Frying conditions	Observations	References
1	Maize Starch (MS)	• Water content of MS – 20%, 40%, 60%, and 80%. • Frying temperature @ 180°C for 20 min.	1. Partial gelatinization and limited swelling 2. Starch modification was related to water content	(Chen et al., 2018)
2	Black rice	• Frying temperature @ 140°C for 5 min	1. Reduced total proanthocyanidins content. 2. Increased apparent amylose content by approx. 5%	(Aalim et al., 2021)
3	Rice	• Frying temperature @ 190°C for 2 min.	1. Significant damage to starch granules observed 2. Stickiness was reduced	(Paesani & Gómez, 2021)
4	Donut (main part is wheat flour)	• Methods of frying: deep fat and hot-air frying • Frying temperature of oil @ 150, 165, and 180 °C.	1. Maximum surface is covered with crust @ high temp (180 °C) using deep fat frying 2. Maximum crust @ lower (150°C) temperature in hot-air frying. 3. Hot-air frying reduced roughness of surface	(Ghaitaranpour et al., 2018)
5	Wheat Starch	• Wheat starch strands were steamed @ 100 °C/4min and fried at 100-160°C with 20°C for 75 sec.	1. Average molar mass reduced to 73.68% with increasing frying temperature. 2. An increase in frying temperature decreases digestion rate and increases resistant starch content	(Yang et al., 2020)
6	You-tiao (a wheat-based product)	• You-tiao was fried @ 110, 130, 150, 170, and 190 °C for 1 min	1. Significant alteration of gluten protein by unfolding and decomposing after frying 2. Observed the aggregation of decomposed protein molecules	(Zhou et al., 2020)
7	Oats	• Stir-frying @ 160 °C for 0, 10, 20, 30, and 40 minutes	1. Significant decrease in peak viscosity 2. Gelatinization degree, colour, fluidity, and particle size were enhanced.	(Qian et al., 2020)

12.3.1.1 Oil Content and Oil Absorption

Frying is a process that involves heat and mass transfer (HMT). Employing this HMT, oil gets transferred from the frying medium to the product that is being fried (called oil uptake – OU), which in turn leads the water inside the food product to escape. By modifying the surface of the food product, this OU can be brought down to a greater extent. The edible coating is one of the best and most effective methods of reducing OU (Salehi, 2020). In cereals, wheat and products based on wheat have been very popular as they are one of the good sources of carbohydrates and energy, and also provide a major amount of protein (if consumed). It has been reported that wheat chips enriched with barley flour reduced the OU during frying (Yuksel et al., 2015).

Multiple factors that affect the oil absorption or uptake include oil quality, frying medium temperature, surface area, etc. Starch gelatinization during frying arrests the moisture movement and thereby reduces the OU (Thanatuksorn et al., 2010). The surface roughness and microstructure (porosity) of wheat gluten-based products have more impact on oil absorption when compared to that of other fruit or vegetable-based food products during frying (Moreno et al., 2010). Edible coatings (EC) have been a go-to option for decreasing the OU in deep-fat fried food products. Water-soluble gums have been used as coating agents, gelling agents, texture modifiers, thickeners, stabilizers, packaging films, and emulsifiers (Salehi, 2020). Other than the above, the chemical composition of cereal products to be fried also plays a major role in oil content and oil uptake (Thanatuksorn et al., 2010).

12.3.1.2 Color

The color of foods after processing, before consumption, plays a major role in consumer acceptance. Sometimes perception is directly or indirectly influenced by factors like socio-economic status, cultural backgrounds, acceptability, and adaptability to a wide range of products and necessary processing technologies involved (Ndlala et al., 2019; Siah and Quail, 2018). Color not only determines the perception of consumers but also sometimes tells about the health of the product (Oyedeji et al., 2017). The frying operation carried out at conditions similar to atmospheric conditions may lead to changes in the color of the product. It was proven that temperature and pressure during frying have a synergistic effect on the color of the product (Mariscal and Bouchon, 2008). The lightness value in yellow-fleshed slices of cassava roots was decreased with the temperature of the frying medium (i.e., the product became darker than that of the untreated case) (Oyedeji et al., 2017). Similar instances and trends were reported in the case of Gulabjamun balls (Kumar et al., 2006). The lightness reduction may be due to an increase in crust formation and intense browning reaction because of exposure to a very high temperature. In the case of low-temperature frying of cassava chips like vacuum frying the reduction in lightness of the product was comparatively less, since vacuum frying employs lower temperature and pressure (Mariscal and Bouchon, 2008). It was also recommended that lower temperature of the cooking medium and oil with comparatively lesser boiling point is to be used for frying operation. Redness in the frying of cassava chips (in general cereal chips) generally represents crust formation and gives a hard feeling when observed from a consumer perspective, which may be due to the Maillard reaction ensuing from the utilization of obtainable reducing sugars. Similarly, the yellowness of cassava chips was due to carotene present, and it was reduced with the rise in the temperature of frying (Oyedeji et al., 2017).

Ndlala et al. (2019) tested the color of both crust and crumb of Magwinya samples with wheat bran (WB) and without WB, and concluded through detailed analysis that samples with WB were darker than that without WB. The reasons were: i). effect of the color of WB was reflected in the final product; ii). increased Maillard reaction and caramelization. All the L*, a*, and b* values of the crust were higher than that of the crumb. It was also seen that these color changes were due to the fermentation time of the Magwinya batter and the frying time of the product (Onipe et al., 2015).

12.3.1.3 Texture

Textural and rheological properties play a significant and decisive role in consumer perception, opinion, and acceptability of any food product, but processed and modified food products in particular. Many factors influence the these parameters such as formulation of food ingredients, conditions of processing, and raw material behavior under certain conditions as well (Ndlala et al., 2019). In general, the texture of a food product during frying is affected by gelatinization of the starch present, the sugar content of the product, and the breakability of the cell walls (Oyedeji et al., 2017). The degree of disruption of the structure of starch is mainly affected by the temperature of the treatment medium or conditions like moisture content of food before being processed and treatment time (Chen et al., 2018). The texture is also altered or developed inside the food product during frying as a result of changes in carbohydrates, proteins, and fats in a similar fashion, as occurs during baking or boiling (Bordin et al., 2013).

Textural properties of cereal-based fried foods such as Magwinya (Ndlala et al., 2019; Onipe et al., 2015) and normal maize starch (NMS) (Chen et al., 2018) are described using hardness representing force applied by teeth for compression of food (in other words, it is the breaking force of a product), cohesiveness representing the extent of deformation undergone by the food material before rupturing food when being bit, chewiness indicating the required number of chews to make the food suitable to be swallowed, and springiness giving the rate of retaining original shape/size once the food product is partially compressed in the mouth (Gwartney et al., 2004; Ndlala et al., 2019).

Oyedeji et al. (2017) observed the textural changes in cassava chips during various frying methods such as atmospheric (fresh) frying, atmospheric (pre-dried chips) frying, and vacuum frying, and found that there was a rapid increase in hardness (breaking force) at lower time of frying for all of the abovementioned methods, since the rapid moisture removal from the surface caused the dehydration in the crust, thus forming a harder crust. After 8–12 min of frying maximum hardness was observed, but there was a consistent dip in the hardness, due to the gelatinization of the starch present in cassava at an extended frying duration. The chips showed a greater increment in breaking force at slightly lower frying times in vacuum frying than that in atmospheric frying as the vacuum was responsible for quicker moisture removal. Reduction of maximum breaking occurrence time to less than 8 min showed reduction in boiling point of water at vacuum conditions, which led to a slower and more uniform removal of moisture throughout frying; similar instances were observed in the work reported by Ndlala et al. (2019) for Magwinya and Ghaitaranpour et al. (2018) for doughnuts. Hardness values of wheat chips were increased with the addition of barley flour due to enhancement of the water-holding ability of mixed flour and thus extended crust formation time and ultimately hardness (Yuksel et al., 2015). The hardness of rice crackers decreased the over-spraying rate and spinning speed during the spray frying method (Udomkun et al., 2020).

12.3.1.4 Size

Deep-fat frying of food products, especially cereal-based products like jilebi (Chakkaravarthi et al., 2014), Gulabjamun balls (Kumar et al., 2006), cassava chips (Oyedeji et al., 2017), magwinya (Ghaitaranpour et al., 2018; Ndlala et al., 2019), and rice crackers (Udomkun et al., 2020) is a kind of HMT process, with the application of heat to frying medium thereby to product and mass of the product gets exchanged i.e., with the application of heat through oil to the abovesaid and many more food products, water present in the raw food product gets attained the heat, gets evaporated and tries to escape out of the food product structure matrix (Chen et al., 2018). Size of the food product plays a crucial role in oil uptake as lower thickness raw products absorb more oil and vice versa (Asokapandian et al., 2020). As in the case of raw products, moisture content plays one of the key roles in maintaining the size and bulkiness of the product. With frying, the size is expanded and the density of the food products is decreased gradually.

Chakkaravarthi et al. (2014) observed that there is a significant change in the size and density of Jilabi during frying before being dipped into sucrose solution, which may be due to moisture loss and oil uptake, expansion of strands because of gelatinization of the residual starch, and steam formation inside leading to expansion. Though there was no conclusive evidence regarding the role of frying medium temperature in the expansion and density reduction, researchers strongly believe that there is a role of temperature too in this action like the other ones (Ghaitaranpour et al., 2018; Melito and Farkas, 2013). Similar instances were seen with papad (Math et al., 2004) and cassava chips (Oyedeji et al., 2017). Udomkun et al. (2020) reported that there was a significant decrease in the density of spray-fried rice crackers when fried at a higher spraying rate and greater spinning speed.

12.3.1.5 Shape

Shape is also one of the key parameters along with size, density, and other surface-related parameters to be kept in mind before commissioning the final product through the frying method. The initial shape of the product is responsible for the quantity of the oil uptake during the frying process. The surface-to-volume ratio holds a linear relationship with the surface area of the product and oil content (Asokapandian et al., 2020). Shrinkage and shape loss of the product happens concerning the initial moisture content of the food product. In the case of cassava chips with yellow flesh frying, shrinkage during the initial stages was rapid and upon the formation of crust, which created a barrier for moisture content to escape out, and gas expansion resulted in the expansion of surface in a haphazard manner. Greater frying temperatures also led cassava slices to undergo maximum shrinkage by inducing higher mass diffusivity and higher moisture loss. Surface characteristics also led to lower shrinkage than expected (Oyedeji et al., 2017). Similar cases were found with doughnuts (Tan and Mittal, 2006).

12.3.1.6 Porosity

Porosity and volume are key physical parameters both in the case of raw as well as processed foods as they influence some of the other structural and mechanical properties of food (Pedreschi, 2012). Porosity (φ) is defined as "the volume fraction of the air or the void fraction in the sample" (Asokapandian et al., 2020). The porosity of the crust formed while frying a food product, especially starchy foods like cereals, is one of the major factors considered to be responsible for greater or enhanced oil uptake during the process (Asokapandian et al., 2020). Oil uptake during the frying process shares a linear relation with porosity before frying or raw product. Net porosity (i.e., porosity by excluding the oil-occupied surface volume) was effectively described the oil uptake relation with the porosity of the food product from time to time (Gaikwad et al., 2021; Pedreschi, 2012).

The size of the pores present also plays a decisive role in oil uptake and moisture content reduction (Zhang et al., 2020). Similar kinds of results were seen in the jilebi-making process where oil uptake was dependent on the diameter of pores (Chakkaravarthi et al., 2014). Crust uniformity, crust coverage percentage, and surface roughness of doughnut were affected by frying medium temperature, treatment time, pore size, and initial porosity of the dough (Ghaitaranpour et al., 2018). In the case of ball-milled wheat flour, a rise in porosity of batter governed the oil uptake of the whole frying process and final product as well (Thanatuksorn et al., 2010).

12.3.1.7 Moisture Content

Frying is a multifaceted process involving HMT between product and frying medium simultaneously. This happens at a frying medium temperature in the range ~150 °C to 180°C or sometimes up to 200°C where the boiling point of water is way less. During the frying process, oil uptake and moisture removal from a food product are affected by multiple factors such as i). Raw material/product parameters such as size, shape, porosity and pore size, surface roughness, density, and initial moisture content; ii). process parameters such as the temperature of frying, time of interaction of food material with frying medium; iii). pre-treatments (Asokapandian et al., 2020; Zhang et al., 2016). In the case of doughnuts, moisture loss followed the typical drying rate curve (i.e., the initial one min faster rate of drying where crust formation took place, followed by a constant rate of moisture removal, which means till the crust became hard), and finally exhibited a falling rate of moisture removal when there were the least possible chances for evaporated water to escape out. Doughnuts with lower initial moisture content absorbed more oil content than those with higher moisture content (Tan and Mittal, 2006). Water is an important constituent of food products that influences the porosity and crunchiness of final fried food products (Shieh et al., 2004).

Frying is a process of dehydration where a considerable amount of moisture is evaporated. In the case of Magwinya as a result of heat transfer through convection from frying medium to the surface of Magwinya and through conduction from surface to inner layers (Ndlala et al., 2019; Onipe et al., 2015).

The moisture content of the final product was affected by the frying time and temperature of the frying medium linearly in the case of papad and puri (Math et al., 2004; Salehi, 2020; Koerten et al., 2017). Other than these, moisture loss was also enhanced by a decrement in pore diameter and surface-to-volume ratio (Koerten et al., 2017).

12.3.2 Effect of Frying on Compositional Properties

Cereals are one of the most important sources of food all over the world. Not only are they a major source of food and energy, but they are also very good sources of proteins, minerals, and vitamins. Consumption of cereals as a whole can help provide protection from diseases like diabetes, cardiovascular diseases, constipation, obesity, and many other disorders (Contreras et al., 2017; Oghbaei and Prakash, 2016; Zhang et al., 2020). In the case of deep-fat frying of any food product, the interaction between the product and the frying medium has a detrimental effect on the nutritive value of the product as opposed to the positie effect on sensory value (Bouchon & Dueik, 2018). The changes during frying may affect the compositional properties of cereal-based food products (Table 12.1). The changes in compositional properties of fried cereal-based food products are discussed below.

12.3.2.1 Changes in Carbohydrates

Carbohydrates are one of the major sources of energy for the human body, and cereals and pulses are abundant sources of carbohydrates. Both the method of processing and the actual sources generally increase the effective availability of carbohydrates (Oghbaei and Prakash, 2016; Poutanen, 2009). Carbohydrates are the most available classes of organic biomolecules present in nature and

Table 12.2 Impact of Frying on Biological Properties of Cereal-Based Food Products

S. No.	Cereal or cereal-based food products	Frying conditions	Observations	References
1	Kabau (cereal)	• 150 °C for 3 min	1. Reduced protein content 2. Amino acids dropped to lowest levels 3. No significant change observed in anti-nutritional factors	(Fitriani et al., 2021)
2	Cereal-based foods	• > 160 °C	1. The formation of acrylamide is not only observed in cereals but also observed in all fried food products.	(Pedreschi et al., 2014)
3	You-tiao (traditional Chinese, fried, twisted dough-roll)	• Frying temperature 225, 200, and 175 °C, and frying time 48, 67, and 86 s.	1. Frying temperature reduced from 200 to 175°C and prolong the frying time (86 s) leads to lower the pH (6.0) of dough and acrylamide formation. 2. Fermentation (yeast) could also reduce the acrylamide content from the fried products.	(Huang et al., 2008)

are synthesized during the photosynthesis process by plants (Zeb, 2019b). They are classified into three major groups based on their structural components: monosaccharides, oligosaccharides, and polysaccharides. Monosaccharides are majorly known in the form of simple sugars that cannot be hydrolyzed into simple compounds, as they are the basic structural units of carbohydrates. Oligosaccharides are generally made of 2 to 10 units of monosaccharides joined by a bond called "glycosidic bond." Maltose is one of the most available oligosaccharides in cereals. Polysaccharides are also known as glycons, and are made of more than 10 units of monosaccharides that are joined by multiple types of glycosidic bonds. Starch and cellulose are some of the common examples of homopolysaccharides, and heparin and pectin are the common examples of heteropolysaccharides (Zeb, 2019b).

In the crust of the fried product, starch and non-starch carbohydrates are broken down during frying because of heavy temperature and interactions between frying medium and carbohydrate groups, and starch-lipid complexes are readily formed. Sucrose is split into glucose and fructose by hydrolysis, which are lost during Maillard reaction and caramelization (Fellows, 2017).

12.3.2.2 Changes in Protein

Cereals and pulses are potential sources of protein in plant-based products for low-income families in regions like Africa and Asia. Digestibility of protein is a key component in satisfying daily protein requirements (Oghbaei and Prakash, 2016). Proteins are the polymers of amino acids, and are made up of a carboxylic group and an amino acid group joined by a carbonyl-amino group defined as a peptide bond or an amide bond (Zeb, 2019b). Some of the essential amino acids like lysine and tryptophan are partially destroyed during the frying process (Fellows, 2017; Oke et al., 2018). Peptides and amino acids are more reactive at greater temperatures. In general, the frying temperatures may go up to 150°C to 200°C; many of the proteins are denatured at temperatures greater than 100 °C. The stability of some of the biologically active proteins (i.e., some enzymes) is also negatively affected by the frying process (Oke et al., 2018; Pokorný & Dostálová, 2016).

Soluble protein fraction in tortilla chips is reportedly reduced during the frying and baking process due to undesirable changes in protein digestibility (Vázquez-Durán et al., 2014). It is found that frying causes a reaction between reducing sugars and amino acids, called the Maillard reaction, which leads to the formation of carcinogenic compounds that are dangerous to human and animal life (Fellows, 2017). The reaction also influences flavor, color, and other nutritional components of food. The reaction, in general, occurs in three stages. The first stage involves the interaction of free amino groups and reducing sugars forming acrolein. The second stage involves Strecker degradation, which results in the degradation of amino acid groups into ammonia and aldehydes under high internal pressure and high temperature. The third and final stage involves the formation of acrylamide involving the combination of acrylic acid and nitrogenous compound with brown color. The same process has been mostly seen in cereals and bread (Baskar & Aiswarya, 2018).

12.3.2.2 Changes in Vitamin

Vitamins are one of the key natural compounds that accomplish multiple biological roles. The most commonly consumed cereals such as rice, wheat, and corn are good sources of vitamins like vitamins A, B_1, B_2, B_3, B_5, B_6, B_9, E, and K. But they are not rich sources of vitamins like B_{12}, C, and D (Garg et al., 2021). Fat-soluble vitamins such as vitamin A, E, D, and K are almost unaffected during the frying of cereals or cereal-based products, whereas water-soluble vitamins such as vitamin B and C are largely destroyed or degraded in quality during frying (Zeb, 2019b). Vitamins are sensitive to higher temperatures and oxidation reactions. In the frying process, the temperature of the frying medium may go up to 200°C, so, it is obvious that there are high chances for the loss of vitamins. Vitamin C is generally the most thermo-sensitive followed by B-group vitamins such as thiamine (B_1), riboflavin (B_2), niacin (B_3), and pyridoxine (B_6) as the most affected vitamins (Bordin et al., 2013; Oke et al., 2018). In the crust of fried food products, fat-soluble vitamins are lost at a comparatively higher rate than that of the inner region. Oxygen-sensitive vitamins are also lost during frying (Fellows, 2017).

12.3.2.3 Changes in Carotenoids

Carotenoids have long been considered as one of the good contributors to the human diet in the segment of micronutrients. These are synthesized by microorganisms and plants but not by animals including humans. They are isoprene derivatives generally found in seeds, fruits, and plants that impart yellow color or orange color to the food product (Siah and Quail, 2018). There are over 600 fat-soluble pigments from plants including non-polar hydrocarbons (HC), carotenes (α and β-carotene, lycopene), and their oxygenized derivatives, xanthophylls. Carotenoids generally interact with reactive-oxygen species and therefore act as antioxidants (i.e., doing scavenging activity of free radicals, singlet-oxygen scavengers, and lipid antioxidants) (Irakli et al., 2011; Luterotti and Kljak, 2010; Maiani et al., 2009; Siah and Quail, 2018).

Deep-fat frying of cereal-based products may reduce the carotenoid retention until the frying temperature reaches 140°C, and after that point, there may be higher degradation, due to the amount of oil uptake and mass transfer between product and frying medium. With the repetition of frying, the carotenoid content of oil is increased (Zeb, 2019c).

12.3.2.4 Changes in Phytosterols

Plant-based sterols (i.e., phytosterols) are secondary metabolites of plants that occur in the form of sterol conjugates and free sterol alcohols in cereals especially. They are triterpene compounds that show similarity in structure to the structure of cholesterol (Zeb, 2019b). The proportion of these sterol compounds vary among different plant parts and a wide range of cereal species (Nystrom et al., 2009). Approximately 40 different compounds fall in the phytosterols category. Cereals, especially rice, wheat, corn, and rye along with some leguminous plants and some oil seeds like sunflower, are very good sources of these plant-based sterols (Dziedzic et al., 2016; Luithui et al., 2018). With temperature rise during frying, polymerization and oxidation of phytosterols are accelerated, thus so is oligomerization and thermo-oxidation of phytosterols (Zeb, 2019d).

12.3.2.5 Changes in Folate

Folate is one of the B-group's vitamins that can be found in various foods and forms that have single-carbon constituents, variable oxidation status (vitamers), and a number of glutamyl residues. Cereals often contain a wide range of vitamers with different formyl and methyl derivatives along with unsubstituted tetrahydrofolate when compared to fruits and vegetables (Cheng et al., 2009; Shewry and Ward, 2010). Various types of thermal processing methods such as frying, baking, boiling, etc., besides other modification activities such as milling, grinding, etc., have substantial effects on the total folate content (Kariluoto and Piironen, 2009).

12.3.2.6 Changes in Total Dietary Fiber

Dietary fiber (DF) is a composite mixture of carbohydrate polymers that are bound with non-carbohydrate components, consisting of non-starch polysaccharides along with protein, lignin, fatty acids, etc. (Singh et al., 2018). DF is considered a major pro-health component of food grains, and is mostly concentrated in the outermost layers of the edible portion of grains. The same layer consists of certain vitamins, phytochemicals, and minerals, which significantly promote human health. DF is classified based on its water-solubility characteristics as water-insoluble and water-soluble. Water-insoluble DF (IDF) includes cellulose, hemicellulose, and lignin. Water-soluble DF (SDF) consists of pectin, gums, and mucilages. Among the above DF, cereals are rich sources of hemicellulose, which are polysaccharides present in the cell wall (Dhingra et al., 2011). The consumption of cereal DF and whole grain have long been considered to be good protection of health and play a pro-active role in reducing the potential risk of diseases like type-2 diabetes and cardiovascular diseases (Poutanen, 2009). Consumption of fiber-rich foods, mainly cereals and millets, has positive effects on the prevention of numerous diseases. DF is particularly insoluble and plays a crucial role in treating diabetes due to its slow and continuously releasing nature (Poutanen, 2009).

During frying, the ratio between soluble DF and insoluble DF (i.e., SDF/IDF) is altered. Ye et al (2016) extensively reviewed the possible changes that happen during different processing methods of foods, and concluded that the thermal treatment can degrade some of the SDF components, and alter their fat absorption capacity and hydration properties as well. Overall SDF degradation may cause oil uptake, which is not a desirable change for a good quality final product.

12.3.2.7 Changes in Phenolic Compound

The most abundant group of phenolic compounds available in cereal grains are phenolic acids. Alkylresorcinols and lignans are the two highly discussed and investigated sections in phenolic compounds. They have been assumed to have a wide range of biological activities both in vitro and in vivo (Poutanen, 2009). In cereals especially, these phenolic acids can occur in three different forms: i) insoluble and bound to the cell wall, ii) soluble conjugated form, and iii) the free acid form (Li et al., 2009). Consumption of cereal-based products subsidizes the intake of phenolic acid only when whole grain cereals are employed for the making of cereal-based products (Oghbaei and Prakash, 2016). Phenolic acids in cereals are considered antioxidants and endorsed to have anti-cancer activity, antioxidative stress, and diseases related to neurodegenerative issues (Luithui et al., 2018; Van Hung, 2016). These phenolic acids are generally found in the outer layer bran of the grain (Li et al., 2009).

Oghbaei and Prakash (2019) carried out a study on roasting (Chapathi) and deep-fat frying (Poorie) of cereal-based food products and identified that the roasting and frying processes reduce phenolic and flavonoid content from Chapathi and Poorie. High heat treatment mainly impacts the availability and nutrient content. Similarly, they observed that the deep-fat fried Poories had a lower content of flavonoids and phenols compared to roasted Chapathi, which was due to the high temperature (170°C) utilized during the frying process. Frying at a lower temperature (≤ 150°C) for a short duration (10–15 s) can control the loss of nutrient and phenolic content from cereal-based food products (Zeb, 2019b).

12.3.2.8 Changes in Fatty Acids and Triacylglycerols

During frying saturated fatty acids (SFAs) are relatively more stable than unsaturated fatty acids (USFAs) resulting in more absorption of oil (in other words, more oil uptake) (Zeb, 2019c). Triacylglycerols (TAGs) are one of the major components of frying mediums such as oil or fat, but are found less in food materials (Zeb, 2019c, 2019b). They are more susceptible to degradation by the high temperatures produced during the frying process (Zeb, 2019c). Free fatty acids (FFAs) are developed during frying because of the interaction of moisture present in the food product and frying medium (Orthoefer and List, 2007). Oligomerization and thermal oxidation are increased when the frying temperatures are high.

Frying of oil at higher temperature (> 180°C) is responsible for oxidation, hydrolysis, and polymerization, which deteriorates the quality and sensorial attributes of fried products (Bhardwaj et al., 2016). While frying, unsaturated oil is oxidized more rapidly compared to less unsaturated oil (Zeb, 2019e). Frying is also responsible for producing trans fatty acids (TFAs) due to hydrogenation, which causes serious health issues in humans. TFAs increase the risk of cardiovascular diseases and reduce the amount of high-density lipoprotein cholesterol. The formation of TFAs is mainly due to the increase in temperature and repeatedly using the same oil for frying (Bhardwaj et al., 2016). The level of TFAs can be reduced by replacing the oil or blending the oils, frying at < 180°C, and avoiding the number of times using the same oil (Zeb, 2019e).

12.3.3 Effect of Frying on Biological Properties

Frying of cereal-based food products at higher temperature (> 120–200°C) for a short duration (5–10 min) causes dehydration, which causes structural rearrangement of biological components and results in an increase in Maillard reaction, formation of acrolein, acrylamide, furans, and changes in anti-nutritional, heterocyclic aromatic amines, aromatic hydrocarbon, and mycotoxins. The changes during frying may also affect the biological properties of cereal-based food products (Table 12.2). The changes in biological properties of fried cereal-based food products are discussed below.

12.3.3.1 Changes in Amino Acid

A peptide is a small chain of amino acids connected with peptide bonds and present in cereal-based foods in distinct quantities. According to Pokorný and Dostálová (2010), the nutritional content of food products is mainly influenced due to the reaction of particular amino acids during frying. Cereal-based food products are comprised of many amino acids such as alanine, aspartic acid, asparagine, cystine, glutamic acid, histidine, leucine, phenylalanine, serine, threonine, tyrosine, and valine (Zeb, 2019b). During the frying of cereal-based foods at high temperatures (> 120–200 °C), Maillard products are generated due to the reaction between carbohydrates and amino acids. The chemical compounds formed during the Maillard reaction are considered to be carcinogenic to human health (Pokorný and Dostálová, 2010).

12.3.3.2 Changes in Antinutritional

Cereals are the major component of carbohydrates, protein, unsaturated fats, vitamins, and minerals. Cereals have several anti-nutritional factors including alkaloids, lectins, oxalates, phytates, protease inhibitors, tannins, trypsin inhibitors, etc. Anti-nutritional factors are biological components that are deleterious to human health and reduce the bioavailability of nutrients by absorbing those nutritive biomolecules (Nagraj et al., 2020; Ram et al., 2020). These compounds are responsible for reducing the nutrient intake in the human body and cause a reduction in gastrointestinal and metabolic functions and several other diseases. Therefore, the reduction of anti-nutritional factors from cereals is of great interest, and several traditional and thermal methods are followed including cooking, de-branning, fermentation, germination, milling, roasting, and soaking (Samtiya et al., 2020).

In most cereals, phytate is present in the form of phosphorous, which reduces the bioavailability of micronutrients to humans. The presence of phytic acid in cereals can also interfere with calcium, magnesium, iron, and zinc due to its divalent cationic ability (Ram et al., 2020; Samtiya et al., 2020). Phytate content from cereals can be reduced utilizing traditional and thermal methods. Oboh et al. (2010) successfully reduced phytate content from different varieties of cereals by roasting at

120–130°C before milling and a significant increase was observed in calcium, carbohydrates, fat, magnesium, and zinc content in the cereals.

Protease inhibitors are present in barley, rye, oats, and wheat at low concentrations but inhibit proteases activity and cause gastric and hyper-genesis problems in humans. Similarly, trypsin inhibitors are found in the embryo of the cereals, and the activity of these inhibitors can be deactivated by heating or dry roasting at above 100°C for 30 min to 1 h (Samtiya et al., 2020).

Tannins are mainly present in the seed coat of cereals and considered to be nutritionally undesirable due to the inhibition of digestive enzymes and reduced intake of minerals and vitamins. They also affect the growth rate, digestibility of protein, and feed intake in humans (Ram et al., 2020; Samtiya et al., 2020). Ram et al. (2020) reported that the amount of tannins in cereal grains ranges between 0.1 to 1.5% and can be reduced using a combination of traditional and thermal methods. The utilization of the dry roasting method can successfully reduce almost 87% of polyphenols from wheat grains (Gunashree et al., 2014).

Alkaloids are active compounds comprised of heterocyclic nitrogen and are considered harmful to human health. They are generally found in barley, oats, rye, and wheat and produced from amino acids such as lysine, tryptophan, and tyrosine (Ram et al., 2020; Samtiya et al., 2020). Sharma et al. (2021) carried out popping of amaranth grains using dry roasting without fat and observed that the amino acid (arginine, cysteine, and lysine) content was damaged owing to starch gelatinization and protein denaturation.

12.3.3.3 Formation of Maillard Reaction

Maillard reaction is also referred to as a non-enzymatic browning of food products. It is a complex chain chemical reaction between reducing sugar and amino acids by the impact of high heat that provides browning and distinct aroma to food products. The formation of the Maillard reaction mainly depends on the product to be fried, time of frying, and temperature of frying. While frying cereal products, Maillard reaction occurs, which reduces the nutritional content from foods by replacing them with harmful compounds such as acrylamide, furans, hydroxymethylfurfural (HMF), and aromatic hydrocarbons (Zeb, 2019b; Morales and Arribas-Lorenzo, 2008). Maillard reaction can be controlled by frying at a higher temperature for a short duration and later processed with a lower (< 120°C) temperature (Pokorný, J., and Dostálová, 2010).

12.3.3.4 Formation of Acrylamide

Acrylamide is a complex chemical compound produced during the Maillard reaction of carbohydrate-rich food products that are baked, fried, and roasted at high temperatures (> 120°C) (Galani et al., 2017). According to International Agency for Research on Cancer (IARC), acrylamide is a cancer-causing compound for humans (Mousavi Khaneghah et al., 2020). Cereals are rich in amino acids and asparagine is one of them that undergoes a Maillard reaction along with reducing sugars such as fructose and glucose, at temperature > 120°C, and produces acrylamide in cereal-based fried food products. According to the European Commission, the level of acrylamide in cereal-based food products ranges between 50–500 mg/kg (European Commission, 2013). Zeb (2019b) reported that the frying medium abundant in polar compounds can also increase the concentration of acrylamide in fried food products. Moreover, the excess consumption of acrylamide products leads to neurological disorders, cellular mutation, and effects on the digestive tract (Khezerolou et al., 2018). The formation of acrylamide in fried products can be reduced using vacuum frying or frying at a temperature < 120°C. (Khezerolou et al., 2018; Pokorný and Dostálová, 2010).

12.3.3.5 Formation of Furans

Furan and furfural are organic volatile compounds formed due to the heating of carbohydrates. Furan is a carcinogenic compound for human health by IARC (Sirot et al., 2019). Hydroxymethylfurfural (HMF) is formed during thermal degradation of carbohydrates (reducing sugar) during caramelization and also through the Maillard reaction. While frying cereals-based food products, caramelization occurs at specific temperatures (Zeb, 2019b). HMF is also known to be cytotoxic for mucous membranes, mutagenicity, and respiratory tract (Shapla et al., 2018). The possible range of HMF is 0.9 mg/kg in caramel products, but an increase in frying time and temperature may increase the formation of HMF (Sirot et al., 2019; Zeb, 2019b). Ramírez-Jiménez et al. (2000) reported that the level of HMF was up to 10.7 mg/kg in fried dough-nut at 160°C for 30 to 40 sec. Göncüoglu and Gökmen (2013) observed the formation of HMF in frying oil while frying wheat flour dough. Mir-Bel et al. (2013) conducted a study of frying Spanish doughnuts in an open fryer and also in a vacuum fryer (100–180°C), and the author reported that the formation of HMF was lower (< 0.05 mg/kg) in the vacuum fryer in comparison to the open fryer.

12.3.3.6 Formation of Acrolein

Acrolein is generally formed during the oxidation of amino acids such as asparagine, methionine, and threonine. Zeb (2019b) reported that at higher temperatures, methionine oxidizes and produces methional that releases acrolein in the fried food products. Acrolein is a toxic compound formed during the Maillard reaction and is mainly found in cereal-based fried foods as well as in frying mediums since they are abundant with carbohydrates and amino acids. Pokorný and Dostálová (2010) reported that during frying cereals, glycerol is condensed to acrolein and the extended heating time and temperature release acrylamide in the product and also in the cooking medium.

12.3.3.7 Formation of Heterocyclic Aromatic Amines

Heterocyclic aromatic amines (HAAs) are formed during the Maillard reaction due to heating, roasting, and frying of amino acids at higher temperatures (Barzegar et al., 2019). The heterocyclic amino acids are histidine, hydroxyproline, proline, and tryptophan. HAAs are classified into several subclasses based on their chemical composition: benzoxazines, furopyridines, indoles, pyridines, pyridoimidazoles, quinolones, quinoxalines, etc. HAAs are also classified based on polarity and temperature such as pyrolytic HAAs and thermic HAAs. Thermic HAAs formed at a temperature range between 100 to 300°C and pyrolytic HAAs formed at a temperature of > 300°C (Özdestan et al., 2014). According to the IARC, most of the HAAs (class 2B) are carcinogenic to human health. Several HAAs are also responsible for increasing the risk of prostate tumors in humans and animals. As per the Council of Europe, the daily intake of HAAs should be lower than 1 μg/day (Barzegar et al., 2019). The formation of HAAs can be reduced by replacing the fat with grapeseed oil or olive oil or sunflower oil (Zeb, 2019b).

12.3.3.8 Formation of Polycyclic Aromatic Hydrocarbons

Polycyclic aromatic hydrocarbons (PAHs) are organic compounds comprised of carbon and hydrogen atoms. PAHs are formed in foods owing to baking, drying, frying, and roasting at a temperature > 160°C (Singh and Agarwal, 2018). PAHs are responsible for the formation of carcinogenic fumes during the frying of cereal-based food products at a higher temperature (Zeb, 2019b).

According to the European Union, the regulatory value of PAHs for cereal-based food products is 5 mg/kg. As per the IARC, PAHs are cariogenic to human and animal health. PAHs are also responsible for tumor, mutagenic, teratogenic, immune toxic effects, etc. (Singh and Agarwal, 2018). The level of PAHs can be reduced in cereal-based fried products by frying at a lower temperature for a short duration.

12.3.3.9 Formation of Mycotoxins

Mycotoxins are formed due to the growth of molds during the storage of cereal-based food products (Nedović et al., 2016). Mycotoxins are also formed by species of fungi including *Aspergillus spp.*, *Fusarium spp.*, and *Penicillium spp.* Fumonisins are one of the groups of mycotoxins formed due to the growth of *Fusarium proliferatum* and *F. verticilliodes* (Pokorný and Dostálová, 2010). Mycotoxins are toxic compounds that cause several types of cancer, hepatic diseases, and immune and neurological disorders (Fletcher and Blaney, 2016). Pokorný and Dostálová (2010) reported that the degradation of fumonisin was not observed at a temperature below 170°C, but rather at temperatures above 180°C. A similar observation was reported by Suman and Generotti (2015) for frying corn chips at 190°C for 15 min.

12.4 EFFECT OF FRYING ON HUMAN HEALTH

The process of frying food results in food preservation through dehydration, deterioration of microbes, and reduced water activity. While frying foods, the oil infuses in the food and results in compositional transformation, which relies on various aspects like physical and chemical composition of oil as well as that of the food, temperature of frying oil, and duration of frying. These aspects impact the nutritive content of the final fried product. Undesirable variations might develop, especially with respect to micronutrients in the course of frying (Oke et al., 2018). It has been reported that there were about 400 kinds of degraded compounds in fried products among which few might cause adverse health effects including carcinogens (Bouchon et al., 2009). Frying of food results in chemical reactions like auto-oxidative at temperatures above 100°C, anaerobic polymerization around 220–320°C, and thermal oxidation around 210°C, and is the most prominent reaction causing adverse health effects among consumers (Clarck et al.,1991). The products of lipid oxidation react with certain amino acids and result in the formation of acrylamide, acrolein, and trans fatty acids that make foods less nutritive (Asokapandian et al., 2020). Frying food at a higher temperature for a prolonged period results in lower protein content, especially the loss of amino acid (lysine), increase in fat, and higher energy uptake that may lead to weight gain (USDA, 2013). However, the mineral loss is small while frying since most minerals are water-soluble and are retained during the short frying time and single use of frying oil. On the other hand, vitamins are heat-labile and destroyed due to oxidative reactions, and thus most vitamins are not retained in fried foods (Oke et al., 2018).

Various reports have shown that the loss of nutrients or formation of compounds postfrying occur in overheated and reused oil/fats and cause adverse health effects. Whereas those fried products from the standard frying process have little to no harmful effects on animals. (Deep frying food is defined as a process where food is completely submerged in hot oil at temperatures above 175°C.) Furthermore, excess eating of fried food may result in health complications due to some toxic compounds produced while frying. Generation of trans fats and depletion of essential fatty acids from repetitively used oil increases the risk of cardiovascular diseases due to the deposition of fat in the arteries. Also, certain harmful substances produced amid the deterioration of fried foods are related to various disorders like Parkinson's, Alzheimer's, and cancers (Guillén et al., 2008). Another study reported that increased blood pressure is linked to the consumption of polar compounds formed in fried products (Soriguer et al., 2003). Thus, it is necessary to control the procedure, quality maintenance of fried products that prevent the detrimental effects on humans (Hosseini et al., 2016).

12.5 IMPORTANCE OF FRYING

In recent times there has been increasing demand for fried food products, which has led to the evolution of various kinds of fried food products. Frying is one of the most adequate and extensively utilized modes of cooking, due to its short cooking time and, chiefly, its texture is broadly accepted globally. It is a quick and convenient mode of food production, comparatively cost-effective, and the resulting attributes like appearance (color/texture), taste, and aroma facilitate its comprehensive utilization. These characteristics are the consequence of physiochemical transformations occurring while frying the food, along with redox reactions, water loss, and polymer formation leading to alteration in the properties of the food. Protein denaturation, starch gelatinization, Maillard reaction, etc., are biochemical changes that occur during cereal frying (Viana et al. 2015). Higher temperature frying oil in cereal-based foods evaporates the moisture from the surface and expands the food, thus leading to desirable characteristics like crisp texture and melting mouthfeel. Frying also aids in the preservation of food through deterioration of microbes as well as deactivation of enzymes by heat and decreases the water activity of the food products. The storage life of fried products is increased through the amount of moisture left, post frying. The products that get dehydrated to a higher extent upon frying (e.g., corn rings) have a longer shelf-life of about a year at ambient conditions (Zaghi et al., 2019).

The composition of food like its protein and fat content leads to characteristic textural changes. The transformation in protein quality is due to the occurrence of the Maillard reaction upon frying. The changes in the nutritive composition of fried food rely on the process parameters of frying. At a higher temperature of frying, oil leads to the quick development of crust that closes the contact surface of the fried product, which reduces the extent of alterations inside the product, and subsequently, conserves food nutrients (Heredia et al., 2014).

Certain nutritive dense cereal oil can be used to enrich fried foods with some beneficial nutrients. Rice bran oil, which has a higher smoking point, could be used in deep-fat frying of foods due to its rich source of vitamin E (32.3 g) and K (24 μg). γ-oryzanol has higher unsaturation that decreases blood cholesterol level. Cereal fried products like corn chips incorporated with stale bread powder result in lower oil content, higher dry matter, ash content, and sensorial acceptability (Yuksel et al., 2017). Wheat flour doughnuts fortified with 10% wheat bran have reduced fat intake with highly acceptable aesthetic values (Kim et al., 2012).

12.6 FUTURE SCOPE

Fried products are more popular and more in demand due to their palatability and flavor due to the process of frying. However, the frying of food results in higher oil uptake, which leads to adverse health effects on regular consumption. Also, fried food product gets either soggy (Gulabjamun or Jalebi or doughnuts) or packed fried food products like corn chips could get rancid if exposed to oxygen during shelf life. Therefore, proper mitigation techniques are required to address the aforementioned frying issues. Since frying food is an outer cooking phenomenon, slightly modified techniques can be employed to the exterior of food products to lower the oil intake. Since most of the oil absorption takes place postfrying (cooling) rather than while frying, intense research is needed to resolve this problem. In addition, the technique of frying itself could be altered in order to retain nutrients along with lower oil absorption. Also, more studies are required to investigate the frying medium or use of other sources of oil like cereal rice bran oil, which is already nutritive dense. Low calorific value heating medium can also be employed. Hence, further research is required to lower oil intake, and also more fried cereal products are encouraged since cereals and cereal products are dense with protective nutrients that could help to combat non-communicable diseases.

REFERENCES

Aalim, H., Wang, D., & Luo, Z. (2021). Black rice (Oryza sativa L.) processing: Evaluation of physicochemical properties, in vitro starch digestibility, and phenolic functions linked to type 2 diabetes. Food Research International, 141, 109898. https://doi.org/10.1016/j.foodres.2020.109898

Achir N, Vitrac O, Trystram G (2008) Heat and mass transfer during frying. In: Sahin S, Sumnu SG (eds) Advances in deep-fat frying of foods contemporary food engineering. CRC Press, Boca Raton, pp 5–32. https://doi.org/10.1201/9781420055597

Asokapandian, S., Swamy, G. J., & Hajjul, H. (2020). Deep fat frying of foods: A critical review on process and product parameters. Critical Reviews in Food Science and Nutrition, 60(20), 3400–3413.

Barzegar, F., Kamankesh, M., & Mohammadi, A. (2019). Heterocyclic aromatic amines in cooked food: A review on formation, health risk-toxicology and their analytical techniques. Food Chemistry, 280, 240–254.

Baskar, G., & Aiswarya, R. (2018). Overview on mitigation of acrylamide in starchy fried and baked foods. In Journal of the Science of Food and Agriculture (Vol. 98, Issue 12). https://doi.org/10.1002/jsfa.9013

Berk, Z (2018). Food Process Engineering and Technology, Frying, Baking, Roasting. Academic press pp 583–589. doi:10.1016/B978-0-12-415923-5.00024-1

Bhardwaj, S., Passi, S. J., Misra, A., Pant, K. K., Anwar, K., Pandey, R. M., & Kardam, V. (2016). Effect of heating/reheating of fats/oils, as used by Asian Indians, on trans fatty acid formation. Food Chemistry, 212, 663–670.

Bordin, K., Kunitake, M. T., Aracava, K. K., & Trindade, C. S. F. (2013). Changes in food caused by deep fat frying – A review. Archivos Latinoamericanos de Nutricion, 63(1), 5–13.

Bouchon, P (2009) Understanding oil absorption during deep-fat frying. In: Taylor, S. (ed) Advances in food and nutrition research. Elsevier Press, pp 209–234

Bouchon, P., & Dueik, V. (2018). Frying of foods. In Fruit Preservation (pp. 275–309). Springer, New York, NY. https://doi.org/10.1007/978-1-4939-3311-2_10

Chakkaravarthi, A., Nagaprabha, P., Kumar, H. N. P., Latha, R. B., & Bhattacharya, S. (2014). Jilebi 3: Effect of frying conditions on physical characteristics. Journal of Food Science and Technology, 51(5), 865–874. https://doi.org/10.1007/s13197-011-0595-6

Chen, L., Tian, Y., Bai, Y., Wang, J., Jiao, A., & Jin, Z. (2018). Effect of frying on the pasting and rheological properties of normal maize starch. Food Hydrocolloids, 77, 85–95. https://doi.org/10.1016/j.foodhyd.2017.09.024

Cheng, Y., Liu, Y., Huang, J., Li, K., Zhang, W., Xian, Y., & Jin, L. (2009). Combining biofunctional magnetic nanoparticles and ATP bioluminescence for rapid detection of Escherichia coli. Talanta, 77(4), 1332–1336. https://doi.org/10.1016/j.talanta.2008.09.014

Clarck WL, Serbia GW (1991) Safety aspects of frying fats and oils. Food Technology, 45:84–94

Contreras, C., Benlloch-Tinoco, M., Rodrigo, D., & Martínez-Navarrete, N. (2017). Impact of microwave processing on nutritional, sensory, and other quality attributes. The Microwave Processing of Foods: Second Edition, 65–99. https://doi.org/10.1016/B978-0-08-100528-6.00004-8

Dhingra, D., Michael, M., Rajput, H., & Patil, R. T. (2011). Dietary fibre in foods: a review. Journal of Food Science and Technology 2011 49:3, 49(3), 255–266. https://doi.org/10.1007/S13197-011-0365-5

Dziedzic, K., Szwengiel, A., Górecka, D., Rudzińska, M., Korczak, J., & Walkowiak, J. (2016). The effect of processing on the phytosterol content in buckwheat groats and by-products. Journal of Cereal Science, 69, 25–31. https://doi.org/10.1016/J.JCS.2016.02.003

European Commission, 2013. Commission Recommendation of 8 November 2013on Investigations into the Levels of Acrylamide in Food. European Commission, Brussels, Belgium.

Fellows, P. J. (2017). Frying. In Food Processing Technology (pp. 783–810). Woodhead Publishing. https://doi.org/10.1016/b978-0-08-100522-4.00018-3

Ferreira, F. S., Sampaio, G. R., Keller, L. M., Sawaya, A. C., Chávez, D. W., Torres, E. A., & Saldanha, T. (2017). Impact of air frying on cholesterol and fatty acids oxidation in sardines: Protective effects of aromatic herbs. Journal of Food Science, 82(12), 2823–2831.

Fitriani, A., Santoso, U., & Supriyadi, S. (2021). Conventional Processing Affects Nutritional and Antinutritional Components and In Vitro Protein Digestibility in Kabau (*Archidendron bubalinum*). International Journal of Food Science, 2021. https://doi.org/10.1155/2021/3057805

Fletcher, M.T. & Blaney, B.J. (2016). Mycotoxins, Editor(s): Colin Wrigley, Harold Corke, Koushik Seetharaman, Jon Faubion, Encyclopedia of Food Grains (Second Edition), Academic Press, pp 290–296. https://doi.org/10.1016/B978-0-12-394437-5.00112-1.

Gaikwad, P. S., Pare, A., & Sunil, C. K. (2021). Effect of process parameters of microwave-assisted hot air drying on characteristics of fried black gram papad. Journal of Food Science and Technology, 1–11.

Galani, J. H., Patel, N. J., & Talati, J. G. (2017). Acrylamide-forming potential of cereals, legumes and roots and tubers analyzed by UPLC-UV. Food and Chemical Toxicology, 108, 244–248.

Garg, M., Sharma, A., Vats, S., Tiwari, V., Kumari, A., Mishra, V., & Krishania, M. (2021). Vitamins in Cereals: A Critical Review of Content, Health Effects, Processing Losses, Bioaccessibility, Fortification, and Biofortification Strategies for Their Improvement. Frontiers in Nutrition, 8, 254.

Ghaitaranpour, A., Koocheki, A., Mohebbi, M., & Ngadi, M. O. (2018). Effect of deep fat and hot air frying on doughnuts physical properties and kinetic of crust formation. Journal of Cereal Science, 83, 25–31. https://doi.org/10.1016/j.jcs.2018.07.006

Göncüoğlu, N., & Gökmen, V. (2013). Accumulation of 5-hydroxymethylfurfural in oil during frying of model dough. Journal of the American Oil Chemists' Society, 90(3), 413–417.

Guillén, M. D., & Goicoechea, E. (2008). Toxic oxygenated α, β-unsaturated aldehydes and their study in foods: A review. Critical Reviews in Food Science and Nutrition, 48(2), 119–136.

Gunashree, B. S., Kumar, R. S., Roobini, R., & Venkateswaran, G. (2014). Nutrients and antinutrients of ragi and wheat as influenced by traditional processes. International Journal of Current Microbiology and Applied Sciences, 3(7), 720–736.

Gwartney, E. A., Larick, D. K., & Foegeding, E. A. (2004). Sensory texture and mechanical properties of stranded and particulate whey protein emulsion gels. Journal of Food Science, 69(9). https://doi.org/10.1111/j.1365-2621.2004.tb09945.x

Heredia, A., Castelló, M. L., Argüelles, A., & Andrés, A. (2014). Evolution of mechanical and optical properties of French fries obtained by hot air-frying. LWT-Food Science and Technology, 57(2), 755–760.

Hosseini, H., Ghorbani, M., Meshginfar, N., & Mahoonak, A. S. (2016). A review on frying: procedure, fat, deterioration progress and health hazards. Journal of The American Oil Chemists' Society, 93(4), 445–466.

Huang, W., Yu, S., Zou, Q., & Tilley, M. (2008). Effects of frying conditions and yeast fermentation on the acrylamide content in you-tiao, a traditional Chinese, fried, twisted dough-roll. Food Research International, 41(9), 918–923. https://doi.org/10.1016/j.foodres.2008.07.023

Irakli, M. N., Samanidou, V. F., & Papadoyannis, I. N. (2011). Development and validation of an HPLC method for the simultaneous determination of tocopherols, tocotrienols and carotenoids in cereals after solid-phase extraction. Journal of Separation Science, 34(12), 1375–1382. https://doi.org/10.1002/JSSC.201100077

Kariluoto, S., & and Piironen, V. (2009). Total folate. Health Grain Methods: Analysis of Bioactive Compounds in Small Grain Cereals, 59–68. https://www.cabdirect.org/cabdirect/abstract/20113091691

Khezerolou, A., Alizadeh-Sani, M., Zolfaghari Firouzsalari, N., & Ehsani, A. (2018). Formation, properties, and reduction methods of acrylamide in foods: A review study. Journal of Nutrition, Fasting and Health, 6(Issue), 52–59.

Kim, B. K., Chun, Y. G., Cho, A. R., & Park, D. J. (2012). Reduction in fat uptake of doughnut by microparticulated wheat bran. International Journal of Food Sciences and Nutrition, 63(8), 987–995.

Kumar, A. J., Singh, R. R. B., Patel, A. A., & Patil, G. R. (2006). Kinetics of colour and texture changes in Gulabjamun balls during deep-fat frying. LWT-Food Science and Technology, 39(7), 827–833.

Li, L., Harflett, C., Beale, M., & and JL Ward. (2009). Phenolic acids. In Health Grain methods: Analysis of bioactive components in small grain cereals (pp. 41–52). https://www.cabdirect.org/cabdirect/abstract/20113091689

Luithui, Y., Baghya Nisha, R., & Meera, M. S. (2018). Cereal by-products as an important functional ingredient: effect of processing. Journal of Food Science and Technology 2018 56:1, 56(1), 1–11. https://doi.org/10.1007/S13197-018-3461-Y

Luterotti, S., & Kljak, K. (2010). Spectrophotometric Estimation of Total Carotenoids in Cereal Grain Products. Acta Chim. Slov, 57, 781–787.

Maiani, G., Castón, M. J. P., Catasta, G., Toti, E., Cambrodón, I. G., Bysted, A., Granado-Lorencio, F., Olmedilla-Alonso, B., Knuthsen, P., Valoti, M., Böhm, V., Mayer-Miebach, E., Behsnilian, D., & Schlemmer, U. (2009). Carotenoids: Actual knowledge on food sources, intakes, stability and bioavailability and their protective role in humans. Molecular Nutrition and Food Research, 53(S2), S194–S218. https://doi.org/10.1002/MNFR.200800053

Manay, N. S., & Shadaksharaswamy, M. (2008). Cereals. In Foods: facts and principles (3rd ed., pp. 219–255). New Age Publishers.

Mariscal, M., & Bouchon, P. (2008). Comparison between atmospheric and vacuum frying of apple slices. Food Chemistry, 107(4), 1561–1569. https://doi.org/10.1016/j.foodchem.2007.09.031

Math, R. G., Velu, V., Nagender, A., & Rao, D. G. (2004). Effect of frying conditions on moisture, fat, and density of papad. Journal of Food Engineering, 64(4), 429–434. https://doi.org/10.1016/j.jfoodeng.2003.11.010

Melito, H. S., & Farkas, B. E. (2013). Effect of infrared finishing process parameters on physical, mechanical, and sensory properties of par-fried, infrared-finished gluten-free donuts. Journal of Food Engineering, 117(3), 399–407. https://doi.org/10.1016/j.jfoodeng.2013.03.012

Mir-Bel, J., Oria, R., & Salvador, M. L. (2013). Reduction in hydroxymethylfurfural content in 'churros', a S panish fried dough, by vacuum frying. International Journal of Food Science and Technology, 48(10), 2042–2049.

Morales, F. J., & Arribas-Lorenzo, G. (2008). The formation of potentially harmful compounds in churros, a Spanish fried-dough pastry, as influenced by deep frying conditions. Food Chemistry, 109(2), 421–425.

Moreno, M. C., Brown, C. A., & Bouchon, P. (2010). Effect of food surface roughness on oil uptake by deep-fat fried products. Journal of Food Engineering, 101(2), 179–186. https://doi.org/10.1016/j.jfoodeng.2010.06.024

Mousavi Khaneghah, A., Fakhri, Y., Nematollahi, A., Seilani, F., & Vasseghian, Y. (2020). The concentration of acrylamide in different food products: a global systematic review, meta-analysis, and meta-regression. Food Reviews International, 1–19.

Nagraj, G. S., Chouksey, A., Jaiswal, S., & Jaiswal, A. K. (2020). Broccoli. In Nutritional Composition and Antioxidant Properties of Fruits and Vegetables (pp. 5–17). Academic Press.

Ndlala, F. N., Onipe, O. O., Mokhele, T. M., Anyasi, T. A., & Jideani, A. I. O. (2019). Effect of wheat bran incorporation on the physical and sensory properties of a South African cereal fried dough. Foods, 8(11). https://doi.org/10.3390/foods8110559

Nedović, V., Raspor, P., Lević, J., Šaponjac, V. T., & Barbosa-Cánovas, G. V. (Eds.). (2016). Emerging and traditional technologies for safe, healthy and quality food (pp. 257–268). Cham, Germany: Springer International Publishing.

Nystrom, L., Nurmi, T., Anna-Maija Lampi, A., & Piironen, V. (2009). Sterols. In Health Grain methods: analysis of bioactive compounds in small grain cereals (pp. 7–14). https://www.cabdirect.org/cabdirect/abstract/20113091693

Oboh, G., Ademiluyi, A. O., & Akindahunsi, A. A. (2010). The effect of roasting on the nutritional and antioxidant properties of yellow and white maize varieties. International Journal of Food Science and Technology, 45(6), 1236–1242

Oghbaei, M., & Prakash, J. (2016). Effect of primary processing of cereals and legumes on its nutritional quality: A comprehensive review. Cogent Food and Agriculture, 2(1). https://doi.org/10.1080/23311932.2015.1136015

Oghbaei, M., & Prakash, J. (2019). Bioaccessible phenolics and flavonoids from wheat flour products subjected to different processing variables. Cereal Chemistry, 96(6), 1068–1078.

Oke, E. K., Idowu, M. A., Sobukola, O. P., Adeyeye, S. A. O., & Akinsola, A. O. (2018). Frying of food: a critical review. Journal of Culinary Science & Technology, 16(2), 107–127. https://doi.org/10.1080/15428052.2017.1333936

Onipe, O. O., Jideani, A. I. O., & Beswa, D. (2015). Composition and functionality of wheat bran and its application in some cereal food products. International Journal of Food Science and Technology, 50(12), 2509–2518. https://doi.org/10.1111/ijfs.12935

Oreopoulou, V., Krokida, M., & Marinos-Kouris, D. (2006). Frying of foods. Handbook of industrial drying. CRC Press, Boca Raton, 1204–1223.

Orthoefer, F. T., & List, G. R. (2007). Dynamics of Frying. In Deep Frying: Chemistry, Nutrition, and Practical Applications: Second Edition (Second Edi). AOCS Press. https://doi.org/10.1016/B978-1-893997-92-9.50018-9

Oyedeji, A. B., Sobukola, O. P., Henshaw, F., Adegunwa, M. O., Ijabadeniyi, O. A., Sanni, L. O., & Tomlins, K. I. (2017). Effect of frying treatments on texture and colour parameters of deep fat fried yellow fleshed cassava chips. Journal of Food Quality, 2017(May). https://doi.org/10.1155/2017/8373801

Özdestan, Ö., Kaçar, E., Keşkekoğlu, H., & Üren, A. (2014). Development of a new extraction method for heterocyclic aromatic amines determination in cooked meatballs. Food Analytical Methods, 7(1), 116–126.

Paesani, C., & Gómez, M. (2021). Effects of the pre-frying process on the cooking quality of rice. LWT-Food Science and Technology, 140, 110743. https://doi.org/10.1016/j.lwt.2020.110743

Pedreschi, F. (2012). Frying of Potatoes: Physical, Chemical, and Microstructural Changes. Drying Technology, 30(7), 707–725. https://doi.org/10.1080/07373937.2012.663845

Pedreschi, F., Mariotti, M. S., & Granby, K. (2014). Current issues in dietary acrylamide: formation, mitigation and risk assessment. Journal of the Science of Food and Agriculture, 94(1), 9–20. https://doi.org/10.1002/JSFA.6349

Pokorný, J., & Dostálová, J. (2010). Changes in nutrients, antinutritional factors, and contaminants at frying temperatures. In Frying of food: Oxidation, nutrient and non-nutrient antioxidants, biologically active compounds and high temperatures (pp. 71–104). CRC Press Boca Raton, FL.

Pokorný, J., & Dostálová, J. (2016). Changes in nutrients, antinutritional factors, and contaminants at frying temperatures. In Frying of Food: Oxidation, Nutrient and Non-Nutrient Antioxidants, Biologically Active Compounds and High Temperatures, Second Edition (pp. 71–104). https://doi.org/10.1201/b10437-8

Poutanen, K. (2009). Cereal foods in diet and health. In Health Grain Methods: Analysis of Bioactive Components in Small Grain Cereals (pp. 1–5). https://doi.org/10.1016/B978-1-891127-70-0.50004-2

Qian, X., Sun, B., Zhu, C., Zhang, Z., Tian, X., & Wang, X. (2020). Effect of stir-frying on oat milling and pasting properties and rheological properties of oat flour. Journal of Cereal Science, 92, 102908. https://doi.org/10.1016/j.jcs.2020.102908

Ram, S., Narwal, S., Gupta, O. P., Pandey, V., & Singh, G. P. (2020). Anti-nutritional factors and bioavailability: approaches, challenges, and opportunities. In Wheat and Barley Grain Biofortification (pp. 101–128). Woodhead Publishing.

Ramírez-Jiménez, A., García-Villanova, B., & Guerra-Hernández, E. (2000). Hydroxymethylfurfural and methylfurfural content of selected bakery products. Food Research International, 33(10), 833–838.

Salehi, F. (2020). Effect of coatings made by new hydrocolloids on the oil uptake during deep-fat frying: A review. Journal of Food Processing and Preservation, 44(11), 1–12. https://doi.org/10.1111/jfpp.14879

Samtiya, M., Aluko, R. E., & Dhewa, T. (2020). Plant food anti-nutritional factors and their reduction strategies: An overview. Food Production, Processing and Nutrition, 2(1), 1–14.

Shapla, U. M., Solayman, M., Alam, N., Khalil, M. I., & Gan, S. H. (2018). 5-Hydroxymethylfurfural (HMF) levels in honey and other food products: effects on bees and human health. Chemistry Central Journal, 12(1), 1–18.

Sharma, K. D., Sharma, B., & Saini, H. K. (2021). Processing, value addition and health benefits. In Millets and Pseudo Cereals (pp. 169–184). Woodhead Publishing.

Shewry, P., & Ward, J. (2010). Healthgrain methods: Analysis of Bioactive Components in Small Grain Cereals. (First Edit, Vol. 1). AACC International.

Shieh, C. J., Chang, C. Y., & Chen, C. S. (2004). Improving the texture of fried food. Texture in Food, 2, 501–523. https://doi.org/10.1533/978185538362.3.501

Siah, S., & Quail, K. J. (2018). Factors affecting Asian wheat noodle color and time-dependent discoloration – A review: Cereal Chemistry, 95(2), 189–205. https://doi.org/10.1002/CCHE.10035
Singh, A., Kaur, V., & Kaler, R. S. S. (2018). A review on dietary fiber in cereals and its characterization. Journal of Applied and Natural Science, 10(4), 1216–1225. https://doi.org/10.31018/JANS.V10I4.1894
Singh, L., & Agarwal, T. (2018). PAHs in Indian diet: Assessing the cancer risk. Chemosphere, 202, 366–376.
Sirot, V., Rivière, G., Leconte, S., Vin, K., Traore, T., Jean, J., ... & Hulin, M. (2019). French infant total diet study: Dietary exposure to heat-induced compounds (acrylamide, furan and polycyclic aromatic hydrocarbons) and associated health risks. Food and Chemical Toxicology, 130, 308–316.
Soriguer, F., Rojo-Martínez, G., Dobarganes, M. C., García Almeida, J. M., Esteva, I., Beltrán, M., ... & González-Romero, S. (2003). Hypertension is related to the degradation of dietary frying oils. The American Journal of Clinical Nutrition, 78(6), 1092–1097.
Suman, M., & Generotti, S. (2015). Transformation of mycotoxins upon food processing: masking, binding and degradation phenomena. Masked mycotoxins in food: formation, occurrence and toxicological relevance. RSC Publishing, Cambridge, 73–89.
Tan, K. J., & Mittal, G. S. (2006). Physicochemical properties changes of donuts during vacuum frying. International Journal of Food Properties, 9(1), 85–98. https://doi.org/10.1080/10942910500473947
Thanatuksorn, P., Kajiwara, K., & Suzuki, T. (2010). Characteristics and oil absorption in deep-fat fried batter prepared from ball-milled wheat flour. Journal of The Science of Food and Agriculture, 90(1), 13–20. https://doi.org/10.1002/jsfa.3766
Udomkun, P., Tangsanthatkun, J., & Innawong, B. (2020). Influence of process parameters on the physicochemical and microstructural properties of rice crackers: A case study of novel spray-frying technique. Innovative Food Science and Emerging Technologies, 59, 102271. https://doi.org/10.1016/j.ifset.2019.102271
USDA. (2013). United States department of agriculture. Retrieved from http://ndb.nal.usda.gov/ndb/foods/show/6572
Van Hung, P. (2016). Phenolic compounds of cereals and their antioxidant capacity. Critical Reviews in Food Science and Nutrition, 56(1), 25–35.
Van Koerten, K. N., Somsen, D., Boom, R. M., & Schutyser, M. A. I. (2017). Modelling water evaporation during frying with an evaporation dependent heat transfer coefficient. Journal of Food Engineering, 197, 60–67. https://doi.org/10.1016/j.jfoodeng.2016.11.007
Vázquez-Durán, A., Gallegos-Soto, A., Bernal-Barragán, H., López-Pérez, M., & Méndez-Albores, A. (2014). Physicochemical, Nutritional and sensory properties of deep fat-fried fortified tortilla chips with broccoli (Brassica oleracea L. convar. italica Plenck) flour. Journal of Food and Nutrition Research, 53(4), 313–323.
Viana, R. D., de Carvalho Oliveira, F., Monte, M. J. S., Pereira, L. M. R., & de Carvalho, J. O. (2015). Ação de antioxidantes no reaproveitamento de óleos vegetais. Revista Interdisciplinar, 7(4), 13–21.
Wrigley, C. W. (2018). Cereals. In Swainson's Handbook of Technical and Quality Management for the Food Manufacturing Sector (pp. 457–479). Woodhead Publishing. https://doi.org/10.1016/B978-1-78242-275-4.00018-6
Yang, Y., Li, T., Li, Y., Qian, H., Qi, X., Zhang, H., & Wang, L. (2020). Understanding the molecular weight distribution, in vitro digestibility and rheological properties of the deep-fried wheat starch. Food Chemistry, 331, 127315. https://doi.org/10.1016/j.foodchem.2020.127315
Ye, F., Tao, B., Liu, J., Zou, Y., & Zhao, G. (2016). Effect of micronization on the physicochemical properties of insoluble dietary fiber from citrus (Citrus junos Sieb. ex Tanaka) pomace. Food Science and Technology International, 22(3), 246–255.
Yuksel, F., Karaman, S., & Kayacier, A. (2015). Barley flour addition decreases the oil uptake of wheat chips during frying. Quality Assurance and Safety of Crops and Foods, 7(5), 621–628. https://doi.org/10.3920/QAS2014.0472
Yuksel, F., Karaman, S., Gurbuz, M., Hayta, M., Yalcin, H., Dogan, M., & Kayacier, A. (2017). Production of deep-fried corn chips using stale bread powder: Effect of frying time, temperature and concentration. LWT-Food Science and Technology, 83, 235–242.
Zaghi, A. N., Barbalho, S. M., Guiguer, E. L., & Otoboni, A. M. (2019). Frying process: From conventional to air frying technology. Food Reviews International, 35(8), 763–777.

Zeb, A. (2019a). Food frying. In Food Frying: Chemistry, Biochemistry, and Safety (pp. 3–24). https://doi.org/10.1002/9781119468417.

Zeb, A. (2019b). Chemistry of fried foods. In Food Frying: Chemistry, Biochemistry, and Safety (pp. 115–174). https://doi.org/10.1002/9781119468417.

Zeb, A. (2019c). Chemistry of Interactions in Frying. In Food Frying (pp. 175–205). John Wiley & Sons, Ltd. https://doi.org/10.1002/9781119468417.

Zeb, A. (2019d). Frying Techniques. In Food Frying (pp. 23–64). John Wiley & Sons, Ltd. https://doi.org/10.1002/9781119468417.

Zeb, A. (2019e). Chemistry of the frying foods. In Food Frying (pp. 73–114). John Wiley & Sons, Ltd. https://doi.org/10.1002/9781119468417.

Zhang, T., Li, J., Ding, Z., & Fan, L. (2016). Effects of Initial Moisture Content on the Oil Absorption Behavior of Potato Chips During Frying Process. Food and Bioprocess Technology, 9(2), 331–340. https://doi.org/10.1007/s11947-015-1625-6

Zhang, X., Zhang, M., & Adhikari, B. (2020). Recent developments in frying technologies applied to fresh foods. Trends in Food Science & Technology, 98, 68–81. https://doi.org/10.1016/j.tifs.2020.02.007.

Zhou, R., Sun, J., Qian, H., Li, Y., Zhang, H., Qi, X., & Wang, L. (2020). Effect of the frying process on the properties of gluten protein of you-tiao. Food Chemistry, 310, 125973. https://doi.org/10.1016/j.foodchem.2019.125973

CHAPTER 13

Baking of Cereals

Amit Kumar Tiwari[a], Reetu[a], Kawaljit Singh Sandhu[a], Maninder Kaur[b], and Manisha Bhandari[c]
[a]Department of Food Science & Technology, Maharaja Ranjit Singh Punjab Technical University, Bathinda, Punjab, India
[b]Department of Food Science & Technology, Guru Nanak Dev University, Amritsar, Punjab, India
[c]Department of Food Science & Technology, Punjab Agricultural University, Ludhiana, Punjab, India

CONTENTS

DOI: 10.1201/9781003242192-17

13.1 INTRODUCTION TO BAKING

Baking is a centuries-old process with a thriving global industry. It is a unit operation in which heated air is used to change the eating quality of various types of food products. It is considered as a cooking form essentially performed in an oven. The action of heat transforms a semi-solid piece of dough into an edible product during baking. The secondary objective of baking is to preserve food items by killing bacteria and removing moisture from the surface to reduce water activity (a_w) (Arepally et al., 2020). Baking is a highly complex process that involves both heat and mass transfer. During baking, dough generally expands depending upon oven condition, method of preparation of dough, and recipe used. Food undergoes numerous physical, chemical, and biochemical changes at baking time. These changes include starch gelatinization, protein denaturation, formation of porous structure, expansion in volume, and release of carbon dioxide as a result of leavening (Yolacaner et al., 2017). Bakery products can be formulated with different methods. Straight dough method, Chorleywood method, and sponge and dough method are the most popular among all methods. Baking necessitates extremely high temperatures, often between 160 and 250°C, and thus a significant amount of thermal energy is required (Bredariol et al., 2019).

The dough experiences various physical and biological changes as it enters the baking oven (Cappelli et al., 2021). Physical changes include oven spring and crust formation. When dough is heated, it expands significantly in the first few minutes. This quick increase in volume is known as oven spring. Several variables have been blamed for oven spring. Heated gases, carbon dioxide, and water causes increase in the internal pressure of the dough, causing it to rise quickly during preliminary phases of the baking process. The second physical change includes crust formation. When dough is subjected to high temperatures in the oven, it develops a skin and creates a crust as moisture evaporates quickly from the surfaces. The strength is provided to the loaf by crust formation (Khatkar, 2011).

There are various chemical changes that occur in dough during baking. The high oven temperature causes a series of physicochemical reactions that leads to the development of the crust layer, the formation of the crumb's alveolar structure, and the expansion of the product (Cappelli et al., 2021). Starch gelatinization, protein coagulation, and browning reaction are prominent changes in dough during baking. Starch begins to gelatinize at about 60°C. The small quantity of water present in dough causes a limited amount of gelatinization, which helps in creating bread texture. Starch gelatinization and gluten denaturalization are opposite to each other in terms of water, as one is associated with water absorption and other with water removal. As temperature proceeds to 74°C,

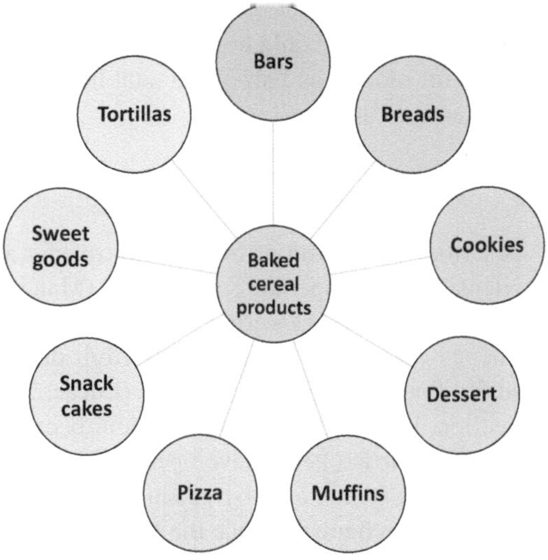

Figure 13.1 Various products from baked cereals.

gelatinization sets in and the gluten matrix that surrounds the individual cells is converted into a semi-rigid film structure (Khatkar, 2011). Roughly at 160°C, the browning reaction begins. During baking the Millard type browning reaction plays a significant role. It is the interaction of aldehyde group present on sugars and amino groups present in proteins. Condensation of amino groups and aldehyde group from reducing sugars produces intermediate chemical compounds, which go through a series of reactions to produce brown-colored pigments called melanoidines (Arepally et al., 2020). These significantly impart colour and flavor to the bakery product.

The baked product can be categorized into three different categories depending upon the water activity of the product (i.e., low moisture, intermediate moisture, and high moisture content baked products). The examples of low baked products (Figure 13.1) are crackers, biscuits, wafers, nuts, and baked potato crisps with water activity of 0.2±0.3. The intermediate moisture products are pastries, cakes, soft cookies, chapattis with water activity of 0.5 ± 0.8. High moisture content baked products with water activity 0.9–0.99 can be further categorized into alkaline, low acid, and acidic food. Crumpets and tortillas are alkaline; breads, rolls, muffins, cheesecake, pizzas, meat pies, sausage rolls, pasties, filled cakes, quiches, baked potatoes, and roasted meats are low acid; and fruit tarts, pies, and sourdough bread are acidic baked products (Fellows, 2009).

13.2 MODE OF HEAT TRANSFER DURING BAKING

Both transport phenomena and structural changes must be considered in order to gain a complete understanding of the baking process. Conduction, convection, and radiation are the three basic modes of heat flow in dough-to-bread transformation (Lucas, 2014).

13.2.1 Conduction

Conduction is the transfer of heat from a source to a substance that absorbs it. During conduction heat flows from the source to the material absorbing it. Surfaces like stones, trays, grates, and conveyor bands can be baking support for conduction heat transfer. It is extremely important that there should be as much contact area as possible between a hot surface and another surface for efficient conduction. Therefore, the bases of utensils should be broad and flat (Purlis, 2014). Overall, during pan baking, conduction accounts for more than half of the total heat delivered to the final product, and in some cases, two-thirds (Lucas, 2014).

13.2.2 Convection

The term "convection" describes the natural or forced heat transfer as a result of the air circulation and steam caused by density and temperature variations. (Manay, 2001). When heating of liquid or air occurs, the parts closest to heat get warmer and less dense. They rise and are replaced by the denser material and hence convection currents flow from denser to less dense areas. In oven convection currents usually flow in a vertical direction (Lucas, 2014). Most ovens use both conduction and natural convection, but the latter is less efficient because heat is spread unevenly throughout the product. Smith and Harris (1974) developed an air jet impingement oven, one of the most recent convection heating systems, as a by-product of microwave cooking research. It was based on the forced convention mechanism. Since the mid-1970s, forced convection has been broadly adopted in tiny units created for pizza shops (Yin and Walker, 1995). The principle of forced convection has been applied to industrial scale ovens as it provides uniform heat throughout the product (Walker, 2016).

13.2.3 Radiation

Another key mechanism of heat transfer in ovens is radiation. Radiation waves can propagate through glass, liquids, and gases without heating them. A medium is not needed in this instance because of usage of energy in the EM (electromagnetic) spectrum (Arepally et al., 2020). The most effective wavelengths lie in the infrared (IR) range, which is undetectable to the naked eye. Radiation is emitted by the oven's heating components (burner flames) and all hot metal sections. The radiation's contribution to heat transfer at the top surfaces of food may be evaluated using h-monitor, and it has been shown to be substantially larger in direct ovens (70–80%) than in indirect ovens (45–60%) (Baik et al., 1999). During baking, the surface of food is being heated when radiation approaches the food (Manay, 2001). As a result, interior heating is negligible. Energy from IR received at the surface is transmitted inwardly through methods such as conduction and to some extent by convection. Because the surface heats up quickly and becomes brown before the heat can penetrate to the center, the process cannot be used for thicker products like bread loaves. It is also difficult to bake irregular-shaped food using IR since any exposed portions would be baked before the remaining. Although radiant energy plays an essential role in practically all ovens, it must be used in combination with conduction and convection, as is the case today (Walker, 2016).

13.3 DIFFERENT TYPES OF BAKING OVENS

Bakery products are consumed by people of all ages in different parts of the world. At present, different types of baked food items such as cookies, bread, muffins, biscuits, cakes, rolls, wafers, tortillas, and pies are available on the market. There are many different steps in the process of making different types of baked products, one of which is baking. A variety of thermal reactions occur during baking, including non-enzymatic browning, protein denaturation, starch gelatinization, and so on, resulting in high-quality texture, flavor, and color. Predominantly in baking, under the influence of high temperatures, a dough turns into a lightly porous and easily digestible tasty product (Manhiça et al., 2012; Jerome et al., 2019). At present, a variety of simple to sophisticated ovens are available on the market to bake cereal-based products. After several years of research and development done in the field of cereal-based products various types of baking ovens have been constructed. Traditional furnace, air-forced convection rotary ovens, continuous and batch ovens, traditional French bread ovens, microwave ovens, steam ovens, infrared radiating ovens, etc., are examples (Paton et al., 2013). Different types of ovens have been summarized in Table 13.1.

Table 13.1 Different Modes of Heating and Types of Ovens and Their Salient Features in the Baking Process (Davidson, 2016)

Type of Heating	Type of oven	Salient features
Radiant Heating	Direct Gas-Fired Ovens	1. It provides radiant heat over a wide range of heat inputs. 2. Responsive and rapid control of heat input. 3. Accurate temperature control. 4. Suitable for different types of baked products including cookies, crackers, and various biscuits.
	Electric Ovens	1. Electrical energy is easily controlled. 2. Thyristor units control the electric ovens and also provide controlled heat input to the baking chamber. 3. Suitable for different types of cookies, biscuits, and crackers. 4. In baking chamber, it provides a very dry atmosphere, which is why requirement of steam injection remains in the first zone for some products. 5. Operating cost is high.
	Indirect Radiant Ovens	1. It provides a very stable, radiant heat, which is why it is preferred by many bakers. 2 Hot gases recirculate from the burner and stay in closed circulation system providing better fuel efficiency. 3. Burner provides sufficient heat to maintain the desired baking temperature. 4. Suitable for all types of products, except soda crackers where direct heating is required. 5. Suitable for different types of fuel along with diesel oil.
Convection Heating	Direct Convection Ovens	1. Circulating fan blows the hot gases from burner to top and bottom part of baking chamber. 2. From the baking chamber wet gases and hot convection gases removed by extraction system. 3. Convection air rapidly dry the pieces of dough. 4. Lack of heat recovery system. 5. Diesel oil is not suitable. 6. Construction cost of this oven is low. 7. Suitable for baking of Danish Butter Cookies. 8. Hot air blows on the surface of product where structure of biscuit and volume are optimizes.
	Indirect Convection Ovens	1. Multipass heat exchanger connected through a burner tube where burner fires. 2. Within heat exchanger, combustion products are being circulated that do not enter the baking chamber rather than thefan provide hot air from heat exchanger to baking chamber. 3. Unsuitable for baking hard sweet biscuits and crackers. 4. It dries the dough pieces to low moisture contents. 5. Product surface rapidly dries and may contribute to an excessive moisture gradient in the biscuit and consequent 'checking' or cracking of the packaged biscuits. 6. Without contrast it provides bland flavor and even coloring to the biscuit. 7. It utilizes diesel oil or gas fuels.

(*continued*)

Table 13.1 (Continued)

Type of Heating	Type of oven	Salient features
Conduction Heating	Used in different types of oven designs, indirect radiant ovens, convection ovens and direct gas-fired ovens	1. Heavy mesh bands and steel baking bands conduct heat rapidly via bottom of the dough pieces. 2. These types of bands are versatile and can be used in many types of oven designs, such as indirect radiant ovens, direct gas-fired ovens, and convection ovens. 3. Carbon steel is used for making of steel bands, usually 1.2 mm thick. 4. Suitable for baking of high sugar and fat content cookies. 5. Steel bands are also used for the baking of 'Marie' biscuits. 6. CB5 heavy mesh baking bands are woven with a tight (herring bone) pattern to provide thick, solid, and heavy mesh 7. Preheating temperature of bands is 120–150°C. Heat immediately conducts into the base of dough pieces as soon as they are deposited onto the band. 8. Widely used throughout the industry. Suitable for baking of rotary-molded products, salted and soda crackers, and hard sweet biscuits.

13.3.1 Tunnel Ovens

Tunnel ovens have been used in various types of food industries over the years. Many researchers around the world have found that tunnel ovens consume less energy and have the potential of mass production of bakery products. A tunnel type multizone oven has relatively good energy utilization efficiency of 60 to 70% (for electric oven) and 50 to 70% (for gas oven) (Stear, 1990). As compared to a batch oven, the main feature of a tunnel oven is that the entire baking chamber is divided into several zones along with the length of the oven. The temperature of the lower and upper baking chambers in each zone can be controlled independently using sensors and sophisticated instruments, in which application of a temperature sequence is possible (Matz, 1988). It is flexible to make several specific baked food items and all the parameters and conditions can be adjusted according to the product. In fact, its flexibility also attracts bakers' attention since it can be used to make different types of food products and also according to bakers' interest. However, predicting baking events in tunnel-type ovens is more difficult than in batch type ovens, which usually operate at a constant temperature. The baking chamber of a tunnel-type multizone oven shows a wavy temperature profile along with its length. In this oven heat transfer is influenced by geometry, size, heating mode, and wall properties of the oven, and also by geometry, size, and the physicothermal properties of the baked products (Baik et al., 1999).

13.3.2 Jet Impingement Oven

This is a special type of hot-air heating system in which jets of hot air at high velocity impinge and transfer heat rapidly over the food item (Wahlby et al., 2000). In this oven-forced convection heating is generally used at high velocity (5 to 50 m/s) of hot air (Kocer et al., 2007). Today, the impinging jet is used for various types of food applications like baking, cooking, toasting and roasting, etc., due to their uniform and rapid heat transfer rate. The high velocity of the jet lowers the boundary layer between the heating medium and the heating surface and provides uniform rapid heat transfer (Jambunathan et al., 1992). Jet impingement baking is more advantageous than conventional baking due to its uniform heating, high rate of heat transfer, higher efficiency, shorter baking time, rapid moisture removal, reduced moisture loss, lower oven temperature, and comparable quality parameters (Li and Walker, 1996; Sarkar and Singh, 2004; Ovadia and Walker, 1998).

13.3.3 Infrared Radiating Ovens

A well-known technology, infrared radiation heating has become prevalent in the food industry for the past few decades. Its working ability heats the surface rapidly and hence it is used in various food processing operations. It is used for drying fruits, vegetables, fish, pasta, and rice, as well as meat frying, biscuits baking, grilling, food, boiling, cooking, etc. (Krishnamurthy et al., 2008; Wang and Sheng, 2004; Kumar et al., 2009). Infrared is electromagnetic radiation with wavelengths that fall between visible light and microwaves (i.e., 0.78 to1000 μm). On the basis of wavelength, these infrared waves are classified into three different types: near infrared (0.78–1.4 μm), mid infrared (1.4–3.0 μm), and far infrared (3.0–1000 μm). Most of the food components absorb the energy from the far infrared wave region (Sandu, 1986; Krishnamurthy et al., 2008; Sakai and Hanzawa, 1994; Nowak and Lewicki, 2004). Infrared heating offers more advantages than conventional heating as it takes less startup and processing time, less energy consumption, and provides efficient uniform heating. (Krishnamurthy et al., 2008). However, there are some disadvantages of infrared heating including coating reflective properties make it insensitive and prolonged contact with biological material causes fractured food and low penetration power. Infrared heating is also affected by the surface characteristics of food product and the penetration power of radiations. (Krishnamurthy et al., 2008; Datta and Rakesh, 2013).

13.3.4 Electric Resistance Oven

The electric resistance oven (ERO) works on the principles of Ohm's law and Joule's first law. In this oven heating occurs due to its electrical resistance generated during passage of electrical current through a material.

Conventional heating ovens have been used since ancient times and are effective in making various baked food items. Ohmic heating technology is going to prove to be a boon in the food industry sector in the coming years, allowing us to bake high-quality food in less time and at lower cost. Heat is distributed very quickly and evenly, as heating take place volumetrically and does not rely on conventional heat transfer based on conduction, convection, or radiation (Varghese et al., 2014; Sakr and Liu, 2014; Jaeger et al., 2016).

Electrical conductivity and electrical field strength of the food material are the main parameters that affect the heat generation during ohmic heating. Both determine the electric current and temperature rise depending on the applied total specific energy input and the material's specific thermal capacity. Heating rate is affected by applied power. Electrical conductivity depends on the specific properties of food such as concentration, composition, and ion mobility and increases with salt content, water, and temperature (Kumar et al., 2014).

Some studies have focused on using ohmic heating to produce crustless breads (by using water addition up to 50–60%). It has been found that foods baked by ohmic heating have better shelf life and quality than conventional baked products (Derde et al., 2014; Hayman et al., 1998; Gally et al., 2016; Masure et al., 2018; Chhanwal et al., 2012). Based on the ohmic heating characteristics this type of oven provides shorter heating time as well as higher heating rate with larger heat output and also reduced baking time.

The main problem with this oven is that the moisture is unevenly distributed during baking so that there is no crust formation in the baked products (Derde et al., 2014; He and Hoseney, 1991; Martin et al., 1991).

13.3.5 Microwave Ovens

At present, microwave heating is used in various food industries and in households for baking, heating, cooking, and thawing because it allows instant and rapid heating and precise control over all processes (Campañone and Zaritzky, 2005; Campañone and Zaritzky, 2010; Campañone et al. 2012; Chavan and Chavan, 2010; Cha-um et al., 2011; Datta and Rakesh, 2013; Sakiyan et al., 2011; Datta and Anantheswaran, 2001; Turabi et al., 2008). The main advantage of a microwave oven is that the radiation in it easily passes through the dry crust and is absorbed by the wet interior (Walker, 2016), although it does not develop brown crust and often has toughening. Domestic and industrial microwaves operate at 2450 MHz and 915 MHz, respectively. The penetration power of longer-wavelength and lower-frequency ovens is much better than ovens that work on convection and conduction heating (Walker, 2016; Lucas, 2014).

13.4 EFFECT OF BAKING

13.4.1 Physical Changes

During the bread-baking operation, the physicochemical characteristics of the bread dough may alter dramatically. When proofed bread dough is heated in an oven, changes in various quality attributes like texture, crust color, appearance, and aroma occur, all of which are extremely important to consumers and can be visually noticed (Ahrné et al., 2007; Purlis and Salvadori, 2007). Oven spring, surface drying, crust formation, and browning of crust are the important physical changes that occur in dough during baking.

13.4.1.1 Oven Spring

The phenomena of rapid expansion of proofed dough is known as oven spring. This is an irreversible process. Several factors contribute to oven spring. Firstly, rise in temperature and yeast activity affects the rapid expansion of dough, secondly aqueous carbon dioxide becomes less soluble and expands in air bubbles due to rise in temperature, and finally vaporization of the water-ethanol azeotrope takes place at 78°C. Generally, the oven spring stage completes in less than eight minutes (Delcour and Hoseney, 2010). The initiation step in the baking process is to raise the temperature of the external crumb layers. As the temperature of the product rises, chemical reactions begin, significantly increasing CO_2 gas production and continuing to expand the volume of the product (HadiNezhad and Butler, 2010). Baking can be divided into three stages: first stage, second stage, and final stage. During the first stage, quick expansion in volume occurs and out of total time required for baking, this stage takes up 25% of time. This is due to the strong heat absorption at the start of the process, when the food product is at room temperature and exposed to the oven's high temperature (Therdthai et al., 2002). When dough surface is exposed to the oven atmosphere, formation of crust starts along with drying of dough surface due to dry air at high temperature. Furthermore, the crust remains cooled due to vaporization of water from the outer surfaces. The volume growth continues to its maximum volume mostly during the second stage of the baking process as the crumb temperature increases. All chemical reactions including protein coagulation, evaporation of moisture, and gelatinization of starch remain maximum at this stage. At the final stage of baking slight shrinkage occurs in the product due to the release of gases in the forms of bubbles. Also, dough structure strengthens due to further increase in temperature. (Sani et al., 2014).

Cake volume expansion plays an important role in the evaluation of cake quality criteria (HadiNezhad and Butler, 2010). High-quality moist cakes should have uniform moisture and consistent volume, and both depend on important factors including baking time-temperature combination and baking condition along with ingredients used in batter preparation. Azmi et al. (2019) evaluated the volume expansion of oven-baked cakes and air fryer-baked cakes, where different temperature and time combinations were considered for baking of cakes such as baking temperature 150 ^{0}C, 160 ^{0}C, and 170 ^{0}C for both convection oven and air fryer oven and baking time 35, 40, and 45 min for convection oven and 25, 30, and 35 min for air fryer oven. The study showed that heights of convection oven-baked cakes and air fryer-baked cakes ranged from 21–50% and 38–66%, respectively. The air fryer-baked cake showed more height than convection oven-baked cake due to convective heat transfer and enhanced air flow in air fryer oven chamber leading to greater volume expansion. During baking longer heating time increases the evaporation of water, which creates air bubbles in the cake to expand and form porous structure further yielding a larger cake. Baking in a convection oven for a longer time reduces the height of the cake, which is due to shrinkage increased at the end of the baking process (Im et al., 2003; Shahapuzi et al., 2012; Alifak and Sakıyan, 2017; Sevimli et al., 2005).

13.4.1.2 Crust Formation

As dough is subjected to oven temperature, it develops a skin and produces a crust when moisture evaporates quickly from the surface. Maximum water evaporation occurs at the time of crust formation. The development of crust and non-enzymatic browning during baking are major contributing factors in bread flavor development (Zanoni et al., 1995). Browning of the crust is mostly influenced by oven temperature, and has a link with weight loss at the time of baking. Another key characteristic of baked goods is surface color, which can be used as a critical baking index. Maillard reaction plays a vital role in the development of color and aroma in bread crust, but it is also linked to the formation of toxic compounds that can be carcinogenic (Brathen and Knutsen, 2005).

13.4.2 Chemical Changes

When dough is subjected to high temperature various chemical changes including starch gelatinization, protein coagulation, and browning occur. Heat constraints affect dough ingredients and microorganisms, but the magnitude of the effects is determined by their thermal stability.

13.4.2.1 Starch Gelatinization

Starch is the most prominent ingredient in baked foods such as snacks, weaning foods, and breakfast cereals. Two processes, gelatinization and retrogradation, require attention during cereal cooking to ensure optimal quality. Raw starch is difficult for humans to digest. However, due to partial gelatinization and fragmentation caused by shear and temperature, the digestibility of starch may be increased by the baking process (Butterworth et al., 2011). Factors such as temperature and availability of water held are responsible for changes in starch granules. (Valamoti et al., 2008). Baking generates the most diverse changes to starch grains among all the cooking methods used. The appearance of baked starch is distinct from that of other cooking methods and remains easier to distinguish. Wheat starch gelatinization happens at temperatures ranging from 50 to 75°C, depending on the water concentration in the system and the temperature used at the time of heating (Vanin et al., 2009). The limited water content present in dough prevents starches from gelatinizing completely, which helps in texture setting and gas retention (Khatkar, 2011). Bread and its many sections may contain fully gelatinized, partially gelatinized, or ungelatinized starches, depending on the hydrothermal conditions of the baking process. In food industries and various research institutes wide ranges of baking conditions are being used, despite of that effect of these baking conditions on gelatinization is not being evaluated. Generally, heat treatment reduces the content of rapidly digestible starch while increasing the content of resistant starch in baked foods (Chen et al., 2015).

13.4.2.2 Protein Coagulation

During baking various physicochemical changes occur in starch–protein matrix. Mixing is one of the most important phases in determining the mechanical qualities of the dough, which have a direct impact on the ultimate quality of baked product. During mixing, protein particles are fragmented and aligned, resulting in a viscoelastic structure with gas-retaining capabilities. Gliadin and glutenin are the parts of proteins with the ability to form gluten. Gluten is made up of two primary subfractions: glutenins, which provide strength and elasticity to the dough, and gliadins, which are responsible for dough viscosity. Hydrophobic amino acids named glutamine predominate in gluten proteins and have a significant ability to develop hydrogen bonds between protein strands (Rosell, 2016). Gluten structural characteristics demonstrates the importance of intra- and intermolecular disulfide bonds in the formation of the gluten matrix in dough (Lavelli et al., 1996). Flour proteins also transform considerably by baking. Increased temperature is believed to enhance the formation of cross-links between proteins, leading to loaf setting at the time of baking (Rosell, 2016). When protein denaturation occurs, the gluten strands that surround individual gas cells release water and transform into a semirigid structure that gives rise to the bread crumb.

13.4.2.3 Browning Reactions

Browning occurs due to formation of intermediate chemicals compounds that polymerize to produce brown pigments. Three processes, namely Millard reaction, caramelization, and dextrinization, are mainly responsible for development of browning in different bakery products (Nguyen et al., 2016). These reactions help in development of desirable characteristics such as taste, color, and flavor in baked products. Millard reaction is the interaction between aldehyde groups of reducing sugars and amino groups of proteins, and/or other nitrogen-containing substances whereas caramelization refers to a series of reactions that happens when carbohydrates, particularly sucrose and reducing sugars, are heated directly (Purlis, 2010). The Maillard type browning reaction does not produce a single or small number of end products, but rather follows a complex chemical pathway that produces a huge variety of structures (Somoza, 2005). The favorable conditions for Millard type browning is a temperature over 50°C or a pH is between 4–7; on the other hand, caramelization necessitates greater heat treatment conditions, such as higher temperature over 120°C and relatively lower water activity (Gokmen et al., 2008). The reactions result in the formation of ring-shaped organic molecules, which darken the surface of baked dough. Toasty and savory smells and flavor components are also produced by Maillard reactions. Studies show that when foods containing starches are cooked the amino acids, namely asparagine, react with sugars and produce acrylamide. Acrylamide is a probable carcinogen that has been linked to neurotoxicity in humans as well as in animals. Studies conducted by the U.S. Environmental Protection Agency in 1985 found that acrylamide toxicity may cause paresthesias in the fingers, weakness in the feet, numbness in the lower limbs, and coldness.

13.4.3 Nutritional Changes

Baking is the main technique used in the commercial processing of foods. It is known from many studies that one or more nutrients are lost during baking. Generally, in bread the loss of nutrients in the crust is greater than that of the interior crumbs that normally cannot reach the high oven temperature, so there is less loss of nutrients. Factors other than temperature also adversely affect nutrients including moisture, time, pH, oxygen, oxidants, metals, enzymes, and other additives (Ranhotra and Bock, 1988). Formation of anti-nutritional factor and loss of nutrients are not the only consequences of baking; it also improves the nutrition of food products such as destruction of protease inhibitors, amylase, undesirable microorganisms, and certain anti-nutrients. The effects of baking on nutritional, functional, and biological properties are summarized in Table 13.2.

Table 13.2 Effect of Baking on Nutritional, Functional, and Biological Properties of Cereals

Properties	Component	Product	Effect of baking	Reference
Nutritional property	Mineral (Na, P, K, Ca, Mg, S, Cu, Fe, Min, Zn)	Wheat bread	- The higher the baking temperature, the lower the macromolecule content	Bredariol et al. (2020)
	Protein	White bread	- Decrease of protein content in outer crust is higher than inner content	Westurlund et al. (1989)
	Vitamins	Chapati	- Reduction in the content of vitamins	Yadav et al. (2008)
	Sugar	White bread	- Decrease in sugar content in crumb in the series of maltose, fructose, and glucose	Westurlund et al. (1989)
Functional property	Spread ratio	Cookies	- The higher the baking time and water content, the lower the spread ratio - Incorporation of fiber led to decreased spread ratio	Mudgil et al. (2017)
	Hardness	Cookies	- Increase in baking time led to increase in hardness of the cookie - Incorporation of fiber led to increase in the hardness of the cookie	Mudgil et al. (2017)
	Volume expansion	Cake	- Increase in carbon dioxide thus resulting in expansion of volume - Increase in volume expansion on higher rate of airflow	Sani et al. (2014); Swami et al. (2015)
	Color	Cookies	- Browning of baked product due to acrylamide formation and sugar caramelization	Icoz et al. (2004); Isleroglu et al. (2012); Palazoglu et al. (2015)
Biological property	Protein digestibility	Wheat bread	- Increase in time and temperature of baking results in increased protein digestibility until the breakdown temperature	Swieca et al. (2013); Bredariol et al. (2020)
	Oxalate	Wheat bread	- The higher the baking temperature and time, the lower the oxalate content in the product	Bredariol et al. (2020)
	Phytate	Whole wheat flour bread	- Higher baking temperature and time led to reduction in phytate content	Turk et al. (1992); Qazi et al, (2003); Helou et al, (2016)
	Total phenols	Cookies	- Low-temperature baking had higher phenolic content	Zilic et al. (2016)
	Anthocyanins	Cookies	- Lower-temperature baking for longer time had higher anthocyanin content	Zilic et al. (2016)
	Antioxidant activity	Cookies	- High-temperature baking for higher time interval had higher antioxidant activity	Lindenmeier and Hofmann (2004); Zilic et al. (2016)

13.4.3.1 Carbohydrates and Fats

Bread is a rich source of carbohydrate in the human diet. On baking, a sequence of changes occurs in the starch as it endures gelatinization. Baking causes a reduction in different sugars in the following order – maltose, fructose, and glucose. A higher amount of glucose is found in the crust when compared to crumb, which is attributed to the hydrolysis of glucans at high temperature that prevails in the crust. The maltose content has been found to be lower in the crust of bread than other fractions of the bread, which is attributed to the longer baking times of bread. This decrease of maltose is accompanied by increase in maltulose, which indicates epimerization of maltose during baking. The presence of maltulose in crust can be considered as an indicator of the Maillard reaction, which gives a volatile aroma and flavor that is more visible in crust than crumb (Westerlund et al., 1989). Normally the hydrolysis of fats and carbohydrates occurs during baking. For example, when the starch is gelatinized during baking, its digestibility increases. Maillard reaction is common during baking, in which the participation of simple and hydrolyzed carbohydrates is important. It adversely affects the carbohydrate content of baked products. Linoleic acid and other fatty acids commonly form unstable hydroperoxides through lipoxygenase activity at high baking temperatures. This leads to the oxidation of lipids and vitamins, especially fat-soluble vitamins, which affects the nutritional value of the baked products (Ranhotra and Bock, 1988).

13.4.3.2 Vitamins

Loss of vitamin during the transformation of grain to bread is a nutritional issue to be considered. Levels of vitamins in baked products are influenced by various factors such as time of baking, temperature of baking, pH, and the process of mixing. Vitamin C and thiamine are the heat-labile nutrients susceptible to baking losses. A study by Hallberg et al. (1982) showed vitamin C losses of 76% and 87% in two hamburger-based meals warmed for 4 hours at 75°C. Bread dough has an acidic pH during the fermentation process, due to which the losses of thiamine are reduced because it is stable at acidic pH levels. Generally, thiamine losses during baking do not exceed 25%. Maleki and Daghir (1967) also reported losses of thiamine in the range of 7 to 24% in enriched white Arabic bread. Furthermore, Bednarcyk (1978) reported the average loss of thiamine was not more than 20.3% in graham crackers made with fortified flour. Thiamine is completely lost when the pH level of the baked product rises above 6. The losses of thiamine are lower in products baked at lower temperature. The point to be noted here is that the loss of thiamine as well as the loss of other vitamins in baked products vary from product to product and should not be considered as universal. However, some vitamins such as pantothenic acid, vitamin C, and folic acid are, of course, less stable compared to others (Ranhotra and Bock, 1988). Losses of folic acid during bread making were reported by Gujska and Majewska (2005). This study shows that the total losses of folic acid depends on baking process and ranges from 12 to 21% in wheat and rye breads. Baking probably entails minimal adverse effects on fat-soluble vitamins, but under extreme baking conditions fat is broken down and oxidation of tocopherols takes place (Ranhotra and Bock, 1988). Yeast is a good source of riboflavin and thus bread made with yeast results in higher riboflavin content (approximately 30% enrichment). Baking leads to reduction in pyridoxine levels confirming that pyridoxine (especially pyridoxal and pyridoxamine forms) is easily degraded by the presence of heat and oxygen. Also, baking results in reduction of folate content by 25% (Dewettinck et al., 2008).

13.4.3.3 Minerals

Heat treatment affects the absorption of certain minerals initially through cleavage of complexes, which makes these minerals less absorbable for physiological needs. Minerals are divided into macro- (S, Mg, Ca, K, P, Na) and micronutrients (Zn, Mn, Fe, Cu). Bredariol et al. (2020) focused on baking parameters and their consequences on nutritional aspects. The study shows that wheat breads after baking result in macrominerals ranging from 65.0–165.0 mg/100 g for sulfur, 50.0– 120.0 mg/100 g for magnesium, 255.0–587.5 mg/100 g for calcium, 245.0–540.0 mg/100 g for potassium, 115.0–170.0 mg/100 g for phosphorus, and 1750.0–2072.5 mg/100 g for sodium. The values ranged from for microminerals from 2.3–5.6 mg/100 g for zinc, 3.2–4.9 mg/100 g for the manganese, 3.7–4.6 mg/100 g for iron, and 0.2–0.6 mg/100 g for copper. In this study data showed that content of macrominerals in bread decreases with higher baking temperature. Baking conditions have a profound effect on macrominerals. As far as microminerals are concerned, the baking temperature and timeframe have significantly less effect on their levels (Rybicka and Gliszczynska Swiglo, 2017; Tuncel et al., 2014). Baking at high temperature results in lower amounts of macromolecule contents in bread while the effect of longer baking time either lowers the macromolecules or sometimes has no variation in the macromolecule content in the bread. For micromolecules, high temperature and longer baking time had no significant effect. The introduction of steam during the beginning of baking had no significant effect on the content of macro- and microminerals. Stability of macromolecules during baking is governed by several conditions. Lower temperatures of baking and lower baking times are favorable in providing bread with higher amounts of macromolecules. However, baking done at high temperature for longer time (20°C for 20 min) results in reduction of phytate and oxalate content therefore favoring higher availability of macrominerals such as potassium, calcium, and magnesium. For enhanced macromolecules, baking is done at lower temperature for longer time intervals so that the phytase enzyme receives the optimum temperature for longer time intervals (Bredariol et al., 2020).

13.4.3.4 Protein

Heat application during baking causes the denaturation of protein, which in turn enhances the protein digestibility of the cereal grain. When this denaturation occurs in the presence of reducing sugars such as maltose, lactose, and fructose, non-enzymatic browning occurs (i.e., Maillard reaction). This reaction destroys all the amino acids present in the cereals amongst which lysine is the most significant. Although baking results in reduced lysine content it can be improved with the addition of materials having high lysine content (i.e., products made of whole wheat flour) (Ranhotra and Bock, 1988). During the process of baking, either dextrin or damaged starch impacts the interaction between protein and polysaccharides and the distribution of water. These interactions between proteins, polysaccharides, and water molecules result in modification of shelf-life period through an unclear mechanism that shows that water from the starch granules migrates to the protein and a starch-dextrin-protein interaction is formed (Delcour and Hoseney, 2010). Baking leads to reduction of 80% ethanol-extractable protein. With increase in the time of baking, reduction in the amount of protein occurs due to its denaturation on heating for longer time durations. The ethanol-extracted protein is present in larger amount in the outer crusts of bread therefore the possibility of depolymerization of protein thermally is higher in the outer crust. However, an increase in the protein content in bread could be due to the formation of products of Maillard reactions, which are nitrogenous (Westurlund et al., 1989).

13.4.4 Functional Changes

13.4.4.1 Expansion during Baking

Baking results in expansion of bread and cake irrespective of the treatment given. Expansion always occurs in three stages. In the first stage, starting at 4–6 min, there is a little expansion followed by the second stage in which rapid expansion of the batter to the maximum volume occurs. Finally, contraction occurs at the end of the baking (HadiNezhad and Butler, 2010). For high- and medium-temperature bakes, peak volume is seen at 16–17 min while at low-temperature bake it occurs at 20 min. At first, heat penetrates into the batter of the cake therefore resulting in a little expansion while in the second stage expansion of bubbles takes place due to the increase in vapor pressure, hence causing rapid expansion. In the third stage, further increase causes swelling of starch followed by agglomeration of protein unless the structure of cake sets and strengthens it and increase in pressure causes slight contraction of the cake (Whitaker and Barringer, 2004). With increase in the temperature of baking, a chemical reaction takes place resulting in production of carbon dioxide thus causing volume expansion. High airflow results in the slightly more expansion of volume than in cakes without airflow (Sani et al., 2014).

13.4.4.2 Spread Ratio

Spread ratio is one of the important parameters for good quality product as it indicates the dough's viscous property and is influenced by recipe, procedures, ingredients, and conditions that are used in the production of biscuit. Cookie spreads until there is an increase in the sudden viscosity during baking. The reason of sudden increase of viscosity is still unknown as the starch is not gelatinized properly. This increase in the viscosity could be attributed to the property of protein of flour (Pareyt and Delcour, 2008). The degree to which cookie spreads is controlled by the rate of spread and the set time. The set time is dependent on the water content in dough, and it acts both as a solvent and provides the strength to the cookie dough. Although spread ratio is widely dependent on the components of the dough that binds water (Tekle, 2009), an increase in spread ratio of cookies was reported with increase in the baking time. Incorporation of dietary fibers such as water-soluble pentosans, pectins, water-insoluble hemicelluloses, lignin, and cellulose results in reduction of spread ratio and tenderness of the cookies (Mudgil et al., 2017).

13.4.4.3 Color

Color development is an important factor for acceptability of any final product. Color is generally developed toward the end stages of baking, which helps to determine whether baking is completed or not. Non-enzymatic browning (Maillard reaction) is a major factor for the development of color on the surface of bakery products. On baking, low-moisture and high-temperature conditions are responsible for color development and acrylamide formation in these products (Palazoglu et al., 2015). On increasing baking time and temperature, lightness value decreases, which means the sample becomes darker. This could be attributed to the browning of casein and sugar solutions at higher baking time and temperature (Icoz et al., 2004). Once the surface of product dehydrates, color development begins due to Maillard reaction and caramelization of sugar. These changes depend on factors such as time, air velocity, rate of heat transfer, temperature, and relative air humidity (Ureta et al., 2014). Increase in the temperature of the oven also significantly affects the rate of color development. The formation of brown pigment increases with increased time and temperature.

13.4.4.4 Hardness

Baking temperature is highly responsible for the hardness of the baked product. The higher the temperature of baking, the higher the hardness of the product. This is attributed to the existence of the crystalline sugar in the dough, which melts as the temperature of baking increases, and this melted sugar is responsible for the strengthening of the structure of the cookie (Kawai et al., 2016). Hardness of the baked products also depends on the time of baking. With increase in baking time, there is an increase in the hardness of products (Mudgil et al., 2017). Difference in the hardness of various products also depends on the solid fat index (Devi et al., 2018). Cookies baked with conventional methods were harder when compared against cookies baked by the use of vacuum combined methods (Palazoglu et al., 2015).

13.4.5 Biological Changes

13.4.5.1 Phytate

Phytate is commonly known as an anti-nutrient factor that forms complexes with minerals, protein, and starch and make them unavailable for bioavailability. During baking, phytate is hydrolyzed and is easily absorbed in the body as no complex is formed between the phytate and phosphorus content of the grain. On increasing the amount of yeast and time of fermentation, there is an increased amount of hydrolysis of phytates in the final product. On baking, the destruction of phytic acid occurs as the enzyme phytase (55–70°C) is activated since the baking temperature is the optimum temperature for the activity of the enzyme. The enzyme phytase is activated once the temperature is above 55°C, and this temperature is achieved in the process of baking (Bredariol and Vanin, 2021). Gontzea and Sutzescu (1968) reported that the phytase enzyme of rye is more active than that of wheat. Several other factors responsible for the reduction of phytate are particle size of the flour, temperature, time of fermentation, pH, and water content. Lower pH favors the reduction of phytate (De Angelis et al., 2003). During the process of baking, phytic acid is partially or completely hydrolyzed to myoinositols, which are not capable of forming stable complexes with any cations, therefore avoiding formation of complexes with zinc, magnesium, iron, and calcium (Belitz et al., 2009).

13.4.5.2 Oxalate

Oxalate is an anti-nutritient that forms complexes with minerals such as iron, zinc, and many other thus reducing their bioavailability. Baking leads to the reduction of oxalate content in baked cereal goods, especially near the surface areas as the temperature rise here is fast (Vanin et al., 2009). As during baking, high temperature is required (220°C) that surpasses the point of volatization of oxalates (189–191°C) causing a reduction in the oxalate content of the baked products. The two main factors affecting the oxalate content during baking are baking time and temperature. The higher the time and temperature of baking, the higher the reduction in the amount of oxalate content in the baked product. Baking temperature is more powerful when initially no use of steam is done. Bredariol et al (2020) reported a reduction in the content of oxalate on baking bread from wheat flour.

13.4.5.3 Protein Digestibility

Baking leads to an improvement in the protein digestibility of cereals when subjected to heat until a breakdown value is reached where it starts to decrease. Digestion of protein and the level of hydrolysis are highly dependent on the rate of heating during baking. The introduction of steam during the start of baking greatly favors the increment of protein digestibility (Bredariol et al., 2020). Increase in the protein digestibility of pita bread on baking was observed by Abdel-Aal and Hucl (2002). Also an increase in the protein digestibility of bread, cookies, and biscuits was observed by Abdel-Aal (2008). Although bread is not a food product with a high protein content, it has higher protein digestibility as baking avoids the undigestible proteins. On baking, protein digestibility is enhanced until a breakdown temperature is reached, due to the proteolysis level and structure of protein digestion, which are greatly dependent on the rate of heating. Increase in protein digestibility could be due to the greater surface area of bread flour providing more accessibility to enzymes to digest molecules of starch and protein. Higher protein digestibility depends on controlling of temperature, time, steam, etc. Baking temperature lower than 220°C, baking time less than 15 min, and steam at the beginning of baking are responsible for enhancing the protein digestibility of flour (Bredariol et al., 2020).

13.4.5.4 Starch Digestibility

Higher baking temperature and lower baking time are favored for enhanced starch digestibility. Longer duration of baking also results in increased extent of starch gelatinization as the core temperature of the cookies also rises. This results in the enhancement of retrogradation of starch during cooling and limits the rate at which starch is hydrolyzed. On prolonged baking, starch is modified by heat-moisture due to which the bifringence property of starch is lost partially leading to specific changes in the digestibility of the starch because the native bonds of starch are broken down and resistant ones are reformed (Naseer et al., 2021).

13.4.5.5 Phenolic Substances

Today, there is an increasing interest in foods high in bioactive compounds for their benefits to human health such as anti-cancer activity and free radical scavenging properties (Dziki et al., 2014; Falcinelli et al., 2018). Food is usually processed (thermal treatment: baking, boiling, steaming, roasting) with the goal of increasing the bioavailability of their nutrients (Rashmi & Negi, 2020; Ou et al., 2019). In particular, baking of breads promotes the formation of antioxidants in bread crust (Deshou et al. 2009). Studies conducted on the effect of baking on free and bound phenolic acid in whole grain muffin, cookie, and bread showed that free phenolic compound increases due to phenolic acids liberated after baking of einkorn (El-Sayed et al., 2013). Cheng et al. (2006) also reported that free phenolic acids in wheat increases due to degradation of conjugated polyphenolic compounds after heat treatment. Effect of baking on phenolic compound depends on a number of factors such as baking recipe, heating conditions, and type of phenolic compounds (El-Sayed et al., 2013). On baking of rye bread, an increase in the total phenolic content was observed in whole bread. The high increase in phenols was observed in the crust of the baked bread. This could be attributed to the melonoidin products formed during the baking having high phenolic content (Michalaska et al., 2008).

13.4.5.6 Antioxidant Activity

Baking of cereal products results in increased antioxidant activity due to the formation of melanoidins as a result of Maillard reaction (Lindenmeier and Hofmann, 2004). Yael et al. (2012) reported an increase in antioxidant activity of red and yellow quinoa seeds. During baking, many antioxidants seemed to be formed in the crust because of Maillard reaction that takes place and results in the formation of melanoidins. Rye bread on baking showed an increased amount of antioxidants in the crust of the bread (Michalaska et al., 2008).

13.5 CONCLUSION

Demand for and interest in bakery products are increasing with time. Due to their nutritional adequacy and convenience these products are widely consumed. Bakery products have a significant position in the market due to their desirable properties such as extended shelf life, diversified taste and texture, and broad use. Conduction, convection, and radiation are the three basic modes of heat flow in dough-to-bread transformation. In most ovens, both conduction and natural convection remain the mode of heat transfer, but the latter is less efficient because heat is spread unevenly throughout the product. When dough is subjected to high temperature various chemical changes such as starch gelatinization, protein coagulation, and browning occur. The development of crust and non-enzymatic browning during baking are major contributing factors in flavor development in various products. Generally, heat treatment reduces the content of rapidly digestible starch while increasing the content of resistant starch in baked foods.

REFERENCES

1. Abdel-Aal, E. S., & Hucl, P. (2002). Amino acid composition and in vitro protein digestibility of selected ancient wheats and their end products. *Journal of Food Composition and Analysis, 15*(6), 737–747.
2. Abdel-Aal, E. S. (2008). Effects of baking on protein digestibility of organic spelt products determined by two in vitro digestion methods. *LWT-Food Science and Technology, 41*(7), 1282–1288.
3. Ahrne, L., Andersson, C.-G., Floberg, P., Rosén, J., Lingnert, H. (2007). Effect of Crust Temperature and Water Content on Acrylamide Formation during Baking of White Bread: Steam and Falling Temperature Baking. *LWT- Food Science & Technology, 40*(10), 1708–1715. https://doi.org/10.1016/j.lwt.2007.01.010
4. Alifak, Y. O., & Sakıyan, O. (2017). Dielectric properties, optimum formulation and microwave baking conditions of chickpea cakes. J*ournal of Food Science & Technology, 54*, 944–953. https://doi.org/10.1007/s13197-016-2371-0
5. Arepally, D., Reddy, R. S., Goswami, T. K., & Datta, A. K. (2020). Biscuit baking: A review. *LWT-Food Science & Technology, 131*, 109726. https://doi.org/10.1016/j.lwt.2020.109726
6. Azmi, M. M. Z., Taip, F. S., Kamal S. M. M. & Chin N. L. (2019) Effects of temperature and time on the physical characteristics of moist cakes baked in air fryer. *Journal of Food Science & Technology, 56*(10), 4616–4624. https://doi.org/10.1007/s13197-019-03926-z
7. Baik, O. D., Grabowski, S., Trigui, M., Marcotte, M., & Castaigne, F. (1999). Heat transfer coefficients on cakes baked in a tunnel type industrial oven. *Journal of Food Science 64*(4), 688–94. http://dx.doi.org/10.1111/j.1365-2621.1999.tb15111.x
8. Bednarcyk, N. E. (1978). *Nutritional value of biscuits and crackers*. 53rd Ann. Tech. Conf. Biscuit and Crackers Manuf. Assoc., San Francisco. 5–9.
9. Belitz, H. D., Grosch, W., & Schieberle, P. (2009). Cereals and cereal products. *Food chemistry*, 670–745.
10. Brathen, E., & Knutsen, S. H. (2005). Effect of temperature and time on the formation of acrylamide in starch based and cereal model system, flat breads and bread. *Food Chemistry, 92* (4), 693–700. https://doi.org/10.1016/j.foodchem.2004.08.030
11. Bredariol, P., Carvalho, R. A. & Vanin, F. M. (2020). The effect of baking conditions on protein digestibility, mineral and oxalate content of wheat breads. *Food Chemistry, 1*, 332,127399. https://doi.org/10.1016/j.foodchem.2020.127399
12. Bredariol, P., Spatti, M., & Vanin, F. M. (2019). Different baking conditions may produce breads with similar physical qualities but unique starch gelatinization behaviour. *LWT- Food Science & Technology, 111*, 737–743. https://doi.org/10.1016/j.lwt.2019.05.094
13. Bredariol, P., & Vanin, F. M. (2021). Bread baking Review: Insight into Technological Aspects in order to Preserve Nutrition. *Food Reviews International*, 1–18.
14. Butterworth, P. J., Warren, F. J., & Ellis, P. R. (2011). Human α-amylase and starch digestion: An interesting marriage. *Starch-Stärke, 63*(7), 395–405. https://doi.org/10.1002/star.201000150
15. Campañone, L. A., & Zaritzky, N. E. (2005). Mathematical analysis of microwave heating process. *Journal of Food Engineering, 69*(3), 359–368. https://doi.org/10.1016/j.jfoodeng.2004.08.027

16. Campañone, L. A., & Zaritzky, N. E. (2010). Mathematical modeling and simulation of microwave thawing of large solid foods under different operating conditions. *Food & Bioprocess Technology*, *3*(6), 813–825. https://doi.org/10.1007/s11947-009-0249-0
17. Campañone, L. A., Paola, C. A., & Mascheroni, R. H. (2012). Modeling and simulation of microwave heating of foods under different process schedules. *Food & Bioprocess Technology*, *5*, 738–749. https://doi.org/10.1007/s11947-010-0378-5
18. Cappelli, A., Lupori, L., & Cini, E. (2021). Baking technology: A systematic review of machines and plants and their effect on final products, including improvement strategies. *Trends in Food Science & Technology*, *115*, 275–284. https://doi.org/10.1016/j.tifs.2021.06.048
19. Cha-um, W., Rattanadecho, P., & Pakdee, W. (2011). Experimental and numerical analysis of microwave heating of water and oil using a rectangular wave guide: Influence of sample sizes, positions, and microwave power. *Food & Bioprocess Technology*, *4*(4), 544–558. https://doi.org/10.1007/s11947-009-0187-x
20. Chavan, R. S., & Chavan, S. R. (2010). Microwave baking in food industry: A review. *International Journal of Dairy Science*, *5*(3), 113–127. https://dx.doi.org/10.3923/ijds.2010.113.127
21. Chen, X., He, X., Fu, X., & Huang, Q. (2015). In vitro Digestion and Physicochemical Properties of Wheat Starch/flour Modified by Heat-moisture Treatment. *Journal of Cereal Science*, *63*, 109–115. https://doi.org/10.1016/j.jcs.2015.03.003
22. Cheng, Z., Su, L., Moore, J., Zhou, K., Luther, M., Yin, J.-J., & Yu, L. (2006). Effects of postharvest treatment and heat stress on availability of wheat antioxidants. *Journal of Agricultural and Food Chemistry*, *54*, 5623e5629. https://doi.org/10.1021/jf060719b
23. Chhanwal, N., Tank, A., Raghavarao, K. S. M. S., & Anandharamakrishnan, C. (2012). Computational fluid dynamics (CFD) modeling for bread baking process—A review. *Food & Bioprocess Technology*, *5*(4), 1157–1172. https://doi.org/10.1007/s11947-012-0804-y
24. Datta, A. K., & Anantheswaran, R. C. (2001). *Handbook of microwave technology for food applications*. M. Dekker.
25. Datta, A. K., & Rakesh, V. (2013). Principles of microwave combination heating. *Comprehensive Reviews in Food Science & Food Safety*, *12*(1), 24–39. https://doi.org/10.1111/j.1541-4337.2012.00211.x
26. Davidson, I., (2016). *Biscuit Baking Technology (Second Edition), Oven Designs*, Academic Press, 73–91.
27. De Angelis, M., Gallo, G., Corbo, M. R., McSweeney, P. L., Faccia, M., Giovine, M., & Gobbetti, M. (2003). Phytase activity in sourdough lactic acid bacteria: purification and characterization of a phytase from Lactobacillus sanfranciscensis CB1. *International journal of food microbiology*, *87*(3), 259–270.
28. Delcour J.A., & Hoseney R.C. (2010). *Principles of cereal science and technology Third Edition,* AACC International.
29. Derde, L. J., Gomand, S. V., Courtin, C. M., & Delcour, J. A. (2014). Moisture distribution during conventional or electrical resistance oven baking of bread dough and subsequent storage. *Journal of Agricultural & Food Chemistry*, *62*(27), 6445–6453. https://doi.org/10.1021/jf501856s
30. Deshou, J., Christopher, C., Pranav, M., Sandeep, K., & Peterson, D.G. (2009) Identification of hydroxycinnamic acid-maillard reaction products in low-moisture baking model systems. *Journal of Agricultural and Food Chemistry*, *57*, 9932–9943. https://doi.org/10.1021/jf900932h
31. Devi, A., & Khatkar, B. S. (2018). Effects of fatty acids composition and microstructure properties of fats and oils on textural properties of dough and cookie quality. *Journal of food science and technology*, *55*(1), 321–330.
32. Dewettinck, K., Van Bockstaele, F., Kühne, B., Van de Walle, D., Courtens, T. M., & Gellynck, X. (2008). Nutritional value of bread: Influence of processing, food interaction and consumer perception. *Journal of Cereal Science*, *48*(2), 243–257.
33. Dziki, D., Różyło, R., Gawlik-Dziki, U., & Świeca, M. (2014). Current trends in the enhancement of antioxidant activity of wheat bread by the addition of plant materials rich in phenolic compounds. *Trends in Food Science & Technology*, *40*, 48–61. https://doi.org/10.1016/j.tifs.2014.07.010
34. El-Sayed, M., Abdel-Aal, & Iwona, R. (2013). Effect of baking on free and bound phenolic acids in wholegrain bakery products. *Journal of Cereal Science*, *57*(3), 312–318. https://doi.org/10.1016/j.jcs.2012.12.001

35. Falcinelli, B., Calzuola, I., Gigliarelli, L., Torricelli, R., Polegri, L., Vizioli, V., Benincasa, P., & Marsili, V. (2018). Phenolic content and antioxidant activity of wholegrain breads from modern and old wheat (*Triticum aestivum* L.) cultivars and ancestors enriched with wheat sprout powder. *Italian Journal of Agronomy*, *13*, 297–302. https://doi.org/10.4081/ija.2018.1220
36. Fellows, P. (2009). *Baking and roasting. Food Processing Technology*. Woodhead Publishing. 978-1-84569-216-2
37. Gally, T., Rouaud, O., Jury, V., & Le-Bail, A. (2016). Bread baking using ohmic heating technology; a comprehensive study based on experiments and modelling. *Journal of Food Engineering*, *190*, 176–184. https://doi.org/10.1016/j.jfoodeng.2016.06.029
38. Gokmen, V., Açar, O. C., Serpen, A. & Morales, F. J. (2008). Effect of leavening agents and sugars on the formation of hydroxymethylfurfural in cookies during baking. *European Food Research & Technology*, *226*(5),1031–1037 http://dx.doi.org/10.1007/s00217-007-0628-6
39. Gontzea, I., & Sutzescu, P. (1968). Natural antinutritive substances in foodstuffs and forages. *Natural antinutritive substances in foodstuffs and forages.*
40. Gujska, E. & Majewska, K. (2005). Effect of Baking Process on Added Folic Acid and Endogenous Folates Stability in Wheat and Rye Breads. *Plant foods for human nutrition*, *60*(2), 37–42. https://doi.org/10.1007/s11130-005-5097-0
41. HadiNezhad, M., & Butler, F. (2010). Effect of flour type and baking temperature on cake dynamic height profile measurements during baking. *Food and Bioprocess Technology*, *3*, 594–602. https://doi.org/10.1007/s11947-008-0099-1
42. Hallberg, L., Rossander, L., Persson, H., & Svahn, E. (1982). Deleterious effects of prolonged warming of meals on ascorbic acid content and iron absorption. *American Journal of Clinical Nutrition*, *36*, 846–850. https://doi.org/10.1093/ajcn/36.5.846
43. Hayman, D. A., Hoseney, R. C., & Faubion, J. M. (1998). Effect of pressure (Crust Formation) on bread crumb grain development. *Cereal Chemistry*, *75*(5), 581–584. https://doi.org/10.1094/CCHEM.1998.75.5.581
44. He, H., & Hoseney, R. C. (1991). A critical look at the electric resistance oven. *Cereal Chemistry*, *68*(2), 151–155.
45. Helou, C., Gadonna-Widehem, P., Robert, N., Branlard, G., Thebault, J., Librere, S., Jacquot, S., Mardon, J., Piquet-Pissaloux, A., Chapron, S., & Tessier, F. J. (2016). The impact of raw materials and baking conditions on Maillard reaction products, thiamine, folate, phytic acid and minerals in white bread. *Food & function*, *7*(6), 2498–2507.
46. Içöz, D., Sumnu, G., & Sahin, S. E. R. P. İ. L. (2004). Color and texture development during microwave and conventional baking of breads. *International Journal of Food Properties*, *7*(2), 201–213.
47. Im, J. S, Huff, H. E, & Hsieh, F. H. (2003). Effects of processing conditions on the physical and chemical properties of buckwheat grit cakes. *Journal of Agricultural & Food Chemistry*, *51*, 659–666. https://doi.org/10.1021/jf0259157
48. Isleroglu, H., Kemerli, T., Sakin-Yilmazer, M., Guven, G., Ozdestan, O., Uren, A., & Kaymak-Ertekin, F. (2012). Effect of steam baking on acrylamide formation and browning kinetics of cookies. *Journal of food science*, *77*(10), E257–E263.
49. Jaeger, H., Roth, A., Toepfl, S., Holzhauser, T., Engel, K.-H., Knorr, D., Vogel, R. F., Bandick, N., Kulling, S., Heinz, V., & Steinberg, P. (2016). Opinion on the use of ohmic heating for the treatment of foods. *Trends in Food Science & Technology*, *55*, 84–97. https://doi.org/10.1016/j.tifs.2016.07.007
50. Jambunathan, K., Lai, E., Moss, M. A., & Button, B. L. (1992). A review of heat transfer data for single circular jet impingement. *International Journal of Heat & Fluid Flow*, *13*(2), 106–115. https://doi.org/10.1016/0142-727X(92)90017-4
51. Jerome, R .E., Singh, S.K., & Dwivedi, M. (2019). Process analytical technology for bakery industry: A review, *Journal of Food Process Engineering*, e13143. https://doi.org/10.1111/jfpe.13143
52. Kawai, K., Hando, K., Thuwapanichayanan, R., & Hagura, Y. (2016). Effect of stepwise baking on the structure, browning, texture, and in vitro starch digestibility of cookie. *LWT-Food Science and Technology*, *66*, 384–389.
53. Khatkar, B. S. (2011). *Bread making process. In Baking Science & Technology*. Arihant Publication.

54. Kocer, D., Nitin, N., & Karwe, M. (2007). *Application of CFD in jet impingement oven. In D. W. Sun (Ed.), Computational fluid dynamics in food processing*. Boca Raton: CRC Press.
55. Krishnamurthy, K., Khurana, H. K., Soojin, J., Irudayaraj, J., & Demirci, A. (2008). Infrared heating in food processing: An overview. *Comprehensive Reviews in Food Science & Food Safety*, 7(1), 2–13. https://doi.org/10.1111/j.1541-4337.2007.00024.x
56. Kumar, C. M., Appu Rao, A. G., & Singh, S. A. (2009). Effect of infrared heating on the formation of sesamol and quality of defatted flours from *Sesamum indicum* L. *Journal of Food Science*, *74*(4), H105–H111. https://doi.org/10.1111/j.1750-3841.2009.01132.x
57. Kumar, P., Ramanathan, M., & Ranganathan, T. V. (2014). Ohmic heating technology in food processing – A review. *International Journal of Food Engineering Research & Technology*, *3*, 1236–1241.
58. Lavelli, V., Guerrieri, N., & Cerletti, P. (1996). Controlled reduction study of modification induced by gradual heating of gluten protein. *Journal of Agricultural and Food Chemistry*, 44(9), 2549–2555. https://doi.org/10.1021/jf960019e
59. Li, A., & Walker, C. E. (1996). Cake baking in conventional, impingement and hybrid ovens. *Journal of Food Science*, *61*(1), 188–191. https://doi.org/10.1111/j.1365-2621.1996.tb14756.x
60. Lindenmeier, M., & Hofmann, T. (2004). Influence of baking conditions and precursor supplementation on the amounts of the antioxidant pronyl-L-lysine in bakery products. *Journal of Agricultural and Food Chemistry*, *52*(2), 350–354.
61. Lucas, T. (2014). *Baking. Bakery products science & technology*. Wiley Publisher, 9781119967156.
62. Maleki, M. & Daghir, S. (1967). Effect of baking on retention of thiamine, riboflavin, and niacin in Arabic bread. *Cereal Chemistry*, *44*,483–487.
63. Manay, N. S. O. (2001). *Cooking of food. Food: facts & principles*. New Age International.
64. Manhiça, F. A., Lucas, C., & Richards, T. (2012). Wood consumption and analysis of the bread baking process in wood-fired bakery ovens. *Applied Thermal Engineering*, *47*, 63–72. https://doi.org/10.1016/j.applthermaleng.2012.03.007
65. Martin, M. L., Zeleznak, K. J., & Hoseney, R. C. (1991). A mechanism of bread firming. I. Role of starch swelling. *Cereal Chemistry*, *68*, 498–503.
66. Masure, H. G., Wouters, A. G. B., Fierens, E., & Delcour, J. A. (2018). Electrical resistance oven baking as a tool to study crumb structure formation in gluten-free bread. *Food Research International*, *116*, 925–931. https://doi.org/10.1016/j.foodres.2018.09.029
67. Matz, S. A. (1988). *Equipment of bakers*. Pan-Tech International, McAllen TX: Pan-American International.
68. Michalska, A., Amigo-Benavent, M., Zielinski, H., & del Castillo, M. D. (2008). Effect of bread making on formation of Maillard reaction products contributing to the overall antioxidant activity of rye bread. *Journal of cereal science*, *48*(1), 123–132.
69. Mudgil, D., Barak, S., & Khatkar, B. S. (2017). Cookie texture, spread ratio and sensory acceptability of cookies as a function of soluble dietary fiber, baking time and different water levels. *LWT*, *80*, 537–542.
70. Naseer, B., Naik, H. R., Hussain, S. Z., Zargar, I., Bhat, T. A., & Nazir, N. (2021). Effect of carboxymethyl cellulose and baking conditions on in-vitro starch digestibility and physico-textural characteristics of low Glycemic Index gluten-free rice cookies. *LWT*, *141*, 110885.
71. Nguyen, H. T., Peters, R. J. & Van Boekel, M.A. (2016). Acrylamide and 5-hydroxymethylfurfural formation during baking of biscuits: Part I: Effects of sugar type. *Food Chemistry*, *192*(2), 575–585. https://doi.org/10.1016/j.foodchem.2015.07.016
72. Nowak, D., & Lewicki, P. P. (2004). Infrared drying of apple slices. *Innovative Food Science & Emerging Technologies*, *5*(3), 353–360. https://doi.org/10.1016/j.ifset.2004.03.003
73. Ou, J., Wang, M., Zheng, J., & Ou, S. (2019) Positive and negative effects of polyphenol incorporation in baked foods. *Food Chemistry*, *284*, 90–99. https://doi.org/10.1016/j.foodchem.2019.01.096
74. Ovadia, D. Z., & Walker, C. E. (1998). Impingement in food processing. *Food Technology*, *52*(4), 46–50.
75. Palazoğlu, T. K., Coşkun, Y., Tuta, S., Mogol, B. A., & Gökmen, V. (2015). Effect of vacuum-combined baking of cookies on acrylamide content, texture and color. *European Food Research and Technology*, *240*(1), 243–249.

76. Pareyt, B., & Delcour, J. A. (2008). The role of wheat flour constituents, sugar, and fat in low moisture cereal based products: a review on sugar-snap cookies. *Critical reviews in food science and nutrition, 48*(9), 824–839.
77. Paton, J., Khatir, Z., Thompson, H., Kapur, N., & Toropov, V. (2013). Thermal energy management in the bread baking industry using a system modelling approach. *Applied Thermal Engineering, 53*(2), 340–347. https://doi.org/10.1016/j.applthermaleng.2012.03.036
78. Purlis, E. (2010). Browning development in bakery products–A review. *Journal of Food Engineering, 99*(3), 239–249. https://doi.org/10.1016/j.jfoodeng.2010.03.008
79. Purlis, E. (2014). Optimal design of bread baking: Numerical investigation on combined convective and infrared heating. *Journal of Food Engineering, 137*, 39–50. https://doi.org/10.1016/j.jfoodeng.2014.03.033
80. Purlis, E., & Salvadori, V. O. (2007) Bread Browning Kinetics during Baking. Journal of Food Engineering, *80*(4), 1107–1115. DOI: 10.1016/j.jfoodeng.2006.09.007
81. Ranhotra, G. S. & Bock, M. A. (1988). *Effects of Baking on Nutrients. In: Karmas, E., Harris, R.S. (eds) Nutritional Evaluation of Food Processing*. Springer, Dordrecht. https://doi.org/10.1007/978-94-011-7030-7_13
82. Rashmi, H.B. & Negi, P.S. (2020). Phenolic acids from vegetables: A review on processing stability and health benefits. *Food Research International, 136*, 109298. https://doi.org/10.1016/j.foodres.2020.109298
83. Rosell, C.M. (2016). Bread: Chemistry of Baking. *Encyclopedia of Food and Health*. 484–489. hhttp://dx.doi.org/10.1016/B978-0-12-384947-2.00088-X
84. Rybicka, I., & Gliszczynska-Swiglo, A. (2017). Minerals in grain gluten-free products. The content of calcium, potassium, magnesium, sodium, copper, iron, manganese, and zinc. *Journal of Food Composition & Analysis, 59*, 61–67. https://doi.org/10.1016/j.jfca.2017.02.006
85. Sakai, N., & Hanzawa, T. (1994). Applications and advances in farinfrared heating in Japan. *Trends in Food Science & Technology, 5*(11), 357–362. https://doi.org/10.1016/0924-2244(94)90213-5
86. Sakiyan, O., Sumnu, G., Sahin, S., Meda, V., Koksel, H., & Chang, P. (2011). A study on degree of starch gelatinization in cakes baked in three different ovens. *Food & Bioprocess Technology, 4*(7), 1237–1244. https://doi.org/10.1007/s11947-009-0210-2
87. Sakr, M., & Liu, S. (2014). A comprehensive review on applications of ohmic heating (OH). *Renewable & Sustainable Energy Reviews, 39*, 262–269. https://doi.org/10.1016/j.rser.2014.07.061
88. Sandu, C. (1986). Infrared radiative drying in food engineering: A process analysis. *Biotechnology Progress, 2*(3), 109–119. https://doi.org/10.1002/btpr.5420020305
89. Sani, N. A., Taip, F. S., Kamal, S. M., & Aziz, N. (2014). Effects of temperature and airflow on volume development during baking and its influence on quality of cake. *Journal of Engineering Science and Technology, 9*(3), 303-313.
90. Sarkar, A., & Singh, R. P. (2004). Air impingement technology for food processing: Visualization studies. *LWT – Food Science & Technology, 37*(8), 873–879. https://doi.org/10.1016/j.lwt.2004.04.005
91. Sevimli, K. M., Sumnu, G., & Sahin, S. (2005). Optimization of halogen lamp—microwave combination baking of cakes: a response surface methodology study. *European Food Research & Technology, 221*, 61–68. https://doi.org/10.1007/s00217-004-1128-6
92. Shahapuzi, N. S. M., Taip, F. S., & Aziz, N. A. (2012). *Experimental studies on the effect of airflow on oven temperature and product qualities in cake baking process*. In: International conference on agricultural and food engineering for life.
93. Smith, D. P., & Harris, H. H. (1974). *Factors in design and construction of a device for heating and dispensing food components*. ASSOCIATED FOOD EQUIPMENT CO DALLAS TX.
    Somoza, V. (2005). Five years of research on health risks and benefits of Maillard reaction products: An update. *Molecular Nutrition & Food Research, 49*(7), 663–672. https://doi.org/10.1002/mnfr.200500034
94. Stear, C. A. (1990). *Hand book of bread making technology*. London, New York: Elsevier Applied Science.
95. Swami, S. B., Thakor, N. J., & Murudkar, P. R. (2015). Effect of yeast concentration and baking temperature on quality of slice bread. *Journal of Food Research and Technology, 3*(4), 131–141.

96. Świeca, M., Gawlik-Dziki, U., Dziki, D., Baraniak, B., & Czyż, J. (2013). The influence of protein–flavonoid interactions on protein digestibility in vitro and the antioxidant quality of breads enriched with onion skin. *Food Chemistry*, *141*(1), 451–458.
97. Tekle, A. (2009). The effect of blend proportion and baking condition on the quality of cookies made from taro and Wheat flour blend. *Thesisi Addis Ababa University, Ethiopia.*
98. Therdthai, N., Zhou, W., & Adamczak, T. (2002). Optimisation of the temperature profile in bread baking. *Journal of Food Engineering*, *55*(1), 41–48. https://doi.org/10.1016/S0260-8774(01)00240-0
99. Tuncel, N. B., Yilmaz, N., Kocabiyik, H., & Uygur, A. (2014). The effect of infrared stabilized rice bran substitution on B vitamins, minerals and phytic acid content of pan breads: Part II. *Journal of Cereal Science*, *59*(2), 162–166. https://doi.org/10.1016/j.jcs.2013.12.005
100. Turabi, E., Sumnu, G., & Sahin, S. (2008). Optimization of baking of rice cakes in infrared–microwave combination oven by response surface methodology. *Food & Bioprocess Technology*, 1(1), 64–73. https://doi.org/10.1007/s11947-007-0003-4
101. Türk, M., & Sandberg, A. S. (1992). Phytate degradation during breadmaking: effect of phytase addition. *Journal of Cereal Science*, *15*(3), 281–294.
102. Ureta, M. M., Olivera, D. F., & Salvadori, V. O. (2014). Baking of muffins: Kinetics of crust color development and optimal baking time. *Food and bioprocess technology*, *7*(11), 3208–3216.
103. Valamoti S M, Samuel, D, Bayram M, Marinova E. (2008). Prehistoric cereal foods from Greece and Bulgaria: Investigation of starch microstructure in experimental and archaeological charred remains. *Vegetation History & Archaeobotany*, *17*, 265–276
104. Vanin, F. M. Lucas, T. & Trystram, G. (2009). Crust Formation and Its Role during Bread Baking. *Trends in Food Science & Technology*, *20*(8), 333–343. https://doi.org/10.1016/j.tifs.2009.04.001
105. Varghese, K. S., Pandey, M. C., Radhakrishna, K., & Bawa, A. S. (2014). Technology, applications and modelling of ohmic heating: a review. *Journal of Food Science & Technology*, *51*(10), 2304–2317. https://doi.org/10.1007/s13197-012-0710-3
106. Wahlby, U., Skjoldebrand, C., & Junker, E. (2000). Impact of impingement on cooking time and food quality. *Journal of Food Engineering*, *43*(3), 179–87. https://doi.org/10.1016/S0260-8774(99)00149-1
107. Walker, C. E. (2016). Oven technologies. *Encyclopedia of Food Grains*, *3*(2), 325–334.
108. Wang, J., & Sheng, K. C. (2004). Modeling of muti-layer far-infrared dryer. *Drying Technology*, *22*(4), 809–820. https://doi.org/10.1081/DRT-120034264
109. Westurlund, E., Theander, O., & Åman, P. (1989). Effects of baking on protein and aqueous ethanol-extractable carbohydrate in white bread fractions. *Journal of Cereal Science*, *10*(2), 139–147.
110. Whitaker, A. M., & Barringer, S. A. (2004). Measurement of contour and volume changes during cake baking. *Cereal chemistry*, *81*(2), 177–181.
111. Yael, B., Liel, G., Hana, B., Ran, H., & Shmuel, G. (2012). Total phenolic content and antioxidant activity of red and yellow quinoa (Chenopodium quinoa Willd.) seeds as affected by baking and cooking conditions. *Food and Nutrition Sciences*, *2012.*
112. Yin, Y., & Walker, C. E. (1995). A quality comparison of breads baked by conventional versus nonconventional ovens: a review. *Journal of the Science of Food & Agriculture*, *67*(3), 283–291. https://doi.org/10.1002/jsfa.2740670302
113. Yolacaner, E. T., Sumnu, G., & Sahin, S. (2017). *Microwave-assisted baking. In The Microwave Processing of Foods.* Woodhead Publishing. https://doi.org/10.1016/B978-0-08-100528-6.00006-1
114. Zanoni, B., Peri, C., & Bruno, D. (1995). Modelling of browning kinetics of bread crust during baking. *LWT- Food Science & Technology*, *28* (6), 604–609. https://doi.org/10.1016/0023-6438(95)90008-X
115. Žilić, S., Kocadağlı, T., Vančetović, J., & Gökmen, V. (2016). Effects of baking conditions and dough formulations on phenolic compound stability, antioxidant capacity and color of cookies made from anthocyanin-rich corn flour. *LWT-Food Science and technology*, *65*, 597–603.

CHAPTER 14

Parboiling of Cereals

Gargi Ghoshal
Dr. S. S. Bhatnagar University Institute of Chemical Engineering & Technology,
Panjab University, Chandigarh, Punjab, India

CONTENTS

DOI: 10.1201/9781003242192-18

14.1 INTRODUCTION

In the parboiling process, paddy is subjected to hydrothermal processing for gelatinization of starch prior to milling. Parboiling consists of three steps: soaking, steaming, and drying. Parboiling results in physical changes in grain, chemical modifications, and also transformations in organoleptic properties of the grain to significantly manipulate the milling properties, storage duration, cooking as well eating attributes of rice (Aykroyd et al., 1940; Ballogou et al., 2013; Wang et al., 2021, Meresa et al., 2020; Parnsakhorn and Noomhorm, 2012; Bhattacharya et al., 2013; da Fonseca, 2011; Luh and Mickus, 1991; Oli et al., 2014; Kimura, 1983). Worldwide ~130 million tons of parboiled paddy are manufactured per annum. Approximately 3–4 million tons of high-priced milled parboiled rice is being traded globally (Kowfi and Nagdi, 2017, 2016a, b). Parboiling was initiated in India and is largely practiced in Bangladesh, Sri Lanka, and in various Sub-Saharan African countries such as Benin, Cameroon, Ghana, Nigeria, and Senegal (Kapur, 1997; Kapur, 1996; Kowfi and Nagdi, 2016a, 2016b, Roy et al., 2006; Ahiduzzaman and Sadrul, 2009; Velupillai, 1980; Ndindeng et al., 2015). Approximately 1.2 tons of parboiled rice per hour are produced (Kowfi and Nagdi, 2016a, 2016b, Roy et al., 2006). India exported 1.97 million MT of rice in December, 2021 up by 31% year on year and by 33% month on month. Non-Basmati exports rose by 36% on year to 1.63 million MT, while Basmati exports increased by 3% to 343,346 MT. This reflects the competitiveness of India's non-Basmati exports throughout December amid kharif crop arrivals, especially for parboiled and broken rice. (www.ers.usda.gov/topics/crops/rice/rice-sector-at-a-glance/)

The rural technique is lengthy, time consuming, laborious, and energy intensive. In conventional parboiling practices the soaking process needs 48 h whereas drying of soaked, steamed paddy is done until 6.9–9.7% (wb) moisture content, thus 6.9–9.7% MJ/kg additional energy is required compared to raw rice processing (Araullo et al., 1985; Billiri and Siebenmorgen, 2014). Raw as well as parboiled rice compositions are listed in Table 14.1. As a requirement for development of high- quality nutritious parboiled rice from local varieties of paddy, it is essential to optimize the

Table 14.1 Composition of Raw and Parboiled Rice

Rice type→ Properties↓	Raw brown	Raw milled	Parboiled brown	Parboiled milled
Moisture (%)	12.60 ± 0.54	11.10 ± 1.16	12.07 ± 0.74	10.83 ± 0.64
Protein (%)	6.85 ± 0.34	6.66 ± 0.34	6.76 ± 0.20	6.36 ± 0.32
Crude fat (%)	2.65 ± 0.20	0.5 ± 0.07	2.69 ± 0.13	0.38 ± 0.06
Ash (%)	1.21 ± 0.06	0.47 ± 0.12	1.18 ± 0.17	0.55 ± 0.04
K (mg/100g)	181.71 ± 9.27	65.46 ± 5.57	152.89 ± 8.77	143.21 ± 8.86
P (mg/100g)	61.27 ± 2.08	41.98 ± 5.40	56.42 ± 1.84	58.85 ± 5.05
Mg (mg/100g)	16.88 ± 0.57	15.06 ± 2.34	15.95 ± 0.28	15.43 ± 2.66
Ca (mg/100g)	6.85 ± 0.43	6.70 ± 0.42	6.23 ± 0.41	4.61 ± 0.85
Zn (mg/100g)	1.98 ± 0.11	2.09 ± 0.09	1.90 ± 0.10	1.15 ± 0.31
Fe (mg/100g)	0.57 ± 0.35	0.40 ± 0.29	0.55 ± 0.47	0.43 ± 0.35
Na (mg/100g)	0.54 ± 0.20	0.53 ± 0.06	0.44 ± 0.14	0.59 ± 0.07
Mn (mg/100g)	0.36 ± 0.05	0.45 ± 0.06	0.42 ± 0.07	0.28 ± 0.02
Cu (mg/100g)	0.16 ± 0.07	0.18 ± 0.04	0.15 ± 0.06	0.17 ± 0.03
Se (mg/100g)	0.04 ± 0.00	0.04 ± 0.00	0.03 ± 0.00	0.03 ± 0.00
Σ	270.35	235.02	224.75	132.88

Source: Lanfer-Marquez et al., 2005; Kwofie & Ngadi, 2017.

design to make it more economical. Again, with the current inclination towards rising global energy utilization likely to arrive at 630 quadrillion Btu by 2025 (IEO, 2013) due to globalization and population expansion, every attempt must be made to provide sustainable approaches to energy use. The Canadian Department of Foreign Trade Affairs, in collaboration with McGill University, Canada, made parboiled rice using African Rice and also practiced energy supply and developed equipments such as improved stoves along with briquetting and innovative gasification) in Sub-Saharan Africa (Kowfi and Nagdi, 2017, 2016 a, b; Ndindeng et al., 2015). Table 14.2 summarizes the research and development of optimization of rice parboiling steps done at different institutions and research centres.

Table 14.2 Research and Development for Optimization of Rice Parboiling Steps Done in Different Institution and Research Centres

Process	Soaking	Steaming	Drying
Schule	Batch system in medium-temperature water followed by a second stage in high-temperature water under pressure in the same tank.	Steaming is not required. Starch gelatinization obtained by soaking in high-temperature water under pressure	In high-temperature air followed by medium-temperature air.
CFTRI	Batch system in hot water 60–75°C for 2–3.5h.	Steam is pressure injected through the perforated pipes.	Sun drying or mechanical drying by medium-temperature air.
Jadavpur University	Batch system in hot water 60–70°C for 2.5–3h.	Steam is pressure injected through perforated pipes. Alternatively continuous system with steam at ambient pressure in autoclave equipped with screw conveyor	Cooling before drying. This is done by using high-temperature air followed by medium-temperature air
Avorio	Continuous system in medium-temperature water	Continuous steaming under pressure in an autoclave Cooling before drying. This is done by using medium-temperature air. Equipped with mechanical conveyors	Cooling before drying. This is done by using high-temperature air followed by medium-temperature water
Crystal rice	Batch system in high-temperature water under vacuum, followed by hydrostatic pressure	Continuous system in a rotary autoclave under steam pressure	By high-temperature air, followed by medium-temperature air
Malek	Batch system in high-temperature water	Continuous steaming under pressure in a medium vertical by high-temperature air, followed by medium-temperature air [5] Stationary autoclave	may be done after milling
CRGA parboiling	Batch system in medium-temperature water, medium followed by higher-temperature water	Continuous system in a horizontal cooker under high steam pressure for a short time	In high-temperature air, followed by medium-temperature air
Thermik Fluide Parboiling	Batch system in hot water soaking (65–75°C)	Paddy is stored in tubes place in a cylinder with a circulating heated thermic fluid	Air drying in shade
Infrared heating	Batch system in hot water soaking (55°C)	No steaming	Simultaneous parboiling and drying with IR heating (2100 W) and vibration

Source: Luh & Mickus, 1991; Pillaiyar et al., 1996; Likitrattanaporn & Noomhorm, 2011.

Parboiled rice is rice cooked in husk that is dried and milled. Though raw and parboiled rice are intrinsically the same there are some extrinisic differences. Although biologically alike the two forms are function wise very dissimilar and thus can be considered as two different cereal varieties. The following describes how raw rice is different from parboiled rice.

1. Raw rice is comparatively white and opaque as compared to parboiled rice, which is translucent, glassy, hard, and yellowish/amber in color.
2. Parboiled rice is relatively broader and slightly shorter than raw rice grain.
3. In parboiled rice starch packing properties are different. Grains with chalky/white belly are absent during the parboiling process since gelatinization grains are continuous and well packed.
4. Flow properties are also different in parboiled rice.
5. The percentage of broken rice is reduced and eventually very high yield is obtained.
6. After parboiling vitamin B content in milled rice is higher than in raw rice.
7. The cooking behavior and consumption quality of prehydrated parboiled rice is different from raw rice.
8. Cooked parboiled rice is harder, more swollen, and has less adhesiveness than cooked raw rice.
9. After parboiling the manufacturing properties of rice are enhanced. Raw rice is not suitable for making some products but parboiled rice is appropriate.
10. The yield, quality, and extraction efficiency of parboiled rice bran oil manufacturing are better than for raw rice.

There is greater dissimilarity in product with higher degree of parboiling conditions in terms of temperature, time of soaking, and steaming pressure. After being introduced in South Asian countries, parboiling was primarily used only by local people. Subsequently, more than centuries ago, definite technological transformation in milling practices and food habits convey the superior vitamin retention in milling exercise according to the observation of WHO. In the 19th century with the acceptance of machine milled head rice, health authorities were concerned that people eating raw (machine-) milled rice become extremely prone to the attack of beriberi. People eating parboiled rice (brown raw rice or unpolished or partially polished raw rice) were protected from various diseases caused by the deficiency of Vitamin B1 (thiamine) in high degree of polished parboiled and raw rice, respectively. Consumption of rice made by parboiling instead of raw rice is thus supported by the WHO (Bhattacharya et al., 2004, 2013, www.coek.info).

14.2 FUNDAMENTALS OF PARBOILING PROCESS AND MACHINERY

14.2.1 Types of Parboiling Processes

Parboiled rice can be classified by the nature of the changes in rice depending on the processing conditions (Bhattacharya, 2011).

These different processes are:

i. Conventional Parboiling
ii. Pressure Parboiling
iii. Dry heat Parboiling

14.2.1.1 Conventional Parboiling

In conventional parboiling the Soak-Drain-Cook-Dry method of parboiling is the most common as well as extensively utilized practice. The following steps are the normal practice.

Soaking of paddy is done in water at an appropriate temperature either at room temperature for 2 or 3 days or hot soaking water at 60–75°C for 24 h or until the rice attains moisture content of about 30% (wet basis) followed by draining of paddy. Then paddy is steamed, then dried. After steaming, paddy is subjected to microwave or infrared or several means of thermal processing for starch gelatinization. Atmospheric steaming or pressure steaming under enhanced pressure (0.5 to 2 kg/cm^2 gauge pressure) is done.

14.2.1.2 Pressure (Low Moisture) Method

This is is alternative method of parboiling. In this method paddy is not soaked to saturation at all but only partially soaked or even simply wetted with water then paddy is subjected to high-pressure steaming at 13 Kg/cm^2 gauge for compulsory gelatinization. Steaming of paddy at atmospheric pressure is common in this technique, but it is frequently carried out in traditional parboiling. Its exceptional attribute is essentially the low moisture percentage for cooking of grain. This practice did well for very short periods in the state of Punjab, India but vanished afterwards, but was very useful to know the fundamental character of parboiling.

14.2.1.3 Dry Heat Parboiling

A third alternative is dry heat parboiling. Soaked paddy is subjected to conduction heating instead of steaming. After soaking, paddy is concurrently gelatinized by steaming, followed by drying. This practice is not extensively applied, but it did facilitate understanding of the basics of the parboiling practice (Khan et al., 1974).

The above mentioned diverse categories of parboiling lead to different products and different quality of products. Soaking can be done either at ambient or at high temperature (about 70°C). Correspondingly steam treatment type of cooking is carried out in gentle or in rigorous circumstances. Consequently there can be an extensive range in the rigorousness of the practice among each of the three types. Severe parboiling practice has powerful consequences on features of the end products (Ying-dan, 2021).

Soaking is generally the most time-intensive process during parboiling treatment, ranging from approximately 12–48 h in conventional parboiling practice to 2–4 h in recent practices. With increasing temperature from 25 to 80°C, time of soaking reduces from 60 h to 1.5 h (Kapur, 1997). Usually soaking is done in concrete masonry tanks (Figure 14.1).

After steaming, paddy is subjected to drying in the sun in local rice mills. During sun drying, paddy is poured in layers on the concrete floor or spread on tarpaulin to diminish foreign impurities and blemishes. Vacuum drying, drying using hot-air, fluidized bed drying, and drying using superheatedsteam are also used (Swasdisevi et al., 2010). The fluidized bed drying method has gained interest because of its efficiency. In fluidized bed drying hotair or superheated steam is used as drying medium. Making of air, solid particles mix by vigorous shaking to evaporate water. Industrial parboiling in developed countries like the United States is entirely automatic using steaming coils with built-in properties in soaking vessels. In industrial parboiling paddy is soaked in hot water at 80–90°C and distributed for 15 min and sustained at 65 °C for 4/5 h; then it is drained off and steam commences by steaming coils for 10–20 min followed by transport through conveyor belts for drying. Different types of driers are used including rotary-drum dryer in combination with steam heat-exchangers, furnaces with husk-fired devices, bin driers, and rotary hot-air driers (Figure 14.1) (Luh and Mickus, 1991).

While numerous parboiling methods have been introduced by different scientists, parboiling in some regions is practiced using small earthen pot or using the boiler method. The parameters of parboiling systems appreciably influenced the quality attributes of parboiled paddy. Soaking

Figure 14.1 Equipment used for parboiling of rice (collected from parboiling plant in Punjab, India).

water temperature and pressure of steam are the essential factors influence quality of rice. Although, higher water temperature produces better quality of rice, but temperatures at higher than the gelatinization temperature will spoil ultimate quality of rice. Steaming enhances yield of rice after milling and also expand the storage quality but is energy exhaustive. The main energy requirement during soaking and steaming steps during parboiling has been principally completed using fuel wood and straight burning of husk, the by-product of rice industry, in African and South Asian countries, correspondingly. The Sun light energy is the major foundation for local parboiling firm for paddy drying. Nonetheless, a combination of partial Sun drying and partial mechanical drying arrangement is prevalent in parts of Asia at medium scale capacity parboiling firm. Extensive disparity between energy demand for theoretical with the actual laboratory parboiling and plant quantity is a sign of incompetent arrangement applied for limited parboiling thus it is the need of hour for the development of uncomplicated but energy proficient parboiling arrangements.

14.3 IMPACT OF PROCESSING TREATMENT ON NUTRITIONAL CHARACTERISTICS (MACRO- AND MICRONUTRIENTS)

14.3.1 Changes after Parboiling

Numerous modifications occur inside the grain during parboiling, which are discussed as follows.

14.3.2 Enzymatic Action

Rice grain is live kernel, but since it is no longer active, numerous enzymes are most likely hypothetically active in kernel. If rice is dipped in aqueous media, an excellent arrangement of biological action is to be assessed. Principal achievement is sprouting of dormant grain. Starting with germination of germs moistening hence during soaking in parboiling should absolutely be an elicit aim at the commencement of the germination procedure.

Irrespective of the method the accessibility of light, air, temperature of soaking, etc., determine the outcome. If paddy soaking is done at ambient temperature for longer duration, conditions may change to anaerobic and consequently seed would die; on the other hand enzymatic changes would still likely happen but if soaking is done at higher (>70 °C) temperature, the seed would expire rapidly comparatively. Enzymes are denatured, but activity of microbes occurs at a slow rate. These alterations have not been explored in depth. However, the wide enzymatic alteration in sugars, for example, have been studied. Ali and Bhattacharya (1980) confirmed the substantial enzymatic alteration to reducing sugars from sucrose at the time of soaking, as well as denovo manufacturing of simple mono, di-saccharides, and amino acids. It could be observed that the degree of alterations was associated with the soaking temperature. This sugar and amino acid reactions were also partially the reasons for the yellowish color of rice post parboiling. Other reported changes include enhancement of phenolic compounds, enzyme activities, and discharge of chloride in soaking water (Anthoni Raj et al., 1980, 1996). Ramalingam and Anthoni Raj (1996) considered the bacteriological inhabitants and the amount of diverse organic constituents in the effluent of soaking water which affect usual bacteriological action throughout soaking. When polished rice grains were kept in a tube fitted with a cap followed by heating in an oven at different temperatures, substantial extra yellowish coloring occurred in parboiled rice grains, as compared to raw rice grains. This discoloration obviously illustrates the incidence of additional sugars and amino acids in polished parboiled rice obviously from enzymatic and chemical reaction during soaking. Of major significance is the level of yellowing in polished parboiled rice upon its exposure to high heat during soaking and steaming, which is associated (besides the temperature of heating) with the temperatures in which the paddy is soaked . This type of yellow color development obviously the enzymatic development of sugars and amino acids vary based on soaking temperatures. While extra heat discoloration in paddy occurs the extent of discoloration is smaller when the paddy soaking is done at room temperature (as enzyme action is reduced). Similarly soaking above 80°C causes less yellowing as enzymes are deactivated. Though it was significant at midway (50–70°C) soaking temperatures, at 60°C maximum yellowing occurs. These results can be described based on enzymatic activity basis only, including generation of sugars and amino acid non-enzymatic reaction. It is therefore not hard to predict the reason that milled parboiled rice grains have minimum discoloration as compared to polished raw rice grains. Potential resettlement of minute molecules limiting potential transformation throughout soaking is the probable interior relocation of small molecules soluble in water from the bran layer ultimately to inner endosperm. Polished parboiled rice has higher quantity of B-vitamins, sugars, and essential minerals than polished raw rice. Internal movement of soluble constituents (e.g., B vitamins, sugars in the kernel) occurs throughout soaking period. Depending on the husk opening there may be other features associated with yellowing of rice. When soaking is carried out this way the husk splits open (at about ≥ 70°C temperature maintained in soaking water), the pigments present are leached out which causes discoloration. (Bhattacharya and Subba Rao, 1966b).

14.3.3 Modifications during Steaming

While the transformations that occur during soaking do not affect the quality of product, some staining does occur, which may affect consumer acceptance. The essential features of parboiled rice are conditional as parboiled rice is actually precooked rice inside husk. Consequently it is the steaming/cooking stage that results in the most changes to parboiled rice as compared to raw rice.

Many researchers have studied the effect of soaking and steaming to understand the effects properly (Bhattacharya and Subba Rao, 1966a, 1966b). The entire range of their consequences was considered independently by researchers. Soaking of paddy has been carried out at diverse temperatures (from room temperature to 80°C) for various durations (at under soaking, optimal soaking, and oversoaking). At 0–1.4 kg/cm^2, gauge pressure steaming was done with increasing pressure for diverse time periods (10–60 min). Finally, the paddy was soaked and a combination of soaking and steaming was done in addition to drying and milling. The gelatinization generally happens at steaming stages along with other steps and by other signature features of the product (e.g. superior yield of head rice, reduced cooking time with intact grain as well as better thiamine retention after milling. Consequenses for soaking at above the gelatinization temperature (e.g., 70°C and oversoaking at 80°C) are observed. Though soaking alone brought considerable changes, but not completely, in the properties of rice. Steaming followed by soaking changes all three components, starch, protein and fat, in rice.

14.3.4 Recent Research

Martins et al. (2021) studied red paddy (*Oryza sativa* L.) parboiling and modelling of the collective features of parboiled rice and its cooking behaviors. Rice parboiling is generally used to enhance its strength and nutrients. Though parboiling results in modification of consistency after cooking, textural quality also has a key effect on user satisfaction. Thus, through quantitative detection and modelling the correlation among parboiling parameters and postcooking quality is meaningful on the basis of process control. Martins et al. (2021) found that postparboiling resulted in insignificant modification in color, composition, and density of rice, whereas the texture of cooked rice was noticeably influenced by its cooking time. A second-order kinetic rate equation was fitted during the development of textural features at cooking time. They tried to fit different kinetic equations to check the kinetic behaviour on textural properties and found that second order equation is best fitted.

Prom-u-thai (2009) studiedthe effect of iron enrichment on organoleptic properties such as appearance and sensory and cooking features of parboiled rice. They found that iron (Fe) fortification increased iron content and its bioavailability as well in milled white rice. The objective of their research was to assess iron fortification on sensory and cooking behaviour of parboiled rice with two different quantities of iron 250 and 450 mg kg^{-1} paddy (www.onlinelibrary.wiley.com). Effects were observed in terms of appearance, cooking properties, basic sensory properties and overall approval contrasting with controlled parboiled rice in Thailand and Bangladesh.

When iron enrichment using 250 mg was done for kg^{-1} paddy considerably elevated Fe content in polished rice was obtained (about 19.1 mg iron kg^{-1} in comparison to 6.2 mg Fe kg^{-1} for unfortified parboiled polished rice), without issues relating to consumer approval based on the existing preassessment method. Enriched Fe was preserved after cooking, and panelists in Thailand and Bangladesh could not tell the difference between this paddy as compared with controlled and regional varieties of parboiled rice. However, 450 mg Fe kg^{-1} paddy had more yellowness in grain and altered texture, flavor, and flakiness after cooking. Thus, panellists found this level not suitable for enrichment in Thailand and Bangladesh. But 250 mg Fe level kg^{-1} paddy dose is effectual as well as adequate to rice eaters. End user satisfaction is robustly linked with pre-cooking appearance, cooking, and sensory qualities.

Wu et al. (2020) studied the effect of elevated-pressure presoaking on consistency and retrogradation characteristics of parboiled rice. Reduced taste, low digestibility, and objectionable yellowish parboiled rice colors hindered consumer acceptance. Thus a novel technique for manufacturing of premium quality parboiled rice is needed.

In another study influence of high hydrostatic pressure (HHP) at presoaking temperatures on textural, color properties, and the degree of retrogradation of parboiled rice was explored. With HHP ranges from 100 to 500 MPa, the rate of absorption of water improves along with reduction of cooking time. Parboiled rice samples with high lightness scores (L) also had inferior intensity of color (B). Compared with control group, parboiled rice sample with presoaking at high-pressure conditions demonstrated a diminution of textural properties from 0.69% to 32.99% (www.pubfacts.com). The thermal properties using (DSC) signify that the enthalpy values of parboiled rice samples reduced following presoaking at high pressure. Molecular configuration of parboiled rice distinguished by FTIR confirmed that HHP presoaking could reduce retrogradation intensity.

Taleon et al. (2021) studied influences of parboiling parameters on zinc and iron biofortified and control polished rice. This paper deals with biofortification of zinc and iron in parboiled rice to serve people with mineral deficiencies. Retention of zinc and iron in polished rice at small scale and at industrial scale parboiling methods was estimated. Conditions similar to commercial and household parboiling techniques resulted in 52.2–59.7% and 70.7–79.6% zinc, 55.4–79.1% and 78.2–119.8% iron retentions, respectively. On the other hand, in polished rice after removal of 8–16% bran, zinc retention was 50.6–66.8% whereas in soaking of rough rice at 20°C and 29.9–56.0% zinc retention was found. When soaking of rough rice was carried out at 65°C zinc retention was found but it was less than in raw rice (58.0–80.6%). But iron retention was comparable. Parboiled rice at room temperature soaking could contribute more zinc (Taleon et al., 2021).

14.4 IMPACT OF PROCESSING TREATMENT ON FUNCTIONAL CHARACTERISTICS (RHEOLOGICAL, THERMAL, STRUCTURAL, ETC.)

14.4.1 Modification of Starch

During parboiling starch modification occurs, which is accountable for various features. Research on starch modification during parboiling of rice has proven a number of positive physicochemical characteristics. Nonetheless parboiled rice is a precooked grain inside husk and thus requires lengthier cooking time. On the contrary, parboiled rice absorbs water slowly in boiling temperature but hydrates quickly at higher degree at low temperature than raw rice.

It was confirmed that hydration of polished rice at room temperature simultaneously until it reaches the equilibrium moisture content. Amylose solubility, change of viscosity followed by swelling of rice flour in water occur usually. Slurry made of parboiled-rice flour at ambient temperature shows dissimilar viscous consistency as compared to raw-rice flour slurry (Biswas and Juliano, 1988; Biliaderis et al., 1992, 1993; Unnikrishnan and Bhattacharya, 1981, 1983). An earlier study reported that nevertheless parboiled rice exhibited poorer viscosity in the Brabender viscograph at cooking temperatures (Ali and Bhattacharya, 1980; Biswas and Juliano, 1988). A similar pattern was seen in gelatinization with alkali at room temperature, which exhibited superior sediment or gel volume (Bhattacharya and Ali, 1976; Pillaiyar, 1984a, 1984b, 1985) in parboiled rice compared to raw rice. Gel mobility made with dilute alkali gel was higher in parboiled than raw rice (Unnikrishnan and Bhattacharya, 1988; Igathinathane, 2005). Chinnaswamy and Bhattacharya (1986) arranged various sets of a wide range of moisture followed by steaming at diverse pressures and then conduction heating using hot sand parboiled rice samples. The lower temperature hydratability, gelatinzing at equilibrium moisture, and elevated ($\geq$ 30%) moisture in paddy consistently resulted in comparatively lower EMC-S, signifying large number of starch molecule reassociation (retrogradation). Conversely, dry-heat parboiled rice, at low final moisture, produced an elevated EMC-S. It was found that before soaking if the parboiled rice is heated the MCs increase and if wetted and tempered the EMCs decreases due to starch modification during parboiling. After parboiling birefringence of raw rice and A-type XRD patterns were lost in parboiled rice (Raghavendra Rao and Juliano, 1970). This was apparently because of some sort of reassociation or re crystallization of starch, but no evidence was found to elucidate the cooking (and altered viscosity, etc.) retardation. XRD pattern for parboiled rice was demonstrated with the emergence of V-type XRD pattern, which indicates development of helical amylose made of lipid–amylose complex, whereas B pattern due to retrogradation was not formed (Charbonnier, 1975; Priestley, 1976). Priestley (1976) suggested that lipid–amylose complex was formed but starch retrograded product was not found, which resulted in delayed heat treatment with firm texture of parboiled rice. Mahanta et al. (1989) made traditional, dry-heat, and "pressure" parboiled rice and both XRD and DSC were done. The results showed that V pattern was absent but several weak B and A patterns appeared. According to Biliaderis and Galloway (1989) two types of lipid–amylose complexes are produced (L-Am I and L-Am II). L-Am I was produced at lesser temperature and is amorphous in nature with lost birefringence, and detached at about 90–100°C. L-Am II was crystalline and at approximately 110–120°C melting occurred (Ong et al, 1995). Another study by Mestres et al. (1988) observed the incidence or deficiency of starch that is retrograded. It was established that if complexed amylopectin is present, B-type endotherm was absent (at high moisture) or explained as fragile and inequitable A and B- or B and V- or A+ and V- type. Irrespective of rate it melted at approximately 55°C, but posed little problem during cooking. Ultimately, under severe circumstances of heating, retrograded amylose endotherm arises and melts at $\geq$140°C (Biliaderis et al., 1993). The maximum moisture percentages during parboiling reach 35–40% during steaming, usual crystallite liquefying occurs approximately at 70°C, and some unmelted annealed starch left behind melts around 80–110°C. Several residual A X-ray patterns have been observed that cause cooking resistance. If moisture amount allowed, when the parboiled rice subsequently was dampened and might or might not show a clear B-type starch crystallinity dotherm but considered amylopectin would definitely be present. Nevertheless as it melts at about 55°C, it causes no difficulty in cooking but only affects the hydration during soaking at reduced temperature and lower viscosity. Ultimately L-Am I and/or L-Am II of lipid–amylose complex formation depends on two factors: moisture content and temperature. L-Am I provides mild resistance to cooking whereas L-Am II provides severe cooking resistance. Parboiling at high steam pressure of 110–130°C in traditional or "pressure"parboiling arrangement would result in with significant cooking resistance. Conflict is higher in parboiled paddy made with "pressure steaming" process as pre steaming paddy moisture is maintained too small. Derycke et al. (2005) confirmed this using two condemnable different types of brown rice of 12% and 24% amylose, respectively, after steaming at 112 and 121°C of presoaked brown rice. They effectively achieved the same results as above, and also confirmed using TR-WAXS to learn the thermal properties along with X-ray properties of the particular flours. Selected amounts of moisture (66%, 40%, and 25%) were chosen, and they found that A-type pattern noticeably affected diffraction melting temperature. Crystallinity index (CI) appeared to diminish at about 65–70°C and disappeared at 90 and 105°C for both samples of 66% and 40% moisture content. On the other hand, for 25% moisture content sample, the CI diminished at extremely slow rate and vanished at 140°C. Thus, it was confirmed that starch is greatly influenced by the amount of moisture in the sample. From thermograms of DSC one endotherm was present at 65% moisture content, whereas two endotherms were found at the level 40% but at 25% moisture level no peak was found. L-Am II pattern was not found at 66% moisture sample, but at 40% and 25% moisture similar endotherms existed at 100–130°C. Parboiled rice samples exhibited A, B, and V patterns from static XRD study. Manful et al. (2008) observed the consequences of soaking and steaming parameters on X-ray and DSC patterns of parboiled rice samples and concluded that X-ray pattern slowly diminished whereas V pattern was seen and weakly visible (Figure 14.2).

14.4.2 Recent Research

Liu et al. (2020) scrutinized effect on the pasting, color, configuration, and rheology of germinated brown rice (GBR) after mild-parboiling treatment (MPT). The results were also matched with white rice (WR). They confirmed that several cracks emerged on the outer surface of the cross-section of GBR, temporarily, and the area of cross-section adjacent to the surface turned dense and even smooth post MPT. Partial gelatinization of GBR happened at some point in MPT, as confirmed by the reduction of comparative crystallinity from 27.51% to 16.87%. They also found conversion from A-type crystalline to jumble of A-type and V-type crystalline structure. Significant ($p < 0.05$) dark color as well higher total color (ΔE) of GBR was found following MPT. Apart from breakdown, other pasting restrictions of GBR enhanced considerably ($p < 0.05$), similar or even greater than those of WR after MPT. WR flours, GBR flours, and MP-GBR flours showed non-Newtonian, visco-elastic, gel-like, shear-thinning, and thixotropic behavior. Furthermore, it was found that dynamic modulus like elastic (G'), viscous (G'') modulus, loss tangent ($\tan \theta$), steady shear viscosity (η), and hysteresis loop area of GBR were enhanced similarly or somewhat higher than those of WR while the shear stress (τ) of GBR diminished similarly as WR after MPT (Iqbal et al., 2021).

14.5 IMPACT OF PROCESSING TREATMENT ON BIOLOGICAL CHARACTERISTICS (BIOACTIVE PROFILE, ANTI-NUTRITIONAL FACTOR, NUTRIENT DIGESTIBILITY, ETC)

14.5.1 Nutrition

Vitamins B and other minor ingredients provide greater nutritional properties to polished parboiled rice. Thus, many countries globally give importance to parboiled rice. Several earlier reports confirmed enhancement of some Bvitamins like thiamin and nicotinic acid were retained in parboiled rice as compared to polished raw rice (www.aaccipublications.aaccnet.org). Due to leaching in soaking water initial insignificant loss of thiamin results but could be higher after husk is split open during soaking. Followed by part of the vitamin was damaged throughout steaming due to high heat but it is insignificant under ordinary circumstances, but it is significant when done with pressure. Similarly nicotinic acid loss may occur but can be considered systematically. The higher vitamin content in polished parboiled rice occurs merely as a negligible fraction of the remaining vitamin which was separated (with bran) at the time of polishing than the raw rice as maximum amount during soaking transfer to endosperm in parboiled rice. According to Bolling and El Baya

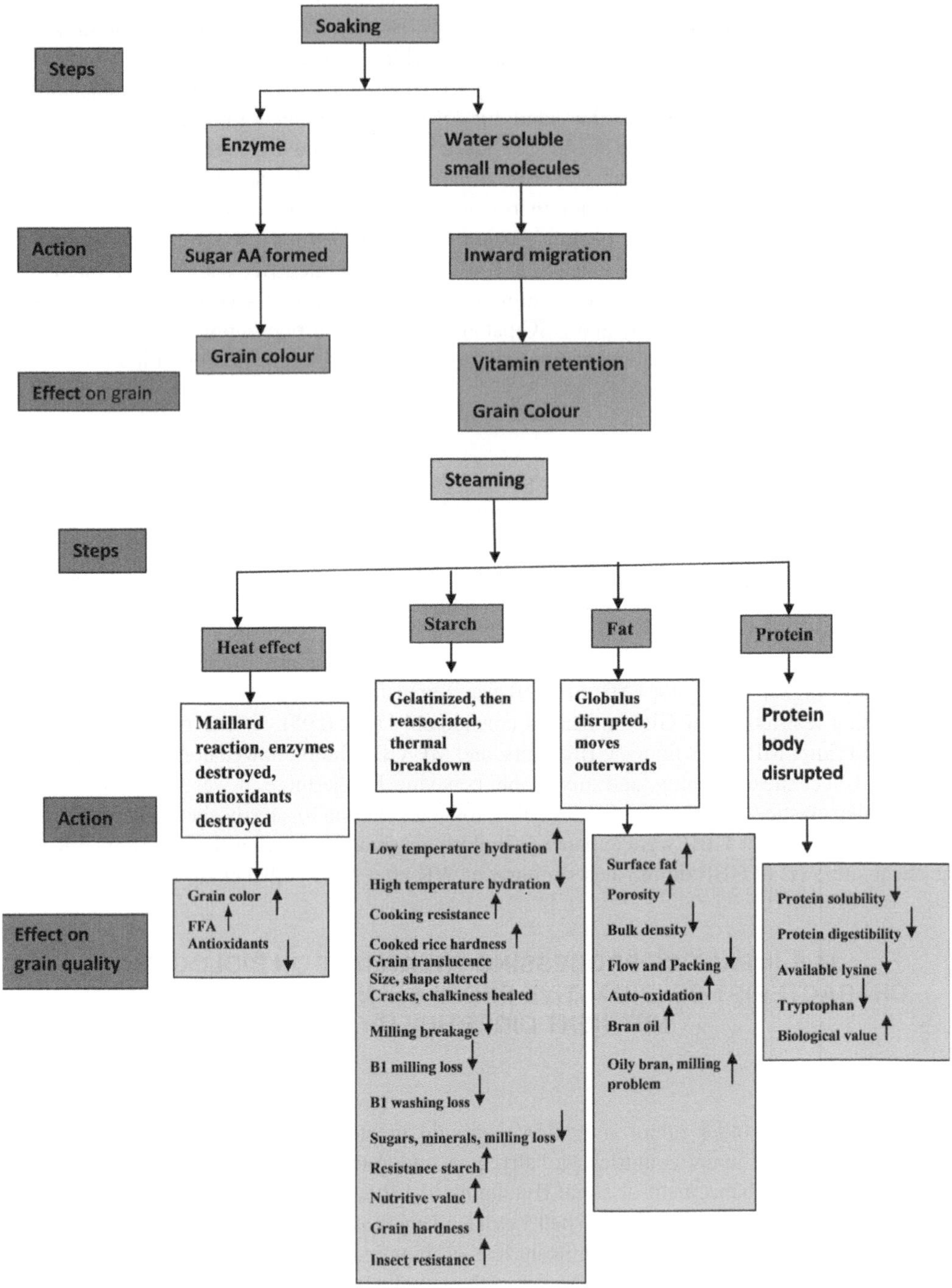

Figure 14.2 Changes occurring in the rice grain during parboiling and their effect on the product quality. FFA is free fatty acid and AA amino acid. ↓ indicates decreasing and ↑ indicates increasing.

(1975) and Ocker et al. (1976) the parboiled milled kernel almost certainly has a higher quantity of B vitamins (Kik, 1955), except for riboflavin. Thus, parboiling causes reduction of loss of vitamins during milling, and it is suitable to note the comparable conditions concerning sugars, amino acids, and minerals. Another report revealed that polished parboiled rice retains higher sugars than polished raw rice (Williams and Bevenue, 1953). According to Ali and Bhattacharya (1980) due to its transpiration, the sugar content was also an active end product. Significant enzymatic transformation of both non-reducing to reducing sugars, further generation of sugars happen; as a considerable portion of sugars was leached out into soaking water and also a slight decline during soaking at hot water because of Maillard reaction. Nonetheless, the key features were significantly reduced sugar loss during polishing of parboiled rice compared to that of raw rice. The degree of enzymatic manufacturing and discharging of sugars varies according to the dispensation circumstances, and the final sugar concentrations in milled rice may differ extensively.

The compositions of minerals of parboiled rice have been evaluated by many authors. It has been demonstrated that there are small transformations as a result of all parboiling methods, but the substance of various minerals (P, Ca, Fe, Mn, Mb, and Cr) are superior in milled parboiled rice in comparison to milled raw rice (Bhattacharya and Ali, 1985). In other words, the milling loss of constituents like B vitams and sugars was prevented after parboiling. But the amount of other minerals (Mg, Zn, Cu) were not influenced after milling of parboiled rice. Similar phenomenon was imperative for riboflavin. Yang and Cho (1995) studied again to confirm the consequences of parboiling on chemical composition of polished parboiled rice. The mechanism of diminished loss of components all through polishing was proposed to correlate lower damage of water-soluble nutrients (sugars, amino acids, B vitamin, mineral) during polishing of brown rice after parboiling treatment. It can be seen that water-soluble components can diffuse into the endosperm at the time of soaking during parboiling. Only a few soluble components have been found to percolate out into the soaking water throughout soaking (Ramalingam and Anthoni Raj, 1996), including thiamine if husk is split open in soaked water. According to Subba Rao and Bhattacharya (1966) retention of thiamine in rice postpolishing was consistently lower in all samples that were only soaked but not steamed, irrespective of the soaking period. If interior movement of B1 all through soaking was the cause of reduced polishing loss in parboiled rice than polishing followed by soaking only, particularly long-lasting soaking, should also have higher maintenance of thiamine in the polished rice. Conversely, the milling retention of vitamin B1 enhanced sharply and quickly and was eventually similar to soaked paddy when steamed for a few minutes. Generally stability increased and thus high retention was seen when starch was gelatinized. Correspondingly retention of vitamin B1 after polishing was increased in postsoaking without steaming. If elevated temperatures at ≥70°C soaking was carried out eventually gelatinization occured. The trends were indistinguishable in case of polishing loss of the constituents like sugars and amino acids (Ali and Bhattacharya, 1980). On the contrary, it was found that gelatinization was the probable reason for the larger postpolishing preservation of the kernels. They concluded that the lowering of polishing loss of the minute aquous media-soluble molecules was not based principally on the inner movement at the time of soaking but chiefly by their entrapment because of holding of the gelatinized starchy endosperm at the time of steaming.

Polished parboiled rice contains lesser amounts of B vitamins, but reduction of the remaining vitamins throughout storing period of rice or at the time of washing before cooking was also lower during parboiled rice cooking (Bhattacharya and Ali, 1985). It also retained minerals, particularly Ca, P, Fe, of larger quantity. The mineral retention enhanced its nutrititional value. Protein content was unchanged in parboiled polished rice. It was reported previously that the nutritional value of the protein was better. Though protein digestibility was diminished. It was found that better biological value was obtained (Eggum, 1979). Consequently the total compatible protein became unchanged. Conversely Benedito de Barber et al. (1977) established that parboiling causes reduction of lysine in small to significant decline in amino acids like tryptophan and methionine among existing amino acids. Lamberts et al. (2008) found epsilon-amino group present in protein-bound lysine was more unstable than open lysine. Tetens et al. (1997) recommended due to the presence of the complexed polymorphs of starch in parboiled rice that inhibited further the resistant starch.

14.5.2 Effect of Soaking Time on Starch Gelatinization

Miah et al. (2002) studied outcome of soaking duration in hot water to the extent of starch gelatinization. They explored the influence of level of gelatinization on the polishing qualities by altering the soaking duration in hot water throughout soaking in parboiling and to create the correlation among these parameters.

Influence of soaking on gelatinization enthalpy and the midpoint temperature T_m can be seen from the explanatory DSC graph of raw and parboiled rice flour samples at 30% concentration at 10° C min^{-1} heating rate. All the samples generated one endotherm on temperature increase. The heat scan for all the samples, representing dissimilar extents of soaking in hot water, were practically indistinguishable in shape, along with alteration due to heating of raw rice (considering fully ungelatinized) happening at about 50° C prior to other treatments. Nonetheless, the peak heights slowly reduced by means of enhancement of soaking duration for raw rice to 120 min soaking duration for parboiled rice. The maximum degree of gelatinization enthalpies were 6.03 J g^{-1} and 0.86 J g^{-1} for raw and parboiled rice flour, respectively, representing that 14% of the starch lingered ungelatinized after soaking for 120 min at scanning rate 1 °C min^{-1}. Using 30% concentration of rice flour, the midpoint transition temperatures (T_m), extent of starch gelatinization, and gelatinization enthalpies (ΔH) for diverse hot soaking handling were determined. Parboiled rice flour generated T_m values in the range 83–84.2 °C, characterized by linear adequacy of fitting insignificant slope of (+) 0.0099. Whereas raw rice of the same variety had a substantially lesser T_m of »78.5 °C. Normand & Marshall (1989) reported comparable T_m (75.9 ° C) was checked for rice variety of identical amylose content (25%) (ifst.pericles-prod.literatumonline.com). Normand & Marshall (1989) reported gelatinization enthalpy, ΔH of 10.3 J g^{-1}, which was larger than the raw rice though the age of rice postharvesting was not described. However, great influences were observed on enthalpy with improvement in gelatinization enthalpy with rising age.

Rice samples soaked at 80 °C were scanned in the DSC for 15, 30, 45, 60, and 120 min, and it was found that 86% higher gelatinization values were reached when soaked for120 min. At the time of inequitable gelatinization it was found that re-orientated subtle amylose molecules enclosed the previously swollen granules of amylopectin (water diffusion), consequently causing retrogradation to take place during cooling as well as during storage. This technique transformed unique arrangement of the unbroken particle afterwards, on reheating; this amylose structure exterior of the granules hinders subsequent swelling of starch in water at temperature where unparboiled rice starch swelled. Biliaderis et al. (1986) proposed a method of incomplete melting then again crystallization during cooling and following storage. This elucidates the late starting of gelatinization (T_m) in the case of all the four parboiled rice samples.

14.5.3 Effect on Heat Transfer

Heating rates are dependent on gelatinization temperature and gelatinization enthalpy. Transition-midpoint (t_m) for gelatinization temperatures and enthalpies both were extrapolated to zero scan rate to eradicate the influence of heat transfer. Modulation or alteration for the raw as well as for parboiled rice, melting temperature t_m at 30% w/w concentration diminish slowly with reducing heating rate, with considerable deviation of between them in similar scan rates. As designated in the DSC plot, parboiled rice generated high t_m at 4 °C when scan rate was zero, and t_m» 73.4 and 77.1 °C for raw and parboiled rice, respectively, when soaking was done in hot water for 15 min. The consequent enthalpy conversion was also enhanced during reduction of heating rate for both, as confirmed in a previous report by Normand & Marshal (1989). There were considerable differences found in scan rates of raw rice, which showed 1.75 times larger slope than parboiled rice, giving extrapolated enthalpy values of 8.16 and 5.76 J g^{-1} for raw and parboiled rice respectively, at zero scan rates. Raw rice produced a large slope, which was obviously due to discharge of amylose succeeding enlargement of perturbed amylopectin, which is dependent on heating rate. Starch concentrations also affect gelatinization enthalpy and t_m. Starch-to-water ratio plays an imperative role in performance of gelatinization. Though initiation of gelatinization occurs at similar temperature it was concluded that t_m enhances with rising concentration of starch. This performance confirmed the results of BeMiller (1993). Gelatinization enthalpy reduces with rising ratio of starch and water in linear mode. Possible justification is through endothermic convention, smaller amount of obtainable starch particle along with additional prediction to soak up heat through the diffusion technique. On the contrary, at small level of starch, amylose leaching takes place. Extra liberally allowed diffusion of water to amylopectin resulted automatically in high consumption of heat (eventually, high enthalpy). At higher concentration, conversely, starch granules cannot take up heat because of the overcapacity of grains, consequently leading to a poorer enthalpy value. The general reason is that with improvement of soaking period during hot water soaking throughout parboiling, single grains gelatinized further followed by retrogradation, where the grains changed to resistant starch, generating a linear slope of »0.6. Extent of soaking, gelatinization, and proportion of transparent kernels in raw and parboiled rice were determined. During starch gelatinization, if moisture is more than 57% subsequent to 45 min of soaking, opaque kernel lose their cloudiness changed to translucent kernels. The maximum 26% of transparent kernel was achieved from parboiled paddy after soaking for 120 min; at 45-min soaking nil transparent grain was found. Marshall et al. (1993) found similar results, and additionally described the reason for pasting properties of starch slurry as dense on incomplete gelatinization but due to extra heat processing it transformed to clear consistency until complete gelatinization (Eliasson, 1986).

14.5.4 Influence of Gelatinization on Milled Rice Color

Parboiled rice color is based on consumer preference and market demand of the product. Quick cooling of the paddy after steaming is essential for color as lower rate of cooling would generate rice of darker color (Bhattacharya & Subba Rao, 1966b). Hot soaking period has a distinct effect on color improvement. The color-inducing effect is principally due to non-enzymatic browning due to Maillard reaction. Sugar liberated all through the parboiling interacts with amino acids present in the grain, resulting in development of darker color in polished rice. The pigments present in bran and husk diffuse into the endosperm supply during soaking (Patil et al., 1982). Correlation between extent of gelatinization and the color of polished rice has been established. During parboiling in diverse degrees of soaking (eventual gelatinization), the comparative darkness slowly enhances. It was found that 80 ° C for 45-min soaking condition was sufficient for suitable darkness, which is equivalent to L* value 34 where maximum value of L* is 36. This was very close to comparative darkness of L* value 35, obtained from parboiled rice available in the market manufactured by Sainsbury's company under the brand name Uncle Ben's.

Ezzudin (2021) estimated the influence of amylose percentage on enrichment of total phenolic content when butterfly pea extract was used during parboiling of three types (high amylose, low amylose, and waxy) of paddy. Taking parboiled and raw rice as positive and negative control, respectively, the rices were evaluated on the basis of parameters such as total phenolic content (TPC), total flavonoid content (TFC), antioxidant activities, physiochemical properties, histology, and microstructure. Due to parboiling processing antioxidant activities, TFC and TPC of all varieties of rice considerably decreased during processing but the author found that all the above mentioned properties were reinstated after addition of butterfly pea extract during parboiling treatment. Increase of the properties exhibited a powerful negative Pearson relationship (range of r is 0.839 to 0.996 at $p < 0.05$) with amylose amount. The microstructure exhibited noticeable porous structure in parboiled as well as fortified rice grains, and it was discovered that decrease in amylose amount signifying amylopectin plays a role in consumption of flavonoid complex. Phenolic compounds also produce complex structure because of gelatinization, as confirmed from peak in FTIR spectra in the range 3400–3200 cm^{-1} that exhibited dependence on amylose percentage. Cooking time is inversely proportional to the amylose amount, bulk density, and water absorption capacity, and cooked rice texture increased after parboiling treatment.

Kumar et al. (2022) evaluated microstructure, digestibility, rheology, starch hydrolysis, and particle size distribution of parboiled and raw rice throughout imitation of gastrointestinal digestion in an in vitro and semi-dynamic way. Some conflicting studies on the insignificant results of parboiling process on digestion of rice and glycemic index (GI) values have been performed. Kumar et al. (2021). found that though fat and moisture contents were not much affected, protein content was enhanced considerably from 4.04 to 5.26 g/100 g, postparboiling. The apparent viscosity and dynamic moduli for parboiled rice were larger than that of the raw rice before and during in vitro digestion. Microstructures were observed through optical microscope as well confocal microscope, which confirmed compactness of gel networks with a lower number of uneven granules or cluster in parboiled rice throughout digestion.

14.6 CONCLUSION

To determine milled parboiled rice quality, the extent of gelatinization in starch and the physicochemical and biological properties must be determined. It has been established that the extent of gelatinization is enhanced with increasing hot soaking period during parboiling. Also, the heat flow during endothermic heating of parboiled rice starch (measured by DSC) is influenced by soaking time during parboiling. Gelatinization enthalpy values are dependent on heating rate and starch quantities in rice and are inversely proportional with gelatinization enthalpy. The gelatinization during parboiling changes the strength of the grain and thus the yield of rice during milling . The essential factor for end user acceptance is organoleptic properties such as overall appearance, especially opacity or translucency, which was significantly enhanced after longer hot soaking periods producing the 57% or more grain gelatinization. The color, especially reduction of level of brightness or increase of darkness, another parameter for customer acceptance, also enhanced by lengthening the hot soaking period during parboiling. Ultimately, it may be concluded that minimum 45-min soaking duration at 80 °C after 10 min steaming time at 1 atmosphere extra pressure is needed to ensure all the essential virtues of rice. Therefore, rice millers should meet the above conditions during practice of parboiled rice to enhance milling quality and increase the yield and quick completion by diminishing soaking duration of the range 24–48 h to 45–120 min.

REFERENCES

Ahiduzzaman, M. and Sadrul Islam, A. K. M. (2009). Energy utilization and environmental aspects of rice processing industries in Bangladesh. *Energies*, 2(1), 134–149.

Ali, S. Z. and Bhattacharya, K. R. (1976). Comparative properties of beaten rice and parboiled rice. *Lebensm Wiss Technol*, 9, 11–13.

Ali, S. Z. and Bhattacharya, K. R. (1980). Changes in sugars and amino acids during parboiling of rice. *J Food Biochem*, 4, 169–179.

Ali, S. Z. and Bhattacharya, K. R. (1982). Studies on pressure parboiling of rice. *J Food Sci Technol*, 19, 236–242.

Anthoni Raj, S. and Singaravadivel, K. (1980). Influence of soaking and steaming on the loss of simpler constituents in paddy. *J. Food Sci. Technol*, 17, 141–143.

Anthoni Raj, S., Singaravadivel, K., and Subramaniyan, V. (1996). Excretion of chlorides during soaking of paddy. *J Food Sci Technol*, 33, 57–59.

Araullo, E. V., de Padua, D. B., Graham, M. (1985). *RICE postharvest technology*. Ottawa, Canada: International Development Research Centre.

Aykroyd, W. R., Krishnan, B. G., Passmore, R., and Sundararajan, A. R. (1940). The rice problem in India. *Indian Med Res Memoirs*, 32, 84.

Ballogou, V. Y., Sagbo, F. S., Soumanou, M. M., Manful, J. T., Toukourou, F., and Hounhouigan, J. D. (2013). Effect of processing method on physico-chemical and functional properties of two fonio (*Digitatia xillir*) landraces. *J Food Sci Technol*, 1–8.

BeMiller, J. N. (1993). Starch-based gums. In: *Industrial Gums: Polysaccharides and Their Derivatives*, 3rd edn (edited by Whistler, R.L. & BeMiller, J.N.). New York: Academic Press, 99–122.

Benedito de barber, C., Martinez, J. and Barber, S. (1977). Effects of parboiling processes on the chemical composition and nutritional characteristics of rice bran. In Barber, S. and Tortosa, E. (Eds) *Rice By-products Utilization*, Vol. IV. Valencia, Spain: Inst Agroquimica Technol Alimentos, 121–130.

Bhattacharya, K. R. (2004). Parboiling of rice. In Champagne, E. T. (Ed) *Rice Chemistry and Technology*, 3rd edn. St. Paul, MN: American Association of Cereal Chemists, 329–404.

Bhattacharya, K. R. (2011). *Chapter–8. Effect of parboiling on rice quality in Rice quality: A Guide to Rice Properties and Analysis*. Woodhead Publishing, 247–297.

Bhattacharya, K. R. (2013). Erratum: improvements in technology for parboiling rice (*Cereal Foods World* (2013) 58(1), 23–26). *Cereal Foods World.*, 58(2), 110.

Bhattacharya, K. R. and Ali, S. Z. (1985). Changes in rice during parboiling, and properties of parboiled rice. In Pomeranz, Y. (Ed.) *Advances in Cereal Science and Technology*, Vol. VII, St. Paul, MN: American Association of Cereal Chemists, 105–167.

Bhattacharya, K. R. and Subba Rao, P. V. (1966a). Processing conditions and milling yields in parboiling of rice. *J Agric Food Chem*, 14, 473–475.

Bhattacharya, K. R and Subba, R. P. V. (1966b). Effect of processing conditions on quality of parboiled rice. *J Agric Food Chem*, 14, 476–479.

Biliaderis, C. G. (1992). Structures and phase transitions of starch in food system. *Food Technol*, 46, 98–109, 145.

Biliaderis, C. G. and Galloway, G. (1989). Crystallization behavior of amylose-V complexes: structure–property relationships. *Carbohydr Res*, 189, 31–48.

Biliaderis, C. G., Page, C. M., Maurice, T. J. and Juliano, B. O. (1986). Thermal characterization of rice starches: a polymeric approach to phase transitions of granular starch. *J Agric Food Chem*, 34, 6–14.

Biliaderis, C. G., Tonogai, J. R., Perez, C. M. and Juliano, B.O. (1993). Thermophysical properties of milled rice starch as influenced by variety and parboiling method. *Cereal Chem*, 70, 512–516.

Billiris, M. A. and Siebenmorgen, T. J. (2014). Energy use and efficiency of rice-drying systems in commercial, cross-flow dryer measurements. *Am. Soc. Agric. Biol. Eng*, 30(2),12.

Biswas, S. K. and Juliano, B. O. (1988). Laboratory parboiling procedures and properties of parboiled rice from varieties differing in starch properties. *Cereal Chem*, 65, 417–423.

Bolling, H. and El Baya, A. W. (1975). Einfluss von Parboiling auf die physikalischen und chemischen Eigenschaften des Reises (Effect of parboiling on the physical and chemical properties of rice). *Getreide Mehl*, 29, 230–233.

Charbonnier, R. (1975). Structure physique de l'amidon du riz natif, étuvé, précuit, appertisé, retrogradé (Physical structure of native, parboiled, precooked, canned, and retrograded rice starch). *Bull Inf Rizicul France*, 160, 14–21.

Chinnaswamy, R. and Bhattacharya, K. R. (1986). Pressure-parboiled rice: a new base for making expanded rice. *J Food Sci Technol*, 23, 14–19.

da Fonseca, F. A., Soares, J. M. S., Caliari, M., Bassinello, P. Z., da Costa, E. E., and Garcia, D.M. (2011). Changes occurring during the parboiling of upland rice and in the maceration water at different temperatures and soaking times. *Int J Food Sci Technol* 46(9), 1912–1920.

Derycke,V., Vandeputte, G. E., Vermeylen, R., de man, W., Goderis, B., Koch, M. H. J. and delcour, J. (2005). Starch gelatinization and amylose-lipid interactions during rice parboiling investigated by temperature resolved wide angle X-ray scattering and differential scanning calorimetry. *J Cereal Sci*, 42, 334–343.

Eggum, B. O. (1979). The nutritional value of rice in comparison with other cereals. In *Chemical Aspects of Rice Grain Quality*. Los Baños, Laguna, Philippines: International Rice Research Institute, 91–111.

Eliasson, A. C. (1986). Viscoelastic behaviour during the gelatinization of starch. *J Texture Stud*, 17, 253–265.

Ezzudin, M. R. (2021). Influence of amylose content on phenolics fortification of different rice varieties with butterfly pea (*Clitoriaternatea*) flower extract through parboiling. *LWT Food Science and Technology*, 147, 111493.

IEO (International Energy Outlook). (2013). In *Administration USEI*, editor. U.S. Department of Energy Washington, DC 20585: U.S. Department; 2013. p. 300.

Igathinathane, C., Chattopadhyay, P.K., Pordesimo, L.O. (2005). Combination soaking procedure for rough rice parboiling. *Trans. Am. Soc. Agric. Eng*. 48(2), 665–671.

Iqbal, S., Zhang, P., Peng Wu, P., Ge, A., Ge, F., Deng, R., and Chen, X.D. (2021). Evolutions of rheology, microstructure and digestibility of parboiled rice during simulated semi-dynamic gastrointestinal digestion. *LWT Food Science and Technology*, 148, 11700.

Kapur, T., Kandpal, T. C., and Garg, H. P. (1996). Solar energy paddy parboiling India: financial Feasibility. Anal, 21(10), 931–937.

Kapur, T., Kandpal, T.C., and Garg, H.P. (1997). Rice processing in India: a generalized framework for energy demand estimation. *Int J Energy Res*, 21(4), 309–325.

Khan, AU, AmilhussinA, Arboleda, J.R., Manalo, A.S., and Chancellor, W.J. (1974). Accelerated drying of rice using heat-conduction media.; 17(5)

Kik, M. C. (1955). Influence of processing on nutritive value of milled rice. *J. Agric. Food Chem*., 3, 600–603.

Kimura, T. (1983). Properties of parboiled rice produced from Japanese paddy. *AMA, Agricultural Mechanization in Asia, Africa and Latin America*, 14(2): 31–33.

Kumar, A., Lal, M.K., Sarangadhar Nayak, S., Sahoo, U., Behera, A., Bagchi, T.B., Parameswaran, C., Swain, P., Sharma, S. (2022). Effect of parboiling on starch digestibility and mineral bioavailability in rice (Oryza sativa L.), LWT156, 113026. https://doi.org/10.1016/j.lwt.2021.113026.

Kwofie, E.M., Ngadi, M., Mainoo, A. (2016a). Local rice parboiling and its energy dynamics in Ghana. *Energy Sustain Dev*. 34:10–9.

KwofieEM, NgadiM. (2016b). Sustainable energy supply for local rice parboiling in West Africa: the potential of rice husk. *Renew Sustain Energy Rev*, 56,1409–1418.

Kwofie, E.M., Ngadi, M. (2017). A review of rice parboiling systems, energy supply, and consumption, Renewable and Sustainable Energy Reviews.

Lamberts, L., Rombouts, I., Brijs, K., Gebruers, K. and Delcour, J. (2008). Impact of parboiling conditions on Maillard precursors and indicators in long-grain rice cultivars. *Food Chem*, 110, 916–922.

Liu, Q., Kong, Q., Li, X., Lin, J., Chen, H., QingbinB., Yuan, Y. (2020). Effect of mild-parboiling treatment on the structure, colour, pasting properties and rheology properties of germinated brown rice. *LWT–Food Science and Technology*. 130, 109623

Luh, B., Mickus, R. (1991). Parboiled rice. In: Luh, B, editor. *Rice*. US: Springer;. p. 470–507.

Mahanta, C. L., Ali, S. Z., Bhattacharya, K. R. and Mukherjee, P. S. (1989). Nature of starch crystallinity in parboiled rice. *Starch/Stärke*, 41, 171–176

Manful, J. T., Grimm, C. C., Gayin, J., and Coker, R. D. (2008). Effect of variable parboiling on crystallinity of rice samples. *Cereal Chem*, 85, 92–95.

Marshall, W.E., Wadsworth, J.I., Verma, L.R. &Velupillai, L. (1993). Determining the degree of gelatinisation in parboiled rice: comparison of a subjective and objective method. *Cereal Chemistry*, 70, 226–230.

Martins, G.M.V., SeverinaD.B., Maria, E.M.D, M′ario, Eduardo, ERMC, Hugo, M.L.O. (2021). Modeling the combinatory effects of parboiling and cooking on red paddy rice (*Oryza sativa* L.) properties. *LWT–Food Science and Technology*, 147, 111607

Meresa, A., Ayalew, D., Seifu,Y., Getu, T., and Kiber, T. (2020). Effect of Parboiling Conditions on Physical and Cooking Quality of Selected Rice Varieties. *International Journal of Food Science*. Volume, Article ID 8810553, 9 pages https://doi.org/10.1155/2020/8810553

Mestres, C., Colonna, P., and Buleon, A. (1988). Characteristics of starch networks within rice fl our noodles and mungbean starch vermicelli. *J Food Sci*, 53, 1809–1812.

Miah, M.A.K. (2002) Parboiling of rice. Part II: Effect of hot soaking time on the degree of 7 1% 8 1% 9 1% Exclude quotes On Exclude bibliography On Exclude matches < 1% starch gelatinization. *International Journal of Food Science and Technology*, 6/2002.

Miah, M.A.K., Anwarul, H., Paul, D. M. and Brian, C. (2002). Parboiling of rice. Part I: Effect of hot soaking time on quality of milled rice. *International Journal of Food Science and Technology*, 37, 527–537.

Ndindeng, S.A., Manful, J., Futakuchi, K., Mapiemfu-Lamare, D., Akoa-Etoa, J.M., Tang, E.N., et al. (2015). Upgrading the quality of Africa's rice: a novel artisanal parboiling technology for rice processors in sub-Saharan Africa. *Food Sci Nutr*.

Normand, F.L. and Marshall, W.E. (1989). Differential scanning calorimetry of whole grain milled rice and milled flour. *Cereal Chemistry*, 66, 317–320.

Ocker, H. D., Bolling, H., and El Baya, A. W. (1976). Effect of parboiling on some vitamins and minerals of rice: thiamine, ribofl avin, calcium, magnesium, manganese and phosphorus. *Riso*, 25, 79–82.

Oli, P., Ward, R., Adhikari, B., Torley, P. (2014). Parboiled rice: understanding from a materials science approach. *J Food Eng*, 124:173–83.

Ong, M. H. and Blanshard, M. V. (1995). The significance of starch polymorphism in commercially produced parboiled rice. *Starch/Stärke*, 47, 7–13.

Parnsakhorn, S. and Noomhorm, A. (2012). Effects of storage temperature on physical and chemical properties of brown rice, parboiled brown rice and parboiled paddy, *Thai J. Agric. Sci*., 45(4), 221–231.

Patil, R.T., Bal, S. and Mukherjee, R.K. (1982). Effect of parboiling treatments on rice quality. *J. Agric. Engg*. 19, 91–97.

Pillaiyar, P. (1984a). Applicability of the rapid gel test for indicating the texture of commercial parboiled rices. *Cereal Chem*, 61, 255–256.

Pillaiyar, P. (1984b). A rapid test to indicate the texture of parboiled rices without cooking. *J Texture Studies*, 15, 263–273.

Pillaiyar, P. (1985). A gel test to parboiled rice using dimethyl sulphoxide. *J Food Sci Technol*, 22, 1–3.

Pillaiyar, P., Sabarathinam, P.L., Subramaniyan, V., Sulochana, S. (1996). Parboiling of paddy using thermic fluid. *J Food Eng*, 27(3):267–78.

Priestley, R. J. (1976). Studies on parboiled rice. I. Comparison of the characteristics of raw and parboiled rice. *Food Chem*, 1, 5–14.

Priestley, R.J. (1976). Studies on parboiled rice: part 1—Comparison of the characteristics of raw and parboiled rice. *Food Chem*, 1(1):5–14.

Prom-u-thai, C., Rerkasem, B., Fukai, S., Huang, L. (2009). Iron fortification and parboiled rice quality: appearance, cooking quality and sensory attributes. DOI 10.1002/jsfa.3753

Raghavendra, R.S.N. and Juliano, B.O. (1970). Effect of parboiling on some physicochemical properties of rice. *J Agric Food Chem*, 18, 289–294.

Ramalingam, N., and Anthoni Raj, S. (1996). Studies on the soak water characteristics in various paddy parboiling methods. *Bioresource Technol*, 55, 259–261.

Roy, P., Shimizu, N., Shiina, T., Kimura, T. (2006). Energy consumption and cost analysis of local parboiling processes. *J Food Eng*; 76(4): 646–55

Subba Rao, P. V. and Bhattacharya, K. R. (1966). Effect of parboiling on thiamine content of rice. *J Agric Food Chem*, 14, 479–482.

Swasdisevi, T., Sriariyakula, W., Tia, W., Soponronnarit, S. (2010). Effect of pre-steaming on production of partially-parboiled rice using hot-air fluidization technique. *J Food Eng*, 96(3):455–62

Taleon, V., MdZakiul, H., Roelinda, J., Rita, W., and MdKhairul, B. (2021). Effect of parboiling conditions on zinc and iron retention in biofortified and non-biofortified milled rice. *Journal of the Science of Food and Agriculture*. DOI 10.1002/jsfa.11379

Tetens, I., Biswas, S. K., Glitso, L.V., Kabir, K. A., Thilsted, S. H. and Choudhury, N. H. (1997). Physico-chemical characteristics as indicators of starch availability from milled rice. *J Cereal Sci*, 26, 355–361.

Unnikrishnan, K. R. and Bhattacharya, K. R. (1981). Swelling and solubility behaviour of parboiled rice flour. *J Food Technol*, 16, 403–408.

Unnikrishnan, K. R. and Bhattacharya, K. R. (1983). Cold-slurry viscosity of processed rice flour. *J Texture Stud*, 14, 21–30.

Unnikrishnan, K. R. and Bhattacharya, K. R. (1988). Application of gel consistency test to parboiled rice. *J Food Sci Technol*, 25, 129–132.

Velupillai, L., Verma, L.R., Wratten, F.T. (1980). Study of parboiling techniques used in Sri Lanka. Paper–American Society of Agricultural Engineers.

Yang, M. O. and Cho, E. J. (1995). The effect of milling on the nutrients of raw and parboiled rices. *J Korean Soc Food Sci*, 11, 51–57.

Ying-dan, Z., Yong, W., Dong, L., Li-jun, W. (2021). The effect of dry heat parboiling processing on the short-range molecular order structure of highland barley. *LWT–Food Science and Technology,* 140, 110797.

Wang, Z., Zhang, M., Liu, G., Deng, Y., Zhang, Y., Tang, X., Li, P., Wei., Z. (2021). Effect of the degree of milling on the physicochemical properties, pasting properties and in vitro digestibility of Simiao rice. *Grain and Oil Science and Technology*, 4, 45–53.

Williams, K.T., Bevenue, A. (1953). A note on the sugars in rice. *Cereal Chem*, 30, 267–269.

Wu, Z., He,Y., Yan,W., Zhang, W. Liu, X. (2020). Ailing Hui,aHaiyanWanga and HonghongLia. Effect of high-pressure pre-soaking on texture and retrogradation properties of parboiled rice. DOI 10.1002/jsfa.11058

CHAPTER 15

Popping/Puffing of Cereals

Emi Grace Mary Gowshika R.
Women's Christian College, Chennai, Tamil Nadu, India

CONTENTS

DOI: 10.1201/9781003242192-19

15.1 INTRODUCTION

One of the most important and relished products in the food industry are snack items. The art of producing a snack is a challenging task, as it has to meet customer expectations and taste preferences, and at the same time should be innovative and unique to capture the market trend from already existing snacks. One of the important and age-old methods used to meet these targets and criteria are puffing and popping. It is considered one of the most cost-effective, simplest, and traditional forms of dry heat methods that is used to prepare ready-to-eat snacks, weaning foods for infants, etc. This technique has been used for hundreds of year, and is still being utilized by employing advanced technologies. The most widely used technique in the puffing industry is explosion puffing, which works by suddenly releasing pressure through water vapor expansion ideology (Sullivan and Craig, 1984).

The popularity of the products obtained through puffing or popping is due their its desirable aroma and acceptable taste. Cereal grains can be used as a pre-cooked, ready-to-eat snack food-base for some specialty foods or be developed into supplementary foods. A notable example is rice grain puffed and popped products like expanded parboiled rice (Hoke et al., 2005) or parboiled rice flour (Lai and Cheng, 2004). Many convenience snack foods like popped and puffed rice, popcorn, popped sorghum, puffed soybean, popped roasted wheat and other grains, and legumes can be prepared using this technology to produce popular Indian delicacies that are famous worldwide (Jaybhaye et al., 2014; Anderson, 1971).

There are many methods for puffing or popping different food products, but most include frying in hot oil, hot air, gun puffing, hot sand roasting, and microwave heating. The oil from frying is absorbed and can easily get rancid, whereas roasting has a risk of producing defects and burning. The sand roasting method uses wood chips or husk fired furnace and is based on heat through conduction (Hoke et al., 2005). This technique has harmful effects such as silica contamination and other environmental hazards. A better method in comparison to the other methods is fluidized bed air puffing through high temperature short time (HTST), which also provides better puffing quality as the product is more evenly exposed to the medium of heat (Brito-De La Fuente and Tovar, 1995)). Electromagnetic waves like microwaves are used at present to avoid these limitations by using conventional popping and puffing methods. This process provides better energy efficiency in very limited time. Microwave is the worldwide recognized method for producing popcorn.

Many varieties of cereals are popped or puffed, but only a few pop or puff well. This has been found to be influenced by various factors such as varietal differences; seasons; grain contents, which includes moisture content, bran content, and bran thickness; physical characteristics of the grain, which includes type of endosperm; and also the method of popping or puffing (Mirza et al., 2014; Joshi et al., 2014a). Rice with medium content of amylose has been found to produce better puffed quality when compared with rice having high content of amylose (Shuh and Tsai, 1995).

Morphology and composition of the kernel also determines the effect of puffing or popping of each cereal grain (Mariotti et al., 2006). Certain other factors having an influence on the quality of puffing or popping are surface tension, initial size of the micropore, yield stress, popping temperature, and rupture stress (Henry et al., 1995). Expansion volume is affected by kernel properties of each cereal grain including shape, size, and density (Joshi et al., 2014a; Hoke et al., 2005). Large kernels in corn have been found to produce lower-quality popcorn, whereas small corn kernels produce good-quality popcorn due to increased content of soft endosperm (Pajic and Babic, 1991). This chapter provides an overview of previously conducted studies relating to various methods of puffing and popping on cereal grains to understand the processing treatments and to study the impact of processing on nutritional, functional, physical, and biological characteristics of puffed or popped cereal grains.

15.2 PROCESSING TREATMENTS

15.2.1 Principle

Puffing involves heating the cereal grains in a pressure chamber between 7 to 14 kg/cm^2 pressure and then releasing this pressure instantaneously by opening the puffing gun or the chamber.. When the pressure is suddenly released, water vapor expansion occurs causing the cereal pellets or grains to blow up to multiple times their original size (e.g., 8 to 16 fold for wheat and 6 to 8 fold for rice). The moisture content in the final product is drastically reduced to 3% to achieve the desired crispiness. While processing wheat, an initial step to remove much of the bran from the grain is done before the puffing process (Cattaneo, et al., 2015). Whereas rice is parboiled and cooked with sugar syrup and is dried to a moisture content of 25 to 30% through rotating dryers and are toasted and puffed (Kent-Jonas and Singh, 2021).

15.2.2 Technology

In the process of puffing, after the expansion of the cereal grains, a part of the starch turns into sugar and dextrin. The material state and properties of the cereal grains are changed due to puffing and new products are produced. The major aim of puffing is to change the chemical properties from the raw cereal grain to the puffed finished product. The action of amylase is taken by the puffing technology and it causes the breakage of starch before the consumption and digestion of the food. This means that the puffing technology imitates a human body. This attribute improves the digestion of puffed products, which makes this technology ideal for food processing. Certain technologies in the past led to incomplete gelatinization of starch, which eventually causes improper shape, hard consistency, deteriorated taste, and lowered digestibility due to aging. This is completely avoided in puffing technology as complete gelatinization of starch occurs in which the alpha starch cannot reverse itself to beta starch after reaction, leading to softness, high digestibility, and good flavor far superior to other processing techniques.

15.2.3 Methods of Popping and Puffing

Various dry heat methods for popping and puffing cereal grains can be adopted, including salt roasting, sand roasting, hot oil frying, gun puffing, and usage of hot medium such as microwave radiation or hot air (Jaybhaye et al., 2014). Puffing in hot sand (250°C hot sand) or in oil (200° to 220°C) is common in India (Hoke et al., 2005).

15.2.3.1 Sand Roasting

Roasting is a dry heat method that is done in an oven or on a split and includes the addition of fat or oil. A combination of convection and radiation is used in oven roasting and radiant heat for a split (Metzger, et al. 1989). Cereals that are pre-gelatinized are exposed to hot sand at about 250°C. A sudden thermal gradient makes the moisture content in the grain vaporize and it escapes through the micropores, which leads to the expansion of the starchy endosperm (Chinnaswamy and Bhattacharya, 1983a).

15.2.3.2 Gun Puffing

In the process of gun puffing milled grains after pre-heating are introduced into a high pressure chamber or gun. A superheated steam is passed into the closed rotating chamber (Luh, 1991). The steam pressure determines the final quality and texture of the finished product; very low pressure results in less crispiness, whereas very high pressure shatters the grain. Adequate time is provided for the superheated steam to cook the cereal grain, and towards the end, the steam is suddenly released to get a crispy puffed grain (Mohammed, Hill and Mitchell, 2010).

15.2.3.3 HTST Fluidized Bed Popping and Puffing

Fluidization is the process in which heat is increased and mass transfer of the product happens, and the surface area is exposed uniformly to the medium of heat. The High-Temperature Short-Time (HTST) Fluidized Bed method has been shown to be more efficient than conduction or hot air roasting puffing processes (Zapotoczny, et al. 2006). The range of temperature is from 240°C to 270°C with an exposure time of 7 to 9.7 seconds for better color quality and higher expansion ratio (8.5 to 10) (Chandrasekhar and Chattopadhyay, 1989). In this process popping is not only influenced by the moisture content of the grain, but also by the moisture present in the heating media (Konishi et al., 2004).

15.2.3.4 Microwave Popping and Puffing

Microwave has been used to pop and puff cereal grains globally. The driving force in microwave expansion is the moisture content. Microwave makes the starchy grains expand by reducing moisture (Ernoult et al., 2002; Boischot et al., 2003). The moisture content and degree of gelatinization are the two major factors in determining the expansion bulk volume, shape, and density of popped and puffed microwave products (Lee et al., 2000). The energy of the microwave heats the cereal grains through a vibrational energy produced on moisture (Mohapatra and Das, 2011). The moisture in the grain produces superheated steam, which induces expansion and creates a local pressure (Dhumal, et al., 2014). After the moisture is lost from the matrix, and when the microwave heating halts, the matrix cools and sets the final structure. When the moisture content is very high, the matrix becomes soft and collapses. The major popping and puffing determinants in the microwave includes: microwave power density, microwave power level, and residence time (Maisont and Narkrugsa, 2010, ; Joshi et al., 2014b, Sweley et al., 2012; Sharma et al., 2014).

15.2.4 Some Puffed and Popped Cereal Grains

15.2.4.1 Puffed Rice

A very popular cereal breakfast consumed in many countries is puffed rice. Pre-gelatinized, milled rice processed through hydrothermal treatment, by heating in high temperature oil, air and

Paddy (rice) 14% mc
↓
Soak overnight in just boiled water (45% mc)
↓
Drying
↓
Parboiled Paddy
↓
Milling
↓
Milled Parboiled Rice
↓
Roast with fine sand (1:2) (240°- 250°C, 22 sec, 15% mc)
↓
Sieving
↓
Preheat in iron pan (80°C, 10.5% mc)
↓
Tempering (30 min)
↓
Mixed with salt solution (11% mc)
↓
Roast with sand (1:10) (240°- 250°C for 11-13 sec)
↓
Expanded rice (1-2% mc)
↓
Sieving
↓
Puffed Rice

Figure 15.1 Puffed Rice Processing Flow.

Source: Chinnaswamy and Bhattacharya, 1983b.

sand or gun puffing method. Puffed rice is an easily digestible, light, and ready-to-consume food (Bhat, Bhat and Salimath, 2008). Its major applications are as cereal drinks, snacks, infant foods, and ready-to-eat breakfast cereals (Chinnaswamy and Bhattacharya, 1983a). Rice kernel increases in volume multiple times during puffing and a fully heat-treated porous, crisp, and ready-to-eat cereal product is processed (Pardeshi and Chattopadhyay, 2014) (Figure 15.1)

15.2.4.2 Popcorn

This food product is made from maize (corn) and is a highly popular product worldwide (Farahnaky, Aliopour and Mazoobi, 2013). Different ingredients are used to coat popcorns such as hydrogenated oil, butter, salt, sugar syrup, and other flavors (Helm and Zuber, 1972; Dofing, Thomas-Compton and Buck, 1990). Factors such as moisture content, kernel size and shape, popping temperature, pericarp thickness, variety, kernel damage, and kernel density affect the final product quality (Mohamed et al., 1993; Richardson, 1959).

15.2.4.3 Popped Sorghum

A very traditionally consumed popular Central Indian snack is popped sorghum (Parker, et al., 1999). It is mostly mixed with spice and oil for taste and flavor or even sweetened according to

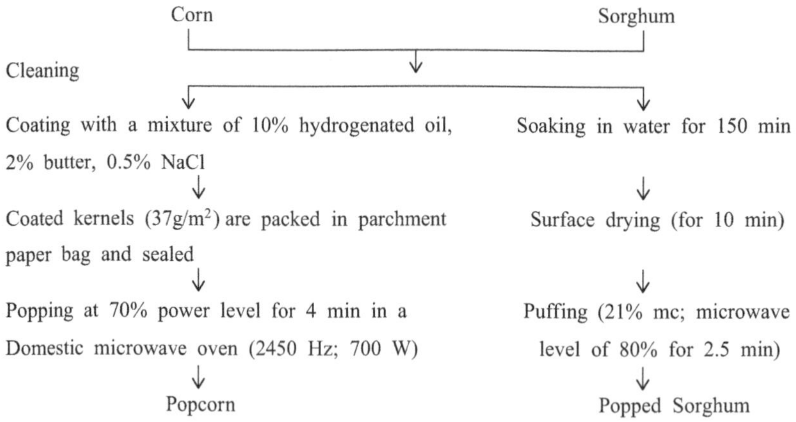

Figure 15.2 Popcorn and Popped Sorghum Processing Flowchart.

Source: Singh and Singh, 1999.

preference (Gundboudi, 2006). Its applications are seen as ready-to-eat food and as weaning food formulation (Thorat et al., 1988). Popping characteristics are affected by the varietal differences in sorghum grains (Haikerwal and Mathieson, 2010). Microwave and conventionally popped sorghum is affected by certain grain characteristics like protein content, amylose content, moisture content, and certain physical aspects like thousand grain weight, grain size, hardness, and bulk density (Thorat et al., 1988; Murthy et al., 1982; Gaul and Rayas-Duarte, 2008). Different methods of popping sorghum affect the popping quality of the popped product differently (Gupta, Ratna and Srivastava, 1995). Moist heat, dry heat, and usage of oil as heating medium provide high popping yield, whereas only 60% was the yield in microwave (Yenagi, 2005) (Figure 15.2).

15.2.4.4 Popped Finger Millet

Popped finger mullet is very rich in fiber content in the raw state. After popping of finger millet, its total dietary fiber, soluble dietary fiber, and insoluble dietary fiber are reduced (Kulkarni, 1990). The removal of seed coat and other characteristics such as degree of removal of seed coat, type of grain, and exclusion of shoots and roots of sprouts in the malted forms have an effect on the fiber content by reducing their levels in the popped finger millet (Burgos and Armada, 2015).

15.2.4.5 Puffed Proso Millet

Different varieties of proso millet have varied puffing quality. The variations and differences in puffing characteristics are attributed to moisture content, pressure under which puffing is performed, and other optimum conditions (Hadimani, 1994). Puffed expansion and yield is better when moisture contents are 15–18% and when gun puffed between 140 to 160 psi pressure. An increase in protein content and decrease in total dietary fiber and ash content was noted when puffed with 12% moisture content and gun puffed at 120 psi pressure.

15.2.4.6 Popped Quinoa

There are increased nutritional and health benefits when popped quinoa is consumed, leading to its importance among the people. It contains high-quality protein, calcium, dietary fiber, zinc, phosphorous, and iron contents. At 260°C, maximum puffing quality (87%) is noted with an expansion ratio (4.36) and 18.34% moisture content (Mandhare, More and Nagulwar, 2020).

15.3 IMPACT OF PROCESSING TREATMENT ON NUTRITIONAL CHARACTERISTICS

15.3.1 Starch

The major component of a grain is starch. The glycoside bonds and hydrogen bonds present between the starch molecules might be destroyed during puffing, and this in turn aids the process of gelatinization during explosion. After the explosion, a mixture of gelatinized starch, melted starch, and degraded starch are present. Starch can be broken down into much smaller molecules of glucose, maltose, etc., which could decrease the content of starch by increasing the sugar content.

15.3.2 Protein

The total protein content is not affected by the process of explosion puffing, but its solubility is decreased. This decrease was observed in the contents of globulin, glutenin, albumin, and gliadin. The sodium dodecyl sulfate (SDS) protein content was found to be increased in a few studies (Huang et al., 2018; Ummadi et al., 1995). During extrusion, protein aggregation led to a decrease in protein solubility. The process of shearing and heating in the expansion process can denature the protein and alter its structure. The non-polar groups inside the protein would thereby be exposed and would increase the formation of protein aggregates and the surface hydrophobicity, which leads to a decrease in protein solubility (Chavan, McKenzie and Shahidi, 2001).

15.3.3 Lipids

After the process of puffing, the fat content in the grains is significantly decreased. The amylose-fat complex appearance, which is formed with the combination of amylose and fat free, is attributed to the decrease in fat content. Under high pressure and temperature, the fat is decomposed into monoglycerides and fatty acids, which ultimately leads to the decreased fat content in puffed and popped cereal products.

15.3.4 Sugar

The sugar content is highly increased due to gelatinization of the starch, which breaks down these complex molecules into smaller sugar molecules like glucose, sucrose, etc. Sugar adds many beneficial characteristics to finished products such as flavor, binding, and browning. It also controls the mouthfeel and texture of puffed and popped grain products.

15.3.5 Micronutrients

The stability of various B vitamins like riboflavin, thiamin, and niacin during puffing and popping of various cereal grains have been studied. There is a loss of thiamine due to increased moisture content (<15%), long residence time (<1 min), and high barrel temperature (<153°C). The retention average of thiamine (54%) and riboflavin (92%) has also been observed. The effect of the stability of thiamine, riboflavin, pyridoxin, cyanocobalamin, and folic acid has been improved with an increase in moisture content. The Nutritional Composition of Grains before and after Puffing is given in Table 15.1.

Table 15.1 Nutritional Composition of Grains before and after Puffing

	Variety	Moisture	Protein	Fat	Starch	Total Sugar
Before Puffing	Millet	10.4%	10.5%	3.8%	74.3%	83.09%
	Barley	12.3%	12.4%	1.7%	70.4%	80.4%
	Black Rice	14.02%	8.2%	2.2%	78.07%	72.8%
	Rice	13.8%	6.8%	1.2%	75.4%	80.5%
	Glutinous Rice	13.9%	7.3%	1.6%	73.9%	70.2%
	Wheat	10.9%	11.1%	1.1%	67.2%	73.8%
After Puffing	Millet	6.3%	10.3%	3.5%	43.4%	88.09%
	Barley	8.9%	12.2%	1.2%	38.2%	85.7%
	Black Rice	7.1%	8.05%	1.6%	51.5%	81.08%
	Rice	8.5%	6.8%	0.9%	56.8%	86.6%
	Glutinous Rice	7.07%	7.01%	0.8%	55.1%	75.5%
	Wheat	7.7%	11.8%	0.5%	49.1%	79.9%

Source: (Huang et al., 2018).

15.4 IMPACT OF PROCESSING TREATMENT ON FUNCTIONAL CHARACTERISTICS

15.4.1 Flavor

The flavor or the volatile components of cereal grain after puffing or popping is drastically changed and may be reduced due to the sudden pressure and temperature changes involved in processing. Usually flavor is added to the puffed or popped product so as to increase its palatability and taste. Flavor is also affected due to the presence of starch and protein content, which gets altered during processing.

15.4.2 Color

Puffed and popped cereal grains have a lighter color based on certain factors including salt and moisture content, salt solution, moisture content, and microwave power. Redness of the product is rendered by moisture content, salt solution, salt and moisture content, and microwave power. The yellowness is based on moisture content and salt solution. It has been further observed that 10–13% of increase in moisture content leads to lightness in colors, and when 16–19% of moisture content is increased, no difference in color is observed. The browning reaction is increased due to longer heating time and lower moisture content, which eventually leads to the product appearing to be light, yellow, and red in puffed and popped cereal grain products (Brennan, 2006).

15.4.3 Puffed Yield

Under different processing conditions, the puffing ability of cereal grain products is measured. Puffed yield is affected by the moisture content, salt content, and microwave power. Cereal grains soaked in salt solution having higher moisture content have higher puffing quality when compared to soaking in water. It was found that cereal grains puffed at 800 watts and soaked in salt solution have increased puffing rates than those puffed under 600 and 700 watts (Maisont and Narkrugsa, 2010).

15.4.4 Expansion Volume

Degree of expansion is used to express the expansion volume of puffed and popped cereal grains. Expansion volume has been found to be affected by moisture content, salt solution, microwave power, and interaction between salt solution, microwave power, and moisture content of the products. Grains soaked in salt solution have more expansion volume when compared to ones soaked in water. The moisture content in grains is superheated into vapor, which is a driving factor for product expansion volume (Simsrisakul, 1991).

15.5 IMPACT OF PROCESSING TREATMENT ON PHYSICAL CHARACTERISTICS

15.5.1 Water Absorption Index (WAI)

The ability of cereal grain to bind itself to water and utilize the level of gelatinization of starch after swelling in excess water is referred to as water absorption index (WAI). The WAI for most cereal grains including white rice, black rice, glutinous rice, millet, and wheat almost double after puffing, and puffed barley WAI was 1.5 times high when compared to the raw product (Rampersad et al., 2003). The high amount of starch contains damaged polymer chains. This was formed at higher pressure and temperature and is the reason for high WAI . The factors known to affect WAI are amount of fiber, type of protein, and denaturation degree in the processed grains (Figure 15.3).

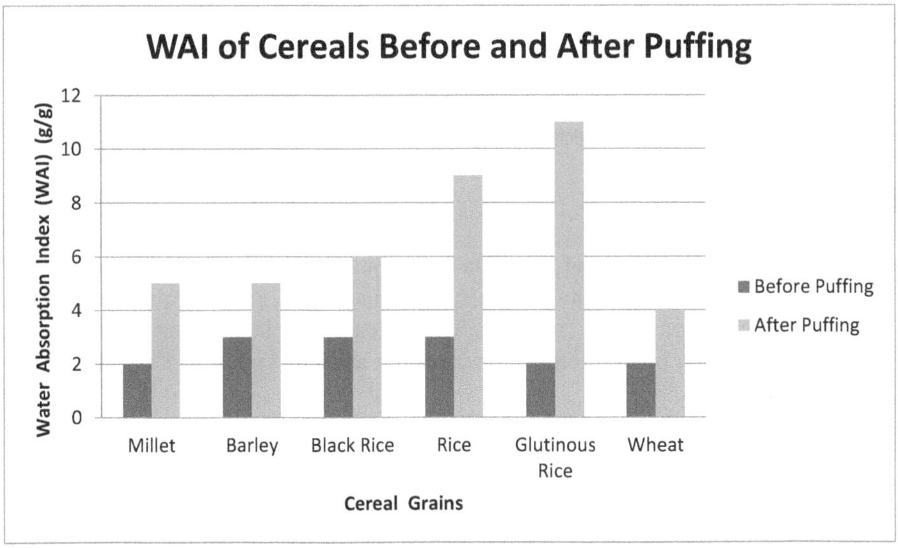

Figure 15.3 WAI of Cereals Before and After Puffing.

Source: Huang et al., 2018.

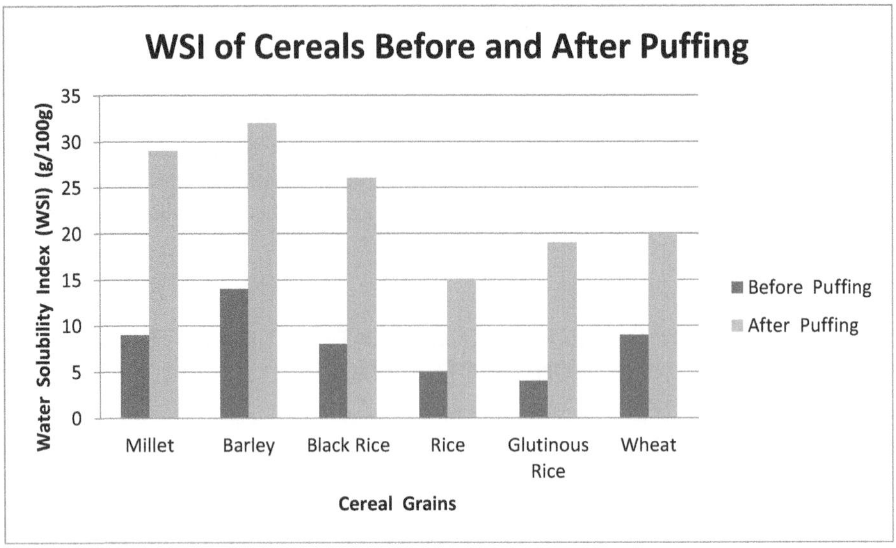

Figure 15.4 WSI of Cereals Before and After Puffing.

Source: Huang et al., 2018.

15.5.2 Water Solubility Index (WSI)

The degree of degradation of macromolecules is indicated by WSI. The amount of soluble polysaccharides obtained after puffing from starch macromolecules is also measured in this index. The degradation of starch in grains during puffing into various micromolecules like soluble polysaccharides and dextrin has been observed. The destruction of both tertiary and quaternary protein structures also increase WSI in grains (Guerrero et al., 2012) (Figure 15.4).

15.5.3 Gelatinization Degree

The changes in the morphology of starch causes gelatinization. Puffed grains have a considerable increase in degree of gelatinization. The structural changes, gelatinization, and degradation of starch are the main reasons for the increase. The increase in gelatinization degree enhances the hydrolysis of starch, which proves helpful for digestion and absorption of the food in the human body (Tang, et al., 2006).

15.6 IMPACT OF PROCESSING TREATMENT ON BIOLOGICAL CHARACTERISTICS

15.6.1 Starch and Protein Digestibility

The digestibility of protein increases mostly due to the opening of complex structures, which serve as a focus place for digestive enzymes (Van, Van and Boom, 2010). The stability of tertiary and quaternary protein structures maintained by a force drastically reduced when the grain was puffed or popped (Axtell, et al. 1981). The exposure of many hydrolytic sites due to irreversible denaturation of protein molecules further increases the opportunity to contact proteins by enzymes, which also results in improved protein digestibility (Rathod and Annapure, 2017). The puffing effects on starch have been investigated and it has been found that with the decrease of resistant starch and slowly digestible starch, rapidly digestible starch increases. This shows that starch digestibility is enhanced due to puffing. There is a change in the structure of the grain after puffing; it becomes porous and loose and the hindrance by steric is reduced, which in turn provides more area for hydrolyzation of starch grains by enzymes. Therefore, puffing and popping is an effective process to enhance the digestibility and absorption of starch in the human body (Roopa and Premavalli, 2008).

15.6.2 Phenolic Profile

After puffing of cereal grains such as quinoa and kiwicha, a notable quantitative and qualitative difference in phenolic compounds is observed. Total hydroxycinnamic acids are decreased in kiwicha by 22% and in quinoa by 35%. The negatively affected compounds are caffeoylquinic acid, feruloylquinic acid, and caffeic acid derivatives in kiwicha, whereas, in quinoa rans-p-coumaric acid is reduced. A significant decrease in total phenolic content has been observed in kiwicha due to puffing and a significant increase is seen in quinoa. Thermal processing further increases the bioavailability of phenolic compounds.

15.6.3 Anti-nutrients

It has been observed that popped cereals have less anti-nutrient content compared with raw cereal grains. The outer layers (bran) of the cereals are the residing places for most of the anti-nutritional factors. This is reduced in processing like popping and puffing as the outer layers of the cereal grains are removed. Among the anti-nutrients present, in millet varieties, phytic acid and tannin content was on the higher side in little and finger millet than in pearl millet. It was also inferred that heat also reduces the anti-nutrient content in the popped and puffed cereals (Ai, et al., 2016).

15.7 CONCLUSION

Puffing and popping processing of cereal grains have a high impact on both quality and nutritional contents when compared with raw cereal grains. Many cereals are now commonly popped or puffed due to its characteristics like lightness when consumed, increased macronutrients such as protein and fiber, and more micronutrient such as calcium, zinc, iron, phosphorous, and other trace minerals. Puffing and popping methods are used to create products like ready-to-eat breakfast cereals, infant formulation and weaning food formulation, snacks, cereal grains, etc. Certain puffing and popping qualities like moisture content, starch digestibility, protein digestibility, and other factors have been found to be increased making the puffed and popped grains palatable. This method is considered as one of the most traditional ways of processing cereal grains and is also preferred by people worldwide.

REFERENCES

Ai, Y., et al. (2016). Effects of Extrusion Cooking on the Chemical Composition and Functional Properties of Dry Common Bean Powders. *Food Chemistry*; 211: 538–545.

Anderson, W.T. (1971). Identifying the convenience-oriented consumer. *Journal of Marketing Research*, 8: 179–183.

Axtell, J.D., et al. (1981). Digestibility of sorghum proteins. *Proceedings of the National Academy of Sciences (USA)*; 78: 1333–1335.

Bhat Upadya, V.G., Bhat, R.S & Salimath, P.M. (2008). Physico-chemical characterization of popping-Special Rice Accessions. *Karnataka Journal of Agricultural Sciences*; 21(2): 184–186.

Boischot, C., Moraru, C.I & Kokini, J.L. (2003). Expansion of glassy amylopectin extrudates by microwave heating. *Cereal Chemistry*; 80(1): 56–61.

Brennan, J.G. (2006). *Food Processing Handbook*. Wiley-VCH Publishers, Weinheim, Germany.

Brito-De La Fuente, E & Tovar, L.R. (1995). Popping and cleaning of amaranth seeds in a fluidized bed. *Food and Bio-products Processing*; 73(4): 56–61.

Burgos, V.E & Armada, M. (2015). Characterization and Nutritional Value of Precooked Products of Kiwicha Grains (Amaranthus caudatus). *Food Science & Technology*; 35 (ahead).

Cattaneo, S., et al. (2015). Heat damage and in-vitro, starch digestibility of puffed wheat kernels. *Food Chemistry*; 188: 286–293.

Chandrasekhar, P.R & Chattopadhyay, P.K. (1989). Studies on micro-structural changes of parboiled and puffed rice. *Journal of Food Processing and Preservation*; 14: 27–37.

Chavan, U.D., McKenzie, D.B & Shahidi, F. (2001). Functional properties of Protein Isolates from Beach Pea (Lathyrus maritimus L.). *food Chemistry*; 74(2): 177–187.

Chinnaswamy, R & Bhattacharya, K.R. (1983a). studies on expanded rice: optimal processing condition. *Journal of Food Science*; 48: 1604–1608.

Chinnaswamy, R & Bhattacharya, K.R. (1983b). studies on expanded rice. Physic-chemical basis of varietal differences. *Journal of Food Science*; 48: 1600–1603.

Dhumal, C.V., et al. (2014). Optimization of process parameters for development of microwave puffed product. *Journal of Ready to Eat Food*; 1(3): 111–119.

Dofing, S.M., Thomas-Compton, M.A & Buck, J.S. (1990). Genotype-popping method interaction for expansion volume in popcorn. *Crop Science*; 30: 62–65.

Ernoult, V., Moraru, C.I & Kokini, J.L. (2002). Influence of fat on expansion of glassy amylopectin extrudes by microwave heating. *Cereal Chemistry*; 79(2): 265–273.

Farahnaky, A., Aliopour, M & Mazoobi, M. (2013). Popping properties of corn grains of two different varieties at different moistures. *Journal of Agricultural Science and Technology*; 15: 771–780.

Gaul, J.A & Rayas-Durate, P. (2008). Effect of moisture contents and tempering method on the functional and sensory properties of popped sorghum. *Cereal Chemistry*; 85(3): 344–380.

Guerrero, P., et al. (2012). Extrusion of Soy Protein with Gelatin and Sugars at Low Moisture Content. *Journal of Food Engineering*; 110(1): 53–59.

Gundboudi, Z.A. (2006). Nutritional and processing qualities of pop sorghum cultivars and value addition (Doctoral dissertation, University of Agricultural Sciences, Dharwad).

Gupta, A., Ratna, S.S & Srivastava, S. (1995). Effect of storage conditions on popping quality of sorghum. *Journal of Food Science and Technology*; 32(3): 211–213.

Hadimani, N.A. (1994). *Pearl millet (Pennisetum americannum) grain quality with special emphasis on milling and popping of millet*. Ph.D. Thesis, CFTRI, Mysore, India.

Haikerwal, M & Mathieson, A.R. (2010). Extraction and Fractionation of Proteins of Sorghum Kernels. *Journal of the Science of Food & Agriculture*; 22(3): 142–145.

Helm, J.L & Zuber, M.S. (1972). Inheritance of pericarp thickness in dent corn. *Crop Science*; 12: 428.

Henry, G., et al. (1995). Modeling deformation and flow during vapor-induced puffing. *Journal of Food Engineering*; 25(3): 357–362.

Huang, R., et al. (2018). Effects of Explosion puffing on nutritional composition and digestibility of grains. *International Journal of Food Properties*; 21(1): 2193–2204.

Jaybhaye, R.V., et al. (2014). Processing and technology for millet based food products: a review. *Journal of Ready to Eat Food*; 1(2): 32–48.

Joshi, N.D., Mohapatra, D & Joshi, D.C. (2014a). varietal selection of some indica rice for production of puffed rice. *Food and Bioprocess Technology*; 7(1): 299–305.

Joshi, N.D., Mohapatra, D & Joshi, D.C. (2014b). puffing characteristics of parboiled milled rice in a domestic convection-microwave oven and process optimization. *Food and Bioprocess Technology*; 7(6): 1678–1688.

Kent-Jones, D. W. and Singh, R. P. (2021, April 1). *Cereal Processing. Encyclopedia Britannica.* Retrieved from: www.britannica.com/technology/cereal-processing

Konishi, Y., et al. (2004). Effect of moisture content on the expansion volume of popped amaranth seeds by hot air and superheated steam using a fluidized bed system. *Bioscience, Biotechnology and Biochemistry*; 68(10): 2186–2189.

Kulkarni, L.R. (1990). *Chemical composition and protein quality evaluation of minor millets.* M.H.Sc. Thesis (Unpublished), University of Agricultural Science, Dharwad.

Lai, H & Cheng, H.H. (2004). Properties of pre-gelatinized rice flour made by hot air or gun puffing. *International Journal of Food Science and Technology*; 39: 201–212.

Lee, E., et al. (2000). Effect of gelatinization and moisture content of extruded starch pellets on morphology and physical properties of microwave-expanded products. *Cereal Chemistry*; 77(6): 769–773.

Luh, B.S. (1991). Rice utilization. Springer Books, 2nd edition.

Maisont, S & Narkrugsa, W. (2010). Effects of salt, moisture content and microwave power on puffing qualities of puffed rice. *Kasetsart Journal (Natural science)*; 44(2): 251–261.

Mandhare, L.L., More, D.R & Nagulwar, M.M. (2020). Effect of popping methods on popping characteristics of quinoa seed. *Journal of Pharmacognosy and Phytochemistry*; 9(5): 1943–1945.

Mariotti, M., et al. (2006). Effect of puffing on ultra-structure and physical characteristics of cereal grains and flours. *Journal of Cereal Science*; 43(1): 47–56.

Metzger, D.D., et al. (1989). Effect of moisture content on popcorn popping volume for oil and hot air popping. *Cereal Chemistry*; 66: 247–248.

Mirza, N., et al. (2014). Variation in popping quality related to Physical, Biochemical and Nutritional properties of Finger Millet geneotypes. *Proceedings of the National Academy of Sciences, India Section B: Biological Sciences*; 1–9.

Mohamed, A.A., Ashman, R.B & Kirleis, A.W. (1993). Pericarp thickness and other kernel physical characteristics relate to microwave popping quality of popcorn. *Journal of Food Science*; 58: 342–346.

Mohammed, Z.H., Hill, S.E & Mitchell, J.R. (2010). Covalent crosslinking in heated protein systems. *Journal of Food Science*; 65(2): 221–226.

Mohapatra, M & Das, S.K. (2011). Effect of process parameters and optimization on microwave puffing performance of rice. *Research Journal of Chemistry and Environment*; 15(2): 454–461.

Murty, D.S., et al. (1982). A note on screening the India sorghum collection for popping quality. *Journal of Food Science and Nutrition*; 19: 79–80.

Pajic, Z & Babic, M. (1991). Interrelation of popping volume and some agronomic characteristics in popcorn hybrids. *Genetica*; 23: 137–144.

Pardeshi, I.L & Chattopadhyay, P.K. (2014). Whirling bed hot air puffing kinetics of rice-soy ready-to-eat (RTE) snacks. *Journal of Ready to Eat Food*; 1(1): 1–10.

Parker, M.L., et al. (1999). Effects of popping on the endosperm cell walls of sorghum and maize. *Journal of Cereal Science*; 30(3): 209–216.

Rampersad, R., Badrie, N & Comissiong, E. (2003). Physic-chemical and sensory characteristics of flavored snacks from extruded Cassava/Pigeonpea flour. *Journal of Food Science*; 68(1): 363–367.

Rathod, R.P & Annapure, U.S. (2017). Physicochemical properties, protein and starch digestibility of lentil based noodle prepared by using Extrusion Processing. *LWT- Food Science and Technology*; 80: 121–130.

Richardson, D.L. (1959). Effect of certain endosperm genes on popping ability in popcorn. *Agronomy Journal*; 51: 631–635.

Roopa, S & Premavalli, K.S. (2008). Effect of Processing on Starch Fractions in different varieties of Finger Millet. *Food Chemistry*; 106(3): 875–882.

Sharma, V., Champawat, P.S & Mudgal, V.D. (2014). Process development for puffing of sorghum. *International Journal of Current Research and Academic Review*; 2(1): 164–170.

Shuh, M.C & Tsai, L.C. (1995). The characteristics of explosion puffing rice products with different amylose contents. *Food Science Taiwan*; 22: 465–478.

Simsrisakul, M. (1991). Important factors affecting puffing quality of Paddy and properties of Puffed rice flour. M.Sc. Thesis, Chulalongkorn University, Bangkok.

Singh, J & Singh, N. (1999). Effects of different ingredients and microwave power on popping characteristics of popcorn. *Journal of Food Engineering*; 42: 161–165.Sullivan, J.F & Craig, J.D Jr. (1984). The development of explosion puffing. *Food Technology*; 38(2): 52–55, 131.

Sweley, J.C., Meyer, M.C & Rose, D.J. (2012). Effects of hybrid, environment, oil addition and microwave wattage on popped popcorn morphology. *Journal of Cereal Science*; 56: 276–281.

Tang, H., et al. (2006). Molecular arrangement in blocklets and starch granule architecture. *Carbohydrate Polymers*; 63(4): 555–560.

Thorat, S.S., et al. (1988). Effect of various grain parameters on popping quality of sorghum. *Journal of Food Science and Technology*; 25(8): 361–363.

Ummadi, P., Chenoweth, W.L & Pkw, N. (1995). Changes in solubility and distribution of Semolina proteins due to extrusion processing. *Cereal Chemistry*; 72(6): 564–567.

Van, D.E.R.M ., Van, D.G.A.J & Boom, R.M. (2010). Understanding Molecular Weight Reduction of Starch during Heating-shearing process. *Journal of Food Science*; 68(8): 2396–2404.

Yenagi, N.B., et al. (2005). Suitable process for popping of sorghum cultivars for value addition, *Maharashtra Journal of Agricultural Extension Education*; 16: 366–368.

Zapotoczny, P., et al. (2006). Effect of Temperature on the physical, functional and mechanical characteristics of hot-air puffed amaranth seeds. *Journal of Food Engineering*; 76(4): 469–476.

CHAPTER 16

Microwave Processing of Cereals

Ranjana Verma,[1] Nilakshi Chauhan,[2] Farhan M. Bhat,[2] and Preeti Choudhary[3]
[1]Department of FSNAT, COCS, CSK HPKV, Palampur, Himachal Pradesh, India
[2]Department of FSNAT, COCS, CSK HPKV, Palampur, Himachal Pradesh, India
[3]College of Horticulture & Forestry, Neri, Hamirpur, Dr YS Parmar UHFS, Solan, Himachal Pradesh, India

CONTENTS

DOI: 10.1201/9781003242192-20

16.1 INTRODUCTION

The world's population is reliant on agriculture for feed and food. The major food crops of the world are rice, wheat, and maize, which contribute a considerable amount of energy, protein, fat, vitamin, and mineral content in the diet of mankind. Cereal grains containing higher amount of starch content, which is the storage carbohydrate, play a major role in the human diet and nutrition (as a slow-release source of energy that is gradually used over time). Unprocessed cereals contain a good amount of fibre. Processing methods alter quality parameters of food products such as the structure, nutritional, and organoleptic properties. Microwave technology is widely used in food processing due to high heating/penetration rate potential, with significantly reduced food preparation time, uniform distribution of heating, handling precision, ease of use, and low maintenance (Gonzalez, 2002). Furthermore, when compared to conventional heating, microwave processing modifies the flavour and nutritional qualities of food to a lesser extent during cooking or reheating (Witkiewicz and Nastaj, 2010).

The combination of electrical energy and electromagnetic fields in the microwave cavity, as well as its dispersion in the product, are the fundamental mechanisms implicated inmicrowave heating. The phenomena of microwave processing promote an instantaneously rapid increase in the temperature within the product, whereas in traditional processing the transfer of energy from the surface of material, with time is not constant. This remains considerably very high with slow penetration temperatures (Goldblith, 1966). Intermolecular friction is caused by the dipole rotation of polar molecules in microwave processing, which produces energy within the substance, resulting in a decreased temperature gradient in the product and an increase in heat to an instantaneous degree. The presence of water and polar ions in food products and their corresponding solids have a direct impact on heat generation; microwave generates heat continuously inside the food material and the temperature of the food rises rapidly when subjected to the radiation. In food products containing high moisture content subjected to processing, the moisture content will evaporate and the temperature will be maintained at 100°C. The temperature of the product might rapidly rise immediately after the water evaporates causing the risk of scaling (Sale, 1976).

Senise and Jermolovicius (2004) investigated the role of microwave processing in potential technological areas in food processing industries/segments such as in pasta drying, drying and dehydration, pasteurization, blanching, sterilisation, cooking, and defrosting. Today microwave processing is attracting attention because this technology is being used for chemical processes, pest control and germination, improved grain or tuber production, and other purposes (Sacharow and Schiffmann, 1992; Hoffman and Zabik, 1985). The aim of this chapter is to discuss the effect of microwave processing on the nutritional, functional, and biological quality of cereals and its products.

16.1.1 Microwave Processing: Basic Principles and Mechanisms

1) The coupling of electric energy from an electromagnetic field in a microwave cavity and its dissipation within the product are the fundamental phenomena involved in microwave heating. In contrast to traditional heating procedures that transmit energy from the surface and have significant thermal time constants and delayed heat penetration, this results in an instant temperature increase inside the product (Goldblith, 1966). Intermolecular friction, mostly induced by bipolar rotation of polar molecules, creates heat internally in the material, resulting in a lower heat gradient in the product and an instant temperature increase in microwave processing. Foods' aqueous and polar ionic contents, as well as their associated solid constituents, have a direct impact on how the heating is carried out (Buffler, 1992; Marsaioli Jr. 1991; Schiffmann, 1986). In contrast to traditional methods, microwave cooking produces heat continuously, resulting in a continual and

rapid temperature rise when food is exposed to this radiation. The heat is used to evaporate water during the heating of a moist product, and the temperature is maintained at 100°C. The electromagnetic spectrum of microwaves varies between frequencies of 300 MHz and 300 GHz. Microwave heating is a special form of radiative heat transfer and differs from thermal heat transfer processing. When radiation is absorbed by a body, it is transformed to heat and generates thermal effects. The microwaves are electromagnetic radiations having wavelength range of 0.1to 1 m, whichresembles a frequency range of 0.3to 3 GHz. Microwaves are partly reflected, transmitted, and absorbed by the food material in which the absorbed waves generate heating. Unlike other radiations, a microwave penetrates deeper into food materials and thus results in inside heating of the product that extends throughout the material irrespective of their thermal conductivity. Microwave treatment does not produce any non-thermal effects in food material (Mertens and Knorr, 1992).

The chemical makeup of the food substance to be irradiated affects microwave heating. Microwaves mostly interact with polar compounds and charged particles in food. The most important interaction in food materials with respect to polar molecules is the water. The electromagnetic field associated with microwaves changes continuously at high frequency, with the water molecules acting as dipoles alternates in response to electromagnetic field of microwaves resulting in generation of frictional heat transfer. Food materials contain polar molecules like water that generally has a random orientation. Upon application of an electrical field, these water molecules orient themselves according to the polarity of the field that alternates rapidly at 2.45×10^{9} cycles per second. This energy of repeated oscillations generates heat due to intermolecular friction. Thus, the rotation of dipole water molecules leads to friction within themselves and the surrounding medium resulting in generation of heat. Very high speed heating as obtained by microwave treatment does not allow desirable physical and biochemical reactions to occur in food materials. The industrial applications of microwave processing of foods are mainly limited for tempering of frozen foods by increasing the temperature of frozen items from -4 to -20°C (Schiffman, 1986).

16.1.2 Effect of Microwave Processing on the Nutritional Properties of Cereals

Microwave processing hasa great impact on the nutritional and structuralproperties of cereals (Plagemann et al., 2014). Researchersare drawn tothe significant effect of microwave treatment on protein degradation and accelerating reactions associated with protein content. These effects may be accomplished with the structure changes. Yang et al. (2014) observed the effect of length of time of microwave processing on grains and found that 5 min of microwave treatment can lower the crude protein subfraction content from 45.00 to 6.00% and increase the rate of degradation of protein into amino acids as compared to raw grains. It was observed that shorter period timeframe (5 min) with lower energy consumption can enhance the valuable nutrients and conserve the protein quality. Quitain et al. (2006) reported that microwave processing enhanced the quality of protein by induced irradiation of microwave that degraded the protein content to amino acids, thereby significantly increasing the quantity of amino acids as compared to traditional processing methods. Microwave processing leads to denaturation of proteins and thus improves their digestibility (Boye et al., 2010).

Cereals such as maize, rice, and wheat are staple foods of the world's population and play an important role in the human diet as sources of energy. Starches are macromolecules that are naturally renewable and biodegradable carbohydrates and widely findapplications in the food, textile, pharmaceutical, and chemical sectors. Starch consists of two type of molecules: amylose and amylopectin. Amylose haslinear chain linked together by 1,4 α -D glucopyranosyl units, whilethe amylopectin consists of branched chain molecules linked together by 1,4 and 1,6 α -D glucopyranosyl units.The crystallinity of starch is commonly analyzed through the X-ray diffraction diagram. Microwave processing affects the structural composition and the properties of starch, such as polarity, energy, viscosity, gelatinization, molecular weight, particle size, and so on, are all impacted by the crystallinity of starch (due to molecular vibration). (Szepes et al., 2005). Modification of starch plays an important role in different segments of the food industry for product formulation. Lewandowiczet al. (1997) observed the effect of microwave processing on the structural composition and crystallinity of cereal starch (wheat, corn, and waxy corn) and found that cereal starches can be modified from B to A pattern after being treated with microwave radiation. Jiang et al. (2011) investigated the structural changes following microwave heating using the water–starch system as a model (frequencies: 27, 915, and 2450 MHz). It was observed that the crystallinity of starch changed from type A to type B and the crystallinity also increased after processing with microwave heating (frequencies: 27, 915, and 2450 MHz). Furthermore, the authors also foundthat microwave heating affectsthe molecular weight of dried starch and degree of polymerization at varied frequencies such as 27 MHz, 915 MHz, and 2450 MHz. Processing results in biochemical changes through molecular transformation and chemical reactions within the processed cereal products. These changes include gelatinization of starch, denaturation of proteins, modification of lipids, changes in texture, enzyme inactivation, browning reactions, and increase in soluble dietary fiber (Nayak et al., 2015). These changes are associated with increasing mouth feel, structure, and texture profile besides increase in nutrient availability, organoleptic properties, and inactivation of heat-labile toxic compounds (Ragaee et al., 2014).

16.1.3 Effect of Microwave Processing on Functional Properties of Cereals

16.1.3.1 Starch

Starch is the most important carbohydrate in all cereal crops, accounting for 56 to 80% of grain dry matter (Eliasson and Larsson, 1993). It is an important source of energy and storage polysaccharides in most cereal crops (Shevkani et al., 2017). Cereal starches differ in shape, size, and distribution of granular properties. The origin of the starch in the plant species depends upon various factors, including modification in starch particle size (1–100 lm in diameter), morphology such as shape (circular, oval, triangle), size (unimodal or bimodal), and association as independent (simple) or aggregated (complex). Microwave processing affects the structural configuration/composition of the starch, amorphous, and semi-crystalline structure. The molecular structure of amylose (AM) and amylo-pectin (AP), as well as the structural configuration of starch molecules, determines the functional qualities of starch, which have important characteristics in the formulation/preparation of food products (Madruga et al., 2014). The effectsof microwave processing on functional properties such as rheological properties structural and morphological of cereal starch are described as follows.

16.1.3.1.1 Rheological Properties

Mechanical and rheological properties play a pivotal role in governing the quality of food products. Cereal flour contains visco-elastic gluten, which gives specific rheological properties to the dough and in turn influences the final quality of food products. The viscosity of starch aqueous systems is considered a major macroscopic feature that is vital tothe formulation and development of products and a slight change in its molecular conformation and structure can result in significant variations in rheological characteristics/properties. Microwave heating can significantly alter the rheological properties of starch, specifically by lowering its rigidity. Microwave processing at 0.5 watt/gm, for specific period of time (i.e., 60 min), resulted in a rise in the pasting properties and decreased the consistency of wheat starch and maize starch with 30% moisture content (Lewandowicz et al., 2000). Swinkels (1985) observed the effect of microwave processing on rheological characteristics such pasting properties like temperature and expansion characteristics in the starch of wheat, maize, and waxy maize and observed that the microwave-treated waxy corn contained negligible amount of amylose as compared to other cereal starches, whereas the content of amylose in conventionally treated wheat and maize was almost 28%. Several other studies have shown that amylose is more responsive to microwave heating than amylopectin.

The effect of microwave processing on different varieties of maize including normal maize, waxy maize, and amylo maize V starches was investigated by Luo et al. (2006). Treatment of different varieties of maize with microwave radiation at 1 W/g for specific period of time (i.e., 20 min) with 30% moisture resulted in a rise in pasting temperature and lowering consistency. Furthermore, the authors also analyzed the effect of microwave irradiation on rapid viscoanalysis profile of all maize starches and found that after irradiation the maximum viscosity, trough, and cold paste viscosity were reduced after holding phase for 30 min. Microwave treatment increases the intermolecular and intramolecular attraction between the hydrogen bonds due to starch chain aggregation.

16.1.3.1.2 Gelatinization Properties

One of the most essential functions of starch, especially in food processing, is gelatinization. Cereal starch treated with microwave processing causes structural reconfiguration in the starch macromolecule, which results in higher gelatinization temperatures and reduces the process enthalpy.

A significant increase in gelatinization was observed for wheat, normal corn, and waxy corn starch at a set temperature (i.e., 13.0°C, 11.0°C, and 6.0°C, respectively). However, a significant reduction in gelatinization enthalpy of microwave-processed cereal starches such as wheat and corn was observed and the enthalpy of gelatinization remainedunchanged in waxy corn (Lewandowicz et al., 2000). Luo et al. (2006) observed that microwave with advanced technological parameters and controlled condition had a smaller effect on the quality attributes of maize starch containing 30% moisture content, which reflects the importance of controlled parameters during microwave processing. They also foundthat microwave treatment of normal maize, waxy maize, and amylo-maize V starches for 20 min reduced the enthalpy of gelatinization in amylo-maize V (B-type) as compared to normal maize and waxy maize (A-type), but increased the gelatinization transition temperatures and generalization temperature range. They further reported that the reduction in DH values showed that double helices structure in the crystalline and non-crystalline portions of the starch granule was destructed by the microwave processing conditions. At the same time, the enhancement of the gelatinization temperature range led to rearrangement of starch molecule, which resulted in the creation of crystallites with varying stabilities.

16.1.3.1.3 Digestibility Properties

When waxy and normal rice starches containing the moisture content of about 20% were subjected to hydrothermal microwave treatment at range of about approximately 270–1300W for 60 min, the digestibility of waxy and normal starches slightly increased as power of microwave increased (Anderson and Guraya, 2006). Similarly, when other cereals starches such as native corn, wheat, RS2, and RS1 were subjected to microwave processing, the digestibility of these starches did not alter the attributes of kinetic energy curve (Gi et al., 2012). Microwave processing of non-waxy, high amylose, and waxy varieties of barley enhanced the starch digestibility, with significantly higher levels of RDs starch and lower levels of slowly digestible starch and resistant starches compared to controls that is unprocessed (Emami et al., 2012).

Furthermore, it was shown that increasing starch digestibility by microwaving at a higher power range or keeping tempered barley samples for a particular length of time prior to microwaving can improve and increase the starch digestibility. Microwave heat-moisture treatment at the frequency of 1000W for approximately 30 min with a moisture level of near about 20% increased the content of resistant starch in Canna starch (Zhang et. al, 2010), resulting in doubling the content of resistant starch in modified starch as compared to native starch content. Microwave processing of starches of cereals, resulted in 50% reduction of rapidly digestible starch content, while the content of slowly digestible starch remained unaltered. These findings contradict the earlier research done by Palav and Seetharaman (2007), who investigated that microwave heat-moisture treatment, reduces resistant starch content, which may be due to low moisture content utilized during the processing.

16.1.3.1.4 Granule Morphology

Identification of genus and species of particular grain is dependent upon the granule morphology. The morphology of starch granule variescrop to crop. Wheat and barley consist of A-type starch having lenticular shape while B-type starches are spherical in shape. Microwave processing is dependent on various parameters such as biological origin of starch, water availability, and handling parameters. For instance, the starch granule of lentils (smooth surface) with 25% moisture content when subjected to 85°C for 6 min and 650 W did not modify the morphological characteristics whereas morphological characteristics at the same conditions of roughed surfaced grains might be affected by the microwave processing (Gonzalez et al., 2002). Microwave radiation at 450W for 15 min depending upon the biological origin of starch affects the micromorphology. The starch granule of cereals like maize having 6.8% moisture content get easily deformed and lose the capacity of agglutination, but there is no noteworthy change in the size and distribution of the particles. Microwave radiation does not affect cereal starch granule with 30% moisture at 1W for 20 min (Luo et al., 2006). However, the internal morphology of the starch is degenerated as it becomes porous and tissue organization is fragile so the center cavity becomes visible.

16.1.3.1.5 Crystallinity

Crystallinity shows the structural composition of agrain. The crystallinity of starch granules is around about 14 to 45%, depending on arrangement of double helical structure of amylopectin in a molecule and other variables such as origin, structure, and moisture content. These variables are affected partially or completely bythe microwave processing conditions. The effects occur in the form of demolition of the crystalline organization due to increase in temperature and vaporization ofwater content of granules. It has been observed that there is a difference in the deformation of the starch in waxy corn with 30% moisture and normal starch as revealed by X-ray diffraction studies (Lewandowicz et al., 2000). Microwave radiation at 0.5W for 1 hour decreased the crystallinity of normal wheat starch whereas other starches (waxy corn and corn) remained unaffected but at 450W for 15 min the starch crystallinity of the maize decreased immensely from 84.6 to 30% (Szepes et al., 2005). Microwave radiation yields more well-organized starch crystal than in the inherent form due to shift in crystallites and formation of double helix chains. These shifts may be attributed to vaporization of water molecule from the center and consequent swing of double helix toward the center (Zhang et al., 2010; Zhang et al., 2009).

16.1.3.1.6 Microstructural Characteristics of Starches

Microwave heating does not affect the granule morphology, but heated starch granules show slightly more aggregation than non-heated granules. Similar results were revealed in the case of native and microwave processed non-waxy starch. This finding shows that internal recrystallization mechanisms may not result in changes in the physical structure of starch granule.

16.1.3.2 Protein

Proteins are highly complex entities present in all life forms. They have vital roles in maintaining body functioning including some of the major processes such as catalysis of metabolic reaction and replication of DNA (Burgess et al., 2009). Cereal cropsarea keydietary source of the world's population, and contribute around about 60% of total protein requirement in the human diet. Most proteins haveviscoelastic properties that play an important role in various segments of food processing industries. Proteins are essential ingredients in a lot of food processing operations, where they provide specific functional activities. Functional properties intricate relationships between protein composition, structure, conformation, and physicochemical properties in which these are associated or assessed (Kinsella, 1976). Various functional properties and effect of microwave processing on themare described in the following.

16.1.3.2.1 Solubility

Solubility is one of the essential parameters in relation to the functional quality of proteins, as it has a considerable impact on other functional properties such as foaming and emulsification properties (Vojdani, 1996). During the creation and testing of new protein components, solubility is considered as the essential attribute in relation to the functional property of specific food products (Zayas, 1997). Protein–water solubility is linked to emulsifying and foaming characteristics (Kinsella, 1976). Emulsifying properties of protein-rich food products are dependent upon emulsifying capacity and emulsion stability, the two key features of functional attributes (Wang and Zayas, 1992). The emulsifying ability of a protein emulsifier is based on its ability to create adsorption sheets surrounding the oil globules and lower the surface tension at the oil–water interfaces. The tendency of emulsion droplets to remain distributed in the absence of creaming, coalescence, or flocculation is referred to as emulsion stability. Ashraf et al. (2012) studied the effect of microwave processing on functionality of cereal and legume and observed that microwave processing for 50s can enhance the emulsifying capacity of wheat and *haleem* wheat flour due to the presence of hydrophobic and hydrophilic amino acids in both the cereal flour, which leads to increase the Emulsifying capacity property.

16.1.3.2.2 Forming Capacity

Protein-foaming properties influence food quality in various food products. Two parameters are commonly used to describe foaming properties: foam ability and foam stability. Instant protein adsorption and reconfiguration at the air-water interface define foam ability, whereas the creation of a ductile, coherent film at the surface determines foam stability. Sparging the gas at a steady flow rate into the protein solution through a sintered glass disc is the most repeatable approach for foam generation in the laboratory, and it provides more stable bubbles than other techniques (Halling, 1981). The functional properties of proteins are affected by various parameters such as concentration of the solution, the negative log of hydrogen ion concentration (pH), solubility, condition (temperature) as well as the apparatus and method used to measured efficiency of functional properties (Kinsella, 1976; McWatters, 1979; Tornberg and Hermansson et al., 1977) . Microwave-baked products have been shown to last longer than conventionally baked products (Ozmutlu et al., 2001; Yin and Walker, 1995). Ashraf et al. (2012) reported that the maximum foaming capacity was attained at300s at the frequency of 2450 MHz in cereal flour. The increase in forming capacity by utilizing microwave

Table 16.1 Effect of Microwave Processing on Functional Properties of Cereal Crops

Crops	Variables of Microwave processing	Time Length	Effects on various parameters
Corn Starch	450 W	15 min	Destructed Granules of corn Starch
	1W/g	20–25min	Reduction in functional properties of starch lower the solubility index also decrease the amount of heat required for gelatinization
	300 and 650 W	½ hour and 150–155s	Lower the rheological characteristics
Waxy Corn Starch	0.5–.06W/g and 1W/g	1 hour and 20 min	Little changes were observed in functional properties of cereal such as lower the solubility, gelatinization, etc.
Wheat	1.0 W/g, 300W	1 hour and 30s	Lower the solubility index, degree of crystallinity, and amount of heat required for gelatinization
Rice	250–1300 W	1 hour	Enhance the pasting properties and digestibility index of flour Does not alter the morphological characteristics
Barley	200w –1.2 kw	7min	The structural changes observed such as crystallinity of starch changes from type A to B Enhance the water absorption and solubility index

processing is mainly due to increase in the concentration of hydrophobic amino acids in cereal flour. The matrix of hydrophobic amino acids affects the foaming capacity (Kong et al., 2007).

16.1.3.2.3 Water Absorption

The functional qualities of protein derived from rice bran using microwave processing were investigated, and it was discovered that water absorption was substantially reduced as a result of a possiblechange in configuration of protein of rice bran due to the microwave treatment (Khan et al., 2011). Ashraf et al. (2012) observed that microwave treatment increases the water-holding capacity in wheat and *haleem* wheat flour. The cereal flour treated with microwave at time length of approximately 300 s had higher water-holding capacity, which may be due to exposure of protein containing hydrophilic bound responsible for increase in the uptake the water content in cereal flour. The effect of microwave processing on functional properties of cereal such as corn, wheat, rice and barley is presented in Table 16.1

16.1.4 Effect of Microwave Processing on Biological Properties of Cereal

Processing of cereals by any method alters its nutritional quality depending upon the type and hardness of grains. Due to non-uniform distribution of nutrients and phytonutrients in cereal grains, the processing can have a significant influence on the composition and quality of resulting finished products. Processing techniques enhances colour, shelf life, and flavour and lead to reduction of anti-nutritional factors. Anti-nutritional factors like trypsin inhibitors, hemagglutinins, tannins, and phytates inhibit protein digestibility, thus processing aids in protein digestibility. The presence of these anti-nutritional factors decreases the bioavailability of micronutrients due to chelation of minerals in the gastrointestinal tract of the human body). Anti-nutritional compounds are heat sensitive and microwave processing is one of the tools that can be used for degradation of anti-nutrients in cereal crops. Microwave treatment, in particular, may partially diminish phytic acid due to its heat-sensitive nature and the development of insoluble complexes between phytate and other components

(Kakati et al., 2010). According to Kala and Mohan (2012), microwave processing generates free radicals, which results in the formation of lower amount of inositol and inositol phosphate, which further lower the phytic acid concentration. The inactivation of trypsin inhibitors is caused by the denaturation of heat-sensitive proteins (Yang et al., 2014), as well as the degradation of sulfhydryl and disulphide (-S-S-) groups that are vulnerable to destruction by microwave processing (Embaby, 2010; Zhong et al., 2015). Furthermore, tannins are water-soluble and have a thermally volatile nature that can be reduced by microwave heating (Yang et al., 2014).

16.1.4.1 Phytic Acid

Phytic acid is commonly known as phytates and is naturally found in cereal and grains (Bora, 2014; BuadesFuster et al., 2017). The content of phytic acid increases in cereal crops with the maturity of the grain and approximately 60–90% of phosphorus phytate is found in mature grain/kernel of cereals (Kumar et al., 2010). Phytic acid is usually negatively charged and forms complexes with positively charged metal ions such as iron, zinc, calcium, and magnesium, lowering their bioavailability and slowing their rate of absorption (Samtiya et al., 2020). Because of these chelating actions, phytic acid acts as anti-nutrientand cause mineral deficiencies in human and animal nutrition (Bora, 2014; Grases et al., 2017). The phytic acid content of various cereal crops has been effectively reduced by microwave processing. The content of phytate was reduced approximately 33% after microwave processing of buckwheat grains at 850 W for 30 min (Table 16.2) (Deng et al., 2015). The fact that phytic acid is heat sensitive lends credence to this notion. Formation of insoluble complexes between phytic acid and other constituents such as phytate-protein and phytate-protein-mineral complexes lower concentration of phytic acid. As a result, microwave processing may be an effective way to decreasephytic acid in many cereal grains and increase nutrient absorption. Soaking before microwave processing increases the ability to lower the content of phytic acid.

16.1.4.2 Tannins

The molecules of tannins are water-soluble with molecular weights higher than 500 Da (Nikmaram et al., 2017). Tannins are secondary metabolites found primarily in fruits, barks, and leaves. Barley, sorghum, and millets are examples of cereal grains that contain them (Timotheo and Lauer, 2018; Morzelle et al., 2019; Samtiya et al., 2020). Hydrolysable and condensed are the two forms of tannins. Millets and peanuts contain a lot of condensed tannins (De Camargo & Lima, 2019). Tannins cause hinderence in absoption and digestion of proteins by inactiving the enzyme present in the digestive track via formation of hydrogen bound between the hydroxyl group of tannin and carbonyl group of the protein resulting in irreversible and reversible tannin–protein complexes (Raes et al., 2014; Joye, 2019). Tannin content in millet crops (pearl and finger millet) are comparatively high as compared to other cereal grains. Microwave processing of pearl millet, finger millet, and sorghum at various frequencies (900 W, 2450 MHz for approximately 40–100 s resulted in tannin content reductions of approximately 45%, 45%, and 77%, respectively (Table 16.2) indicating that microwave processing has the potential to lower the content of tannins to 50% (Singh et al., 2017). Deng et al. (2015) reported that microwave processing reduced the tannin content to 28% in buckwheat grains. They further reported that the content of tannin was reduced through microwave processing by breaking down of tannin–protein aggregation between the molecules.

16.1.4.3 Trypsin Inhibitors

Anti-nutritional factors such as trypsin (protease) inhibitors are thought to prevent the full use of a food's nutritious value. Inhibitors of trypsin bind to the protease enzyme, reducing protein breakdown in the small intestine while encouraging quicker protein release from the body. As a result, sulfur-containing amino acids like methionine and cysteine in bean have decreased bioavailability (Nikmaram et al., 2017). Time-temperature combinations in microwave processing affect the activity of trypsin (protease) inhibitors in cereal bran and as the heating temperature along with time increase, the activity of trypsin reduces. Microwave processing reduced the trypsin inhibitor activity in cereal bran to 77, 77.29, and 78% when subjected to the frequency of 2450 MHz for 1.5, 2.0, and 2.5 min, respectively (Kaur et al., 2012). The authors focused their research on three types of grain brans, wheat, barley, and oat, and compared dry heating (hot air oven) and microwave heating. Anti-nutritional compounds such as trypsin (protease) inhibitors, phytic acid, oxalate, and saponins were effectively detoxified through microwave processing for 2.5 min. The heat-sensitive characteristic was responsible for the significant decrease in the activity of trypsin in cereals. Irakli et al. (2020a,2020b) observed a 28 to 30% reduction in trypsin inhibitors in rice bran and reported that the thermal processing principally involves the splitting of covalent bond and breakdown of disulfide bonds might be responsible for inactivation of anti-nutritional compound. Suhag et al. (2021) reported that microwave processing comparatively reduced the amount of trypsin as compared to other anti-nutritional compounds in food grains.

16.1.4.4 Oxalate

Due to its proclivity for forming connections with divalent metal cations, oxalate can lower the bioavailability of minerals including calcium, iron, and magnesium in the human diet. Consumption of food containing high amount of oxalate may lead to the formation of calcium oxalate crystals, which can obstruct renal tubules and lead to urinary calculus and hypocalcemia in humans (Nikmaram et al., 2017). According to Kaur et al. (2012) microwave processing for the time periods of 1.0, 2.0, and 3.0 min reduced the oxalate content in cereal brans upto 44, 57, and 65%, respectively. Similarly, Irakli et al. (2020a,2020b) reported that microwaveprocessing was one of the most successful methods for lowering the oxalate concentration in rice bran to approximately 34% among the other heating methods (i.e., dry heating and infrared heating). As a result, microwave treatment can be a simple, quick, and effective way to remove oxalate content from cereals. Microwave heating reduces oxalate levels, which could be attributed to heat stress, which degrades total oxalate (Kala and Mohan, 2012). The effects of microwave processing on functional properties (Table 16.1) such as rheological properties structural and morphological of cereal starch are described as follows.

Table 16.2 Effect on Microwave Processing on Biological Properties of Cereal Crops

Grains	Variables of Microwave processing	Time Length	Reduction in anti-nutritional compound
Cereal bran	Microwave processing at 2450 MHz	3 min	Reduction in phytic acid content (53.80%) Reduction in oxalate content (60–65%) Reduction in saponin content (90–100%) Reduction in tannin content (8.0%)
Peral millet	Microwave processing at 900 W, 2450 MHz	40–100s	Reduction in phytic acid content (71%) Reduction in tannin content (45%)
Finger millet	Microwave processing at 900 W, 2450 MHz	40–100s	Reduction in phytic acid content (58%) Reduction in tannin content (45%)
Sorghum	Microwave processing at 900 W, 2450 MHz	40–100s	Reduction in phytic acid content (11%) Reduction in tannin content (77%)
Buckwheat	Microwave processing at 850 W and 2450 MHz	30 min	Phytic acid reduction (33.0%) Reduction in trypsin inhibitor activity (12–13%) Reduction in saponin content (20%) Reduction in tannin content (28%)

16.1.5 Effect of Microwave Processing on Cereal-based Food Products

Microwaves are used for a wide range of applications, including cereal preparation, frying, cooking, and pasteurization. Higher yields with shorter processing periods and improved end product quality, both in terms of organoleptic and nutritional qualities, are advantages of using microwaves over traditional processing procedures.

16.1.5.1 Pre-cooking

Pre-cooking of food helps to reduce preparation time. During the formulation of cereal-based products, these operations primarily affect the starch content of grains by reducing its gelatinization time. Wang et al. (1993) and Caballero-Córdoba et al. (1994) studied the effect of toasting on cereal flours.In this study cereal flour such as wheat and rice flour was pre-cooked in a microwave oven for 11 min at the frequency of 2450 MHz containing moisture content 13.0 and 14.0%, respectively. It wasfound that microwave-treated flour had good nutritional and sensory properties as comparedto cereal flour treated with conventional methods. They further investigated that pre-cooked cereal flours (rice and wheat) in different proportion with soy flour processed with microwave heating for approximately 10 min was found best in formulation of different food products like porridge and soup, respectively.Martinez-Bustos et al. (2000) investigated the effect of microwave processing on nixtamalized maize flour (NMF) for the formulation of masa and tortillas and reported that microwave processing of maize flour for 10,15, and 20 min at 2450 MHz frequency reduced the 50% processing time and liquid waste discharge. The wholegrain NMF sample heated in a microwave oven with 70 g.kg^{1}water and 2.0 g.kg^1 Ca $(OH)_2$ for 20 min was statistically similar to the control sample in terms of hardness. A protocol developed by Velupillai et al. (1989) for parboiling of rice with microwave processing noted that: (1) soaking the coarse grain; (2) soaked grain subjected to first level microwave processing for partial gelatinization of starch content in grain enhanced the water absorption properties of microwave-treated grain; (3) From the treated soaked rice, remaining water content is discharged, if any; (4) The drained and processed coarse rice grain further subjected for the second-level of microwave processing to make the rice completely gelatinized with reduced water content of the rice.

16.1.5.2 Cooking

The cooking and thawing processes in amicrowave oven have been well documented in many previous studies. There are various studies suggesting that thawing or cooking affects the organoleptic properties such as taste, flavour, and consistency of finished product. Li (1995) observed that the cooking or reheating (microwaved) of food leads to formation of changes in organoleptic properties such as loss of taste and aroma, which is a major disadvantage due to imbalance of sensory characteristics in the final product. Clarke & Farrell (2000) studied the effects of microwave processing on textural changes in frozen pizza base and observed that the frozen pizza base containing water binders (pea fiber, potato starch, locust bean gum and oat fiber) chemical emulsifying agents and protein hydrolyzing enzymes (fungi associated proteases) does not alter the textural profile of the pizza after the microwave processing for 120, 150, and 180 s.

16.1.5.3 Baking

There are limited studiesavailable in the context of utilization of microwave processing in bakery segments because microwave processing produces goods with inferior quality attributes,which may be due to differences in heat generation and mass transfer mechanism as compared to products developed using conventional heating methods. Many researchers faced challenges in microwave processing to meet the quality characteristics in baked products. Microwave processing affects organoleptic properties like taste, aroma, colour, and textural profile in commonly consumed bakery goods such as bread, cookies, biscuits, and cake. Many researchers have recommended overcoming the effects of microwave processing on finished products by modifyingthe cooking practice.

16.1.5.4 Bread

Baking under controlled conditions produces bread with desired quality characteristics. The desired quality characteristics required for production of baked goods involve heating rate, the amount of heat required, moisture condition in oven, and time and temperature combination. Bread when heated in a microwave oven loses moisture rapidly, and after microwaving, the durability mechanics of bread is increased significantly (Jahnke, 2003; Chavan & Chavan, 2010). Yin & Walker (1995) studied the effect of microwave processing on quality characteristics of toast and reported that various variables of microwave processing such as mechanism of heat and mass transfer, inadequate starch gelatinization due to shorter time period of microwave cooking, changes in gluten content induced by microwave radiation, and generation of gases and water vapours causes inferior quality of toast in microwaves.

Park (2001) investigated the interaction of microwave processing with moisture content in bread and created aninterrelated graph for loss of moisture and input of energy. It was further reported that despite having the same moisture content, microwave processed bread had significantly lowered quality parameters as compared to conventionally heated bread. In microwave heating the moisture content of the bread was reduced, which further disturbed the structural properties of bread thereby lowering its quality. Microwave-treated breads havelesser water absorption properties as compared to non-microwaved breads. Scientists have also worked on the development of stabilizers in baking through microwave technique with gelling agents, resins, and enzymatic constituents so as to control migration of water and recrystallization of starch in glazed baked goods cooked with microwave irradiation (Jahnke, 2003).

16.1.5.5 Cakes

In the cake-making process, the batter prepared after mixing the ingredients is put into cake molds followed by subsequent transferring to the pre-heated oven. Besides baking temperature and baking time other factors that influence the optimum baking conditions include the sweetener proportion of the cake formula, the quantity of milk used in making the batter, the consistency of the batter, size of the pan, etc. (Young &Cauvain, 2007; Bennion& Bamford, 1997. In a study conducted by Stinson (1986), slight sensory differences were observed in the cakes baked in microwave and conventional ovens. The study revealed that quality cake can be baked approximately 20% faster in a microwave oven as compare to conventional oven.

Modification of ingredients used in cake affect the dielectric behaviours of the cake batters. Tsoubeci (1994) formulated the different type of cakes with sucrose, whey protein isolate, maltodextrin blends, and fat with special fiber in a model cake system by utilizing both traditional as well as microwave ovens and observed that in microwave processing sucrose substitution influenced the dielectric properties of the batters and increased absorption power.

Many attempts have beentaken by researchers to standardize the methodology of microwave-baked cake using surface response that have revealed that cake batter containing wheat starch, polysorbate (0.3%), shortening (45.20%), and water (133.70%) baked in microwave oven (100% power) for 6 min produced good quality cakes comparable with those baked conventionally (Şummu et al.,2000).

16.1.5.6 Cookies

The quality of cookies is influenced by a number of factors like type of ingredients and their proportion, weight, diameter, and other quality attributes. According to the literature, generally cookies baked using traditional methods in ovens of approximately 80 to 120°C for 3.5 to 4.0 min have a final moisture content of 5.0%. If the moisture content variation between the outer sides of cookies and in the center is higher than approximately 1.5%, this leads to the formation of cracks during storage, due to expansion and contraction of the cookie (Bernussi et al., 1998; Manley, 2000).

Bernussi et al. (1998) studied the effects of microwave heating along with the traditional method of baking on moisture gradient and quality attributes of cookies. They reported that cookies pre-baked in a traditional oven at 140°C for 4 to 5 min and then baked with a microwave oven at frequency of 2450 MHz with energy capacity of 617 W for 30s hadsignificantly decreased the moisture gradient, from 2.18 to 0.90%, reducing the occurrence of cracking, from 41.70 to nil. As a result, both baking techniques reduced cookie cracking by lowering the moisture gradient and maintaining the quality attributes of finished products such as textural profile and sensory characteristics.

16.1.5.7 Pasta Products

Drying of pasta is a crucial factor that determines the quality and storage stability of pasta as the moisture content in dry pasta must be reduced from 30% to 12–13%. It is difficult to dry pasta products because of slow migration of moisture to the outer surface. Dry air is, by itself, quite effective in reducing the water content at or near the surface, while the bounded water/moisture content requires time to move to the outer surface. Microwave heating can efficiently provide a solution to this problem as it provides the movement of water content towards the outer surface of food product. Effective drying of pasta using hot air and microwave drying was suggested by a study conducted by Decareau (1986). The effect of traditional methodsand microwave heating on the nutritional, textural,and sensory attributes of cooked and raw macaroni was evaluated by Altan and Maskan (2005). It was observed that macaroni samples dried with energy generated through microwave processing (60 and 220 W) led to significant reduction in the drying time when compared with those dried by using only conventional hot air, but there was incomplete gelatinization of starch during drying.

The household microwave oven was used to study the effect of microwave-assisted fluidized bed drying on macaroni beeds, and a considerable decrease in drying time in macaroni beeds with an increase in microwave power and air temperature was found (Goksu et al., 2005). Berteli & Marsaioli Jr. (2005) investigated the potential of microwave processing/irradiation to enhance the air drying of Penne-type short cut pasta and observed that microwave processing significantly reduced the average drying time without affecting the nutritional and sensory attributes of the final product when compared to conventional air drying. The most common advantages of utilizing microwaves to dry pasta products are the minimization of drying time and the preservation of quality attributes in the final products.

16.1.5.8 Popcorn

Popcorn is a nutritious snack prepared from whole grain of special variety of corn and has great demand in day-to-day life. With the popularity of household microwave ovens, microwave popcorns have become more popular due to ease of preparation. Ready-to-use popcorns are readily available with different flavours and seasoning in small convenient packs. According to reports, the quality of popcorn depends on its water content, temperature of popcorn, grain magnitude and contour, variability, peel width, grain bulk, and grain injury (Richardson 1959; Mohamed et al., 1993). This is one of the common and popular uses of microwave in many countries particularly in the United States (Moraru and Kokini, 2003). The expansion in popcorn is mainly due to the starch present in corn (Schwartzberg et al., 1995). Popping above 75% can be obtained in kernels seasoned with10% hydrogenated oil, butter 2%, 0.5% salt, and by using microwave oven with the 2450MHz and 660W power (Singh and Singh, 1999).

16.2 CONCLUSION

Microwave processing has widened itshorizons and is now used to prepare cereal and its products as an efficient technique to save energy and time. Microwave processing is responsible for micro- and macrochanges in food structure leading to changes in rheological properties such as viscosity, gelatinization, etc. There is an increase in the electrical conductivity of microwave-treated flours, which is valuable in the production of creamy soups and salad dressings. Foods with higher water content are more prone to lose nutritional content during microwave treatment. The various reactions that occur in food components during heating by microwave due to nutritional loss are little debatable. Though microwave radiations are extensively used in food industries,it is necessary to improve certain points. For example,non-uniform microwave heating can lead to formation of cold and hot spots on the food. Future research should be focused on the combination of microwave with other techniques to overcome its negative effects. Although there has been great improvement in the production of many high-throughput methods for drying of grains, including controlling water content in biscuits, modification of starch, etc., still research is required to improve the quality of products.

REFERENCES

Altan, A. &Maskan, M. (2005). Microwave assisted drying of short-cut (ditalini) macaroni: Drying characteristics and effect of drying processes on starch properties. *FoodResearch International,* 38(7), 787–796.

Anderson, A. K. &Guraya, H. S. (2006). Effects of microwave heat-moisture treatment on properties of waxy and non-waxy rice starches. *Food Chem.*97, 318–323.

Ashraf, S., S.M.G. Saeed, S.A. Sayeed&R. Ali, .(2012). Impact of microwave treatment on the functionality of cereals and legumes. *International Journal of Agriculture and Biology,*14: 356–370

Bennion, E.,B. &Bamford, G.,S.,T. 1997. The Technology of Cake Making. Chapman & Hall, 6th ed., ISBN-10-0751403490, London.

Bernussi, A. L. M.;Chang, Y. K.&Martinez-Bustos, F. (1998). Effect of Production by Microwave Heating After Conventional Baking on Moisture Gradient and Product Quality of Biscuits (cookies). *Cereal Chemistry.*75(5), 606–611.

Berteli, M. N. &Marsaioli Jr., A. (2005). Evaluation of short cut pasta air dehydration assisted by microwaves as compared to the conventional drying process. *Journal of Food Engineering.* 68(2), 175–183.

Bora, P. (2014). Anti-nutritional factors in foods and their effects. *Journal ofAcademia and Industrial Research,* 2(6), 1–14.

Boye, J., Zare, F., Pletch, A. (2010). Pulse protein: Processing, characterization, functional properties and application in food and feed. *Food Research International,* 43, 414–431.

BuadesFuster, J. M.,SanchísCort´es, P., Perell´oBestard, J., &GrasesFreixedas, F. (2017). Plant phosphates, phytate and pathological calcifications in chronic kidney disease. *Nefrologia,* 37(1), 20–28.

Buffler, C. R. (1992). Microwave cooking and processing.*New York Van Nostrand Reinhold,* ISBN-10-0442008678, New York.

Burgess, R. R. &Deutscher, M. P. (2009). Guide to Protein Purification.2nd ed.Academic Press, New York.

Caballero-Córdoba, G., M., Wang, S., H.&Sgarbieri, V., C. (1994). Característicasnutricionais e sensoriais de sopacremosa semi-instantâneaàbase de farinhas de trigo e soja desengordurada. *Pesquisa AgropecuáriaBrasileira*, 29(7), 1137–1143.

Chavan, R.S. &Chavan, S.R. (2010). Microwave baking in food industry: A review. *International Journal of Dairy Science*.5(3), 113–127.

Clarke, C.I. &Farrell, G.M. (2000). The effects of recipe formulation on the textural characteristics of microwave-reheated pizza bases. *Journal of the Science of Food and Agriculture*.80 (8), 1237–1244.

De Camargo, A. C., &Lima, R. da S. (2019). A perspective on phenolic compounds, their potential health benefits, and international regulations: The revised Brazilian normative on food supplements. *Journal of Food Bioactives*.7, 7–17.

Decareau, R. V.&Peterson, R. (1986). Microwave food processing equipment through the world. *Food Technology,*40, 99–105.

Deng, Y., Padilla-Zakour, O., Zhao, Y., &Tao, S. (2015). Influences of high hydrostatic pressure, microwave heating, and boiling on chemical compositions, antinutritional factors, fatty acids, in vitro protein digestibility, and microstructure of buckwheat. *Food and Bioprocess Technology,*8, 2235–2245.

Eliasson, A.,C. &Larsson, K. (1993). *Cereal in Bread Making*. New York, Marcel Dekkar, Inc., 376 pp.

Emami, S., Perera, A., Meda, V., &Tyler, R. T. (2012). Effect of microwave treatment on starch digestibility and physico-chemical properties of three barley types. *Food Bioprocess Technology*, 5, 2266–2274.

Embaby, H. E. S. (2010). Effect of soaking, dehulling, and cooking methods on certain antinutrients and in vitro protein digestibility of bitter and sweet lupin seeds. *Food Science and Biotechnology,*19, 1055–1062.

GI, H., D., M., Ja´mbor, A´ ., Juha´sz, E., Gergely, S. (2012). Effects of microwave heating on native and resistant starches. *Acta Alimentary*, 41, 233–247.

GoksuE. I.; Summu, G. &Esin, A. (2005). Effect of microwave on fluidized bed drying of macaroni beads. *Journal of Food Engineering*, 66(4), 463–468.

GoldblithS.A. (1966). Basic principles of microwaves and recent developments. *Advances Food Research*. 15, 277–301.

Gonzalez, Z., Perez, E.(2002). Evaluation of lentil starches modified by microwave irradiation and extrusion cooking. *Food Res. Int.* 35, 415–420.

Grases, F., Prieto, R. M., &Costa-Bauza, A. (2017). Dietary phytate and interactions with mineral nutrients. In *Clinical aspects of natural and added phosphorus in foods* New York, NY: Springer.

Halling, P.J. (1981). Protein-Stabilized Foams and Emulsions. *Critical Reviews of Food Science and Nutrition, 15*,155–203.

Hoffman, C.J. &Zabik, M.E. (1985). Effects of microwave cooking/reheating on nutrients and food systems: a review of recent studies. *Journal of the American Dietetic Association*. 85(8), 922–6.

Irakli, M., Lazaridou, A., &Biliaderis, C. G. (2020a). Comparative evaluation of the nutritional, antinutritional, functional, and bioactivity attributes of rice bran stabilized by different heat treatments. *Foods, 10*(1), 1–18.

Irakli, M., Lazaridou, A., &Biliaderis, C. G. (2020b). Comparative evaluation of the nutritional, antinutritional, functional, and bioactivity attributes of rice bran stabilized by different heat treatments. *Foods, 10*(1), 57. https://doi.org/10.3390/foods10010057

Jahnke, M. (2003). *Method of making microwavable yeast-leavened bakery product containing dough additive*. United States Patent 6,579,546.

Jiang, Q., Xu, X., Jin, Z., Tian, Y., Hu, X., &Bai, Y. (2011). Physico-chemical properties of rice starch gels: Effect of different heat treatments. *Journal of Food Engineering, 107*,353–357.

Joye, I. (2019). Protein digestibility of cereal products. *Foods, 8*(6), 1–14.

Kakati, P., Deka, S. C., Kotoki, D., &Saikia, S. (2010). Effect of traditional methods of processing on the nutrient contents and some antinutritional factors in newly developed cultivars of green gram [Vigna radiata (L.) Wilezek] and black gram [Vigna mungo (L.) Hepper] of Assam, India. *International Food Research Journal,*17, 377–384.

Kala, B., K.,& Mohan, V., R. (2012). Effect of microwave treatment on the antinutritional factors of two accessions of velvet bean, Mucuna pruriens (L.) DC. Var. Utilis (wall. Ex wight) bak. Ex burck. *International Food Research Journal,* 19(3), 961–969.

Kaur, S., Sharma, S., Dar, B. N., & Singh, B. (2012). Optimization of process for reduction of antinutritional factors in edible cereal brans. *Food Science and Technology International,*1850), 445–454.

Khan, S. H.,Butt, M. S.,Sharif, M. K., Sameen, A., Mumtaz, S. and Sultan,M. T. (2011). Functional properties of protein isolates extracted from stabilized rice bran by microwave, dry heat, and parboiling. *Journal of Agriculture and Food Chemistry*. 59,2416–2420.

Kinsella, J.E. (1976. Functional Properties of Proteins in Foods: A Survey. *Critical Reviews of Food Science and Nutrition,7,* 219–280.

Kong, W.-L., Bao, J.-B., Wang, J., Hu, G.-H., Xu, Y., & Zhao, L. (2007). Preparation of open-cell polymer foams by CO_2 assisted foaming of polymer blends. *Polymer*, 90, 331–341.

Kumar, V., Sinha, A. K., Makkar, H. P. S., &Becker, K. (2010). Dietary roles of phytate and phytase in human nutrition: A review. *Food Chemistry,* 120(4), 945–959

Lewandowicz, G., Fornal, J., & Walkowski, A. (1997). Effect of microwave radiation on potato and tapioca starches. *Carbohydrate Polymers*, 34, 213.

Lewandowicz, G., Jankowski, T., Fornal, J. (2000). Effect of microwave radiation on physico-chemical properties and structure of cereal starches. *Carbohydrate. Polymer*. 42, 193–199.

Li, H.-C. (1995). *Identification of aroma compounds and evaluation of volatile losses during the frying and microwave reheating of a flour-based batter system (pancake)*. Ph.D. dissertation, University of Minnesota, United States -- Minnesota. Retrieved June24, 2010, from Dissertations & Theses: A&I. (Publication No. AAT 9534129).

Luo, Z., He, X., Fu, X., Luo, F., Gao, Q. (2006). Effect of microwave radiation on the physicochemical properties of normal maize, waxy maize and amylo-maize V starches. *Starch/Sta¨rke* 58, 468–474.

MadrugaMS, de AlbuquerqueFSM, SilvaIRA, do AmaralDS, MagnaniM, NetoVQ. (2014) Chemical, morphological and functional properties of Brazilian jackfruit (Artocarpus heterophyllusL.) seeds starch. *Food Chemistry* 143:440–445

Manley, D. J. R. (2000). *Technology of biscuits, crackers and cookies*–Third Edition. CRC Woodhead Publishing Limited, ISBN- 1855735326, Abington.

Marsaioli Jr., A. (1991). Desenvolvimento de um protótipo de secadorcilíndrico–rotativo a microondas e a arquente para a secagemcontínua de produtossólidosgranulados. 197p. Tese (DoutoremEngenharia de Alimentos) –Faculdade de Engenharia de Alimentos,UniversidadeEstadual de Campinas.

Martinez-Bustos, F. M.; Garcia, M.; Chang, Y. K.;Sanchezsinencio, F. &Figueroa, C. (2000). Characteristics of nixtamalised maize flours produced with the use of microwave heating during alkaline cooking. *Journal of the Science of Food and Agriculture*. 80(6), 651–656.

McWatters, K.,H. & Holmes, M.,R. (1979). Influence of pH and Salt Concentration on Nitrogen Solubility and Emulsification Properties of Soy Flour. *Journal of Food Science, 44,*770–773.

Mertens, B. & Knorr, D. (1992). Developments of nonthermal processes for food preservation . *Food Technology.* **46**(5): 124–133 .

MohamedAA, AshmanRB and KirleisAW .(1993). Pericarp thickness and other kernel physical characteristics relate to microwave popping quality of popcorn. *Journal of Food Science,*58: 342–346.

Moraru, C.I. &Kokini, J.L. (2003). Nucleation and expansion during extrusion and microwave heating of cereal foods, *Comprehensive reviews in Food Science and FoodSafety*. 2(3), 120–138.

Morzelle, M. C., Salgado, J. M., Massarioli, A. P., Bachiega, P., Rios, A. de O., Alencar, S. M.,. (2019). Potential benefits of phenolics from pomegranate pulp and peel in alzheimer's disease: Antioxidant activity and inhibition of acetylcholinesterase. *Journal of Food Bioactives*. 5, 136–141.

Nayak, B., Liu, R., H. &Tang, J. (*2015*). Effect of processing on phenolic antioxidants of fruits, vegetables, and grains– A review. *Critical reviews in Food Science and Nutrition,*55(7),887–918

Nikmaram, N., Leong, S. Y., Koubaa, M., Zhu, Z., Barba, F. J. &Greiner, R. (2017). Effect of extrusion on the anti-nutritional factors of food products: An overview. *Food Control,*79, 62–73.

Ozmutlu, O., Sumnu, G., Sahin, S. 2001. Effects of different formulations on the quality of microwave baked breads. *European Food Research Technology,*213, 38–42.

Palav, T. &Seetharaman, K. (2007). Impact of microwave heating on the physico-chemical properties of a starch–water model system. *Carbohydrate. Polymer*, 67, 596–604.

Park, T. (2001). A study of microwave-water interactions in bread system. Ph.D. dissertation, Rutgers. The State University of New Jersey–New Brunswick, United States -- New Jersey. Retrieved June 24, 2010, from Dissertations & Theses: A&I. Publication No. AAT 9973982

Plagemann, R., Langermann, von J. &Kragl, U. (2014). Microwave assisted covalent immobilization of enzymes on inorganic surfaces. *Eng. Life Sci.*14(5), 493–499.

Quitain, A. T., Daimon, H., Fujie, K., Katoh, S. &Moriyoshi, T. (2006). Microwave-assisted hydrothermal degradation of silk protein to amino acids. *Industrial and Engineering Chemical Research*, 45,4471–4474.

Raes, K., Knockaert, D., Struijs, K., &Van Camp, J. (2014). Role of processing on bio accessibility of minerals: Influence of localization of minerals and anti-nutritional factors in the plant. *Trends in Food Science and Technology,*37(1), 32–41.

Ragaee, S., Seetharaman, K. & Abdel-Aal, E-S., M. (2014).The impact of milling and thermal processing on phenolic compounds in cereal grains. *Critical Reviews in Food Science and Nutrition.*; 54(7):837–849.

Richardson DL. (1959). Effect of certain endosperm genes on popping ability in popcorn. *Agronomy Journal*, 51: 631–635.

Sacharow, S. & Schiffmann, R. (1992). *Microwave Packaging*. Pira International, *Leatherhead* UK, 155p, *Safety*, 2(3),120–138.

Sale, A. J. H. (1976). A review of microwave for food processing. *International Journal of Food Science & Technology*.11 (4), 319–329.

Samtiya, M., Aluko, R. E., & Dhewa, T. (2020). Plant food anti-nutritional factors and their reduction strategies: An overview. *Food Production, Processing and Nutrition,*2(6), 1–14

Schiffmann, R. F. (1986). Food product development for microwave processing. *Food Technology Journal.* 40(6), 94–98

Schwartzberg, H.,G., Wu, J.,P.,C., Nussinovitch, A. & Mugerwa, J. (1995). Modelling deformation and flow during vapor-induced puffing. *Journal of Food Engineering*,25, 329–372.

Senise, J. T. & Jermolovicius, L. A. (2004). Microwave Chemistry–A Fertile Field For Scientific Research And Industrial Applications. *Journal of Microwaves andOptoelectronics*. 3(5), 97–112.

Shevkani, K., Singh, N., Bajaj, R. & Kaur, A. (2017). Wheat starch production, structure, functionality and applications: a review. *International Journal of Food Science and Technology,* 52,38–58.

Singh, A., Gupta, S., Kaur, R., & Gupta, H. R. (2017). Process optimization for anti-nutrient minimization of millets. *Asian Journal of Dairy and Food Research,*36(4), 322–326.

Singh, J. & Singh, N. (1999). Effect of different ingredients and microwave power on popping characteristics of popcorn. *Journal of Food Engineering,*42(3), 161–165.

Stinson, C. T. (1986). A Quality Comparison of Devil's Food and Yellow Cakes Baked in a Microwave/ Convection Versus a Conventional Oven. *Journal of Food Science*. 51(6), 1578–1579.

Suhag, R., Dhiman, A., Deswal, G., Deswal G., Thakur, D., Sharanagat, V. S., Kumar, K.T. & Kumar V. (2021). Microwave processing: A way to reduce the anti-nutritional factors (ANFs) in food grains. _LWT - Food Science and Technology. 150 (2021) 111960:111960

Şummu, G., Ndife, M., K. & Bayindirli, L. (2000). Optimization of microwave baking of model layer cakes. *European Food Research and Technology*, 211(3), 169–174.

Swinkels, J., J., M. (1985). Composition and properties of commercial native starches. *Starch/Stärke,*37, 1–5

Timotheo, C. A. &Lauer, C. M. (2018). Toxicity of vegetable tannin extract from Acacia mearnsii in Saccharomyces cerevisiae. *International Journal of Environmental Science and Technology, 15*(3), 659–664.

Tornberg, E., & Hermansson, A.,M. (1977). Functional characterization of protein stabilized emulsions: effect of processing. *Journal of Food Science*, 42,468–472

Tsoubeci, M. N. (1994). Molecular interactions and ingredient substitution in cereal based model systems during conventional and microwave heating. Ph.D. dissertation, University of Minnesota, United States -- Minnesota. Retrieved June 24, 2010, from *Dissertations,& Theses*: A&I.(Publication No. AAT 9514682).

Velupillai, L., Verma, L., R. & Tsangmuichung, M. (1989). *Process for parboiling rice*. United States Patent 4,810,511.

Vojdani, F. (1996). Solubility. In: HallGM (ed) Methods of testing protein functionality. Chapman, London, pp 11–60.

Wang, C., R. &Zayas, J., F. (1992). Emulsifying capacity and emulsion stability of soy proteins compared with corn germ protein flour. *Journal of Food Science*, 57,726–731.

Wang, S. H.,Clerici, M., T., P., S. &Sgarbieri, V., C. (1993). Característicassensoriaisnutricionais de mingau de preparorápidoà base de farinhas de arroz e soja desengordurada e leiteempó. *Alimentos e Nutrição*, 5(1), 77–86.

WitkiewiczK, NastajJF. (2010). Simulation Strategies in Mathematical Modeling of Microwave Heating in Freeze-Drying Process. *Drying Technology,*28, 1001–1012.

Yang, H. W.,Hsu, C. K.&Yang, Y. F. (2014). Effect of thermal treatments on anti-nutritional factors and antioxidant capabilities in yellow soybeans and green-cotyledon small black soybeans. *Journal of the Science of Food and Agriculture, 94*(9), 1794–1801.

Yin, Y., &Walker, C., E. (1995). A quality comparison of breads baked by conventional versus nonconventional ovens: A review. *Journal of Science and Food Agriculture*, 67, 283–291.

Young, L. & Cauvain, S.P. (2007). Technology of Breadmaking. Springer; 2 Ed., ISBN-10-0751403458, Germany.

ZayasJF.(1997). Functionality of proteins in foods. *Springer*. Berlin, Heidelberg, New York.

Zhang, J., Chen, F., Liu, F. &Wang, Z. W. (2010). Study on structural changes of microwave heat-moisture treated resistant Canna edulis Ker starch during digestion in vitro. *Food Hydrocolloids* 24, 27–34.

Zhang, J., Wang, Z. W., &Shi, X. M. (2009). Effect of microwave heat/moisture treatment on physicochemical properties of *Canna edulis* Ker starch, *Journal of Science Food and Agriculture* 89,653–664.

Zhong, Y., Wang, Z., &Zhao, Y. (2015). Impact of radio frequency, microwaving, and high hydrostatic pressure at elevated temperature on the nutritional and antinutritional components in black soybeans. *Journal of Food Science, 80*(12), 2732–2739.

CHAPTER 17

Infrared Heating of Cereals

Shulin Yang,[1] Zhenui Cao,[2] Xin Ying,[2] Xiaoming Wei,[2] Lisa F.M. Lee Nen That,[3] Jessica Pandohee,[4] and Bo Wang[5]
[1]COFCO Grains Holdings Limited, Beijing, China
[2]COFCO Nutrition & Health Research Institute, Beijing, China
[3]School of Science, RMIT University, Bundoora, Victoria, Australia
[4]Centre for Crop and Disease Management, School of Molecular and Life Sciences, Curtin University, Bentley, WA, Australia
[5]School of Behavioural and Health Sciences, Australian Catholic University, New South Wales, Australia

CONTENTS

DOI: 10.1201/9781003242192-21

17.1 INTRODUCTION

Infrared (IR) is a small section of the electromagnetic spectrum with characteristic wavelength in the range of 0.75 to 1,000 µm, frequency between 300 GHz–430 THz, and photon energy between 1.7 eV to 1.24 meV (Figure 17.1). IR is further divided into three regions based on their wavelength. These include near-infrared (NIR, 0.75–1.40 µm), mid-infrared (MIR, 1.40–3.00 µm), and far-infrared (FIR, 3.00–1000 µm) regions (Sakai & Hanzawa, 1994). It is well-accepted that IR is responsible for the thermal transmission in most heat and light. This heating effect of IR has been used to naturally heat treat food by harnessing IR from sunlight since ancient times. The use of IR as a traditional method of food drying was aimed at reducing water content in food in order to increase the storage time of food for winter or for times when hunting and gathering was not possible.

Today IR is used in the food processing industry for much more than just food drying and plays an important role in cereal and grain production. One of the major purposes of IR irradiation is as a decontamination procedure in post-harvesting of cereals (Los et al., 2018). Cereals are at risk of bacterial, spore, yeast, and fungal infections once they have been harvested. The irradiation of wheat harvest using IR for 60 s has been shown to significantly reduce the bacterial population from 5.7 × 104 cfu/g to 0.73 × 103 cfu/g (Hamanaka et al., 2000) and thus reduce agricultural losses due to microbial spoilage. Enzyme inactivation is another important application of IR irradiation as it results in the denaturation of lipases and amylases present in cereals, which causes deterioration of lipids and starch leading to rancidity and off-flavours (Sawai et al., 2003). In order to optimise the effect of IR on wheat germs, Li et al. (2016) showed that irradiating wheat samples at 90°C for 20 mins was a compromise between inactivating lipase and lipoxygenase and keeping free fatty acid and peroxide values low for 60 days. Other benefits of IR irradiation in cereal post-processing include drying, roasting, blanching, and taste enhancing.

Cereals are complex samples made of macromolecules (such as starch, protein, lipids), small biochemical molecules (comprising phytochemical and bioactive compounds), and other micronutrients, inorganic salts, and water (Zamaratskaia et al., 2021). While IR irradiation extends the shelf-life of cereals and their safety for future consumption, it also affects the nutritional and functional characteristics of cereals. Other organoleptic qualities of cereals that are less studied but are essential for consumer food acceptance are also changed post IR treatment (Dogu-Baykut & Kirkin, 2021; Semwal & Meera, 2021). The aim of this chapter is to summarise and discuss the effect of IR irradiation on the nutritional, functional, and biological characteristics of cereals.

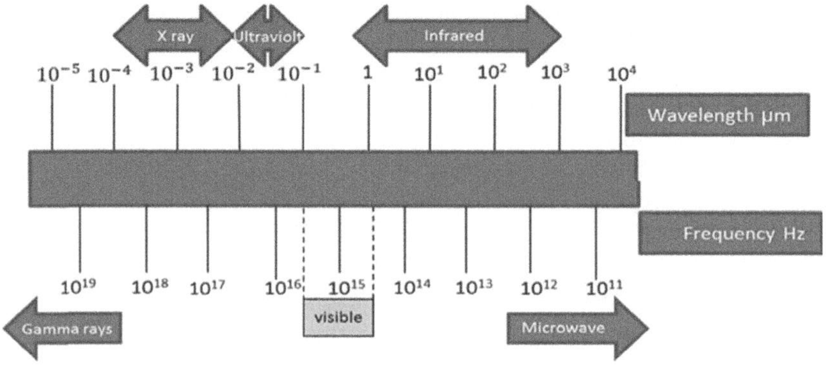

Figure 17.1 Electromagnetic radiation spectrum.

17.2 PROCESSING TREATMENT

17.2.1 Fundamentals of Treatment

In general, when the cereals are treated with IR, due to the difference in the physicochemical properties and surface condition of the cereal and the wavelength of used IR, some radiation is absorbed, some is reflected to the atmosphere while some penetrates through the material (as shown in Figure 17.2). Once the IR is absorbed by the food, the molecules and atoms in the cereal that have the same natural radio frequency as the radiation frequency of IR produce strong resonance to absorb the radiation energy. This energy is then predominantly used to induce the thermal motion of these molecules and atoms (Zhang et al., 2017). Subsequently, the motion of the molecules is accelerated to change their vibrational energy, and this finally leads to an increased temperature in the internal region of the cereal. Meanwhile, due to the moisture evaporation on the surface, the temperature of the external region of the cereal is lower than its internal parts so this temperature gradient drives heat transfers along using the material as the medium. Meanwhile, the moisture content in the internal region of the raw cereal grains is usually higher than its external region (i.e., the surface). The IR treatment also facilitates the formation of a humidity gradient. Therefore, the heating or drying efficiency of cereal using IR treatment is higher compared to contemporary techniques.

The absorption capacity of different compounds to IR and the absorption of the same compound to IR with different wavelengths varies. In general, the effectiveness of IR treatment of cereal grains is usually positively correlated with the IR absorption capacity of the materials (Sakai & Hanzawa, 1994). In cereals, polyatomic molecules such as organic compounds and macromolecules have a wide absorption band for IR, so they can effectively absorb IR energy during the treatment (Van Kranendonk, 1959). Briefly, the absorption of IR by water occurs in the wavelength range of 2–11 μm, which covers the entire far IR range. Carbohydrates have an IR absorption wavelength range of 3 and 7–10 μm. On the other hand, lipids have a broad absorption band that ranges almost the entirety of the far IR spectrum, with strong absorption bands at 3–4, and 9–10 μm (Aboud et al., 2019). Amino acids, polypeptides, and proteins show two strong absorption bands at 3–4 and 6–9 μm (Sandu, 1986). As such, among the three regions of the IR, the far-infrared region is the most widely used in food applications.

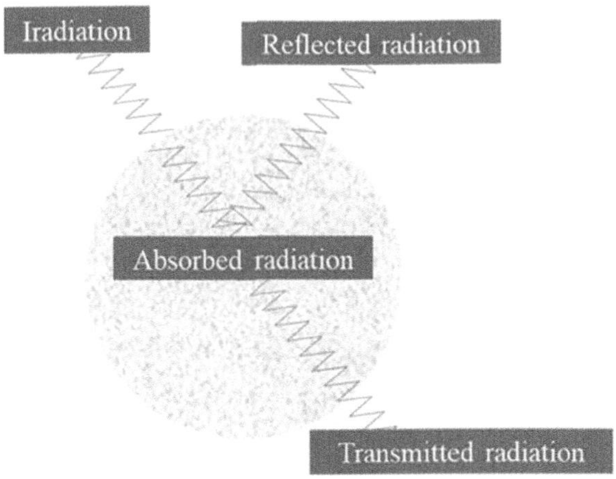

Figure 17.2 Process of radiation.

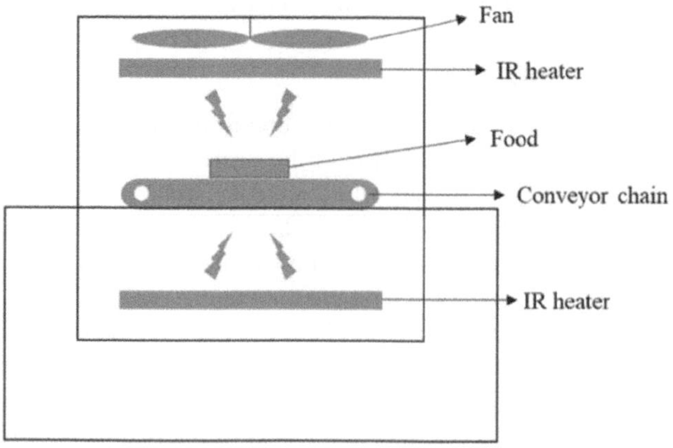

Figure 17.3 Infrared heating and drying equipment.

17.2.2 Process and Machinery Involved

The structure of IR heating or drying equipment is shown in Figure 17.3. Briefly, cereals are treated with IR radiation during their movement on the conveyor chain. Among all the components in the system, the IR heating element is the core one. Generally, based on the heating mode, this element can be categorised as gas and electric heating radiators. The latter one can be further divided into tubular/flat metallic heaters (long waves), ceramic heaters (long waves), quartz tube heaters (medium, short waves), and halogen tube heaters (ultra-short waves) (Rastogi, 2012).

To date, three basic radiation laws are used to describe the distribution and quantity of IR energy during the IR treatment of samples: Stefan–Boltzmann's, Planck's, and Wien's displacement law. The first are used to determine the intensity of emitted radiation and the spectral distribution, while the last one determines the maximum emission wavelength of the IR heating source (Sakare et al., 2020). Based on the advance of theory and overcome of the technical bottlenecks, the application of IR technology at the industrial scale has been explored. For example, Pan et al. (2015) reported the successfully application of pilot-scale IR dry system in tomato peeling facilities and the catalytic IR technology has been commercialised by Catalytic Drying Technology Inc. in Kansas, U.S.A.

17.3 IMPACT OF PROCESSING TREATMENT ON NUTRITIONAL CHARACTERISTICS

Cereals have high nutritional value as they contain both macronutrients and micronutrients including proteins, carbohydrates, dietary fibre, vitamins, and minerals (McKevith, 2004). Several studies have demonstrated that IR radiation has an impact on the nutritional value of cereals. The effects of IR treatment on these nutrients in cereals are discussed in this section.

17.3.1 Impact of IR on Starch in Cereals

The total carbohydrate content determined in maize, rice, and sorghum is 73.1 ± 2.5%, 80.1 ± 2.3%, 75.3 ± 2.0%, respectively. A brief IR treatment at 22,000°C for 0.5 min and samples exiting the radiator at 140 °C increased the levels of maize, rice, and sorghum to 82.5 ± 4.6%, 81.9 ± 1.5%, 78.4 ± 4.0%, respectively (Keya & Sherman, 1997). A reduction in moisture could have led to the increase in levels of these cereals. Rousta et al. (2014) also observed this increase in sorghum due to loss of moisture after IR treatment.

The effect of IR and microwave heating were tested on starch in three types of Canadian barley: normal starch barley (NB), high amylose barley (HAB), and waxy starch barley (WB) (Emami et al., 2011). The initial total starch (% dry weight basis) for NB, HAB, and WB were 65.1 ± 3.1, 58.2 ± 1.9, and 46.3 ± 1.4 and the amylose content (% of starch) for NB, HAB, and WB were 28.7 ± 0.4, 46.5 ± 0.3, 0.7 ± 0.1, respectively. The IR treatment resulted in a higher gelatinised starch percentage compared to microwave radiation in all three types of barley. The starch gelatinisation % was dependent on the type of barley, which had different ratio of amylose to amylopectin. The highest % of starch gelatinisation was determined from WB followed by NB and HAB. Swaminathan et al. (2015) reported that in sorghum flour, the starch gelatinisation was determined by testing the peak viscosity, which is an indication of gelatinisation of starch for a cereal undergoing heat treatment. As both temperature (80–120°C) and time of IR treatment (5–15 min) increased, the peak viscosity decreased, which would be ideal for preparation of calorie-dense foods.

17.3.2 Impact of IR on Proteins in Cereals

Crude protein content in maize, rice, and sorghum increased after IR treatment due to a loss in moisture (Keya & Sherman, 1997; Rousta et al., 2014). Moreover, following a combined IR-vacuum treatment, the level for crude protein in rice bran increased from 12.26 ± 0.05 g/100g dry basis to 13.08 ± 0.14 g/100g dry basis (Kreungngern et al., 2021). Tuncel et al. (2010) also reported minute differences in crude protein content of corn samples by comparing untreated sample and the three drying methods: Infrared radiation treatment, hot air treatment, and combined IR and hot air treatment. A study by Özer et al. (2018) showed that the substitution of corn flour with IR stabilised immature rice grain flour (IRGF) in gluten-free bread led to a 7-fold rise in protein content. The protein level (%, d.w.) increased from 2.52 ± 0.69 in gluten-free bread with corn starch to 17.49 ± 0.49 in bread with IRGF. Protein quality of hull-less barley flour was also investigated where the content of the essential amino acids was not affected in untreated samples and samples treated with tempering moisture and IR heating (Bai et al., 2018).

A study was carried out on the comparison of far-infrared radiation (FIR) and hot air drying (HA) in polished and unpolished pigmented rice (Ratseewo et al., 2020). Three varieties of pigmented Thai rice, namely Mali Dang (red), Hom Nil (purple), and riceberry (purple), were tested and ten amino acids were detected in both pigmented and unpigmented rice and included threonine, isoleucine, tryptophan, phenylalanine, histidine, methionine, arginine, valine, leucine, and lysine. Both drying methods resulted in a decline in the amino acid content in the pigmented rice varieties. The polished Mali Dang rice had the biggest decrease in isoleucine (99%), lysine (80%), and phenylalanine (75%) for FIR-dried rice. The loss in methionine content 89% was the highest in HA-treated rice. The findings suggest that the drying methods have a more detrimental effect on polished rice than unpolished rice as the external layer protects the inner parts of the grain.

17.3.3 Impact of IR on Lipids in Cereals

Storage stability is an issue in cereals as they are susceptible to rancidity caused by lipase activity (Tuncel et al., 2014). Several studies have looked at the effect of IR on lipids and storage stability in cereals, especially rice (Irakli et al., 2018; Yılmaz et al., 2018; Yılmaz et al., 2013). The disruption of cells during the milling process releases lipase, which hydrolyses fat to free fatty acids (FFA) and glycerol. Stabilisation is an essential step to inactivate enzymes and prolong shelf life. FFA content is a good indicator of lipase activity (Irakli et al., 2018; Yılmaz et al., 2018). The effect of IR treatment has resulted in the decline of FFA content and improved storage stability. Studies in rice have shown that in untreated samples, FFA content in rice bran and immature rice grain increased from 4–5% to 35–49% after 20 days to 6 months of storage (Yan et al., 2020; Yılmaz et al., 2018; Yılmaz et al., 2013). However, the treatment of IR caused a significant decline in FFA content (10–14%) as IR power increased. Yılmaz (2016) reported that the highest increase of FFA content was observed in the first month of storage followed by a gradual raise, and a short processing time did not satisfactorily inhibit the lipase activity. Although IR treatment lowered the FFA content, rice containing >5% FFA content was not suitable for human consumption (Yılmaz et al., 2013). Optimisation of IR conditions is crucial and Irakli et al. (2018) suggested that optimal IR treatment for rice bran was for 15 mins at 140°C. Optimal combined IR-vacuum treatment conditions determined by response surface methodology and central composite design consisted of IR power (999W), vacuum strength (650 mmHg), and processing time (562 s) (Kreungngern et al., 2021). Comparison of three different techniques, namely continuous ambient air drying (AAD), combined effect of IR drying and tempering (IRD), and combined effect of hot air drying (HAD) in brown and rough rice showed that during the ten months of storage, IRD treatment had the lowest increase in FFA content for rough rice (5.12 ± 0.05%) and brown rice (6.91 ± 0.07%) (Ding et al., 2015). The highest increase in FFA content was obtained by AAD for rough rice (10.05 ± 0.09%) and brown rice (13.16 ± 0.04%). This suggests that rough rice is more stable and IRD treatment supports the highest storage stability and lipase inactivation.

Swaminathan et al. (2015) reported that a decrease in FFA content (88%) was observed as temperature (120°C) and processing time (10 mins) were increased. An increase in IR radiation intensity and processing time and a decrease in distance between the wheat germ samples and emitters led to a decrease in FFA content (from 1.55 to 1.81–1.83 g oleic acid/100 g oil) during the 90 days of storage (Gili et al., 2017).

The most abundant fatty acids in rice are palmitic acid, oleic acid, and linoleic acid (Irakli et al., 2018; Kreungngern et al., 2021; Yan et al., 2020; Yılmaz et al., 2018; Yılmaz et al., 2013). Studies looking at the effect of IR showed that there were no significant differences in the fatty acid profile (Irakli et al., 2018; Yılmaz et al., 2018; Yılmaz et al., 2013). Yan et al. (2020) reported that there were minute differences in the fatty acid profile from the effect of IR. In IR-treated samples to test the stability of fatty acids, the levels of saturated fatty acids, monounsaturated fatty acids, and polyunsaturated fatty acids varied from 17.9 mg/g, 35.83 mg/g, 37.5 mg/g to 19.1 mg/g, 38.4 mg/g, and 40.6 mg/g at day 0 of storage. These levels remained higher in the 20 days of storage as compared to untreated samples. In untreated samples, there was significant decrease in the saturated fatty acids (8.0 mg/g), monounsaturated fatty acids (18.5 mg/g), and polyunsaturated fatty acids (19.4 mg/g).

17.3.4 Impact of IR on Vitamins in Cereals

IR treatment has had contradicting effects on vitamins in cereals. Tuncel et al. (2014) showed that the substitution of wheat flour with IR stabilised rice bran (SRB) at increasing levels (2.5, 5.0, and 10.0%) caused an augmentation in different forms of vitamins B content, namely thiamine, riboflavin, niacin, pyridoxamine, and pyridoxine. Three types of bread were tested including white bread, wheat bran added to white bran, and whole grain wheat bread. Niacin had the highest increase and the lowest rise in content was riboflavin. This is due to their reported known amounts in rice as niacin is the most abundant in rice bran while riboflavin is present in small amounts in rice. The highest content of niacin was also observed in wheat bran bread due to its presence in aleurone layer and using both wheat bran and SRB exacerbated its content. The study looking at the substitution of corn starch with IR-stabilised immature rice grain flour at increasing levels (30, 50, 70, and 100%) in gluten-free breads tested only thiamine and riboflavin content (Özer et al., 2018). There was no significant difference in thiamine content while riboflavin had a significant increase in breads with

IRGF. Although there is a higher content of thiamine than riboflavin in the bran, the heat treatment caused by IR treatment (1600W) may have a negative effect on thiamine content.

IR power, temperature, and processing time are the main factors that affect the rate of degradation in tocopherols and tocotrienols. Several studies have demonstrated that IR treatment increased the tocopherol content in treated samples compared to untreated samples (Özer et al., 2018; Ratseewo et al., 2020; Wanyo et al., 2014; Yılmaz et al., 2018; Yılmaz, 2016). In some studies, there was no γ-tocopherol and δ-tocopherol in untreated samples, but it was present in all IR-treated samples (Wanyo et al., 2014; Yılmaz et al., 2018). The IR treatment causes the release of bound tocopherols to high molecular weight compounds in rice grain (Yılmaz et al., 2018). It is important to optimise the IR processing conditions as a high temperature could lead to degradation of heat-sensitive tocopherol (Kreungngern et al., 2021). Studies by Irakli et al. (2018) in rice reported a decrease in tocopherol content in treated samples and untreated samples. However, IR treatment stabilised the decline in tocopherols as the percentage loss in untreated samples was higher than treated samples indicating that IR had stabilised the samples. A similar trend was observed in a study by Kreungngern et al. (2021). Similar results were observed in the tocotrienol content, which decreased in both treated and untreated samples (Irakli et al., 2018; Kreungngern et al., 2021).

Gili et al. (2017) demonstrated in wheat germ studies that during the 90 days of storage, the total tocopherol content for untreated wheat germ was between 4133 and 4181 while IR-treated samples had a significant decrease between 3395 and 3899. IR-treated samples still had sustained high levels of tocopherols. Three factors, namely IR radiation intensity, treatment time, and distance between the wheat germ samples and emitters, played a role in the effect of IR treatment, and it was not possible to determine a relationship among the factors.

17.3.5 Impact of IR on Minerals in Cereals

Minerals are usually abundant in the bran layer of cereals and the most dominant minerals in rice bran are phosphorus and potassium (Saunders, 1985; Tuncel et al., 2014). A study looked at the effect of substituting wheat flour with SRB at increasing levels of 2.5, 5.0, and 10.0% in three types of bread, namely white bread, bread containing 40% of wheat bran, and whole grain wheat bread (Tuncel et al., 2014). The mineral content, namely zinc, iron, potassium, calcium, and phosphorus, was also investigated in the three types of breads. Levels of zinc, iron, potassium, and phosphorus were augmented at increasing doses of SRB. However, the amount of calcium declined as SRB was increased. A higher content of calcium is usually observed in wheat flour and the amount of SRB substituted in the breads led to its decrease.

Corn starch was substituted with IRGF at increasing levels of 30, 50, 70, and 100% in gluten-free bread (Özer et al., 2018). IRGF substitution had a positive effect on the levels of potassium, calcium, magnesium, phosphorus, iron, zinc, manganese, and sulphur. The levels of magnesium and manganese had the highest incline in bread with 100% level of IRGF. Sodium content was the only mineral that decreased in bread with IRGF.

17.4 IMPACT OF PROCESSING TREATMENT ON FUNCTIONAL CHARACTERISTICS

17.4.1 Hydration Properties

The hydration property is one of the most important properties of cereal products. This is because the hydration step is commonly used to process different grains for many reasons, such as cooking, extraction, fermentation, germination, malting, etc. To evaluate the swelling behaviour of the starch component of cereal grains, water absorption index (WAI) and water solubility index (WSI) are commonly used. Briefly, cereal products are soaked in an excessive amount of water for a certain time, followed by the centrifugation and separation of the precipitate from the supernatant. The ratio between the original dry cereal and precipitate sample is called the WAI. Subsequently, the separated supernatant can be dehydrated and WSI can be calculated as the ratio between the weight of dried supernatant and original dry cereal (Yu et al., 2017). In the study by Altan (2014), barley grains were processed by IR heating, followed by microwave irradiation. The authors observed that the WAI of processed grain was increased by 220% by the combined treatment. Similarly, Mapengo & Emmambux (2020) investigated the functional properties of IR-treated maize meal complexed with stearic acid. The results indicated that with the increase of temperature caused by IR treatment, both WAI and WSI were significantly increased.

The improved hydration property of cereal product via IR treatment was due to the structure change of the starch component. During the IR treatment, the crystalline structure of starch granules is disrupted and the exposed hydroxyl groups in starch molecules facilitate water binding, which further causes an enhanced granular swelling and hence increases the WAI and WSI (Ye et al., 2018).

17.4.2 Rheological Properties

IR treatment also affects rheological properties of cereal meal, starch, flour, etc., in terms of altering their pasting profile. In the study by Mapengo et al. (2021), maize starch and maize starch-stearic acid complex were processed using IR-based heat-moisture treatment. The results showed both processed starch and starch-fatty acid complex exhibited longer pasting time, increased the pasting temperature, and decreased the peak viscosity during pasting, compared with native maize starch. The authors also observed that the firmness of the gel formed by the processed starch increased with the increase of the duration of IR treatment (1 to 2 h). This trend was also reported in the IR-based heat-moisture treated sorghum flour (Swaminathan et al., 2015), corn starch (Ismailoglu & Basman, 2015), and wheat starch (Ismailoglu & Basman, 2016) .

The mechanism of the reduced pasting viscosity and increased pasting temperature of cereal flour or cereal starch was partially due to the increased hydrogen bonding between starch chains. It was also hypothesised that after IR-based moisture treatment, the denatured protein matrix partially prevented the hydration, dispersibility, and molecular entanglement of embedded starch during the pasting, finally causing a reduction in pasting viscosity (Ogundele & Kayitesi, 2019).

17.4.3 Color Profile

IR treatment can also introduce colour change (browning) in cereals (Bhinder et al., 2019). The intensity of the browning is proportional to the extent of the IR treatment in terms of temperature and time. Bhinder et al. (2019) treated buckwheat grain using IR at 130–170°C for 10 min and investigated the colour change. Browning index (BI), defined by the absorbance of flour extract at 420 nm, was measured before and after the treatment. Before the IR treatment, the original BI of eight types of buckwheat varied in the range of 0.30–0.39 and after 10 min treatment at 130°C, this index was increased by nearly 50% (0.57–0.69). The browning with similar intensity of buckwheat using the conventional heating was only observed when grains were heated at 160°C for 40 min (Małgorzata et al., 2016). Swaminathan et al. (2015) also reported the effect of IR treatment on the colour change of sorghum during the processing. The authors observed the lightness of the grains started to decrease once the temperature was beyond 120°C in the time range of 5–15 min.

It was proposed that the browning of the grain is due to Maillard reaction and polymerisation of phenolic compounds during IR treatment. Žilić et al. (2013) observed that the amount of 5-hydroxymethyl furfural (HMF) in the maize flour was increased from a "not detectable level" to

21.25 and 46.88 μg/g, after processing the sample at 120 and 140°C for 85-–100 s, respectively. In the study by Bhinder et al. (2019), the authors reported the browning of the IR-treated buckwheat was positively correlated to the content of Maillard reaction products. Bhinder (2019) and Žilić (2013) reported the correlation between the colour change and the total phenolic content in the sample. It was proposed that the oxidation and polymerisation of the phenolic compounds during IR treatment could also contribute to the browning.

Additionally, IR treatment can also induce the fading of colour in some cereal grains. For example, Žilić et al. (2013) reported the fading of deep yellowish colour of maize after IR treatments at 110°C for 50 s. The "L^*" value of the IR-treated sample increased from 57.60 to 61.54 while the "b^*" value decreased from 36.69 to 31.15, suggesting decline of the darkness and yellowness. This is probably due to the degradation of carotenoids during the treatment.

17.4.4 Thermal Properties

During the IR treatment of cereals, physical or chemical modification of the materials may occur (Aboud et al., 2019). As such, the thermal properties of grain are usually affected. In general, the extent of the thermal properties change is proportional to the degree of modification.

In the study by Ding et al. (2018), hot air, IR, and ambient air were used to dry brown rice. Subsequently, the effect of dehydration method on the thermal property of the rice flour was investigated. These authors observed that compared with the flour from the brown rice dried using hot and ambient air, the one made from IR-treated sample had lower onset (T_o) and conclusion (T_c) temperatures, range of gelatinisation temperature (W_p), and gelatinisation enthalpy (ΔH). This could be due to the high temperature and high initial moisture content of starch granules during the IR drying, which favoured the annealing of the starch granules located in the outer layer of rice kernels and altered the thermal and pasting properties of starch.

IR treatment can also affect the thermal properties of cereals by hydrolysing amylopectin. In the study by Su et al. (2020), germinated barley grains were dried using hot air or IR at 60°C for 12–36 h and the starch was isolated to evaluate its thermal properties. Compared with germinated and hot air-dried barley starch, the germinated and IR dried one exhibited a significantly lower gelatinisation enthalpy (ΔH). This was due to the more significant starch hydrolysis in the barley grains during the drying using IR than the one processed with hot air (55.89 vs. 32.72% as the proportion of smaller amylopectin with chain length ≤12). Compared with hot air drying, the IR treatment can provide more energy to destroy the non-reducing ends of the amylopectin chain and reduce the relative crystallinity in the starch. Therefore, a partially hydrolysed sample with smaller amylopectin chains can be produced, with a lower amount of energy required for gelatinisation.

17.4.5 Structural Properties

During the IR treatment, the rapid increase of the water vapour pressure within the cereals can cause cotyledon rupture, leading to potential fissures and cavities on the surface structure of cereals (Ogundele & Kayitesi, 2019). In the study by Andrejko et al. (2011), after the wheat grain was processed with IR at 180°C for 120 s, dramatic cracks from the centre to the surface of the grain were observed (shown in Figure 17.4). These defects were found to be correlated with a compromised compression resistance of the grain. The authors also observed the gelatinisation of starch in the grain after the IR treatment, due to the melting of crystallites.

The microstructure of the starch granule in the cereal grain can also be affected by the IR treatment. In the study by Ismailoglu & Basman (2016), wheat starch was tempered to 30% moisture content and treated with IR at 750 W. The authors observed clear rupture of some starch granules using bright light microscopy. Under polarised light, a clear polarisation cross was observed at the

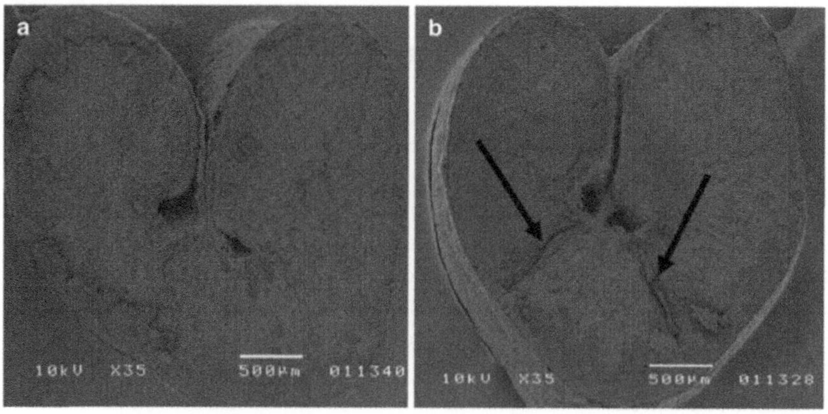

Figure 17.4 Transverse section of wheat grain: a) intact grain, b) grain after IR processing.

Source: Adapted from Andrejko et al., 2011.

center of native starch granule but the birefringence at the center of some wheat granules disappeared after IR treatment. Similar results have been reported in corn starch (Ismailoglu & Basman, 2015) and maize meal (Mapengo & Emmambux, 2020).

In the study by Mapengo & Emmambux (2020), the starch granules formed aggregates after the maize meal was processed by IR-based heat moisture treatment. Additionally, the authors reported that the size of these aggregates increased with the increased duration of the IR treatment. This phenomenon could be due to the reassociation of starch chain with each other in the weak region during the IR treatment. Su et al. (2020) compared the morphology of the starch granules in the barley flour treated by IR and hot-air drying, after 36 h germination. The result suggested the starch granules in the IR-dried sample were more separated (i.e., without being surrounded by protein and other components) (as shown in Figure 17.5). This was possible because the IR energy directly reached the surface of the cereal grain, which made the enzyme more active to hydrolyse the flour, resulting in the release of granules more efficiently.

17.5 IMPACT OF PROCESSING TREATMENT ON BIOLOGICAL PROPERTIES

17.5.1 Anti-nutrients

Anti-nutrients or anti-nutritional factors are biological compounds that reduce nutrient availability and digestibility by binding with nutrients and/or other molecules providing nutritional benefits. The bran of cereal grains (wheat, oat, rice, and millet) is well known to contain phytates, oxalates, and tannins (Nadeem et al., 2010; Saleh et al., 2013). While phytic acid is a natural antioxidant, improves blood glucose and cholesterol, and has anticancer properties, it also reduces starch digestibility and mineral bioavailability. Tuncel et al. (2014) showed that rice bran stabilized with IR at 700 W for 3 min had a significantly higher amount of phytic acid whereas Rousta et al. (2014) reported a significant decrease in phytate and tannin content in sorghum irradiated with IR using a 1000 W lamp for 60, 90, and 120 s with longer treatments showing a larger decrease of anti-nutritional factors. Moreover, the content of phytic acid in IR-treated immature rice grain was shown to be significantly higher than in untreated samples (Özer et al., 2018).

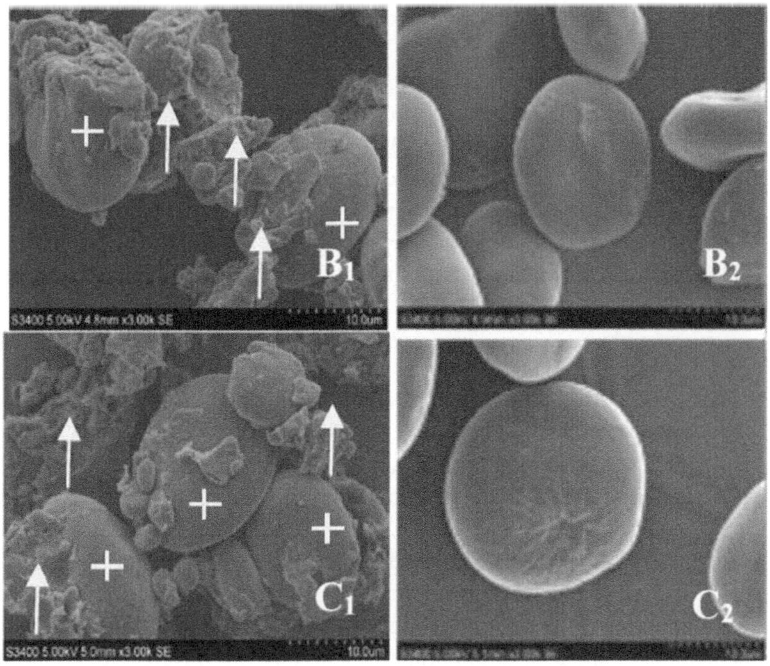

Figure 17.5 Scanning electron micrographs (1500×) of naked barley flour (1) ("+" represent starch granules and "↑" represent protein and fibre structures); scanning electron micrographs (1700×) of naked barley starch (2); polarized light micrographs of naked barley starch (×40).

Source: Partially adapted from Su et al. (2020).

Phenolic compounds, which include phenolic acids (such as p-hydroxybenzoic, protocatechuic, and vanillic acids) and cinnamic acids (such as p-coumaric, caffeic, ferulic, sinapic acids), are also known as phytochemical compounds. In cereals phenolic acids are mainly stored in the cell wall material in bran and possess bioactive properties and health-promoting properties (Dykes & Rooney, 2007). Infrared irradiation has had varied effects on phenolic compound content in cereals including an increase in pigmented rice and a reduction in hull-less barley flours, but a decrease in rice bran, husks, and ground rice husk (Bai et al., 2018; Ratseewo et al., 2020; Wanyo et al., 2014). To date, no comprehensive study of the effect of IR irradiation on all the anti-nutritional factors in cereals has been reported. Other important but yet understudied compounds such as gossypol, lectins, protease inhibitors, and amylase inhibitors in cereals reduce the digestibility of proteins and mineral absorption (Gilani et al., 2019; Kaur et al., 2015).

17.5.2 Phytochemical Profile

As a staple food, cereals are known to be rich in starch, proteins, and lipids but their phytochemical content is less discussed. Phytochemicals are small bioactive molecules (also known as specialised metabolites) that are manufactured by plants for fungal, bacterial, or viral resistance (Godstime et al., 2014). In humans phytochemical compounds bring health benefits such as anti-inflammatory, antioxidant, and antibacterial properties by neutralising reactive oxygen species in the body (Dillard & German, 2000). Interestingly, the largest group of bioactive compounds and phytochemicals in cereal grains is phenolic compounds including phenolic acids, alkylresorcinols, and flavonoids amongst less abundant carotenoids and β-glucan (Belobrajdic & Bird, 2013; Ward et al., 2008).

Antioxidant properties of cereals are increased post irradiation of IR waves. FIR of unpolished and polished pigmented rice increased the level of flavonoid content, anthocyanin, and antioxidant capacity in pigmented rice (Ratseewo et al., 2020). Nam et al. (2004) also reported an increase in antioxidant properties of rice hull extract after FIR accompanied by fewer volatile aldehydes. Interestingly the antioxidant properties of the rice hull extract only lasted for three days. The application of short waves of IR has been shown to not have significant effects on the total phenolic and antioxidant amount in rice bran, with the only loss of nutrient being thiamine (Yılmaz & Tuncel, 2015).

17.6 CONCLUSION

Infrared heating is a technology that has many applications in post-harvest cereal production. Through its heating and drying effects, IR radiation is today commonly applied to cereals to increase shelf-life and improve their safety by destroying spores, bacteria, or fungus as well as deactivating enzymes that would otherwise cause rancidity or off-flavours. This chapter discussed how IR treatment affects the nutritional, functional, and biological characteristics of cereals. While IR treatments have been reported to have negative impacts on the digestibility of cereals, they also improve their phytochemical and antioxidant content. It is important to note that the wavelength, duration, and power used for the treatments have different effects on cereal contents. Moreover, the chemical compounds and contents in various cereals have been shown to react differently to IR treatment, a behaviour that could be attributed to the complex matrix they are embedded in. Therefore, for the cereal industry it is crucial the IR strategies be optimised for the desired health benefits.

REFERENCES

Aboud, S. A., Altemimi, A. B., R. S. Al-Hilphy, A., Yi-Chen, L., & Cacciola, F. (2019). A Comprehensive Review on Infrared Heating Applications in Food Processing. *Molecules, 24*(22), 4125. www.mdpi.com/1420-3049/24/22/4125

Altan, A. (2014, 2014/02/01/). Effects of pretreatments and moisture content on microstructure and physical properties of microwave expanded hull-less barley. *Food Research International, 56*, 126–135. https://doi.org/https://doi.org/10.1016/j.foodres.2013.12.029

Andrejko, D., Grochowicz, J., Goździewska, M., & Kobus, Z. (2011, 2011/11/01). Influence of Infrared Treatment on Mechanical Strength and Structure of Wheat Grains. *Food and Bioprocess Technology, 4*(8), 1367–1375. https://doi.org/10.1007/s11947-009-0238-3

Bai, T., Nosworthy, M. G., House, J. D., & Nickerson, M. T. (2018). Effect of tempering moisture and infrared heating temperature on the nutritional properties of desi chickpea and hull-less barley flours, and their blends. *Food Research International, 108*, 430–439. https://doi.org/10.1016/j.foodres.2018.02.061

Belobrajdic, D. P., & Bird, A. R. (2013, 2013/05/16). The potential role of phytochemicals in wholegrain cereals for the prevention of type-2 diabetes. *Nutrition Journal, 12*(1), 62. https://doi.org/10.1186/1475-2891-12-62

Bhinder, S., Singh, B., Kaur, A., Singh, N., Kaur, M., Kumari, S., & Yadav, M. P. (2019, 2019/07/01/). Effect of infrared roasting on antioxidant activity, phenolic composition and Maillard reaction products of Tartary buckwheat varieties. *Food Chemistry, 285*, 240–251. https://doi.org/https://doi.org/10.1016/j.foodchem.2019.01.141

Dillard, C. J., & German, J. B. (2000). Phytochemicals: nutraceuticals and human health. *Journal of the Science of Food and Agriculture, 80*(12), 1744–1756. https://doi.org/https://doi.org/10.1002/1097-0010(20000915)80:12<1744::AID-JSFA725>3.0.CO;2-W

Ding, C., Khir, R., Pan, Z., Wood, D. F., Venkitasamy, C., Tu, K., El-Mashad, H., & Berrios, J. (2018, 2018/10/30/). Influence of infrared drying on storage characteristics of brown rice. *Food Chemistry, 264*, 149–156. https://doi.org/https://doi.org/10.1016/j.foodchem.2018.05.042

Ding, C., Khir, R., Pan, Z., Zhao, L., Tu, K., El-Mashad, H., & McHugh, T. H. (2015). Improvement in shelf life of rough and brown rice using infrared radiation heating. *Food and Bioprocess Technology, 8*, 1149–1159. https://doi.org/10.1007/s11947-015-1480-5

Dogu-Baykut, & Kirkin, C. (2021). Impact of Organoleptic and Consumer Acceptability for Non-Thermally Processed Grain-Based Food Products. In M. Selvamuthukumaran (Ed.), *Non-Thermal Processing Technologies for the Grain Industry* (Vol. 1, pp. 304). CRC Press.

Dykes, L., & Rooney, L. W. (2007). Phenolic compounds in cereal grains and their health benefits. *Cereal Foods World, 52*, 105–111.

Emami, S., Meda, V., & Tyler, R. T. (2011). Effect of micronisation and electromagnetic radiation on physical and mechanical properties of Canadian barley. *International Journal of Food Science and Technology, 46*, 421–428. https://doi.org/10.1111/j.1365-2621.2010.02505.x

Gilani, G. S., Cockell, K. A., & Sepehr, E. (2019). Effects of Antinutritional Factors on Protein Digestibility and Amino Acid Availability in Foods. *Journal of AOAC INTERNATIONAL, 88*(3), 967–987. https://doi.org/10.1093/jaoac/88.3.967

Gili, R. D., Palavecino, P. M., Penci, M. C., Martinez, M. L., & Ribotta, P. D. (2017). Wheat germ stabilization by infrared radiation. *Journal of Food Science and Technology, 54*(1), 71–81. https://doi.org/10.1007/s13197-016-2437-z

Godstime, O., Felix, E., Augustina, J., & Christopher, E. (2014). Mechanisms of Antimicrobial Actions of Phytochemicals against Enteric Pathogens – A Review. *Journal of Pharmaceutical, Chemical and Biological Sciences, 2*(2), 77–85.

Hamanaka, D., Dokan, S., Yasunaga, E., Kuroki, S., Uchino, T., & Akimoto, K. (2000, 01/01). The sterilization effects of infrared ray on the agricultural products spoirage microorganisms. *2000 ASAE Annual Intenational Meeting, Technical Papers: Engineering Solutions for a New Century, 2*, 971–979.

Irakli, M., Kleisiaris, F., Mygdalia, A., & Katsantonis, D. (2018). Stabilization of rice bran and its effect on bioactive compounds content, antioxidant activity and storage stability during infrared radiation heating. *Journal of Cereal Science, 80*, 135–142. https://doi.org/10.1016/j.jcs.2018.02.005

Ismailoglu, S. O., & Basman, A. (2015). Effects of infrared heat-moisture treatment on physicochemical properties of corn starch. *Starch – Stärke, 67*(5–6), 528–539. https://doi.org/https://doi.org/10.1002/star.201400266

Ismailoglu, S. O., & Basman, A. (2016). Physicochemical properties of infrared heat-moisture treated wheat starch. *Starch – Stärke, 68*(1–2), 67–75. https://doi.org/https://doi.org/10.1002/star.201500160

Kaur, S., Dar, B. N., Pathania, S., & Sharma, S. (2015). Reduction of Antinutritional Factors in Cereal Brans for Product Development. *Journal of Food Processing and Preservation, 39*(3), 215–224. https://doi.org/https://doi.org/10.1111/jfpp.12223

Keya, E. L., & Sherman, U. (1997). Effects of a brief, intense infrared radiation treatment on the nutritional quality of maize, rice, sorghum, and beans. *Food and Nutrition Bulletin, 18*(4), 1–6.

Kreungngern, D., Kongbangkerd, T., & Ruttarattanamongkol, K. (2021). Optimization of rice bran stabilization by infrared-vacuum process and storage stability. *Journal of Food Processing Engineering, 44*, e13668. https://doi.org/10.1111/jfpe.13668

Li, B., Zhao, L., Chen, H., Sun, D., Deng, B., Li, J., Liu, Y., & Wang, F. (2016). Inactivation of Lipase and Lipoxygenase of Wheat Germ with Temperature-Controlled Short Wave Infrared Radiation and Its Effect on Storage Stability and Quality of Wheat Germ Oil. *PLOS ONE, 11*(12), e0167330. https://doi.org/10.1371/journal.pone.0167330

Los, A., Ziuzina, D., & Bourke, P. (2018). Current and Future Technologies for Microbiological Decontamination of Cereal Grains. *Journal of Food Science, 83*(6), 1484–1493. https://doi.org/https://doi.org/10.1111/1750-3841.14181

Małgorzata, W., Konrad, P. M., & Zieliński, H. (2016, 2016/04/01/). Effect of roasting time of buckwheat groats on the formation of Maillard reaction products and antioxidant capacity. *Food Chemistry, 196*, 355–358. https://doi.org/https://doi.org/10.1016/j.foodchem.2015.09.064

Mapengo, C. R., & Emmambux, M. N. (2020, 2020/09/30/). Functional properties of heat-moisture treated maize meal with added stearic acid by infrared energy. *Food Chemistry, 325*, 126846. https://doi.org/https://doi.org/10.1016/j.foodchem.2020.126846

Mapengo, C. R., Ray, S. S., & Emmambux, M. N. (2021, 2021/06/01/). Structural and digestibility properties of infrared heat-moisture treated maize starch complexed with stearic acid. *International Journal of Biological Macromolecules, 180*, 559–569. https://doi.org/https://doi.org/10.1016/j.ijbiomac.2021.03.100

McKevith, B. (2004). Nutritional aspects of cereals. *Nutrition Bulletin, 29*(2), 111–142. https://doi.org/10.1111/j.1467-3010.2004.00418.x

Nadeem, m. k., Anjum, F., Amir, R. A. I., Khan, M. R., Hussain, S., & Javed, M. S. (2010, 01/01). An overview of anti-nutritional factors in cereal grains with special reference to wheat-A review. *Pakistan Journal of Food Sciences, 20*, 54–61.

Nam, K. C., Kim, J.-H., Ahn, D. U., & Lee, S.-C. (2004, 2004/01/01). Far-Infrared Radiation Increases the Antioxidant Properties of Rice Hull Extract in Cooked Turkey Meat. *Journal of Agricultural and Food Chemistry, 52*(2), 374–379. https://doi.org/10.1021/jf035103q

Ogundele, O. M., & Kayitesi, E. (2019, 2019/04/01). Influence of infrared heating processing technology on the cooking characteristics and functionality of African legumes: a review. *Journal of Food Science and Technology, 56*(4), 1669–1682. https://doi.org/10.1007/s13197-019-03661-5

Özer, M., Yılmaz Tuncel, N., & Tuncel, N. B. (2018). The effects of infrared stabilized immature rice grain flour in gluten-free bread preparation. *Cereal Chemistry, 95*, 527–535. https://doi.org/10.1002/cche.10056

Pan, Z., Li, X., Khir, R., El-Mashad, H. M., Atungulu, G. G., McHugh, T. H., & Delwiche, M. (2015, 2015/09/01/). A pilot scale electrical infrared dry-peeling system for tomatoes: Design and performance evaluation. *Biosystems Engineering, 137*, 1–8. https://doi.org/https://doi.org/10.1016/j.biosystemseng.2015.06.003

Rastogi, N. K. (2012). Chapter 13 – Infrared Heating of Fluid Foods. In P. J. Cullen, B. K. Tiwari, & V. P. Valdramidis (Eds.), *Novel Thermal and Non-Thermal Technologies for Fluid Foods* (pp. 411–432). Academic Press. https://doi.org/https://doi.org/10.1016/B978-0-12-381470-8.00013-X

Ratseewo, J., Meeso, N., & Siriamornpun, S. (2020). Changes in amino acids and bioactive compounds of pigmented rice as affected by far-infrared radiation and hot air drying. *Food Chemistry, 306*, 125644. https://doi.org/10.1016/j.foodchem.2019.125644

Rousta, M., Sadeghi, A. A., Shawrang, P., Aimn Afshar, M., & Chamani, M. (2014). Effect of gamma, electron beam and infrared radiation treatment on the nutritional value and anti-nutritional factors of sorghum grain. *Iranian Journal of Applied Animal Science, 4*(4), 723–731.

Sakai, N., & Hanzawa, T. (1994, 1994/11/01/). Applications and advances in far-infrared heating in Japan. *Trends in Food Science & Technology, 5*(11), 357–362. https://doi.org/https://doi.org/10.1016/0924-2244(94)90213-5

Sakare, P., Prasad, N., Thombare, N., Singh, R., & Sharma, S. C. (2020, 2020/09/01). Infrared Drying of Food Materials: Recent Advances. *Food Engineering Reviews, 12*(3), 381–398. https://doi.org/10.1007/s12393-020-09237-w

Saleh, A. S. M., Zhang, Q., Chen, J., & Shen, Q. (2013). Millet Grains: Nutritional Quality, Processing, and Potential Health Benefits. *Comprehensive Reviews in Food Science and Food Safety, 12*(3), 281–295. https://doi.org/https://doi.org/10.1111/1541-4337.12012

Sandu, C. (1986). Infrared Radiative Drying in Food Engineering: A Process Analysis. *Biotechnology Progress, 2*(3), 109–119. https://doi.org/https://doi.org/10.1002/btpr.5420020305

Saunders, R. M. (1985). Rice bran: Composition and potential food uses. *Food Reviews International, 1*(3), 465–495. https://doi.org/10.1080/87559128509540780

Sawai, J., Sagara, K., Hashimoto, A., Igarashi, H., & Shimizu, M. (2003). Inactivation characteristics shown by enzymes and bacteria treated with far-infrared radiative heating. *International Journal of Food Science & Technology, 38*(6), 661–667. https://doi.org/https://doi.org/10.1046/j.1365-2621.2003.00717.x

Semwal, J., & Meera, M. (2021). Infrared Radiation: Impact on Physicochemical and Functional Characteristics of Grain Starch. *Starch – Stärke, 73*(3–4), 2000112. https://doi.org/https://doi.org/10.1002/star.202000112

Su, C., Saleh, A. S. M., Zhang, B., Feng, D., Zhao, J., Guo, Y., Zhao, J., Li, W., & Yan, W. (2020, 2020/12/15/). Effects of germination followed by hot air and infrared drying on properties of naked barley flour and starch. *International Journal of Biological Macromolecules, 165*, 2060–2070. https://doi.org/https://doi.org/10.1016/j.ijbiomac.2020.10.114

Swaminathan, I., Guha, M., Hunglur, U. H., & Rao, D. B. (2015). Optimization of infrared heating conditions of sorghum flour using central composite design. *Food Science and Biotechnology, 24*(5), 1667–1671. https://doi.org/10.1007/s10068-015-0216-7

Tuncel, N. B., Yılmaz, N., Kocabıyık, H., & Uygur, A. (2014). The effect of infrared stabilized rice bran substitution on B vitamins, minerals and phytic acid content of pan breads: Part II. *Journal of Cereal Science, 59*, 162–166. https://doi.org/10.1016/j.jcs.2013.12.005

Tuncel, N. B., Yılmaz, N., Kocabıyık, H., Öztürk, N., & Tunçel, M. (2010). The effects of infrared and hot air drying on some properties of corn (*Zea mays*). *Journal of Food, Agriculture & Environment, 8*(1), 63–68. https://doi.org/10.1234/4.2010.1441

Van Kranendonk, J. (1959, 1959/01/01/). Induced infrared absorption in gases calculation of the ternary absorption coefficients of symmetrical diatomic molecules. *Physica, 25*(1), 337–342. https://doi.org/https://doi.org/10.1016/S0031-8914(59)93671-7

Wanyo, P., Meeso, N., & Siriamornpun, S. (2014). Effects of different treatments on the antioxidant properties and phenolic compounds of rice bran and rice husk. *Food Chemistry, 157*, 457–463. https://doi.org/10.1016/j.foodchem.2014.02.061

Ward, J. L., Poutanen, K., Gebruers, K., Piironen, V., Lampi, A.-M., Nyström, L., Andersson, A. A. M., Åman, P., Boros, D., Rakszegi, M., Bedő, Z., & Shewry, P. R. (2008, 2008/11/12). The HEALTHGRAIN Cereal Diversity Screen: Concept, Results, and Prospects. *Journal of Agricultural and Food Chemistry, 56*(21), 9699–9709. https://doi.org/10.1021/jf8009574

Yan, W., Liu, Q., Wang, Y., Tao, T., Liu, B., Liu, J., & Ding, C. (2020). Inhibition of lipid and aroma deterioration in rice bran by infrared heating. *Food and Bioprocess Technology, 13*, 1677–1687. https://doi.org/10.1007/s11947-020-02503-z

Ye, J., Hu, X., Luo, S., Liu, W., Chen, J., Zeng, Z., & Liu, C. (2018). Properties of Starch after Extrusion: A Review. *Starch – Stärke, 70*(11–12), 1700110. https://doi.org/https://doi.org/10.1002/star.201700110

Yılmaz, F., Yılmaz Tuncel, N., & Tuncel, N. B. (2018). Stabilization of immature rice grain using infrared radiation. *Food Chemistry, 253*, 269–276. https://doi.org/10.1016/j.foodchem.2018.01.172

Yılmaz, N. (2016). Middle infrared stabilization of individual rice bran milling fractions. *Food Chemistry, 190*, 179–185. https://doi.org/10.1016/j.foodchem.2015.05.094

Yılmaz, N., & Tuncel, N. B. (2015). The effect of infrared stabilisation on B vitamins, phenolics and antioxidants in rice bran. *International Journal of Food Science and Technology 50*, 84–91.

Yılmaz, N., Tuncel, N. B., & Kocabıyık, H. (2013). Infrared stabilization of rice bran and its effects on γ-oryzanol content, tocopherols and fatty acid composition. *Journal of the Science of Food and Agriculture, 94*(8), 1568–1576. https://doi.org/10.1002/jsfa.6459

Yu, C., Liu, J., Tang, X., Shen, X., & Liu, S. (2017). Correlations between the physical properties and chemical bonds of extruded corn starch enriched with whey protein concentrate [10.1039/C6RA26764E]. *RSC Advances, 7*(20), 11979–11986. https://doi.org/10.1039/C6RA26764E

Zamaratskaia, G., Gerhardt, K., & Wendin, K. (2021, 2021/01/01/). Biochemical characteristics and potential applications of ancient cereals – An underexploited opportunity for sustainable production and consumption. *Trends in Food Science & Technology, 107*, 114–123. https://doi.org/https://doi.org/10.1016/j.tifs.2020.12.006

Zhang, M., Chen, H., Mujumdar, A. S., Tang, J., Miao, S., & Wang, Y. (2017, 2017/04/13). Recent developments in high-quality drying of vegetables, fruits, and aquatic products. *Critical Reviews in Food Science and Nutrition, 57*(6), 1239–1255. https://doi.org/10.1080/10408398.2014.979280

Žilić, S., Ataç Mogol, B., Akıllıoğlu, G., Serpen, A., Babić, M., & Gökmen, V. (2013, 2013/07/01/). Effects of infrared heating on phenolic compounds and Maillard reaction products in maize flour. *Journal of Cereal Science, 58*(1), 1–7. https://doi.org/https://doi.org/10.1016/j.jcs.2013.05.003

PART V

Non-Thermal Processing of Cereals and Its Impact

CHAPTER 18

High Hydrostatic Pressure Processing of Cereals

Rajat Suhag,[1,3] Chandrakala Ravichandran[2,3], and Ashutosh Upadhyay[3]
[1]Faculty of Science and Technology, Free University of Bozen-Bolzano, Bolzano, Italy
[2]Department of Food Technology, Rajalakshmi Engineering College, Chennai, Tamil Nadu, India
[3]Department of Food Science and Technology, National Institute of Food Technology Entrepreneurship and Management, Kundli, Sonipat, Haryana, India

CONTENTS

DOI: 10.1201/9781003242192-23

18.1 INTRODUCTION

Cereal grains are a major source of energy, carbohydrates, fiber, proteins, vitamins, and minerals in animal and human diets. The cereal group is an important component of a balanced diet. They provide important essential vitamins and minerals such as vitamin A, vitamin B_{12}, iron, and calcium for the growth and maintenance of the human body. Cereals include wheat, rice, maize, rye, oat, barley, sorghum, millet, etc. Among these, wheat and rice account for more than half of all staple foods across many countries. Numerous studies have discovered that cereal grains possess functional compounds like β-glucan, dietary fibre, inulin, resistant starch, and phenolics with proven health advantages. Scientific evidence also reveals the fact that consumption of whole grains may reduce the incidence of risk associated with diseases such as stroke, hypertension, diabetes, metabolic syndrome, and cancer (Borneo & León, 2012). Furthermore, cereals are processed in a variety of ways like milling, germination, fermentation, extrusion, and microwave heating, which cause alteration in physicochemical, cooking, and functional properties. Novel non-thermal technologies are capturing the food processing sector in recent years due to their green approach, ensuring food safety, enhanced shelf life, and natural products with fewer implications on sensory and nutritional quality. High hydrostatic pressure processing (HHP) is one such technique with quasi-instantaneous uniform distribution of pressure throughout the sample of any shape and size delivering products with enhanced shelf life and high quality.

HHP is a novel non-thermal processing method that employs high hydrostatic pressure, normally around or above 100 MPa. This process inactivates various microbes and enzymes, without affecting the flavor or nutrients of the food (Balakrishna et al., 2020). It is commonly utilized in the food industry to make a variety of products. During HHP a food product experiences equal pressure on all sides in a short amount of time. Foods with porous matrix are not appropriate for HHP processing because the food matrix may collapse or distort (Maqsood-ul-Haque & Kamal, 2021). HHP may also significantly reduce the pathogen/bacterial count and can cause denaturation of proteins without altering the molecular bonding. Reduction in bacterial count, prolongs the shelf life and improves the safety and quality of food products. Furthermore, HHP can reduce the impact of changes that result in the loss of vitamins and minerals, as well as the formation of off-flavours. Working principles of HHP are (i) Le Chatelier's principle; (ii) isostatic pressing; and (iii) microscopic-ordering (Ravash et al., 2020; Serna-Hernandez et al., 2021).

1. ***Le Chatelier's Principle***: It states that an equilibrium chemical system will experience a change in reaction and a decrease in volume when pressured and vice versa.
2. ***Isostatic Pressing***: It is also called Pascal's Principle. According to this the pressure is distributed evenly in all the dimensions. The substance reverts to its original shape upon decompression.
3. ***Microscopic Ordering Principle***: Asserts that increasing the pressure improves the degree of ordering of a substance's molecules at a constant temperature. As a result, pressure and temperature have opposing effects on molecule structure.

The major proportion of cereals is starch. Starch is a natural, biodegradable polymer consisting of mixture of linear amylose and branched amylopectin. The primary goal of using HHP to process starch has been to produce starches with good functional characteristics. Native starch suffers from limitations in the food industry due to their instability under temperature, pH, shear, etc. Starch is commonly used as thickening agent, stabilizer, gelling agent in beverages, creams, desserts, sauces, etc. Through modification, its application could be extended due to varying functional attributes.

18.2 MECHANISM OF HHP ON STARCH PROCESSING

High-pressure application on cereals involves use of wide pressure ranging from 100 to 800 MPa. Based on the desired objective, the pressure employed varies. As per H. Liu, Fan, et al. (2016), pressure application on starches is categorized based on final intended application. When the static pressure of 200–8000 MPa is applied onto the starch suspension containing excess water to achieve gelatinization is called high-pressure gelatinization. In contrast, low moisture starch exposed to static pressure of 200–1100 MPa to modify their internal and external structure is called high-pressure compression. When a dynamic pressure of 20–100 MPa is applied to achieve modification of starch it is called high-pressure homogenization.

The basic objective to process starch under high pressure is to achieve gelatinization at lower temperatures and cause alteration in native structure and its properties. Starches are divided into three types based on their X-Ray diffraction patterns: A-type (representing cereal sources, e.g., rice, peanut, barley and wheat); B-type (high amylose starches and tuber sources, e.g., corn and potato starch); and C-type (representing legume, root, and fruit sources, e.g., lentil, tapioca, and arrowroot starch) (Maniglia, Castanha, Rojas, & Augusto, 2021; Y. Hibi et.al., 1993). Generally, A-type starches are the most susceptible to HHP, due to their short amylopectin branches and lower degree of polymerization (6–15). Whereas B-type are the most resistant as they are highly polymerized glucose chains with polymerization degree of 40–50. Thirdly, C-type starches are a combination of both A- and B-type and amylopectin with non-reducing ends demonstrating medium sensitivity (Pei-Ling et.al., 2010; Alcázar-Alay & Meireles, 2015). Influence of HHP on starch gelatinization has been extensively researched recently (Raghunathan et al., 2021). Heat treatment and HHP both cause gelatinization, but the mechanisms are distinct. Temperature and pressure are the critical parameters for changing starch structures in food matrix, as in the case of protein denaturation (Knorr et al., 2006).

Gelatinization is the process causing irreversible changes in starch due to molecular disruption within the starch granule. The rupture of starch granules caused by heat is known as gelatinization. It increases the amount of starch available for amylase degradation. As a result, starch gelatinization is essential during food processing to make starch digestible or to thicken/bind water in food products like sauce or soup. Gelatinization can be achieved through thermal and pressure treatment. HHP affects the temperature at which starch gelatinizes (Jiang et al., 2015). Furthermore, after the swelling stage, pressured granule dispersion has lower viscosity than thermally processed starch granules. The lack of amylose solubilization due to unbroken or incomplete breakdown of starch granules is the cause of reduced viscosity (Fukami et al., 2010), as shown in Figure 18.1. Heat-induced gelatinization leads to an increase in amylose solubilization through disintegration, as demonstrated in Figure 18.2. Since amylopectin is more stable in the granules, amylose solubility is increased, resulting in decreased melting of crystalline regions. Van der Waals and hydrogen bonding both become more stable under pressure. As a result, it prevents helix unwinding and side-by-side dissociation. This causes the crystalline area to disintegrate partially and the helical structure to unwind. Overall, the mechanism behind HHP-induced gelatinization of starch is due to stabilization effect produced by the linkage of large amount of water molecules to the structure of helix via Van der Waals interactions, changing the crystallinity characteristics of the starch from type A to type B (Knorr et al., 2006). Liu et al. (2020) reported that starch crystallites are disturbed sequentially during heat gelatinization, starting with the less stable crystallites and progressing to the more stable crystallites. Pressure, on the other hand, causes non-preferential crystallite disruption. Therefore, the internal structure of granules after pressure and heat treatment is expected to differ, altering the functional properties including swelling power, thermal transitions, gel texture, and viscoelasticity.

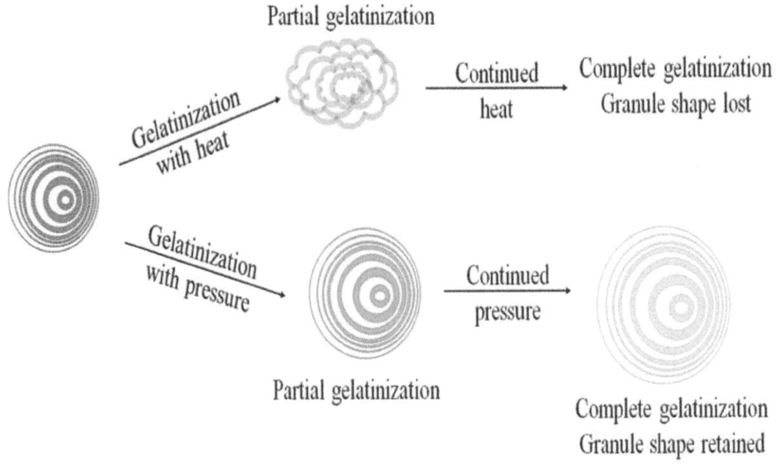

Figure 18.1 Schematic representation of heat and pressure gelatinization (reprinted from Balakrishna et al. (2020) under open access Creative Common CC BY license).

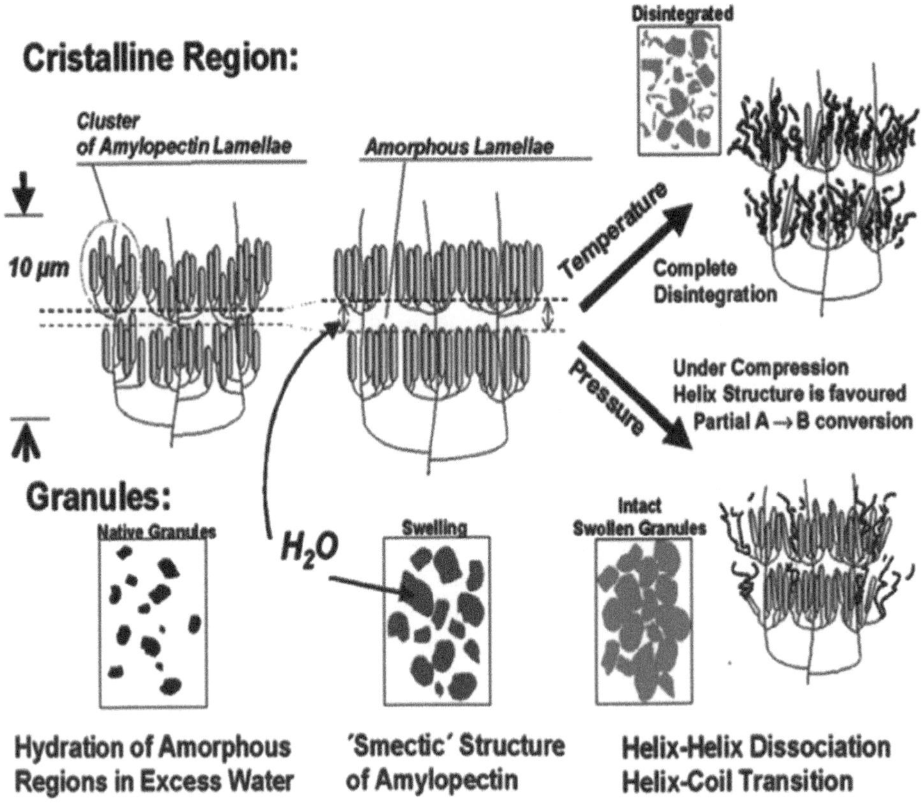

Figure 18.2 Mechanism of HHP-induced gelatinisation (reprinted from Knorr et al. (2006) with permission from Elsevier).

Morphological characteristics studied by scanning electron microscopy (SEM) revealed irregular shapes with smooth surfaces for native starch granules of rice, sorghum, and high amylose maize (Maniglia et al., 2021). HHP treatment at 200 and 400 MPa caused partial gelatinization showing rough surfaces with cracks, but at 600 MPa it produced irreversible swelling. Amylose and amylopectin chains interact strongly during HHP treatment, resulting in a dense structure with pits and perforations on the surface (Raghunathan et al., 2021; Shen et al., 2018; Zhang et al., 2019).

HHP treatment results in partial gelatinization of rice, wheat, corn, and waxy corn starches at constant pressure and temperature greater than 300 MPa, but complete gelatinization at 600 MPa (Kawai et al., 2012). HHP can gelatinize most types of starches at temperatures under 0°C if the treatment pressure is adequate. If heat is used in addition to pressure, then low working pressure is sufficient to cause starch gelatinization. Rice starch was fully gelatinized by HHP treatments at 500 MPa and 600 MPa (Pei-Ling et al., 2010). Furthermore, the nature of starch and processing duration have an impact on the physical features of gels formed from starch suspensions. Wheat, corn, and rice starch-based hydrogels prepared using HHP treatment exhibited cream-like structure, while hydrogels prepared using tapioca starch had a denser structure. Hydrogels made from tapioca and rice starch had greater viscosity, stiffness, and G' (storage modulus) values after a 15 min HHP treatment at 600 MPa, indicating overall structural reinforcement (Larrea-Wachtendorff et al., 2020). In addition, Deng et al. (2014) reported two-cycle (two 15-min cycles at 200 and 600 MPa) HHP treatment is more beneficial than continuous HHP treatment. In comparison to continuous 30 min HHP treatment, they found that HHP treatment of rice starch for two 15 min cycles resulted in more gelatinization and structural disturbances, elimination of surface protrusion, and reduced resistant starch at the same pressure.

Finally, the application of HHP has generated intriguing outcomes in the case of starch properties, resulting in extended applications. However, more research is needed on the interplay and performance of modified starch with other substances found in industrial formulations, like proteins, fats, salt, sucrose, and other polysaccharides. It is also worth remarking that the results for each type of starch may vary, necessitating more research to expand the existing information.

18.3 EFFECT OF HHP ON NUTRITIVE VALUE OF CEREALS

18.3.1 Protein

Changes in plant proteins caused by HHP have been a matter of discussion in the literature in recent times. In addition, emphasis has been placed on how pressure can be utilized to tune specific functional characteristics, like the ability to form gels or aggregates, or the potential to stabilize nano-emulsions and foams for various food products (Queirós et al., 2018). The ability to induce breakage, conformational changes, and aggregate of protein molecules in solution by causing a volume change in the protein molecules acts as the mechanisms of protein modification by HHP. This concept is grounded on Le Chatelier's principle (Akharume et al., 2021; Messens et al., 1997).

Pressure affects protein denaturation and final configuration by causing structural changes at several stages (Gharibzahedi & Smith, 2021). In general, HHP affects only non-covalent linkages, such as hydrogen, hydrophobic, and ionic linkages, while covalent bonds, which have a limited compressibility under high pressure, are mostly unaffected (Considine et.al., 2008). Due to covalent bonds, the core structure of the protein is essentially unaffected, but the secondary, tertiary, and quaternary structures are susceptible to disruption due to ionic, hydrophobic, hydrogen bonds, and electrostatic interactions (Yang & Powers, 2016). Pressures above 2 GPa generally affect the primary structure, while pressures of >400 and 100–200 MPa, mostly compromise the hydrogen, hydrophobic, disulphide, and ionic bonds in secondary and tertiary/quaternary structures, respectively (Knorr et al., 2006).

In germinated foxtail millet flour after HHP, protein content changed insignificantly. The protein content reduced non-significantly from 13.65% to 13.11% post HHP at 600 MPa, 120 min at 60°C (N. Sharma et al., 2018). The creation of a protein–starch complex or hydrogen/covalent/ionic interaction amid starch and protein molecules could be responsible for the decrease in protein content. Furthermore, in comparison to other processing methods such as soaking and germination, Kakati, Deka, Kotoki, & Saikia (2010) found that pressure cooking slightly reduced protein content. In another study, Angioloni & Collar (2012) evaluated the effect of incorporating HHP-treated wheat, oat, sorghum, and finger millet flour on protein content of bread. The protein level of bread made with HHP-treated flours was significantly reduced ($p < 0.05$). After HHP treatment of flours, the protein content of wheat, oat, finger millet, and sorghum bread decreased by 26.0%, 26.95%, 28.66%, and 28.48%, respectively.

Despite technological advancement, the detailed mechanism behind changes in protein confirmation, functionality, and protein content of cereals has yet to be discovered further. As a result, more research is needed for construction of protein structures with customized functionality and tailored applications.

18.3.2 Fat

Thermal treatment of foods induces oxidation processes and reduces overall acceptability. Fat oxidation is a thermodynamically controlled process (Medina-Meza et.al., 2013). High-pressure treatment can alter the thermodynamic balance of biochemical processes. This is true in the case of fat oxidation, where the dynamics are impacted by HHP (Medina-Meza et al., 2014). HHP-induced fat oxidation can change the product's color, flavor, nutritional value, and functional qualities, and result in hazardous by-products such as lipid hydroperoxides, carbonyl compounds, oxysterols, etc. (Kubow, 1992; Schaich, 2005). HHP treatment has been reported to induce fat oxidation in high-fat products such as chicken breast muscles (Wiggers et al., 2004), turkey thigh muscles (Tuboly et al., 2003), mackerels (Senturk & Alpas, 2013), ham (Clariana & García-Regueiro, 2011), etc.

Fat oxidation begins in the presence of heat, light, or metal ions and is catalyzed by enzymes already present in the food matrix. Further, it involves formation of free alkyl groups by transfer of a hydrogen ion from a fatty acids α-methylene group. Oxygen-sensitive free radical produces a lipid peroxyl radical when combined with molecular oxygen. As it combines with other fatty acids, it produces hydroperoxide and added radicals. Only the available free radical resumes the reaction with the help of other fatty acids. When two free fatty acid radicals react and generate a non-radical, the reaction chain comes to a halt, which can transpire after 10–100 cycles (Barbhuiya et al., 2021; Min & Boff, 2002).

Few research works have been undertaken to study the impact of HHP on cereal fat content and fat oxidation. Wang, Zhu, Ramaswamy, Hu, & Yu (2018) investigated the influence of HHP parameters (pressure and holding duration) on fat acidity of brown rice. The effects of HHP treatment on fat acidity of brown rice were found to be insignificant at lower pressures (below 400 MPa). HHP treatment at 400 MPa increased the levels of fat acidity from 24.9 to 35.5 (0 min) and 33.0 (10 min) mg KOH/100 g, respectively. This shows that HHP treatment of brown rice at 400 MPa enhanced the hydrolytic rancidity of lipids. Furthermore, holding duration had no effect on fat acidity among HHP parameters, although treatment pressure was a key factor. HHP treatment at 200 MPa resulted in the minimum fat acidity level, and with increase in pressure there was gradual increase in fat acidity levels (H. Wang et al., 2018). In addition, HHP treatment has been shown to lower fat content in various cereal flours. Fat content in wheat, oat, ragi, and sorghum breads produced using HHP-treated (350 MPa, 10 min) flour reduced from 3.85 to 3.49, 9.07 to 8.67, 4.60 to 4.40 and 4.42 to 4.22 g, respectively (Angioloni & Collar, 2012). The explanation for the decrease in fat content after

HHP could be related to oxidation of fat and formation of other compounds. To better understand the mechanism behind the reduction in fat content and fat oxidation caused by HHP treatment of cereals and other food products, more research is needed.

18.3.3 Dietary fiber

Fibers classified as non-starch polysaccharides have been gaining attention in recent years due to their significant anti-diabetic, anti-obesity, and cardio-protective properties. Through physical transformation, low-cost and devalued fiber sources can be turned into natural, useful, and nutritional food ingredients. Various types of fibers like fruit fibers, gums, and inulin have found a number of applications as thickening and gelling agents in food products. Fibers can substitute flour, sugar, or fat in food preparations, improving oil and water retention and assisting in the production of stable emulsions (Elleuch et al., 2011). Dietary fibers refer to carbohydrate polymer consisting of 10 or more monomers that are not degraded in the small intestine by endogenous enzymes of humans (McCleary et al., 2011) and can be categorized as follows: (a) Naturally occurring edible carbohydrate polymer in foods and is consumed as such; (b) carbohydrate polymer obtained after physical, chemical, or enzymatic processing of raw material and exhibiting physiological health benefits; (c) synthetic carbohydrate polymers exhibiting physiological health benefits. Cellulose, hemi-cellulose, gums, pectin, inulin, resistant starch, etc., are some naturally occurring dietary fibers (Chiewchan, 2017). Furthermore, dietary fibers are classified as soluble or insoluble dietary fibers based on their solubility. Soluble dietary fibers can dissolve in water and form gel network, whereas insoluble dietary fibers are not dissolved in water and have strong hygroscopic and swelling characteristics (Thebaudin et al., 1997).

Several physical processing methods such as high shear homogenization, extrusion, ultrasound, micro-fluidization, HHP etc., have been used to modify the physicochemical and functional properties of the native structure of dietary fibers. Angioloni & Collar (2012) developed breads using HHP-treated flours of wheat, oat, ragi, and sorghum and found decrease in the content of soluble and insoluble dietary fiber in comparison to bread prepared using conventional flours. Insoluble dietary fiber content reduced from 1.28 to 1.16, 6.95 to 6.64, 6.85 to 6.55, and 4.74 to 4.53 g/100 g bread for HHP-treated flours of wheat, oat, ragi, and sorghum, respectively. Similarly, soluble dietary fiber content reduced by 8.70%, 4.42%, 4.39%, and 4.12% for HHP-treated flours of wheat, oat, ragi, and sorghum, respectively. A primary factor for the changing of the fiber level is assumed to be the intensive instantaneous heat energy released from direct pressure treatment and its associated forces (cavitation, friction, heat, impact compression, or the electrostatic pressure treatment) (Mateos-Aparicio et al., 2010). The findings suggest that bread prepared using HHP-treated flours oat, ragi, and sorghum can be marked as high fiber breads (>6 g total dietary fiber/100 g bread) according to nutritional claims for dietary fiber foods (European Commission, 2006). In addition, peels of orange, mango, and prickly pear showed insignificant reductions in total dietary fibers after HHP treatment (Tejada-Ortigoza et.al., 2017).

18.3.4 Ash

The inorganic residue left post combustion or complete oxidation of organic materials in a food product is referred to as ash. The minerals in the food sample make up the majority of the inorganic residue.

In comparison to the control, HHP enhanced the ash content of germinated brown rice at all pressures (100, 300, and 500 MPa) (Xia et al., 2017). HHP at 300 MPa for 10 min resulted in the highest ash content of 12.82 g/Kg, compared to 8.67 g/Kg before the treatment (control). HHP treatment at 100 and 500 MPa for 10 min yielded similar ash content of 9.43 and 9.47 g/Kg, with no significant reduction. In contrast, Angioloni & Collar (2012) reported a decrease in the ash content of breads manufactured using HHP-treated (350 MPa, 10 min) wheat, oat, ragi, and sorghum flours. Ash content reduced from 0.64 to 0.58, 1.45 to 1.39, 1.12 to 1.07, and 0.96 to 0.92 g/100 g for wheat, oat, ragi, and sorghum bread. Similarly, 18.02 and 3.77% increase in ash content was reported for *Prosopis chilensis* seed (Briones-Labarca et.al., 2011) and Granny Smith apple (Briones-Labarca et.al., 2011), respectively, post HHP. These findings suggest that various foods with distinct matrixes react to HHP treatment in different ways. To determine the impacts of HHP on cereal ash content, more studies are necessary.

18.3.5 Energy

Human energy demand is compensated by oxidation of glucose, protein, fat, and occasionally alcohol. Indirect calorimetry measures oxygen input and/or carbon dioxide output, and direct calorimetry measures generation of heat from nutrient oxidation; both are used to determine energy expenditure. Energy expenditure is determined by three factors: (a) body composition, which determines resting energy expenditure (REE); (b) food intake, which determines the energy cost of processing food, and; (c) body movement, which determines activity-induced energy expenditure (AEE). Through a greater REE, energy expenditure and hence energy need are generally higher in bulkier people, such as obese and overweight individuals. Negative energy balance subjects have lower energy expenditure, which is mostly due to a decrease in AEE. Physical movement is not a basis for energy balance but rather a function. Energy intake is the most critical consideration in determining energy balance. To account for energy losses in urine and feces, nutrient composition charts have typically employed the *Atwater factors* to translate gross nutritional intake to metabolizable energy intake. Carbohydrate, protein, fat, and alcohol have *Atwater* values of 4, 4, 9, and 7 kcal/g, respectively (Westerterp, 2020).

Angioloni & Collar (2012) reported the energy values of breads prepared by replacing conventional wheat flour by 60% oat, 40% ragi, and 40% sorghum flour before and after HHP treatment (350 MPa, 10 min). The energy value of bread made with HHP-treated wheat flour (275 kcal/100 g) was similar to that of normal (276 kcal/100 g). Similarly, HHP treatment of ragi flour had no effect on the energy values of ragi bread (264 kcal/100 g). Breads made using HPP-treated oat and sorghum flour, on the other hand, showed a difference. For oat and sorghum breads, after HHP treatment energy values decreased significantly from 282 to 262 and 271 to 267 kcal/100 g, respectively (Angioloni & Collar, 2012). These findings imply that HHP treatment alters cereals in distinct ways.

18.4 EFFECT OF HHP ON FUNCTIONAL PROPERTIES OF CEREALS

18.4.1 Water-holding Capacity

The ability of the hydrated fiber matrix to retain water in the form of coupled water, hydrodynamic water, and physically trapped water is known as water-holding capacity, with the latter contributing the most to this attribute (Alfredo et.al., 2009). The ability of fiber to hold water indicates its physiological relevance in intestine function and blood sugar regulation (Wolever, 1990). The water-holding capacity of maize starch increased linearly as the HHP treatment time increased. The water-holding capacity of HHP-treated maize starch at 300 MPa for 60 min was found to be 0.0488 g water/g dry starch, which was approximately 2.7 times that of HPP treatment at 300 MPa for 5 min (Santos et al., 2014). Because pressure favorably impacts the helix shape with the stabilization of hydrogen bonds, it is acknowledged that starch pressurization in surplus of water can produce granule enlargement and incomplete disintegration (Pei-Ling et al., 2010). Despite the starch granule surface's resistance to alteration when pressure is applied, the inside portion is packed with a gel-like structure with void space, modifying the starch characteristics (Błaszczak et al., 2005; Pei-Ling et al., 2010).

According to Marti et al. (2014), regardless of processing duration, HHP treatment considerably enhanced the water-holding capacity of wheat bran. When compared to untreated samples, water-holding capacity increased significantly following HHP treatment at 600 MPa for 5 and 15 min, but there was no significant difference in water-holding capacity of samples treated for 5 and 15 min. Furthermore, coarse wheat bran showed a higher value of water-holding capacity than fine bran (Marti et al., 2014). This could be linked to the biochemical and structural characteristics of fiber, which has a significant role in water-holding and swelling characteristics (Figuerola et al., 2005). A change in the arrangement of some fiber components after HHP treatment could explain the increase in water-holding capacity.

Lee & Koo (2019) studied the influence of HHP (150 and 300 MPa for 30 min) on the water-holding capacity of low protein wheat flour and low protein wheat-oat flour blends (20% and 40% oat flour concentrations). Water-holding capacity of low protein wheat increased from 92.76% to 97.38% following HHP treatment at 150 MPa, but reduced to 93.46% when pressure was elevated to 300 MPa. Water-holding capacity of low protein wheat-oat flour blends showed similar trends of increasing at 150 MPa and decreasing at 300 MPa at both concentrations (Lee & Koo, 2019). The structural damage caused by higher treatment pressure may be linked to the decrease in water-holding capacity.

18.4.2 Water Solubility Index

The water solubility index measures the solubility of biomolecules both prior and after processing in surplus of water (C. Sharma et al., 2017). Several studies have looked at how HHP affects the water solubility of various cereals. The water solubility index of white, black, and red whole grain quinoa flour increased after treatment with HHP (F. Zhu & Li, 2019). For black, white, and red quinoa flour, HHP at 600 MPa and 55°C increased the water solubility index from 15.8 to 40.2%, 15.4 to 34.0%, and 16.5 to 34.6%, respectively. Furthermore, the water solubility of all quinoa flour varieties showed an increase with an increase in temperature at a specific HHP pressure. For example, when the temperature was elevated from 55°C to 95°C at 600 MPa, the water solubility index of black, white, and red quinoa flour increased from 40.2 to 47.5%, 34.0 to 40.7%, and 34.6 to 42.5%, respectively. Moreover, the type of quinoa flour had an effect on the magnitude of increases in water solubility index (F. Zhu & Li, 2019).

Hang Liu and group reported the solubility of common buckwheat starch (CB) (H. Liu, Wang, et al., 2016) and tartary buckwheat starch (TB) (H. Liu, Guo, et al., 2016) under HHP. CB solubility increased as HHP pressures increased at 50 and 60°C, but reduced as HHP pressures increased from 120 to 600 MPa at 70, 80, and 90°C (H. Liu, Wang, et al., 2016). Similarly, for TB at temperatures of 50 and 60°C solubility increased with pressure and reached maximum value at 600MPa. At 70, 80, and 90°C the solubility decreased with increase in pressure with maximum value for native TB (H. Liu, Guo, et al., 2016). Furthermore, for both CB and TB, solubility increased with increasing processing temperatures at a particular pressure. The solubilization of amylose was restricted by partially or totally degraded starch granules, reducing CB and TB solubility (Stolt et al., 2000). Furthermore, HPP caused the development of amylose–lipid complexes in CB and TB granules decreased the fluidity of soluble amylose, lowering its solubility (Oh et al., 2008).

S. M. Zhu, Lin, Ramaswamy, Yu, & Zhang (2017) reported the solubility of HHP-treated rice bran proteins. They found that solubility increased from 47.8% to 60.3% and 65.2% for HHP treatment at 100 and 200 MPa, respectively. Further increase in HHP levels to 300, 400, and 500

MPa consistently reduced the solubility to 41.7%, 39.9%, and 25.9%, respectively. It is evident that lower pressures lead to partial opening of protein structures enabling greater solubility, whereas higher pressures lead to structure formation causing reduction in solubility.

18.4.3 Oil Absorption Capacity (OAC)

OAC is essential and affects the texture, flavor, mouthfeel, and product yield of food products (Seena & Sridhar, 2005). It is important in the formation of several food products such as pancakes, desserts, baked products, beverages, confectioneries, etc. (Farooq & Boye, 2011). The physical trapping of oil within proteins, as well as non-covalent connections like electrostatic, hydrophobic, and H-bonding as forces implicated in lipid–protein interactions, have been attributed to OAC. The binding of oil's hydrocarbon chains to the nonpolar side chains of amino acids causes oil retention (Farooq & Boye, 2011; N. Wang et al., 2020).

S. M. Zhu et al. (2017) investigated the OAC of HHP-treated rice bran protein. OAC showed a sharp increase from 2.63 to 7.39 and 7.57 g/g for rice bran protein HHP treated at 100 and 200 MPa, respectively. Further increase in HHP treatment pressure to 300, 400, and 500 MPa reduced the OAC of rice bran protein to 5.12, 4.22, and 4.85 g/g, respectively. Increase in OAC may be attributed to unfolding of rice bran protein post HHP, exposing the hydrophobic groups and allowing them to interact with oil. These results indicate that using a moderate HHP pressure of 100– 200 MPa to increase the OAC of rice bran protein is an efficient technique to make it usable as a component in high oil binding food products.

Furthermore, several investigations described the OAC of cereals starched after HHP. OAC of wheat starch and corn starch increased from 153% to 350% and 188% to 424% after HHP at 600 MPa, 20 min (Heydari et al., 2021). Whereas OAC of sorghum starch was found to decrease consistently after HHP treatment, native sorghum starch has OAC of 0.98 g/g which reduced to 0.92, 0.85, 0.76, 0.70, and 0.55 g/g, respectively, after HHP treatment at 120, 240, 360, 480, and 600 MPa for 20 min (H. Liu et.al., 2016). Because different cereals have shown different outcomes, more research is needed to fully understand the mechanism of HHP on OAC.

18.4.4 Emulsifying and Foaming Properties

Protein emulsifying and foaming capabilities are important in determining the structural, stabilization, and sensory characteristics of foods like toppings, mousses, margarine, drinks, ice cream, etc. (Aryee et al., 2018). Emulsifiers are capable of forming stable films and dispersions by wringing oil droplets into an environment. These films can then be made to form foams or emulsions (Bessada et al., 2019). The emulsifying activity index (EAI) and emulsifying stability index (ESI) must be established to identify the emulsifying capabilities of proteins. The EAI determines the amount of emulsified oil per gram of protein, whereas the ESI determines the emulsion's resistance over time (Burger & Zhang, 2019).

Plant proteins, as previously stated, have foaming qualities that are frequently measured in terms of foaming capacity (FC) and foaming stability (FS). FC is concerned with the volume (percentage) of incorporated air upon whipping, while FS is concerned with the foam stability (volume) over time (Shevkani et.al., 2019). Plant proteins' emulsifying and foaming abilities, as well as their behavior resembling that of some of the more well-known emulsifiers, make them promising candidates for application in the food sector.

Some of the functional properties discussed above are directly linked to the emulsifying and foaming properties of emulsions/foams containing proteins. As previously stated, HHP can alter protein structure, hydrodynamic volume, and surface hydrophobicity. Therefore, modifying its solubility, the tendency to adsorb on interfaces, and possibility of interactions with itself or other medium components. As a result, HPP looks to be a good technique for manipulating proteins' surface-active characteristics and their ability to promote the creation of dispersed food production systems like emulsions and foams, as well as to stabilize them.

EAI of rice bran protein increased from 17.5% to 25.4% after HHP at 100 MPa for 10 min. Further increase in HHP pressure showed no significant improvement in EAI (S. M. Zhu et al., 2017). HHP-induced improvement in EAI may be linked to the unfolding of rice bran protein and exposure of more hydrophobic groups. Furthermore, HHP treatment improved the ESI of rice bran protein. After HHP treatment from 100 to 400 MPa increased the ESI several times from 18.2 min, a slight reduction was reported at 500 MPa (S. M. Zhu et al., 2017). The unfolding of rice bran protein structure exposes lipophilic and hydrophilic groups, enhancing protein–solvent interaction and avoiding oil droplet coalescence. Furthermore, due to its severity, a greater HHP treatment of 500 MPa may reduce molecular flexibility and ESI.

HHP treatment also improved the FC of rice bran protein with pressure. HHP treatment at 100, 200, 300, 400, and 500 MPa for 10 min increased the FC from 14.8% to 23.8, 30.5, 34.5, 37.8, and 39.0%, respectively, whereas FS showed variable trend, showing an increase of 47.2, 35.1, and 40.1% at 100, 400, and 500 MPa, respectively. But no significant difference at 200 and 300 MPa HHP treatment (S. M. Zhu et al., 2017).

18.4.5 Swelling Power

In the broadest sense, starch gelatinization refers to the heat disordering of crystalline forms in native starch granules, but it also refers to phenomena such as granular swelling and soluble polysaccharide leaching. The temperature at which starch gelatinizes is less relevant in most food systems than the qualities that are dependent on swelling (Richard F & William R, 1990). Researchers have now discovered that the swelling behavior of HHP-treated starch and heat-treated starch resulted in different characteristics. For example, when compared to heat-treated starch, HHP-treated starch showed lower enzyme susceptibility. As a result, studying the swelling behavior of HHP-treated starch will help us better comprehend the gelatinization qualities generated by HHP. Recently, C. Wang, Xue, Yousaf, Hu, & Shen (2020) conducted a thorough investigation on the influence of HHP on swelling behavior of rice starch.

Some previous studies have also investigated the impact of HHP on swelling power of cereals starch. Swelling power of TB increased with increase in HHP pressure at 50–60°C, whereas a decrease in swelling power was reported with increase in HHP pressure at 70–90°C (H. Liu, Guo, et al., 2016). Similar results of increase in swelling power at lower processing temperatures (50–60°C) and decrease at higher processing temperatures (70–90°C) with increase in HHP pressure were reported for CB (H. Liu, Wang, et al., 2016). At lower temperatures, the HHP-modified TB and CB samples had a larger swelling power, indicating that amylose molecules are aggregating under pressure. The drop in swelling power of HHP-modified samples at higher temperatures could be attributed to starch molecule rearrangement, which hindered hydration and swelling of CB and TB. Similarly, for HHP-treated black, white, and red quinoa flour swelling power was found to increase with increase in HHP pressure at lower processing temperatures (55 and 65°C), but a decrease in swelling power was observed at high temperatures (75, 85, and 95°C) with increase in HHP pressure (F. Zhu & Li, 2019). These findings imply that HHP treatment has a favorable effect on swelling power at lower processing temperatures, but that it has a negative effect at higher temperatures. As a result, HHP treatment should be used at lower processing temperatures to improve cereal swelling power.

18.4.6 Thermal Properties

Thermal characteristics are essential to determine since many processes of food processing and preservation include heat transfer. Food thermal characteristics are significant in the designing of cold stores and cooling systems, and also in assessing time for heating, refrigerating, drying, or freezing foods. As the thermal characteristics of the food depend heavily on biochemical configuration and temperature, it is feasible to anticipate such thermal characteristics with mathematical patterns that take account of the chemical composition and temperatures impacts.

Thermal analysis of three types of quinoa flours (black, white, and red) after HHP treatment showed no significant difference in the peak gelatinization temperatures (T_p) up to 400 MPa and transition enthalpy (ΔH) decreased at all pressure levels of HHP treatment. For white and red quinoa flours T_p increased at 500 MPa. After HPP treatment at 600 MPa, ΔH decreased from 12.0 to 1.7 J/g for black quinoa flour, whereas no endothermic peaks were reported for white and red flour, suggesting complete gelatinization (F. Zhu & Li, 2019).

In addition, the thermal characteristics of HHP-treated TB and CB were examined. T_p of native TB and CB were practically identical at 77.0 and 76.5°C, respectively, after HHP at 120, 240, 360, 480, and 600 MPa, which decreased consistently. T_p of TB and CB were determined to be 68.5 and 69.5°C, respectively, at 600 MPa. Similarly, after HHP treatment, ΔH decreased consistently for both TB and CB. ΔH of native CB (22.5 J/g) was slightly higher than that of native TB (19.8 J/g) and which reduced to 8.6 and 6.6 J/g for CB and TB, respectively, when HHP-treated at 600 MPa (H. Liu, Guo, et al., 2016; H. Liu, Wang, et al., 2016). Furthermore, similar results of decrease in T_p and ΔH after HHP treatment were reported for starches of wheat, normal corn, and waxy corn starch (Heydari et al., 2021).

18.4.7 Gelation Properties

Another important functionality of food products is the capability to form gels under specific settings, which is important for several food products like jellies, puddings, meat products, etc., where elasticity, water retention, and texture are important. Gelation is a generic method of converting a liquid to a solid that has been used to create a spectrum of products with different textures. Simpler food gels include gelatin-based confectionery, which are made up of a water–gelatin gel with added sweetness, flavor, and color. Gelation can be caused by a variety of physical or chemical techniques, including heat, pressure, chemical, and enzymatic treatments.

Gel formed by both TB and CB showed a progressive reduction in hardness, gumminess, adhesiveness, and chewiness after HHP treatment. For CB, cohesiveness varied from 0.415 (600 MPa) to 0.652 (control), hardness reduced significantly from 148.7 N (control) to 22.3 N (600 MPa), and springiness showed no significant difference after HHP treatment (H. Liu, Wang, et al., 2016). Hardness of gels formed by TB ranged from 83.6 N (control) to 6.8 N (600 MPa) indicating CB formed harder gels than TB. Cohesiveness of TB gels ranged from 0.488 (600 MPa) to 0.664 (control) (H. Liu, Guo, et al., 2016). Gel hardness may be reduced due to reduction in the amount of leached amylose (Puncha-Arnon & Uttapap, 2013). Furthermore, inadequate/poor water–starch and starch–starch molecular interactions may also contribute to weak textural characteristics of HHP-treated gels (Vittadini et.al., 2008). In contrast, gel formed by brown rice showed an increase in the hardness from 83 N (control) to 107 N after HHP treatment at 400 MPa and no significant difference was observed for cohesiveness and springiness (H. Wang et al., 2020).

18.5 EFFECT OF HHP ON BIOLOGICAL PROPERTIES

Measurement of digestibility is a method of determining the bioavailability of nutrients. Digestion is a complex mechanism that provides organisms with essential nutrients and energy they need for maintenance and growth. As a result, numerous researchers have shown interest in modeling the digestion and absorption in its entirety or in part utilizing a number of *in vivo* and *in vitro* approaches.

The *in vivo* digestibility of nutrients can be assessed both indirectly and directly. For direct method, meticulous tracking of feed consumption and fecal excretion of an animal exposed to nutritional intervention is done over a set length of time. The downside of this procedure is the possibility of bodily excretions and urine contamination, as well as the fact that the collection is not always correct. The indirect method of determining digestibility does not include quantifying consumption; nevertheless, a marker such as fecal excretion could be added to or incorporated in the meal (Bovera et al., 2013).

Nutritionists have used the *in vitro* approach to try to replicate the physiological response of the digestive system, with varying degrees of accuracy, the compartments and working conditions present in the digestive system of the target organisms to gain some insight into how nutrients or food are metabolized or digested. Various sorts of bioreactors and operational characteristics have been actively developed in order to achieve the two primary aims of such studies: (a) to assess the presumed biological efficiency after digestive process in relation to the possible bioavailability of a specified nutritional or chemical compound (predictive models) or (b) to gain knowledge of the different mechanisms, and interactions throughout the digestion activity (explanatory models) (Moyano et al., 2015).

F. Zhu & Li (2019) explored the effect of HHP on the *in vitro* starch digestibility of three types of quinoa flour. HHP decreased the *in vitro* starch hydrolysis for all the flour types with increase in HHP pressure levels. At 600 MPa, *in vitro* starch hydrolysis reduced from 950 to 861 g/kg, 954 to 862 g/kg, and 967 to 888 g/kg for black, white, and red quinoa flour, respectively (F. Zhu & Li, 2019). Similarly, for TB and CB *in vitro* starch hydrolysis reduced significantly with increase in pressure. Furthermore, after HHP treatment, rapidly digestible starch content reduced significantly and slowly digestible starch and resistant starch content increased for both TB and CB. At 600 MPa, rapidly digestible starch, slowly digestible starch and resistant starch content were 29.0%, 49.1%, and 8.2%, respectively for CB and 17.0%, 50.5%, and 6.7% for TB, respectively (H. Liu, Guo, et al., 2016; H. Liu, Wang, et al., 2016). Increase in the content of resistant starch suggests stronger amylose–amylose and amylose–amylopectin interactions after HHP. Also, resistant starch has health benefits like controlling of serum cholesterol levels, reducing insulinemic responses and glycemic index, and protecting against colorectal cancer (Asp et al., 1996).

Furthermore, HHP treatment has been reported to modify the total phenolic content (TPC) of food grains. TPC of germinated foxtail millet increased by approximately 17% with increase in HHP pressure and duration (N. Sharma et al., 2018). TPC was also found to increase after HHP treatment in three varieties of quinoa flours (black, white, and red). TPC of control black quinoa flour was 7.46 mg GAE/g, with a maximum value of 7.68 mg GAE/g at 200 MPa, while maximum values of 7.97 mg GAE/g at 400 MPa and 7.43 mg GAE/g at 600 MPa were found for white quinoa flour (control = 6.46 mg GAE/g) and red quinoa flour (control = 6.84 mg GAE/g) (F. Zhu & Li, 2019). Increased rate of mass transfer due to HHP, which improves the grains' solvent extraction capacity by disrupting cellular walls and hydrophobic interactions, may be the reason for increased TPC (Kim et al., 2017).

According to N. Sharma et al. (2018), antioxidant activity of HHP-treated germinated foxtail millet grain flour improved with increase in pressure and temperature determined through Ferric reducing antioxidant potential (FRAP). At all pressure levels maximum FRAP values were reported at temperature of 40°C. FRAP values increased with increase in pressure levels and maximum values were found to be 7.85 μmol/g at 600 MPa and 40°C (N. Sharma et al., 2018). Turbulence and shear due to HHP may disrupt the cell walls and organelles, enhancing the extractability of antioxidants and improving their availability (Xia, Wang, et al., 2017).

Cereals being a primary source of energy and nutrients also contain many anti-nutrients (ANs). ANs are natural or artificial components that prevent nutrients from being released or absorbed, lowering their bioaccessibility and bioavailability in food. ANs play a role in plants chemical defense system against pathogens (Suhag et al., 2021). In germinated foxtail millet grain flour, HHP has been demonstrated to efficiently minimize ANs. Phytic acid and tannin content in control sample were found to be 0.5421% and 0.0016%, respectively, which reduced to 0.1870% and 0.0006%, respectively, after HHP treatment at 600 MPa and 60°C (N. Sharma et al., 2018). The heat labile and water-soluble properties of phytic acid and tannin may account for this reduction. Furthermore, formation of insoluble complexes amid ANs and other compounds may cause similar reductions (Suhag et al., 2021).

18.6 CONCLUSION AND FUTURE SCOPE

This chapter congregated the information about the modifications produced in different nutritional components, functional, and biological properties of various cereals as result of HHP treatment (Table 18.1). HHP treatment has been shown to improve hydration, emulsification, gelation, and

Table 18.1 Effect of HHP on Properties of Cereals

Cereal	Treatment conditions	Major findings	References
Brown rice	Pressure: 200–600 MPa Holding time: 0–20 min	• Reduced cooking time from 34 to 14 min. • Reduced hardness after HHP treatment. • Pressure above 500 MPa decreased springiness and gumminess.	(Yu et al., 2015)
Brown rice	Pressure: 200–500 MPa Holding time: 5–15 min	• Improved rice hydration with limited swelling and intact starch granules. • HHP treatment of 400 MPa for 10 min lead to gelatinization degree of 41.2%, 50.2%, and 39.9% in brown rice, medium milled rice, and fully milled rice, respectively.	(Zhu et al., 2016)
Brown rice	Pressure: 100–500 MPa Holding time: 15 min Temperature: 20°C Germination conditions: 37°C and 36 h	• HHP treatment enhanced formation of alcohols, aldehydes, and ketones, improving flavor in germinated black rice.	(Xia, Mei, et al., 2017)
Brown rice	Pressure: 100–500 MPa Holding time: 10 min Temperature: 18°C Germination conditions: 36 h and 95% germination rate	• Starch granules gelatinized at 300 MPa. • Reduction in hardness, cohesiveness, gumminess, and resilience of germinated black rice due to starch gelatinization.	(Xia, Tao, et al., 2017)
Paddy rice	Pressure: 100–600 MPa Holding time: 30 min Temperature: 25°C	• Reduction is starch enthalpy. • Elimination of white-core phenomenon due to uniform distribution of water in rice grain after HHP treatment.	(Xu et al., 2019)
Normal and waxy rice	Pressure: 300 and 400 MPa Holding time: 25 min Temperature: 55°C	• Decreased hardness, increased springiness, and cohesiveness. • Reduction in retrogradation of cooked rice.	(Tian et al., 2014)
Black rice	Pressure: 200–400 MPa Holding time: 15 min Temperature: 25°C	• Improvement in water absorption and texture properties of cooked black rice, while reduction in leached amylose. • Decreased retrogradation degree from 91 to 71% when stored for 21 days.	(Meng et al., 2018)

Table 18.1 (Continued)

Cereal	Treatment conditions	Major findings	References
Black rice	Pressure: 200–500 MPa Holding time: 15 min Temperature: 25°C	• Increased total content of key flavor components alcohols, aldehydes, and ketones in cooked black rice after HHP treatment. • HHP reduced the retrogradation degree of cooked black rice.	(Meng et al., 2020)
Rice	Pressure: 100–500 MPa Holding time: 30 min Temperature: 25°C	• Increased water adsorption and decreased cooking time. • Reduction of hardness values from 0.69 to 32.99%. • Decreased enthalpy of parboiled rice.	(Wu et al., 2021)
Foxtail millet	Pressure: 200–600 MPa Holding time: 30–120 min Temperature: 20–80°C	• Improved water absorption of germinated grains. • Increased starch gelatinization degree of flour with maximum value = 64.93%.	(Sharma et al., 2018)
Brown rice	Pressure: 100–500 MPa Holding time: 10 min Temperature: 18°C Germination conditions: 37°C and 36 h	• Increased *in vitro* bioavailability of Ca and Cu from 12.59 to 52.17% and 2.87 to 23.06%, respectively. • Increased total antioxidant activity and starch anti-enzymatic capacity.	(Xia, Wang, et al., 2017)
Brown rice	Pressure: 0.1–100 MPa Holding time: 24 h Temperature: 37°C Germination conditions: 37°C and 48 h	• Maximum total polyphenol content reported at 30 MPa. • Reduction of phytic acid content from 6.1 to 3.0 mg/g.	(Kim et al., 2015)
Rough rice	Pressure: 0.1–100 MPa Holding time: 24 h Temperature: 37°C Germination conditions: 37°C and 2 days	• Increase in total phenolic content from 85.37 μg/g (at 0.1 MPa) to 183.52 μg/g (at 100 MPa).	(Kim et al., 2017)
Buckwheat protein	Pressure: 100–600 MPa Holding time: 1–300 min	• Decrease in the IgE binding of buckwheat protein hydrolyzed with alkaline protease from 83.3 to 100%.	(Lee et al., 2017)
Wheat	Pressure: 100–300 MPa Holding time: 10–600 s Temperature: 30°C	• HPP treatment at 300 MPa for 10 min, resulted in mycotoxin content under detection limit (0.5 mg/kg) analyzed after storage (0 and 6 weeks).	(Schmidt et al., 2019)

thermal characteristics. It also increases antioxidant properties and *in vitro* digestibility, as well as lowers the AN concentration. However, cereals research is limited, and there are many more aspects to be discovered. Furthermore, the findings stated above should be exploited in the development of novel and nutritional food products in order to deliver the health benefits cereals provide.

REFERENCES

Akharume, F. U., Aluko, R. E., & Adedeji, A. A. (2021). Modification of plant proteins for improved functionality: A review. *Comprehensive Reviews in Food Science and Food Safety*, 20(1), 198–224. https://doi.org/10.1111/1541-4337.12688

Alcázar-Alay, S. C., & Meireles, M. A. A. (2015). Physicochemical properties, modifications and applications of starches from different botanical sources. *Food Science and Technology*. https://doi.org/10.1590/1678-457X.6749

Alfredo, V. O., Gabriel, R. R., Luis, C. G., & David, B. A. (2009). Physicochemical properties of a fibrous fraction from chia (Salvia hispanica L.). *LWT – Food Science and Technology*. https://doi.org/10.1016/j.lwt.2008.05.012

Angioloni, A., & Collar, C. (2012). Effects of pressure treatment of hydrated oat, finger millet and sorghum flours on the quality and nutritional properties of composite wheat breads. *Journal of Cereal Science*. https://doi.org/10.1016/j.jcs.2012.08.001

Aryee, A. N. A., Agyei, D., & Udenigwe, C. C. (2018). Impact of processing on the chemistry and functionality of food proteins. In *Proteins in Food Processing: Second Edition*. https://doi.org/10.1016/B978-0-08-100722-8.00003-6

Asp, N.-G., van Amelsvoort, J. M. M., & Hautvast, J. G. A. J. (1996). Nutritional Implications Of Resistant Starch. *Nutrition Research Reviews*. https://doi.org/10.1079/nrr19960004

Balakrishna, A. K., Abdul Wazed, M., & Farid, M. (2020). A review on the effect of high pressure processing (HPP) on gelatinization and infusion of nutrients. *Molecules*. https://doi.org/10.3390/molecules25102369

Barbhuiya, R. I., Singha, P., & Singh, S. K. (2021). A comprehensive review on impact of non-thermal processing on the structural changes of food components. *Food Research International*. https://doi.org/10.1016/j.foodres.2021.110647

Bessada, S. M. F., Barreira, J. C. M., & Oliveira, M. B. P. P. (2019). Pulses and food security: Dietary protein, digestibility, bioactive and functional properties. *Trends in Food Science and Technology*. https://doi.org/10.1016/j.tifs.2019.08.022

Błaszczak, W., Valverde, S., & Fornal, J. (2005). Effect of high pressure on the structure of potato starch. *Carbohydrate Polymers*. https://doi.org/10.1016/j.carbpol.2004.10.008

Borneo, R., & León, A. E. (2012). Whole grain cereals: functional components and health benefits. *Food Funct.*, *3*(2), 110–119. https://doi.org/10.1039/C1FO10165J

Bovera, F., Lestingi, A., Piccolo, G., Iannaccone, F., Attia, Y. A., & Tateo, A. (2013). Effects of water restriction on growth performance, feed nutrient digestibility, carcass and meat traits of rabbits. *Animal*. https://doi.org/10.1017/S1751731113001146

Briones-Labarca, V., Muñoz, C., & Maureira, H. (2011). Effect of high hydrostatic pressure on antioxidant capacity, mineral and starch bioaccessibility of a non conventional food: Prosopis chilensis seed. *Food Research International*. https://doi.org/10.1016/j.foodres.2011.01.013

Briones-Labarca, V., Venegas-Cubillos, G., Ortiz-Portilla, S., Chacana-Ojeda, M., & Maureira, H. (2011). Effects of high hydrostatic pressure (HHP) on bioaccessibility, as well as antioxidant activity, mineral and starch contents in Granny Smith apple. *Food Chemistry*. https://doi.org/10.1016/j.foodchem.2011.03.074

Burger, T. G., & Zhang, Y. (2019). Recent progress in the utilization of pea protein as an emulsifier for food applications. *Trends in Food Science and Technology*. https://doi.org/10.1016/j.tifs.2019.02.007

Chiewchan, N. (2017). Microstructure, constituents, and their relationship with quality and functionality of dietary fibers. In *Food Microstructure and Its Relationship with Quality and Stability*. https://doi.org/10.1016/B978-0-08-100764-8.00010-1

Clariana, M., & García-Regueiro, J. A. (2011). Effect of high pressure processing on cholesterol oxidation products in vacuum packaged sliced dry-cured ham. *Food and Chemical Toxicology*. https://doi.org/10.1016/j.fct.2011.03.027

Considine, K. M., Kelly, A. L., Fitzgerald, G. F., Hill, C., & Sleator, R. D. (2008). High-pressure processing – Effects on microbial food safety and food quality. *FEMS Microbiology Letters*. https://doi.org/10.1111/j.1574-6968.2008.01084.x

Deng, Y., Jin, Y., Luo, Y., Zhong, Y., Yue, J., Song, X., & Zhao, Y. (2014). Impact of continuous or cycle high hydrostatic pressure on the ultrastructure and digestibility of rice starch granules. *Journal of Cereal Science*. https://doi.org/10.1016/j.jcs.2014.06.005

Elleuch, M., Bedigian, D., Roiseux, O., Besbes, S., Blecker, C., & Attia, H. (2011). Dietary fibre and fibre-rich by-products of food processing: Characterisation, technological functionality and commercial applications: A review. *Food Chemistry*. https://doi.org/10.1016/j.foodchem.2010.06.077

European Commission. (2006). EU register on nutrition and health claims made on foods. *Official Journal of the European Union*.

Farooq, Z., & Boye, J. I. (2011). Novel Food and Industrial Applications of Pulse Flours and Fractions. In *Pulse Foods*. https://doi.org/10.1016/B978-0-12-382018-1.00011-3

Figuerola, F., Hurtado, M. L., Estévez, A. M., Chiffelle, I., & Asenjo, F. (2005). Fibre concentrates from apple pomace and citrus peel as potential fibre sources for food enrichment. *Food Chemistry*. https://doi.org/10.1016/j.foodchem.2004.04.036

Fukami, K., Kawai, K., Hatta, T., Taniguchi, H., & Yamamoto, K. (2010). Physical Properties of Normal and Waxy Corn Starches Treated with High Hydrostatic Pressure. *Journal of Applied Glycoscience*, *57*(2), 67–72. https://doi.org/10.5458/jag.57.67

Gharibzahedi, S. M. T., & Smith, B. (2021). Effects of high hydrostatic pressure on the quality and functionality of protein isolates, concentrates, and hydrolysates derived from pulse legumes: A review. *Trends in Food Science & Technology*, *107*, 466–479. https://doi.org/10.1016/j.tifs.2020.11.016

Heydari, A., Razavi, S. M. A., Hesarinejad, M. A., & Farahnaky, A. (2021). New Insights into Physical, Morphological, Thermal, and Pasting Properties of HHP-Treated Starches: Effect of Starch Type and Industry-Scale Concentration. *Starch/Staerke*. https://doi.org/10.1002/star.202000179

Jiang, B., Li, W., Shen, Q., Hu, X., & Wu, J. (2015). Effects of high hydrostatic pressure on rheological properties of rice starch. *International Journal of Food Properties*. https://doi.org/10.1080/10942912.2012.709209

Kakati, P., Deka, S. C., Kotoki, D., & Saikia, S. (2010). Effect of traditional methods of processing on the nutrient contents and some antinutritional factors in newly developed cultivars of green gram (Vigna radiata (L.) Wilezek) and black gram (Vigna mungo (L.) Hepper) of Assam, India. *International Food Research Journal*.

Kawai, K., Fukami, K., & Yamamoto, K. (2012). Effect of temperature on gelatinization and retrogradation in high hydrostatic pressure treatment of potato starch-water mixtures. *Carbohydrate Polymers*. https://doi.org/10.1016/j.carbpol.2011.07.046

Kim, M. Y., Lee, S. H., Jang, G. Y., Li, M., Lee, Y. R., Lee, J., & Jeong, H. S. (2015). Influence of Applied Pressure on Bioactive Compounds of Germinated Rough Rice (Oryza sativa L.). Food and Bioprocess Technology, 8(10), 2176–2181. https://doi.org/10.1007/s11947-015-1565-1

Kim, M. Y., Lee, S. H., Jang, G. Y., Li, M., Lee, Y. R., Lee, J., & Jeong, H. S. (2017). Changes of phenolic-acids and vitamin E profiles on germinated rough rice (Oryza sativa L.) treated by high hydrostatic pressure. *Food Chemistry*. https://doi.org/10.1016/j.foodchem.2016.08.069

Knorr, D., Heinz, V., & Buckow, R. (2006). High pressure application for food biopolymers. *Biochimica et Biophysica Acta – Proteins and Proteomics*. https://doi.org/10.1016/j.bbapap.2006.01.017

Kubow, S. (1992). Routes of formation and toxic consequences of lipid oxidation products in foods. *Free Radical Biology and Medicine*. https://doi.org/10.1016/0891-5849(92)90059-P

Larrea-Wachtendorff, D., Sousa, I., & Ferrari, G. (2020). Starch-Based Hydrogels Produced by High-Pressure Processing (HPP): Effect of the Starch Source and Processing Time. *Food Engineering Reviews*. https://doi.org/10.1007/s12393-020-09264-7

Lee, C., Lee, W., Han, Y., & Oh, S. (2017). Effect of Proteolysis with Alkaline Protease Following High Hydrostatic Pressure Treatment on IgE Binding of Buckwheat Protein. *Journal of Food Science*. https://doi.org/10.1111/1750-3841.13627

Lee, N. Y., & Koo, J. G. (2019). Effects of high hydrostatic pressure on quality changes of blends with low-protein wheat and oat flour and derivative foods. *Food Chemistry*. https://doi.org/10.1016/j.foodchem.2018.07.171

Liu, H., Fan, H., Cao, R., Blanchard, C., & Wang, M. (2016). Physicochemical properties and in vitro digestibility of sorghum starch altered by high hydrostatic pressure. *International Journal of Biological Macromolecules*. https://doi.org/10.1016/j.ijbiomac.2016.07.088

Liu, H., Guo, X., Li, Y., Li, H., Fan, H., & Wang, M. (2016). In vitro digestibility and changes in physicochemical and textural properties of tartary buckwheat starch under high hydrostatic pressure. *Journal of Food Engineering*. https://doi.org/10.1016/j.jfoodeng.2016.05.015

Liu, H., Wang, L., Cao, R., Fan, H., & Wang, M. (2016). In vitro digestibility and changes in physicochemical and structural properties of common buckwheat starch affected by high hydrostatic pressure. *Carbohydrate Polymers*. https://doi.org/10.1016/j.carbpol.2016.02.028

Liu, Y., Chao, C., Yu, J., Wang, S., Wang, S., & Copeland, L. (2020). New insights into starch gelatinization by high pressure: Comparison with heat-gelatinization. *Food Chemistry*. https://doi.org/10.1016/j.foodchem.2020.126493

Maniglia, B. C., Castanha, N., Rojas, M. L., & Augusto, P. E. (2021). Emerging technologies to enhance starch performance. *Current Opinion in Food Science*, 37, 26–36. https://doi.org/10.1016/j.cofs.2020.09.003

Maqsood-ul-Haque, S., & Kamal, N. A. S. (2021). High pressure process treatment (HPP) as an alternative food preservation method on fruits and vegetables: A brief review. *Malaysian Journal of Chemical Engineering and Technology (MJCET)*. https://doi.org/10.24191/mjcet.v4i1.11233

Marti, A., Barbiroli, A., Bonomi, F., Brutti, A., Iametti, S., Marengo, M., ... Pagani, M. A. (2014). Effect of high-pressure processing on the features of wheat milling by-products. *Cereal Chemistry*. https://doi.org/10.1094/CCHEM-07-13-0133-CESI

Mateos-Aparicio, I., Mateos-Peinado, C., & Rupérez, P. (2010). High hydrostatic pressure improves the functionality of dietary fibre in okara by-product from soybean. *Innovative Food Science and Emerging Technologies*. https://doi.org/10.1016/j.ifset.2010.02.003

McCleary, B. V., DeVries, J. W., Rader, J. I., Cohen, G., Prosky, L., Mugford, D. C., ... Okuma, K. (2011). Collaborative study report: Determination of insoluble, soluble, and total dietary fiber (Codex Definition) by an enzymatic-gravimetric method and liquid chromatography. *Cereal Foods World*. https://doi.org/10.1094/CFW-56-6-0238

Medina-Meza, I. G., & Barnaba, C. (2013). Kinetics of Cholesterol Oxidation in Model Systems and Foods: Current Status. *Food Engineering Reviews*. https://doi.org/10.1007/s12393-013-9069-0

Medina-Meza, I. G., Barnaba, C., & Barbosa-Cánovas, G. V. (2014). Effects of high pressure processing on lipid oxidation: A review. *Innovative Food Science and Emerging Technologies*. https://doi.org/10.1016/j.ifset.2013.10.012

Meng, L., Zhang, W., Hui, A., & Wu, Z. (2020). Effect of high hydrostatic pressure on pasting properties, volatile flavor components, and water distribution of cooked black rice. *Journal of Food Processing and Preservation*. https://doi.org/10.1111/jfpp.14900

Meng, L., Zhang, W., Wu, Z., Hui, A., Gao, H., Chen, P., & He, Y. (2018). Effect of pressure-soaking treatments on texture and retrogradation properties of black rice. *LWT*. https://doi.org/10.1016/j.lwt.2018.03.079

Messens, W., Van Camp, J., & Huyghebaert, A. (1997). The use of high pressure to modify the functionality of food proteins. *Trends in Food Science and Technology*. https://doi.org/10.1016/S0924-2244(97)01015-7

Min, D., & Boff, J. (2002). *Lipid Oxidation of Edible Oil*. https://doi.org/10.1201/9780203908815.pt3

Moyano, F. J., Saénz de Rodrigáñez, M. A., Díaz, M., & Tacon, A. G. J. (2015). Application of in vitro digestibility methods in aquaculture: Constraints and perspectives. *Reviews in Aquaculture*. https://doi.org/10.1111/raq.12065

Oh, H. E., Hemar, Y., Anema, S. G., Wong, M., & Neil Pinder, D. (2008). Effect of high-pressure treatment on normal rice and waxy rice starch-in-water suspensions. *Carbohydrate Polymers*. https://doi.org/10.1016/j.carbpol.2007.11.038

Pei-Ling, L., Xiao-Song, H., & Qun, S. (2010). Effect of high hydrostatic pressure on starches: A review. *Starch/Staerke*. https://doi.org/10.1002/star.201000001

Puncha-Arnon, S., & Uttapap, D. (2013). Rice starch vs. rice flour: Differences in their properties when modified by heat-moisture treatment. *Carbohydrate Polymers*. https://doi.org/10.1016/j.carbpol.2012.08.006

Queirós, R. P., Saraiva, J. A., & da Silva, J. A. L. (2018). Tailoring structure and technological properties of plant proteins using high hydrostatic pressure. *Critical Reviews in Food Science and Nutrition*. https://doi.org/10.1080/10408398.2016.1271770

Raghunathan, R., Pandiselvam, R., Kothakota, A., & Mousavi Khaneghah, A. (2021). The application of emerging non-thermal technologies for the modification of cereal starches. *LWT*. https://doi.org/10.1016/j.lwt.2020.110795

Ravash, N., Peighambardoust, S. H., Soltanzadeh, M., Pateiro, M., & Lorenzo, J. M. (2020). Impact of high-pressure treatment on casein micelles, whey proteins, fat globules and enzymes activity in dairy products: a review. *Critical Reviews in Food Science and Nutrition*. https://doi.org/10.1080/10408398.2020.1860899

Richard F, T., & William R, M. (1990). Swelling and gelatinization of cereal starches. I. Effects of amylopectin, amylose, and lipids. *Cereal Chemistry*.

Santos, M. D., Ferreira, J., Fidalgo, L. G., Queirós, R. P., Delgadillo, I., & Saraiva, J. A. (2014). Changes in maize starch water sorption isotherms caused by high pressure. *International Journal of Food Science and Technology*. https://doi.org/10.1111/ijfs.12273

Schaich, K. M. (2005). Lipid Oxidation: Theoretical Aspects. In *Bailey's Industrial Oil and Fat Products*. https://doi.org/10.1002/047167849x.bio067
Schmidt, M., Zannini, E., & Arendt, E. K. (2019). Screening of post-harvest decontamination methods for cereal grains and their impact on grain quality and technological performance. *European Food Research and Technology*. https://doi.org/10.1007/s00217-018-3210-5
Seena, S., & Sridhar, K. R. (2005). Physicochemical, functional and cooking properties of under explored legumes, Canavalia of the southwest coast of India. *Food Research International*. https://doi.org/10.1016/j.foodres.2005.02.007
Senturk, T., & Alpas, H. (2013). Effect of High Hydrostatic Pressure Treatment (HHPT) on Quality and Shelf Life of Atlantic Mackerel (Scomber scombrus). *Food and Bioprocess Technology*. https://doi.org/10.1007/s11947-012-0943-1
Serna-Hernandez, S. O., Escobedo-Avellaneda, Z., García-García, R., Rostro-Alanis, M. de J., & Welti-Chanes, J. (2021). High hydrostatic pressure induced changes in the physicochemical and functional properties of milk and dairy products: A review. *Foods*. https://doi.org/10.3390/foods10081867
Sharma, C., Singh, B., Hussain, S. Z., & Sharma, S. (2017). Investigation of process and product parameters for physicochemical properties of rice and mung bean (Vigna radiata) flour based extruded snacks. *Journal of Food Science and Technology*. https://doi.org/10.1007/s13197-017-2606-8
Sharma, N., Goyal, S. K., Alam, T., Fatma, S., Chaoruangrit, A., & Niranjan, K. (2018). Effect of high pressure soaking on water absorption, gelatinization, and biochemical properties of germinated and non-germinated foxtail millet grains. *Journal of Cereal Science*. https://doi.org/10.1016/j.jcs.2018.08.013
Shen, X., Shang, W., Strappe, P., Chen, L., Li, X., Zhou, Z., & Blanchard, C. (2018). Manipulation of the internal structure of high amylose maize starch by high pressure treatment and its diverse influence on digestion. *Food Hydrocolloids*, 77, 40–48. https://doi.org/10.1016/j.foodhyd.2017.09.015
Shevkani, K., Singh, N., Chen, Y., Kaur, A., & Yu, L. (2019). Pulse proteins: secondary structure, functionality and applications. *Journal of Food Science and Technology*. https://doi.org/10.1007/s13197-019-03723-8
Stolt, M., Oinonen, S., & Autio, K. (2000). Effect of high pressure on the physical properties of barley starch. *Innovative Food Science and Emerging Technologies*. https://doi.org/10.1016/S1466-8564(00)00017-5
Suhag, R., Dhiman, A., Deswal, G., Thakur, D., Sharanagat, V. S., Kumar, K., & Kumar, V. (2021). Microwave processing: A way to reduce the anti-nutritional factors (ANFs) in food grains. *LWT*. https://doi.org/10.1016/j.lwt.2021.111960
Tejada-Ortigoza, V., García-Amezquita, L. E., Serna-Saldívar, S. O., & Welti-Chanes, J. (2017). The dietary fiber profile of fruit peels and functionality modifications induced by high hydrostatic pressure treatments. *Food Science and Technology International*. https://doi.org/10.1177/1082013217694301
Thebaudin, J. Y., Lefebvre, A. C., Harrington, M., & Bourgeois, C. M. (1997). Dietary fibres: Nutritional and technological interest. *Trends in Food Science and Technology*. https://doi.org/10.1016/S0924-2244(97)01007-8
Tian, Y., Zhao, J., Xie, Z., Wang, J., Xu, X., & Jin, Z. (2014). Effect of different pressure-soaking treatments on color, texture, morphology and retrogradation properties of cooked rice. *LWT – Food Science and Technology*. https://doi.org/10.1016/j.lwt.2013.09.020
Tuboly, E., Lebovics, V. K., Gaál, Ö., Mészáros, L., & Farkas, J. (2003). Microbiological and lipid oxidation studies on mechanically deboned turkey meat treated by high hydrostatic pressure. *Journal of Food Engineering*. https://doi.org/10.1016/S0260-8774(02)00260-1
Vittadini, E., Carini, E., Chiavaro, E., Rovere, P., & Barbanti, D. (2008). High pressure-induced tapioca starch gels: Physico-chemical characterization and stability. *European Food Research and Technology*. https://doi.org/10.1007/s00217-007-0611-2
Wang, C., Xue, Y., Yousaf, L., Hu, J., & Shen, Q. (2020). Effects of high hydrostatic pressure on swelling behavior of rice starch. *Journal of Cereal Science*. https://doi.org/10.1016/j.jcs.2020.102967
Wang, H., Hu, F., Wang, C., Ramaswamy, H. S., Yu, Y., Zhu, S., & Wu, J. (2020). Effect of germination and high pressure treatments on brown rice flour rheological, pasting, textural, and structural properties. *Journal of Food Processing and Preservation*. https://doi.org/10.1111/jfpp.14474
Wang, H., Zhu, S., Ramaswamy, H. S., Hu, F., & Yu, Y. (2018). Effect of high pressure processing on rancidity of brown rice during storage. *LWT*. https://doi.org/10.1016/j.lwt.2018.03.042

Wang, N., Maximiuk, L., Fenn, D., Nickerson, M. T., & Hou, A. (2020). Development of a method for determining oil absorption capacity in pulse flours and protein materials. *Cereal Chemistry*. https://doi.org/10.1002/cche.10339

Westerterp, K. R. (2020). Energy metabolism. In B. P. Marriott, D. F. Birt, V. A. Stallings, & A. A. B. T.-P. K. in N. (Eleventh E. Yates (Eds.), *Present Knowledge in Nutrition* (pp. 3–14). https://doi.org/10.1016/B978-0-323-66162-1.00001-9

Wiggers, S. B., Kröger-Ohlsen, M. V., & Skibsted, L. H. (2004). Lipid oxidation in high-pressure processed chicken breast during chill storage and subsequent heat treatment: Effect of working pressure, packaging atmosphere and storage time. *European Food Research and Technology*. https://doi.org/10.1007/s00217-004-0931-4

Wolever, T. M. S. (1990). Relationship between dietary fiber content and composition in foods and the glycemic index. *American Journal of Clinical Nutrition*. https://doi.org/10.1093/ajcn/51.1.72

Wu, Z., He, Y., Yan, W., Zhang, W., Liu, X., Hui, A., Wang, H., & Li, H. (2021). Effect of high-pressure presoaking on texture and retrogradation properties of parboiled rice. *Journal of the Science of Food and Agriculture*, 101(10), 4201–4206. https://doi.org/10.1002/jsfa.11058

Xia, Q., Mei, J., Yu, W., & Li, Y. (2017). High hydrostatic pressure treatments enhance volatile components of pre-germinated brown rice revealed by aromatic fingerprinting based on HS-SPME/GC–MS and chemometric methods. *Food Research International*. https://doi.org/10.1016/j.foodres.2016.12.001

Xia, Q., Tao, H., Huang, P., Wang, L., Mei, J., & Li, Y. (2017). Minerals in vitro bioaccessibility and changes in textural and structural characteristics of uncooked pre-germinated brown rice influenced by ultra-high pressure. *Food Control*. https://doi.org/10.1016/j.foodcont.2016.07.018

Xia, Q., Wang, L., Xu, C., Mei, J., & Li, Y. (2017). Effects of germination and high hydrostatic pressure processing on mineral elements, amino acids and antioxidants in vitro bioaccessibility, as well as starch digestibility in brown rice (Oryza sativa L.). *Food Chemistry*. https://doi.org/10.1016/j.foodchem.2016.07.114

Xu, X., Yan, W., Yang, Z., Wang, X., Xiao, Y., & Du, X. (2019). Effect of ultra-high pressure on quality characteristics of parboiled rice. *Journal of Cereal Science*. https://doi.org/10.1016/j.jcs.2019.03.014

Y. Hibi, T. Matsumoto, & S. Hagiwara. (1993). Effect of high pressure on the crystalline structure of various starch granules. *Cereal Chem.*

Yang, J., & Powers, J. R. (2016). Effects of High Pressure on Food Proteins. In *Food Engineering Series* (pp. 353–389). https://doi.org/10.1007/978-1-4939-3234-4_18

Yu, Y., Ge, L., Zhu, S., Zhan, Y., & Zhang, Q. (2015). Effect of presoaking high hydrostatic pressure on the cooking properties of brown rice. *Journal of Food Science and Technology*. https://doi.org/10.1007/s13197-015-1901-5

Zhang, K., Ma, X., Dai, Y., Hou, H., Wang, W., Ding, X., … Dong, H. (2019). Effects of high hydrostatic pressure on structures, properties of starch, and quality of cationic starch. *Cereal Chemistry*. https://doi.org/10.1002/cche.10132

Zhu, F., & Li, H. (2019). Effect of high hydrostatic pressure on physicochemical properties of quinoa flour. *LWT*. https://doi.org/10.1016/j.lwt.2019.108367

Zhu, S. M., Hu, F. F., Ramaswamy, H. S., Yu, Y., Yu, L., & Zhang, Q. T. (2016). Effect of High Pressure Treatment and Degree of Milling on Gelatinization and Structural Properties of Brown Rice. *Food and Bioprocess Technology*. https://doi.org/10.1007/s11947-016-1770-6

Zhu, S. M., Lin, S. L., Ramaswamy, H. S., Yu, Y., & Zhang, Q. T. (2017). Enhancement of Functional Properties of Rice Bran Proteins by High Pressure Treatment and Their Correlation with Surface Hydrophobicity. *Food and Bioprocess Technology*. https://doi.org/10.1007/s11947-016-1818-7

CHAPTER 19

Ultrasonication Processing of Cereals

Balmeet Singh Gill,[1] Sukriti Singh,[2] Harpreet Kaur,[1] Dilpreet Singh,[1] and Manisha Bhandari[3]
[1]Department of Food Science and Technology, Guru Nanak Dev University, Amritsar, Punjab, India
[2]Department of Food Technology, Uttaranchal University, Dehradun, India
[3]Department of Food Science and Technology, Punjab Agricultural University, Ludhiana, Punjab, India

CONTENTS

19.1 INTRODUCTION

The properties of food products have been under evaluation for the last few decades. The demand of the food sector is for limited processing of foods, which requires alterations in the processing techniques because the application of various processing methods under critical conditions tends to reduce the nutritive significance and bioavailability of foods by causing chemical and physical changes, and as a result, lowers the organoleptic acceptability of foods (Majid et al., 2015).

DOI: 10.1201/9781003242192-24

Today, newer techniques involving physical modifications have been developed to improve nutritional significance (Kaur and Gill, 2019). Ultrasonication is an emerging technique of modification that involves numerous advantages such as enhanced quality, limited processing, and minimized use of chemicals, thereby retaining the safety of food products (Knorr et al., 2011; Gill and Kaur, 2019). Ultrasonication plays an important role in the research and development sector of the food industry. It is based on acoustic energy and has a frequency higher than is audible by human beings (>16 kHz), and there are two frequency ranges (i.e., low- and high-energy sound waves). Low-energy sound waves utilize low power and low intensity and have a frequency range above 100 kHz at intensity less than 1 Wcm^{-2}. On the other hand, high-energy sound waves use high power and high intensity at frequency range 20–500 kHz and intensity above 1 Wcm^{-2} (Mason et al., 2011). High-energy ultrasound waves are an important technique used to evaluate the physicochemical characteristics of food such as firmness, ripeness, acidity, sugar content, etc. Low-energy ultrasound waves are needed to alter the chemical and physical characteristics of food (Soria and Villamiel, 2010) by applying shear, pressure, and temperature variation in the solution by which they propagate and produce cavitations to inactivate microorganisms present in food products (Majid et al., 2015). Ultrasonication has been successfully applied in processing of cheese; to retain the quality of fresh fruits and vegetables in either pre-harvest or post-harvest; in cooking oils, food gels, cereal products, frozen and aerated foods; and in food products based on emulsified fat. Other applications involve the assessment of type, state, and size of protein and also the detection of adulteration of honey.

This non-thermal technique is widely accepted in foods that are heat sensitive due to its specific characteristics and ability to enhance shelf life, prevent microbial spoilage, and its usefulness in bacterial biofilms (Majid et al., 2015). It is a successful tool that is used in the production and processing of nanoparticles of polymers. It is highly effective as it breaks up the aggregates resulting in reduced size and polydispersity of nanoparticles (Tang et al., 2003). When compared to conventional thermal processing treatments, it has several benefits over them such as enhanced yield of production at better rates of processing, preservation is better of heat labile foods, easy application and operation, better aseptic conditions of processing, enhanced quality and functionality of the foods that are processed, cost-effective technology, consumes less energy, and chances of equipment contamination are low (Gharibzahedi and Smith, 2020). Ultrasonication is referred to as a green technology as it has the potential to produce products that are free from environmental pollutants. Ultrasonication works via modification of matrix, texture of foods, reformation in the kinetic and thermodynamic parameters of enzymes for inclusive functionality, crystallization of lactose in the dairy products, enhanced process of extraction, reduction in the microbial load, and alteration in the micelles of casein with no addition of chemical preservatives (Rao et al., 2021). This technique demands less energy when compared to other processing treatments, especially the thermal ones, since there is no large energy requirement (mechanical or electrical). This technique not only focuses on improving and retaining the quality and safety of food products but also contributes to developing novel food products (Kaur and Gill, 2019). The advantages of ultrasonication are , higher efficiency of mixing and extraction application, lower temperature of working, fast transfer of energy reduced floor space, and energy requiredg (Kehinde et al., 2021). Ultrasonication can be widely used in the food applications due to its chemical and mechanical effects those are antifoaming action, aggregates dispersion, equipment's sterilization, viscosity modification, destruction of cell, microorganism's inactivation, particle size reduction, and cell growth modulation (Feizollahi et al., 2021).

19.2 MODE OF ACTION

Ultrasonication is a non-thermal technology that offers maximum quality and safety to food products and helps achieve the objective of sustainable "green" chemistry and extraction.

Ultrasonication found its utility in the food industry as it can be used for various processing techniques such as homogenization, extraction, pasteurization, activation/deactivation of enzymes, particle size reduction, and emulsification (Alex and Bates, 2011). Ultrasound waves are used in combination with other treatments such as temperature (ultrasonication), heat (thermosonication), pressure (manosonication), and heat and pressure (manothermosonication) (Ravi Kumar and Sachin., 2017). Factors that influence the ultrasonication process are treatment time, temperature, sonication power frequency, and energy input (Grgić et al., 2019). When ultrasound is applied on a liquid medium it causes acoustic cavitation that results in generation of hotspots that have a short life span having intense heating at a very high temperature (5000 °C) (1000 atm); the heating and cooling is done at the rate of 10^{10} K/s (Tang et al., 2003). This is a physical phenomenon that leads to the formation, growth, and eventual collapse of bubbles (or cavities) leading to minimum radius generating a very high shear energy inside a liquid (Majid et al., 2015; Gallo et al., 2018). Ultrasonication therefore forms stable emulsions and dispersions with particles of smaller size at much lower ratio of mechanical mixing. These lower ratios of mixing are converted to supplementary efficiencies that require lower amount of energy providing similar dispersion or homogenization quantity leading to lower energy consumption for cooling of the processed solutions (Gharibzahedi and Smith, 2020). The ultrasound waves when propagates in the liquid system causing the bubbles to oscillate, which causes mechanical, thermal, and chemical changes in the medium and leads to the generation of high temperature (5000 K) and pressure (1000 atm) (Soria and Villamiel, 2010). The cavitation results in micro-streaming and enhances the transfer of heat and mass, and the collapse of cavitation bubbles is rapid and rigorous, which leads to drastic changes in pressures (>1000 atm) and temperatures (up to 500°C) (Cui and Zhu, 2020). Formation of free radical, sheer stress, turbulences, increase in cell permeability, cell membrane thinning, and selectivity loss are some of the mechanical, chemical, and biological effects of ultrasonication (Yusaf and Al-Juboori, 2014; Lateef et al.,2007; Sams and Feria, 1991). Ultrasound is non-toxic, eco-friendly, safe, has high yield, is an effective means of inactivating microbes resulting in minimum loss of flavor, can be expanded in food industries, and can improve other properties such as emulsification, crystallization, and homogenization (Kentish and Ashok, 2011; Vercet et al., 2002; Chouliara et al., 2010; Chemat et al., 2011).

In the food processing industry, ultrasonication processing is done in the range of 20 KHz to 10 MHz, and is further characterized into three ranges:

1. 20–100 kHz – Causes variation in physicochemical characteristics or structure of food (Sharma et al., 2020).
2. 100 kHz–1 MHz – Causes chemical reaction due to free radical formation in food.
3. 1 kHz–10 MHz – No changes in food but is used for analytical research such as measuring food composition, adulteration, physical characteristics of batter, and uniformity (Tobergate and Curtis, 2013).

Ultrasonication can be done in a liquid suspension by applying ultrasound directly on the product (immersing probe into suspension), coupling ultrasound with the device, and by submerging the product in an ultrasonic bath (immersing sample container in which sample is suspended in suspension liquid into a bath containing a liquid through which ultrasonic waves are propagated). Direct sonication yields high effective energy output in comparison to indirect sonication.

A sonicator consists of a sonicator probe (acoustic element) or horn of various shape and diameter responsible for conducting the acoustic energy from a transducer to the suspension. The amount of acoustic energy transferred to the suspension will depend on various factors such as applied power, total amount of suspension time, mode (pulse or continuous), immersion depth in the suspension (2 to 5 cm), container geometry (flat, round, or conical bottom flask), and distance from the bottom (no more than 1 cm). The effective acoustic energy delivered can be calculated by recording the increase in temperature in liquid at a given device output power setting and using the following formula:

$$P = \frac{dT}{dt} MCp$$

where

P – Delivered acoustic power (W)
T – Temperature (K)
t – Time (s)
Cp- Specific heat of the liquid (J/g·K)
M – Mass of liquid (g) (Taurozzi et al., 2012)

19.3 IMPACT OF ULTRASONICATION ON FUNCTIONAL CHARACTERISTICS

Ultrasonication is a physical treatment that affects and improves the composition, structure, and qualities of food products in a novel way. It can be used efficiently in the food sector for a variety of search-related food applications to produce desired food items with higher yields, shorter processing times, reduced environmental impact, and lower energy usage. It alters food characteristics such as starch, protein, matrix of protein-starch, etc., and can be used as a functional ingredient in food products. As a result, ultrasonication can be used to generate and improve starch-based products' functionality and stability (Kaur and Gill, 2019). The amorphous area of starch granules is disrupted by ultrasonication; their shape and size remain unchanged, but their surface becomes porous, which enhances physicochemical qualities such as swelling power, solubility, and pasting parameters. In ultrasonication, shear pressures are generated when bubbles collapse forcefully, fracturing polymer chains such as starch. Also, when the solvent molecules dissociate to create radicals, the polymer degrades (Sujka and Jamroz, 2013). Impacts of ultrasonication on the functional properties of cereals are summarized in Table 19.1.

Ultrasonication improved the water absorption capacity (WAC) and oil absorption capacity (OAC) of black gram-extracted proteins. There was a significant increase in the WAC with a slight increase in the OAC of the modified protein (Kamani et al., 2021). This enhanced functional property could be due to the smaller particle size and improvement in the solubility due to the processing treatment. However, increased time period for protein ultrasonication can reduce the WAC ascribing to the denaturation of the molecular structure of the protein and increment in the hydrophobic surface of the protein. The stearic factor that is responsible for the determination of the WAC is the amino acid's hydrophobic–hydrophilic balance of the molecule of the protein. Also, increased ultrasonication time is responsible for the enhanced re-aggregation of the protein molecule that inhibits the ability of retaining the water. During the ultrasonication, the buried hydrophobic regions of the proteins were exposed, which allowed the modified protein to be entrapped in the oil (Biswas and Sit, 2020).

Ultrasonication results in increased emulsification activity that could be due to certain characteristics such as solubility, surface charge, and hydrophobicity. Also, an increase in zeta can also be a possible reason for the enhanced emulsification activity. A change in the tertiary and quaternary structure of proteins was observed due to the major pressure that was generated during the sonication technique that increased the potential of protein molecules to absorb oil at the oil–water interface. This alteration is responsible for the dissociation of the hydrophobic interactions in the

Table 10.1 Effect of Ultrasonication on the Functional Properties of Cereals

Cereal	Dose	Time period (min)	Effect on functional property	References
Quinoa	20 KHz (500 W)	5, 15, 25, and 35	• Increase in the emulsion activity and stability • Decrease in the turbidity • Improvement in the oil- and water-binding capacities	Mir et al., 2019
Black gram by product	20 KHz	5, 10, 20, and 30	• Increase in the water and oil absorption capacities • Increase in the emulsion capacity and stability • Increase in the foaming capacity and decrease in the foaming stability	Kamani et al., 2021
Proso millet	20 KHz (10 W)	5, 12.5, and 20	• Solubility increased • Increase in the foaming capacity and stability with increase in the amplitude • Highest emulsion activity and stability at 12.5 min	Nazari et al., 2018
Wheat grains	30 KHz (100 W)	15 and 30	• Increase in swelling power and solubility • Increase in oil absorption capacity	Karwasra et al., 2020
Rice flour	24 KHz (180 W)	60	• Increase in the water absorption capacity, water absorption index, and swelling power (these decreased with the increase in temperature duing ultrasonication) • Decrease in the water solubility index	Vela et al., 2021
Maize	100 W	30 min	• Increase in swelling power and solubility	Luo et al., 2008
Wheat bran	24 KHz (400 W)	5, 10, and 15 min	• Increase in the swelling power	Habus et al., 2021
Wheat gluten protein	28, 40, and 80 KHz	10 min	• Increase in water- and oil-holding capacity • Enhanced protein solubility	Zhang et al., 2020
Oat grains	150, 250, and 350 W	10 and 20 min	• Increase in swelling power and solubility • Increased water absorption capacity • Increased oil absorption capacity • Increased syneresis	Falsafi et al., 2019

interior parts that leads to diffusion of oil molecules on the surface of small-size soluble particles that have favorable absorption of the oil droplets on the surface (Wu et al., 2020). However, similar to WAC, increase in the time period for ultrasonication results in reduced emulsification activity due to the disruption of the structure of the protein molecules and reduced interfaces of oil and water or aggregration of the protein molecules that have been denatured. Ultrasonication of the protein molecules resulted in reduced emulsification activity contributing to the alteration of the protein molecule's orientation due to the stability treatment that resulted in the limiting of the stability of the protein film in the emulsion system at the oil–water interface (Mir et al., 2019).

Foaming capacity is increased in ultrasonication treatment leading to the partial denaturation of protein that contributes to higher diffusion of the air and water interface due to the increased cohesiveness and flexibility of foam. Ultrasonication treatment tends to expose more of the hydrophobic groups, which leads to enhanced function of the protein at the air–water interface (Zuniga-Salcedo et al., 2019). An increase in the swelling power on ultrasonication was observed due to the breakdown of intermolecular bonds in starches resulting in reduced compression of the arrangements of the starch granules (Karwasra et al., 2020). With increase in time of ultrasonication, an increase in the starch molecule's disorganization occurs along with the lixiviation of the amylose component that allows the penetration of water in the granules of the starch leading to increased swelling power.

During ultrasonication, the protein layer that covers the starch granules is removed causing subsequent leaching of the amylose in the water (Chan et al., 2010).

19.4 RHEOLOGY

The rheology of any food product is the major factor that affects the textural or structural properties. When starch from the cereal source is used as a functional ingredient in any food product, the properties of that food product are modified. It is important to understand the properties of a particular starch (source) and the effect of the starch on the rheological properties of a particular food product. Ultrasound treatment increases the cavitation pressures that cause starch granules to degrade, making them more accessible to water during the heating step. The viscosity of a starch suspension is dependent on the duration of ultrasonic treatment of a starch sample. It was found that ultrasound treatment for increasing power and time causes more disruption of starch granules, which weakens the crystalline region of the molecule, causing the molecules to entrap more water, resulting in high viscosity which might be due to the shear pressures generated by ultrasonic treatment causing significant damage to the starch granules, resulting in the straightening out of amylose molecules and a reduction in shear action within the fluid layers, resulting in a drop in viscosity (Kaur and Gill, 2019). Thus, it can be concluded that ultrasound-treated starch is used to get high apparent viscosity or modified rheological properties as compared to untreated starch. The rheological properties are also influenced by the swelling power of the starch used in a particular food. For instance, rice has the greatest values of swelling power and solubility among native starches, which might be ascribed to the greater phosphate group concentration on amylopectin. Phosphate groups reduce the amount of bonding inside the crystalline domain, resulting in increased granule hydration. This might be due to starch granules being physically and chemically disrupted, resulting in increased water absorption and retention (Jambrak et al., 2010. Therefore, it is the cheapest non-thermal technique for starch modification and can be utilized as a promising processing technique in the food industry.

19.5 STRUCTURE

The structure of food components, including texture and consistency, is the most significant predictor of the sensory quality of food items, and contributes to the palatability and enjoyment of having food (Moelants et al., 2014). The structure of starch from a variety of botanical sources, including potato, maize, wheat, rice, mung bean, cassava, and sago, have all been demonstrated to be altered by ultrasonication. Ultrasonic waves cause physicochemical changes in these starches, which are linked to disturbances in the swelling granule, causing amylose and amylopectin to escape. This illustrates how the sonication technique may be used to physically alter powdered starch to give it better characteristics (Bonto et al., 2020a). When ultrasonic waves break cell walls, a huge surface area is exposed, allowing mass transport of cellular fragments into the aqueous medium. The surface becomes porous as a result of the empty cell walls. Sonication aids the development of micropores on rice's surface as well as cracks and fissures within the grain (Toma et al., 2001). The development of fissures as a result of sonication is common on rice, independent of the amylose concentration or presence or absence of the bran layer. Ultrasound vibrations (40–53 kHz) were applied to the kernel for 5 min, causing fractures and fissures that kept the head rice grain unbroken. The degree of fractured kernels rises as the sonication period is increased to 30 and 60 min. Increasing the frequency of sound waves (e.g., 280, 360, or 273 kHz) causes the grains to easily fracture and the starch granules to be damaged (Gallant et al., 1997. A study investigated the effect of ultrasonication on the morphology of a rice grain. The morphological alterations of rice grain seem to reduce the hardness of the treated grain. The microporous surface allows for simple water absorption, while the development of surface fissures and internal fractures allows for water diffusion into the inner endosperm during cooking, resulting in rupture and amylose leaching (Bonto et al., 2020c). It can be concluded that a controlled frequency of ultrasonication can enhance the functional value of food as well as improve the cooking quality and milling yield of cereal grains.

19.6 PROTEIN

In recent years, ultrasound has gained a lot of interest for its use in protein modification, either as a pre-treatment to improve protein modification or as a chemical reaction by modifying physical and functional properties including gelation, foamability, emulsification, and solubility (Majid et al., 2015). Various ultrasonic powers (0–1020 W) have been widely used to induce structural changes in plant-derived renewable proteins and lead to protein unfolding, resulting in improved protein functional characteristics. Peanut protein isolates, soy protein isolates, black bean protein isolates, and jackfruit seed protein isolates are only a few of the proteins available (Wen et al., 2019. In the modification of the protein, there are several mechanisms involved relating to the impact of ultrasonication on the -SH group present in the protein. (i) The intermolecular break down of S-S bonds is responsible for causing acoustic cavitation, which reduces particle size in protein dispersions. (ii) The -SH groups are formed by decreasing disulfide bonds (S-S) with the help of cavitation. (iii) Incomplete protein molecules' unfolding is influenced by intense ultrasonic shear waves increases surface-exposed buried -SH groups (Gharibzahedi and Smith, 2020). Without the use of chemicals or heat processes, this technique may effectively alter the structure of legumes and cereal proteins, which may be an effective strategy to enhance consumer acceptance due to the labeling of products as "green and cleaner."

19.7 THERMAL PROPERTIES

Functional properties such as starch and proteins are highly affected by their thermal characteristics. On application of probe ultrasonication at for 20 min 63 W/cm^2, modification of oat's starch enthalpy and gelatinazion temperature occurred; however, the thermal characteristics of starch granules treated by ultrasound waves were not significantly changed. With increasing the ultrasound intensity, the gelation temperature is shifted to higher values (Falsafi et al., 2019). Generally, two thermal transitions are there in soy protein isolates, which corresponds to the denaturation temperatures of two protein fraction that are 7S (-conglycinin, 77°C) and 11S (glycinin, 92°C) (Karki et al., 2010. A low-power ultasonication treatment for 1 min results in a considerable reduction in temperature of denaturation, but a high power sonication for 2 min at high power might increase thermal denaturation temperature. In most processing periods, the enthalpy of protein fractions treated with low-power sonication did not change, while high-power sonication decreased this thermodynamic characteristic. This is due to conformational variations in soy protein isolate structure as a result of various cavitation intensities. A greater denaturation rate is usually linked to lower thermal stability or a lower amount of organized secondary structure (Wani et al., 2015. It was reported that to improve the functioning of cereal and legume proteins, operational parameters involved in ultrasonication should be optimized at high intensities (Gharibzahedi and Smith, 2020).

19.8 IMPACT OF ULTRASONICATION ON NUTRITIONAL CHARACTERISTICS

Traditional food preservation methods rely on the inactivation of pathogenic or spoilage bacteria and enzymes using either heat removal (e.g., freezing), heat addition (e.g., canning), or chemical additives (e.g., chlorine) to assure food product safety and shelf life (Ojha et al., 2018). The thermal processing of food material leads to the destruction of several nutritional components, which reduces nutritive value and thus consumer acceptance. Non-thermal techniques are more reliable because most of the nutrients are preserved. Therefore, ultrasound treatment has gained popularity in food processing.

Different risks related to the deficiency of micro- and macronutrients can be overcome by adequate intake of micro- and macronutrients on a daily basis. The "macro" nutrients such as carbohydrates, proteins, and lipids are taken in large quantity. However, the "micro" nutrients are those that are required in small quantities for the proper functioning of the body, and include vitamins, minerals, and bioactive components like antioxidants. Ultrasonication is not only used in food preservation. It it also being used to enrich particular food products and to increase the nutritive value of food such as cereal. Ultrasound waves of high frequency are used, which produce pores on the surface of the grain. Due to the porous nature of the surface of the grain, its absorption capacity is increased and the retention capability of the grain is also improved (Bonto et al., 2018). Moreover, ultrasonication also alters the native characteristics of starch, and proteins are changed. This will improve the functional value of particular nutrients in the cereal. Table 19.2 summarizes the impact of ultrasonication on the nutritional characteristics of cereals.

19.9 MACRONUTRIENTS

The functional value of cereal foods can be improved by the modification of starch and protein structures. Due to the exposure of rice starch to ultrasound waves, the starch molecules get degraded to a small size. The smaller starch molecules are more susceptible to enzymatic breakdown. Therefore, the digestibility of some cereal grains is increased, along with the glycemic index (Bonot et al., 2020b). The increased glucose content in dehulled rice has also been reported (Ding et al., 2018a). The sonication technique also improves the functional value of cereal protein, by reducing the size of protein molecules (alpha-helix and beta fold), which increases the solubility of the protein (e.g., the wheat protein is reduced to 200 nm). Treated proteins are used for food fortification, additives for extrusion as flour products, and encapsulation for bioactive compounds, and ultrasonication increased the protein interaction with other nutrients, which improves the nutritive value of food (Wen et al., 2019).

On treatment of rice protein with ultrasonication, a decrease in the composition of amino acids was observed due to the release of some of the proteins into the medium during the process (Li et al., 2016). Ultrasonication results in decreased starch content since during this processing, the structure of the starch granules changes, which promotes the decomposition of starch, and then there is an increase in the reducing sugar content (Xia et al., 2020.

On ultrasonication of polished rice, a reduction in starch content was observed from 80.8 to 76.3 g/100 g while an increase in the protein and lipid content occurred from 8.63 to 9.22 g/100 g and 0.42 to 1.96 g/100 g, respectively. Ultrasonication decreased insoluble dietary fibre from 3.22 to 2.29 and increased soluble dietary fibre from 0.51 to 1.15 g/100 g, with a slight decrease in the total dietary fibre content from 3.73 to 3.44 g/100 g (Geng et al., 2021).

Ultrasonication of sorghum grains at a frequency of 20 KHz and power of 750 W resulted in higher fatty acid content of palmitic, arachidic, and stearic acid when the treatment was for a time

Table 19.2 Effect of Ultrasonication on the Nutritional Properties of Cereals

Cereal	Dose	Time	Effect	References
Brown rice noodles	600 W	20 min	• Increase in the protein content • Increase in the lipid content • Decrease in the starch content • Increase in the total, insoluble, and soluble dietary fibre	Geng et al., 2021
Yellow dent corn	• 200 W • 130, 160, and 200 W	• 15, 30, 45, and 60 min • 15 min	• Loose distribution of the starch granules with smaller amount of protein matrix • No change in the particle size of starch granules except for 60 min treatment • Lower gelatinization enthalpy • The sulfhydryl component of the protein increased • The content of disulfide bonds decreased	Liu et al., 2021
Wheat	700 W	2, 4, and 6 min	• Loss in protein content	Ertas, 2013
Brown rice	16 KHz (2000 W)	30 min	• No significant variation in starch, protein, fat, fibre, and vitamin content	Cui et al., 2010
Non-waxy and waxy rice variety	40 KHz (130 W)	5 and 15 min	• Loss of iron and phosphorus • Reduction in thiamin and niacin	Bonto et al., 2020
White wheat	22 KHz (630 W)	8 h	• Increased protein and lipid content • Decreased starch content • No effect on the ash content	Naumenko et al., 2022
Soybean	100, 200, and 300 W	30 min	• Increase in the protein • Decrease in the lipid content and the total sugar content • Increase in the moisture content	Yang et al., 2015

period of 10 min. Also, ultrasonication increased the amount of linolenic acid and reduction in the amount of eicosenoic acid. The increase in these contents is due to the cavitation that occurs during the ultrasonication process. Cavitation results in rupturing of the biological membranes and the cell wall (Hassan et al., 2017).

On increasing the power of ultrasonication of soyabean, an increase in the protein content of 47.66% at the power of 300 was seen while the decrease in the lipid and total sugar content was 27.41 and 21.66%, respectively. The ultrasonication of soybean also results in enhancing the moisture content of the soybean from 81.42 to 87.26%. The following result indicates that degradation of the total sugar and lipids provides the energy for the synthesis of the proteins. Also, an increase in the amino acid content was observed to be 18.28% at 300 W. When soybean was ultrasonicated, damage of the cell membrane occurs promoting the enzymatic activity in the processed sample, thus increasing the amino acid profile (Yang et al., 2015).

19.10 MICRONUTRIENTS

The iron content of rice was increased by direct soaking of ultrasonicated white rice in a solution of iron. The treated rice absorbs more water and with water it absorbs iron. Due to this absorption, the iron content of rice grain is increased, reported to be 28-fold more uptake of iron in sonicated (321 ± 13.4 mg kg^{-1}) rice as compared with untreated rice (283 ± 4.24 mg $^{kg-1}$) (Bonto et al., 2020b). The fortification of rice with Vitamin B5 is also possible with pre-treatment of rice with ultrasounds waves followed by soaking the sample in vitamin B5 solution. The result reported in the literature for sonicated rice was 447.59 μg/g and 186.16 μg/g for non-sonicated rice (Bonto et al., 2018).

A decrease in the iron, phosphorus, thiamin, and niacin content occurred in the ultrasonicated uncooked rice kernels. This could be due to the production of microjets during the collapse of the bubble that further generate the process of cavitation at the interface of water and rice that resulted in the leaching of the minerals in the water (Bonto et al., 2020c).

Thus, it can be concluded that ultrasonication is most suited for the improvement of the nutritive quality of cereal-based food.

19.11 IMPACT OF ULTRASONICATION ON BIOLOGICAL CHARACTERISTICS

Extraction of various bioactive compounds like flavonoids, phenolics, anthocyanins, and other antioxidant components from different plant and plant seeds using novel technologies such as ultrasound-assisted treatments, microwave-assisted treatment, high-pressure processing, and supercritical fluids has gained interest among researchers around the world as traditional methods such as solvent extraction, leaching, and percolation require long processing time and have low efficiency. The impacts of ultrasonication on the biological properties of cereals are listed in Table 19.3.

19.12 POLYPHENOLIC COMPOUNDS AND ANTIOXIDANT ACTIVITY

Ultrasonication-assisted extraction of bioactive compounds has been found to reduce processing time, temperature, and solvent consumption (Hossain et al., 2012). As a non-thermal processing technology, ultrasonication has been used to enhance the production of bioactive compounds (primary and secondary metabolites) (Hasan et al., 2017). Wang et al. (2020) optimized the ultrasound-assisted extraction of phenolic compounds in red rice bran, and the result showed an increase in the anthocyanins and vitamin C content. Lohani and Muthukumarappan (2017) studied the effect of continuous ultrasonication after fermentation and found improved extraction of phenolic content in sorghum flour. Ultrasonication initiated the release of bounded phenolics from the starch–protein matrix of fermented sorghum flour and hence increased free phenolic content. An increase in total phenolic content (41%) and antioxidant activity (42.4%) in ultrasonicated sorghum flour was observed in comparison to the control sorghum flour. Ultrasonication-assisted extraction (240W frequency 50/60 Hertz) of phenolic contents in brown rice using different solvents methanol, ethanol, and isopropanol was studied by Muhammad et al. (2018). There was a significant increase in total phenolic acid and antioxidant activity of brown rice after ultrasonication. Hassan et al. (2020) studied the effect on phytochemical characteristics of ultrasound-processed sorghum sprouts and observed that ultrasonication at 40% amplitude for 5 min showed the improved radical scavenging activity, phenolic profile, decrease in antinutritional activity (alkaloids, saponins, tannins, and phytates), and a higher percentage of *in vitro* protein digestibility. Sullivan et al. (2018) studied the effect of ultrasonication on the digestibility and biological properties of sorghum gluten-like flour and observed that ultrasonication (40% amplitude, 10 min) significantly improved the oxygen and nitric oxide radical scavenging activities of flour.

During the process of wheat germination, ultrasonication treatment leads to the stimulation of production of secondary metabolites in plants and grains such as phenolic compounds and thus increases the antioxidant activity (Naumenko et al., 2022). An increase in the antioxidant activity by DPPH (2,2-diphenyl-1-picryl-hydrazyl-hydrate), TROLOX (6-hydroxy-2,5,7,8-tetramethylchroman-2-carboxylic acid), and ORAC (Oxygen Radical Absorbance Capacity) was observed for sorghum

Table 19.3 Effect of Ultrasonication on the Biological Properties of Cereals

Cereal	Dose	Time	Effect	References
Wheat	20 KHz (750 W)	2, 4, 8, 16, and 20 h	• The content of rapidly digestible starch increased • The content of the resistant starch decreased • Degradation in the amount of phenolic compounds • *In vitro* antioxidant activity decreased • Degradation in anthocyanins	Cui and Zhu, 2020
Brown rice noodles	600 W	20 min	• Digestible starch content pf the starch decreased	Geng et al., 2021
Wheat	700 W	2, 4, and 6 min	• Reduction in the phytic acid	Ertas, 2013
Sorghum	20 KHz (500 W)	2, 3, and 4 min	• Increase in the total phenolic content	Lohani and Muthukumarappan, 2021
Wheat bran	24 KHz (400 W)	5, 10, and 15 min	• Reduction of polyphenol oxidase • Slight lowering of the total phenolic content and the antioxidant activity	Habus et al., 2021
Tartary buckwheat	20 KHz (90 W)	5 to 60 min	• Increase in the total phenolic content, flavonoid content, and antioxidant activity	Dzah et al., 2020
Maize starch	40 KHz (600 W)	10, 20, and 30 min	• Lower content of slowly digestible starch and resistant starch	Chan et al., 2021
White variety	22 KHz (630 W)	8 h	• Increase in the total flavonoid content and antioxidant activity	Naumenko et al., 2022
Sorghum	20 KHz	5 and 10 min	• Increase in the protein digestibility	Beazley, 2017
Buckwheat flour	20 KHz	5, 10, 15, 20, and 30 min	• Increase in the *in vitro* protein digestibility up to 10 min and further ultrasonication results in decreased digestibility	Jin et al., 2021

flour after 10 min of ultrasonication. This increase in the antioxidant capacity is due to the increased capacity of the solvent to extract polyphenolic compounds (Sullivan et al., 2018).

A reduction in DPPH activity was observed due to the generation of low molecular weight peptides. Yadav et al. (2021) observed that ultrasound amplitude (66%, 26 min) and grain-to-water ratio of 1:3 showed reduction in phytate (by 66.98%) and tannin contents (by 62.83%) of finger millet. Ultrasonication alone or in combination with other techniques can be used to improve the antioxidant activity in cereals and reduce the antinutritional factors and can be used as a substitute to traditional extraction methods. The antioxidant activity of ultrasonicate cereal grain was improved as compared to the untreated one. Iftikhar et al. (2020) reported DPPH value of treated rice bran was 72.97% and for the untreated sample it was 49.54%.

19.13 GAMMA-AMINOBUTYRIC ACID (GABA)

Ding et al. (2018a) applied ultrasonication (25 kHz, 16 W/L, 5 min) on germinated red rice (66 h) and metabolomics analysis, and found a significant increase in γ-aminobutyric acid (GABA), riboflavin, O-phosphoethanolamine, and glucose-6-phosphate in comparison to the untreated germinated rice. GABA shunt-related metabolites such as alanine, succinic, and glutamic acid were also increased after ultrasonic treatment. In another study conducted by Ding et al. (2019) reported that on ultrasonicating (25 kHz) germinated oats (96 h), the glutamic acid, free sugars, GABA, alanine, and phenolic compounds including 2p, 2f avenanthramides were enhanced. Yang et al. (2015) studied the effect of ultrasonic treatment (40 kHz, 30 min) in soaked soybean sprouts and observed a 43.4% increase in GABA. Ding et al. (2018b) studied the effect of ultrasonication of sprouted soft wheat for 72 h and found that there was an increase in GABA by 30.7%.

19.14 ANTINUTRITIONAL FACTORS

Ertas (2013) reported a decrease in the phytate content of various legumes and cereals. The decrease could be due to the leaching and solubilization of these anti-nutritional phytic acid salts in the water, and this effect of leaching and solubilization is more effective if the water is hot. Increase in the temperature and time of soaking results in greater reduction of the phytic salts. Ultrasound disrupts the integrity of the surface components that are treated, which results in better leaching of the phytate into the solvent thus reducing it (Feizollahi et al., 2021).

With increase in ultrasonication power, trypsin inhibitors are decreased due to the high unfolding of the proteins on higher power. The buried hydrophobic regions that contain phenylalanine, tryptophan, and tyrosine are exposed and the formation of aggregates takes place. This aggregation of protein is the main reason for the inactivation of the trypsin inhibitors as the activity of enzyme is decreased due to the obstruction in the active sites (Xia et al., 2020).

Ultrasonication of the soybean results in decreased activity of lipoxygenase isozyme by 36.22% at 300 W. This is due to the collapse of the small voids in the cell or of small bubbles present and the severe stress that is produced due to the shear. It decreased the trypsin inhibitor content by 98.78% at 300 W as ultrasonication causes it to be used as the source of the energy during the process of germination. This could be due to the alterations in the conformations of structure and inducing a decrease in the disulfide bonds (Yang et al., 2015).

19.15 STARCH/PROTEIN DIGESTIBILITY

An increase in the rapidly digested starch content was observed in cereals such as wheat, rice, maize, and barley. This increase in the starch digestibility could be attributed to the increased porosity of the starch granules on ultrasonication resulting in increased areas for attack of enzymes. Ultrasonication disrupts the double helix structure of the starch molecules making various sites susceptible to attack by digestive enzymes that degrade the amylopectin chain (Kaur and Gill, 2019). When compared to polished rice, ultrasonicated rice had reduced starch digestibility, which could be due to the covering of the starch granules by lipids, dietary fibres, and proteins or altered structure of the starch that is difficult to digest (Geng et al., 2021).

On ultrasonication of sorghum flour for 10 and 20 min, a reduction in the soluble protein was observed when hydrolyzed by proteolytic and pancreatin enzymes. This could be due to the smaller size of soluble proteins obtained on ultrasonication as the components that have lower molecular weights are not measured (Sullivan et al., 2018).

Ultrasonication of sorghum resulted in lower amount of protein concentration that expresses greater amounts of peptides or amino acids chain. Ultrasonication increased the digestibility and the solubilization of the protein with the increase in time and amplitude of the treatment (Beazley, 2017). Ultrasonication of buckwheat protein isolate increases the *in vitro* protein digestibilty. The digestibility was increased up to 41%, which could be due to effect of ultrasound on the structural conformation and microstructure of the protein isolates, which in turn increases the digestibility by increasing the access of enzymes to the protein. After 10 min of ultrasonication, protein renatures buries the protein inside the cavity resulting in reduction in digestibility of protein, and ultrasonication for longer periods may result in the formation of aggregates of the proteins (Jin et al., 2021).

19.16 CONCLUSION

Ultrasound is an emerging technology that is considered green technology due to its non-toxicity and being eco-friendly and the fact that it saves a huge amount of energy and therefore maximizes production. Ultrasonication has been employed to evaluate the composition of foods and to detect contamination. An adequate amount of research has been carried out on ultrasonication techniques applied in food technology; however, considerable future research is required to develop highly automated ultrasound systems that reduce cost, labor, energy, and maximize the production of safe and value-added food products.

REFERENCES

Alex, P., & Bates, D. (2011). Industrial applications of high power ultrasonics. In Ultrasound technologies for food and bioprocessing, Springer, New York, pp. 599–616.

Beazley, S. F. (2017). Effects of ultra-sonication process on digestibility of kafirin in sorghum.

Biswas, B., & Sit, N. (2020). Effect of ultrasonication on functional properties of tamarind seed protein isolates. *Journal of food science and technology*, *57*(6), 2070–2078.

Bonto, A. P., Camacho, K. S. I., & Camacho, D. H. (2018). Increased vitamin B5 uptake capacity of ultrasonic treated milled rice: A new method for rice fortification. *LWT*, *95*, 32–39.

Bonto, A. P., Jearanaikoon, N., Sreenivasulu, N., & Camacho, D. H. (2020a). High uptake and inward diffusion of iron fortificant in ultrasonicated milled rice. *LWT- Food Science & Technology*, *128*, 109459.

Bonto, A. P., Tiozon Jr, R. N., Rojviriya, C., Sreenivasulu, N., & Camacho, D. H. (2020b). Sonication increases the porosity of uncooked rice kernels affording softer textural properties, loss of intrinsic nutrients, and increased uptake capacity during fortification. *Ultrasonics Sonochemistry*, *68*, 105234.

Bonto, A. P., Tiozon Jr, R. N., Sreenivasulu, N., & Camacho, D. H. (2020c). Impact of ultrasonic treatment on rice starch and grain functional properties: A review. *Ultrasonics Sonochemistry*, *71*, 105383.

Chan, H. T., Bhat, R., & Karim, A. A. (2010). Effects of sodium dodecyl sulphate and sonication treatment on physicochemical properties of starch. *Food Chemistry*, *120*(3), 703–709.

Chan, C. H., Wu, R. G., & Shao, Y. Y. (2021). The effects of ultrasonic treatment on physicochemical properties and in vitro digestibility of semigelatinized high amylose maize starch. *Food Hydrocolloids*, *119*, 106831.

Chemat, F., Zill-e-Huma, & Khan, M. K. (2011). Applications of ultrasound in food technology: Processing, preservation, and extraction. *Ultrasonics Sonochemistry, 18*, 813–835.

Chouliara, E., Georgogianni, K. G., Kanellopoulou, N., & Kontominas, M. G. (2010). Effect of ultrasonication on microbiological, chemical, and sensory properties of raw, thermalized, and pasteurized milk. *International Dairy Journal, 20*, 307–313.

Cui, L., Pan, Z., Yue, T., Atungulu, G. G., & Berrios, J. (2010). Effect of ultrasonic treatment of brown rice at different temperatures on cooking properties and quality. *Cereal chemistry*, *87*(5), 403–408.

Cui, R., & Zhu, F. (2020). Effect of ultrasound on structural and physicochemical properties of sweetpotato and wheat flours. *Ultrasonics sonochemistry*, *66*, 105118.

Ding, J., Hou, G. G., Nemzer, B. V., Xiong, S., Dubat, A., & Feng, H. (2018a). Effects of controlled germination on selected physicochemical and functional properties of whole-wheat flour and enhanced γ-aminobutyric acid accumulation by ultrasonication. *Food Chemistry*, *243*, 214–221.

Ding, J., Ulanov, A. V., Dong, M., Yang, T., Nemzer, B. V., Xiong, S., Zhao, S., & Feng, H. (2018b). Enhancement of gamma-aminobutyric acid (GABA) and other health-related metabolites in germinated red rice (*Oryza sativa* L.) by ultrasonication. *Ultrasonics Sonochemistry*, *40*(Pt A), 791–797.

Ding, J., Johnson, J., Chu, Y. F., & Feng, H. (2019). Enhancement of γ-aminobutyric acid, avenanthramides, and other health-promoting metabolites in germinating oats (*Avena sativa* L.) treated with and without power ultrasound. *Food Chemistry*, *283*, 239–247.

Dzah, C. S., Duan, Y., Zhang, H., Boateng, N. A. S., & Ma, H. (2020). Ultrasound-induced lipid peroxidation: Effects on phenol content and extraction kinetics and antioxidant activity of Tartary buckwheat (Fagopyrum tataricum) water extract. *Food Bioscience*, *37*, 100719.

Ertas, N. (2013). Dephytinization processes of some legume seeds and cereal grains with ultrasound and microwave applications. *Legume Research*, *36*(5), 414–421.

Falsafi, S. R., Maghsoudlou, Y., Rostamabadi, H., Rostamabadi, M. M., Hamedi, H., & Hosseini, S. M. H. (2019). Preparation of physically modified oat starch with different sonication treatments. *Food Hydrocolloids*, *89*, 311–320.

Feizollahi, E., Mirmahdi, R. S., Zoghi, A., Zijlstra, R. T., Roopesh, M. S., & Vasanthan, T. (2021). Review of the beneficial and anti-nutritional qualities of phytic acid, and procedures for removing it from food products. *Food Research International*, *143*, 110284.

Gallant, D. J., Bouchet, B., & Baldwin, P. M. (1997). Microscopy of starch: evidence of a new level of granule organization. *Carbohydrate Polymers*, *32*, 177191.

Gallo, M., Ferrara, L., & Naviglio, D. (2018). Application of Ultrasound in Food Science and Technology: A Perspective. *Foods (Basel, Switzerland)*, *7*, 164.

Geng, D. H., Lin, Z., Liu, L., Qin, W., Wang, A., Wang, F., & Tong, L. T. (2021). Effects of ultrasound-assisted cellulase enzymatic treatment on the textural properties and in vitro starch digestibility of brown rice noodles. *LWT*, *146*, 111543.

Gharibzahedi, S. M. T., & Smith, B. (2020). The functional modification of legume proteins by ultrasonication: A review. *Trends in Food Science & Technology*, *98*, 107–116.

Grgić, I., Ačkar, Đ., Barišić, V., Vlainić, M., Knežević, N., & Medverec 494 Knežević, Z. (2019). Non-thermal methods for starch modification-A review. In *Journal of Food Processing and Preservation*, *43*, e14242.

Habuš, M., Novotni, D., Gregov, M., Štifter, S., Čukelj Mustač, N., Voučko, B., & Ćurić, D. (2021). Influence of particle size reduction and high-intensity ultrasound on polyphenol oxidase, phenolics, and technological properties of wheat bran. *Journal of Food Processing and Preservation*, *45*(3), e15204.

Hasan, M., Bashir, H. and Bae, H. (2017). Use of ultrasonication technology for the increased production of plant secondary metabolites. *Molecules*, 22, 1046.

Hassan, S., Imran, M., Ahmad, N., & Khan, M. K. (2017). Lipids characterization of ultrasound and microwave processed germinated sorghum. *lipids in health and disease*, *16*(1), 1–11.

Hassan, S., Imran, M., Ahmad, M. H., Khan, M.I., Changmou X. U., Khan, M. K. & Muhammad, N. (2020). Phytochemical characterization of ultrasound-processed sorghum sprouts for use in functional foods. *International Journal of Food Properties*, *23*, 853–863

Hossain, M. B., Brunton, N.P., Patras, A., Tiwari, B., O'Donnell, C.P., Martin-Diana, A.B (2012). Optimization of ultrasound-assisted extraction of antioxidant compounds from marjoram (*Origanum majorana* L.) using response surface methodology. *Ultrasonics Sonochemistry*, *19*, 582–590.

Iftikhar, M., Zhang, H., Iftikhar, A., Raza, A., Begum, N., Tahamina, A., & Wang, J. (2020). Study on optimization of ultrasonic-assisted extraction of phenolic compounds from rye bran. *LWT-Food Science & Technology*, *134*, 110243.

Jambrak, A. R., Herceg, Z., Šubarić, D., Babić, J., Brnčić, M., Brnčić, S. R., & Gelo, J. (2010). Ultrasound effect on physical properties of corn starch. *Carbohydrate Polymers*, *79*, 91–100.

Jin, J., Okagu, O. D., Yagoub, A. E. A., & Udenigwe, C. C. (2021). Effects of sonication on the in vitro digestibility and structural properties of buckwheat protein isolates. *Ultrasonics sonochemistry*, *70*, 105348.

Kamani, M. H., Semwal, J., & Meera, M. S. (2021). Functional modification of protein extracted from black gram by-product: Effect of ultrasonication and micronization techniques. *LWT*, *144*, 111193.

Karki, B., Lamsal, B. P., Jung, S., van Leeuwen, J. H., Pometto III, A. L., Grewell, D., & Khanal, S. K. (2010). Enhancing protein and sugar release from defatted soy flakes using ultrasound technology. *Journal of Food Engineering*, *96*, 270–278.

Karwasra, B. L., Kaur, M., & Gill, B. S. (2020). Impact of ultrasonication on functional and structural properties of Indian wheat (Triticum aestivum L.) cultivar starches. *International Journal of Biological Macromolecules*, *164*, 1858–1866.

Kaur, H., & Gill, B. S. (2019). Effect of high-intensity ultrasound treatment on nutritional, rheological, and structural properties of starches obtained from different cereals. *International Journal of Biological Macromolecules, 126*, 367–375.

Kehinde, B. A., Sharma, P., & Kaur, S. (2021). Recent nano-, micro-and macrotechnological applications of ultrasonication in food-based systems. *Critical reviews in food science and nutrition, 61*(4), 599–621.

Kentish, S., & Ashokkumar, M. (2011). The physical and chemical effects of ultrasound. In H. Feng, G. V. Barbosa-Canovas, & J. Weiss (Eds.), *Ultrasound Technologies for Food and Bioprocessing* (pp. 1–12). London: Springer.

Knorr, D., Froehling, A., Jaeger, H., Reineke, K., Schlueter, O., & Schoessler, K. (2011). Emerging technologies in food processing. *Annual Review of Food Science and Technology, 2*, 203–235.

Lateef, A., Oloke, J. K., & Prapulla, S. G. (2007). The effect of ultrasonication on the release of fructosyltransferase from *Aureobasidium pullulans* CFR 77. *Enzyme and Microbial Technology, 40*, 1067–1070.

Li, S., Yang, X., Zhang, Y., Ma, H., Liang, Q., Qu, W., … & Mahunu, G. K. (2016). Effects of ultrasound and ultrasound assisted alkaline pretreatments on the enzymolysis and structural characteristics of rice protein. *Ultrasonics Sonochemistry, 31*, 20–28.

Liu, J., Yu, X., & Liu, Y. (2021). Effect of ultrasound on mill starch and protein in ultrasound-assisted laboratory-scale corn wet-milling. *Journal of Cereal Science, 100*, 103264.

Lohani, U. & Muthukumarappan, K. (2017). Modeling of continuous ultrasonication to improve total phenolic content and antioxidant activity in sorghum flour: A comparison between response surface methodology and artificial neural network. *International Journal of Food Engineering, 13*, 20160086.

Lohani, U. C., & Muthukumarappan, K. (2021). Study of continuous flow ultrasonication to improve total phenolic content and antioxidant activity in sorghum flour and its comparison with batch ultrasonication. *Ultrasonics Sonochemistry, 71*, 105402.

Luo, Z., Fu, X., He, X., Luo, F., Gao, Q., & Yu, S. (2008). Effect of ultrasonic treatment on the physicochemical properties of maize starches differing in amylose content. *Starch-Stärke, 60*(11), 646–653.

Majid, I., Nayik, G. A., & Nanda, V. (2015). Ultrasonication and food technology: A review. *Cogent Food & Agriculture, 1*, 1071022.

Mason, T. J., Chemat, F., & Vinatoru, M. (2011). The extraction of natural products using ultrasound or microwaves. *Current Organic Chemistry, 15*, 237–247.

Mir, N. A., Riar, C. S., & Singh, S. (2019). Structural modification of quinoa seed protein isolates (QPIs) by variable time sonification for improving its physicochemical and functional characteristics. *Ultrasonics Sonochemistry, 58*, 104700.

Moelants, K. R., Cardinaels, R., Van Buggenhout, S., Van Loey, A. M., Moldenaers, P., & Hendrickx, M. E. (2014). A review on the relationships between processing, food structure, and rheological properties of plant-tissue-based food suspensions. *Comprehensive Reviews in Food Science and Food Safety, 13*, 241–260.

Muhammad, Z, Farooq, A, Muhammad, Z, Muhammad, N, Abdullah, H and Ehsan, M (2018). Effect of Ultrasonic Extraction Regimes on Phenolics and Antioxidant Attributes of Rice (*Oryza sativa* L.) Cultivars. *Iranian Journal of Chemistry & Chemical Engineering-International English Edition, 37*, 109–119.

Naumenko, N., Potoroko, I., & Kalinina, I. (2022). Stimulation of antioxidant activity and γ-aminobutyric acid synthesis in germinated wheat grain Triticum aestivum L. by ultrasound: Increasing the nutritional value of the product. *Ultrasonics Sonochemistry, 86*, 106000.

Nazari, B., Mohammadifar, M. A., Shojaee-Aliabadi, S., Feizollahi, E., & Mirmoghtadaie, L. (2018). Effect of ultrasound treatments on functional properties and structure of millet protein concentrate. *Ultrasonics Sonochemistry, 41*, 382–388.

Ojha, K. S., Tiwari, B. K., & O'Donnell, C. P. (2018). Effect of ultrasound technology on food and nutritional quality. In *Advances in Food and Nutrition Research, 84*, 207–240.

Rao, M. V., Sengar, A. S., Sunil, C. K., & Rawson, A. (2021). Ultrasonication-A green technology extraction technique for spices: A review. *Trends in Food Science & Technology, 116*, 975–991.

Ravikumar, Madhusudan & Sachin, A. J. (2017). Ultrasonication: An Advanced Technology for Food Preservation. *International Journal of Pure & Applied Bioscience, 5*, 363–371

Sams, A. R., & Feria, R. (1991). Microbial effects of ultrasonication of broiler drumstick skin. *Journal of Food Science, 56*, 247–248.

Sharma, S., Pradhan, R., Manickavasagan, A., & Dutta, A. (2020). Characterization of ultrasonic-treated corn crop biomass using imaging, spectral and thermal techniques: a review. *Biomass Conversion and Biorefinery*, 1–16.

Soria, A. C., & Villamiel, M. (2010). Effect of ultrasound on the technological properties and bioactivity of food: A review. *Trends in Food Science & Technology, 21*, 323–331.

Sujka, M., & Jamroz, J. (2013). Ultrasound-treated starch: SEM and TEM imaging, and functional behavior. *Food Hydrocolloids, 31*, 413–419.

Sullivan, A. C., Pangloli, P., & Dia, V. P. (2018). Impact of ultrasonication on the physicochemical properties of sorghum kafirin and in vitro pepsin-pancreatin digestibility of sorghum gluten-like flour. *Food Chemistry, 240*, 1121–1130.

Tang, E. S. K., Huang, M., & Lim, L. Y. (2003). Ultrasonication of chitosan and chitosan nanoparticles. *International Journal of Pharmaceutics, 265*(1–2), 103–114.

Taurozzi, J., Hackley, V. and Wiesner, M. (2012), Preparation of Nanoparticle Dispersions from Powdered Material Using Ultrasonic Disruption, Special Publication (NIST SP), National Institute of Standards and Technology, Gaithersburg, MD, 1200(2), 1200–2.

Tobergte D. R. & Curtis, S. (2013) Ultrasound technologies for food bioprocessing. Springer, New York.

Toma, M., Vinatoru, M., Paniwnyk, L., & Mason, T. J. (2001). Investigation of the effects of ultrasound on vegetal tissues during solvent extraction. *Ultrasonics Sonochemistry, 8*, 137–142.

Vela, A. J., Villanueva, M., & Ronda, F. (2021). Low-frequency ultrasonication modulates the impact of annealing on physicochemical and functional properties of rice flour. *Food Hydrocolloids, 120*, 106933.

Vercet, A., Sánchez, C., Burgos, J., Montanés, L., & Lopez Buesa, P. L. (2002). The effects of manothermosonication on tomato pectic enzymes and tomato paste rheological properties. *Journal of Food Engineering, 53*, 273–278.

Wang, Y, Zhao, L, Zhang, R, Yang, X., Sun, Y., Shi, L., & Xue, P. (2020). Optimization of ultrasound-assisted extraction by response surface methodology, antioxidant capacity, and tyrosinase inhibitory activity of anthocyanins from red rice bran. *Food Science & Nutrition, 8*, 921–932.

Wani, I. A., Sogi, D. S., Shivhare, U. S., & Gill, B. S. (2015). Physico-chemical and functional properties of native and hydrolyzed kidney bean (*Phaseolus vulgaris* L.) protein isolates. *Food Research International, 76*, 11–18.

Wen, C., Zhang, J., Yao, H., Zhou, J., Duan, Y., Zhang, H., & Ma, H. (2019). Advances in renewable plant-derived protein source: The structure, physicochemical properties affected by ultrasonication. *Ultrasonics Sonochemistry, 53*, 83–98.

Wu, D., Tu, M., Wang, Z., Wu, C., Yu, C., Battino, M., ... & Du, M. (2020). Biological and conventional food processing modifications on food proteins: Structure, functionality, and bioactivity. *Biotechnology advances, 40*, 107491.

Xia, Q., Tao, H., Li, Y., Pan, D., Cao, J., Liu, L., ... & Barba, F. J. (2020). Characterizing physicochemical, nutritional and quality attributes of wholegrain Oryza sativa L. subjected to high intensity ultrasound-stimulated pre-germination. *Food Control, 108*, 106827.

Yadav, S., Mishra, S., & Pradhan, R. C. (2021). Ultrasound-assisted hydration of finger millet (Eleusine Coracana) and its effects on starch isolates and antinutrients. *Ultrasonics Sonochemistry, 73*, 105542.

Yang, H., Gao, J., Yang, A., & Chen, H. (2015). The ultrasound-treated soybean seeds improve edibility and nutritional quality of soybean sprouts. *Food research international, 77*, 704–710.

Yusaf, T., & Al-Juboori, R. A. (2014). Alternative methods of microorganism disruption for agricultural applications. *Applied Energy, 114*, 909–923.

Zhang, H., Chen, G., Liu, M., Mei, X., Yu, Q., & Kan, J. (2020). Effects of multi-frequency ultrasound on physicochemical properties, structural characteristics of gluten protein and the quality of noodle. *Ultrasonics Sonochemistry, 67*, 105135.

Zuniga-Salcedo, M. R., Ulloa, J. A., BAUTISTA-ROSALES, P. U., Rosas-Ulloa, P., Ramírez-Ramíez, J. C., Silva-Carrillo, Y., ... & Hernández, C. (2019). Effect of ultrasound treatment on physicochemical, functional and nutritional properties of a safflower (Carthamus tinctorius L.) protein isolate. *Italian Journal of Food Science, 31*(3).

CHAPTER 20

Ozonation of Cereals

Devina Vaidya, Manisha Kaushal, Anil Gupta, and Anupama Anand
Department of Food Science and Technology,
Dr. YS Parmar University of Horticulture and Forestry, Nauni, Solan, Himachal Pradesh, India

CONTENTS

DOI: 10.1201/9781003242192-25

20.1 INTRODUCTION

Cereals such as rice, wheat, and maize play an important role as staple food crops in many areas of the world. Whole grains, as well as their products, contribute to a major component in the daily diet, providing a wholesome amount of carbohydrates, proteins, dietary fibres, vitamins as well as various bioactive phytochemicals including carotenoids, lignans, phenolics, sterols, β-glucans, phytates, etc. Various studies have indicated the protective role of whole grain foods against several diseases associated with a sedentary and unhealthy lifestyle such as diabetes, cardiovascular diseases, and certain types of cancers (Nadeem et al., 2010). However, for the production and maintenance of high-quality grains and cereals many postharvest quality and quantity loss factors should be considered such as temperature, humidity, contamination and degradation by insects, rodents, microorganisms, toxin production, and storage conditions (Granella et al., 2018). Most of the current methods involving chemical, physical, and biological processes are not practical at large scale, due to time consumption, nutrition losses, toxic effects, or low detoxification efficiency, thus there is a constant search for novel, eco-friendly methods, which not only help in proper disinfestation but also maintain the nutritional quality of the grains as well as their value-added products (Chen et al., 2014).

Today, ozone (a triatomic form of oxygen), which has the Generally Recognized as Safe (GRAS) status from the US Food and Drug Administration (FDA), having strong anti-microbial and sanitising properties and is being used in various applications in the food industry. It is bluish at ambient temperature, soluble in water at acidic pH values, and degrades after producing numerous free radical species, specifically hydroxyl (OH^-) radical. Ozone may be used to decontaminate a variety of foods, including fruits, vegetables, spices, herbs, drinks, meat, and fish, in both gaseous and aqueous forms (Pandiselvam et al., 2019). Due to their high amount of production and importance as a staple food item, the use of ozone has been broadened to cereals and cereal products. Ozonation is now generally utilised as a sanitising technique, mostly to extend product shelf life, which boosts production profitability by lowering trade losses in the food processing industry. The use of ozone in the food sector is regulated in numerous nations throughout the world, including the United States, Japan, Australia, Canada, and the European Union (Agriopoulou et al., 2022). Moreover, as a viable green insecticide against a range of stored grain insects and pests as well as mycotoxin degradation, the treatment enhances both the microbiological safety and the qualitative shelf life of food goods. Synthetically, ozone has a very unstable form that may be made from oxygen in the air at low concentrations (0.3 ppm) using 185 nm wavelength light generated by high transmission UV lamps or using the more common corona discharge technique (Sivaranjani et al., 2021). Various studies have reported the effect of ozonation on the nutritional, sensory, rheological, and functional qualities of cereals and their products. Thus, in comparison to other procedures, the current chapter will offer an overview of the effectiveness of ozonation on grain processing, as well as limits and future views.

20.2 FUNDAMENTALS OF OZONATION AND MACHINERY

20.2.1 History of Ozonation

Ozone was discovered by Schonbein in 1840, and in 1906, the first commercial-scale disinfection of potable water with ozone was put into practice in Nice (France) (Graham, 1997). Various cities in France, the Netherlands, Germany, Austria, Switzerland, and other nations have employed ozonation as a standard method for water treatment and disinfection (Overbeck, 2000). The majority of bottled water currently involves ozone treatment, which began in 1982 when the FDA declared ozone to be GRAS for this product (Majchrowicz, 1998). The US Department of Agriculture (USDA) approved ozone in its gaseous form for the storage of meat in 1957 as well as for the reconditioning of recycled poultry chilling water in 1997 (Güzel-Seydim et al., 2004). Ozone is an alternative disinfectant for high moisture foods such as strawberries and raspberries. In the same year, an independent panel of experts convened by the Electric Power Research Institute (EPRI) declared ozone to be a GRAS chemical for use as a food disinfectant and sanitiser when used in line with good manufacturing standards (Graham, 1997). Ozone has now been permitted for use as a disinfectant or sanitiser in foods and food processing in the United States after the FDA did not object to the expert panel's conclusions. It was authorised by the FDA in 2001 for use as an anti-bacterial agent in the treatment, storage, and processing of meats and their products (Novak and Yuan, 2007).

20.2.2 Machinery Involved

Ozone is a gaseous chemical found in the atmosphere that is formed when lightning or high-energy UV rays impact it. The equipment for ozonation consists of five basic components: an oxygen cylinder, ozone generators, ozone analyser, a flow rate meter, and an ozone destructor. The turbulence is maintained by air fans, which help in the circulation of ozone gas throughout the product. The generation of ozone by oxygen is the most important step, which is further continuously used in the product, analysed by the ozone analyser, and measured by the flow rate meter. There are two methods commonly used to produce ozone for food applications: corona discharge (CD), which is generally preferred commercially, and with ultraviolet (UV) light method (Cullen et al., 2012; O'Donnell et al., 2012). The oxygen molecule is divided into two oxygen atoms by an electrical discharge. Excess electrons in these unstable oxygen atoms cause them to join with other oxygen molecules, lowering their energy state. This combination forms ozone as a product. This highly unstable ozone reacts with the product and causes changes in its molecular structure.

20.2.2.1 Oxygen Generators

Oxygen generators make it simple to create oxygen-enriched air (>90 % O2). These machines take in ambient air, filter it (removing dust particles), then separate and eliminate nitrogen, leaving air that is significantly more oxygenated. The airflow is pressurised and passed through a molecular sieve (microscopic porous bead) bed, which adsorbs or retains nitrogen and moisture, supplying oxygen-enriched air to the concentrator's supply output. Due to the increased amounts of ozone generation today, the majority of ozone-generating systems employ oxygen as the input gas (Tiwari et al., 2010).

20.2.2.2 Ozone Generators

20.2.2.2.1 Corona Discharge Method

To generate ozone by the corona discharge method, oxygen molecules are passed through the electrical field where they split into oxygen free radicals. The free radical reacts with diatomic oxygen to form a triatomic ozone molecule (O_3). As depicted in Figure 20.1, there are two electrodes through which an electric discharge of approximately 30kV/s is passed, which splits the oxygen into free radicals. A ceramic dielectric medium separates them, resulting in a small discharge gap. Voltage, current frequency, dielectric material property and thickness, discharge gap, and absolute

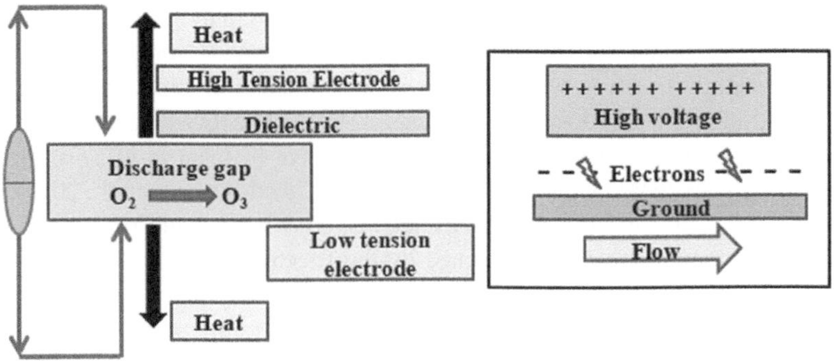

Figure 20.1 Corona discharge method of ozonation.

pressure within the discharge gap all affect ozone generation (Khadre et al., 2001). The operating conditions of these generators can be subdivided into low-frequency (60 Hz), high voltage (>20,000 V)), medium frequency (600 Hz), medium voltage (1,000 Hz), and low voltage (10,000V). Corona discharge generators are capable of creating significant ozone concentrations. Although there are many different types of generating cells, the basic principle stays the same. Electrons are accelerated over an air gap to provide enough energy to break the oxygen-oxygen double bond, resulting in atomic oxygen. Hence, this phenomenon is also called "The miniature lightening in a completely controlled and enclosed environment."

Corona discharge has the following advantages:

- High ozone concentrations
- Best for water applications
- Quick odour elimination
- Equipment may operate for years without maintenance

20.2.2.3 UV-lamp Method

Although sunlight produces ozone in the upper atmosphere, the photochemical principle can be applied to produce ozone where it is needed. Both black lights and UV bulbs generate wavelengths necessary to produce ozone from oxygen sources where UV lamps of 185 nm are used to produce ozone. Air (generally ambient) is passed over a UV light, which splits oxygen (O_2) molecules in the gas and the resultant oxygen atoms (O_1) connect to other oxygen molecules for the goal of finding stability (O_3). The ozone in the output gas is then fed into the water, where it inactivates pollutants by rupturing the organisms' cell walls. A UV ozone generator with a reaction chamber composed of a high-reflectivity material is built to protect wiring, electrical connections, and other electronic components from the impacts of UV radiation, heat, and ozone. Using UV lamps ozone concentration of about 0.1–0.001 % by weight can be produced, which is 14 times less than the corona discharge method. The process of ozone ventilation is more easily accomplished at low temperatures. Ozone is mostly produced in the air by high-voltage electrical discharges or pure oxygen. This radiation generates a rupture in a common oxygen molecule in the atmosphere, causing free oxygen atoms to separate. Ozone polymers are produced when these atoms collide with other oxygen molecules (Duguet, 2004). An oxygen molecule absorbs enough energy to split into two oxygen atoms.

$$O_2 + hv \rightarrow O + O$$

Each of these atoms binds to an oxygen molecule to produce another molecule of ozone.

$$O + O_2 \rightarrow O_3$$

Finally, the ozone molecule is destroyed again absorbing more UV radiation.

$$O_3 + h\nu\ O \rightarrow O_2$$

UV radiation is absorbed in the closed cycle of ozone formation and degradation. The ozone produced dissipates spontaneously in oxygen over a period of time (Gonçalves, 2009).

UV light has the following advantages:

- Simple construction
- Lower cost than corona discharge
- Output not impacted by humidity
- Fewer by-products than corona discharge

However, this method is still not applicable commercially due to the low generation of ozone, as it does not fulfill the amount of ozone required to sanitise high-capacity silos of grains ranging from 5 to 10,000 tons.

20.2.3 Other Components

A treatment chamber is an air-tight and leakproof collector where the ozone enters from the top and returns through the bottom outlet after completing the exposure time. The dimension and type are based on the mode of application of ozone and the type of product being processed. The outlet is connected to the ozone analyser and then to ozone destructor. The ozone concentration in % by weight or g/L that enters and leaves the treatment chamber is measured by the ozone analyser. While a flow meter is used to measure the flow rate of ozone in g/hour that enters and leaves the treatment chamber (Janex et al., 2000).

Excess ozone is eliminated at the end of the treatment chamber, and the treated and decontaminated air is recirculated or expelled to the ambient environment via an ozone destructor. To prevent any adverse effects on workers, excess ozone may be converted down into oxygen and released into the atmosphere. The excess ozone in this device is allowed to escape into the air, allowing for regulated interaction between the two. Because ozone is unstable in the air, it decomposes and oxygen is released through the outlet (Prabha et al., 2015).

Furthermore, since air preparation systems contain air filters that must be replaced frequently and tube-type ozone generators must be shut down for annual tube cleaning and other general maintenance, the ozonation system's maintenance cost includes periodic cleaning, repair, and replacement of equipment parts. Depending on the size and quantity of ozone generators in the system, cleaning tubes may take several days. The broken tubes during cleaning or which deteriorate after years of operation at high voltages must be replaced periodically. Thus, the instrumentation of ozone treatment is of a high maintenance cost, which should be reduced for maximum output in the future.

20.3 MECHANISM OF ACTION OF OZONE ON CEREALS

Ozone has a strong anti-bacterial effect in cereals and cereal products for a wide variety of microorganisms such as *Fusarium*, *Cladosporium*, *Mucor*, *Rhizopus Aspergillus*, *Penicillium*,

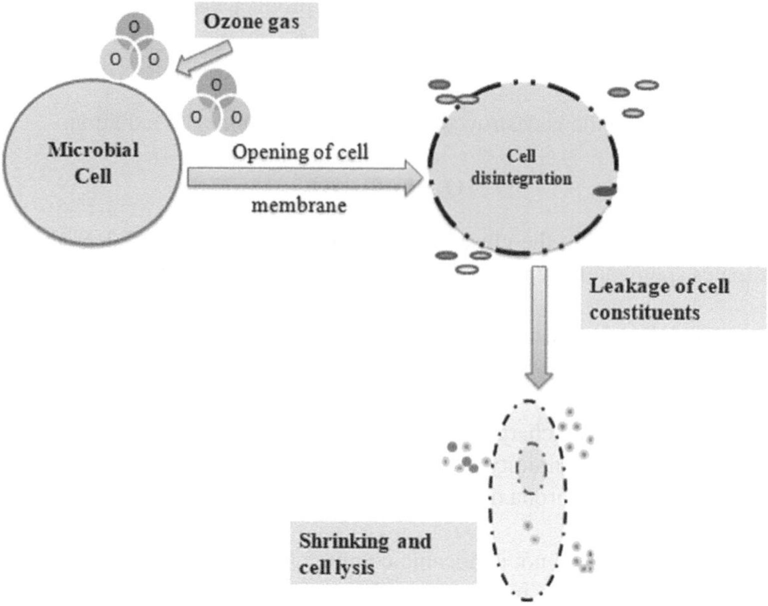

Figure 20.2 Microbial deactivation by ozonation.

Curvularia, etc. The oxidation of sulfhydryl and amino acid groups of protein and enzymes in microorganisms is followed by the oxidation of polyunsaturated fatty acids as the mode of action by ozone for microbial inactivation (Brodowska et al., 2018; Afsah-Hejri et al., 2020).

A schematic example of microbial inactivation due to ozonation is shown in Figure 20.2.

Ozone penetrates to the kernels to inactivate microorganisms that enter through seed coat diffusion or microscopic pores in the grains (White et al., 2013). The time required to reduce the initial concentration of ozone to half its half-life falls as the moisture content of grains increases. The ozone degradation rate constantly rises as the moisture content of rice grains rises, and it also changes with grain bed thickness (Pandiselvam et al., 2019). Various microorganisms have different sensitivity to ozone treatment, which is determined by a variety of criteria, including the selection of the optimum concentration without compromising product quality. The effect of ozone may not immediately control microbial growth, but the total plate count tends to decrease over the storage period after ozonation.

20.4 IMPACT OF OZONE ON DIFFERENT ASPECTS OF CEREALS INDUSTRY

20.4.1 Insects and Rodents

Due to the massive losses during storage, insect infestation in grains is a huge challenge to world food security. Various environmental conditions such as humidity, temperature, and light, may promote an infestation. The eradication of moulds and insect pests while retaining the nutritional value of the food is a key difficulty in grain storage (Srivastava and Mishra, 2021). Annual losses in quantity and quality of stored food harvests are severe due to the usage of underdeveloped storage facilities. Because infection levels vary from year to year, the estimation of quantitative losses varies. However, the presence of insect pests and moulds during handling and storage is responsible for over 30% of crop loss (FDA, 2017).

The majority of stored wheat insects preferentially devour grain embryos and feed on the starchy endosperm, decreasing the end-use quality of flour. Chemical fumigants, which enter the body of the insects through spiracles and destroy the hemolymph components, are the most common method of control for these insects (Glowacz et al., 2015). To address environmental issues and limit human health threats, the use of insecticides and chemical fumigants such as methyl bromide, phosphine, malathion, and fenitrothion has been prohibited and substituted in other ways. By acting as a powerful oxidising agent and reacting with proteins, DNA, and double bonds of polyunsaturated fatty acids, ozone has the potential to reduce the harmful effect of insecticides on food items.

Subramanyam et al. (2017) investigated *Rhyzopertha dominica* mortality in stored wheat with ozone treatments of 0.42 and 0.84 g/L for up to 36 and 38 hours, respectively, and discovered that the time for 99 % mortality of *Rhyzopertha dominica* adults was 1.6 times slower at an ozone concentration of 0.42 g/L for 38 hours. The changes in lethal time may be due to the rate of ozone penetration into wheat kernels at various concentrations. Gas diffusion, air seepage velocity within the grain layer, initial ozone concentration, and adsorption on the grain surface all play a role in the ozone penetration rate in wheat. Moreover, it could also be possible due to the strong oxidising properties of ozone, which in greater concentrations enhanced depolymerisation and structural alteration of chemical components (Obadi et al., 2016).

Trombete et al. (2016) investigated the effects of ozone on the chemical characteristics, mineral profile, and rheological properties of wheat at an ozone concentration of 60 mg/L and found no significant changes in wheat quality. Mishra et al. (2018) also improved the treatment conditions for ozone fumigation in stored wheat grains, using a 2.5 g/L ozone concentration for an 8-hour treatment period. Adults, pupae, larvae, and eggs of *Rhyzopertha dominica* died at a rate of 97, 100, 99, and 100 %, respectively. They also obtained rheological characteristics using FTIR peaks and scanning electronic microscopic pictures, as well as minor alterations at the molecular level.

These findings resulted in information on the interaction of ozone with grain characteristics, as ozone at optimal concentrations may be used on a wide scale in grain silos for insect control. However, exceeding the recommended treatment dose may have a deleterious effect on grain quality, since rheological characteristics may change leading to oxidation, resulting in poor grain quality (Goze et al., 2017). As a result, the treatment technique must be optimised to achieve 100 % insect death with minimal grain quality alterations.

20.4.2 Degradation of Toxins

The application of ozone in food processing has been emerging under investigation for the degradation of toxins present in food grains. Mycotoxins are naturally occurring compounds generated in food by fungus as secondary metabolites, and mycotoxicosis is induced by consuming mycotoxin-contaminated food or feed. In either case, the sickness is caused by direct eating of infected food or by inhalation of spores and skin contact. Mycotoxin can have acute or chronic effects on human health depending on the species, kind, and dose. Ozone has been frequently used as an anti-bacterial agent in food processing to remove or decrease bacterial and fungal cells due to its strong oxidative potential (Alwi and Ali, 2014). The same oxidising ability is included in the molecular decomposition of mycotoxins, with the effectiveness of this decontamination depending on parameters such as the kind of food, ozone concentration, and exposure period.

Aflatoxins (AF), which are produced mostly by two species of the genus *Aspergillus, A. flavus,* and *A. parasiticus*, are one of the most significant classes of mycotoxins due to their great toxic potential. Further, AFB_1, AFB_2, AFG_1, and AFG_2 chemicals are classified as genotoxic and carcinogenic by the International Agency for Research on Cancer. *Fusarium* toxins, such as deoxynivalenol (DON), are the most prominent mycotoxins in wheat grains and are frequently detected in high concentrations in grains and derived products (El-Desouky et al., 2012). DON is not classed as a genotoxic substance but is linked to serious side effects including protein synthesis inhibition, anorexia, endocrine dysfunction, and immune system changes. Deoxynivalenol and total aflatoxins demonstrated the most significant decreases, with 64.3 and 48.0 %, respectively, according to Trombette et al. (2016). Furthermore, applying 60 mg/L of O_3 for 300 min and 2 kg samples, the total fungal count (TFC) was lowered by approximately 3.0 log cfu/g of wheat grain. When maize grits were treated with 60 mg/L of O_3 for 480 min, Porto et al. (2019) found the most significant reduction in aflatoxin levels when compared to other treatments. This may be explained by the fact that corn grits had a higher surface area than kernels, and the moisture level of the grain mass also played a part in detoxifying. Thus, ozone treatment may be beneficial.

20.4.3 Microbial Reduction

Ozone, either in a gaseous or aqueous form, has been shown to reduce natural microflora as well as bacterial, fungal, and mould contamination in cereals and grain products, including *Bacillus, Coliform, Micrococcus, Flavobacterium, Alcaligenes, Serratia, Aspergillus, and Penicillium* spores. (Naito and Takahara, 2006). The mechanism of action of ozone on the microbial cell lysis has already been illustrated in Figure 20.2. Various studies have shown the effect of ozone treatment on the microbial reduction of different species. Allen et al. (2007) found that ozone was extremely successful in inactivating fungus associated with barley, regardless of whether the fungi were spores or mycelia; however, mycelia were less resistant to ozone. Applying 0.16 and 0.10 mg/L ozone for 5 min resulted in 96 % inactivation of spores and mixes of spores and tiny amounts of mycelia, respectively. While, White et al. (2010) conducted a study of ozone-treated maize at 2.4 and 4.8 mg /L for 24 hours and reported that more fungi growth was observed along the sides of the storage containers with ozone-deactivated fungi genera *Rhizopus, Fusarium, Aspergillus, Mucor*, and *Penicillium* in order (greatest to least). This could be the result of airflow restrictions caused by maize contact with the storage container wall, allowing less ozone to interact with kernels along the wall. According to De Alencar et al. (2012), the degradation of pigments in fungal colonies was correlated with a disorder in the structure of fungus owing to the oxidation of microbial pigments by free radical oxygen molecules, which resulted in a colour shift from green to white.

20.4.4 Starch Modification

Ozonation is used to study the effects of native starch on rheological, physico-chemical, and thermal characteristics. The oxidation of starch molecules involves two primary reactions where the first step involves converting a hydroxyl group to carboxyl and carbonyl groups by oxidation and then depolymerising starch molecules by cleaving glycosidic linkages (Pandiselvam et al., 2019). When native starch molecules do not possess desirable functional qualities, they can be improved through physical, chemical, enzymatic, and genetic methods. Chemical methods (acetylation, cross-linking, hydroxypropylation, oxidation, and etherification) are generally effective in modifying starch, as modified starch using chemical methods has several advantages in the food industry, which include low retrogradation and gel syneresis, as well as improved gel texture, paste clarity, and film adhesion. These properties are influenced by ozonation as per the reports of various researchers.

According to Chan et al. (2011), ozonation of 5 mg/L for 10 min reduced the molecular mass of maize starch (dry form) by 22.6 % compared to native starch. This may be because chain lengths are getting smaller. However, the granular form changed, and surfaces of corn starch solution exposed to ozone for 1 hour resulted in increased granule size due to starch granule swelling. Lee et al. (2017) found similar results in wheat flour suspensions, indicating that ozone treatment reduced starch granules and subsequent starch degradation resulted in enhanced water absorption, producing starch molecule swelling. In wheat flour, the effect of ozone on thermal parameters such as gelatinisation temperature and enthalpy was not significantly different from control samples (Goze et al., 2016). In contrast, waxy rice starch's thermal characteristics decreased, possibly due to polyphenols, proteins, and non-starch polysaccharides that prevented starch from oxidising (Ding et al., 2015). Obadi et al. (2018) looked into the treated pasting characteristics of whole wheat flour, finding a decrease in the peak, trough, final, setback, and breakdown viscosities, demonstrating the weakening of starch granules due to ozone oxidation. The effects of ozone treatment on different aspects of cereal industry viz. including insects, degradation of toxins, microbial reduction, and starch modification are depicted in Table 20.1.

Table 20.1 Effect of Ozone Treatment on Different Aspects of the Cereal Industry

Parameter	Food grains	Processing conditions (Ozone dose and time)	Impact of ozone treatment	References
		Effect of insect mortality		
Sitophilus granarius	Wheat	135 mg/L for 8 days	100 % mortality (Adults)	Hansen et al., 2013
Tribolium castaneum	Corn	47.8 mg/L for 6 min	100 % mortality (Adults)	McDonough et al., 2011
Rhizopertha dominica	Wheat	70 mg/L for 4 days	97 % mortality (Adults)	Mason and Woloshuk, 2011
		Toxin degeneration		
Aflatoxin	Corn	47.8 mg/L for 1.8 min	30% reduction	Campbadal et al., 2011
Deoxynivalenol	Wheat	75 mg/L for 90 min	53.48% reduction	Wang et al., 2016
Total aflatoxin and deoxynivalenol	Wheat	60 mg/L for 300 min	64.3% and 58.0% reduction, respectively	Trombette et al., 2016)
Aflatoxin B_1, B_2 G_1 and G_2	Corn	60 mg/L for 480 min	54.6, 57.0, 36.1, and 30.0% decline, respectively	Porto et al., 2019
		Microbial degradation		
Aspergillus spp. Penicillium spp	Rice	10.13 mg/L for 60 hours	100% efficacy	Santos et al., 2016
Aspergillus spp. Penicillium spp.	Corn	2.14 mg/L for 50 hours	78.5% and 98.0% reduction, respectively	Brito Junior et al., 2018
Fusarium spp.	Wheat	60 mg/L for 300 min	58.0% reduction	Trombette et al., 2016)
		Starch modification		
	Wheat	500 mg/L for 60 min	Swelling power increased	Li et al., 2012
	Wheat flour	1500 mg/L for 45 min	Amylopectin fractions and molecular weight decreased Swelling power increased	Sandhu et al., 2012
	Corn starch	4.2 mg/L for 1 hour	Gelatinisation temperature and pasting temperature increased	Catal and Ibanoglu, 2012
	Rice starch	0.006 mg/L for 2 hours	Carbonyl and carboxyl content increased Swelling power decreased	Ding et al., 2015
	Wheat flour	120 mg/L for 60 min	Water absorption index and pasting properties increased	Lee et al., 2017

(continued)

Table 20.1 (Continued)

Parameter	Food grains	Processing conditions (Ozone dose and time)	Impact of ozone treatment	References
		Seed germination		
	Maize kernels	0.048 mg/L for 5 min	Increase in total root length by 0.5 meters	Normov et al., 2019
	Wheat grains	60 mg/L for 120 min	No significant changes	Savi et al., 2014
	Wheat grains	60 mg/L for 180 min	Decrease in germination capacity by 12.5%	Alexande et al., 2018

20.5 EFFECT ON NUTRITIONAL PARAMETERS, FUNCTIONAL, AND BIOLOGICAL PARAMETERS OF CEREALS AND CEREAL PRODUCTS

Whole grains are high in fibre, vitamins, minerals, and bioactive phytochemicals (phenolics, carotenoids, vitamin E, lignans, β-glucan, inulin, resistant starch, sterols, and phytates), some of which act as immediate free radical scavengers, while others act as cofactors of antioxidant enzymes or indirect antioxidants, all of which provide health benefits beyond basic nutrition by lowering the risk of chronic diseases (Zhu , 2014). Cereals, in addition to being a source of bioactive components, have been shown to enhance satiety, reduce calorie intake, and improve meal satiating ability, and are inversely related to body mass index (BMI) and weight gain in a variety of populations due to their lower energy density (Tiwari et al., 2010). Whole grains include all of the critical components and naturally existing nutrients in their original quantities, which may be affected by ozonation and impact the overall quality of the finished product. Aside from the various benefits of ozonation on cereals and their derivatives, excessive usage may result in the oxidation degradation of chemical elements in the grains, surface oxidation, discoloration, or the formation of undesirable modifications. Hence, the effect of ozonation on various nutritional, functional, and biological parameters of cereals and cereal products is discussed in the following.

20.5.1 Impact of Ozonation on the Nutritional Profile of Cereals

20.5.1.1 Macronutrients

Ozonation has been shown to significantly affect the properties of carbohydrates like starch, as well as other macronutrients like proteins and lipids. Ozonated maize starches have different visual features due to a change in starch granule form from smooth to rough and fibrous surface, as well as lower swelling power due to depolymerisation and decreased hydrophobicity of amylopectin due to increased oxidation of hydrophilic areas. Furthermore, ozone treatment of starch resulted in the partial breakdown of glycosidic bonds, resulting in smaller starch granules and higher peak and final viscosities (Alvani et al., 2011). The water-holding capacity of oxidised starches in excess water was reported to increase due to strong and complex network formation between hydroxyl groups and carboxyl groups of starch constituents by interchain hydrogen bonding, as indicated by higher L^* values, with decreased chroma and yellowness due to the oxidation of flour pigments (Catal and Ibanoglu, 2012).

The impact of ozonation on proteins causes the oxidation of the polypeptide back-chain and the modification of the side chain amino acids by breaking the proteinprotein cross-linking. It has been observed that ozone treatment increases the protein content of cereal grains. Zhou et al. (2015) found a significant increase in protein content, total and essential amino acids in ozone-treated hybrid rice grains. Higher wet gluten content was found in ozone oxidised wheat flour, which decreased as ozone concentration increased, suggesting that at a certain ozone concentration, gluten structure begins to deteriorate, as seen by reduced wet gluten values at high ozonation levels (Mei et al., 2016). In contrast to β-pleated sheets, the influence of ozonation on wheat protein shape indicated a decrease in the α-helix and β-turn structural components, lowering the α-helix to β-pleated ratio due to reduced molecular mass of glutenin units. In cakes made from ozonated wheat, oxidation of the -SH group to generate S-S bonds increased the molecular weight distribution of the polymeric proteins (Gozé et al., 2017). The decrease in the quantity of β-turn may be due to oxidation or cross-linking of remaining amino acids.

Furthermore, ozonation has been demonstrated to affect lipid function. Unsaturated lipids had their characteristics altered by ozone treatment for usage in the food, pharmaceutical, and cosmetic sectors. Obadi et al. (2018) studied the effects of ozonation on the fatty acid profile of whole grain flours and discovered that palmitic acid rose considerably, while linoleic, linolenic, and stearic acid decreased with treatment intensity, and oleic acid remained unaltered. As a result of the biochemical alteration, the acid value and lipase activity dropped, while polyphenols and relative antioxidant potential rose. However, the amounts of peroxide and p-insidine were successfully increased.

20.5.1.2 Micronutrients

Micronutrients including vitamin E and B complex vitamins like thiamine, niacin, and riboflavin are abundant in cereals. Except for yellow corn, which includes -carotene as a pigment, vitamin C, vitamin A, and vitamin B12 are the limiting vitamins. Vitamin B5 (pantothenic acid) and B6 (pyridoxine) were marginally affected by ozone, whereas vitamin E, B1 (thiamin), B9 (folic acid), and nicotinamide remained intact regardless of ozone concentration (Dubois et al., 2006). The elements including potassium, iron, zinc, and magnesium, as well as trace minerals like selenium, are abundant in whole grain grains. Several minerals (K, Ca, Zn, Cu, Mn, P, and Mg) have shown a significant increase in concentration after ozonation, although their yield has been significantly affected (Broberg et al.2015). In the Fe, Na, and S, there was no noticeable change. Due to ozonation, the concentrations of Fe, Cu, and Mn in wheat grains increased, but the concentration of Zn in wheat grains decreased (Yabo et al., 2017). In contrast, Mishra et al. (2019) found a modest decrease in Cu, Zn, and Se concentrations with no significant effect on Fe, Mn, Mg, K, P, and Ca. P, Ca, and Mg levels in wheat kernels increased after ozonation at a greater level than the ambient concentration.

20.5.2 Impact of Ozonation on Functional Properties of Cereals

20.5.2.1 Rheological and Pasting Properties

The most essential indicators determining the quality of wheat flour are dough qualities. When utilised in gaseous form, ozone has been shown to improve bread qualities such as whiter crumb and increased specific volume. The strength of the dough was increased after minimal exposure to ozone, but prolonged treatment produced a substantial decline, and the peak time of dough growth was also altered, indicating a decrease in extensibility and an increase in resistance to an extension following wheat flour ozonation. After treatment, there was an increase in gluten strength (W) and P/L (P-toughness and L-extensibility) values (El-Desouky et al., 2012). Sandhu et al. (2011 b) concluded that ozonated wheat flour has a high water retention capacity due to the increased amount of water required to generate acceptable dough. In contrast, no variation in the water absorption of wheat flour was observed by Li et al. (2015), while influencing the development time, stability time, and farinograph quality number. After ozone treatment, Mendez et al. (2002) found that the extensibility, dough strength, and bread-making qualities of wheat flour did not alter significantly. However, ozonation improved viscoelasticity, which contributed to a better rheological profile and stability of the dough, as a result of oxidation of the -SH groups to form disulfide linkages (Li et al., 2012), whereas lower dough strength at high levels of ozone treatment may be due to overoxidation anddepolymerisation of proteins and amino acids (Zhu, 2018).

The pasting temperature of wheat flour following ozonation remained unchanged, but peak viscosity, which correlates with starch swelling power, rose at lower doses of ozone treatment and thereafter fell (Mei et al., 2016). Setback viscosity, on the other hand, was found to decrease with elevation as the ozone concentration increased. While Sandhu et al. (2012) found a rise in pasting temperature, peak, breakdown, setback, and final viscosities due to the oxidation of starch hydroxyl groups, which resulted in the creation of carboxyl and carbonyl groups.

20.5.2.2 Thermal Properties and Gelatinisation

A Differential Scanning Calorimeter (DSC) was used to assess the thermal characteristics of ozonated flour. By increasing the gliadin denaturation transition temperature, ozonation caused major changes in the thermal characteristics of wheat flour (Obadi et al., 2018). However, no differences in gelatinisation temperature profiles or enthalpies were found for ozonated maize, tapioca, or sago starches (Chan et al., 2011). Since gelatinisation qualities are determined by the crystallinity and microstructure of starch granules, it is assumed that oxidation had little effect on the structural orientation of starch. Sandhu et al. (2012) also found that ozonation had no effect on wheat starch transition temperature or gelatinisation enthalpy. However, Çatal and İbanoğlu. (2014) found that the gelatinisation temperature of wheat starch increased with treatment time. Ozonated starches had better shear thinning, reduced gelatinisation, and lower retrogradation potential, and thus may be used to change flours for applications that require more cooking stability and less retrogradation (Pandiselvam et al., 2019).

20.5.2.3 Colour

Ozonation influenced the colour profile of treated flour in a slight but considerable way in which the L* (brightness) and a* values increased, while the b* value declined (László et al., 2008). L* and a* values remained stable after 19 weeks of storage,although b* value declined significantly. Mei et al. (2016) found an increase in the whiteness of ozonated wheat flour due to the oxidation of carotenoids that cause the yellow hue of untreated flour. Wang et al. (2016) reported a rise in the L* value as well as a decrease in the b* values of ozone-treated wheat flour. The application of ozone treatment to wheat products will enhance consumer acceptance since consumers prefer bright and white colours. The easily available conjugated double bond of carotenoids (carotene, flavones, and xanthophylls) to oxidation by ozone has been attributed to the decline in the b* value. There were no significant changes in colour profile when whole wheat grains were treated with ozone, which may be attributed to the outer layers of the grain protecting pigment in the endosperm (Trombete et al. 2016).

20.5.3 Impact of Ozonation on Biological Properties of Cereals and Overall Functionality of Cereal and Cereal Products

20.5.3.1 Polyphenols and Antioxidants

Due to oxidation, ozone treatment (0.06 L/min for 30 min) reduced the tannin content of sorghum flour by % (Yan et al., 2012). In another investigation, ozone treatment did not affect the concentration of ferulic acid in wheat kernels (8 kg of wheat kernels under 89 g of O_3/m^3 O_2) (Dubois et al., 2006). As a result, it was established that ozone treatment did not affect flour polyphenols, due to the protection provided by the flour/kernel matrix. In all systems, the cohesive matrix of cell wall, starch, and protein tends to prevent ozone from quickly entering to interact with polyphenols.

The auto decomposition of ozone, which produces several free radical species such as hydroperoxyl (HO_2), hydroxyl (OH-), and superoxide (O_2) radicals, may explain the decrease in total phenolic content with prolonged ozone exposure time. The antioxidant ability of phenolic compounds can protect the body by scavenging free radicals or interacting with ozone, lowering polyphenol levels. According to Shaghaghian et al. (2014), an increase in ozone concentration resulted in a non-significant antioxidant loss, with the maximum antioxidant capacity drop of wheat germ samples being 0.113 0.023 mg/mL after 4 hours of 6000 ppm ozonation.

20.5.3.2 Germination

Biological grain activation is a biological processing and traditional method for improving the functional, nutritional, and sensory characteristics of cereal grains, as well as boosting their micronutrient content (Sujyashree et al., 2021). The minimal environmental factors necessary for grain activation include optimal humidity, oxygen availability for aerobic respiration, and an adequate temperature and duration for the various metabolic processes (Rakcejeva et al., 2014). The biological activity of grains increases nutritional value by increasing the bioavailability of nutritional components, vitamins, bio-elements, and other physiologically active substances by partially hydrolyzing starch, proteins, hemicelluloses, and even celluloses. Biological activation increases enzymatic activity in grains, causing dormant hydrolytic enzymes to break down starch, fibres, and proteins, resulting in more digestible substances and increased functional qualities without the use of chemicals. Furthermore, several anti-nutritional factors (enzyme inhibitors, haemaglutinins, and anti-vitamins) lose their function after activation, allowing for complete valorisation of biological substances in grains (Sangronis and Machado, 2007; Elkhalifa and Bernhardt, 2010; Donkor et al., 2012).

20.5.3.3 Effect of Ozonation on Functionality and Microbial Spoilage on Cereal-based Products

Various items manufactured from flour or kernels treated with ozone have resulted in better functionality as compared to those made from untreated flour. Bread prepared with ozone-treated flour (1500 mg/L) for 4.5 min and a flour blend with ozone-treated flour had bigger specific volumes, more crumb cells, and a whiter look as compared to the bread made with regular flour. The bread treated with ozone for longer than 9 min (45 min) had lower quality (lower specific volume) as per the study conducted by Sandhu et al. (2011a) and Sandhu et al. (2011b). While, bread prepared from mildly ozone-treated wheat flour showed enhanced quality attributes such as improved specific volume and crumb cell counts, and reduction in staling rates, bread made from overly treated wheat flour had deteriorated quality. The findings mentioned above mainly consist of the findings of the influence of ozone treatment on protein and starch characteristics. Moderate ozone treatment in products strengthens the gluten network and dough, whereas severe treatment weakened the dough and degraded the gluten structure. A GC-MS study revealed that bread prepared with ozone-treated flour (up to 45 min) has a different volatile content than bread made with untreated flour (Obadi et al., 2018). This is most likely related to lipid oxidation in flour. Moreover, the oxidation of carotenoids and other pigments increased the lightness of bread colour while decreasing the yellowness (Sandhu et al., 2011b). Further, ozone-treated flour bread had the same specific volume as bread made from potassium bromate-oxidised flour under similar conditions (Sandhu et al., 2011a).

A probable carcinogen that has been outlawed in several countries for use in bread making known as potassium bromate is still used as an oxidant in others. The use of ozone as a substitute for potassium bromate in bread production has a lot of potential.

Gluten-free baking items are becoming increasingly popular, and gluten-free bread made with sorghum flour treated with ozone enhanced the brightness, cell volume, and diameter of the bread slices while decreasing its stiffness (Marston et al., 2015). Ozonation only had a minor impact on the specific volume of bread. The increase in the area of air cells of the bread slice could be due to the higher viscosity of starch paste containing the gas produced during fermentation contributing to the bread loaf volume. However, the decrease in the firmness could be due to a more open structure of crumb, while the impact of ozone processing on non-starch components such as proteins in gluten-free grains and gluten-free food quality remains an area of future research.

Steaming-fermented wheat dough produces a bread known as Chinese stream bread, which is made from medium hard flour treated with ozone in a gaseous form (Mei et al., 2016). Chinese steamed bread prepared from ozone-treated flour at a concentration of 5 mg/L ozone with a flow rate of 3.3 L/min for 2 hours had more chewiness than that made from untreated flour while the hardness and flexibility of the bread were improved by increasing the treatment period to 1 hour. The sensory study revealed that 1-hour ozone treatment resulted in the best overall sensory acceptability of the bread, whereas longer treatment times resulted in lower sensory quality. Further, the elasticity of Chinese steamed bread prepared from ozonated flour had a significantly lower sensory score than the control as per reported by Mei et al. (2016). The effect of ozone treatment on protein and starch characteristics of Chinese streamed bread prepared from ozone-treated flour could account for a major part of these findings. Moderate/mild ozone therapy enhanced protein quality, whereas severe treatment degrades protein quality.

Fresh and semi-dried noodles were made with ozone-treated wheat and buckwheat flour at 5 g/hour with a 5 L/min flow rate for 1 hour (ozone gas) by Li et al. (2012) and Bai et al.(2017), respectively. The fresh noodles prepared from ozone-treated flour had a much lower amount of germs and were brighter in colour throughout the storage period of up to 10 days as compared to control noodles. The hardness and chewiness of the cooked noodles were unaffected by ozone treatment, while a reduction in the adhesiveness of the cooked noodles was observed (Li et al., 2012; Li et al., 2013). The microbial count of the noodles containing buckwheat was lowered up to log1.8 cfu/g by using an aqueous ozone treatment (2.21 mg/L). (Bai et al., 2017). The buckwheat noodles' storage life was extended to 9 days due to modified atmospheric packaging containing a 30:70 proportion of N_2 and CO_2 along with ozone water, which improved the physico-chemical properties of the noodles such as texture, sensory attributes, etc. The ozone's microbial cleansing effect may easily be attributed to the noodles' enhanced microbiological shelf life. Foods manufactured with ozone-treated flour such as bread and Chinese steamed bread have a higher resistance to microorganisms for a longer storage period. In a study conducted by Bai et al. (2017), the microbial load of buckwheat noodles prepared from buckwheat flour was reduced to 1.8 log cycles cfu/g ,which increased the shelf life of the noodles, which was mainly due to the antimicrobial effect of ozone. Foods like bread and Chinese steamed bread are made using ozone-treated flour should have a longer microbiological shelf life.

Further, wheat flour treated with ozone was compared with chlorine treated wheat flour (procured for market) for the preparation of high-ratio cakes by Chittrakorn et al. (2014). The HPLC data showed that ozonation caused an increase in the molecular weight of the polymeric protein, which significantly positively affected the cake volume whereas chlorination did not have any effect on the molecular weights. The reason for the polymerisation may be due to the oxidation of the sulfhydryl groups, which formed disulphide linkages between glutenin molecules. The chlorination of soft wheat flour is done frequently to avoid the cakes from collapsing after baking and to maintain the colour of the product. On the other hand, chlorine is harmful, which makes ozone a better alternative

Table 20.2 Effect of Ozonation on the Quality Parameters of Grains and Products

Food grain/products	Processing conditions	Effect of Ozonation on quality parameters	References
Wheat, maize and rice starch	60 mg/L for 1 hour	Colour values were affected in terms of increase in L* and b* value in rice starch and decrease in b* value in wheat and maize starch	Catal and Ibanoglu, 2012
Wheat	70 mg/L for 6 days	Amino acids and baking characteristics remained unaltered; no effect on saturated and unsaturated fatty acids	Coste et al., 2009
Rice starch	60 mg/L for 5 days	Retrogradation as well as swelling of starch reduced	An and King, 2009
Maize	50 mg/L for 30 days	Polyunsaturated fatty acids, amino acids remained unaltered	Mendez et al., 2002
Noodles	5 g/L for 1 hour	Brightened the colour; Slightly affected the hardness and chewiness, while decreased the adhesiveness of the cooked noodles	Li et al., 2012
Cake	20 mg/L for 30 min	Decrease in colour values, i.e., (*L** and *a**) of the crust of the cake The cake height was affected slightly along with nutritional parameters	Sui et al., 2016

to enhance the shelf life of the products. High ratio cakes manufactured from ozone-treated flour (0.006 mg/L for 36 min) had identical nutritional qualities to those prepared from chlorinated wheat flour, according to comparative tests. Therefore, ozone can be used an effective "greener" alternative to chlorine for the modification of soft wheat flour.

The research community and consumers are becoming more interested in gluten-free bread items these days. A gluten-free cake was prepared from sorghum flour treated with ozone and compared to untreated flour. The cake prepared from ozone-treated flour comprised a higher volume index per slice, but the diameter and cell volume were lower. In addition, cake made with treated flour exhibited a lower hardness and a higher lightness. The increase in the volume of the cake may be due to the increase in the viscosity of the starch paste as well as the batter due to ozone treatment. Increased viscosity pastes/batters can better hold gas while mixing ingredients. Moreover, the increase in the volume of the cake caused a loss of stiffness (Marston et al., 2015). Another prominent breakfast cereal becoming popular these days is muesli, which consists of a variety of materials including cereal flakes, seeds, dried fruits, nuts, etc. The effect of ozone treatment at 35 mg/L for 30 min was reported by Lignicka et al.(2021), showing that the total plate count of microbes effectively reduced from 2.68 to 2.15 log cfu /g, while the mould and yeast count decreased from 2.56 to 2.10 log cfu / g and 5.04 to 4.54 log cfu /g, respectively. The above-mentioned studies have efficiently shown the effect of ozone treatment on different cereal-based products. However, the optimisation of dosage and time parameters for every product still needs to focus. Moreover, further studies on the effect of ozonation on the overall nutritional and functional qualities of products should be focused on in the future for the development of quality products. Some of the effects of ozonation on the quality parameters of grains and some of the above-mentioned products are listed in Table 20.2.

20.6 CONCLUSION AND FUTURE PERSPECTIVES

Ozonation is a non-thermal, eco-friendly technology, which has widespread applications in the food industry due to its immense oxidation potential. The application of ozone in the cereal industry helps tackle various aspects, including the potential post-harvest quality losses during storage such as infestations of insects and rodents, toxin production, microbial contamination, and their effects on nutritional and functional parameters. This technology has been growing rapidly as compared to other thermal technologies due to minimum losses of nutritional properties and the fact that it leaves no chemical residues after the reaction. However, limiting factors such as high cost and non-specific process parameters still need to be worked on. Various combination techniques such as ozonation with modified air packaging and ultrasonication have been used to reduce the high cost of ozonation in an effective manner. Hence, ozonation has great potential in the cereal industry as it may help in overcoming various post-harvest losses and improve the shelf life of grains as well as produce high-quality and safe products.

REFERENCES

Afsah-Hejri, L., Hajeb, P., & Ehsani, R. J. (2020). Application of ozone for degradation of mycotoxins in food: A review. *Comprehensive Reviews in Food Science and Food Safety*, *19*(4), 1777–1808.

Agriopoulou, S., Sachadyn-Król, M., Stamatelopoulou, E., & Varzakas, T. (2022). Effect of Ozonation and Plasma Processing on Food Bioactives. In *Retention of Bioactives in Food Processing* (pp. 547–577). Cham: Springer International Publishing.

Alexander, L., Yuri, S., Mikhail, P., Olga, S., Sergey, K., & Irina, L. (2018). Treatment of spring wheat seeds by ozone generated from humid air and dry oxygen. *Research in Agricultural Engineering*, *64*(1), 34–40.

Allen, B., Wu, J., & Doan, H. (2007). Inactivation of fungi associated with barley grain by gaseous ozone. *Journal of Environmental Science and Health, Part B*, *38*(5), 617–630.

Alvani, K., Qi, X., Tester, R. F., & Snape, C. E. (2011). Physico-chemical properties of potato starches. *Food Chemistry*, *125*(3), 958–965.

Alwi, N. A., & Ali, A. (2014). Reduction of Escherichia coli 0157, Listeria monocytogenes and Salmonella enterica sv. Typhimurium populations on fresh-cut bell pepper using gaseous ozone. *Food Control*, *46*, 304–311.

An, H. J., & King, J. M. (2009). Using ozonation and amino acids to change pasting properties of rice starch. *Journal of Food Science, 74(3),* 278–283.

Bai, Y. P., Guo, X. N., Zhu, K. X., & Zhou, H. M. (2017). Shelf-life extension of semi-dried buckwheat noodles by the combination of aqueous ozone treatment and modified atmosphere packaging. *Food Chemistry*, *237*, 553–560.

Brito Júnior, J. G. D., Faroni, L. R. D. A., Cecon, P. R., Benevenuto, W. C. A. D. N., Benevenuto Júnior, A. A., & Heleno, F. F. (2018). Efficacy of ozone in the microbiological disinfection of maize grains. *Brazilian Journal of Food Technology, 21*. https://doi.org/10.1590/1981-6723.02217.

Broberg, M. C., Feng, Z., Xin, Y., & Pleijel, H. (2015). Ozone effects on wheat grain quality–A summary. *Environmental Pollution*, *197*, 203–213.

Brodowska, A. J., Nowak, A., & Śmigielski, K. (2018). Ozone in the food industry: Principles of ozone treatment, mechanisms of action, and applications: An overview. *Critical Reviews in Food Science and Nutrition*, *58*(13), 2176–2201.

Campbadal, C. A., Mason, L. J., Maier, D. E., Denvir, A., & Woloshuk, C. (2011). Ozone application in a modified screw conveyor to treat grain for insect pests, fungal contaminants, and mycotoxins. *Journal of Stored Products Research, 47(3)*, 249–254.

Çatal, H., & İbanoğlu, Ş. (2012). Effect of aqueous ozonation on the pasting, flow and gelatinization properties of wheat starch. *LWT-Food Science and Technology*, *59*(1), 577–582.

Chan, H. T., Leh, C. P., Bhat, R., Senan, C., Williams, P. A., & Karim, A. A. (2011). Molecular structure, rheological and thermal characteristics of ozone-oxidized starch. *Food Chemistry, 126(3)*, 1019–1024. https://doi.org/10.1016/j. foodchem.2010.11.113

Chen, R., Ma, F., Li, P. W., Zhang, W., Ding, X. X., Zhang, Q. I., & Xu, B. C. (2014). Effect of ozone on aflatoxins detoxification and nutritional quality of peanuts. *Food Chemistry*, *146*, 284–288.

Chittrakorn, S., Earls, D., & MacRitchie, F. (2014). Ozonation as an alternative to chlorination for soft wheat flours. *Journal of Cereal Science*, *60*(1), 217–221.
Coste, C., Dubois, M., & Pernot, A. G. (2009). *U.S. Patent Application No. 12/282,032.*
Cullen, P. J., & Norton, T. (2012). Ozone sanitisation in the food industry. *Ozone in food processing*, 163–176.
De Alencar, E. R., Faroni, L. R. D. A., Soares, N. D. F. F., da Silva, W. A., & Da Silva Carvalho, M. C. (2012). Efficacy of ozone as a fungicidal and detoxifying agent of aflatoxins in peanuts. *Journal of the Science of Food and Agriculture, 92*(4), 899–905. https://doi.org/10.1002/jsfa.4668
Ding, W., Wang, Y., Zhang, W., Shi, Y., & Wang, D. (2015). Effect of ozone treatment on physicochemical properties of waxy rice flour and waxy rice starch. *International Journal of Food Science & Technology*, *50*(3), 744–749.
Donkor, O. N., Stojanovska, L., Ginn, P., Ashton, J., & Vasiljevic, T. (2012). Germinated grains–Sources of bioactive compounds. *Food Chemistry*, *135*(3), 950–959.
Dubois, M., Coste, C., Despres, A. G., Efstathiou, T., Nio, C., Dumont, E., & Parent-Massin, D. (2006). Safety of Oxygreen®, an ozone treatment on wheat grains. Part 2. Is there a substantial equivalence between Oxygreen-treated wheat grains and untreated wheat grains? *Food Additives and Contaminants*, *23*(1), 1–15.
Duguet, J. P. (2004). Basic concepts of industrial engineering for the design of new ozonation processes. *Ozone News*, *32*(6), 15–19.
El-Desouky, T. A., A. M. A. Sharoba, A. I. El-Desouky, H. A. El-Mansy, and K. Naguib. (2012): Effect of ozone gas on degradation of aflatoxin B1 and Aspergillus flavus fungal. *Journal of Analytical Toxicology* 2, no. 1 128.
Elkhalifa, A. E. O., & Bernhardt, R. (2010). Influence of grain germination on functional properties of sorghum flour. *Food Chemistry*, *121*(2), 387–392.
FDA. (2017). Food and Drug Administration (FDA). FDA Food Code 2013. *Federal Register* *62*:17237–17164.
Glowacz, M., Colgan, R., & Rees, D. (2015). The use of ozone to extend the shelf-life and maintain quality of fresh produce. *Journal of the Science of Food and Agriculture*, *95*(4), 662–671.
Goncalves, A. A. (2009). Ozone: an emerging technology for the seafood industry. *Brazilian archives of Biology and Technology*, *52*(6), 1527–1539.
Goze, P., Rhazi, L., Lakhal, L., Jacolot, P., Pauss, A., & Aussenac, T. (2017). Effects of ozone treatment on the molecular properties of wheat grain proteins. *Journal of Cereal Science*, *75*, 243–251.
Goze, P., Rhazi, L., Pauss, A., & Aussenac, T. (2016). Starch characterization after ozone treatment of wheat grains. *Journal of Cereal Science*, *70*, 207–213.
Graham, D. M. 1997. Use of ozone for food processing. *Food Technology*. *51*(6):72–75.
Granella, S. J., Christ, D., Werncke, I., Bechlin, T. R., & Coelho, S. R. M. (2018). Effect of drying and ozonation process on naturally contaminated wheat seeds. *Journal of Cereal Science*, *80*, 205–211.
Guzel-Seydim, Z. B., Greene, A. K., & Seydim, A. C. (2004). Use of ozone in the food industry. *LWT-Food Science and Technology*, *37*(4), 453–460.
Hansen, L. S., Hansen, P., & Jensen, K. M. V. (2013). Effect of gaseous ozone for control of stored product pests at low and high temperature. *Journal of Stored Products Research*, *54*, 59–63.
Janex, M. L., Savoye, P., Roustan, M., Do-Quang, Z., Laine, J. M., & Lazarova, V. (2000). Wastewater disinfection by ozone: influence of water quality and kinetics modeling. *Ozone: Science & Engineering*, *22*(2), 113–121.
Khadre, M. A., Yousef, A. E., & Kim, J. G. (2001). Microbiological aspects of ozone applications in food: a review. *Journal of Food Science*, *66*(9), 1242–1252.
László, Z., Hovorka-Horváth, Z., Beszédes, S., Kertesz, S., Gyimes, E., & Hodur, C. (2008). Comparison of the effects of ozone, UV and combined ozone/UV treatment on the color and microbial counts of wheat flour. *Ozone: Science and Engineering*, *30*(6), 413–417.
Lee, M. J., Kim, M. J., Kwak, H. S., Lim, S. T., & Kim, S. S. (2017). Effects of ozone treatment on physicochemical properties of Korean wheat flour. *Food Science and Biotechnology*, *26*(2), 435–440.
Li, M. M., Guan, E. Q., & Bian, K. (2015). Effect of ozone treatment on deoxynivalenol and quality evaluation of ozonised wheat. *Food Additives & Contaminants: Part A*, *32*(4), 544–553.

Li, M., Peng, J., Zhu, K. X., Guo, X. N., Zhang, M., Peng, W., & Zhou, H. M. (2013). Delineating the microbial and physical–chemical changes during storage of ozone treated wheat flour. *Innovative Food Science & Emerging Technologies*, *20*, 223–229.

Li, M., Zhu, K. X., Wang, B. W., Guo, X. N., Peng, W., & Zhou, H. M. (2012). Evaluation the quality characteristics of wheat flour and shelf-life of fresh noodles as affected by ozone treatment. *Food Chemistry*, *135*(4), 2163–2169.

Lignicka, I., Balgalve, A., Ābelniece, K., & Zīdere-Laizāne, A. M. (2021). Comparison of the effect of ultra-violet light, ozone and heat treatment on muesli quality. *Agronomy Research 19(2)*, 531–539

Majchrowicz, A. (1998). Innovative technologies could improve food safety. *Food Review/National Food Review*, *22*(6), 16–20.

Marston, K., Khouryieh, H., & Aramouni, F. (2015). Evaluation of sorghum flour functionality and quality characteristics of gluten-free bread and cake as influenced by ozone treatment. *Food Science and Technology International*, *21*(8), 631–640.

Mason, L. J., & Woloshuk, C. P. (2011). Susceptibility of stored product insects to high concentrations of ozone at different exposure intervals. *Journal of Stored Products Research, 47(4)*, 306–310.

McDonough, M. X., Mason, L. J., & Woloshuk, C. P. (2011). Susceptibility of stored product insects to high concentrations of ozone at different exposure intervals. *Journal of Stored Products Research*, *47*(4), 306–310.

Mei, J., Liu, G., Huang, X., & Ding, W. (2016). Effects of ozone treatment on medium hard wheat (Triticum aestivum L.) flour quality and performance in steamed bread making. *CyTA-Journal of Food*, *14*(3), 449–456.

Mei, J., Liu, G., Huang, X., & Ding, W. (2016). Effects of ozone treatment on medium hard wheat (Triticum aestivum L.) flour quality and performance in steamed bread making. *CyTA-Journal of Food*, *14*(3), 449–456.

Mendez, F.,Maier, D. E.,Mason, L. J., &Woloshuk, C. P. (2002). Penetration of ozone into columns of stored grains and effects on chemical composition and processing performance. *Journal of Stored Product Research*, *39*, 33–44

Mishra, G., Palle, A. A., Srivastava, S., & Mishra, H. N. (2019). Disinfestation of stored wheat grain infested with Rhyzopertha dominica by ozone treatment: process optimization and impact on grain properties. *Journal of the Science of Food and Agriculture*, *99*(11), 5008–5018.

Mishra, G., Srivastava, S., Panda, B. K., & Mishra, H. N. (2018). Sensor array optimization and determination of Rhyzopertha dominica infestation in wheat using hybrid neuro-fuzzy-assisted electronic nose analysis. *Analytical Methods*, *10*(47), 5687–5695.

Nadeem, M., Anjum, F. M., Amir, R. M., Khan, M. R., Hussain, S., & Javed, M. S. (2010). An overview of anti-nutritional factors in cereal grains with special reference to wheat-A review. *Pakistan Journal of Food Sciences*, *20*(4), 54–61.

Naito, S., & Takahara, H. (2006). Ozone contribution in food industry in Japan. *Ozone: science and Engineering*, *28*(6), 425–429.

Normov, D., Chesniuk, E., Shevchenko, A., Normova, T., Goldman, R., Pozhidaev, D., & Trdan, S. (2019). Does ozone treatment of maize seeds influence their germination and growth energy? *Acta Agriculturae Slovenica*, 114(2), 251–258.

Novak, J.S., & Yuan, J. T. (2007). The Ozonation Concept: Advantages of Ozone Treatment and Commercial Developments. *Advances in thermal and non-thermal food preservation, 7*(2), 185–197.

O'Donnell, C., Tiwari, B. K., Cullen, P. J., & Rice, R. G. (Eds.). (2012). *Ozone in food processing*. John Wiley & Sons.

Obadi, M., Zhu, K. X., Peng, W., Ammar, A. F., & Zhou, H. M. (2016). Effect of ozone gas processing on physical and chemical properties of wheat proteins. *Tropical Journal of Pharmaceutical Research*, *15*(10), 2147–2154.

Obadi, M., Zhu, K. X., Peng, W., Noman, A., Mohammed, K., & Zhou, H. M. (2018). Characterization of oil extracted from whole grain flour treated with ozone gas. *Journal of Cereal Science*, *79*, 527–533.

Overbeck, P. K. (2000). WQA ozone task force- an update. *Water Conditioning & Purification March*, 76–78.

Pandiselvam, R., Manikantan, M. R., Divya, V., Ashokkumar, C., Kaavya, R., Kothakota, A., & Ramesh, S. V. (2019). Ozone: An advanced oxidation technology for starch modification. *Ozone: Science & Engineering*, *41*(6), 491–507.

Porto, Y. D., Trombete, F. M., Freitas-Silva, O., De Castro, I. M., Direito, G. M., & Ascheri, J. L. R. (2019). Gaseous ozonation to reduce aflatoxins levels and microbial contamination in corn grits. *Microorganisms*, *7*(8), 220.

Prabha, V., Barma, R. D., Singh, R., & Madan, A. (2015). Ozone technology in food processing: A review. *Trends in Biosciences* *8*(16), 4031–4047.

Rakcejeva, T., Zagorska, J., & Zvezdina, E. (2014). Gassy ozone effect on quality parameters of flaxes made from biologically activated whole wheat grains. *International Journal of Nutrition and Food Engineering*, *8*(4), 396–399.

Sandhu, H. P., Manthey, F. A., & Simsek, S. (2011a). Quality of bread made from ozonated wheat (*Triticum aestivum L.*) flour. *Journal of the Science of Food and Agriculture*, *91*(9), 1576–1584.

Sandhu, H. P., Manthey, F. A., & Simsek, S. (2012). Ozone gas affects physical and chemical properties of wheat (*Triticum aestivum L.*) starch. *Carbohydrate Polymers*, *87*(2), 1261–1268.

Sandhu, H. P., Manthey, F. A., Simsek, S., & Ohm, J. B. (2011b). Comparison between potassium bromate and ozone as flour oxidants in breadmaking. *Cereal Chemistry*, *88*(1), 103–108.

Sangronis, E., & Machado, C. J. (2007). Influence of germination on the nutritional quality of Phaseolus vulgaris and Cajanus cajan. *LWT-Food Science and Technology*, *40*(1), 116–120.

Santos, R. R., Faroni, L. R., Cecon, P. R., Ferreira, A. P., & Pereira, O. L. (2016). Ozone as fungicide in rice grains. *Revista Brasileira de Engenharia Agrícola e Ambiental*, *20*, 230–235.

Savi, G. D., Piacentini, K. C., Bittencourt, K. O., & Scussel, V. M. (2014). Ozone treatment efficiency on Fusarium graminearum and deoxynivalenol degradation and its effects on whole wheat grains (Triticum aestivum L.) quality and germination. *Journal of Stored Products Research*, *59*, 245–253.

Shaghaghian, S., Niakousari, M., & Javadian, S. (2014). Application of ozone post-harvest treatment on Kabkab date fruits: effect on mortality rate of Indian meal moth and nutrition components. *Ozone: Science & Engineering*, *36*(3), 269–275.

Sivaranjani, S., Prasath, V. A., Pandiselvam, R., Kothakota, A., & Khaneghah, A. M. (2021). Recent advances in applications of ozone in the cereal industry. *LWT*, 111412.

Srivastava, S., & Mishra, H. N. (2021). Ecofriendly nonchemical/nonthermal methods for disinfestation and control of pest/fungal infestation during storage of major important cereal grains: A review. *Food Frontiers*, *2*(1), 93–105.

Subramanyam, B., Xinyi, E., Savoldelli, S., & Sehgal, B. (2017). Efficacy of ozone against Rhyzopertha dominica adults in wheat. *Journal of Stored Products Research*, *70*, 53–59.

Sui, Z., Yao, T., Zhong, J., Li, Y., Kong, X., & Ai, L. (2016). Ozonation treatment improves properties of wheat flour and the baking quality of cake. *Philippine Agricultural Scientist*, *99*(1), 50–57.

Sujayasree, O. J., Chaitanya, A. K., Bhoite, R., Pandiselvam, R., Kothakota, A., Gavahian, M., & Mousavi Khaneghah, A.(2021). Ozone: An Advanced Oxidation Technology to Enhance Sustainable Food Consumption through Mycotoxin Degradation. *Ozone: Science & Engineering*, 1–21.

Tiwari, B. K., Brennan, C. S., Curran, T., Gallagher, E., Cullen, P. J., & O'Donnell, C. P. (2010). Application of ozone in grain processing. *Journal of Cereal Science*, *51*(3), 248–255.

Trombete, F., Minguita, A., Porto, Y., Freitas-Silva, O., Freitas-Sá, D., Freitas, S. & Fraga, M. (2016). Chemical, technological, and sensory properties of wheat grains (Triticum aestivum L) as affected by gaseous ozonation. *International Journal of Food Properties*, *19*(12), 2739–2749.

Wang, L., Shao, H., Luo, X., Wang, R., Li, Y., Li, Y., & Chen, Z. (2016). Effect of ozone treatment on deoxynivalenol and wheat quality. *PloS one*, *11*(1), e0147613.

White, S. D., Murphy, P. T., Leandro, L. F., Bern, C. J., Beattie, S. E., & van Leeuwen, J. H. (2013). Mycoflora of high-moisture maize treated with ozone. *Journal of Stored Products Research*, *55*, 84–89.

Yabo, W., Siyu, W., Yue, S., Wei, M., Tingting, D., Weiqin, Y.,& Xiaozhi, W. (2017). Elevated ozone level affects micronutrients bioavailability in soil and their concentrations in wheat tissues. *Plant, Soil and Environment*, *63*(8), 381–387.

Yan, S., Wu, X., Faubion, J., Bean, S. R., Cai, L., Shi, Y. C., & Wang, D. (2012). Ethanol-production performance of ozone-treated tannin grain sorghum flour. *Cereal Chemistry*, *89*(1), 30–37.

Zhou, X., Zhou, J., Wang, Y., Peng, B., Zhu, J., Yang, L., & Wang, Y. (2015). Elevated tropospheric ozone increased grain protein and amino acid content of a hybrid rice without manipulation by planting density. *Journal of the Science of Food and Agriculture*, *95*(1), 72–78.

Zhu, F. (2018). Effect of ozone treatment on the quality of grain products. *Food chemistry*, *264*, 358–366.

CHAPTER 21

Cold Plasma Treatment of Cereals

Anusha Mishra,[1] Ranjitha Gracy T. Kalaivendan,[1] Gunaseelan Eazhumalai,[1] and Uday S. Annapure[1,2]
[1]Department of Food Engineering Technology, Institute of Chemical Technology, Mumbai, Maharashtra, India
[2]Institute of Chemical Technology, Marathwada Campus, Jalna, Maharashtra, India

CONTENTS

DOI: 10.1201/9781003242192-26

21.1 INTRODUCTION

Food product stabilization is one of the utmost priorities in the food processing sector. This is achieved by the inactivation of microorganisms directly or limiting conditions of their growth. It needs to be attained without compromising the nutritional quality of the food product. Non-thermal technology has recently gained dominance over thermal technology due to its sustainability and prevention of adverse impacts of heat on the nutritional profile, flavor, and appearance of foods (Barbosa-Canovas, Pothakamury, & Swanson, 1995). Consumers today are focused on conscious, chemical-free eating behavior, which poses an immense challenge and opportunity for food technologists and academicians all over the globe to fulfill the changing consumer and environmental requirements.

The advent of the thermal processing of food dates back to the 1860s with the introduction of pasteurization in the food fraternity. Later, blanching, sterilization, canning, and drying added cherry to the cake. These technologies helped in delivering safe food to consumers by inactivating microorganisms and spores. Since the progression in technological evolution, the term non-thermal technology has been introduced in the food domain. It offers a substitute for conventional thermal technology to the consumer need for minimally processed foods and preserving the biological ingredients. The commercial application of non-thermal preservation has been successfully flourishing for 15 years and is still growing widely.

Cereals are staples and are consumed in large amounts by the majority of people either directly or in modified form as flour, bran, and various additional ingredients used in the manufacture of other foods. Several modification techniques are applied to cereals that include gelatinization, heat-moisture treatment, and annealing. However, the application of non-thermal technology (i.e., cold plasma technology) is still being researched.

Plasma is known as the fourth state of matter produced energizing the gaseous molecules. It is a mixture of high-velocity electrons, ionized atoms and molecules, free radicals, atoms in their ground and excited state, and UV radiation (Ramanan, Sarumathi, & Mahendran, 2018). When the gaseous atoms are excited by high energy inputs like microwave, radio frequency (Park, Henins, Herrmann, Selwyn, & Hicks, 2001), or electric field (Chen et al., 2004), the electrons will expel out of the atomic orbits. This results in the formation of ions and free electrons. The free electrons on impact with other electrons, positive ions, and stable molecules lead to the production of radicals and non-radical species. The source gas used for plasma generation determines the composition of reactive species. Cold plasma is an ultra-fast sterilization technique that does not alter the nutritional composition of the food being treated as it operates at ambient temperature, and thus it is a favorable technology to be efficiently exploited in the food processing sector.

There are several types of plasma: low-pressure plasma, atmospheric plasma, plasma-activated water, glow plasma, and jet, microwave-assisted, and dielectric barrier discharge as illustrated in Figure 21.1. The most common, dielectric barrier discharge and jet plasma, are used in the food industry, because of their simple design, easy configurations, and accessibility (Pankaj, S. K., Wan, Z., & Keener, K. M.., 2018)

Dielectric barrier discharge is a cold atmospheric plasma that has two metal electrodes wrapped with a dielectric material of varying widths at a discharge distance starting from 0.1 mm to, with the discharge being insulated by the dielectric barrier layer. A working gas moves between the two electrodes resulting in the formation of plasma as the two electrodes ionize. Two concentric electrodes are present in the plasma jet where the inner electrode is connected to a radiofrequency power at a high frequency leading to ionization of the gas present. The plasma exits the nozzle and gives a jet-like appearance (Tendero, 2006).

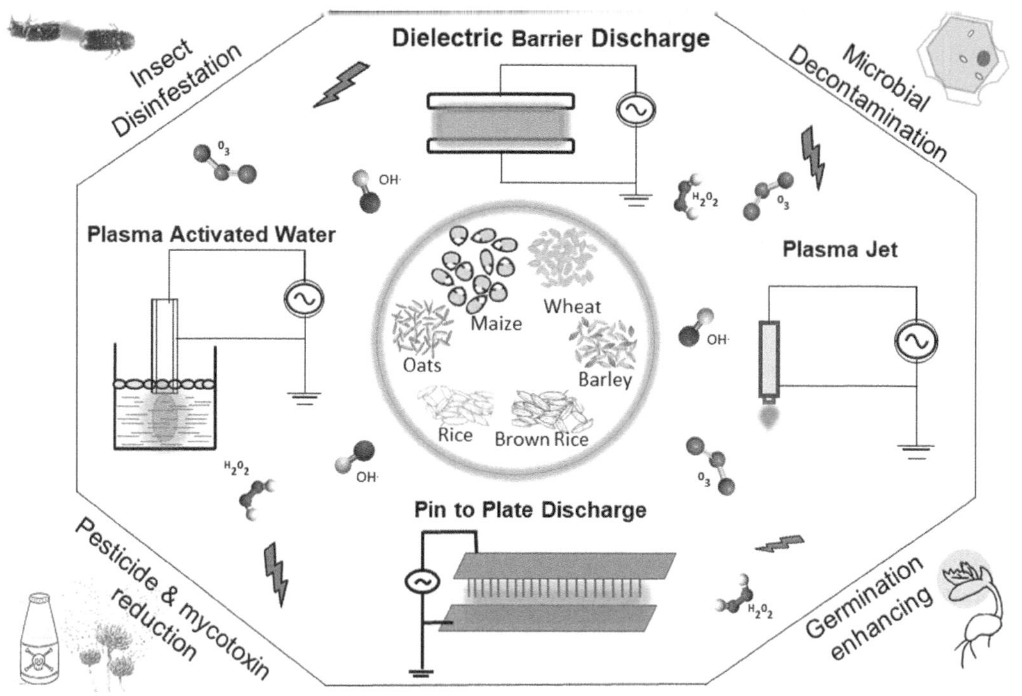

Figure 21.1 Application of cold plasma for cereal processing.

Modification of starch molecules present in food components occurs due to cross-linking, depolymerization, and plasma etching due to interaction with the plasma species. During cross-linking, a C-O-C linkage forms between the two polymeric chains (C-OH) leading to the elimination of water molecule that occurs due to the cleavage of glycosidic bonds. This phenomenon occurs due to the cleavage of glycosidic bonds brought about by hydroxyl radicals formed from the decomposition of the water molecule and gas occurring during the process of generation of plasma. The depolymerization of starch components (amylose and amylopectin) happens due to the collision of plasma ions resulting in the formation of smaller fragments. Another important process taking place due to cold plasma technology is surface etching. It is the physical phenomenon occurring due to physical sputtering, which in turn, increases the surface energy of the substrate leading to easier penetration of water through the surface, which has been etched, resulting in improved hydrophilicity (Thirumdas, Kadam, & Annapure, 2017).

The numerous applications of cold plasma in agriculture include microbial inactivation, shelf-life extension, potent pesticide, and increase in the germination rate. There has been advancement in research on the application of cold plasma technology in surface modification and improvement in functional properties of cereal grains. These include reduction in cooking time of parboiled rice (Sarangapani et al., 2015), and modification in the surface morphology, cooking, textural properties, and increase in brightness and whiteness index of brown rice (Thirumdas et al., 2016). The reason behind these actions is an etching of the rice surface leading to the enhanced penetration of water during soaking/cooking resulting in increased germination and reduction in cooking time and hardness (Chen, 2014;).

The current chapter deals with the detailed application of cold plasma treatment of cereals and its effect on the nutritional, functional, and biological properties of cereals.

21.2 APPLICATIONS OF COLD PLASMA IN CEREALS

Cold plasma is one of the potential novel technologies utilized for various applications in cereal processing such as microbial decontamination in seed surfaces (like wheat, maize, rice, etc.), mycotoxin and allergen elimination to enhance seed germination, and reduction of anti-nutrition components, pesticide dissipations, and insect disinfestations as listed in Table 21.1.

21.2.1 Microbial Decontamination in Cereals

Cereal grains are contaminated by various contaminants at different stages of the food supply chain (farm to fork). Microbial contamination and spoilage must be controlled to ensure food safety to the consumer. Factors like air, water, soil, dust, animals, rodents, birds, and animal feces could be potential sources of microbial contamination. Other environmental conditions like poor storage conditions, improper handling, and processing equipment also contribute to a greater extent. In general, microbial contamination may happen due to bacterial and fungal growth in grains whereas better storage condition avoids the possibilities of bacterial growth as the stored grains have lower moisture content. The most prevalent bacteria among cereal grains are *Lactobacillaceae, Psedomonadaceae, Bacillaceae, Micrococcaceae, Escherichia coli, and salmonella species.* Microbial contamination in grains is more concentrated in grain surface but some studies have reported contamination inside the grains. The ideal method of decontamination for cereal grains must be cost-effective and should retain the sensory and nutritional qualities of grains. Non-thermal technologies have been explored with keen interest due to their ability to decontaminate food products with minimal effect on quality and nutritional profile. Cold plasma (CP) is one of the non-thermal technologies focused on due to its efficiency in microbial decontamination on cereal grains. In CP, the efficacy of decontamination depends on various parameters like product composition, environmental temperature and relative humidity (RH), food properties such as moisture content, pH, and surface properties, and processing parameters such as voltage, frequency, gas composition, flow rate, treatment time, electrode type, inter-electrode gap, headspace, and exposure pattern time (Feizollahi, Iqdiam, Vasanthan, Thilakarathna, & Roopesh, 2020).

The proposed mechanisms for bacterial inactivation are cell wall rupture and intracellular damage by reactive species, which cause cell death. In atmospheric CP systems, atmospheric gas molecules break down into reactive species such as reactive nitrogen species (RNS) and reactive oxygen species (ROS) and induce cell wall rupture via lipid peroxidation, enzyme inactivation, protein denaturation, and DNA damage. The gram-negative bacteria are more susceptible to CP as compared to gram-positive bacteria due to the difference in cell wall thickness. The factors that induce bacterial spore inactivation are UV photons, reactive oxygen and nitrogen species. In spores, the inactivation induced due to DNA damage are caused by UV radiation. ROS and UV photons penetrate the cell wall and cause protein denaturation and lipid oxidation in the spore core. It also denatures the dipicolinic acid responsible for spore resistance. Acids like nitric and nitrous acid produced from the reaction of RNS and free water in the core cause denaturation of the inner membrane of the spore (Laroque, Seó, Valencia, Laurindo, & Carciofi, 2022). The mold and fungal decontamination are much more important in the case of low-moisture cereal grains. The mechanism of action for fungal decontamination are protein denaturation, cell leakage, mycelial tip deformation, DNA fragmentation and release, and apoptosis. The reactive species also cause etching and erosion of fungal spores. The reactive species at low density also causes cell apoptosis leading to lipid accumulation in spore structure (Panngom et al., 2014).

Table 21.1 Applications of Cold Plasma Processing on Cereal Grains and Products

Microbial Decontamination on Cereals

S. No	Cereal Grain	Bacterial/fungal Strains	Parameters	Results	Reference
1	Wheat	*Bacillus atrophaeus*, *Aspergillus flavus*	DBD atmospheric system, 0–120 kV, 50 Hz. Treated at 80 kV, 5 or 20 min. Direct exposure, indirect exposure	Anaverage reduction of 2.57 ± 0.46 log10 and 2.47 ± 0.28 log 10 units were obtained for cells of *B. atrophaeus* after direct and indirect exposure, respectively. For spores of *A. flavus*, the average reduction factors were 1.60 ± 0.41 and 1.53 ± 0.45 log10 units afterdirect and indirect modes of plasma exposure, respectively.	(Los, Ziuzina, Boehm, & Bourke, 2020)
2	Maize	*Aspergillus flavus*, *A. alternata*, *F.culmorum*	Diffuse Coplanar Surface Barrier Discharge (DCSBD), Power 400 V.	The plasma treatment showed reduction of 3.79 log (CFU/g) in *F. culmorum*after a 60-s plasma treatment, 4.21 log (CFU/g) in *A. favus* and 3.22 log (CFU/g) in *A. alternata* after a 300-s plasma treatment.	(Zahoranová et al., 2018)
3	Brown Rice	*Bacillus cereus*, *Escherichia coli*	DBD, 15 kHz voltage, input power 250 W, treatment time 20 min	The bacterial count reduced by 2.01 log CFU/g in cooked rice.	(K. H. Lee et al., 2018)
4	Rice	Total plate count	Atmospheric DBD, power 250 W, 20 min	The plasma treatment reduced the microbial concentration from 4.08–4.11 log CFU/g to 2.68–2.84 log CFU/g.	(J. H. Lee et al., 2019)
5	Barley, wheat	*Escherichia coli*, *B. atrophaeus*, *P. verrucosum*	Atmospheric DBD, 80 kV, 20 min	The maximal reductions of barley background microbiota were 2.4 and 2.1 log10 CFU/g and of wheat – 1.5 and 2.5 log10 CFU/g for bacteria and fungi.	(Los, Agata; Ziuzina, Dana; Akkermans, Simen; Boehm, 2014)

Mycotoxin Reduction in cereals

S. No	Cereal Grain	Mycotoxin	Process Parameter	Results	Reference
1	Barley	Deoxynivalenol (DON)	Atmospheric DBD, 300 W, 34 kV, 0–10 min	The results showed that ACP treatment for 6 and 10 min reduced DON concentration by 48.9% and 54.4%, respectively	(Feizollahi, Iqdiam, Vasanthan, Thilakarathna, & Roopesh, 2020)
2	Corn	Aflatoxin B1 (AFB1)	Atmospheric DBD, 1–30 min, 200 W, 50 HZ, 90 kV, 1–30 min, 40–80% RH%	Aflatoxin in corn was degraded by 62% and 82% by 1 and 10 min HVACP treatment in RH 40% air, respectively.	(Shi, Ileleji, Stroshine, Keener, & Jensen, 2017)

(continued)

Table 21.1 (Continued)

Microbial Decontamination on Cereals

S. No	Cereal Grain	Bacterial/fungal Strains	Parameters	Results	Reference
3	Rice	Deoxynivalenol, zearalenone, enniatins, fumonisin B1, and T2 toxin	cold atmospheric pressure plasma (CAPP), 38 kV, 17 kHz, 5–60 s	All pure mycotoxins exposed to CAPP were degraded almost completely within 60 s.	(ten Bosch et al., 2017)

Insect Disinfestation

S. No	Cereal Grain	Insect	Process Parameter	Results	Reference
1	Wheat flour	*Triboliumcastaneum*	ACP, 80 $kV_{RMS,}$ 5–20 min	Cold plasma increased mortality of T.castaneum 2.6%, from 2.0% (control) when treated for 20 min.	(Los, Ziuzina, van Cleynenbreugel, Boehm, & Bourke, 2020)
2	Wheat flour	*Triboliumcastaneum*	Atmospheric DBD, 80 kV, 0–20 min	Direct treatment for 5 min showed mortality of 66.7% (0 h PTRT), whereas 100% (24% h PTRT mortality in adult insects. Larvae were completely inactivated after 1 min of treatment, whereas eggs, pupae, and adult insects were inactivated after 2, 5, and 10 min of treatment.	(Ziuzina, van Cleynenbreugel, Tersaruolo, & Bourke, 2021)
3.	Whole wheat grains	*Rhyzopertha dominica*	DBD, 44–47 kV, 4–7 min, in-package treatment	The maximum mortality was 88.33% at 47 kV for 7 min achieved. The protein content of wheat is increased to 16.07%	(Madathil, Thirugnanasambandan Kalaivendan, Paul, & Radhakrishnan, 2021)

Pesticide Dissipation

S.No	Cereal Grain	Pesticide	Process Parameter	Results	Reference
1	Maize	Dichlorvos, Omethoate	Radio frequency (13.56 MHz) plasma; 30–120W power; O_2 flow rate 20–40 cm^3/min; 30–120 s treatment time; distance of plasma discharge zone 20–80 cm	The maximum percentage of reduction at 120W power with 40 cm^3/min flow rate at 20 cm discharge zone exposed for 120 s: Dichlorvos – 88%, Omethoate – 94%.	(Bai, Chen, & Mu, 2009)

2	Maize	Chlorpyrifos, Carbaryl	Dielectric Barrier Discharge plasma; 5–25 W; Argon gas flow rate 150–1500 ml/min; 20–60 s treatment time; Frequency 100–1200 Hz; Initial pesticide concentration 2, 4 ppm	The maximum percentage of reduction at 1200 Hz and 25W with 1000 ml/min for 60 s of treatment time. Chlorpyrifos – 91.5%, Carbaryl – 73.1%.	(Feng, Ma, Liu, Xie, & Fan, 2019)

Seed Germination

S. No	Cereal Grain	Parameters	Key findings	Results	Reference
1	Rice	Atmospheric plasma jet, 15 kV at 60 Hz, 0–3 s,	Average germ length increased with increase in exposure time	The plasma treatment increases the wettability without affecting the seed germination count	(Penado, Mahinay, & Culaba, 2018)
2	wheat	DBD, 13.0 kV, air flow 1.5 L min^{-1}, 0–13 min, 1.50 W	the highest root length and shoot length was at 7 min and 4 min treatment, respectively, on and reduced on further increase in time.	The germination potential, germination rate, germination index, and vigor index increased by 26.7,9.1, 16.9, and 46.9% after 7 min DBD plasma treatment,	(Li et al., 2017)
3	Rice	Atmospheric microcorona discharge,	90% of the treated germinated seeds had a seedling length longer than 12 cm, while-66% of the non-treated seedlings had a length longer than 12 cm.	The germination percentage increased to 98% in plasma treated whereas 90% for untreated rice	(Khamsen et al., 2016)
3	Brown Rice	Argon (18, 24 mL/min) gas DBD, 100– 200 W power, 25 s -300 s treatment time	Theplasma treatment increased the root length from 4.5 mm (control) to 7.1 mm, while the seedling height increased from 12.5 to 21.1 mm.	The germination percentage, root length, seedling height increased by 84%, 57%, and 69%, respectively.	(Yodpitak et al., 2019)

21.2.2 Seed Germination Enhancement

Germination takes place during favorable conditions and seeds exhibit a dormancy nature during unfavorable conditions. Thus, the seeds need to be triggered for better germination during dormancy state. The most common methods for improvement of seed germination are using fertilizers or pesticides, besides the traditional methods such as hormone-induced and controlled environment (humidity, temperature, salinity). Though these methods were successful at different times, research works have been conducted to improve seed germination with a novel technology like CP to reduce the limitations of conventional methods (Rifna, Ratish Ramanan, & Mahendran, 2019).

In CP-based systems, the free radicals and reactive species produced by the plasma system can penetrate the seed pericarp and activate cells for enhanced germination. The plasma is also known for surface modification properties by etching on the surface of the seed and improve the hydrophilicity, water absorption capacity, oxygen, and moisture permeability, which influence the germination of seeds including cereals. It also affects the surface energy and hydrophilicity of seeds after the treatment by surface erosion resulting in high and improved germination. The seed surface oxidation resulting from radical activity increases the water contact angle, hence improving water absorption and helping germination. Another method of plasma application is germination enhancement of seeds by plasma-activated water, in which the active plasma radical is exposed directly into the water or applied on the surface of the water. This produces the nitrogen and hydrogen peroxide radicals, which possess the ability to penetrate through the cell membrane and induce the germination process (Ganesan, Tiwari, Ezhilarasi, & Rajauria, 2021).

21.2.3 Mycotoxin Elimination

Cereals are more prone to fungal contamination in the presence of favorable conditions like high moisture, humid, and heat. The fungal contamination and development of mycotoxin are more common under improper storage conditions. The fungal species well known for their mycotoxins are *Aspergillus, Penicillium,* and *Fusarium*. Mycotoxins like deoxynivalenol (DON), zearalenone (ZEN), fumonisins, and patulin produced by these fungal species are highly toxic and have severe health effects in humans and animals (Shanakhat et al., 2018). Various studies have reported CP as a potential method for decontamination of fungal species as well as the destruction of mycotoxin in the food matrix. The mechanism for degradation of mycotoxin is based on the interaction of free radicals (oxygen radicals, nitrogen radicals) produced in the plasma system with mycotoxins. The functionality of the mycotoxin changes and the toxicity is lost as the structure (furan ring deformation) was damaged and changed by the addition of bonds and other components like -OH, H+, etc. This could be a result of the generation of UV photons, ozone, and reactive ions. The free radicals also involve processes like ionization, epoxidation, and oxidation reaction with furan rings, which cause destruction. Thus, CP could be potentially used for the reduction of mycotoxin in cereals (Misra, Barun, Roopesh, & Jo, 2019).

21.2.4 Insect Disinfestation

Besides microbial contamination, insect infestation is another big problem that needs to be addressed as it causes massive losses during production and storage. The major sources of infestation in cereal grains are infestation from farms while harvesting, packaging, unhygienic transportation, and improper storage. The traditional methods of controlling insects and pests are by the application of pesticides, insecticides, and fumigants used at various phases of plant growth, storage, and distribution. These chemical-based pesticides are highly toxic and harmful for humans. The insects infested and stored along with the grains feed on grain endosperm resulting in the deterioration of quality, loss of weight, and viability of seeds, and ultimately reduced shelf life (Ziuzina, van Cleynenbreugel, Tersaruolo, & Bourke, 2021). Various technologies have been used to substitute chemical pesticides with plant-based extracts and insect growth regulators. Investigations are also being done on the potential of non-thermal technologies in cereals to eradicate the insects and pests in the grains. Cold plasma is a promising technology for eliminating insects to 100% mortality (Sarangapani, Patange, Bourke, Keener, & Cullen, 2018). Cold plasma produces free radicals, reactive species, and UV radiation, which react with the insect and engender mortality. The mechanism of disinfestation refers to the fact that reactive species interact with the insect body and cause denaturation of multiple enzymes, which inhibit the systematic insect metabolism, also the oxygen species such as ozone and oxygen radicals reduces respiration rate and cause asphyxiation of insects. Moreover, the other radicals penetrate through the insect and transmit radical reactions with cell constituents. To explicate further, ozone and its derivative radicals are transported into the aphid's body through spiracles, thereafter via the trachea, and finally reach the cells to act on the proteins and organelles (e.g., nucleus), resulting in fatality. Also, on the surface of the insects, lesions caused by etching cause rupture and leakage of cells (Madathil, Thirugnanasambandan Kalaivendan, Paul, & Radhakrishnan, 2021).

21.2.5 Pesticide Dissipation

Agricultural produce retains the pesticide residue applied during agricultural practices. The common pesticides used in agricultural practices are diazinon, chlorpyrifos, cypermethrin, parathion, paraoxon, omethoate, dichlorvos, malathion, azoxystrobin, cyprodinil, and fludioxonil. These residues are harmful to human health and the ecosystem. Cold plasma was also studied for its potential to degrade various organochlorine and organophosphorus pesticide residues in a food matrix. The reactive species and electrons produced from plasma discharge cause causing chemical bond cleavage and various oxidation reactions and degrade the residues in the matrix (Bourke, Ziuzina, Boehm, Cullen, & Keener, 2018). Due to the structural difference, the degradation effect with CP treatment changes with each pesticide residue (Gavahian & Cullen, 2020). The mechanism of action of CP responsible for the observed outcomes in cereals is shown in Figure 21.2.

Mechanism	Effect	↑ or ↓
Surface etching	Germination (water absorption capacity, oxygen,& moisture permeability) Hydration Microbial load Enzymatic hydrolysis Resistant starch	↑ ↑ ↓ ↑ ↑
Oxidative reactions with reactive oxygen and nitrogen species	Denaturation of enzymes Percentage of pesticide concentration Cleavage of bonds Water solubility index Depolymerisation Viscosity	↑ ↑ ↑ ↑ ↑ ↑ or ↓
Polymeric reactions with the radicals and plasma reactive species	Cell wall rupture Erosion of fungal spores	↑ ↑

Figure 21.2 Mechanism of action of plasma and its effects.

21.3 EFFECT OF COLD PLASMA ON THE NUTRITIVE PROPERTIES OF CEREALS

Cereals have a versatile nutritional profile of both major nutrients including carbohydrates, proteins, and lipids as well as minor nutrients such as vitamins and minerals. During each processing operation, a diminishing of the nutrients occurs due to their susceptibility to thermal, chemical, and mechanical extremities. While CP treatment is thermally benign, the mechanical modifications by etching and chemical interactions of the reactive species and free radicals with food constituents may result in noticeable changes in the nutritional profile. This necessitates the exploration of non-thermal CP treatment on the nutritional content of grains. Regardless of the few studies exploring the proximate composition comprised of moisture, protein, lipid, ash, and carbohydrate of plasma-treated cereals and cereal products, they explain the potential modifications in the nutritional profile as nutritional properties considerably govern the functional and rheological properties.

Having reduced moisture content, cereals are considered to be durable commodities as moisture content is an important factor affecting their shelf life. Cold plasma treatment has been found to reduce the moisture content of treated grains in the case of brown rice (Chen, Chen, & Chang, 2012), wheat (Selcuk, Oksuz, & Basaran, 2008), and also in parboiled rice (Sarangapani, Devi, Thirundas, Annapure, & Deshmukh, 2015), which helps storage stability. When the CP species strike the grain surface, the formation of pores and fissures on the surface promotes water evaporation, which in turn may bring reduce moisture. Also, the disassociation of the water molecules into hydroxyl radicals on the application of high-energy CP may be the cause of the moisture reduction. On the contrary, a study of CP disinfestation of in-package whole wheat grains showed an increase in moisture content due to the condensation of vapourised moisture, thereby increasing the unbound moisture content of the grains (Madathil, Thirugnanasambandan Kalaivendan, Paul, & Radhakrishnan, 2021). Nevertheless, the moisture content was found to be in the required range for safe storage conditions (Karunakaran, Muir, Jayas, White, & Abramson, 2001).

Not many studies report quantitative changes in the protein, fat, carbohydrates, and ash content, except a few stating an increase in protein and a reduction in fat and carbohydrate proportions. Incision of peptide chains in proteins by plasma radicals and species gives rise to the formation of free amino acids, which may cause an increase in the protein content (Tiwari et al., 2010). Moreover, the structure and molecular weight fractions of the wheat flour protein were modified significantly after CP treatment due to changes in α-helices and β-sheets proportions in the protein secondary structure (Misra et al., 2015) and the increase in higher molecular weight protein fractions (Bahrami et al., 2016). Strengthening the disulfide bonds between the glutenin subunits on the interaction with plasma species provoked such changes in the protein content of wheat. Correspondingly, CP influences specific amino acids content present in the grains, particularly are in relation to germination. Proline, an osmotic balance controlling amino acid of wheat seedlings, was increased significantly in CP-treated grains compared to untreated grains (Li et al., 2017). Likewise, a surge in the naturally occurring neurotransmitter amino acid Gamma-Amino Butyric Acid (GABA) in germinated brown rice (Chen et al., 2016) and *Indica* rice (Zargarchi & Saremnezhad, 2019) was affirmed after low-pressure plasma treatment. This must be attributed to the germination intensifying effect of CP.

The lipid content of most cereals is substantially low compared to carbohydrate and protein content. Yet, unfavorable reactions with lipids may affect the ultimate keeping quality. Since CP predominantly causes oxidative interaction with the commodity, lipid oxidation may also rely on the process intensities, which may also reduce the total lipid content as observed by Madathil et al. (2021). Further, oxidative indicators of lipid including peroxides, free fatty acids, n-hexanal, and malondialdehyde may increase as they are the products of lipid oxidation. Also, some of the oxidative labile fatty acids were reduced in the plasma-treated wheat flour, which elucidates the CP oxidation of the cereal lipid despite the need for supplementary investigations (Bahrami et al., 2016). Depolymerization and starch modification may also cause a reduction in the carbohydrate content of parboiled rice (Sarangapani et al., 2015), while an increase in crude fiber content of plasma-treated wheat was seen (Madathil et al., 2021). Thus, plasma process conditions such as operating pressure, treatment mode, gas used, the gas flow rate, plasma power, and exposure time along with product attributes such as the particle size, moisture content, and surface homogeneity influence the nutritive properties along with functional and biological properties.

21.4 EFFECT OF COLD PLASMA ON THE FUNCTIONAL PROPERTIES OF CEREALS

Microbial inactivation using the most recent CP technology has been widely explored. This technology has immense potential to improve the functional characteristics of cereal grain. Cold plasma is a highly efficient technology leading to a microbial reduction in food commodities and other biological materials. However, it has a remarkable ability to modify the functional properties of starch, protein, and hydrocolloids, which can be assessed through hydration properties, viscosity, FTIR, degree of gelatinization, crystallinity, and surface morphology through SEM.

21.4.1 Gel Hydration Properties

Hydration properties are used to identify the gelatinization of starch and protein denaturation during the thermic treatment of food components (de la Hera., 2013). Even a slight modification in the fine structure of amylopectin can lead to a significant change in the functional properties of the starch component. A significant difference in the values of water-holding capacity has been found in plasma-treated rice starch (Thirumdas et al., 2016). Variation in the plasma power and duration causes significant modification in the gel hydration properties of the starch.

21.4.1.1 Water-holding Capacity

The extent of a material to hold water against gravity is referred to as the water-holding capacity of a substance. It is the total amount of water that can be absorbed per gram of a substance. This encompasses bound water, hydrodynamic water, physically entrapped water, and capillary water (Damodaran, 2017). In the study conducted by Chaple et al. (2020), there was an increment in the water-holding capacity of the plasma-treated sample with an increase in time duration. This was attributed to the hydrolytic depolymerization of starch. Plasma treatment leads to alteration (due to etching) in the characteristics of the surface leading to easier penetration of water molecules.

21.4.1.2 Water Solubility Index

Water-holding capacity and water solubility index (gel hydration properties) govern the intake of water during thermal treatment, which affects the gelatinization of starch. Solubility of biomolecules (starch, fibers, proteins, or sugars) before or after processing in the presence of an excess of water is known as the water solubility index. At different plasma power and treatment time of CP, there is an increase in the percent solubility of the starch. Higher solubility of treated starch has been observed compared to untreated starch. This is due to the formation of smaller fragments due to depolymerization or partial breakdown leading to higher swelling power. Bie et al. (2016) studied the degradation of molecules and oxidation of starch molecules by the application of plasma technology, which showed an increase in solubility. In a study conducted by Thirumdaset et al. (2016) it was noted that depending on the nature of the gas used the hydrophilic nature of the starch molecule is altered (i.e., either it may increase or decrease). In their study on rice starch, the incorporation of the OH molecule leads to the development of hydrophilicity. This was confirmed using FTIR analysis.

21.4.2 Cooking Properties

Cooking is an important parameter governing the end consumption of any foodstuff. The technology-driven world demands a reduction in the time and cost of the product being consumed. Therefore, CP offers an innovative strategy to reduce the cooking time of cereals. Thirumdas et al. (2016) studied the influence of CP on cooking parameters of brown rice. A significant reduction in the cooking time was observed with the increase in power and time of treatment of the brown rice. The outermost layer of brown rice consists of the layer of bran, which is etched due to the formation of reactive species formed by the plasma. The easier penetration of water occurs due to the opening of the bran layer, thus leading to the reduction in cooking time. The reduction in cooking time from 29 to 21 min was observed with an increase in the power to 50 W and 10 min duration. This phenomenon can be attributed to the addition of polar groups between the starch molecules. In the case of bamboo rice studied by Potluri et al. (2018), there was a reduction in cooking time to 12 min, and the soaking rate was increased by 15%.

Cooking time was inversely related to water uptake. A higher water uptake ratio was responsible for the least cooking time of the rice. Reduction in the time of cooking also led to an increase in the cooking loss due to easier leaching of the fragmented starch molecule.

21.4.3 Degree of Gelatinization

Gelatinization refers to the irreversible swelling of the granules due to hydration leading to destruction of the molecular order and solubilization of starch (Bauer and Knorr, 2004; Zhang et al., 2015). When starch is heated up to 80°C in the presence of an excess of water the process of gelatinization is initiated. The amylopectin loses its crystalline structure and the amylose leaches out starch granules upon gelatinization (Svihus, Uhlen, & Harstad, 2005). The CP treatment leads to an increase in the degree of gelatinization after CP treatment. It was observed in the study conducted by Thirumdas et al. (2016) that in the CP-treated brown rice sample, the degree of gelatinization increased by 40%. The increment in the degree of gelatinization illustrates the reduction in cooking time and increment in the water uptake. Cold plasma can either lead to an increase or decrease in the gelatinization temperature. As observed by Bie et al. (2016) in their study, there was a decrease in enthalpy and an increase in the temperature of gelatinization upon treatment of starches with an oxygen and helium glow plasma. Numerous advantages occur due to CP treatment (i.e., lead to increase in the surface energy, meaning disruption of intermolecular bonds that occurs when a surface is created), the addition of new functional groups, depolymerization, and modification in the hydrophilicity of the component being treated (Wongsagonsup et al., 2014). The thermal properties of CP-treated whole wheat grain and wheat flour have been analyzed. In the study by Chaple et al. (2020) on the CP treatment of the sample, all gelatinization temperatures (onset, peak, and conclusion) were found to have increased. This means a higher amount of energy is needed to begin the gelatinization of starch (Wang et al., 2010). There was a slight reduction in gelatinization enthalpy (ΔH) post CP treatment by 1.3 J/g. They concluded that the enthalpy (ΔH) leads to a loss in the double-helical structure of the molecules of the starch. This indicates that the CP-treated starch consumes less energy for gelatinization. The decrease in crystallinity of starch after the plasma treatment denotes the decrease in the enthalpy (ΔH). In the study conducted by Thirumdas et al. (2016) on the gelatinization of rice starch at 60 W for 10 min, there was a reduction in gelatinization temperature in comparison to the untreated sample. This decrease can be attributed to depolymerization or a change in the amylose/amylopectin ratio of starch granules due to the presence of plasma strains.

21.4.4 Rheological Attributes

Pasting refers to heating and cooling (temperature range of 50–95°C) of starch–water mixtures to a programmed cycle under a constant shearing force (Zhu, 2015). Pasting performance of flours during cooking and cooling includes many processes, namely swelling, deformation, fragmentation, and solubilization, which occur in a very complex media whose viscoelastic properties in the pasted and gelled states are governed primarily by the volume occupied by the swollen particles (Collar, 2016). The instant increase in the value of viscosity through a constant heating process occurring at a specified temperature is known as pasting temperature. It decreased with an increase in plasma voltage and time in the study conducted by Thirumdas et al. (2016). Chaple et al. (2020) observed that with the increase in the duration of the CP treatment the peak viscosity and final viscosity increased in the case of both plasma-treated whole wheat and whole wheat flour.

The cross-linking of starch occurs due to oxidation leading to an increase in the peak viscosities (Lee, Hong, Lee, Chung, & Lim, 2015). The starch granules become loose leading to destruction by the action of plasma reactive species, which results in greater swelling and therefore higher peak viscosity (Pal et al., 2016). Breakdown viscosity increased from 644 cP to 1077 cP and final viscosity from 501 cP to 2096 cP during the cooling process as a result of re-formation of bonds between molecules of chains. The viscosity also increases due to the disruption of hydrogen bonds due to the application of plasma. This attribute indicates that the depolymerization of starch occurs in wheat flour due to plasma thus leading to alteration in the degree of retrogradation of starches.

The higher power and time duration of plasma leads to a significant change in the dough strength. This is due to the alteration in the functionality of the flour due to modification in protein and lipid content. This was studied by Misra et al. (2015) through the formation of disulfide bonds and impact on the volume of the loaf (Menkovska, Mangova, & Dimitrov, 2014). However, there is still a need to optimize conditions to improve the dough strength to be used industrially for new product development. Cold plasma offers one of the most promising measures for the modification of starch resulting in lower final viscosity and higher pasting temperature (Potluri et al., 2018). There was a reduction in viscosity from 373 cP to 306 cP due to chemical modification happening in the rice starch. Zou et al. (2004) justified that the reactive species generated by plasma results in excitation of the chemical groups lead to the necessary modification. And the cross-linking of starch combined with decreased retrogradation and increase in gelatinization temperature lead to a reduction in the mobility of amorphous chains due to the formation of intermolecular bridges. The increase in the time and treatment intensity of the CP leads to an increase in the degree of cross-linking, resulting in lower solubility and swelling power of the granules of the starch. This phenomenon also leads to the reduction of the number of granules of starch available for gelatinization, resulting in an increased pasting profile for the plasma-treated samples in comparison to the non-plasma-treated starch.

21.4.5 Crystallinity

This technique is employed to analyze the crystalline organization of starch (Zobel et al., 1988). It provides information about the form of starch present, quality, and size of the crystallites (Muhrbeck, 1991). X-ray diffraction techniques give information about the polymorphic form of starch present, and it is also possible to obtain information about the quality and size of the crystallites. The structure of starch is described as A, B, C type crystalline in nature depending on the branching and structure of the amylopectin side chain (Zhang et al., 2014). In the study conducted by Thirumdas et al. (2016), the intensity of the peak in the case of plasma-treated brown rice was reduced with the similar A-type (present at the outside of the branched molecule and possess only an α (1–6) linkage to B_1 chain pattern (Hizukuri, 1986). This peak reduction can be associated with the depolymerization of starch due to plasma treatment. The low-pressure energy that is employed in plasma treatment results in the modification of only surface characteristics, which is restricted only to some molecular layers deep and does not alter the bulk properties of the substance being treated (Wolf and Sparavignak, 2010). From studies, it has been concluded that the effect of CP is system dependent and the process parameters applied like treatment time, mode of treatment, and the substrate.

21.4.6 Surface Morphology by Scanning Electron Microscope

The surface morphology of CP-treated samples can be examined through scanning electron microscopy (SEM). It is one of the most promising techniques to study the surface characteristics of a sample modified through thermal or non-thermal technology. As in the case of other non-hermal techniques, CP also led to the creation of fissures and cracks (Lii et al., 2003). Etching caused by plasma is the most important attribute contributing to modification in the surface morphology of the grain or starch structure. The process of removal of material from the surface of the cereal or any substance is known as etching. It was found in the study conducted by Liu et al. (2021) that the surface morphology of untreated rice kernels remains unchanged and was smooth and regular. However, after the CP treatment of 120 W for the 60 s, some irregular structures appeared on the surface of the rice kernel. When the same kernel was treated at higher intensity of 220 W and 60 s, the formation of cavity and cracks was observed with the adherence of smaller particles to form bigger piles on the surface. Thus, from the study it can be concluded that the morphology of the substance can be altered with the increase in power. After being treated at 220 W CP for 60 s, cavities and cracks appeared, and some small particles aggregated together to form bigger piles on the surface. These results demonstrated that the surface structure of an intact kernel may be altered by different power strengths of helium plasma, and the aggregation of small particles led to the rough surface of the rice kernel. Surface morphology characterization of bamboo rice was conducted by Potluri et al. (2018), which revealed that the untreated rice had a homogeneous surface in comparison to CP treatment rice. At 25 W/cm^2 there was a higher etching of the surface leading to increased water absorption (formation of cracks and pits) resulting in reduced cooking time. Thirumdas et al. (2016) examined the influence of CP on the surface morphology of brown rice. Plasma-treated brown rice has fissures and shallow depressions and increased grain roughness. This is due to the breakage of covalent bonds of the surface leading to the formation of volatile monomers that can be vaporized and removed by the vacuum. Preferential etching occurs due to the difference in etching rates occurring on the surface (Taylor and Wolf, 1980).

21.4.7 FTIR

Fourier Transform Infrared (FTIR) spectroscopy is a technique used to obtain the infrared spectrum of absorption, emission, and photoconductivity of solid, liquid, and gas molecules. In the CP application of cereal grains, FTIR can be utilized to study the reduction in cooking time of cereal grains using the absorption spectra of peak intensity. Moisture content can be assessed from absorptions occurring due to vibrations of water molecules at 1630 cm^{-1} bending (Burneau, et al., 1990). The FTIR spectra of Thirumdas et al. (2016) peak at 1640 cm^{-1}, which shows the increase in the absorbance intensities after the plasma treatment of rice. The increase in peaks around that wavenumber may be due to O-H stretching occurring because of the vibration of water molecules.

Deeyal et al. (2013) reported that the peak 1630 cm^{-1} was a result of vibrations of water molecules absorbed in the non-crystalline region of starch molecules after the atmospheric argon plasma treatment.

21.5 EFFECT OF COLD PLASMA ON THE BIOLOGICAL PROPERTIES OF CEREALS

Despite the ideal application potential of CP such as microbial decontamination, insect disinfestation, pesticide dissipation, mycotoxin reduction, germination, and drying enhancement, the changes in the bioactive profile must be optimal. Digestibility of the cereal and the presence of antagonistic substances known as anti-nutritional factors determine the availability of nutrients on consumption. The morphology and molecular structures of the nutrients greatly influence the biological characteristics of cereals. Since CP majorly causes modifications in the surface morphology, a substantial impact on the biological attributes can be expected. Thus, in this section, the effect of CP treatment on biological characteristics of cereals such as nutrient digestibility and anti-nutritional factors are discussed.

21.5.1 Effect on Nutrient Digestibility

Whole grain cereals contain an amalgam of macro- (carbohydrates, protein, and fat) and micronutrients (vitamins and minerals). However, the selected cereals for everyday consumption are widely processed into various forms to enhance convenience as well as palatability, thereby diminishing the nutrients that are present in microproportions and making the cereals carbohydrate-rich as they comprise almost 75% (McKevith, 2004). As starch is an abundant carbohydrate that exists in cereals and starch digestibility plays a vital role in determining the application of the cereal for specific product development, most of the CP studies on cereals focused on investigating the effect on starch digestibility. Starch in cereals is enzymatically digested or hydrolyzed to glucose, which yields energy to the body via the glycolytic pathway (Bhalla, Thakur, Seth, Pratush, & Summerhill, 2009). Based on the structure of the starch, the rate of hydrolysis may change, and accordingly, starch is fractionated into rapidly digestible (RDS) starch that can be hydrolyzed within 20 min and slowly digestible (SDS) starch that can take up to 120 min for digestion (Aarathi, Urooj, & Puttaraj, 2003). The portion of the starch inaccessible to digestion in the small intestine is called the resistant starch (RS), which is the recent component of interest as it promotes gut health and helps in the development of low glycemic products.

Considering the micrometer penetration and surface-level modification of CP, to delineate the impact of CP on the digestibility of cereal starches, most of the existing studies carried out the plasma modifications on the starch isolates from different cereal sources such as wheat, maize, barley, sorghum, buckwheat, and quinoa along with other cereals and cereal products like brown rice and wheat flour. When the cereal or the cereal product is subjected to CP treatment, the dynamic plasma species including the free radicals, reactive oxygen, and nitrogen species interact with the starch granules resulting in two different aftereffects such as the enhancement and the retardation of digestibility. The renowned surface etching phenomenon of the CP causes scorching of the starch granular surface and promotes the leaching out of the polymeric units amylose and amylopectin facilitating increased enzymatic hydrolysis (Lee et al., 2016) consequently the digestibility. This may be expressed with the rise in rapidly digestible starch (RDS) and a corresponding reduction in the resistant starch content (Gao et al., 2019). The surface deformation may also reduce the hardness of the cereal. Moreover, looking into the crystallinity attributes in relation to the amylopectin-to-amylose ratio as they influence the digestibility significantly, it can be seen that oxidative plasma species depolymerize the starch components bringing in the reduction in the amylose content and an increase in the crystallinity (Thirumdas, Kadam, & Annapure, 2017). This tends to give a more ordered lamellar structure of the starch and enables pronounced enzymatic hydrolysis.

However, the depolymerized starch yields low molecular weight oligomeric fragments of both amylose and amylopectin. These depolymerized compounds undergo intramolecular interactions; for example, amylose–amylose, amylopectin–amylopectin, and amylose–amylopectin engender double-helical complex structures, which thereby shield or inhibit the enzymatic starch hydrolysis. In such a case, the rapidly digestible starch may decrease with an increment in the slowly digestible starch and resistant starch content. Also, the depolymerized starch compounds may aggregate in conjunction with one another causing an anomaly in the morphology and escalating the recrystallization resulting in an increase in resistant starch. Dielectric barrier discharge plasma (DBD) treatment of brown rice for 10–20 min was observed with increased rate of glucose release (Lee et al., 2016). Similarly, plasma modification of buckwheat, sorghum, wheat, and quinoa starch with DBD of 20 kV for just 30 s delivers easily digestible attributes to the treated starches (Gao et al., 2019). The SEM of the studied starches have shown the eroded structure after plasma treatment. Owing to this evidence, the authors explicated that the eroded starch leaches out the enzyme activation sites of the polymer and thus enhances the digestibility. Nevertheless, considering some of the other works of literature on the same cereal with different plasma reactors or with different compounding physical or chemical modifications, the obtained digestibility results are quite contrary to this study. Gao et al. (2021) reported a rise in the resistant starch content of buckwheat starch after plasma treatment with the same DBD system; however, the starch was pre-gelatinized before plasma application and complexed with polyphenolic quercetin. Similarly (Shi, Wang, Ji, Yan, & Liu, 2022) observed an increase in wheat-resistant starch content of wheat flour after the plasma-activated water treatment along with heat moisture modification. The presence of other macronutrients such as protein and lipids tends to easily form a complex with the surface-eroded starch molecule and resist the hydrolysis giving rise to higher resistant starch content. Studies on waxy and normal maize (Yan, Feng, Shi, Cui, & Liu, 2020) as well as barley starch (Shen et al., 2021) exhibit parallel results of reduced rapidly digestible starch and marked-up resistant starch fractions. Thus, in the case of applications requiring increase in resistance starch (i.e., reduction in starch hydrolysis), plasma treatment can be combined with other mild physical or chemical modifications like dry heat, heat, and moisture, and modified citrate, respectively. Yet cooking can also be considered as a heat and moisture treatment incurring a protrusion in the resistant starch content. Applying plasma treatment to ready-to-eat cereals for the diabetic population may not be suitable as the reduction in starch hydrolysis is inconsistent. Depending on the type of plasma reactor, mode of application, treatment time, plasma power, operating pressure, and also the physical form and structure of the cereal, the nutrient digestibility may differ. This necessitates further research on optimizing the process parameters as per the requirement.

21.5.2 Effect on Bioactive Profile

Along with the major nutritive components, minor nutrients such as vitamins and minerals including the secondary metabolites, phenolics, and anthocyanins present in some of the whole grains enrich the bioactive profile of cereals (Kaur, Jha, Sabikhi, & Singh, 2014). Being free radical scavengers, these secondary metabolites contribute to the antioxidant capacity of the food. Nevertheless, because of their presence in trivial quantity in cereals and vulnerability to severe processing conditions, the effect of cereal processing on these bioactive compounds has not been explored much. Special rice varieties include, but are not exclusively, brown rice (Pandey, Lijini, & Jayadeep, 2017), Indica rice, and Thai aromatic rice, which have studied for their medicinal and nutraceutical significance (Bhat et al., 2020) considering their bioactive profile. Additionally, the role of germination is remarkable in enhancing bioactivity through metabolic and enzymatic modifications (Cho & Lim, 2016). Since CP has been a known technology for boosting seed germination in terms of their root and shoot length by surface modification of the grains, which results in increased wettability, water, and nutrient uptake, it has also been investigated for its effect on the bioactivity of the selected germinated grain varieties, namely rice (Yodpitak et al., 2019; Park, Puligundla, & Mok, 2020; Zargarchi & Saremnezhad, 2019; Sookwong et al., 2014) wheat, and oat (Sera, Spatenka, Šerý, Vrchotova, & Hruskova, 2010).

The total phenolic content of plasma-treated germinated grains was found to be higher compared to the untreated germinated grains, which in turn engenders higher antioxidant activity indicated by ABTS radical scavenging activity as well as DPPH radical scavenging activity. There are three possible mechanisms for the increase in the total phenolic components after CP application: i) when the seed grains are subjected to CP treatment, the turbulent plasma species collide on the seed surface ensuing fissures and erosions, which tend to help the cell-bound metabolites emanate freely thereby increasing their content; ii) the interaction of these reactive species with the seed cell elements causes modifications in the metabolisms involved while germination thus increases the total phenolic content; and iii) also, in response to the plasma-induced physiological stress reaction to plant tissue, specific secondary metabolites are synthesized, which could also account for the increase in total phenolic content. In comparison with the untreated germinated grain, CP pre-treatment increases the phenolic content in fewer days of germination, and coupled with its stability as in untreated grains, drastic reduction in the phenolics was observed after 1.5 days of germination (Yodpitak et al., 2019). Moreover, vitamin E analogs were also increased in the plasma-treated germinated grains. However, there were no significant increments in the total anthocyanin and ɤ-oryzanol content in the same study concerning plasma treatment. In contrast, plasma treatment of raw rice berry rice flour improved the total anthocyanin content; notably specific anthocyanins like peonidin 3-glucoside and cyanidin 3-glucoside were increased (Settapramote, Laokuldilok, Boonyawan, & Utama-ang, 2021). Similarly, plasma-treated raw bamboo rice flour exhibited increased phytosterol (β-sitosterol) content, which is due to the morphological as well as molecular changes occurring (Potluri, Sangeetha, Santhosh, Nivas, & Mahendran, 2018).

Due to the surface etching phenomenon, CP pre-treatment expedites better absorption and bioavailability of minerals as observed in the case of iron fortification of rice (Akasapu, Ojah, Gupta, Choudhury, & Mishra, 2020). Plasma-treated rice grains expressed higher iron retention in both acidic and alkaline conditions after more than 30 days of storage and increased in vitro bioavailability of the fortified iron content. Regardless of the discussed studies, more investigations on the interaction of CP with the bioactive complexes are needed.

21.5.3 Effect on Anti-nutritional Factors

Anti-nutritional factors (ANF) or non-nutritive substances are compounds affecting the absorption of the nutrients thereby reducing their bioavailability in the human digestive system. The presence of such compounds hampers the nutritional profile of the food. Some of the anti-nutritional compounds present in cereals, namely α-amylase inhibitors, protease inhibitors (trypsin/chymotrypsin inhibitors), phytates, and lectins (haemagglutinins), are eliminated widely during the milling and other secondary processing operations as these ANFs commonly occur in the exterior layers of the grains (Nadeem et al., 2010). Thus, there have been no extensive studies on the effect of CP on the ANFs in the cereal grains. However, certain studies explain the modifications in their enzymatic counterparts, which could explicate the possible effect on the non-nutritive substances. For example, changes in the α-amylase activity can be inversely correlated to the α-amylase inhibitor content. The α-amylase activity of germinated brown rice was increased by 162% after CP treatment of 10 min (Chen et al., 2016). Normally during seed germination, α-amylase in the aleurone or bran layer is instigated to hydrolyze the reserved starch, which may be exhibited as the increase in α-amylase value of grains during germination (Kaneko, Itoh, Ueguchi-Tanaka, Ashikari, & Matsuoka, 2002). Since CP treatment augments the germination rate by surface modification phenomenon, the fissured morphology of brans also substantially facilitates the α-amylase activity increment. This was further proven in plasma-treated brown rice (Lee et al., 2016) even without the germination process. However, it was observed that after 5 min of treatment, when the treatment time was increased the α-amylase activity was reduced; however, the reason for which has yet to be investigated. Furthermore, a reduction in the activity of trypsin inhibitors was also observed in plasma-treated soybeans (Li et al., 2017). Despite being a study on pulses, this investigation looked at the effect of CP on the inactivation of ANFs. Also, previous subsections on the digestibility and bioactive profile of plasma-treated cereals elucidate the potential of CP for better absorption and bioavailability of nutrients that may be hostile to the mechanism of ANFs. In that way, CP could also play an antagonistic role againist ANFs, although further investigations are needed.

21.6 CONCLUSION

Cold plasma treatment of cereal grains is an underexplored area of research despite the prospective applications, namely microbial decontamination, disinfestation, pesticide reduction, mycotoxin inactivation, and also for enhancing germination. Because of the marginal penetrating potential of CP and the heterogeneous morphology of cereals, the influence on the nutritive, biological, and functional properties is limited yet appreciable. However, the discussed studies showed the ability of CP to significantly enrich the qualitative attributes of cereal and cereal products in addition to ensuring microbial and chemical safety. Also, the involvement of numerous process parameters (i.e., type of plasma reactor and gas used, mode of treatment, gas flow rate, operating pressure, plasma power, and exposure time) and product variables (i.e., particle size, the proportion of constituents, surface homogeneity) facilitate tailoring the outcomes as per the requirements by optimizing the influential factors. Nevertheless, a lack of investigations on various cereal crops and their products makes the data on the interactions ambiguous. Moreover, certain limitations pertaining to CP treatment such as inadequate large-scale operations, need for toxicological assessment, and optimization for the plasma species concentrations concerning the operating conditions need to be addressed in future investigations.

REFERENCES

Aarathi, A., Urooj, A., &Puttaraj, S. (2003). In vitro starch digestibility and nutritionally important starch fractions in cereals and their mixtures. *Starch-Stärke*, 55(2), 94–99.

Akasapu, K., Ojah, N., Gupta, A. K., Choudhury, A. J., & Mishra, P. (2020). An innovative approach for iron fortification of rice using cold plasma. *Food Research International*, 136, 109599.

Bahrami, N., Bayliss, D., Chope, G., Penson, S., Perehinec, T., & Fisk, I. D. (2016). Cold plasma: A new technology to modify wheat flour functionality. *Food Chemistry*, 202, 247–253.

Bai, Y., Chen, J., Mu, H., Zhang, C., & Li, B. (2009). Reduction of dichlorvos and omethoate residues by O2 plasma treatment. *Journal of Agricultural and Food Chemistry*, *57*(14), 6238–6245.

Barbosa-Canovas, G. V., Pothakamury, U. R., & Swanson, B. G. (1995). *State of the art technologies for the stabilization of foods by non-thermal processes: physical methods. Food Preservation by Moisture Control* (GV Barbosa-Ca'novasand J. Welt-Chanes, eds.). Technomic Publishing Co., Lancaster and Basel, 493–532.

Bauer, B. A., & Knorr, D. (2004). Electrical conductivity: A new tool for the determination of high hydrostatic pressure-induced starch gelatinisation. *Innovative Food Science & Emerging Technologies*, 5(4), 437–442.

Bhalla, T., Thakur, N., Seth, A., Pratush, A., & Summerhill, S. (2009). *Cereal based alcoholic beverages.* book: Fundamentals of Food Biotechnology, New Delhi. India.
Bhat, F. M., Sommano, S. R., Riar, C. S., Seesuriyachan, P., Chaiyaso, T., & Prom-u-Thai, C. (2020). Status of bioactive compounds from bran of pigmented traditional rice varieties and their scope in production of medicinal food with nutraceutical importance. *Agronomy*, 10(11), 1817.
Bie, P., Pu, H., Zhang, B., Su, J., Chen, L., & Li, X. (2016). Structural characteristics and rheological properties of plasma-treated starch. *Innovative Food Science & Emerging Technologies*, 34, 196–204.
Bourke, P., Ziuzina, D., Boehm, D., Cullen, P. J., & Keener, K. (2018). The Potential of Cold Plasma for Safe and Sustainable Food Production. *Trendsin Biotechnology*, 36(6), 615–626. https://doi.org/10.1016/j.tibtech.2017.11.001.
Burneau, A., Barres, O., Gallas, J. P., & Lavalley, J. C. (1990). Comparative study of the surface hydroxyl groups of fumed and precipitated silicas. 2. Characterization by infrared spectroscopy of the interactions with water. *Langmuir*, *6*(8), 1364–1372.
Chaple, S., Sarangapani, C., Jones, J., Carey, E., Causeret, L., Genson, A., & Bourke, P. (2020). Effect of atmospheric cold plasma on the functional properties of whole wheat (Triticum aestivum L.) grain and wheat flour. *Innovative Food Science & Emerging Technologies*, *66*, 102529.
Chen, H. H. (2014). Investigation of properties of long-grain brown rice treated by low-pressure plasma. *Food and Bioprocess Technology*, 7(9), 2484–2491.
Chen, H. H., Chang, H. C., Chen, Y. K., Hung, C. L., Lin, S. Y., & Chen, Y. S. (2016). An improved process for high nutrition of germinated brown rice production: Low-pressure plasma. *Food Chemistry*, 191, 120–127.
Chen, H. H., Chen, Y. K., & Chang, H. C. (2012). Evaluation of physicochemical properties of plasma treated brown rice. *Food Chemistry*, 135(1), 74–79.
Chen, Y. S., Zhang, X. S., Dai, Y. C., & Yuan, W. K. (2004). Pulsed high-voltage discharge plasma for degradation of phenol in aqueous solution. *Separation and Purification Technology*, *34*(1–3), 5–12
Cho, D.-H., & Lim, S.-T. (2016). Germinated brown rice and its bio-functional compounds. *Food Chemistry*, 196, 259–271.
Collar, C. (2016). Impact of visco-metric profile of composite dough matrices on starch digestibility and firming and retrogradation kinetics of breads thereof: Additive and interactive effects of non-wheat flours. *Journal of Cereal Science*, *69*, 32–39.
Damodaran, S. (2017). *Food proteins and their applications*: Routledge.
de la Hera, E., Gomez, M., &Rosell, C. M. (2013). Particle size distribution of rice flour affecting the starch enzymatic hydrolysis and hydration properties. *Carbohydrate Polymers*, *98*(1), 421–427.
Deeyai, P., Suphantharika, M., Wongsagonsup, R., & Dangtip, S. (2013). Characterization of modified tapioca starch in atmospheric argon plasma under diverse humidity by FTIR spectroscopy. *Chinese Physics Letters*, *30*(1), 018103.
Feizollahi, E., Iqdiam, B., Vasanthan, T., Thilakarathna, M. S., &Roopesh, M. S. (2020). Effects of atmospheric-pressure cold plasma treatment on deoxynivalenol degradation, quality parameters, and germination of barley grains. *Applied Sciences*, *10*(10), 3530.
Feng, X., Ma, X., Liu, H., Xie, J., & Fan, R. (2019). Argon plasma effects on maize: pesticide degradation and quality changes (June). https://doi.org/10.1002/jsfa.9810
Ganesan, A. R., Tiwari, U., Ezhilarasi, P. N., &Rajauria, G. (2021). Application of cold plasma on food matrices: A review on current and future prospects. *Journal of Food Processing and Preservation*, 45(1), 1–16. https://doi.org/10.1111/jfpp.15070
Gao, S., Liu, H., Sun, L., Cao, J., Yang, J., Lu, M., & Wang, M. (2021). Rheological, thermal and in vitro digestibility properties on complex of plasma modified Tartary buckwheat starches with quercetin. F*ood Hydrocolloids*, 110, 106209.
Gao, S., Liu, H., Sun, L., Liu, N., Wang, J., Huang, Y., Zhang, X. (2019). The effects of dielectric barrier discharge plasma on physicochemical and digestion properties of starch. *International Journal of Biological Macromolecules*,138, 819–830.
Gavahian, M., & Cullen, P. J. (2020). Cold Plasma as an Emerging Technique for Mycotoxin-Free Food: Efficacy, Mechanisms, and Trends Cold Plasma as an Emerging Technique for Mycotoxin-Free Food: Efficacy, Mechanisms, and Trends. *Food Reviews International*, 36(2),193–214. https://doi.org/10.1080/87559129.2019.1630638

Hizukuri, S. (1986). Polymodal distribution of the chain lengths of amylopectins, and its significance. *Carbohydr. Res.*, 147, 342–347.

Kaneko, M., Itoh, H., Ueguchi-Tanaka, M., Ashikari, M., & Matsuoka, M. (2002). The α-amylase induction in endosperm during rice seed germination is caused by gibberellin synthesized in epithelium. *Plant Physiology*, 128(4), 1264–1270.

Karunakaran, C., Muir, W., Jayas, D., White, N., & Abramson, D. (2001). Safe storage time of high moisture wheat. *Journal of Stored Products Research*, 37(3), 303–312.

Kaur, K. D., Jha, A., Sabikhi, L., & Singh, A. (2014). Significance of coarse cereals in health and nutrition: a review. *Journal of Food Science and Technology*, 51(8), 1429–1441.

Khamsen, N., Onwimol, D., Teerakawanich, N., Dechanupaprittha, S., Kanokbannakorn, W., Hongesombut, K., &Srisonphan, S. (2016). Rice (Oryza sativa L.) Seed Sterilization and Germination Enhancement via Atmospheric Hybrid Nonthermal Discharge Plasma. *ACS Applied Materials and Interfaces*, 8(30), 19268–19275. https://doi.org/10.1021/acsami.6b04555

Laroque, D. A., Seó, S. T., Valencia, G. A., Laurindo, J. B., & Carciofi, B. A. M. (2022). Cold plasma in food processing: Design, mechanisms, and application. *Journal of Food Engineering*, 312(July 2021), 110748. https://doi.org/10.1016/j.jfoodeng.2021.110748

Lee, J. H., Woo, K. S., Jo, C., Jeong, H. S., Lee, S. K., Lee, B. W, Khun, J. (2019). Quality Evaluation of Rice Treated by High Hydrostatic Pressure and Atmospheric Pressure Plasma. *Journal of Food Quality*, 2019. https://doi.org/10.1155/2019/4253701.

Lee, K. H., Kim, H.-J., Woo, K. S., Jo, C., Kim, J.-K., Kim, S. H., Kim, W. H. (2016). Evaluation of cold plasma treatments for improved microbial and physicochemical qualities of brown rice. *LWT-Food Science and Technology,* 73, 442–447.

Lee, K. H., Woo, K. S., Yong, H. I., Jo, C., Lee, S. K., Lee, B. W., Kim, H. J. (2018). Assessment of microbial safety and quality changes of brown and white cooked rice treated with atmospheric pressure plasma. *Food Science and Biotechnology*, 27(3), 661–667. https://doi.org/10.1007/s10068-017-0297-6.

Lee, S. J., Hong, J. Y., Lee, E. J., Chung, H. J., & Lim, S. T. (2015). Impact of single and dual modifications on physicochemical properties of japonica and indica rice starches. *Carbohydrate Polymers*, 122, 77–83.

Li, J., Xiang, Q., Liu, X., Ding, T., Zhang, X., Zhai, Y., & Bai, Y. (2017). Inactivation of soybean trypsin inhibitor by dielectric-barrier discharge (DBD) plasma. *Food Chemistry*, 232, 515–522.

Li, Y., Wang, T., Meng, Y., Qu, G., Sun, Q., Liang, D., & Hu, S. (2017). Air Atmospheric Dielectric Barrier Discharge Plasma Induced Germination and Growth Enhancement of Wheat Seed. *Plasma Chemistry and Plasma Processing*, 37(6), 1621–1634. https://doi.org/10.1007/s11090-017-9835-5

Lii, C. Y., Liao, C. D., Stobinski, L., &Tomasik, P. (2003). Effect of corona discharges on granular starches. *Journal of Food Agriculture and Environment*, 1(2), 143–149.

Liu, J., Wang, R., Chen, Z., & Li, X. (2021). Effect of Cold Plasma Treatment on Cooking, Thermomechanical and Surface Structural Properties of Chinese Milled Rice. *Food and Bioprocess Technology*, 14(5), 866–886.

Los, A., Ziuzina, D., Boehm, D., & Bourke, P. (2020). Effects of cold plasma on wheat grain microbiome and antimicrobial efficacy against challenge pathogens and their resistance. *International Journal of Food Microbiology*, 335(April), 108889. https://doi.org/10.1016/j.ijfoodmicro.2020.108889

Los, A., Ziuzina, D., van Cleynenbreugel, R., Boehm, D., & Bourke, P. (2020). Assessing the biological safety of atmospheric cold plasma treated wheat using cell and insect models. *Foods*, 9(7). https://doi.org/10.3390/foods9070898.

Los, Agata; Ziuzina, Dana; Akkermans, Simen; Boehm, D. (2014). Improving microbiological safety and quality characteristics of wheat and barley by high voltage atmospheric cold plasma closed processing. *Paper Knowledge . Toward a Media History of Documents*, 3, 1–49.

Madathil, R. V., ThirugnanasambandanKalaivendan, R. G., Paul, A., & Radhakrishnan, M. (2021). In package control of Rhyzopertha dominica in wheat using a continuous atmospheric jet cold plasma system. *Frontiers in Advanced Materials Research*, 10–25. https://doi.org/10.34256/famr2112.

McKevith, B. (2004). Nutritional aspects of cereals. *Nutrition Bulletin*, 29(2), 111–142.

Menkovska, M., Mangova, M., & Dimitrov, K. (2014). Effect of cold plasma on wheat flour and bread making quality. *Macedonian Journal of Animal Science*, 4(1), 27–30.

Misra, N. N., Barun, Y., Roopesh, M. S., & Jo, C. (2019). Cold Plasma for Effective Fungal and Mycotoxin Control in Foods: Mechanisms, Inactivation Effects, and Applications, 18, 106–120. https://doi.org/10.1111/1541-4337.12398.

Misra, N. N., Kaur, S., Tiwari, B. K., Kaur, A., Singh, N., & Cullen, P. J. (2015). Atmospheric pressure cold plasma (ACP) treatment of wheat flour. *Food Hydrocolloids*, 44, 115–121.

Muhrbeck, P., & Eliasson, A. C. (1991). Influence of the naturally occurring phosphate esters on the crystallinity of potato starch. *Journal of the Science of Food and Agriculture*, 55(1), 13–18.

Nadeem, M., Anjum, F. M., Amir, R. M., Khan, M. R., Hussain, S., &Javed, M. S. (2010). An overview of anti-nutritional factors in cereal grains with special reference to wheat-A review. *Pakistan Journal of Food Sciences*, 20(1–4), 54–61.

Pal, P., Kaur, P., Singh, N., Kaur, A., Misra, N. N., Tiwari, B. K., & Virdi, A. S. (2016). Effect of nonthermal plasma on physico-chemical, amino acid composition, pasting and protein characteristics of short and long grain rice flour. *Food Research International*, *81*, 50–57.

Pandey, S., Lijini, K., & Jayadeep, A. (2017). Medicinal and health benefits of brown rice. In *Brown Rice* (pp. 111–122): Springer.

Pankaj, S. K., Wan, Z., & Keener, K. M. (2018). Effects of cold plasma on food quality: A review. *Foods*, 7(1), 4.

Panngom, K., Lee, S. H., Park, D. H., Sim, G. B., Kim, Y. H., Uhm, H. S., & Choi, E. H. (2014). Non-thermal plasma treatment diminishes fungal viability and up-regulates resistance genes in a plant host. *PloS one*, 9(6), e99300.

Park, H., Puligundla, P., &Mok, C. (2020). Cold plasma decontamination of brown rice grains: Impact on biochemical and sensory qualities of their corresponding seedlings and aqueous tea infusions. *LWT*, 131, 109508.

Park, J., Henins, I., Herrmann, H. W., Selwyn, G. S., & Hicks, R. F. (2001). Discharge phenomenaof an atmospheric pressure radio-frequency capacitive plasma source. *Journal of Applied Physics*, 89(1), 20–28.

Penado, K. N. M., Mahinay, C. L. S., & Culaba, I. B. (2018). Effect of atmospheric plasma treatment on seed germination of rice (Oryza sativa L.). *Japanese Journal of Applied Physics*, 57(1). https://doi.org/10.7567/JJAP.57.01AG08plasma for degradation of phenol in aqueous solution. *Separation and Purification Technology*, 34(1–3), 5–12.

Potluri, S., Sangeetha, K., Santhosh, R., Nivas, G., & Mahendran, R. (2018). Effect of low-pressure plasma on bamboo rice and its flour. *Journal of Food Processing and Preservation*, 42(12), e13846.

Ramanan, K. R., Sarumathi, R., & Mahendran, R. (2018). Influence of cold plasma on mortality rate of different life stages of Triboliumcastaneum on refined wheat flour. *Journal of Stored Products Research*, 77, 126–134.

Rifna, E. J., Ratish Ramanan, K., & Mahendran, R. (2019). Emerging technology applications for improving seed germination. *Trends in Food Science and Technology*, 86(February), 95–108. https://doi.org/10.1016/j.tifs.2019.02.029

Sarangapani, C., Danaher, M., Tiwari, B., Lu, P., Bourke, P., & Cullen, P. J. (2017). Efficacy andmechanistic insights into endocrine disruptor degradation using atmospheric air plasma. *Chemical Engineering Journal*, 326, 700–714.

Sarangapani, C., Devi, Y., Thirumdas, R., Annapure, U. S., & Deshmukh, R. R. (2015). Effect of low-pressure plasma on physico-chemical properties of parboiled rice. *LWT-Food Science and Technology*, 63(1), 452–460.

Sarangapani, C., Patange, A., Bourke, P., Keener, K., & Cullen, P. J. (2018). Recent Advances in the Application of Cold Plasma Technology in Foods. *Annual Review of Food Science and Technology*, 9, 609–629. https://doi.org/10.1146/annurev-food-030117-012517

Selcuk, M., Oksuz, L., &Basaran, P. (2008). Decontamination of grains and legumes infected with Aspergillus spp. and Penicillum spp. by cold plasma treatment. *Bioresource Technology*, 99(11), 5104–5109.

Sera, B., Spatenka, P., Šerý, M., Vrchotova, N., &Hruskova, I. (2010). Influence of plasma treatment on wheat and oat germination and early growth. *IEEE Transactions on Plasma Science*, 38(10), 2963–2968.

Settapramote, N., Laokuldilok, T., Boonyawan, D., & Utama-ang, N. (2021). Optimisation of the dielectric barrier discharge to produce Riceberry rice flour retained with high activities of bioactive compounds using plasma technology. *International Food Research Journal*, 28(2), 386–392.

Shanakhat, H., Sorrentino, A., Raiola, A., Romano, A., Masi, P., &Cavella, S. (2018). Current methods for mycotoxins analysis and innovative strategies for their reduction in cereals: an overview. *Journal of the Science of Food and Agriculture*, 98(11), 4003–4013. https://doi.org/10.1002/jsfa.8933

Shen, H., Ge, X., Zhang, B., Su, C., Zhang, Q., Jiang, H., Li, W. (2021). Understanding the multi-scale structure, physicochemical properties and in vitro digestibility of citrate naked barley starch induced by non-thermal plasma. *Food & Function*.

Shi, H., Ileleji, K., Stroshine, R. L., Keener, K., & Jensen, J. L. (2017). Reduction of aflatoxin in corn by high voltage atmospheric cold plasma. *Food and Bioprocess Technology*, *10*(6), 1042–1052.

Shi, M., Wang, F., Ji, X., Yan, Y., & Liu, Y. (2022). Effects of plasma-activated water and heat moisture treatment on the properties of wheat flour and dough. *International Journal of Food Science & Technology*.

Sookwong, P., Yodpitak, S., Doungkaew, J., Jurithayo, J., Boonyawan, D., &Mahatheeranont, S. (2014). Application of oxygen-argon plasma as a potential approach of improving the nutrition value of pre-germinated brown rice. *Journal of Food and Nutrition Research*, 2(12), 946–951.

Svihus, B., Uhlen, A. K., &Harstad, O. M. (2005). Effect of starch granule structure, associated components and processing on nutritive value of cereal starch: *A review. Animal Feed Science and Technology*, 122(3–4), 303–320.

Taylor, G. N., & Wolf, T. M. (1980). Oxygen plasma removal of thin polymer films. *Polymer Engineering & Science*, 20(16), 1087–1092.

Ten Bosch, L., Pfohl, K., Avramidis, G., Wieneke, S., Viöl, W., & Karlovsky, P. (2017). Plasma-based degradation of mycotoxins produced by Fusarium, Aspergillus and Alternaria species. *Toxins*, 9(3), 1–12. https://doi.org/10.3390/toxins9030097

Tendero, C., Tixier, C., Tristant, P., Desmaison, J., & Leprince, P. (2006). Atmospheric pressure plasmas: A review. *Spectrochimica Acta Part B: Atomic Spectroscopy*, 61(1), 2–30.

Thirumdas, R., Kadam, D., &Annapure, U. (2017). Cold plasma: an alternative technology for the starch modification. *Food Biophysics*, 12(1), 129–139.

Thirumdas, R., Saragapani, C., Ajinkya, M. T., Deshmukh, R. R., & Annapure, U. S. (2016). Influence of low pressure cold plasma on cooking and textural properties of brown rice. *Innovative Food Science & Emerging Technologies*, 37, 53–60.

Tiwari, B., Brennan, C. S., Curran, T., Gallagher, E., Cullen, P., & O'Donnell, C. (2010). Application of ozone in grain processing. *Journal of Cereal Science*, 51(3), 248–255.

Wang, L., Xie, B., Shi, J., Xue, S., Deng, Q., Wei, Y., & Tian, B. (2010). Physicochemical properties and structure of starches from Chinese rice cultivars. *Food Hydrocolloids*, 24(2–3), 208–216.

Wolf, R., &Sparavigna, A. C. (2010). Role of plasma surface treatments on wetting and adhesion. *Engineering*, 2(06), 397.

Wongsagonsup, R., Pujchakarn, T., Jitrakbumrung, S., Chaiwat, W., Fuongfuchat, A., Varavinit, S., and Suphantharika, M. (2014). Effect of cross-linking on physicochemical properties of tapioca starch and its application in soup product. *Carbohydrate Polymers*, *101*, 656–665.

Yan, Y., Feng, L., Shi, M., Cui, C., & Liu, Y. (2020). Effect of plasma-activated water on the structure and in vitro digestibility of waxy and normal maize starches during heat-moisture treatment. *Food Chemistry*, *306*, 125589.

Yodpitak, S., Mahatheeranont, S., Boonyawan, D., Sookwong, P., Roytrakul, S., &Norkaew, O. (2019). Cold plasma treatment to improve germination and enhance the bioactive phytochemical content of germinated brown rice. *Food Chemistry*, 289(September 2018), 328–339. https://doi.org/10.1016/j.foodchem.2019.03.061

Zahoranová, A., Hoppanová, L., Šimončicová, J., Tučeková, Z., Medvecká, V., Hudecová, D., Černák, M. (2018). Effect of Cold Atmospheric Pressure Plasma on Maize Seeds: Enhancement of Seedlings Growth and Surface Microorganisms Inactivation. *Plasma Chemistry and Plasma Processing*, 38(5), 969–988. https://doi.org/10.1007/s11090-018-9913-3.

Zargarchi, S., &Saremnezhad, S. (2019). Gamma-aminobutyric acid, phenolics and antioxidant capacity of germinated indica paddy rice as affected by low-pressure plasma treatment. *LWT*, 102, 291–294.

Zhang, B., Xiong, S., Li, X., Li, L., Xie, F., & Chen, L. (2014). Effect of oxygen glow plasma on supramolecular and molecular structures of starch and related mechanism. *Food Hydrocolloids*, 37, 69–76.

Zhang, J., Cheng, H. T., Xu, H., Xia, Y. J., Liu, C. X., & Xu, Z. J. (2015). Relationship between cooking-eating quality and subspecies differentiation in RILs population from indica and japonica crossing. *Chinese Journal of Rice Science*, 29, 167–173.

Zhu, F. (2015). Composition, structure, physicochemical properties, and modifications of cassava starch. *Carbohydrate Polymers*, *122*, 456–480.

Ziuzina, D., van Cleynenbreugel, R., Tersaruolo, C., & Bourke, P. (2021). Cold plasma for insect pest control: Triboliumcastaneum mortality and defense mechanisms in response to treatment. *Plasma Processes and Polymers* (August 2020), 1–12. https://doi.org/10.1002/ppap.202000178

Zobel, H. F. (1988). Starch crystal transformations and their industrial importance. *Starch-Stärke*, 40(1), 1–7.

Zou, J. J., Liu, C. J., & Eliasson, B. (2004). Modification of starch by glow discharge plasma. *Carbohydrate Polymers*, *55*(1), 23–26.

CHAPTER 22

Irradiation of Cereals

Purba Chakraborty
Dr. S.S. Bhatnagar University Institute of Chemical Engineering and Technology, Panjab University, Chandigarh, Punjab, India

CONTENTS

22.1 INTRODUCTION

The fast growth of the world's population, and the resulting rise in demand for limited food supplies, presents a severe challenge for the food industries with no simple solution. Furthermore, due to insufficient preservation procedures, there are considerable losses in a broad range of essential foods between production and consumption; this makes the discovery of prospective new technologies to remedy this condition an important area in which advancement may be very beneficial. A technique that may enhance the supply of food while remaining economically viable merits careful attention. Irradiation might be one of these processes.

Irradiation is being used effectively to suppress and eliminate dietary allergies and anti-nutritional components such as saponins and tannins (Diehl, 2002; Bhat et al., 2007). Other advantages of irradiation processing include minimal sample preparation, greater penetration level, negligible heat dissipation during processing (1 k Gy of radiation elevates the product temperature by 0.36°C), and absence of a catalyst (Diehl, 2002). Irradiation is currently utilized for three purposes: to control insects in cereals and legumes as an alternative to chemical fumigation; to restrict sprouting or other self-generating degrading processes; and to eliminate vegetative cells, particularly those that may cause human illness. As a consequence, safety, quality, and shelf life are enhanced (Figure 22.1).

DOI: 10.1201/9781003242192-27

Figure 22.1 Why do we need irradiation?

Microbial deterioration is typically not an issue when cereals and cereal products are kept at moisture levels of less than 14%. Insects are responsible for the majority of losses. Cereals may include microbial burdens (pathogens like *Salmonellae*) when used in wet formulations. This affects stock feed formulations since viruses might infect cattle for human consumption if they do not undergo heat treatment. Occasionally, grains and their products with moisture allow mould growth and development of mycotoxins (aflatoxins). Cereal irradiation is quite beneficial in combating all of the aforementioned issues. Cereals have a variety of functional characteristics that are critical to their processing and use. As a result, the final treatments must be determined by dosage levels that do not compromise these functional characteristics.

22.2 TYPES OF IRRADIATION

Only three types of irradiation are allowed to be used commercially for food preservation, according to the Codex Alimentarius Commission: X-rays (electromagnetic beams with a maximum energy of 5MeV), accelerated electrons (electronic beams generated in Van de Graaff generating capacity with a potential output of 10MeV), and radiation from high-energy gamma rays (obtained from radioactive elements such as cobalt-60 with a high penetrating power) (Diehl, 2002; Bhat et al., 2007). Irradiation is categorized depending on the dose amount: radappertization (corresponding to radioactive disinfection), with a dosage level of 30–40 k Gy, mostly used in the canning industry; radurization (used to increase shelf life), with a dosage level of 0.75–2.5 k Gy, mostly used for cereals and pulses; and radicidation (comparable to pasteurization of milk), with a potency level of 2.5–10 (Diehl, 2002; Bhat et al., 2007).

In agriculture, gamma-radiation has been found to have numerous applications, which include minimizing post-harvest losses by suppressing sprouting and contamination, eliminating or regulating infestations, reducing foodborne illnesses and establishing storage periods, and reproducing high-performance, well-adapted, and disordered agronomic varieties (Andress et al., 1994; Emovon, 1996).

22.2.1 Gamma-radiation

Gamma rayskill microorganisms by adversely affecting the cell membrane or injuring significant element in the cell (affecting metabolic enzyme activity or, severely damaging deoxyribonucleic acid (DNA) and ribonucleic acids (RNA)), further affecting their growth and replication. The effect of radiation becomes evident after a certain span, when the DNA helix fails to unfold and the microorganism is unable to reproduce through replication (Diehl, 1995; Yeh & Yeh, 1993; Ciesla et al., 1991).

When irradiating cereals, the rays interact with micro- and macroelements, causing chemical bonds to disperse and the formation of (short-lived) free radicals like hydroxyl, hydrogen atoms (-OH and -H), hydrogen peroxide, and high-energy electrons. The resultant free radicals interact with nucleic acids and the functional groups that bind one nucleotide to another, eventually suppressing microbial and insect growth (Yeh & Yeh, 1993; Ciesla et al., 1991).

22.2.1.1 Gamma-irradiation of Cereals

Currently, the cereal and pulses sector relies primarily on fumigation for postharvest pest management. Fumigation leaves residues, which causes human foodborne illnesses. India established a quasi-barrier in 2004, mandating all incoming cereal imports to be fumigated and disinfected with methyl bromide (MeBr) (Carpenter et al., 2000). However, under the Federal Clean Air Act and the Montreal Protocol, the US Environmental Protection Agency (EPA) phased out most phytosanitary applications of MeBr in 2005. Additionally, MeBr spraying is only practicable at treatment temperatures of 5°C. As a result, a viable alternative to MeBr for integrated pest management in cereals and pulses, as well as for product quality and the environment, is essential. Irradiation of cereals and legumes has emerged as an innovative method of coping with insect and pest problems. The primary benefit of gamma-irradiation is that it may be used after the product has been packaged, removing the risk of cross contamination. The nutritional content, sensory appeal, and other associated quality attributes of cereals and cereal products can be maintained if the dosages are optimized and the product is irradiated at the recommended doses.

22.2.1.2 Factors Influencing the Impact of Irradiation on Food

22.2.1.2.1 On Moisture Content

Moisture acts as a transporter for free radicals, allowing them to associate with other nutrients. As a result, free water is critical in amplifying the irradiation's secondary effects. This has been observed while frozen foods are treated with just minimal side effects. If the moisture content is less than 12%, secondary effects are likewise relatively restricted (Rayas-Duarte & Rupnow, 1994).

22.2.1.2.2 On Temperature

The principal effects of irradiation are unaltered by the temperature at the time of irradiation whereas the secondary effects are temperature sensitive (Olson, 1998). Prior to irradiation, cereals and pulses can be frozen to reduce free radical mobility and the generation of off taste (Carpenter et al., 2000).

22.2.1.2.3 On Irradiation Dosages

The dose of irradiation influences the extent of physicochemical changes in food. At lower dosage levels, the products generated, and the dose have a linear association. However, at greater concentrations, additional interactions between the components occur, resulting in the production of altogether new metabolites. To obtain the intended impact from irradiation, the dosages must be adjusted. The primary process in any application of food irradiation includes chemical changes that eventually determine the quantity of radiation dosage and exposure to be applied on the product (Diehl, 2002; Bhat et al., 2007).

22.2.1.2.4 On Atmosphere during Irradiation

Natural oxygen in the atmosphere functions as a species of free radicals, whose binding with reactive molecules in the food matrix effectively affects the quality (Rayas-Duarte & Rupnow, 1993). Lack of oxygen during irradiation causes several reactions such as decarboxylation, dehydration, and polymerization. Carbon dioxide, carbon monoxide, and aldehydes are some of the most prevalent radiolytic products (Giroux & Lacroix, 1998).

22.2.1.3 Radiation Process

A radiation process involves systematically irradiating a product in order to preserve, alter, or enhance its properties. This process entails subjecting the product to radiation generated by a radioactive source (such as cobalt-60) for a certain amount of time. A portion of this radiation that reaches the product is absorbed. The quantity varies based on the product's bulk and composition, as well as the exposure span. For each specific product, a specific amount of radiation energy is mandatory to achieve the intended impact; the appropriate value is obtained through experimentation.

Cobalt-60 and caesium-137 are the most optimal gamma-radiation sources for irradiation due to their very high gamma ray intensity and relatively long half-life (30.1 years – caesium-137; 5.27 years – cobalt-60). Caesium-137, on the other hand, has been limited to compact self-contained, parched irradiators, which are principally employed for blood irradiation and insect disinfection. Currently, cobalt-60 is used as the gamma-radiation source in all commercial radiation processing industries. The product that has been bombarded with gamma rays, on the other hand, never becomes radioactive and may therefore be treated normally.

22.2.1.4 Effects of Gamma-irradiation

22.2.1.4.1 Effects on Microbial and Insect Infestation

Bacteria, protozoa, slime moulds, yeasts, and fungus are some of the microorganisms that may infect grains and their products. Bacteria and fungus are the most problematic of them. Bacteria are infrequent causes of storage losses, but they are of particular relevance due to pathogenic concerns or potential spoiling in high moisture, unheated goods. Some bacteria can stop germination (Christensen, 1982) (Figure 22.2).

Damage caused by insects during cereal storage includes:

- consumption of the grain itself
- contamination of the product with insect fragments, excreta, etc.

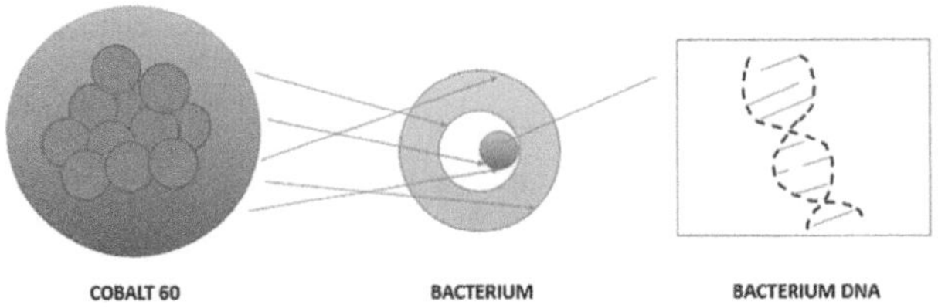

Figure 22.2 Microbial inactivation mechanism through gamma-irradiation: The irradiation source (Co-60) emits gamma rays, which penetrate the structures and microorganisms damaging the microbial DNA, inactivating the cell.

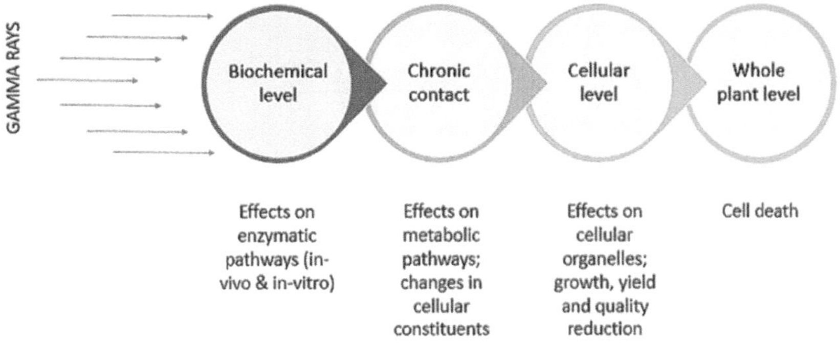

Figure 22.3 Steps of gamma-irradiation on cell inactivation (from biochemical process level to whole plant process level).

- damage to storage structures and containers
- heat and moisture generation in storage
- transfer of human diseases

To manage microbiological issues connected with cereals and their products, gamma-irradiation is used at dose rates around 10 times higher than for insect control (Figure 22.3 and Figure 22.4). It is important to note that combining heat with radiation allows for large dose rate reductions; for example, irradiating bread at 65°C with a dosage of 0.5 k Gy provides similar mould protection to 5 k Gy without heating. A dosage of 7.5 k Gy, which results in a 5-decimal decrease in *Enterobacteriaceae*, has been recommended for the aim of eradicating human infections from stock feed (FAO/ IAEA, 1982). The Codex Alimentarius Commission and the World Health Organization (WHO) have established food irradiation standards, which are used by more than 42 nations. Research on the effects of gamma-irradiation on various cereals are compiled and emphasized in Table 22.1.

22.2.1.4.2 Effects on Cereal Nutritive Components

Through the free radical process, gamma-irradiation alters the physical and chemical characteristics of macro-compounds in cereals. This alteration in turn impacts both the nutritional and sensory acceptability, and other qualitative features of cereals, which can be retained when dosages are monitored, and the product is irradiated at the optimal levels. It can also be used on packaged products to reduce the risk of cross contamination.

Starch: Physicochemical properties of starch comprises of enzymatic digestibility, thermal properties, and chemical composition. Starch has functional qualities in addition to physicochemical features such as solubility, viscosity, gel strength, water-holding capacity, and viscoelastic changes. These qualities can be greatly altered by starch modification processes, impacting starch quality (Zhu, 2018) (Figure 22.5).

In starch the gelatinization temperature, gelatinization enthalpy, and viscosity (maximum, mid, and terminal) reduce with increase in dose, although dispersion mechanism, amylose concentration, and surface splitting of the starch is enhanced (Bashir & Aggarwal, 2019; 2017). The carboxyl content, solubility, freeze-thaw stability, crystalline structure, and water absorption capacity all enhance when the irradiation dosage is lowered, but syneresis, pasting characteristics, and pH are significantly reduced (Lee et al., 2013). Gamma-radiation significantly affects amylose, amylopectin,

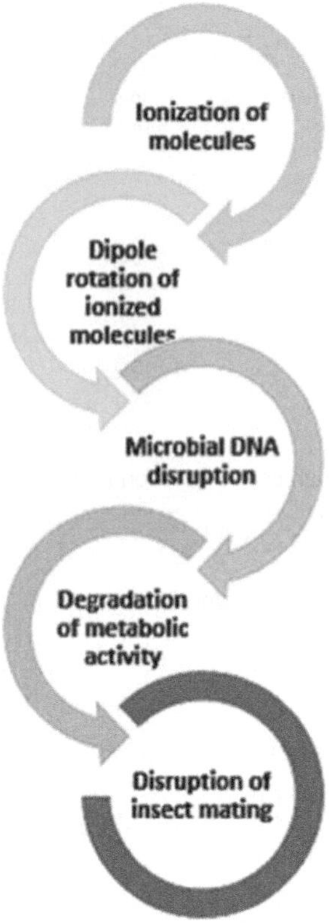

Figure 22.4 Disinfection steps through irradiation in insects.

Table 22.1 Effects of Gamma-irradiation on Properties of Cereals and Legumes

Cereals	Effects on properties	References
Rice	Grain disinfestation and reduced pasting characteristics	Zanao et al. (2009)
Wheat	For 180 seconds, a combination of gamma-radiation (1.5 k Gy) and infrared radiation resulted in 96.0% adult mortalities	Mohamed & Mikhaiel (2013)
	Reduced overall plate count and cysteine concentration; enhanced noodle brightness	Li et al. (2012)
Legumes	Reduced level of anti-nutritional factor, elevated phenolics, antioxidants, and total free amino acid concentration	Singh et al. (2014)
	Increased potential for water and oil absorption, as well as antioxidant properties	Jabeen et al. (2015)
Pearl millet	Increase in water uptake capacity and optical density while decreasing in viscosity	Falade & Kolawole (2013)
Sorghum	Alpha and beta amylase activities were reduced by 22% and 32%, respectively; the microbial load was minimized.	Mukisa et al. (2012)

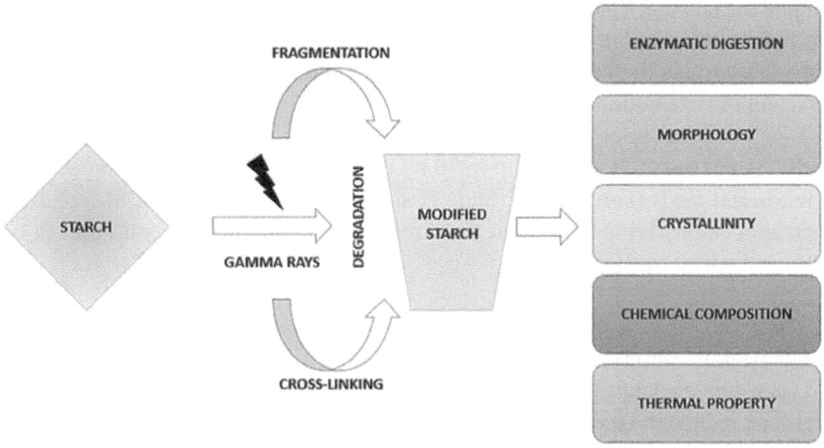

Figure 22.5 Starch irradiation, modifications, and development of modified starch; effects of irradiation on physicochemical properties.

Table 22.2 Effects of Gamma-irradiation on Cereal Components

Component	Dosage	Effect	Reference
Starch	5–10 k Gy	Decreased starch yield and quality from maize	Bashir & Aggarwal, 2019
Protein	≤5 k Gy	Increase in free amino acids and a shift to lower molecular weight in the protein fraction permanent structural change	Mehlo et al., 2013
Lipid	≤10 k Gy	No changes in fatty acid pattern or free fatty acid levels occurred	Jan et al., 2012
Vitamin	≥5 k Gy	Content and activity decreases	Bashir & Aggarwal, 2019; Jan et al., 2012

and phosphorous levels. Amylopectin has been well researched as compared to amylose and is considered to represent concentric rings, blocklets, and double helices. When exposed to gamma-radiation, they tend to split open, relieving tension and generating more sugars while also modifying other physicochemical parameters (Bertoft, 2017).

Proteins: Irradiation of various cereals and legumes causes a change in protein fractions (albumin and globulin). Protein digestibility and solubility alters, which in turn affects the emulsifying and foaming properties (in cases of legumes and oil seeds mostly). It also causes deamination, decarboxylation, disulfide bond reduction, sulfhydryl group oxidation, alteration of amino acid molecules, coordinated metal ion valence alterations, peptide chain dissociation, and protein degradation (Kuan et al., 2013). The effects of gamma-irradiation on cereal components is shown in Table 22.2.

Dosages of 10 kGY or less are preferable in cereal proteins, particularly in case of gluten. Gluten's chemical changes with respect to exposure to irradiation affect principally its structure; the form, and composition of the gluten protein (e.g., native or denatured); wet or dry conditions; and irradiation conditions. Gamma-irradiation can be utilized as an alternative to heat or chemical treatment in the preservation of cereals and legumes without protein profile and functionality degradation.

Lipids: Despite the fact that lipids are among the dietary elements that are more susceptible to irradiation, modest doses of ionizing radiation (<10 k Gy) had no effect on total lipids in wheat. Higher dosages may cause lipids to alter to some extent, depending on the lipid's content, the presence of antioxidants, and the irradiation environment.

Amino acids: As per research, dosage up to 70 k Gy has no significant effect on amino acid levels in various foods and feeds (Farag, 1998; Seda et al., 2002). Hanis et al. (1988) found that 25 k Gy dosage when applied on various cereal meals, it reduced methionine concentrations in oat by 31%, corn by 33%, and wheat by 39%.

Vitamins: The levels of vitamin loss are impacted by a range of parameters, including radiation exposure, cereal grain variety, storage duration and conditions. Minimal losses can be obtained by using oxygen-free packaging and irradiating at relatively low temperature. 10 k Gy to 25 k Gy dosage levels can reduce the levels of thiamin and riboflavin contents in cereals (Aziz et al., 2006).

22.2.1.4.3 Effects on Individual Cereals and Cereal Products

Low radiation doses are efficient at controlling insects at all stages of development. Radiation, on the other hand, can have an impact on the chemical composition and nutritional quality of grains if used in doses greater than those required for pest control of stored grains. Irradiation causes a breakdown of proteins, carbohydrates, lipids, and vitamins, which must be addressed when using this technology to reduce food losses. Ionizing radiation has been demonstrated to impact the milling qualities of grains, the rheological features of dough and batters, and the quality of flours used in breads, cakes, and pasta items. At high doses of radiation, the odor, smell, and taste of cereal grains and cereal grain products become undesirable.

Marathe et al. (2002) conducted storage studies on irradiated whole wheat flour (0.25–1 k Gy) contained in polyethylene packaging and observed that radiation exposure prior to storage of 6 months had no significant adverse effects on proximate composition, vitamin B_1 and B_2 content, free fatty acids, sedimentation value, gelatinization viscosity, dough characteristics, and microbial count. Irradiation at 0.25 k Gy was found to be effective in shelf life enhancement (by up to 6 months) of whole wheat flour.

Abu et al. (2006) investigated the effect of gamma-irradiation and resulting changes on the functional qualities of cowpea and observed a decrease in swelling and pasting ability and increase in oil absorption capacity considerably with dosage. Nene et al. (1975) observed that gamma-irradiation decreased the viscosity during gelation of red gram starch. Rao & Vakkil (1983) investigated the effects of gamma-irradiation on the flatulence-causing oligosaccharides (raffinose, stachyose, and verbascose) in green gram and found that the oligosaccharide concentration was decreased owing to polymeric chain fragmentation. RayasDuarte & Rupnow (1993) also corroborated identical trends for irradiated northern bean starch having increased maltose content due to starch fragmentation and hydrolysis.

Mekkawy (1996) used varied amounts of gamma-irradiation on sorghum grains (10, 50, 100, 150, and 200K Gy). He observed that gamma-irradiation had no effect on sorghum grain total protein, lipid, or ash contents. Even when the highest irradiation dosage was employed, irradiation treatments on sorghum had no detectable effect on tannic acid content (200 k Gy). Tannic acid is a polyphenol with several applications in the food industry (beer, wine clarity, fragrance component, and color stabilizer in soft drinks and juices). The concentration of hemicellulose was found to decrease as the irradiation dose level increased. It was also demonstrated that providing rats with baseline meals supplemented with irradiated sorghum grains increased digestibility coefficient.

Mohamed and Mikhaiel (2013) studied the impact of gamma-radiation (1.5 k Gy) on insects in stored wheat grains. They discovered that combining infrared light with gamma-radiation (1.5 k Gy) resulted in 96.0% adult death for similar insects.

Yu et al. (2005) assessed wheat that had been treated with gamma rays and dehydrated with hot air. They observed that irradiation impacted the drying characteristics and surface temperature of wheat, and that the dehydration rate and surface temperature of wheat increased with increasing irradiation dosage. The cell structure of irradiated wheat grain was analyzed using an electron microscope, and it was determined that the changes in drying characteristics and surface temperature of wheat were caused by the disruption of the cell structure generated by irradiation.

Yu & Wang (2007a) discovered that irradiating dried rice with (0–10 k Gy) had a negative influence on drying time and apparent amylose content while making an impact on average dehydration rate, gel viscosity, and gelatinization conditions. Yu & Wang (2007b) studied the effect of irradiation on the physicochemical and mechanical properties of rice starch (2, 5, 8, and 10 k Gy). The apparent amylose content, gel thickness, and gelatinization temperature were all affected by irradiation pretreatment. The apparent amylose content decreased as dosage increased, yet viscosity of gel and gelatinization temperature increased. These alterations were mostly due to changes in starch structure.

Muramatu et al. (1991) investigated the effects of gamma-irradiation (3, 4, 5, 6, and 7 k Gy) and electron beam (2MeV) on the microbiological load of buckwheat flour and irradiated flour products. They discovered that the peroxide value of lipid in buckwheat flour was enhanced with the amount of gamma rays and electron beams absorbed. Oxygen absorbers, on the other hand, lowered the generation of peroxide value. Irradiation also reduced the viscosity of the dough. The use of an oxygen absorber resulted in a high sensory score of treated buckwheat flour noodles with minor changes in color, taste, and texture.

Zanao et al. (2009) studied the effects of gamma-irradiation (0.5, 1, 3, and 5 k Gy) on the physicochemical (grain disintegration, composition, amylose content, pasting properties, and color) and sensory properties of raw and cooked rice. They observed that gamma-irradiation had no effect on the amount of grain disintegration during irradiation treatment but was detrimental to pest multiplication. Irradiation had no reported discernible effect on the percentage composition or amylase content. As the irradiation dosage escalated, the quality of the paste degraded.

Aziz et al. (2006) investigated the effect of gamma-irradiation on the presence of pathogenic bacteria and the nutritional value of four major cereal grains (maize, sorghum, barley, and wheat). The acidity of grain lipids was lowered by irradiation, with the impact increasing with radiation dosage. Furthermore, peroxide levels increased in direct proportion to the amount of radiation the grains absorbed. The rise in peroxide values are common when peroxide level increases above 100 meq/g fat; which was not observed in this case. Thiamin content, which is considered to be the most radiation-sensitive of the water-soluble vitamins, reduced by 33.3–47.9% after 15 k Gy irradiation. Riboflavin, which is less radiation-sensitive, exhibited a lower reduction of 20.8–32.1% in response to 15 k Gy.

Mukisa et al. (2012) studied the effects of gamma-irradiation (10 and 10+25 k Gy) on inactivation of microorganisms, amylase activity, and sorghum flour properties. Gamma-irradiation lowered the activity of amylases (alpha and beta) by 22% and 32%, respectively. Irradiation led in denser porridge products, which they attributed to gamma-irradiation's depolymerization of starch. They also saw a reduction in microbial population as irradiation dosages were increased.

ElShazali et al. (2011) investigated the anti-nutrients, protein digestibility, and sensory quality of gamma-irradiated (dosage of 2 k Gy) pearl millet flour. According to their research, gamma-irradiation has little effect on individual components like tannin and phytate, but when combined with thermal processing, the amount is significantly decreased. They also observed an increase in the flour's qualitative attributes.

22.2.1.4.4 Effects on Individual Pulses and Legumes

Pulses are abundant in protein, fiber, vitamins, and minerals. These grains augment proteins from other plant sources, such as cereal grains, and offer essential amino acids to many vegetarian diets across the world. Protein content ranges from 15 to 40% depending on the legume, variety, and growth factors (Bashir et al., 2012). Isoflavones and peptides, for example, are bioactive components that can help prevent cancer and cardiovascular disease. Because of their high protein content, legumes blended with an equivalent percentage of water can be utilized as an egg substitute in vegan food products (Bashir & Aggarwal, 2016; Singh et al., 2014).

Jabeen et al. (2015) investigated into the impact of 1 k Gy irradiation on the functional and nutritional properties of soy flour and sprouted soy flour. They noticed that increasing the dose enhanced the flour's functional qualities (water and oil absorption capability). There was a significant increase in antioxidant properties for both regular and sprouted flour with increase in dosages. The effects of specific dosages of gamma-irradiation on cereal and pulses is shown in Table 22.3.

Singh et al. (2014) investigated the nutrient composition of legumes (masur, kabuli chickpea, brown chickpea, and green mung) after gamma treatment (1, 2, 3, 4, and 5 k Gy). Radiation greatly decreased the content of anti-nutritional components in beans in a dose-dependent way. The total free amino acid contents and antioxidants, on the other hand, increased with increase in dose.

Koksel et al. (1998) analyzed barley cultivars, Tokak and Clerine, and irradiated them at two distinct dosage levels (0.05–0.75 and 0.5–5.0 k Gy). Most of the malt quality attributes were adversely affected by moderate irradiation of barley prior to malting. Furthermore, the negative effects of irradiation were reduced at levels up to 0.25 k Gy. Irradiating malt samples resulted in either a minor or no loss of qualitative attributes.

Khattakk & Klopfenstein (1989) observed the effects of gamma-irradiation (0.5, 1, 2.5, and 5 k Gy) on the amino acid profile and accessible lysine in grains and legumes. They found that while some amino acids were lost in the irradiation samples, the accessible lysine amount was increased.

Table 22.3 Effects of Specific Dosages of Gamma-irradiation on Cereal and Pulses

Pulses	Dosage	Irradiation effects	Reference
Rye	0.1–1.0 k Gy	No detectable degradation of linoleic and linolenic acid contents	Vaca & Harms-Ringdahl, 1986
Sorghum grain (Hemaira variety) Sorghum grain (Shahlla variety)	1.0, 3.0, 5.0, 7.0, and 10 k Gy 1.0, 3.0, 5.0, 7.0, and 10 k Gy	Irradiation with 7.0 and 10.0 k Gy lead to significant reduction in tannin content in Shahlla (from 0.35 to 0.26 mg of catechin equiv./100 g) Hemaira's tannin content was not significantly affected by irradiation	Abu-Tarboush, 1998
Wheat grain	0.05–10 k Gy 0.5 and 1 k Gy	At 0.05 k Gy: increase of inhibition activity against *Sitophilus granarius* L. a-amylase whereas at 10 k Gy: decrease of inhibition activity against *Sitophilus granarius* L. a-amylase Irradiated grain had significant increase in inhibition activity against alpha-amylase of *Tribolium confusum* Duv.	Gralik & Warchalewski, 2006
Milled aromatic rice	0.2 k Gy 0.5 k Gy 1 k Gy	Lower sensory appeal for stored irradiated rice was observed compared with those of non-irradiated rice Changes in taste, texture, and odor contributed mainly to the declining overall acceptability of cooked, irradiated rice samples	Sirisoontaralak & Noomhorm, 2007

22.2.1.4.5 *Effects on Organoleptic Properties, Grain Viability, and Biochemistry*

Cereal diets are considered to be the rich source of fiber, and both the production and consumption of diverse cereal-based meals are expanding. These items may get heavily contaminated with numerous microbes throughout cultivation, production, and even marketing (Gharib & Aziz, 1995). Thermal fumigation utilizing ethylene oxide sterilization has been used with varied degrees of effectiveness in worldwide markets. However, there are significant drawbacks to using these technologies for grain sterilization; for example, hazardous residues persist and organoleptic qualities are changed (Campbell et al., 1986). Irradiation has the potential to be a successful alternative technique (Aldryhim & Adam, 1999).

There are few reports on the effect of irradiation on enzyme activity which leads to alteration in organoleptic properties in cereals and pulses. In a study evaluating the effect of gamma-irradiation on barley germination, it was discovered that 0.7 k Gy was the optimal dosage level to suppress radicle and germ formation, lowering malting losses by 13–22% (Marseu & Cojocaru, 1972). However, dosages of 8 k Gy fully blocked a-amylase activity in wheat (Kiss & Farkas, 1977), but doses of 0.3 k Gy diminish the activity of this enzyme in rice (Wang et al., 1983). Gamma treated white rice flour and starch were investigated. At room temperature, the white rice was exposed to 0.5, 1.0, 3.0, 5.0, 7.0, and 9.0 k Gy gamma-radiation. As the irradiation dosage was increased, the viscosity of the rice flour and starch paste reduced constantly. Furthermore, as compared to the native samples, the crystallinity of irradiation starch increased but decreased in irradiated flour (Bao et al., 2005).

Grain viability has apparent and significant implications for both the seed and malting industries. Exposure to levels more than 0.5 k Gy significantly inhibited rice grain germination (Wang et al., 1983). Dose levels of 5.81 k Gy inhibited wheat germination by 80% (Lorenz, 1975). At this dosage level, grain respiration was also reduced, according to the latter.

22.2.2 Regulatories on the Gamma-irradiation Usage

The World Health Organization (WHO), the American Medical Association, the Food Safety and Standards Authority of India (FSSAI), the Institute of Food Technologists, Food Standards Australia New Zealand, the Food and Agriculture Organization (FAO), the International Atomic Energy Agency (IAEA), and Codex have all approved food irradiation. More than 100 nations have implemented irradiation technology, and over 60 distinct irradiated food items are available on the market. According to the American Council on Science and Health, around 8,000 supermarkets and other retail establishments in the United States trade irradiated cereals, pulses, and cereal products. The Codex and FAO suggest gamma-irradiation for cereals and pulses at a maximum of 1 k Gy for disinfestations and a maximum of 5 k Gy for microbial load reduction.

For major commercialization of irradiated food, producers must employ a new technique, merchants must sell the product, and customers must purchase it. It was widely assumed that the main hurdle to food irradiation was customer unwillingness to accept irradiated food. The evidence for this was mostly based on consumer surveys, but some scientists have cautioned that surveys may overestimate the possibility that customers will purchase irradiated food if it is visible and evaluable on store shelves.

22.3 CONCLUSION

Gamma-irradiation is considered to be one of the potential future approaches that will be employed to address the ever-increasing customers' requirements for nontoxic, safe, and enhanced shelf life in order to maintain high food quality. Despite the necessity of disseminating information regarding food irradiation to customers, authorities and manufacturers must also be aware of the overall consumer attitude toward the technique. This involves information regarding risks connected by applying ionizing radiation in food processing and manufacturing. More research is needed to assess the method of interaction of different gamma-irradiation dosages on the most common pathogens detected in cereals and pulses, as well as to optimize the dosages. The overall dose of 10 k Gy, as recommended by the Codex Alimentarius Commission (World Health Organization, 1994), is extremely successful in microbiological decontamination and has no negative impact on nutritive value of cereal grains. Gamma-irradiation up to 5 k Gy enhances the storability qualities of cereals and pulses by significantly reducing fungal growth and the concentration of free fatty acids in the grains. A dose of up to 2 k Gy improves the protein functioning of cereals and pulses. At low dosages, the emulsifying and foaming capabilities are unaffected.

REFERENCES

Abu, J. O., Duodu, K. G., and Minnaar, A. (2006). Effect of gamma irradiation on some physicochemical and thermal properties of cow pea starch. *Food Chemistry*, 95, 386–393.

Abu-Tarboush, H. M. (1998). Irradiation inactivation of some antinutritional factors in plant seeds. *Journal of Agricultural and Food Chemistry*, 46(7), 2698–2702.

Aldryhim, Y. N., and Adam, E. E. (1999). Efficacy of gamma irradiation against *Sitophilus granaries* (L.) (Coleoptera: Curculionidae). *Journal of Stored Products Research*, 35, 225–232.

Andress, E. L., Delaplane, K. S., and Schuler, G. A. (1994). *Food Irradiation. Fact Sheet HE.* Institute of Food and Agricultural Sciences University of Florida, USA 08467.

Aziz, N. H., Souzan, R. M., and Shahin Azza, A. (2006). Effect of γ-irradiation on the occurrence of pathogenic microorganisms and nutritive value of four principal cereal grains. *Applied Radiation and Isotopes*, 64(12), 1555–1562.

Bao, J., Ao, Z., and Jane, J. L. (2005). Characterization of physical properties of flour and starch obtained from gamma-irradiated white rice. *Starch/Starke*, 57, 480–487.

Bashir, K., Aeri, V., and Masoodi, L. (2012). Physicochemical and sensory characteristics of pasta fortified with chickpea flour and defatted soy flour. *Journal of Environmental Science Toxicology and Food Technology*, 1(5), 34–39.

Bashir, K., and Aggarwal, M. (2016). Effects of gamma irradiation on cereals and pulses- A review. *International Journal of Recent Scientific Research*, 7(12), 14680–86.

Bashir, K., and Aggarwal, M. (2019). Physicochemical, structural and functional properties of native and irradiated starch: a review. *Journal of Food Science Technology*, 56, 513–523.

Bashir, K., Swer, T. L., Prakash, K. S., and Aggarwal, M. (2017). Physico-chemical and functional properties of gamma irradiated whole wheat flour and starch. *LWT – Food Science and Technology*, 76, 131–139.

Bertoft, E. (2017). Understanding starch structure: recent progress. *Agronomy*, 7, 56.

Bhat, R., Sridhar, K. R., and Tomita-Yokotanib, K. (2007). Effect of ionizing radiation on anti-nutritional features of velvet bean seeds (Mucunapruriens). *Food Chemistry*, 103, 860–66.

Campbell, G. L., Classen, H. L., and Ballance, G. M. (1986). Gamma irradiation treatment of cereal grains for chick diets. *Journal of Nutrition*, 116, 560–566.

Carpenter, J., Gianessi, L., and Lynch, L. (2000). The Economic Impact of the Scheduled U.S. Phase out of Methyl Bromide, National Centre for Food and Agricultural Policy.

Christensen, C. M. (1982). *Ed. Storage of Cereal Grains and their Products*. A.A.C.C. St. Paul, USA.

Ciesla, K., Zoltowski, T., and Mogilevsky, L. Y. (1991). Detection of starch transformation under gamma irradiation by small angle X-ray scatting. *Starch/ Starke*, 43, 11–16.

Diehl, H. F. (1995). *Safety of Irradiated Foods*, 2nd edn. New York: Marcel Dekker Inc. 283–289.

Diehl, J. F. (2002). Food irradiation- past, present and future. *Radiation Physics and Chemistry*, 63, 211–215.

ElShazali, A. M., Nahid, A. A., Salma, H. A., Isam, A. M. A., and Elfadil, E. B. (2011). Effect of radiation process on antinutrients, protein digestibility and sensory quality of pearl millet flour during processing and storage. *International Food Research Journal*, 18(4), 1401–1407.

Emovon, E. U. (1996). *Keynote Address: Symposium on Irradiation for National Development*, Shelda Science and Technology Complex, SHESTCO, Abuja, Nigeria.

Falade, K. O., and Kolawole, T. A. (2013). Effect of γ-Irradiation on Color, Functional and Physicochemical Properties of Pearl Millet (*Pennisetum glaucum*) Cultivars. *Food and Bioprocess Technology*, 6(9), 2429–2438.

FAO/IAEA. (1982). Training Manual on Food Irradiation Technology and Techniques. 2nd edition. *International Atomic Energy Commission*, Vienna.

Farag, M. Diaa El-Din H. (1998). The nutritive value for chicks of full-fat soybeans irradiated at up to 60 kGy. *Animal Feed Science and Techology*, 73, 319–28.

Gharib, O. H., and Aziz, N. H. (1995). Effect of gamma-irradiation and storage periods on the survival of toxigenic microorganisms and the khapra beetle *Trogoderma gramarium* in crushed corn. *Journal of the Egyptian Society of Toxicology*, 15, 23–28.

Giroux, M., and Lacroix, M. (1998). Nutritional adequacy of irradiated meat: A review. *Food Research International*, 31, 273–276

Gralik, J., and Warchalewski, J. R. (2006). The influence of g-irradiation on some biological activities and electrophoresis patterns of wheat grain albumin fraction. *Food Chemistry*, 99, 289–298.

Hanis, T., Mnukova, J., Jelen, P., Klir, P., Perez, B., and Pesek, M. (1988). Effect of gamma irradiation on survival of natural microflora and some nutrients in cereal meals. *Cereal Chemistry*, 65.

Jabeen, N. M., Kola, M., and Reddy, K. J. (2015). Impact of irradiation on nutritional quality and functional properties of soy flour and sprouted soy flour. *International Journal of Advanced Research*, 3(3), 1120–1129.

Jan, S., Parween, T., Siddiqi, T. O., and Mahmooduz, Z. (2012). Effect of gamma radiation on morphological, biochemical, and physiological aspects of plants and plant products. *Environmental Reviews*, 20(1), 17–39.

Khattak, A. B., and Klopfenstein, C. F. (1989). Effect of Gamma Irradiation on the Nutritional Quality of Grains and Legumes. II. Changes in Amino Acid Profiles and Available Lysine. *Cereal Chemistry*, 66(3), 171–172.

Kiss, I. and Farkas, J. (1977). Storage of wheat and corn at high moisture content as affected by ionizing radiation. *Acta Alimentaria*, 6,193–213.

Koksel, H., Sapirstein, H. D., Celik, S., and Bushuk, W. (1998). Effects of gamma-irradiation of wheat on gluten proteins. *Journal of Cereal Science*, 28, 243–250.

Kuan, Y. H., Bhat, R., Patras, A., Karim, A. A. (2013). Radiation processing of food proteins-A review on the recent development. *Trends Food Science and Technology*, 30, 105–120.

Lee, J. S., Ee, M. L., Chung, K. H., and Othman, Z. (2013). Formation of resistant corn starches induced by gamma-irradiation. *Carbohydrate Polymer*, 97, 614–617.

Li, M., Zhu, K. X., Wang, B. W., Guo, X. N., Peng, W., and Zhou, H. M. (2012). Evaluation the quality characteristics of wheat flour and shelf-life of fresh noodles as affected by ozone treatment. *Food Chemistry*, 135(4), 2163–2169.

Lorenz, K. (1975). Irradiation of cereal grains and cereal grain products. *Critical Reviews in Food Science and Technology*, 6(4), 317–382.

Marathe, S. A., Machaiah, J. P., Rao, B. Y. K., Pednekar, M. D., and Sudha Rao, V. (2002). Extension of shelf-life of whole-wheat flour by gamma radiation. *International Journal of Food Science and Technology*, 37, 163–168.

Marseu, P. and Cojocaru, L. (1972). Effect of ionizing radiation on the germination of barley used in brewing. *Lucari de Ceretare*. 9,193–206.

Mehlo, L., Mbambo, Z., Bado, S., Lin, J., Moagi, S. M., Buthelezi, S., Stoychev, S., and Chikwamba, R. (2013). Induced protein polymorphisms and nutritional quality of gamma irradiation mutants of sorghum. *Mutation Research/ Fundamental and Molecular Mechanisms of Mutagenesis*, 749(1–2), 66–72.

Mehlo, L., Mbambo, Z., Bado, S., Lin, J., Moagi, S.M., Buthelezi, S., Stoychev, S., and Chikwamba, R. (2013). Induced protein polymorphisms and nutritional quality of gamma irradiation mutants of sorghum. *Mutant Research*, 749(1–2), 66–72.

Mekkawy, S. H. (1996). Effect of gamma irradiation on chemical composition and nutritive value of sorghum grains. *Arab Journal of Nuclear Sciences and Applications*, 29 (4), 195–199.

Mohamed, S. A., and Mikhaiel, A. A. (2013). Combined Effect of Infrared and Gamma Radiation on Certain Insects in Stored Wheat Grains and Wheat Flour. *Isotope and Radiation Research*, 45(2), 375–384.

Mukisa, I. M., Muyanja, C. M. B. K., Byaruhanga, Y. B., Schüller, R. B., Langsrud, T., and Narvhus, J. A. (2012). Gamma irradiation of sorghum flour: Effects on microbial inactivation, amylase activity, fermentability, viscosity and starch granule structure. *Radiation Physics and Chemistry*, 81(3), 345–351.

Muramatu, N., Ohinata, H., Karasawa, H., Oike, T., Ito, H., and Ishigaki, I. (1991). Use of gamma irradiation for microbial inactivation of buckwheat flour and products. 8. *Shokuhin Shosha*, 26(1–2), 30–39.

Nene, S. P, Vakil, U. K, and Sreenivasan, A. (1975). Effect of gamma radiation on physicochemical character of red gram (*Cajanus cajan*) starch. *Journal of Food Science*, 40, 943–47.

Olson, D. G. (1998). Irradiation of Food. *Food Technology*, 52, 56–62.

Rao, V. S. and Vakil, U. K. (1983). Effects of Gamma-Irradiation on Flatulence-Causing Oligosaccharides in Green Gram (*Phaseolus areus*). *Journal of Food Science*, 48, 1791–1795.

Rayas-Duarte, P., and Rupnow, J. H. (1993). Gamma-irradiation affects some physical properties of dry bean (*Phaseolus vulgaris*) starch. *Journal of Food Science*, 58, 389–94.

Rayas-Duarte, P., and Rupnow, J. H. (1994). Gamma-irradiatied dry bean (*Phaseolus vulgaris*) starch: Physicochemical properties. *Journal of Food Science*, 59(4), 839–43.

Seda, H. A., Mahmoud, A. A., Ibrahim, A. A., and El-Niely, H. F. G. (2002). Effect of radiation processing on sensory and chemical characteristics of broad beans Giza-2 (*Vicia faba*). *Egyptian Journal of Radiation Science and Application*, 15, 119–139.

Singh, P. K., Sohani, S., Panwar, N., and Bhagyawant, S. S. (2014). Effect of radiation processing on nutritional quality of some legume seeds. *International Journal of Biological and Pharmaceutical Research*, 5(11), 876–881.

Sirisoontaralak, P., and Noomhorm, A. (2007). Changes in physicochemical and sensory properties of irradiated rice during storage. *Journal of Stored Products Research*, 43(3), 282–289.

Vaca, C. E., and Harms-Ringdahl, M. (1986). Radiation-induced lipid peroxidation in whole grain of rye, wheat and rice: effects on linoleic and linolenic acid. *Radiation Physics and Chemistry*, 28(3), 325–330.

Wang, U. P., Lee, C. Y., Chang, J. Y., and Yet, C. L. (1983). Gamma-radiation effects on Taiwan-produced rice grains. *Agricultural and Biological Chemistry*, 47, 461–472.

Yeh, A. I., and Yeh, S. L. (1993). Property differences between cross-linked and hydroxyl propylated rice starches. Cereal Chemistry, 70, 596–601.

Yu, Y., and Wang, J. (2007a). Effect of gamma-irradiation treatment before drying on qualities of dried rice. *Journal of Food Engineering*, 78, 529–536.

Yu, Y., and Wang, J. (2007b). Effect of gamma-ray irradiation on starch granule structure and physicochemical properties of rice. *Food Research International*, 40, 297–303.

Yu, Y., Wang, J., Wang, A., Pang, L., and Fu, J. (2005). Effect of $_{60}$Co gamma irradiation pretreatment on drying characteristic of wheat. *Nongye Gongcheng Xuebao/ Transactions of the Chinese Society of Agricultural Engineering*, 21(5), 145–149.

Zanao, C., Fernanda Pedroso, C. B., Solange, G., Sarmento, S., Bruder, S., and Arthur, V. (2009). Effect of gamma irradiation on physico- chemical and sensorial characteristics of rice (*Oryza sativa* L.) and on the development of *Sitophilus oryzae*. *L. Ciencia e Tecnologia de Alimentos*, 29(1), 46–55.

Zhu, F. (2018). Relationships between amylopectin internal molecular structure and physicochemical properties of starch. *Trends in Food Science and Technology*, 78, 234–242.

CHAPTER 23

Pulse Electric Field Processing of Cereals

Swati Joshi
Department of Food Science and Technology, Punjab Agricultural University, Ludhiana, Punjab, India

CONTENTS

DOI: 10.1201/9781003242192-28

23.1 INTRODUCTION

The food industry has evolved over the past years with the main focus on the processing techniques utilized for new product development. The rise in the market demands for quality food products with fresh-like sensorial characteristics along with improved nutrient content and shelf life has inclined researchers and the food industry to investigate alternative non-thermal processing methods as a substitute for traditional processes like thermal treatments. Despite the benefits of thermal processing, various deteriorative changes in terms of flavor, color, texture, and general appearance take place in the food altering its final quality. This paved the way for the utilization of nonthermal processing technologies, such as oscillating magnetic fields, ultrasound, high-pressure processing, UV radiation, and pulsed electric fields with the aim of microbe-inactivation thereby enhancing the shelf life of the product leading to the manufacture of stable and fresh-like food products that retain bioactive compounds present in the original food.

A pulse electric field (PEF) system consists of an oscilloscope to observe the pulse waveform, high-voltage power source, storage capacitor bank, a charging current limiting resistor, and a treatment chamber. Pulsed electric field utilizes PEF strength between 20–80 kV/cm for short time intervals ranging from nanoseconds to several milliseconds (Koubaa et al., 2016). Thus, it contributes as an alternative to traditional methods to destroy pathogenic microbes and enzyme inactivation with the additional advantage of retaining nutritional and sensorial attributes of liquid food products (Sánchez-Vega et al., 2014).

The application of PEF causes permeabilization of the cellular membrane exposing the cell to an electrical field that causes a movement of free cations and anions charges out and inside the cell along the electric field resulting in the induction of the transmembrane potential. High potential voltage may either alter the membrane permeability or irreversibly destruct the cell membrane structure, leading to faster permeation of the internal molecules (Goettel et al., 2013; Postma et al., 2018; Buchmann et al., 2018).

Under the influence of PEF, the biological membrane is electrically perforated and loses its semi-permeability (Parniakov et al., 2014) allowing the recovery of high value-added compounds from different cell matrices. Also, PEF is an excellent technology to improve freezing and drying processes.

Juices, majorly apple and orange, are among the foods most frequently treated with PEF. The sensorial properties of juices are reported to be well retained, along with extension of the shelf life. Apple sauce, yogurt drinks, and salad dressing have also been shown to preserve their fresh-like nature with extended shelf life after PEF processing. Some major PEF-processed foods include milk (Alirezalu et al., 2020), vegetables (Salehi, 2020), tomato juice (Bresciani et al., 2021), carrot (C. Liu et al., 2020), strawberry juice (Yildiz et al., 2021), and egg and egg products (Baba et al., 2018). PEF has gained considerable attention in the food industry for its application in enzyme inactivation, product sterilization, and meat tenderization due to the negligible temperature induced losses in food products (Lasekan et al., 2017).

Several studies have reported that the application of an electric field with low frequency and low intensity can alter the activity of biosystems, proliferation, differentiation, and growth rate . It has been reported that PEF stimulates the production of secondary metabolites (Morales-de la Peña et al., 2021) and phytosterols and antioxidants from fruits and oil seeds (Shorstkii & Koshevoi, 2019) (Tylewicz et al., 2020). Generally, PEF field intensities range between 0.5 to 2 kV/cm for fresh materials, whereas PEF treatment in case of dry matter requires higher intensities around 20 kV/cm (Wang et al., 2020).

Conventionally, liquid foods are electrical conductors due to the presence of electrolytes and ions; therefore liquid media are desirable for PEF treatment. Current studies have reported that

PEF could be utilized to transform the techno-functional and microstructure properties of solid foods including cereal-based products as well. The effect of PEF with regards to the electric field intensity, residence time, etc., on the nutritional, techno-functional, and antioxidative properties are reviewed in the following. In addition, the modifications in the microstructure of the starch and different protein conformations as a result of electroporation induced by PEF have been explained comprehensively.

23.2 EFFECT OF PEF ON NUTRITIONAL COMPOSITION

23.2.1 Proteins and Amino Acids

Sarkis et al. (2015) studied the influence of PEF on the extraction of protein from sesame oil cake. The pretreatment reduced the concentration of ethanol and elevated temperatures required for protein and peptides extraction. In addition, PEF treatment improved the extraction rate of lignans and polyphenols through various solvents. For diffusion, in the case of different temperatures, PEF showed a positive impact on the extraction of polyphenols and proteins when compared to the control. Hence, the utilization of PEF may reduce the use of excess organic solvents and the requirement of elevated temperatures conventionally used to improve diffusion.

PEF treatment caused an increase in the content of free amino acids; 7.46% and 15.83% increase in the free fatty acids (FAA) for light and dark grains, respectively. An increase in the accessibility of bound amino acids in the matrix caused due to opening up of the protein structure by the PEF treatment led to their interaction with other molecules (Y. F. Liu et al., 2018). However, a 20.2% reduction in the total amino acid content of dark grains was reported following PEF treatment (Kumari et al., 2019). This decline in the total amino acid reduction in the treated samples may be attributed to the higher rate of Maillard reaction due to presence of Maillard intermediaries triggered when the samples were exposed to PEF treatments. This may be justified by the amino-gram test of spent grains, which denoted reduction in the content of glycine and asparagine content conjugation in Maillard reaction (Guan et al., 2010); Wang et al., 2011). The effects of PEF on different nutritional components are listed in Table 23.1.

23.2.2 Starch

After PEF treatment at 50 kV cm^{-1}, starch lost granule shape, and the crystallinity degree were decreased significantly. Meanwhile, the peak, breakdown, and final viscosity of treated starch were decreased with increasing electric field strength. Li et al. (2019) studied birefringence in the native wheat starch granules molecules. Maltese crosses were clearly visible for native granules. The lower field intensities did not significantly affect the clarity of the crosses. PEF was found to alter the inner and surface structures of starch to different levels by disintegration of the intact starch granules as supported by XRD. Duque et al. (2020) reported that the treatment increased the transition temperature and particle size and declined the relative crystallinity and 1042:1020 ratio causing significant modification in the morphology of starch granules.

23.2.3 Fibres

Duque et al. (2020) reported the significance of PEF pretreatment on the ß-glucan content of oat bran and reported an increase up to 94% in the PEF-treated oat bran suspension. This may be attributed to the ability of PEF to facilitate the release of soluble ß-glucan embedded in the aleurone layer of the oat bran, especially at specific elevated energy input levels.

Table 23.1 Effect of PEF on Nutritional Composition

Proteins and amino acids	Sesame cake Rice and gluten protein concentrates Gluten protein	PEF reduced the need of ethanol and high temperatures required for protein extraction. No aggregation or primary structure modifications were induced by moderate intensity pulse electric field treatment (MIPEF). When MIPEF was applied, a significant protein solubility decrease was observed. Gluten rich in highly reactive sulfhydryl groups were highly influenced by PEF treatment, where the disulphide interchange had a great impact in terms of solubility, foaming capacity and water- and oil-holding capacity.	Sarkis et al. (2015) Melchior et al. (2020) Melchior et al. (2020)
Starch	Glutinous Rice Grain Corn starch	The peak viscosity of control and PEF-treated samples showed a corresponding decrease from 1484 to 1407, 1420, 1350, and 1172 Pa·s, respectively when the number of PEF pulses increased from 50 to 300. After the PEF treatment at 50 kV cm^{-1}, starch lost granule shape, and the crystallinity degree were decreased significantly. Meanwhile, the peak, breakdown and final viscosity of treated starch were decreased with increasing electric field strength.	Qiu et al. (2021) Han et al. (2020)
Fibres	Oat bran	Significant (up to 94%) increase in ß-glucan content in the PEF-treated (2 kV/cm) oat suspension, indicating the ability of PEF treatment to enhance the release of soluble ß-glucan from the aleurone layer of the oat bran.	Duque et al. (2020)
Carbohydrates and sugars	Brewers spent grains	PEF pre-treatment resulted in considerably higher amount of total carbohydrate and starch in case of dark brewers spent grains extracts	Kumari et al. (2019)

23.2.4 Carbohydrates and Sugars

The pretreatment increased drastically ($p<0.05$) the extraction rate of reducing sugars and protein when applied to brewer's spent grains (Kumari et al., 2019). The higher amount of protein in the PEF-treated samples may be attributed to the additional extraction of non-protein nitrogen compounds as a result of cell disruption. Higher amounts of total carbohydrate and starch were observed in the case of dark spent grains extracts, but no effect was observed for light spent grains.

23.3 EFFECT OF PEF ON TECHNO-FUNCTIONAL PROPERTIES OF CEREAL PRODUCTS

EFI: electric filed intensity; PF: pulse frequency; PW: pulse width; RT: residence time. WHC: water-holding capacity, OHC: oil-holding capacity; ES: emulsion stability; FC: foaming capacity; FS: foam stability

23.3.1 Water-holding Capacity

Among the PEF-treated samples, WHC increased with the electric field intensity giving a highest a peak value at 10 kV cm−1 (Zhang et al., 2021). The WHC of a protein molecule is closely related to its hydrophilic group distribution, hydration capacity, and 3D network structure. A PEF with larger electric field intensity will cause the extension and unfolding of gluten protein, exposing the hydrophilic groups and increasing the water absorption and interaction between and gluten protein and water molecules (Table 23.2).

Proteins have lipophilic and hydrophilic amphoteric functional groups that impart the emulsifying properties. The emulsifying properties of proteins are associated with protein solubility, surface hydrophobicity, and surface charge distribution. Parameters like pulse frequency and residence time have significant impact on the functional properties of the gluten. However, no significant effect of pulse width on WHC was observed.

Table 23.2 Effect of PFE on Techno-functional Properties of Cereal Products

Water-holding capacity	Vital wheat gluten	improved with increase in electric filed intensity and residence time	Zhang et al. (2021)
Oil absorption capacity	Vital wheat gluten	Improved with electric filed intensity	Zhang et al. (2021)
Emulsion stability	Vital wheat gluten	Improved with electric filed intensity	Zhang et al. (2021)
Foaming capacity	Vital wheat gluten	Improved with EFI and residence time	Zhang et al. (2021)
Foam stability	Vital wheat gluten	Of PEF-treated gluten was higher than control	Zhang et al. (2021)
Solubility	Vital wheat gluten	Improved with EFI and residence time	Zhang et al. (2021)
Thermal stability	Vital wheat gluten	Increased with EFI	Zhang et al. (2021)
Texture	Cooked rice	Hardness, cohesiveness, springiness, gumminess, and chewiness of treated group are significantly higher than the control	Bai et al. (2021)
Volatile components	Cooked rice	PEF causes decrease in the aldehydes, ketones, and hydrocarbons and increase in the esters and acids for treated groups hexanal, that imparts a typical rice odor, in treated group increases by 67.96%, strengthening the flavor of the products.	Bai et al. (2021)
Rheological	Vital wheat gluten Japonica rice	The apparent viscosity of the solution changed slightly with the increase in the EFI compared with the control; The EFI seemed to impose a fluctuating action on the rheological properties. PEF does not significantly impact the PV, TV, BD, and PT, with slight decrease in the SB compared to the control and the electric intensity of PEF has no significantly influence on the SB	Zhang et al. (2021) Wu et al. (2019)
Thermal properties	Cooked rice Japonica Rice	Peak temperature (Tp) of PEF groups increases significantly compared to the control. Conclusion temperature (Tc) decreased in treated samples. Gelatinization enthalpy (ΔH) showed insignificant rise after PEF treatment To, Tp, and Tc increase firstly and then decrease with the increased electric field intensity	Bai et al. (2021) C. Wu, et al. (2019)
Antimicrobial activity	Brewers spent grains	PEF-treated BSG extracts inhibited growth for *Listeria monocytogenes and Salmonella typhimurium*	Kumari et al. (2019)
Immunomodulatory activity	Brewers spent grains	PEF dark BSG extracts showed significantly lower ($p<0.05$) response for chemokines and cytokines production	Kumari et al. (2019)

The OHC was significantly affected ($P<0.01$) by PEF treatment. Significant ($P<0.05$) actions occurred with EFI and PF but not with PW and RT among the PEF treatments.

Gluten has poor emulsification property due to its lower solubility. The unfolding of protein molecules as a result of PEF creates hydrophobic or hydrophilic groups, forming an interface between air and water. The emulsion stability improved with the increase in field intensity under residence time of 1 min. However, no considerable effect of pulse width and residence time was observed. Larger electric intensity (12.5 and 10.0 cm−1) and pulse frequency (900 and 700 Hz) showed significant ($p<0.01$) influence among the treated samples.

Protein foaming capacity describes its ability to reduce the surface tension of the liquid–gas interface to produce bubbles, while foaming stability of proteins refers to its ability to maintain stability of generated foam. The foaming properties are influenced by foam generation method, protein concentration, solubility, hydrophobicity, molecular flexibility, distribution of charged groups, and polar groups. The pulse width showed significant ($p<0.01$) actions on the foaming capacity.

23.3.2 Texture

PEF pretreatment improved chewiness by 33.18–24.60% for PEF-treated samples compared to the control (Bai et al., 2021). Cooked rice exposed to PEF showed higher hardness and much lower gruel solid loss. The improved cohesiveness might be related to enhanced linkage of protein, starch, and other macromolecules of rice. The PEF pretreatment made rice grains porous, as justified by the SEM images, resulting in a uniformly porous structure, thereby increasing the chewiness and springiness of cooked rice.

23.3.3 Volatile Components

Results reported by Bai et al. (2021) revealed that the PEF caused a decline in the ketones, aldehydes, and hydrocarbons and an increase in the acids and esters. A known typical odor in rice, primarily due to hexanal, drastically enhanced by 67.96% from 0.0123 μg kg−1 to 0.0304 μg kg−1, thus strengthening the flavor of the PEF products. Additionally, PEF pretreatment caused a variation in the main content of linoleic acid, oleic acid, linolenic acid, palmitic acid, and stearic acid, improving the essential fatty acids. Also, a significant rise of the peroxide value acid value and p-anisidine value were reported.

23.3.4 Rheololgical Properties

Wu et al. (2019) reported that the PEF did not affect the final viscosity, trough viscosity, breakdown, and peak temperature but there was a slight decline in the setback viscosity compared to the control. However, EFI levels did not play any significant role in any of them. The re-association among starch granules when cooled involves retrogradation or re-ordering of the starch (Schafranski et al., 2021). The setback viscosity is related to the textural attributes and smaller setback viscosity values clearly denote cold water-soluble paste, supporting the profitable utilization of PEF in improving the cold paste stability of starch in cereal-based products. Therefore, variations in setback values for rice starches might be attributed to their changes in molecular structure during PEF exposure.

23.3.5 Thermal Properties

Bai et al. (2021) reported a rise in the peak temperature (Tp), decline in the conclusion temperature, and insignificant rise in the gelatinization enthalpy on PEF exposure to rice. The reduction in the conclusion temperature may be attributed to the destruction of crystalline portions of rice and improving the interaction between water and starch (Wu et al., 2019). The rice exposed to PEF indicated better gelatinization, improved digestibility, and higher glycemic index of rice. For peak temperature, the PEF groups showed better values compared to the control. The increase of Tp and ΔH is probably because the PEF treatment releases starch granules more easily, better exposed starch molecules absorb and bind more water and swelled rapidly on heating. Moreover, the pasted starch restricted the water movement to the internal portion of rice, thereby increasing the gelatinization temperature and cooking time. The unbinding crystalline and amorphous regions can be easily damaged by PEF and leave the binding crystalline regions more stable and uniform (Hong et al., 2018; Sunil, et al., 2018).

23.3.6 Antimicrobial Activity

Application of spent grain extracts treated with PEF light inhibited the growth of Listeria monocytogenes *Salmonella typhimurium* to a better extent (Minimum inhibitory concentration (MIC) value of 50 and 25 mg/mL) compared to the untreated extracts (MIC>50 mg/mL) (Kumari et al., 2019).

23.3.7 Immunomodulatory Activity

The pretreated light colored BSG extracts showed no significant effect on stimulation of immune-modulatory markers after 1 day of incubation period (Kumari et al., 2019). However, PEF pretreated dark colored extracts demonstrated a reduction in the production of cytokines and chemokines in comparison to the untreated extracts.

23.4 EFFECT OF PFE ON BIOLOGICAL AND ANTIOXIDATIVE PROPERTIES

23.4.1 Antioxidative Properties

23.4.1.1 DPPH Radical Scavenging Activity

The results reported by Wang et al. (2020) showed significant loss in the DPPH activity with the highest value of inhibition (48.54%) at the electric field intensity of 10 kV cm−1. However, with increase in field intensity from 10 to 15 kV cm−1, DPPH inhibition declined from 48.54% to 45.83%. A similar trend was reported by Liang et al. (2021). In addition, the lower frequency level (2,000 Hz) positively affected the DPPH inhibition as summarized in Table 23.3.

23.4.1.2 Total Phenolic Content (TPC) and Antioxidant Activity (AA)

Significant differences in the AA and TPC content between treated sorghum flour and control were observed at 1 kV/cm and 3 kV/cm (Lohani & Muthukumarappan, 2016). The AA and TPC of tested sorghum flour significantly ($p<0.05$) increased by 11% and 8.2%, respectively, with change in electric field intensity from 1 kV/cm to 2 kV/cm. Exceptionally, a decline in the TPC as well as AA were reported when electric field intensity was increased further to 3 kV/cm. The greater electric field intensity (at 3 kV/cm) led to the generation of higher energy input (10.8 kJ/kg), which could cause an irreversible loss in the membrane permeability properties of cell and thereby a degradation in the TPC values as well. When the PEF treatment time enhanced from 500 μs to 875 μs, a significant ($p<0.05$) rise of about 5.7% and 6.3% was observed in TPC and AA, respectively.

Table 23.3 Effect of PFE on Biological and Antioxidative Properties

DPPH Total Phenolic Content (TPC) and antioxidant activity (AA) TPC *Benzoic acids* *Cinnamic acids*	Corn peptides Sorghum flour Brewers spent grains Sorghum flour Sorghum flour	With increase in electric field intensity from 10 to 15 kV cm−¹, DPPH inhibition decreased from 48.54% to 45.83%. The highest value (30.50%) was obtained with frequency of 2,000 Hz. However, when the frequency was increased from 2,000 to 2,700 Hz, the DPPH inhibition significantly decreased (p<0.05) TPC and AA of SF significantly (p<0.05) increased by 8.2% and 11%, respectively with the increase in electric field intensity from 1 kV/cm to 2 kV/cm. However, a drop was observed when electric field intensity increased further up to 3 kV/cm Sorghum flour showed a significant (p<0.05) increase of 5.7% and 6.3% in TPC and AA, respectively, when the PEF treatment time enhanced from 500 μs to 875 μs no significant difference between PEF-treated and untreated BSG extracts were observed in TPC Protocatechuic acid and p-hydroxybenzoic acid increased on PEF exposure PEF-treated flour showed higher values for Caffeic acid, p-Coumaric acid, Ferulic acid, and Salicylic acid in comparison to control	Wang et al. (2015) Lohani et al. (2016) Kumari et al. (2019) Lohani et al. (2016) Lohani et al. (2016)
Starch digestibility	Wheat starch Japonica rice	For wheat starch, the RDS increases significantly with electric field intensity except for 2.86 kV cm−1; the SDS in treated samples is significantly less than the control and the RS increases significantly with electric field intensity at lower levels (2.86 and 4.29 kV cm−1) and influences insignificantly among higher levels (5.71, 7.14, and 8.57 kV cm−1) All PEF-treated starches show higher RDS levels but lower SDS levels than those of native starch. The RDS increases by 3.73% at EFI larger than 5.71 kV cm−1, the SDS decreases by 4.31% at EFI of 8.57 kV cm−1	Li et al. (2019) Wu et al. (2019)
Protein digestibiity			
Starch hydrolysis	Cooked rice	PEF-treated samples showed higher hydrolysis than the control sample	Bai et al. (2021)
Glycemic index (GI)	Cooked rice	GI increases by 8.03–4.57% in treated samples	Bai et al. (2021)

With increase in the treatment time, the number of pulse/sec falling on to the sample increased. At 875 μs residence time, metabolic responses were induced owing to the opening of the pores in the cell membrane and subsequently a continuous influx and efflux of the phenolic compounds. Further increase in the treatment time might cause a disruption in the cell wall irreversibly and no further release of the bound phenolics from the samples. The most suitable combination for maximum extraction of TPC and AA from sorghum flour was 2 kV/cm field intensity, 45% (w/v) flour-to-water ratio, and 875 μs resident time. Lohani et al. (2016) concluded that the electroporation as a result of PEF treatment was dominantly influenced by the water-to-flour ratio followed by residence time and the electric field intensity.

23.4.1.3 Benzoic Acids and Cinnamic Acids

Treated sorghum flour had the higher concentration of ferulic, salicylic, and p- caffeic acids compared to control (Lohani et al., 2016). It was concluded that the low and mild intensity PEF intensities can be successfully used to prepare sorghum flour with improved phenolic concentration and antioxidative properties.

23.4.2 Biological Properties

23.4.2.1 Starch Digestibility and Protein Digestibility

Rapid digestible starch (RDS) increases significantly with increase in EFI of the PEF treatment in the case of wheat starch with exception at 2.86 kV cm−1 (Li et al., 2019). The slowly digested starch (SDS) in the treated samples was significantly less than the untreated samples with no significant change caused due to intensity alteration. Also, the resistant starch increased significantly with EFI at lower levels, but the impact was negligible among higher levels (5.71–8.57 kV cm−1).

Similar rise in the RDS levels and decline in the SDS levels were observed. The SDS declined by 4.31% at EFI (8.57 kV cm−1) whereas the RDS increased by 3.73% at EFI greater than 5.71 kV cm−1 (Wu et al., 2019). It was speculated that treatment caused disintegration in starch surface, facilitating the enzymes to act on the inner portions of starch. The RDS content increased at the starting point of enzymic reaction. The left starch grains after PEF treatment were found to be uniform, less compact, and more accessible to the enzymes, which also causes faster hydrolysis of starch and less SDS derived. A significant rise in the protein digestibility was observed in case of peanuts when electric fields were applied for 15 to 45 min. However, no precise literature on the impact of protein digestibility on specifically cereal proteins was found.

23.4.2.2 Starch Hydrolysis and Glycemic Index (GI)

The fermentation-based industries demand an ecological choice that provides higher rate of starch hydrolysis and low production costs. The PEF-treated groups showed higher enzyme hydrolysis in comparison to control attaining after 150 min groups (Bai et al., 2021). The glycemic index is positively corelated to the blood glucose elevation within 180 min of injestion (Osawa & Inoue, 2007). Compared to the control (77.30 ± 0.63), the GI increased significantly by 4.57–8.03% ranging between 80.83–83.51. The higher GI might be beneficial for those persons who need fast energy supply and hypoglycemic population.

23.5 EFFECT OF PFE ON MICROSTRUCTURE, MORPHOLOGY OF STARCH, AND PROTEIN STRUCTURES

23.5.1 Scanning Electron Microscopy (SEM)

The SEM micrographs presented by Wu et al. (2019) showed that the electric energy acted on the superficial structure of the rice starch in solid state, probably due to the dense layer in the outer region that exhibits resistance to the external physical stress (Błaszczak et al., 2005). PEF affected the integrity of cell and starch granules. Compared to the control, treated samples possessed larger pores and irregular arrays with a loose structure (Bai et al., 2021). This porous structure developed as a result of PEF may offer in the future wide opportunities for effective enzymic reactions since it provides easy access to the inner part of grains and exposes more starch that could further be hydrolyzed (Table 23.4).

23.5.2 Secondary and Tertiary Protein Structure as Indicated by FTIR

The number and location of peaks were not altered on exposing with PEF (Wu et al. 2019). Bai et al. (2021) reported an increase in the 1045/1022 ratio indicating higher level of ordered

Table 23.4 Effect of PFE on Microstructure, Morphology of Starch, and Protein Structures

Scanning Electron Microscopy	Vital wheat gluten Cooked rice Sorghum flour	Larger electric field intensity resulted in a more homogeneous structure of the wet gluten Compared to the control, treated group shows larger pores and formation of irregular arrays The SEM images revealed that the electroporation resulted an increase in the porosity of the cellular membrane as starch granules appeared in a loosened structure	Zhang et al. (2021) Bai et al. (2021) Lohani et al. (2016)
Secondary structure	Japonica rice Vital wheat gluten Cooked rice	The number and location of peaks are not altered after PEF exertion The proportion of α-helices and β-sheets decreased and increased with the increase in EFI, respectively. The proportion of random coils changed insignificantly ($p>0.05$). The proportion of β-turns increased at 7.5 kV cm^{-1} and decreased significantly The percentage of α-helix, β-sheet and random coil structure decreases by 5.53–4.71%, 2.74–3.04%, and -4.65–5.42% in the PEF-treated samples. The reduction of β-sheet is less than that of α-helix and random coil	Wu et al. (2019) Zhang et al. (2021) Bai et al. (2021)
Tertiary structure	Vital wheat gluten	The amount of free SH and total SH increased and decreased significantly ($p<0.01$) with an increase in EFI, respectively	Zhang et al. (2021)
Protein surface hydrophobicity	Vital wheat gluten	The hydrophobicity of VWG protein after PEF treatment was significantly ($P<0.01$) reduced compared with the control	Zhang et al. (2021)
X-ray diffraction	Cooked rice	No new peaks emerge or significant location shifts on XRD curves, demonstrating that PEF would not alter the crystalline arrangement of rice	Bai et al. (2021)
Micrographs	Wheat starch	No changes in shape, size or appearance are observed under ordinary light for all three types of starch, indicating that no morphology is damaged by PEF	Li et al. (2019)
XRD pattern	Wheat starch	Compared to the controls, PEF changes the relative crystallinity of the starches significantly at 2.86, 4.29, and 5.71 kV cm−1 for wheat starch	Li et al. (2019)
SSNMR spectra CP/ MAS NMR	Wheat starch Japonica rice	The portions of ordered structure vary the largest at EFI of 2.86 kV cm−1 for wheat starch after treatment PEF does not change the type A crystalline structure of rice starch.	Li et al. (2019) Wu et al. (2019)
Lamellar structure and Molecular weight (SAXS)	Wheat starch Japonica rice Maize starch	Significant influence of PEF on the scatter structure and fractal dimension at all EFI except 4.29 kV cm−1 for wheat starch The increase of AUCAP1 and decrease of AUCAP2 at EFI larger than 5.71 kV cm−1 imply the breakdown of molecular chain: the portion of shorter chain increases and that of the longer chain decreases. PEF does not cause significant difference in Mw of starch at all the EFI tested, which are at the same multiple levels. The Mw of PEF-treated maize starch was decreased significantly with increasing electric field strength, with the lowest values observed at 50 kV/cm. Therefore, under the high intense PEF treatments, the side chains of amylopectin in starches might be destroyed and disconnected, leading to the decrease of molecular weight of maize starches	Li et al. (2019) Wu et al. (2019) Han et al. (2020)
Nuclear Magnetic Resonance (NMR)	Glutinous Rice Grain	The peaks at 101.5, 73.1l, and 62.8 ppm shifted to higher resonance after PFE treatments due to enhancement of starch hydration	Qiu et al. (2021)

structure, suggesting that PEF might disrupt the short-range ordered structures of rice. The 1022/995 decreased from 0.97 (the control) to 0.93–0.94 implying the reduction of order degree, which could lead to higher hydrolysis degree of starch by weakening the ability of rice to resist the action of enzymes.

The changes in band shapes at 1687, 1653, and 1649 cm−1 were observed, which could be related to alteration in the intermolecular α-helices and β-sheets of rice protein, denoting protein unfolding after PEF treatment (Qiu et al., 2021). FTIR results confirmed that 50 to 300 pulses enhanced the water absorption and hydration of rice starch compared to control soaked in KCl solution and concluded that the PEF-induced changes in proteins were similar to the hydration reaction that occurs during starch gelatinization.

Melchior et al. (2020) reported that the effect of the high-intensity PEF treatments on the protein structure is strongly dependent on nature of protein. The amide I region of the gluten protein denoting the protein conformations was analyzed. The results indicated that the high-intensity PEF influenced the strong dipole moment of the gluten protein conformations. β-sheets have hydrogen bonds comparatively weaker than the α-helix and are characterized by small net dipole moment. Transformation occurs from α-helix to β-sheet and random coil as well as to the loss in the β-sheet structure resulting in the loss of order and subsequent destabilization of the secondary protein structure (Giteru et al., 2018); Zhang et al. 2017; W. Zhao & Yang, 2012; Y. M. Zhao et al., 2019).

The amount of free SH and total SH increased and decreased significantly ($P<0.01$) with an increase in EFI when vital wheat gluten was treated with PEF (Zhang et al., 2021). After treatment, the increment of 25.2% for free sulfhydryl group and 46.9% for total sulfhydryl groups was observed at the EFI of 12.5 kV cm^{-1}. This indicates that the PEF extends the gluten and breaks the disulfide linkage.

23.5.3 Protein Surface Hydrophobicity

Protein hydrophobicity of the protein is linked to the conformation and structure of the protein, altering the functional characteristics of it (Nakai, 1983). The hydrophobicity of wheat gluten protein reduced significantly ($P<0.01$) post treatment (Zhang et al., 2021). PEF exposed more tryptophan residues previously embedded within protein molecules (Qian et al., 2016).

23.5.4 X-ray Diffraction

No new peaks on XRD curves emerged nor were any significant location shifts observed (Bai et al., 2021). This showed that PEF did not cause any transformation in the crystalline arrangement in the rice. However, the relative crystallinity declined by 1.03–4.92% in comparison to the control. Similar results indicating the slight damage that PEF would cause to the crystalline structure of rice starch on exposing it with high or low electric field intensity were also reported (Wu et al., 2019).

PEF altered the relative crystallinity of the wheat starch at 5.71, 2.86, and 4.29 kV cm−1 (Li et al., 2019). When exposed to lower intensities, rearrangement of amylose chains in loose amorphous areas might take place, whereas at higher field intensity, damage to crystalline regions occurs.

The XRD peaks of rice samples with 200 to 300 pulses denoting crystallization were found to be much smoother and smaller than the control (Qiu et al., 2021). The greater PEF intensity showed a diminishing XRD curve demonstrating a decline in the peak intensity and hence a loss in the crystallinity of rice. In addition, the XRD curves indicated an instance of partial gelatinization that occurred in rice samples when 200–300 pulses were applied.

23.5.5 Lamellar Structure and Molecular Weight

The knowledge regarding the effect of PEF on the lamellar structure of wheat starch was acquired utilizing small angle X-ray scattering (SAXS). It relates the scattering vector and scattering intensity to analyze the thickness of crystalline layer and its fractal structure. Significant impact of PEF on the fractal and scatter structure for wheat starch was noted except at 4.29 kV cm−1 (Li et al., 2019).

Kumari et al. (2019) reported the peaks in the untreated light and dark BSG extracts denoting a molecular weight of 13.7 and 18.5 kDa, respectively, but peaks of the treated samples denoted a molecular weight of 13.4 and 17.9 kDa. This decline could be attributed to the PEF-induced polarization of the peptides and breakdown of the non-covalent bonds (electrostatic interactions, hydrophobic interactions, hydrogen bonds), dissociation of the subunits and disorganization of the primary and secondary structure of protein, and subsequent release of small peptides and amino acids (Lin et al., 2016).

PEF does not cause any significant change in the molecular weight of starch regardless of the field intensity. AUCAP2 and AUCAP1, referring to the area of AP2 and AP1, denote the concentration of long-chain and short-chain amylopectin. The results implied breakdown of amylopectin at higher intensity (Wu et al., 2019).

The double logarithmic plot of SAXS for rice flour showed diffractive peak at around 0.67 nm−1 pertaining to the semicrystalline region of starch. The results denoted a loss in the semicrystalline lamellae of the starch (Bai et al., 2021). Han et al. (2020) reported a decrease in the Mw of treated maize starch when electric field strength was increased.

23.5.6 Nuclear Magnetic Resonance (NMR)

No impact of PEF over the type A crystalline structure of rice was reported in CP/MAS NMR spectra (Wu et al., 2019). The chemical shifts assigned for regions are usually associated with the crystalline type of starch (Su et al., 2016). The relative amount of starch increased significantly at 5.71 kV cm−1 and declined at 8.57 kV cm−1 in comparison to the control. The findings were even supported by ATR-FTIR, where the short-range structures rearranged at higher intensities (Wu et al., 2019). Similar results were reported by Li et al. (2019) where the crystalline regions decline by 0.7% for wheat.

The peaks at 62.8, 73.1, and 101.5 ppm were shifted to higher resonance post PEF treatment (Qiu et al., 2021). The peak shifts might be the result of starch hydration induced by PEF treatment, showing that higher intensity facilitates the peak shift to greater resonances. Also, PEF induced rearrangement of glutinous rice starch. The shift in the resonance peak may be attributed to the low resonance in case of non-crystalline material. Coinciding with the findings of XRD and ATR-FTIR, it is clear that PEF changes the ordered structure of the three types of starch, with more severity in type B.

REFERENCES

Alirezalu, K., Munekata, P. E. S., Parniakov, O., Barba, F. J., Witt, J., Toepfl, S., Wiktor, A., & Lorenzo, J. M. (2020). Pulsed electric field and mild heating for milk processing: a review on recent advances. *Journal of the Science of Food and Agriculture*, *100*(1), 16–24. https://doi.org/10.1002/jsfa.9942

Baba, K., Kajiwara, T., Watanabe, S., Katsuki, S., Sasahara, R., & Inoue, K. (2018). Low-Temperature Pasteurization of Liquid Whole Egg using Intense Pulsed Electric Fields. *Electronics and Communications in Japan*, *101*(2), 87–94. https://doi.org/10.1002/ecj.12053

Bai, T. G., Zhang, L., Qian, J. Y., Jiang, W., Wu, M., Rao, S. Q., Li, Q., Zhang, C., & Wu, C. (2021). Pulsed electric field pretreatment modifying digestion, texture, structure and flavor of rice. *Lwt*, *138*. https://doi.org/10.1016/j.lwt.2020.110650

Błaszczak, W., Fornal, J., Valverde, S., & Garrido, L. (2005). Pressure-induced changes in the structure of corn starches with different amylose content. *Carbohydrate Polymers*, *61*(2), 132–140. https://doi.org/10.1016/j.carbpol.2005.04.005

Bresciani, A., Giordano, D., Vanara, F., Blandino, M., & Marti, A. (2021). High-amylose corn in gluten-free pasta: Strategies to deliver nutritional benefits ensuring the overall quality. *Food Chemistry*, *353*. https://doi.org/10.1016/j.foodchem.2021.129489

Buchmann, L., Bloch, R., & Mathys, A. (2018). Comprehensive pulsed electric field (PEF) system analysis for microalgae processing. *Bioresource Technology*, *265*, 268–274. https://doi.org/10.1016/j.biortech.2018.06.010

Duque, S. M. M., Leong, S. Y., Agyei, D., Singh, J., Larsen, N., & Oey, I. (2020). Modifications in the physicochemical properties of flour "fractions" after Pulsed Electric Fields treatment of thermally processed oat. *Innovative Food Science and Emerging Technologies*, *64*, 102406. https://doi.org/10.1016/j.ifset.2020.102406

Giteru, S. G., Oey, I., & Azam Ali, M. (2018). The effect of pulsed electric fields on the rheology and microstructure of chitosan-poly(vinyl alcohol) composites. *International Journal of Nanotechnology*, *15*(8–10), 655–662. https://doi.org/10.1504/IJNT.2018.098430

Goettel, M., Eing, C., Gusbeth, C., Straessner, R., & Frey, W. (2013). Pulsed electric field assisted extraction of intracellular valuables from microalgae. *Algal Research*, *2*(4), 401–408. https://doi.org/10.1016/j.algal.2013.07.004

Guan, Y. G., Lin, H., Han, Z., Wang, J., Yu, S. J., Zeng, X. A., Liu, Y. Y., Xu, C. H., & Sun, W. W. (2010). Effects of pulsed electric field treatment on a bovine serum albumin-dextran model system, a means of promoting the Maillard reaction. *Food Chemistry*, *123*(2), 275–280. https://doi.org/10.1016/j.foodchem.2010.04.029

Han, Z., Han, Y., Wang, J., Liu, Z., Buckow, R., & Cheng, J. (2020). Effects of pulsed electric field treatment on the preparation and physicochemical properties of porous corn starch derived from enzymolysis. *Journal of Food Processing and Preservation*, *44*(3). https://doi.org/10.1111/jfpp.14353

Hong, J., Zeng, X. A., Han, Z., & Brennan, C. S. (2018). Effect of pulsed electric fields treatment on the nanostructure of esterified potato starch and their potential glycemic digestibility. *Innovative Food Science and Emerging Technologies*, *45*, 438–446. https://doi.org/10.1016/j.ifset.2017.11.009

Koubaa, M., Barba, F. J., Grimi, N., Mhemdi, H., Koubaa, W., Boussetta, N., & Vorobiev, E. (2016). Recovery of colorants from red prickly pear peels and pulps enhanced by pulsed electric field and ultrasound. *Innovative Food Science and Emerging Technologies*, *37*, 336–344. https://doi.org/10.1016/j.ifset.2016.04.015

Kumari, B., Tiwari, B. K., Walsh, D., Griffin, T. P., Islam, N., Lyng, J. G., Brunton, N. P., & Rai, D. K. (2019). Impact of pulsed electric field pre-treatment on nutritional and polyphenolic contents and bioactivities of light and dark brewer's spent grains. *Innovative Food Science and Emerging Technologies*, *54*, 200–210. https://doi.org/10.1016/j.ifset.2019.04.012

Lasekan, O., Ng, S., Azeez, S., Shittu, R., Teoh, L., & Gholivand, S. (2017). Effect of pulsed electric field processing on flavor and color of liquid foods. *Journal of Food Processing and Preservation*, *41*(3), e12940.

Li, Q., Wu, Q. Y., Jiang, W., Qian, J. Y., Zhang, L., Wu, M., Rao, S. Q., & Wu, C. Sen. (2019). Effect of pulsed electric field on structural properties and digestibility of starches with different crystalline type in solid state. *Carbohydrate Polymers*, *207*, 362–370. https://doi.org/10.1016/j.carbpol.2018.12.001

Liang, R., Cheng, S., Lin, S., Dong, Y., & Ju, H. (2021). Validation of Steric Configuration Changes Induced by a Pulsed Electric Field Treatment as the Mechanism for the Antioxidant Activity Enhancement of a Peptide. *Food and Bioprocess Technology*, *14*(9), 1751–1757. https://doi.org/10.1007/s11947-021-02643-w

Lin, S., Liang, R., Li, X., Xing, J., & Yuan, Y. (2016). Effect of pulsed electric field (PEF) on structures and antioxidant activity of soybean source peptides-SHCMN. *Food Chemistry*, *213*, 588–594. https://doi.org/10.1016/j.foodchem.2016.07.017

Liu, C., Pirozzi, A., Ferrari, G., Vorobiev, E., & Grimi, N. (2020). Effects of Pulsed Electric Fields on Vacuum Drying and Quality Characteristics of Dried Carrot. *Food and Bioprocess Technology*, *13*(1), 45–52. https://doi.org/10.1007/S11947-019-02364-1

Liu, Y. F., Oey, I., Bremer, P., Silcock, P., & Carne, A. (2018). Proteolytic pattern, protein breakdown and peptide production of ovomucin-depleted egg white processed with heat or pulsed electric fields at different pH. *Food Research International*, *108*, 465–474. https://doi.org/10.1016/j.foodres.2018.03.075

Lohani, U. C., & Muthukumarappan, K. (2016). Application of the pulsed electric field to release bound phenolics in sorghum flour and apple pomace. *Innovative Food Science and Emerging Technologies*, *35*, 29–35. https://doi.org/10.1016/j.ifset.2016.03.012

Melchior, S., Calligaris, S., Bisson, G., & Manzocco, L. (2020). Understanding the impact of moderate-intensity pulsed electric fields (MIPEF) on structural and functional characteristics of pea, rice and gluten concentrates. *Food and Bioprocess Technology*, *13*(12), 2145–2155. https://doi.org/10.1007/S11947-020-02554-2

Morales-de la Peña, M., Rábago-Panduro, L. M., Soliva-Fortuny, R., Martín-Belloso, O., & Welti-Chanes, J. (2021). Pulsed Electric Fields Technology for Healthy Food Products. *Food Engineering Reviews*, *13*(3), 509–523. https://doi.org/10.1007/s12393-020-09277-2

Nakai, S. (1983). Structure-Function Relationships of Food Proteins with an Emphasis on the Importance of Protein Hydrophobicity. *Journal of Agricultural and Food Chemistry*, *31*(4), 676–683. https://doi.org/10.1021/jf00118a001

Osawa, M., & Inoue, N. (2007). Studies on the in vitro starch digestibility and the glycemic index in rice cultivars differing in amylose content. *Japanese Journal of Crop Science*, *76*(3), 410–415. https://doi.org/10.1626/jcs.76.410

Parniakov, O., Barba, F. J., Grimi, N., Lebovka, N., & Vorobiev, E. (2014). Impact of pulsed electric fields and high voltage electrical discharges on extraction of high-added value compounds from papaya peels. *Food Research International*, *65*(PC), 337–343. https://doi.org/10.1016/j.foodres.2014.09.015

Postma, P. R., Cerezo-Chinarro, O., Akkerman, R. J., Olivieri, G., Wijffels, R. H., Brandenburg, W. A., & Eppink, M. H. M. (2018). Biorefinery of the macroalgae Ulva lactuca: extraction of proteins and carbohydrates by mild disintegration. *Journal of Applied Phycology*, *30*(2), 1281–1293. https://doi.org/10.1007/s10811-017-1319-8

Qian, J. Y., Ma, L. J., Wang, L. J., & Jiang, W. (2016). Effect of pulsed electric field on structural properties of protein in solid state. *LWT – Food Science and Technology*, *74*, 331–337. https://doi.org/10.1016/j.lwt.2016.07.068

Qiu, S., Abbaspourrad, A., & Padilla-Zakour, O. I. (2021). Changes in the Glutinous Rice Grain and Physicochemical Properties of Its Starch upon Moderate Treatment with Pulsed Electric Field. *Foods 2021, Vol. 10, Page 395*, *10*(2), 395. https://doi.org/10.3390/FOODS10020395

Salehi, F. (2020). Physico-chemical properties of fruit and vegetable juices as affected by pulsed electric field: a review. *International Journal of Food Properties*, *23*(1), 1036–1050. https://doi.org/10.1080/10942912.2020.1775250

Sánchez-Vega, R., Elez-Martínez, P., & Martín-Belloso, O. (2014). Effects of High-Intensity Pulsed Electric Fields Processing Parameters on the Chlorophyll Content and Its Degradation Compounds in Broccoli Juice. *Food and Bioprocess Technology*, *7*(4), 1137–1148. https://doi.org/10.1007/s11947-013-1152-2

Sarkis, J. R., Boussetta, N., Blouet, C., Tessaro, I. C., Marczak, L. D. F., & Vorobiev, E. (2015). Effect of pulsed electric fields and high voltage electrical discharges on polyphenol and protein extraction from sesame cake. *Innovative Food Science and Emerging Technologies*, *29*, 170–177. https://doi.org/10.1016/j.ifset.2015.02.011

Schafranski, K., Ito, V. C., & Lacerda, L. G. (2021). Impacts and potential applications: A review of the modification of starches by heat-moisture treatment (HMT). *Food Hydrocolloids*, *117*. https://doi.org/10.1016/j.foodhyd.2021.106690

Shorstkii, I., & Koshevoi, E. (2019). Extraction Kinetic of Sunflower Seeds Assisted by Pulsed Electric Fields. *Iranian Journal of Science and Technology, Transaction A: Science*, *43*(3), 813–817. https://doi.org/10.1007/s40995-018-0591-z

Su, Y., Du, H., Huo, Y., Xu, Y., Wang, J., Wang, L., Zhao, S., & Xiong, S. (2016). Characterization of cationic starch flocculants synthesized by dry process with ball milling activating method. *International Journal of Biological Macromolecules*, *87*, 34–40. https://doi.org/10.1016/j.ijbiomac.2015.11.093

Sunil, Neelash Chauhan, Jaivir Singh, Suresh Chandra, V. C. and V. K. (2018). "Non-thermal techniques: Application in food industries" A review. *Journal of Pharmacognosy and Phytochemistry*, *7(5)*(5), 1507–1518. www.phytojournal.com/archives/2018/vol7issue5/PartZ/7-4-659-545.pdf

Tylewicz, U., Oliveira, G., Alminger, M., Nohynek, L., Dalla Rosa, M., & Romani, S. (2020). Antioxidant and antimicrobial properties of organic fruits subjected to PEF-assisted osmotic dehydration. *Innovative Food Science and Emerging Technologies*, *62*. https://doi.org/10.1016/j.ifset.2020.102341

Wani, S. P., Anantha, K. H., Sreedevi, T. K., Sudi, R., Singh, S. N., & D'Souza, M. (2011). Assessing the environmental benefits of watershed development: evidence from the Indian semi-arid tropics. *Journal of Sustainable Watershed Science and Management*, *1*(1), 10–20.

Wang, K., Wang, Y., Lin, S., Liu, X., Yang, S., & Jones, G. S. (2015). Analysis of DPPH inhibition and structure change of corn peptides treated by pulsed electric field technology. *Journal of food science and technology*, *52*, 4342–4350.

Wang, L., Boussetta, N., Lebovka, N., & Vorobiev, E. (2020). Cell disintegration of apple peels induced by pulsed electric field and efficiency of bio-compound extraction. *Food and Bioproducts Processing*, *122*, 13–21. https://doi.org/10.1016/j.fbp.2020.03.004

Wu, C., Wu, Q. Y., Wu, M., Jiang, W., Qian, J. Y., Rao, S. Q., Zhang, L., Li, Q., & Zhang, C. (2019). Effect of pulsed electric field on properties and multi-scale structure of japonica rice starch. *Lwt*, *116*. https://doi.org/10.1016/j.lwt.2019.108515

Yildiz, S., Pokhrel, P. R., Unluturk, S., & Barbosa-Cánovas, G. V. (2021). Changes in Quality Characteristics of Strawberry Juice After Equivalent High Pressure, Ultrasound, and Pulsed Electric Fields Processes. *Food Engineering Reviews*, *13*(3), 601–612. https://doi.org/10.1007/S12393-020-09250-Z

Zhang, L., Wang, L. J., Jiang, W., & Qian, J. Y. (2017). Effect of pulsed electric field on functional and structural properties of canola protein by pretreating seeds to elevate oil yield. *Lwt*, *84*, 73–81.

Zhang, C., Yang, Y. H., Zhao, X. D., Zhang, L., Li, Q., Wu, C., Ding, X., & Qian, J. Y. (2021). Assessment of impact of pulsed electric field on functional, rheological and structural properties of vital wheat gluten. *LWT*, *147*, 111536. https://doi.org/10.1016/J.LWT.2021.111536

Zhao, W., & Yang, R. (2012). Pulsed Electric Field Induced Aggregation of Food Proteins: Ovalbumin and Bovine Serum Albumin. *Food and Bioprocess Technology*, *5*(5), 1706–1714. https://doi.org/10.1007/s11947-010-0464-8

Zhao, Y. M., de Alba, M., Sun, D. W., & Tiwari, B. (2019). Principles and recent applications of novel non-thermal processing technologies for the fish industry – a review. *Critical Reviews in Food Science and Nutrition*, *59*(5), 728–742. https://doi.org/10.1080/10408398.2018.1495613

Index

B

D

G

H

I

N

O

P

Q

R

S